st of the Elements

NAME	SYMBOL	ATOMIC NUMBER	ATOMIC WEIGHT
Actinium	Ac	89	227.028
Aluminum	Al	13	26.982
Americium	Am	95	243
Antimony	Sb	51	121.76
Argon	Ar	18	39.948
Arsenic	As	33	74.922
Astatine	At	85	210
Barium	Ba	56	137.327
Berkelium	Bk	97	247
Beryllium	Be	4	9.012
Bismuth	Bi	83	208.980
Bohrium	Bh	107	269
Boron	B	5	10.811
Bromine	Br	35	79.904
Cadmium	Cd	48	112.411
Calcium	Ca	20	40.078
Californium	Cf	98	251
Carbon	C	6	12.011
Cerium	Ce	58	140.115
Cesium	Cs	55	132.905
Chlorine	Cl	17	35.453
Chromium	Cr	24	51.996
Cobalt	Co	27	58.933
Copernicium	Cn	112	285
Copper	Cu	29	63.546
Curium	Cm	96	247
Darmstadtium	Ds	110	281
Dubnium	Db	105	268
Dysprosium	Dy	66	162.5
Einsteinium	Es	99	253
Erbium	Er	68	167.26
Europium	Eu	63	151.964
Fermium	Fm	100	257
Flerovium	Fl	114	289
Fluorine	F	9	18.998
Francium	Fr	87	223
Gadolinium	Gd	64	157.25
Gallium	Ga	31	69.723
Germanium	Ge	32	72.61
Gold	Au	79	196.967
Hafnium	Hf	72	178.49
Hassium	Hs	108	277
Helium	He	2	4.003
Holmium	Ho	67	164.93
Hydrogen	H	1	1.0079
Indium	In	49	114.82
Iodine	I	53	126.905
Iridium	Ir	77	192.22
Iron	Fe	26	55.845
Krypton	Kr	36	83.8
Lanthanum	La	57	138.906
Lawrencium	Lr	103	256
Lead	Pb	82	207.2
Lithium	Li	3	6.941
Livermorium	Lv	116	293
Lutetium	Lu	71	174.967
Magnesium	Mg	12	24.305

NAME	SYMBOL	ATOMIC NUMBER	ATOMIC WEIGHT
Manganese	Mn	25	54.938
Meitnerium	Mt	109	276
Mendelevium	Md	101	259
Mercury	Hg	80	200.59
Molybdenum	Mo	42	95.94
Neodymium	Nd	60	144.24
Neon	Ne	10	20.180
Neptunium	Np	93	237
Nickel	Ni	28	58.69
Niobium	Nb	41	92.906
Nitrogen	N	7	14.007
Nobelium	No	102	259
Osmium	Os	76	190.23
Oxygen	O	8	15.999
Palladium	Pd	46	106.42
Phosphorus	P	15	30.974
Platinum	Pt	78	195.08
Plutonium	Pu	94	244
Polonium	Po	84	209
Potassium	K	19	39.098
Praseodymium	Pr	59	140.908
Promethium	Pm	61	145
Protactinium	Pa	91	231.036
Radium	Ra	88	226.025
Radon	Rn	86	222
Rhenium	Re	75	186.207
Rhodium	Rh	45	102.906
Roentgenium	Rg	111	280
Rubidium	Rb	37	85.468
Ruthenium	Ru	44	101.07
Rutherfordium	Rf	104	265
Samarium	Sm	62	150.36
Scandium	Sc	21	44.956
Seaborgium	Sg	106	270
Selenium	Se	34	78.96
Silicon	Si	14	28.086
Silver	Ag	47	107.868
Sodium	Na	11	22.990
Strontium	Sr	38	87.62
Sulfur	S	16	32.066
Tantalum	Ta	73	180.948
Technetium	Tc	43	98
Tellurium	Te	52	127.60
Terbium	Tb	65	158.925
Thallium	Tl	81	204.383
Thorium	Th	90	232.038
Thulium	Tm	69	168.934
Tin	Sn	50	118.71
Titanium	Ti	22	47.88
Tungsten	W	74	183.84
Uranium	U	92	238.029
Vanadium	V	23	50.942
Xenon	Xe	54	131.29
Ytterbium	Yb	70	173.04
Yttrium	Y	39	88.906
Zinc	Zn	30	65.39
Zirconium	Zr	40	91.224

FIFTH EDITION

Conceptual Chemistry

UNDERSTANDING OUR WORLD
OF ATOMS AND MOLECULES

JOHN SUCHOCKI

Saint Michael's College

PEARSON

Boston   Columbus   Indianapolis   New York   San Francisco   Upper Saddle River
Amsterdam   Cape Town   Dubai   London   Madrid   Milan   Munich   Paris   Montréal   Toronto
Delhi   Mexico City   São Paulo   Sydney   Hong Kong   Seoul   Singapore   Taipei   Tokyo

Editor in Chief: Adam Jaworski
Senior Marketing Manager: Jonathan Cottrell
Assistant Editor: Coleen Morrison
Editorial Assistant: Fran Falk
Marketing Assistant: Nicola Houston
Senior Media Producer: Kristin Mayo
Director of Development: Jennifer Hart
Development Editor: Daniel Schiller
Managing Editor, Chemistry and Geosciences: Gina M. Cheselka
Production Project Manager: Connie Long
Full-Service Project Management/Composition: GEX Publishing Services
Illustrations: Imagineering
Image Lead: Maya Melenchuck
Photo Researcher: Eric Schrader
Text Permissions Manager: Alison Bruckner
Text Permissions Researcher: GEX Publishing Services
Design Manager: Mark Ong
Interior and Cover Design: Wee Design Group
Cover Image Credit: *Fire:* Lijuan Guo/Fotolia; *Sunset:* Muzhik/Shutterstock

Library of Congress Cataloging-in-Publication Data
Suchocki, John.
    Conceptual chemistry : understanding our world of atoms and molecules /
John Suchocki. — Fifth edition.
        pages cm
    Includes index.
    ISBN 978-0-321-80441-9
 1. Chemistry—Textbooks. I. Title.
    QD33.2.S83 2014
    540—dc23
                                                          2012035236

www.pearsonhighered.com   ISBN-10: 0-321-80441-4; ISBN-13: 978-0-321-80441-9

To:

NEIL DEGRASSE TYSON

For Carrying the Candle

John obtained his Ph.D. in organic chemistry from Virginia Commonwealth University. He worked as a postdoctoral fellow in pharmacology at the Medical College of Virginia before moving to Hawaii to become a tenured professor at Leeward Community College, where his interests turned to science education and the development of distance learning programs as well as student-centered learning curricula. In addition to writing *Conceptual Chemistry*, John is the chemistry and astronomy coauthor of the college and high school editions of *Conceptual Physical Science* and *Conceptual Integrated Science* with physicist Paul Hewitt and others. John is currently an adjunct professor at Saint Michael's College in Colchester, Vermont. He also produces science multimedia through his company Conceptual Productions, and writes and illustrates award-winning science-oriented children's books. John is also an avid songwriter who produces music through his recording label CPro Music.

KV 09.28.2018 0732

BRIEF CONTENTS

CONTENTS

1 About Science   2

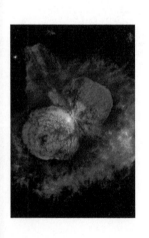

2 Particles of Matter   26

3 Elements of Chemistry 58

4 Subatomic Particles 92

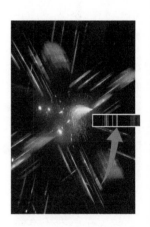

8 How Water Behaves 228

9 How Chemicals React 260

10 Acids and Bases in Our Environment   294

11 Oxidations and Reductions Charge the World   322

12 Organic Compounds 354

13 Nutrients of Life 396

Morphine

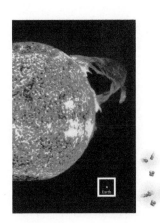

THE CONCEPTUAL CHEMISTRY PHOTO ALBUM

Conceptual Chemistry is a very personal book, as reflected in the many photographs of the author's family and friends that grace its pages. Key to its inception is John's uncle and mentor, Paul Hewitt, author of *Conceptual Physics*, who appears on page 13. On Uncle Paul's lap is John's son Evan Suchocki (pronounced su-HOCK-ee, with a silent *c*), who, as a toddler, sums up the book with his optimistic message.

Taking advantage of water's high heat of vaporization is John's wife, Tracy, who is seen fearlessly walking over hot coals on page 250 and smelling the fragrant-filled balloon on page 27. She is seen again with their oldest child, Ian, on page 100, and again with their second child Evan on page 64. Their third child, Maitreya Rose, is showcased both as a fetus and as a baby on page 430, with her mother on page 410, and as a 2-year-old holding the cellulose- and color-rich Vermont autumn leaves on page 404. She appears yet again in the Chapter 12 opening photograph on page 354 and with her good friend Annabelle Creech as they brush their teeth on page 169. About to enjoy his favorite beverage—by the liter—is son Evan on page 16. He appears again on page 45, using balloons to demonstrate the relationship between the volume of a gas and its temperature. Also, their beloved dog, Sam, shows off his great panting skills on page 240.

Members of John's extended family have also made their way onto the pages of *Conceptual Chemistry*. Nephew Graham Orr is seen on page 76 drinking water both as a child and as a grown-up college student. Exploring the microscopic realm with the uncanny resolution of electron waves is cousin George Webster, who is seen on page 110 alongside his own scanning electron microscope. Cousin Gretchen Hewitt demonstrates her taste for chips on page 410, and brother-in-law Peter Elias smells the camphor of a freshly cut Ping Pong® ball on page 383. Of John's dear friends, we see Rinchen Trashi looking through the spectroscope on page 105 and Nikki Church excited by the carbonation of water on page 219.

The photographs of the children of many of John's friends also grace this book. Ayano Jeffers-Fabro is the adorable girl hugging the tree on page 12. Helping us to understand the nature of DNA in the Chapter 13 opener on page 396 are Daniel and Jacob Glassman-Vinci. Makani Nelson, on page 398, provides us with a fine example of a human body full of cells and biomolecules. Look also for Makani's cameo appearance in the opening montage video at ConceptualChemistry.com.

TO THE STUDENT

Welcome to the world of chemistry—a world where everything around you can be traced to these incredibly tiny particles called *atoms*. Chemistry is the study of how atoms combine to form materials. By learning chemistry, you gain a unique perspective of what things are made of and why they behave as they do.

Chemistry is a science with a practical outlook. By understanding and controlling the behavior of atoms, chemists have been able to produce a broad range of new and useful materials—alloys, fertilizers, pharmaceutical products, polymers, computer chips, recombinant DNA, and more. These materials have raised our standard of living to unprecedented levels. Learning chemistry, therefore, is worthwhile simply because of the impact this field has on society. More important, with a background in chemistry, you can judge for yourself whether available technologies are in harmony with the environment and with what you believe to be proper.

This book presents chemistry conceptually, focusing on the concepts of chemistry with little emphasis on calculations. Although sometimes wildly bizarre, the concepts of chemistry are straightforward and accessible—all it takes is the desire to learn. What you will gain from your efforts, however, may be more than new knowledge about your environment and your personal relation to it—you may improve your learning skills and become a better thinker! But remember, as with any other form of training, you'll only get out of your study of chemistry as much as you put into it. I enjoy chemistry, and I know you can too. So put on your boots and let's explore this world from the perspective of its fundamental building blocks.

Good chemistry to you!

TO THE INSTRUCTOR

As instructors, we share a common desire for our teaching efforts to have a long-lasting, positive impact on our students. We focus, therefore, on what we think is most important for students to learn. For students taking liberal arts chemistry courses, certain learning goals are clear. Those students should become familiar with—and perhaps even interested in—the basic concepts of chemistry, especially those that apply to their daily lives. They should understand, for example, how soap works and why ice floats on water. They should be able to distinguish between stratospheric ozone depletion and global climate change and know what it takes to ensure a safe drinking water supply. Along the way, they should learn how to think about matter in regards to atoms and molecules. Furthermore, by studying chemistry, students should come to understand the methods of scientific inquiry and become better equipped to pass this knowledge along to future generations. In short, these students should become citizens of above-average scientific literacy.

These are noble goals, and it is crucial that we do our best to achieve them. I have come to realize, however, that these feats are not what my former students usually cherish most from having taken a course in chemistry. Rather, it is the personal development they experienced through the process.

As all science educators know, chemistry—with its many abstract concepts—is fertile ground for the development of higher-level thinking skills. Thus, it seems reasonable to share this valuable scientific offering—tempered to an appropriate level—with all students. Liberal arts students, like all other students, come to college not just to learn specific subjects, but also to grow personally. In fact, I would argue that this personal growth is the most vital commodity of a college education. This growth should include improvements in analytical and verbal reasoning skills, along with a boost in self-confidence from having successfully met well-placed challenges. The value of our teaching, therefore, rests not only on our ability to help students learn chemistry but also on our ability to help them learn about themselves.

You will find in this book the standard discussions of the applications of chemistry, as shown in the table of contents. True to its title, this textbook also builds a conceptual base from which students may view nature more perceptively by helping them visualize the behavior of atoms and molecules and showing them how this behavior gives rise to our macroscopic environment. Numerical problem-solving skills and memorization are not stressed. Instead, chemistry concepts are developed in a storytelling fashion, with the frequent use of analogies and tightly integrated illustrations and photographs. Follow-up end-of-chapter questions are designed to challenge students' understanding of concepts and their ability to synthesize and articulate conclusions. Concurrent with helping students learn chemistry, *Conceptual Chemistry* aims to be a tool by which students can learn how to become better thinkers and reach their personal goals of self-discovery.

Explore the world from a chemistry perspective

Chapter-opening features introduce chemistry concepts in a real-world context, setting the stage for the chapter.

The chemical reactions going on in your body are quite similar to those going on within burning wood. In both cases, the products are carbon dioxide, water, and energy.

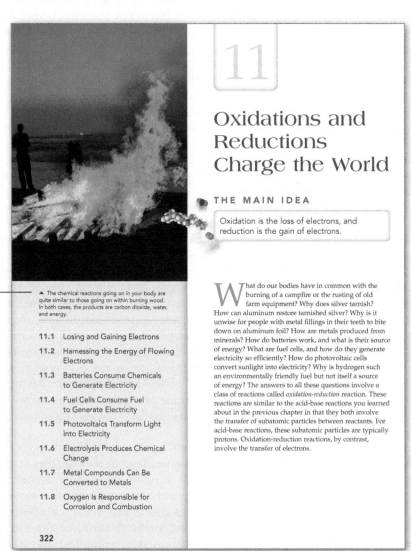

11

Oxidations and Reductions Charge the World

THE MAIN IDEA

Oxidation is the loss of electrons, and reduction is the gain of electrons.

What do our bodies have in common with the burning of a campfire or the rusting of old farm equipment? Why does silver tarnish? How can aluminum restore tarnished silver? Why is it unwise for people with metal fillings in their teeth to bite down on aluminum foil? How are metals produced from minerals? How do batteries work, and what is their source of energy? What are fuel cells, and how do they generate electricity so efficiently? How do photovoltaic cells convert sunlight into electricity? Why is hydrogen such an environmentally friendly fuel but not itself a source of energy? The answers to all these questions involve a class of reactions called *oxidation-reduction* reaction. These reactions are similar to the acid-base reactions you learned about in the previous chapter in that they both involve the transfer of subatomic particles between reactants. For acid-base reactions, these subatomic particles are typically protons. Oxidation-reduction reactions, by contrast, involve the transfer of electrons.

▲ The chemical reactions going on in your body are quite similar to those going on within burning wood. In both cases, the products are carbon dioxide, water, and energy.

322

Chemistry

HANDS ON

The Cool Rubber Band

Predict what happens to the temperature of a rubber band as it is stretched. Predict what happens to the temperature of a stretched rubber band as it relaxes.

PROCEDURE

1. Stretch a rubber band while holding it to your lower lip, which you will find is sensitive to small temperature changes.
2. Relax the stretched rubber band that is in contact with your lower lip.

ANALYZE AND CONCLUDE

1. Does the speed at which you stretch the rubber band make a difference?

2. You touch your hand to the forehead of someone with a fever. You feel that his or her forehead is hot. How does your hand feel to the person with the fever?
3. If the contracting rubber band causes your lip to cool down, what does your lip do to the contracting rubber band?
4. A hammer is hanging by a stretched rubber band. Hot air is then blown over the rubber band with a hair dryer. Is the hammer lifted upward or does it drop downward?
5. True or False: Experiments often raise more questions than they answer.

Hands On Chemistry

Each chapter opens with a hallmark Hands On Chemistry activity that allows the student to experience concepts related to chemistry outside the formal lab setting. Using common household ingredients and equipment, these activities guide students step-by-step through experiments that bring concepts alive.

Engage students with chemistry-related issues

Essays highlight chemistry concepts within the context of our society and lives.

Contextual Chemistry *(continued)*

Shale gas, lower 48 states

Montana Thrust Belt · Bakken · Devonian (Ohio) · Appalachian Basin · Cody · Heath · Gammon · Utica · Hilliard-Baxter-Mancos · Mowry · Michigan Basin · Antrim · Forest City Basin · Denver Basin · New Albany · Marcellus · Monterey Temblor · Lewis · Woodford · Fayetteville · Conasauga · Barnett-Woodford · Barnett · Tuscaloosa · Haynesville-Bossier

Basins

Shale plays
- Current plays
- Prospective plays

Stacked plays
— Shallowest/youngest
— Intermediate depth/age
— Deepest/oldest

▲ Fracking technology was first developed in the Barnett shale on the property of the Dallas-Fort Worth International Airport. The richest gas-bearing shale in the United States is the Marcellus formation, centered over western Pennsylvania.

also toxic materials coming directly from the shale, such as heavy metals and hydrocarbons, as well as unhealthy concentrations of radio-

fracking technology, the United States in 2009 surpassed Russia to become the world's leading producer of natural gas.

In addition to economic benefits, there are also potential environmental benefits to increased production of natural gas. As described in Chapter 17, most of the electricity produced in the United States comes from the burning of coal, which generates large amounts of pollutants, such as particulates, mercury, sulfur dioxides, and carbon dioxide. Natural gas, however, burns with much greater efficiency while generating far fewer pollutants —the output of carbon dioxide, for example, is about half that of coal. Although shale gas is not the ideal environmentally friendly fuel, its development is seen by many in the industry as a responsible way to

supply our energy needs for decades while other more sustainable energy technologies, such as solar energy, are developed.

Fracking technology, however, also involves significant environmental risks. Foremost is the issue of dealing with the large volumes of used fracking fluid. This fluid is usually stored in pools adjacent to the well before being shipped by truck to a disposal site. At each of these stages there is the potential for an accidental spill of the fluid into the environment. Furthermore, if the upper portions of the well are not properly installed, the used fracking fluid coming up the well could potentially seep into the water table, ruining local water supplies. The toxins that could be released include not only cancer-causing agents within the fluid formula but

160

Contextual Chemistry

A SPOTLIGHT ON ISSUES FACING OUR MODERN SOCIETY

Fracking for Shale Gas

We depend greatly on fossil fuels to meet our energy needs. Three major forms of fossil fuels are coal, petroleum, and natural gas. Both coal and petroleum are made of large and complicated carbon-based molecules. The main component of natural gas, by contrast, is methane, CH_4, which is a structurally simple molecule. Traditionally, coal and petroleum have been much more accessible in large quantities than natural gas. A prime reason for this is because much of the world's natural gas remains trapped miles beneath the surface within a type of rock known as shale.

Two recent technological advances are now allowing access to this otherwise difficult to reach natural gas, also called *shale gas*. The first is our ability to drill very deep and then sideways. This is important because shale deposits are laid down horizontally. As shown in the accompanying image, drilling sideways maximizes the surface area of the shale within reach of the bore hole. The second advance is the process of *hydraulic fracturing*, also known as *fracking*, in which channels are punched into the shale using a powerful explosive. A slurry of water and sand with a small amount of other chemicals, such as anticorrosion agents and lubricants, is then injected into these channels. This slurry is known as *fracking fluid*. Under high pressure, this fluid expands into natural cracks, which remain open as grains of sand become lodged within them. After the fracking fluid is removed, large volumes of natural gas, which is lighter than air, escape through the cracks and rise to the surface, where the gas is piped to a storage facility for future use.

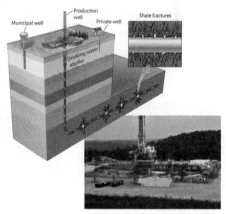

▲ Shale deposits are thousands of feet deep and usually only about 100 feet thick, so the most efficient access is provided by horizontal drilling. Natural cracks in the shale contain large amounts of natural gas. These cracks are forced open by high-pressure fracking fluid.

Millions of gallons of water are used to frack a single well—typically 5 to 15 million gallons, and as much as 400 million gallons for the largest wells. The used fracking fluid is toxic and requires special methods of disposal. Regulations for this disposal, however, vary from state to state. Some states permit local municipal water treatment facilities to accept fracking fluid. In states with more stringent regulations, additional wells are drilled deep below the local water table. The used fracking fluid is pumped to the bottom of these wells where it is pushed into the ground.

There is currently a "gold rush" for shale gas, with thousands of new wells being drilled every year (see accompanying figure). Notably, about 30 percent of the U.S. natural gas supply now comes from fracking operations. This is up from about zero percent over the past decade and is soon expected to reach 50 percent. Hundreds of thousands of jobs have since been created either directly or indirectly from this new industry. Furthermore, land owners are profiting from fees and royalties paid to them by companies who establish wells on their property. With

Contextual Chemistry Essays

Contextual Chemistry essays follow each chapter complete with discussion questions. These essays focus on chemistry-related issues that lend themselves to controversy. An essay can serve as a starting point for a student project or as a centerpiece for in-class student discussion groups. Expanded in this edition, new essay topics include Global Climate Change, Fracking for Shale Gas, and Genetically Modified Foods.

Checkpoints guide the way
to conceptual understanding

Ideas across the chapter are reinforced with checkpoints that help synthesize ideas.

CONCEPT CHECK

Is there gravity on the Moon?

CHECK YOUR ANSWER Yes, absolutely! The Moon exerts a downward gravitational pull on any body near its surface, as evidenced by the fact that astronauts were able to land and walk on the Moon. This NASA photograph shows an astronaut jumping. Without gravity, this jump would have been his last, because he would never have come down.

Boxed Concept Checks

Boxed Concept Checks pose a question followed by an immediate answer. This question-and-answer format reinforces the chemistry concept under discussion, solidifying the ideas presented before the student moves on to new concepts.

---

maximum number of electrons that the shell representing that period can hold.

Notice how electrons in the outermost shell begin to pair only after that shell is half filled. Carbon, for example, has four outer-shell electrons, none of which are paired. This differs from how an energy-level diagram is filled. For example, the second shell consists of the 2s and 2p orbitals, which, as shown in Figure 4.29, have different energy levels. Therefore, you might expect carbon's lower energy 2s orbital to fill with two paired electrons. In an advanced chemistry course,

READING CHECK

When do electrons in the outermost shell of an atom begin to pair with each other?

Reading Checks

In-the-margin Reading Checks flag key points, directing the student to a key sentence within each section of a chapter and remind students to reflect on them at the end of the section.

CALCULATION CORNER SCUBA DIVING AND HOT AIR BALLOONS

We can express Boyle's Law and Charles's Law mathematically as follows:

$$P_1V_1 = P_2V_2 \qquad \frac{V_1}{T_1} = \frac{V_2}{T_2}$$

Boyle's Law   Charles's Law

Here P_1, V_1, and T_1 represent an original pressure, volume, and temperature, respectively, while P_2, V_2, and T_2 represent a new pressure, volume, and temperature, respectively. Each of the preceding equations contains four variables. For such equations, if three of the variables are known we can calculate the fourth (assuming all temperatures are expressed in kelvins).

EXAMPLE

What would be the new volume of a 1.00 liter balloon if it were brought from sea level, where the air pressure is 1.00 atmosphere, to an altitude of 2500 meters, where air pressure is about 0.743 atmospheres? Assume there is no change in temperature. (As we explore in Chapter 16, the "atmosphere" is a common unit of pressure and is equal to the average atmospheric pressure at sea level.)

$$P_1 = 1.00\,\text{atm}$$
$$P_2 = 0.743\,\text{atm}$$
$$V_1 = 1.00\,\text{liter}$$
$$V_2 = ?$$

Use algebra to rearrange the equation for Boyle's Law to solve for V_2:

$$V_2 = P_1V_1/P_2$$
$$= (1.00\,\text{atm})(1.00\,\text{liter})/(0.743\,\text{atm})$$
$$= 1.35\,\text{liters}$$

YOUR TURN

1. A scuba diver swimming underwater in the ocean breathes compressed air at a pressure of 2 atmospheres.

If she holds her breath while returning to the surface, by how much might the volume of her lungs increase?

2. A 5.00-liter rubber balloon is submerged 5.00 meters under ocean water, where its new volume is measured to be 3.38 liters. Show that the pressure at this depth is 1.48 atmospheres.

3. A perfectly elastic 419-liter balloon is heated from 25°C (298 K) to 50°C (323 K). To what new volume does it expand?

4. A hot air balloon 401,000 liters in volume is warmed from 298 K to 398 K. As the air inside the balloon expands, it is unable to stretch the fabric, which is not very elastic. Instead, the expanded air escapes out of a hole placed at the top of the balloon. Show that 135,000 liters of air escapes.

5. Air has a density of 1.18 g/L. Show that the hot air balloon in the previous question is now lighter by 159 kilograms, which helps the balloon to rise.

The answers for Calculation Corners appear at the end of each chapter.

Calculation Corners

Calculation Corners in selected chapters provide practice in the quantitative-reasoning and basic math skills needed to perform chemical calculations. None of the calculations involves skills beyond fractions, percentages, or basic algebra.

Turn classroom concepts
into real-world applications

Features in the text stimulate classroom discussions, engaging students in a real-world understanding of chemistry.

EXPLAIN THIS

What is found between two adjacent molecules of a gas?

NEW! Explain This

Explain This questions activate prior knowledge, illustrate intriguing applications of concepts, and serve as the catalyst for lively classroom discussion. Answers appear in the instructor's manual.

CHEMICAL CONNECTIONS

How is a raindrop connected to a campfire?

NEW! Chemical Connections

In-the-margin Chemical Connections ask students to uncover the link between two seemingly unrelated materials or processes. Answers appear in the instructor's manual.

 FOR YOUR INFORMATION

Astronomers have recently discovered an expired star that has a solid core made of diamond. This star-sized diamond measures about 4000 kilometers wide, which amounts to about 10 billion trillion trillion carats. It has been named "Lucy," after The Beatles song "Lucy in the Sky with Diamonds." In about 7 billion years, our own star, the Sun, is also likely to crystallize into a huge diamond ball.

For Your Information

For Your Information paragraphs in the margins are included in each chapter and highlight interesting information relating to the adjacent chapter content.

MasteringChemistry®

Engaging Experiences

The Mastering platform is the most effective and widely used online homework, tutorial, and assessment system for the sciences. The Mastering system helps instructors maximize class time with easy-to-assign, customizable, and automatically graded assessments that motivate students to learn outside of class and arrive prepared for lecture.

Student Tutorials

Student Tutorials guide students through the toughest topics in chemistry with individualized coaching. These self-paced tutorials coach students with hints and feedback specific to their individual misconceptions.

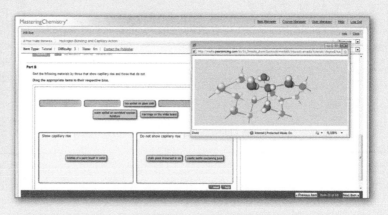

End-of-chapter Problems

End-of-chapter problems from the text are now easily assignable within MasteringChemistry to help students prepare for the types of questions that may appear on a test.

Reading Check Questions

In-the-margin Reading Check questions can be assigned in MasteringChemistry, allowing instructors to assign reading and test students on their comprehension of chapter content.

Easy to get started. Easy to use.

MasteringChemistry® provides a rich and flexible set of course materials to get you started quickly.

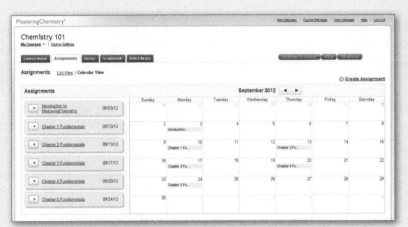

NEW! Calendar Features

The Course Home default page now features a Calendar View displaying upcoming assignments and due dates.

- Instructors can schedule assignments by dragging and dropping the assignment onto a date in the calendar. If the due date of an assignment needs to change, instructors can drag the assignment to the new due date and change the "available from and to dates" accordingly.
- The calendar view gives students a syllabus-style overview of due dates, making it easy to see all assignments due in a given month.

Gradebook

Every assignment is automatically graded. Shades of red highlight vulnerable students and challenging assignments.

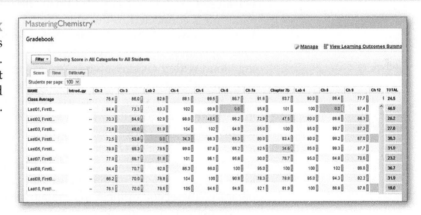

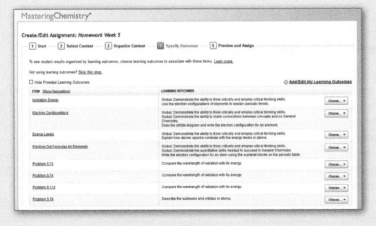

NEW! Learning Outcomes

Let Mastering do the work in tracking student performance against your learning outcomes:

- Add your own or use the publisher provided learning outcomes.
- View class performance against the specified learning outcomes.
- Export results to a spreadsheet that you can further customize and share with your chair, dean, administrator, or accreditation board.

Notably, the organization of content in this fifth edition is similar to that of the fourth edition. Thus, for instructors already using the fourth edition, changes to a course syllabus will be minimal. Under the hood, however, this fifth edition is a major upgrade, especially in terms of its readability, accuracy, and new pedagogical features. Equally as important, the book has been updated to reflect current events, such as the 2011 earthquake and tsunami, and recent advances in science and technology, such as the success of immunotherapies against cancer and the development of hydraulic fracturing for shale gas.

This latest edition sports a new and modern-looking layout. Integrated into the design are learning objectives appearing alongside each chapter section heading. Each learning objective begins with an active verb that specifies what students should be able to do after studying that section, such as "Calculate the energy released by a chemical reaction." These section-specific learning objectives are further integrated into the Mastering Chemistry online tutorial and assessment tool.

Appearing beneath each section heading is an "Explain This" question. These questions would be fairly difficult for students to answer without having read the chapter section. Some require that students recall earlier material. Others reveal interesting applications of chemistry concepts. In all cases, the Explain This question should serve as a launching point for classroom discussion. The answers to these questions appear in the *Instructor Manual*.

Also new to this edition, appearing in the margins of each chapter, are a set of questions that ask students to find the chemical connection between seemingly unrelated materials or processes. Such a "Chemical Connection" question is featured on the back cover of this textbook, where students are asked how a campfire is connected to the Sun. Students tend to struggle with these riddle-like questions, which probe into their understanding of atoms and molecules. These questions are best presented in class, where students will have fun thinking aloud with their classmates. You can expect to hear many creative answers. The author's answers are published in the *Instructor Manual*.

Changes have also been made to the end-of-chapter (EOC) material. Most importantly, each question was reviewed for quality and, as needed, either rewritten or replaced. All questions are now sorted by learning objectives, as shown in a grid appearing at the beginning of the EOC material. The hands-on chemistry activities are now called "Confirm the Chemistry" to highlight their important role in the learning cycle. The more challenging "Think and Explain" questions are now organized by section number to facilitate the assigning of homework.

By popular demand, all 17 chapters are now included in the printed edition. Furthermore, the "Contextual Chemistry" essays have been updated and new ones written so that one of these essays now appears at the end of each chapter.

Content Changes to the Fifth Edition

• New Contextual Chemistry Essays appear throughout the text. One at the end of Chapter 1 uses the topic of global climate change to highlight the ever-important role of science in society. This essay sets the stage for a more thorough discussion of global climate change now appearing in Chapter 16. Another new essay that appears after Chapter 5 discusses the issues involved in hydraulic fracturing and explains how this technology has boosted the United States in becoming the world's leading producer of natural gas. Also, the new essay on genetically modified foods appearing after Chapter 15 was developed out of material that appeared in Chapter 15 of the fourth edition.

- A discussion of atomic orbitals, energy level diagrams, and electron configurations has been included in Chapter 4 to help students understand the structure of the atom as well as atomic behavior. These concepts are then summarized in a revised discussion of Pauling's "argonian" shell model, which is now called the *noble gas shell model* to distinguish it from the traditional shell model used to describe principal quantum states. Placed at the end of Chapter 4 and not discussed in subsequent chapters, this noble gas shell model can easily be skipped, which is advisable for courses designed to prepare students for advanced chemistry. For the non-science-oriented student, however, this model provides valuable insight into the nature of the periodic table, electron-dot structures, and chemical bonding.

Other content changes include the following:

- The discussion of wastewater treatment was moved from Chapter 7 to Chapter 16.
- New material on ocean acidification was written for Chapter 10.
- The section on the greenhouse effect and global climate change was moved to Chapter 16, where it fits with the strong focus on atmospheric chemistry.
- Discussions on angiogenesis inhibitors and other monoclonal antibodies were added to Chapter 14, which also features a more accurate description of the reuptake inhibiting modes of action for amphetamines and cocaine.
- A description of neonicotinoids, now the most widely used insecticides, and their benefits and risks was added to Chapter 15.
- Notably, Chapters 16 and 17 are the most "fact-heavy" chapters. Much research went into updating and cross-referencing the accuracy of the data presented in these chapters. Furthermore, the content of Section 17.3 now focuses on the issues of the nuclear industry, which allowed for removal of the discussion of nuclear fusion that was redundant with the content of Chapter 5. Chapter 17 also features a new subsection on the history and current state of the aging North American power grid.

Changes to the Supplements

The **Conceptual Chemistry Alive! (CCAlive!)** video lecture series, featuring the author's lecture presentations, downloadable worksheets, and many other study resources, continues to be upgraded on a regular basis. A social media component, for example, now lets students work together in study groups that can be moderated by an instructor or a teaching assistant. Also, the author's class journal (blog), where he describes his current classroom activities, is also available to registered instructors. The video lecture series can also be found online in MasteringChemistry and at ConceptualChemistry.com.

The Conceptual Chemistry Laboratory Manual now sports a new lab on Charles' Law. Also, the Mastering Chemistry online homework system has been revised to match the significant changes that were created in the end-of-chapter material of the textbook.

Supplements Available with the Fifth Edition

For the Student

MasteringChemistry® with Pearson eText MasteringChemistry® from Pearson has been designed and refined with a single purpose in mind: to help educators create that moment of understanding with their students. The Mastering platform delivers engaging, dynamic learning opportunities—focused on your course objectives and responsive to each student's progress—that are proven to help students absorb course material and understand difficult concepts. By complementing your teaching with our engaging technology and content, you can be confident that your students will arrive at that moment—the moment of true understanding.

The Conceptual Chemistry Alive! (CCAlive!) video lecture series, featuring the author's lecture presentations is available in MasteringChemistry and at ConceptualChemistry.com along with many other important study resources such as downloadable worksheets and practice quizzes. Each video lecture averages only 7 minutes in length, but there are over 200 of them, spanning the table of contents of the textbook. These video lectures are best thought of as the "talking textbook," in which students get to see and hear the concepts of chemistry. CCAlive! complements the textbook as a means of delivering the content of chemistry. This, in turn, supports the instructor who is seeking to dedicate his or her classes to student-centered learning activities such as Process-Oriented Guided Inquiry Learning (POGIL).

Explorations in Conceptual Chemistry: A Student Activity Workbook (0-321-68172-X) was written by Jeffrey Paradis of California State University, Sacramento. This manual features hands-on activities that help students learn by doing chemistry in a discovery-based team environment. The *Student Activity Manual* is also available in the Pearson Custom Library.

The Laboratory Manual (0-321-80453-8) was written by John Suchocki and Donna Gibson, of Chabot College. The *Laboratory Manual* features experiments tightly correlated to the chapter content. Each lab consists of objectives, a list of materials needed, a discussion, the procedure, and report sheets.

For the Instructor

MasteringChemistry® with Pearson eText MasteringChemistry® from Pearson has been designed and refined with a single purpose in mind: to help educators create that moment of understanding with their students. The Mastering platform delivers engaging, dynamic learning opportunities—focused on your course objectives and responsive to each student's progress—that are proven to help students absorb course material and understand difficult concepts. By complementing your teaching with our engaging technology and content, you can be confident that your students will arrive at that moment—the moment of true understanding.

- By providing answer-specific feedback and coaching, the MasteringChemistry® tutorial system helps students figure out where they are going wrong when problem solving. By offering feedback specific to students' incorrect answers, MasteringChemistry® tutorials coach 92 percent of students to the correct answer.

- The program enables instructors to compare their class performance with the national average on specific questions or topics. At a glance, instructors can see class distribution of grades, time spent, most difficult problems, most difficult steps, and even most common answers.

- Pearson eText gives students access to the text whenever and wherever they can access the Internet. The eText pages look just like the printed text and include powerful interactive and customization functions. This product does not include the bound book.

Instructor Resource Materials (0-321-80450-3) This integrated collection of resources was designed to help you make efficient and effective use of your time. Resources feature art from the text, including figures and tables in JPEG format, as well as three prebuilt PowerPoint® presentations per chapter. The first presentation contains the images, figures, and tables embedded in the PowerPoint® slides, and the second presentation includes a complete lecture outline. The final presentation consists of approximately 15–25 Clicker questions per chapter. A TestGen® version of the Test Bank, which allows you to create and tailor exams to your particular needs, is also offered. All of these resources can be downloaded from the Instructor Resource Center found at www.pearsonhighered.com/chemistry.

The **TestGen® Computerized Test Bank** (0-321-80452-X) Prepared by John Suchocki and his wife, Tracy, the Test Bank contains more than 2100 multiple-choice questions from which to choose in creating your own tests and quizzes. These files are also available in Microsoft® Word format and can be downloaded from the Instructor Resource Center, found at www.pearsonhighered. com/chemistry.

Instructor Manual (0-321-80451-1) Written by John Suchocki, the *Instructor Manual* contains sample syllabi, teaching tips, suggested demonstrations, and answers to all end-of-chapter questions. It is an important resource for the instructor who is seeking to implement student-centered learning techniques such as "student-centered circles" and "minute quizzes."

Acknowledgments

For the creation of this fifth edition, I am most grateful to the many chemistry instructors who offered their time and energy to provide detailed and constructive reviews. This includes reviews of the fourth edition, which set the stage for the fifth edition, as well as reviews of the fifth edition manuscript as it progressed. For these efforts, I thank the following instructors:

Eric Ball, Metropolitan State College Denver
Nathan Bowling, University of Wisconsin–Stevens Point
Charles Carraher, Florida Atlantic University
Richard Delgado, Lindenwood University
Brian Fraser, Genessee Community College
Eric Goll, Brookdale Community College of Monmouth
Mike Maguire, Wayne State University

John Means, University of Rio Grande
Sarah Morse, Bridgewater State University
Gregory Oswald, North Dakota State University
Rill Reuter, Winona State University
Kenneth Traxler, Bimidji State University
Bob Widing, University of Illinois at Chicago

I am grateful to numerous individuals and indebted for their assistance in the development of *Conceptual Chemistry*. Standing at the head of this crowd is my uncle and mentor, Paul G. Hewitt. He planted the seed for this book in the early 1980s and has lovingly nurtured its growth ever since. To my parents, thank you for your continued love and support. To my wife, Tracy, I remain deeply thankful for your endless patience and for the love and time you give to me daily. Tracy's assistance in producing the manuscript and her persistent and creative efforts on the test bank were particularly helpful. To Ian, Evan, and Maitreya, who have grown up knowing only a dad who pores for hours over his computer, thank you for reminding me of the important things in life.

Special thanks to my inspirational high school science teachers, Linda Ford (chemistry) and Edward Soldo (biology) of Sycamore High School, Ohio. Their positive impact on me has been lifelong.

To the faculty and staff of the chemistry and physics departments of Saint Michael's College, I send a grand thank-you for your continued support and friendship. Special thanks are extended to Frank L. Lambert, to whom the fourth edition was dedicated, Professor Emeritus, Occidental College, for his much-appreciated assistance in the development of *Conceptual Chemistry*'s presentation of the second law of thermodynamics. I send a big *mahalo* to the crew that helped in the filming of CCAlive!, including Michael Reese, Peter Elias, Camden Barruga, Ed Nartatez, Kelly Sato, Sharon Hopwood, Patrick Garcia, Irwin Yamamoto, Stacy Thomas, Kai Dodge, and Maile Ventura. Also, I am grateful for the past support of the faculty and staff of Leeward Community College. Many thanks to Bradley Sieve of Northern Kentucky University for his assistance with the lecture PowerPoint® slides in both this edition and the previous edition. I send much thanks to John Singer of Jackson Community College for revising the Clicker questions, as well as Phil Reedy of Delta College of San Joaquin for his valuable feedback on the development of ConceptualChemistry. com and for coordinating the creation of the complete CCAlive! video lecture transcripts, which were created courtesy of Delta College.

To Jeff Paradis of CSU Sacramento, I am thankful for his efforts in creating the activities manual, which complements this textbook so well. For developing the *Conceptual Chemistry Laboratory Manual,* I am forever grateful to Donna Gibson of Chabot College. For past work on the *Conceptual Chemistry* test bank, I am deeply indebted to Bill Centobene of Cypress College as well as Dan Stasko of the University of Southern Maine. Thanks to Kelly Befus of Anoka-Ramsey Community College for checking the accuracy of the fifth edition test bank as well as the lab manual. For the "Wheel of Scientific Inquiry" shown in Figure 1.3, I thank William Harwood and his graduate students at Indiana University.

To the many talented and dedicated folks at Pearson, I send my deepest appreciation. Thanks to President Paul Corey for his longtime support of *Conceptual Chemistry* and to Adam Jaworski for being an inspiration and for organizing and overseeing a mighty team of editors to tend to the development and production of this fifth edition. To Jennifer Hart, thank you for being available as my main channel to the Pearson network—working with you has been a delight, and Pearson is fortunate to have you. To Coleen Morrison, thank you for nimbly receiving the baton when Jennifer was promoted. To Daniel Schiller, my developmental editor for this edition, thank you for bringing the *Conceptual Chemistry* manuscript to that next level of excellence. To Fran Falk, thank you for coordinating the supplements, and to Kristin Mayo, thank you for taking on the details of media production. To Kelly Morrison and her team at GEX Incorporated, thank you for your competence at piecing together the pages of this fifth edition—working with you has been a much-appreciated smooth sail. Special note of thanks to Marianne Miller for her eagle-eye copyediting and sense of humer [sic].

Continued thanks are due to my earlier editors from Benjamin Cummings: Ben Roberts, Jim Smith, Kate Brayton, Hilair Chism, and Irene Nunes. Tremendous thanks go to the reviewers listed here, who contributed immeasurably to the development of this and earlier editions of *Conceptual Chemistry*:

Adedoyin M. Adeyiga, Cheyney University of Pennsylvania
Pamela M. Aker, University of Pittsburgh
Edward Alexander, San Diego Mesa College
Sandra Allen, Indiana State University
Susan Bangasser, San Bernardino Valley College
Ronald Baumgarten, University of Illinois, Chicago
Stacey Bent, New York University
John Bonte, Clinton Community College
Emily Borda, Western Washington University
Richard Bretz, University of Toledo
Benjamin Bruckner, University of Maryland, Baltimore County
Kerry Bruns, Southwestern University
Patrick E. Buick, Florida Atlantic University
John Bullock, Central Washington University
Barbara Burke, California State Polytechnical University, Pomona
Robert Byrne, Illinois Valley Community College

Richard Cahill, De Anza College
David Camp, Eastern Oregon University
Charles Carraher, Florida Atlantic University
Jefferson Cavalieri, Dutchess Community College
William J. Centobene, Cypress College
Ana Ciereszko, Miami Dade Community College
Richard Clarke, Boston University
Natasha Cleveland, Frederick Community College
Cynthia Coleman, SUNY Potsdam
Virgil Cope, University of Michigan–Flint
Kathryn Craighead, University of Wisconsin–River Falls
Jerzy Croslowski, Florida State University
Jack Cummini, Metropolitan State College of Denver
William Deese, Louisiana Tech University
Sara M. Deyo, Indiana University Kokomo
Rodney A. Dixon, Towson University
Jerry A. Driscoll, University of Utah
Melvyn Dutton, California State University, Bakersfield
J. D. Edwards, University of Southwestern Louisiana

Karen Eichstadt, Ohio University
Karen Ericson, Indiana University–Purdue University, Fort Wayne
David Farrelly, Utah State University
Andy Frazer, University of Central Florida
Kenneth A. French, Blinn College
Ana Gaillat, Greenfield Community College
Patrick Garvey, Des Moines Area Community College
Shelley Gaudia, Lane Community College
Donna Gibson, Chabot College
Marcia L. Gillette, Indiana University Kokomo
Palmer Graves, Florida International University
Jan Gryko, Jacksonville State University
William Halpern, University of West Florida
Marie Hankins, University of Southern Indiana
Alton Hassell, Baylor University
Barbara Hillery, SUNY Old Westbury
Chu-Ngi Ho, East Tennessee State University
Angela Hoffman, University of Portland
John Hutchinson, Rice University
Mark Jackson, Florida Atlantic University
Kevin Johnson, Pacific University
Stanley Johnson, Orange Coast College
Margaret Kimble, Indiana University–Purdue University, Fort Wayne
Joe Kirsch, Butler University
Louis Kuo, Lewis and Clark College
Frank Lambert, Occidental College
Carol Lasko, Humboldt State University
Joseph Lechner, Mount Vernon Nazarene College
Robley Light, Florida State University
Maria Longas, Purdue University
David Lygre, Central Washington University
Art Maret, University of Central Florida
Vahe Marganian, Bridgewater State College
Jeremy Mason, Texas Tech University
Irene Matusz, Community College of Baltimore County–Essex
Robert Metzger, San Diego State University
Luis Muga, University of Florida

B. I. Naddy, Columbia State Community College
Donald R. Neu, St. Cloud State University
Larry Neubauer, University of Nevada, Las Vegas
Frazier Nyasulu, University of Washington
Frank Palocsay, James Madison University
Robert Pool, Spokane Community College
Daniel Predecki, Shippensburg University
Britt E. Price, Grand Rapids Community College
Brian Ramsey, Rollins College
Jeremy D. Ramsey, Lycoming College
Kathleen Richardson, University of Central Florida
Ronald Roth, George Mason University
Elizabeth Runquist, San Francisco State University
Kathryn M. Rust, Tennessee Technological University
Maureen Scharberg, San Jose State University
William M. Scott, Fort Hays State University
Francis Sheehan, John Jay College of Criminal Justice
Mee Shelley, Miami University
Anne Marie Sokol, Buffalo State College
Vincent Sollimo, Burlington County College
Ralph Steinhaus, Western Michigan University
Mike Stekoll, University of Alaska
Dennis Stevens, University of Nevada, Las Vegas
Anthony Tanner, Austin College
Joseph C. Tausta, SUNY Oneonta
Bill Timberlake, Los Angeles Harbor College
Margaret A. Tolbert, University of Colorado
Anthony Toste, Southwest Missouri State University
Carl Trindle, University of Virginia
Everett Turner, University of Massachusetts Amherst
Jason K. Vohs, Saint Vincent College
George Wahl, North Carolina State University
M. Rachel Wang, Spokane Community College
Karen Weichelman, University of Southwestern Louisiana
Bob Widing, University of Illinois, Chicago
Ted Wood, Pierce University
David L. Yates, Park University
Sheldon York, University of Denver

To the struggling student, thank you for your learning efforts—you are on the road to making this world a better place.

Much effort has gone into keeping this textbook error-free and accurate. However, some errors or inaccuracies may have escaped our notice. Forwarding such errors or inaccuracies to me would be greatly appreciated. Your questions, general comments, and criticisms are also welcome. I look forward to hearing from you.

JOHN SUCHOCKI
John@ConceptualChemistry.com

▲ Our home is a blue marble of a planet covered mostly with oceans. Land makes up only 30 percent of its surface. The atmosphere we breathe is quite thin compared to the size of our planet—about as thin as an apple skin is compared to an apple.

1

About Science

THE MAIN IDEA

Science is the study of nature's rules

Through science we have learned much about the natural world. For example, we have learned that matter is made of very small fundamental particles called *atoms*. These atoms can then join to form larger fundamental structures called *molecules*. This sort of knowledge has allowed us to create some amazing technologies—from agriculture to medicine to space travel.

Yet science is more than just a body of knowledge. It is also a *method* for exploring nature and discovering the order within it. **Science** is the product of observations, common sense, rational thinking, experimentation, and (sometimes) brilliant insights. It has been built up over many centuries and gathered from places all around the Earth. It is a huge gift to us today from the thinkers and experimenters of the past.

What is so special about science? Why is science such an effective tool for discovery and for solving problems? How is science different from technology? Why is it so important that each of us have an understanding of this eye-opening and creative human endeavor?

Chemistry

The Cool Rubber Band

Predict what happens to the temperature of a rubber band as it is stretched. Predict what happens to the temperature of a stretched rubber band as it relaxes.

PROCEDURE

1. Stretch a rubber band while holding it to your lower lip, which you will find is sensitive to small temperature changes.
2. Relax the stretched rubber band that is in contact with your lower lip.

ANALYZE AND CONCLUDE

1. Does the speed at which you stretch the rubber band make a difference?

2. You touch your hand to the forehead of someone with a fever. You feel that his or her forehead is hot. How does your hand feel to the person with the fever?
3. If the contracting rubber band causes your lip to cool down, what does your lip do to the contracting rubber band?
4. A hammer is hanging by a stretched rubber band. Hot air is then blown over the rubber band with a hair dryer. Is the hammer lifted upward or does it drop downward?
5. True or False: Experiments often raise more questions than they answer.

1.1 Science Is a Way of Understanding the Natural World

EXPLAIN THIS

What is the first step in doing scientific research?

We humans are very good at observing. We are also very good at explaining what we observe. What we recognize today as modern science, however, began not with our powers of observation, nor with our creative explanations. Rather, modern science began when people first became skeptical of their observations and explanations. They wondered whether their observations were accurate. They wondered whether their explanations were correct. To resolve their doubt, they turned to experimentation.

The greatly respected Greek philosopher Aristotle (384–322 B.C.) claimed that an object falls at a speed proportional to its weight. In other words, the heavier the object, the faster it falls. This idea was held to be true for nearly 2000 years, in part because of Aristotle's compelling authority. The Italian physicist Galileo (1564–1642) was doubtful and allegedly showed the falseness of Aristotle's claim with one experiment—demonstrating that heavy and light objects dropped from the Leaning Tower of Pisa fall at nearly equal speeds. You too can refute Aristotle's claim with a simple experiment, as shown in **Figures 1.1** and **1.2**.

As a practical matter, experiments are better at proving ideas wrong than right. For example, is it a truth that all crows are black? Upon seeing millions of black crows, we may become very confident that all crows are black. The moment we see our first white albino crow, however, this once reasonable idea has been proven false. But learning that our ideas are false is useful information. It can prompt us to double-check our thinking. We can then use our experience and creativity to come up with a more encompassing, alternative explanation.

LEARNING OBJECTIVE

Describe the nature of science and the scientific method.

 READINGCHECK

When did modern science begin?

▲ Figure 1.1
Place a half sheet of paper UNDER a heavy book. Lift these up and release them together. Which falls to the floor faster? Is it because the heavy book is pushing down on the light paper?

▲ Figure 1.2
Place the half sheet of paper on TOP of the heavy book. Lift these up and release. The results will surprise you. Which falls to the floor faster? Is it because the light paper is pushing down on the heavy book? Or might it be that, in the absence of wind resistance, all objects fall with the same acceleration?

FOR YOUR INFORMATION

The success of science has much to do with an *attitude* common to scientists. This attitude is one of inquiry and honest experimentation guided by a confidence that all natural phenomena can be explained.

This new explanation may not be perfect, but we can be confident that it is closer to the truth than our previous explanation was. The more experiments we conduct, and the more times we refine our explanations, the closer we get to understanding the actual workings of nature.

The Wheel of Scientific Inquiry

Performing experiments is just one of many activities that scientists use to reach their goal of better understanding nature. As shown in **Figure 1.3**, one of the first activities tends to be the asking of a broad question, such as "Where did the Moon come from?" "Can we efficiently create hydrogen from water using direct solar energy?" or "When did humans first arrive in North America?" All other activities are guided by this broad question. These activities will likely include learning about what is already known, making new observations, narrowing the focus of the research to something manageable, asking specific questions that can be answered by experiment, documenting expectations, performing experiments, confirming the results of experiments, reflecting about what the results might mean, and—perhaps most important—communicating with others.

The order in which these activities are performed is largely up to the scientist. No cookbooks. No algorithms of logic. Just equipment, a blank lab

▶ Figure 1.3
This diagram illustrates essential activities conducted by scientists. Commonly, the first activity is the asking of a broad question that defines the scope of the research. It is usually based upon the scientist's particular interests. The scientist can then move among all the various activities in unique paths and repeat activities as often as necessary.

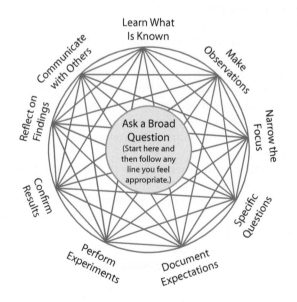

notebook, some self-discipline, a healthy dose of creative curiosity, and a desire to learn about nature for what it is—not for what we might wish it to be. This is the scientific spirit.

1.2 The Discovery of the Buckyball

EXPLAIN THIS

Why does falsifying information discredit a scientist, but not a lawyer?

The scientific process is aptly illustrated by the late 20th century research of chemists Harry Kroto of Florida State University and Rick Smalley and Bob Curl of Rice University in Texas (**Figure 1.4**). Their story began with Harry Kroto's interest in identifying the composition of interstellar dust, which is the dust found in the vast distances between stars.

As will be explored in Chapter 4, it is possible to identify materials in space by studying the light they emit or absorb. Within this light there are patterns that can be matched with known materials. Spectral patterns from our sun, for example, tell us that the Sun is made mostly of hydrogen and helium. The patterns of light coming from interstellar dust, however, are unlike the light patterns coming from any known material. The composition of interstellar dust, therefore, has been a great mystery.

Kroto understood that interstellar dust is created by stars, especially those producing carbon. This led him to the following broad question:

- **Broad Question** Can we reproduce star-like conditions here on Earth to create new carbon-based materials that have the spectral patterns of interstellar dust?

- **Document Expectations** Kroto was visiting his friend and colleague Bob Curl at Rice University. Curl introduced Kroto to Rick Smalley, whose research involved using pulses of laser light to vaporize various materials such as silicon. The laser energy was powerful enough to heat materials to

LEARNING OBJECTIVE

Provide an example of the scientific method in action.

 FOR YOUR INFORMATION

Findings are widely publicized among fellow scientists and are generally subjected to further testing. Sooner or later, mistakes (and deception) are found out; wishful thinking is exposed. This has the long-run effect of compelling honesty. There is little bluffing in a game in which all bets are called. In fields of study where right and wrong are not so easily established, the pressure to be honest is considerably less.

▲ **Figure 1.4**
Rick Smalley and Bob Curl (left) and Harry Kroto (right) together conducted research that led to the discovery of a new form of carbon. Their story illustrates how the scientific process helps us understand nature.

▶ **Figure 1.5**
Smalley's experimental equipment in which lasers vaporized the material to be tested. Connected to this chamber was an instrument called a *mass spectrometer*, which measured the mass of the vaporized molecules.

over 10,000 degrees, which is hotter than the surface of stars. Kroto realized that if Smalley could focus his laser light on carbon, it might produce homemade interstellar dust (**Figure 1.5**). But in order to use Smalley's laser, Kroto needed to describe his expectations to Smalley. Then later, in order to receive grant money, Kroto would need to articulate his expectations on his grant applications.

- **Make Observations**   After Smalley agreed to collaborate, the research team expected that long and massive molecules of many carbon atoms would be the major products. Indeed, such molecules were produced. What caught them by surprise, however, was that the most abundant product was a much smaller molecule consisting of only 60 carbon atoms.

- **Confirm Results**   If the results are real, they must be reproducible. Kroto and his colleagues naturally repeated the experiments to confirm they were generating a molecule consisting of 60 carbon atoms.

- **Narrow the Focus**   That so many of the C_{60} molecules were produced suggested that these molecules must be stable. (Unstable molecules tend to fall apart and are not seen so readily.) What was the identity of this stable molecule? Could it be an undiscovered molecule? Could it be a major component of interstellar dust? These questions prompted the research team to narrow its focus to a study of this particular molecule.

- **Reflect on Findings**   How could they deduce the identity of this molecule? They knew it consisted of 60 carbon atoms. Their next step was to use molecular models to build a reasonable structure. Kroto was familiar with the work of the inventor and architect Buckminster Fuller, who designed geodesic domes. Perhaps C_{60} looked like a geodesic dome. Late one evening, Smalley pieced together what looked like a polyhedron sphere, as shown in **Figure 1.6**.

- **Learn What Is Known**   Assuming Smalley's spherical structure was correct, the research team realized that they were on the verge of discovering a third form of the element carbon. Up to that time, elemental carbon was known to exist in only two forms—diamond and graphite.

- **Communicate with Others**   Kroto, Smalley, and Curl were quick to publish their results and conclusions. With such documentation they were able to lay claim as the discoverers of this new molecule, which they named buckminsterfullerene, in honor of the architect (**Figure 1.7**). But a bold claim requires strong proof. Soon research groups around the world were looking for ways to discount the results. One team suggested that the appearance of C_{60} was merely an artifact of the equipment. In such a case, it wouldn't matter how many times the experiment was repeated, because the procedure was

▲ **Figure 1.6**
The spherical cage of 60 carbon atoms has the structure of a soccer ball consisting of 20 hexagons and 12 pentagons.

◀ **Figure 1.7**
Kroto and his colleagues published their identification of buckminsterfullerene in the widely read journal *Nature*, which featured the molecule on its cover.

flawed. Measuring your weight on a broken scale is a good example of a flawed procedure—no matter how many times you step on the scale, the weight you measure will be wrong every time.

✓ R E A D I N G C H E C K

What is an example of a flawed procedure?

- **Perform Experiments** How could Kroto, Smalley, and Curl prove their structure was correct? A traditional way of figuring out the structure of a molecule is to zap it with a lot of energy, causing the molecule to break apart. By studying the fragments, scientists can figure out the original structure—much as an engineer might be able to figure out the structure of a collapsed bridge by looking at the remains. They used this procedure, and the results were consistent with their soccer ball structure, which they called the *buckyball*.

- **Ask Specific Questions** Assuming their structure was correct, the team should have been able to produce a visible amount of the material. This would allow them to make spectral measurements to further prove the structure and to see if it was related to interstellar dust. But if it was an unknown material, there would be no established procedures for creating larger, more workable quantities. Kroto, Smalley, and Curl spent the next several years trying to produce visible amounts of buckminsterfullerene. They were unsuccessful.

- **Communicate with Others** Theorists persisted by calculating what the spectrum of such a molecule might look like. Reports of these calculations caught the eyes of Don Huffman of the University of Arizona and Wolfgang Kratschmer of the Max Planck Institute in Germany, shown in **Figure 1.8**.

◀ **Figure 1.8**
Don Huffman, left, and Wolfgang Kratschmer, right, as physicists were interested in the identity of interstellar dust. Though they were not trained as professional chemists, they understood chemistry well enough to produce large quantities of the chemical buckminsterfullerene.

FOR YOUR INFORMATION
Because of the great potential for unseen error in any procedure, the results of a scientific experiment are considered valid only if they can be reproduced by other scientists working in similarly equipped laboratories.

These scientists had already used strong electric currents to ignite carbon at star-hot temperatures. In doing so, they created a product whose spectra nicely matched the calculated spectra of buckminsterfullerene. They announced their procedure at a science conference and then soon published a paper describing the isolation of purified buckminsterfullerene, which formed beautiful red crystals.

- **Reflect on Findings** The spectral data for buckminsterfullerene suggested that it was not a major component of interstellar dust. But here was a newly discovered molecule. More than that, it turned out that the buckyball was just one molecule of a whole new class of molecules we now call fullerenes.

Most notable are the fullerenes called nanotubes, which are very long rod-shaped molecules. Shown in **Figure 1.9**, nanotubes can be made into lightweight fibers that are many times stronger than any previously known material. Such fibers are useful in the manufacture of bulletproof clothing, reinforced concrete, and sports equipment. Because they conduct electricity, nanotubes also hold much promise in the field of electronics. They can be used for solar cells, electronic displays, energy storage, and even artificial muscles for robots. Nanotubes can also be used to remove carbon dioxide from the exhausts of power plants, for creating fresh water from ocean water, or potentially as a storage medium for hydrogen gas in hydrogen-powered vehicles. These are but a few of the many potential applications of fullerenes. As we explore in Chapter 3, this area in which we engineer materials by manipulating individual atoms or molecules is known as **nanotechnology.**

CONCEPT CHECK

Kroto, Smalley, and Curl won the Nobel Prize for

a. their identification of the composition of interstellar dust.

b. producing buckminsterfullerene.

c. understanding the implications of a chance discovery.

d. All of the above

CHECK YOUR ANSWER The answer is c: Kroto, Smalley, and Curl initially produced buckminsterfullerene, but only in very small quantities—an ultrasensitive mass spectrometer was required to detect this new material. Their most significant achievement was being open and curious enough to explore their chance discovery and then to understand its implications.

▶ Figure 1.9
Carbon nanotubes, first developed in the early 1990s, provide great strength when embedded within lightweight materials, such as plastics. Nanotubes also conduct heat and electricity, which makes them applicable to a new generation of electronics. As a downside, if inhaled, carbon nanotubes, much like asbestos, may cause certain cancers.

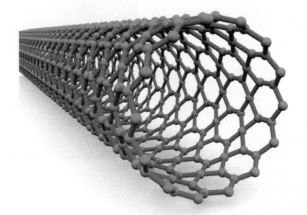

◀ **Figure 1.10**
Kroto, Curl, and Smalley opened up the world of nanotechnology with their discovery of buckminsterfullerene. For this work they were awarded the most prestigious science award, the Nobel Prize.

1.3 Technology Is Applied Science

EXPLAIN THIS

Who generates an idea, who develops it, and who uses it?

Science is concerned with gathering knowledge about the natural world. When we apply this knowledge for practical purposes, we have what we call **technology.** Kroto, Smalley, and Curl were doing science in their discovery of buckminsterfullerene, which led to the discovery of ultrastrong nanotube fibers. From scientific discoveries, an engineer can design new technologies. Nanotubes, for example, hold much promise for useful applications, as described earlier. New technologies, in turn, can be of assistance to scientists in conducting their research. The computer, for example, is technology scientists use on a daily basis. So technology arises from science, but technology also supports progress in science—the two are different from each other but very much related.

Technology is a double-edged sword that can be both helpful and harmful. We have the technology, for example, to extract fossil fuels from the ground and then to burn the fossil fuels for the production of energy. Energy production from fossil fuels has benefited our society in countless ways. On the flip side, the burning of fossil fuels endangers the environment. It is tempting to blame technology itself for problems such as pollution, resource depletion, and even overpopulation. These problems, however, are not the fault of technology any more than a shotgun wound is the fault of the shotgun. It is humans who use the technology, and humans who must decide how to use it responsibly.

Remarkably, we already possess the technology to solve many environmental problems. In this 21st century, we are seeing a switch from fossil fuels to more sustainable energy sources, such as photovoltaics, hydroelectric, wind, solar thermal electric generation, and biomass conversion. Whereas the paper on which the hard copy of this book is printed came from trees, paper will soon come from fast-growing weeds, and less of even these materials will be needed as e-books gain popularity. In some parts of the world, progress is being made to stem the rapid growth of our human population, which is an issue that aggravates almost every problem faced by humans today. We live on a finite planet,

LEARNING OBJECTIVE

Relate technology to the furthering of science and vice versa.

 FOR YOUR INFORMATION

Science is a way of knowing. Technology is a way of doing.

and Earth's population carrying capacity is being acknowledged. The greatest obstacle to solving today's problems lies more with social inertia than with a lack of technology. Technology is our tool. What we do with this tool is up to us. The promise of technology is a cleaner and healthier world. Wise applications of technology *can* lead to a better world.

Risk Assessment

The numerous benefits of technology are paired with risks. When the benefits of a technological innovation are seen to outweigh its risks, the technology is accepted and applied. X rays, for example, continue to be used for medical diagnosis despite their potential for causing cancer. But when risks are perceived to outweigh the benefits, the technology tends to be used very sparingly or not at all.

Risk can vary for different groups. Aspirin is useful for adults, but for young children it can cause a potentially lethal condition known as *Reye's syndrome*. Dumping raw sewage into the local river may pose little risk for a town located upstream, but for towns downstream the untreated sewage is a health hazard. Similarly, storing radioactive wastes underground may pose little risk for us today, but for future generations the risks of such storage are greater if there is leakage into groundwater. Technologies involve different risks and benefits for different people, raising questions that are often hotly debated. Which medications should be sold over the counter to the general public, and how should they be labeled? Should food be irradiated in order to put an end to food poisoning, which kills more than 5000 Americans each year? The risks to all members of society need to be considered when public policies are decided.

People seem to have difficulty accepting the impossibility of zero risk. You cannot go to the beach without risking skin cancer, no matter how much sunscreen you apply. You cannot avoid radioactivity, for it occurs naturally in the air you breathe and the foods you eat. Science, however, can help to determine relative risks. As the tools of science improve, the accuracy of risk assessment improves. Acceptance of risk, on the other hand, is a societal issue. Zero risk is not possible, and a society that accepts no risks receives no benefits.

READINGCHECK

What is the promise of technology?

FORYOUR
INFORMATION

Pharmaceuticals provide measurable benefits. A recently noted risk, however, is that many of these pharmaceuticals are now ending up in our drinking water, lakes, and streams. Although their concentrations are quite low, they are detectable, and these materials are being shown to have a measurable effect on species living in the water. Conclusion: old medicines should be thrown in the trash and never flushed down the drain. Better yet, bring your old prescription drugs to your local pharmacist who will have the means to dispose of these chemicals properly.

CONCEPTCHECK

Does technology come from science or does science come from technology?

CHECK YOUR ANSWER Both! The practical application of knowledge gained through science is technology. In this sense, the science comes first. The tools of technology, however, can be used by scientists to further our understandings of nature. A case in point would be the telescope, which is a tool that permits us to learn about the stars. Science and technology are different, but they complement each other.

1.4 We Are Still Learning about the Natural World

LEARNING OBJECTIVE

Distinguish between scientific facts, hypotheses, laws, and theories.

EXPLAIN THIS

Which is more important for a scientist: critical thinking or creative thinking?

Through science we have already learned so much about nature. That said, as any scientist would tell you, there is still much more that we have yet to learn. Science is a work in progress. The understandings we have one year will not necessarily be the understandings we have the next year.

For example, it is common to think of a fact as something unchanging and absolute. But in science, a **fact** is something agreed upon by competent observers as being true. Interestingly, what humans accept to be factual changes over time as we learn new ideas. It was once an accepted fact that the universe is unchanging and permanent. Today, we recognize the fact that the universe is expanding and evolving.

A hypothesis is a suggested *explanation* for an observable phenomenon. A hypothesis becomes a **scientific hypothesis** when, and only when, it can be tested through experiments. The more tests that the scientific hypothesis passes, the greater the confidence we have that the hypothesis is true. However, if the hypothesis fails even one test, then the hypothesis is taken to be false. A new, more encompassing hypothesis is needed.

READING CHECK

When does a hypothesis become a scientific hypothesis?

CONCEPT CHECK

Which statement is a *scientific* hypothesis?

1. The Moon is made of Swiss cheese.

2. Human consciousness arises from an essence that is undetectable.

CHECK YOUR ANSWER Both statements attempt to explain observed phenomena, so both are hypotheses. Only statement (1) is testable, however, and therefore only statement (1) is a *scientific* hypothesis.

It was once believed that mass is lost as wood burns, because, clearly, ashes weigh less than the wood that burned. The 18th-century French chemist Antoine Lavoisier (1743–1794), however, was skeptical (**Figure 1.11**). To him it appeared that burning wood lost mass because it was losing gases to the atmosphere. He went on to hypothesize that during any chemical change, such as burning, mass transforms from one substance to another, but it is always conserved. This means that the total mass before the reaction is equal to the total mass after the reaction. To test this hypothesis he conducted experiments during which burning took place in a sealed chamber. He found that the chamber and its contents weighed the same before and after the burning. The old hypothesis didn't fit this observation. Lavoisier's new hypothesis was a better alternative.

When a hypothesis has been tested and supported by experimental data over and over again and has not been contradicted, it may become formally stated as a **scientific law,** or *principle*. Lavoisier's conservation of mass hypothesis was repeatedly confirmed over many years, so it became known as the *law of mass conservation*, which we discuss in more detail in Chapter 9. But remember, the goal of science is to describe the rules of nature as accurately as possible. Ideally, a scientific law matches perfectly with the rules of nature. In practice, however, our so-called "laws" are merely our best approximations. If a scientific law is eventually found to be inaccurate through reproducible and verifiable evidence, then the law—in order to be closer to ideal—must be modified or changed. In the early 20th century, for example, it was found that mass actually does change ever so slightly during a chemical reaction. The law of mass conservation, therefore, is not perfectly accurate. The change in mass during a chemical reaction, however, is exceedingly small and not easily measured. The law of mass conservation, therefore, still holds some practical value, which is why it is still widely embraced.

Scientists use the word *theory* in a way that differs from its usage in everyday speech. In everyday speech, a theory is no different from a hypothesis—an explanation for an observable phenomenon. A **scientific theory,** on the other hand, is a well-tested explantion that unifies a broad range of observations within the natural world. Physicists, for instance, speak of the *theory of relativity* and use it to explain how we are held to Earth by gravity and how a strong gravitational field causes

▲ **Figure 1.11**

Antoine Lavoisier, shown here with his wife, Marie-Anne, who assisted him in many of his experiments, was a concerned citizen as well as a first-rate scientist. He established free schools, advocated the use of fire hydrants, and designed street lamps to make travel through urban neighborhoods safer at night.

▲ **Figure 1.12**
The tree Ayano hugs is made primarily from carbon dioxide and water, the very same chemicals Ayano releases through her breath. In return, the tree releases oxygen, which Ayano uses to sustain her life. We are one with our environment down to the level of atoms and molecules.

time to slow down. Biologists speak of the *theory of natural selection* and use it to explain both the unity and the diversity of life. Chemists speak of the *theory of the atom* and use it to explain how mass is seemingly conserved in a chemical reaction and how one material can transform into another.

Theories are a foundation of science, but, like facts, hypotheses, and laws, they are not fixed. Rather, they evolve as they go through stages of redefinition and refinement so as to mirror nature as accurately as possible. Since it was first proposed 200 years ago, for example, the theory of the atom has been repeatedly refined as new evidence about atomic behavior has been gathered. Those who know little about science may argue that scientific theories have little value because they are always being modified. Those who understand science, however, see it differently. Science is self-correcting, and its theories grow stronger as they are modified. A summary of what we mean by a scientific fact, hypothesis, law, and theory is provided in Table 1.1.

The domain of science is restricted to the observable natural world. While scientific methods can be used to debunk various claims, science has no way of verifying testimonies involving the supernatural. The term *supernatural* literally means "above nature." Science works within nature, not above it. Likewise, science is unable to answer such philosophical questions as "What is the purpose of life?" or such religious questions as "What is the nature of the human spirit?" Though these questions are valid and have great importance to us, they rely on subjective personal experience and do not lead to testable hypotheses.

Why Should We Learn Science?

Just as you can't enjoy a ball game, computer game, or party game until you know its rules, so it is with nature. Because science helps us learn the rules of nature, it also helps us appreciate nature. You may see beauty in a tree, but you'll see more beauty in that tree when you realize that it was created from substances found not in the ground but primarily in the air—specifically, the carbon dioxide and water put into the air by respiring organisms such as yourself (**Figure 1.12**). Learning science builds new perspectives and is not unlike climbing a mountain. Each step builds on the previous step, while the view grows evermore astounding (**Figure 1.13**).

There are also many practical reasons for which we should become familiar with science. Consider what the world was like before the advent of scientific thinking, around the time of Galileo in the late 1500s. Since that time, science has revealed much about the workings of the natural universe. With this knowledge arose society-changing technologies. For transportation, we've gone from horses to trains to cars to airplanes to spaceships. With the development of agriculture and advances in medicine, the human population has grown from about 500 million to almost 7 billion. The materials we've developed provide for everything from skyscrapers to computers to the medicines that protect our health and prolong our lives. For better or for worse, science has had and will continue to have a huge impact on society.

TABLE 1.1   A Summary of Scientific Terms

Scientific Fact—An agreed upon truth about the natural world.
Example: Wood burns.

Scientific Hypothesis—A testable explanation for an observable phenomenon.
Example: Wood transforms mostly to gaseous materials as it burns.

Scientific Law—An experimentally confirmed description of the natural world.
Example: The mass of the products of a reaction equals the mass of the reactants.

Scientific Theory—A well-tested explanation that unifies many observations.
Example: Matter is made of tiny particles called atoms that are not destroyed during a chemical transformation.

◀ **Figure 1.13**
"Wow, Great Uncle Paul! Are we like unhatched chicks, ready to poke through our shells to a new environment and new understanding of our place in the universe?"

We are now awakening to the fact that the resources of our planet are limited. Should we pay no attention and consume and pollute as we wish? Should we abandon the understandings of nature we've gained through science and merely hope for the best? Or should we use these understandings wisely and move toward living on this planet in a sustainable fashion? Are the decisions we now and will soon face better met with ignorance or with knowledge?

 **FOR YOUR INFORMATION**

Science helps us to gain an important perspective—that we humans are a part of nature, not apart from nature.

1.5   Chemistry Is Integral to Our Lives

EXPLAIN THIS

How has chemistry influenced our modern lifestyles?

When you wonder what the land, air, or ocean is made of, you are thinking about chemistry. When you wonder how a rain puddle dries up, how a car utilizes the energy of gasoline, or how your body acquires energy from the food you eat, you are again thinking about chemistry. By definition, **chemistry** is the study of matter and the transformations it can undergo. **Matter** is anything that has mass and occupies space. It is the stuff that makes up all material things—anything you can touch, taste, smell, see, or hear is matter. The scope of chemistry, therefore, is very broad.

Chemistry is often described as a central science, because it touches all the other sciences (**Figure 1.14**). It springs from the principles of physics, and it serves as the foundation for the most complex science of all—biology. Indeed, many of the great advances in the life sciences today, such as genetic engineering, are applications of some very exotic chemistry. Chemistry is also the foundation for earth science. It is also an important component of space science. Just as we learned about the origin of the Moon from the chemical analysis of moon rocks in the early 1970s, we are now learning about the history of Mars and other planets from the chemical information gathered by space probes.

Progress in science is made as scientists conduct research. Research is any activity whose purpose is the discovery of new knowledge. Many scientists focus on **basic research,** which leads us to a greater understanding of how

LEARNING OBJECTIVE

Describe chemistry as a central science with an emphasis in applied research.

 **READINGCHECK**

What is the definition of chemistry?

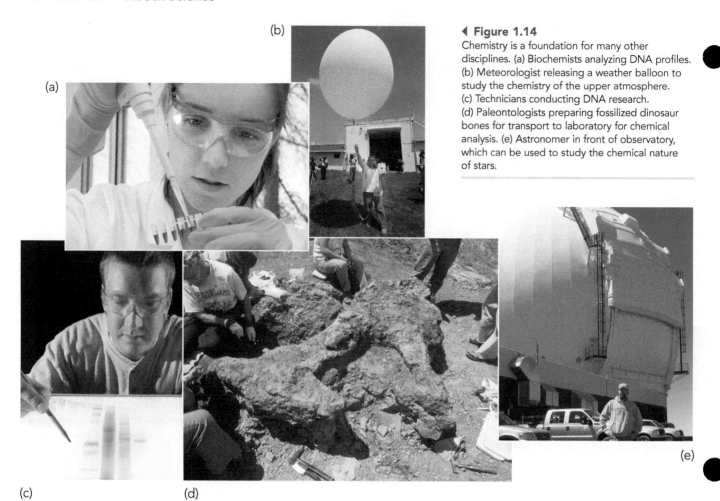

(a)

(b)

(c)

(d)

(e)

◀ **Figure 1.14**
Chemistry is a foundation for many other disciplines. (a) Biochemists analyzing DNA profiles. (b) Meteorologist releasing a weather balloon to study the chemistry of the upper atmosphere. (c) Technicians conducting DNA research. (d) Paleontologists preparing fossilized dinosaur bones for transport to laboratory for chemical analysis. (e) Astronomer in front of observatory, which can be used to study the chemical nature of stars.

FOR YOUR INFORMATION

Industries in the United States employ about 900,000 chemists.

▲ Figure 1.15
The Responsible Care symbol of the American Chemistry Council.

the natural world operates. Basic research in chemistry tells us how atoms combine to form molecules or how the structures of molecules can be determined. **Applied research** focuses on the development of useful applications of the knowledge laid down by basic research. The majority of chemists choose applied research as their major focus. Applied research in chemistry has provided us with medicine, food, water, shelter, and so many of the material goods that characterize modern life. Just a few of the myriad of examples are shown in Figure 1.14.

Over the course of the past century, we became very good at manipulating atoms and molecules to create materials to match our needs. At the same time, however, mistakes were made when it came to caring for the environment. Waste products were dumped into rivers, buried in the ground, or vented into the air without regard for possible long-term consequences. Many people believed that Earth was so large that its resources were virtually unlimited and that it could absorb wastes without being significantly harmed.

Most nations now recognize this as a dangerous attitude. As a result, government agencies, industries, and concerned citizens are involved in extensive efforts to take care of the environment. For example, members of the American Chemistry Council, who, as a group, produce 90 percent of the chemicals manufactured in the United States, have adopted a program called Responsible Care. Through this program, members of this organization have pledged to manufacture their products without causing environmental damage. The Responsible Care program emblem is shown in **Figure 1.15**. Through the wise use of chemistry, waste products can be minimized, recycled, engineered into useful products, or rendered environmentally safe. This is an area of research known as *Green Chemistry*, which we explore in detail in the Contextual Chemistry essay at the end of the next chapter.

Transparent matrix of processed silicon dioxide (Chapter 10)

Chemically disinfected drinking water (Chapter 7)

Caffeine solution (Chapter 14)

Thermoset polymer (Chapter 12)

Prescription medicines stored in refrigerator (Chapter 14)

Chlorofluorocarbon-free refrigerating fluids (Chapter 9)

Electrical energy from a fossil fuel or nuclear power plant (Chapter 17)

Metal alloy (Chapter 6)

Roasting carbohydrates, fats, proteins, and vitamins (Chapter 13)

Natural gas laced with odoriferous sulfur compounds (Chapter 12)

Fertilizer grown vegetables (Chapter 15)

▲ **Figure 1.16**
Most of the material items in any modern house are shaped by some human-devised chemical process.

Chemistry has influenced our lives in many important ways, and it will continue to do so in the future **Figure 1.16**. For this reason, it is in everyone's interest to become familiar with the basic concepts of chemistry.

CONCEPT CHECK

Chemists have learned how to produce aspirin using petroleum as a starting material. Is this an example of basic or applied research?

CHECK YOUR ANSWER This is an example of applied research, because the primary goal was to develop a useful commodity. However, the ability to produce aspirin from petroleum depended on an understanding of atoms and molecules developed from many years of basic research.

1.6 Scientists Measure Physical Quantities

EXPLAIN THIS

What is true about the numerator and denominator of any conversion factor?

LEARNING OBJECTIVE

Convert the units of a known physical quantity.

Science starts with observations. When possible, it is helpful to quantify observations by taking measurements. By quantifying observations, we are able to make objective comparisons, share accurate information with others, or look for trends that might reveal some inner workings of nature.

Scientists measure *physical quantities*. Some examples of physical quantities you will be learning about and using in this book are length, time, mass,

Any measurement of a physical quantity must always include two things. What are they?

▲ Figure 1.17
The metric system is finally making some headway in the United States, where various commercial goods, such as Evan's favorite soda, are now sold in metric quantities.

weight, volume, energy, temperature, heat, and density. Any measurement of a physical quantity must always include a number followed by a unit that tells us not only what was measured but also the scale of the measurement. It would be meaningless, for example, to say that your dog weighs 40, because without a specific unit, no one would know what that meant: 40 ounces, 40 pounds, 40 kilograms? A dog that weighed 40 kg would be more than 35 times heavier than one that weighed 40 oz. Units such as ounces, pounds, and kilograms, or feet, yards, and kilometers are all units that allow us to make meaningful comparisons when we measure physical quantities, and they must be included to complete the description.

There are two major unit systems used in the world today. One is the United States Customary System (USCS, formerly called the British System of Units), used in the United States, primarily for nonscientific purposes.* The other is the Système International (SI), which is used in most other nations. This system is also known as the International System of Units or as the metric system. The orderliness of this system makes it useful for scientific work, and it is used by scientists all over the world, including those in the United States. (And the International System is beginning to be used for nonscientific work in the United States, as **Figure 1.17** shows.) This book uses the SI units given in Table 1.2. On occasion, USCS units are also used to help you make comparisons. One major advantage of the metric system is that it uses a decimal system, which means all units are related to the next smaller or larger units by a factor of 10. Some of the more commonly used prefixes, along with their decimal equivalents, are shown in Table 1.3. From this table, you can see that 1 kilometer is equal to 1000 meters, where the prefix *kilo-* indicates 1000. Likewise, 1 millimeter is equal to 0.001 meter, where the prefix *milli-* indicates 1/1000. You need not memorize this table, but you will find it a useful reference when you come across these prefixes in your course of study.

TABLE 1.2   Metric Units for Physical Quantities and Their USCS Equivalents

PHYSICAL QUANTITY	METRIC UNIT	ABBREVIATION	USCS EQUIVALENT
length	kilometer	km	1 km = 0.621 miles (mi)
	meter	m	1 m = 3.285 feet (ft)
	centimeter	cm	1 cm = 0.3937 inches (in.)
			1 in. = 2.54 cm
	millimeter	mm	none commonly used
time	second	s	second also used in USCS
mass	kilogram	kg	1 kg = 2.205 pounds (lb)
	gram	g	1 g = 0.03528 ounces (oz)
			1 oz = 28.345 g
	milligram	mg	none commonly used
volume	liter	L	1 L = 1.057 quarts (qt)
	milliliter	mL	1 mL = 0.0339 fl oz
	cubic centimeter	cm^3	1 cm^3 = 0.0339 fl oz
energy	kilojoule	kJ	1 kJ = 0.239 kilocalories (kcal)
	joule	J	1 J = 0.239 calories (cal)
			1 cal = 4.184 J
temperature	degree Celsius	°C	(°C × 1.8) + 32 = degrees Fahrenheit, °F
	kelvin	K	°C + 273 = K

*Two other countries that continue to use the USCS are Liberia and Myanmar.

TABLE 1.3   Metric Prefixes

PREFIX	SYMBOL	DECIMAL EQUIVALENT	EXPONENTIAL FORM	EXAMPLE
tera-	T	1,000,000,000,000.	10^{12}	1 **tera**meter (Tm) = 1 trillion meters
giga-	G	1,000,000,000.	10^{9}	1 **giga**meter (Gm) = 1 billion meters
mega-	M	1,000,000.	10^{6}	1 **mega**meter (Mm) = 1 million meters
kilo-	k	1000.	10^{3}	1 **kilo**meter (km) = 1 thousand meters
hecto-	h	100.	10^{2}	1 **hecto**meter (hm) = 1 hundred meters
deka-	da	10.	10^{1}	1 **deka**meter (dam) = ten meters
no prefix	–	1.	10^{0}	1 meter (m) = 1 meter
deci-	d	0.1	10^{-1}	1 **deci**meter (dm) = 1 tenth of a meter
centi-	c	0.01	10^{-2}	1 **centi**meter (cm) = 1 hundredth of a meter
milli-	m	0.001	10^{-3}	1 **milli**meter (mm) = 1 thousandth of a meter
micro-	μ	0.000 001	10^{-6}	1 **micro**meter (μm) = 1 millionth of a meter
nano-	n	0.000 000 001	10^{-9}	1 **nano**meter (nm) = 1 billionth of a meter
pico-	p	0.000 000 000 001	10^{-12}	1 **pico**meter (pm) = 1 trillionth of a meter

CALCULATION CORNER UNIT CONVERSION

Welcome to Calculation Corner! *Conceptual Chemistry* focuses on visual models and qualitative understandings. As with any other science, however, chemistry has its quantitative aspects. In fact, it is only by the interpretation of quantitative data obtained through laboratory experiments that chemical concepts can be reliably confirmed. It is only natural that there are times when your conceptual understanding of chemistry can be nicely reinforced by some simple, straightforward calculations.

Often in chemistry, and especially in a laboratory setting, it is necessary to convert from one unit to another. To do so, you need only multiply the given quantity by the appropriate conversion factor. All conversion factors can be written as ratios in which the numerator and denominator represent the equivalent quantity expressed in different units. Because any quantity divided by itself is equal to 1, all conversion factors are equal to 1. For example, the following two conversion factors are both derived from the relationship 100 centimeters = 1 meter:

$$\frac{100 \text{ centimeters}}{1 \text{ meter}} = 1 \qquad \frac{1 \text{ meter}}{100 \text{ centimeters}} = 1$$

Because all conversion factors are equal to 1, multiplying a quantity by a conversion factor does not change the value of the quantity. What does change are the units. Suppose you measured an item to be 60 centimeters in length. You can convert this measurement to meters by multiplying it by the conversion factor that allows you to cancel centimeters.

EXAMPLE

Convert 60 centimeters to meters.

ANSWER

$$(60 \text{ centimeters}) \frac{(1 \text{ meter})}{(100 \text{ centimeters})} = 0.6 \text{ meter}$$

$$\uparrow \qquad\qquad \uparrow \qquad\qquad \uparrow$$

quantity   conversion   quantity
in centimeters   factor   in meters

You can create two conversion factors for every equality. For example, from Table 1.2 we see that 1 kilometer equals 0.621 miles. To create these two conversion factors, show each unit as a numerator in one conversion factor but as a denominator in the other conversion factor:

1 km/0.621 mi   0.621 mi/1 km

Multiply the quantity provided to you by the conversion factor of choice, which is the one that shows the original

unit in the denominator. This way, the original unit will be canceled, leaving you with the desired new unit. For example,

$$(26.2 \text{ mi}) (1 \text{ km}/0.621 \text{ mi}) = 42.2 \text{ km}$$

quantity in miles   conversion factor   quantity in
                                        kilometers

Always be careful to write down your units. They are your ultimate guide, telling you what numbers go where and whether you are setting up the equation properly. Remember, you should set up a conversion factor so that the desired unit is always in the numerator and the unit to be cancelled is always in the denominator.

Conversion factors can be combined to allow multiple conversions in a single equation.

EXAMPLE

How many pounds are there in 48 ounces? Use data from Tables 1.2 and 1.3.

ANSWER

Start by writing down the given quantity, which in this case is 48 ounces. Multiply by a series of conversion factors to transform this quantity into the desired unit—from ounces to grams, from grams to kilograms, and then from kilograms to pounds:

$$(48 \text{ oz})\left(\frac{28.345 \text{ g}}{1 \text{ oz}}\right)\left(\frac{1 \text{ kg}}{1000 \text{ g}}\right)\left(\frac{2.205 \text{ lbs}}{1 \text{ kg}}\right) = 3 \text{ lbs}$$

Of course, the above conversion could be done in a single step knowing that 16 ounces equals 1 pound. A string of conversion factors in a single equation, however, becomes quite useful when the direct conversion factor is unknown.

YOUR TURN

Multiply each physical quantity by the appropriate conversion factor to find its numerical value in the new unit indicated. You will need paper, a pencil, a calculator, and Tables 1.2 and 1.3.

a. 7320 grams to kilograms

b. 235 kilograms to pounds

c. 4585 milliliters to quarts

d. 100 calories to kilocalories

e. 100 calories to joules

The answers for Calculation Corners appear at the end of each chapter.

Chapter 1 Review

LEARNING OBJECTIVES

Describe the nature of science and the scientific method. (1.1)	→	*Questions 1–3, 19, 20, 27–31, 49*
Provide an example of the scientific method in action. (1.2)	→	*Questions 4–6, 32–35*
Relate technology to the furthering of science and vice versa. (1.3)	→	*Questions 7–9, 21, 25, 26, 36-39, 50*
Distinguish between scientific facts, hypotheses, laws, and theories. (1.4)	→	*Questions 10–12, 24, 40–44*
Describe chemistry as a central science with an emphasis in applied research. (1.5)	→	*Questions 13–15, 45, 46*
Convert the units of a known physical quantity. (1.6)	→	*Questions 16–18, 22, 23, 47, 48*

SUMMARY OF TERMS (KNOWLEDGE)

Applied research A type of research that focuses on developing applications of knowledge gained through basic research.

Basic research A type of research that leads us to a greater understanding of how the natural world operates.

Chemistry The study of matter and the transformations it can undergo.

Fact Something agreed upon by competent observers as being true.

Matter Anything that has mass and occupies space.

Nanotechnology An area where we engineer materials by manipulating individual atoms or molecules.

Science A body of knowledge built from observations, common sense, rational thinking, experimentation, and (sometimes) brilliant insights.

Scientific hypothesis A testable explanation for an observable phenomenon.

Scientific law An experimentally confirmed description of the natural world. Also known as a *principle*.

Scientific theory A well-tested explanation that unifies a broad range of observations within the natural world.

READING CHECK QUESTIONS (COMPREHENSION)

1.1 Science Is a Way of Understanding the Natural World

1. What would have to be done to refute Aristotle's hypothesis that heavier objects fall faster?

2. Are experiments better at proving ideas right or proving them wrong?

3. According to Figure 1.3, what is usually the first step to conducting scientific research?

1.2 The Discovery of the Buckyball

4. What did Kroto, Smalley, and Curl use to heat carbon to star-hot temperatures?

5. How many carbon atoms are there in a buckyball?

6. Who were the first to isolate large quantities of purified buckminsterfullerene?

1.3 Technology Is Applied Science

7. What is technology?

8. What issue aggravates almost every problem faced by humans today?

9. Are medical X rays used because they carry zero risks?

1.4 We Are Still Learning about the Natural World

10. Is it possible for a fact to change?

11. What do we call a suggested explanation for an observable phenomenon?

12. Which works to unify a broad range of observations: a scientific hypothesis or a scientific theory?

1.5 Chemistry Is Integral to Our Lives

13. Why is chemistry often called the central science?

14. What is the difference between basic research and applied research?

15. What pledge has been made by members of the American Chemistry Council through the Responsible Care program?

1.6 Scientists Measure Physical Quantities

16. What are the two major systems of measurement used in the world today?

17. Why are prefixes used in the metric system?

18. A milligram is equal to how many grams?

CONFIRM THE CHEMISTRY (HANDS-ON APPLICATION)

19. Fill a glass with water. Place a card on top of the glass. Holding the card in place, turn the glass upside down. While holding the upside-down glass, predict what will happen when you let go of the card. What happens when you tilt the glass sideways?

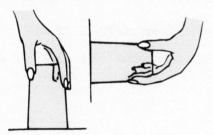

20. Poke a small hole in the bottom of an aluminum can. Hold your finger over the hole and fill the can with water. Remove your finger from the hole and, of course, water comes out in a downward stream. Now predict whether water will still come out of the can through the hole while the can falls to the ground. Try it and see. To avoid a mess, drop the can into a bucket.

THINK AND SOLVE (MATHEMATICAL APPLICATION)

21. We can define a "risk/benefit" ratio as the amount of risk taken divided by the amount of potential benefit:

$$\text{risk benefit ratio} = \frac{\text{risk}}{\text{benefit}}$$

If the risks are equal to the benefits, then the risk/benefit ratio equals one. If an activity offered you a risk/benefit ratio of 100, would you take it? How about if the risk/benefit ratio was 0.01? Is the risk benefit ratio of buying a lottery ticket large or small?

22. Using conversion factors, calculate the age of someone who is exactly 25 years old in units of months. How many days is this? (Assume 1 year = 365 days.)

23. Using conversion factors, show that there are about 63 billion seconds in 2000 years. (Assume 1 year = 365 days.)

THINK AND COMPARE (ANALYSIS)

24. Rank the following in order of believability:

a. Scientific theory

b. Hypothesis

c. Scientific hypothesis

For questions 25–26, rank the activities in order of lowest (least risky) to highest (most risky) risk/benefit ratio. Discuss your rankings with others.

25. A teenager travels 500 miles in

a. a commercial airline.

b. his or her own car while talking on a cell phone.

c. his or her own car while not talking on a cell phone.

26. A sick person with a prescription from his or her medical doctor purchases

a. brand name prescription medicine.

b. generic prescription medicine.

c. cheaper nonprescription herbal remedies.

THINK AND EXPLAIN (SYNTHESIS)

1.1 Science Is a Way of Understanding the Natural World

27. What sort of explanations are given by science?

28. How might you come to the conclusion that the Earth revolves around the Sun?

29. Of the scientific activities listed in Figure 1.3, which do you think would be the top two activities undertaken by older, well-seasoned scientists? How about a younger, less-seasoned scientist?

30. Of the scientific activities listed in Figure 1.3, which is likely the most time consuming?

31. During which of the scientific activities listed in Figure 1.3 does the scientist come up with a hypothesis?

1.2 The Discovery of the Buckyball

32. Why is reproducibility such a vital component of science?

33. Some politicians take pride in maintaining a particular point of view. They think that changing their point of view would be seen as a sign of weakness. How is changing one's view considered in science?

34. How might the demand for reproducibility in science have the long-run effect of compelling honesty?

35. Why weren't Don Huffman and Wolfgang Kratschmer included in the Nobel Prize for the discovery of buckminsterfullerene?

1.3 Technology Is Applied Science

36. What technologies tend to be hotly debated? Name some examples.

37. In the 1950s both the United States and the Soviet Union conducted numerous above-ground tests of nuclear bombs. What were some of the risks and benefits of these nuclear tests? Comment on the fact that the full risks of a technology are not always immediately apparent.

38. A scientific paper published in the late 1990s suggested that certain vaccines were responsible for increased rates of autism. Peer review found that the author falsified evidence. The paper was retracted and the author lost his medical license. To this day, however, many parents refuse to vaccinate their children due to this discounted research. As a consequence, the rates of preventable diseases, such as whooping cough, have risen. Why is it so difficult for many of us to distinguish between perceived and actual risk?

39. Why are the benefits of vaccination difficult to perceive?

1.4 We Are Still Learning about the Natural World

40. Which of the following are scientific hypotheses?
   a. Stars are made of the lost teeth of children.
   b. Albert Einstein was the greatest scientist ever.
   c. The planet Mars is coated with cotton candy.
   d. Tides are caused by the Moon.
   e. You were Abraham Lincoln in a past life.
   f. A human remains self-aware while sleeping.

41. What kinds of questions is science unable to answer?

42. In response to the question "When a plant grows, where does the material come from?" the ancient Greek philosopher Aristotle (384–322 B.C.) hypothesized that all material came from the soil. What experiment might be performed to test this hypothesis?

43. What happens to a scientific theory that cannot be adequately modified in the face of contradictory evidence?

44. Chemically speaking, how is your breath connected to paper?

1.5 Chemistry Is Integral to Our Lives

45. Of physics, chemistry, and biology, which is the most fundamental science and why?

46. Why might biology be considered a more complex science than chemistry?

1.6 Scientists Measure Physical Quantities

47. A major advantage of the metric system is that it uses the easy to work with decimal system. What is the major advantage of using the United States Customary System?

48. Examine Table 1.3 carefully. What is the relationship between the number of zeros in the decimal equivalent and its exponential form?

THINK AND DISCUSS (EVALUATION)

49. There are nine activities listed over the circumference of the wheel of inquiry in Figure 1.3. Which is most commonly employed? Explain your reasoning with others.

50. Medicines, such as pain relievers and antidepressants, are being found in the drinking water supplies of many municipalities. How did these medicines get there? Does it matter that they are there? Should something be done about it? If so, what?

READINESS ASSURANCE TEST (RAT)

If you have a good handle on this chapter, then you should be able to score at least 7 out of 10 on this RAT. Check your answers online at www.ConceptualChemistry.com. If you score less than 7, you need to study further before moving on.

Choose the BEST answer to the following.

1. Scientific thinking began when
   a. people began to communicate their explanations of the natural world to others.
   b. ancient Greek thinkers began pondering philosophical questions.

c. mysterious forces were proposed to explain the creation of Earth.

d. people became skeptical of their explanations and turned to experimentation.

2. Kroto, Smalley, and Curl discovered

a. that diamond can be transformed into graphite.

b. a third form of the element carbon.

c. that star-hot temperatures can be produced in the laboratory.

d. the mathematical structure of the soccer ball.

e. Two of the above

3. A flawed procedure is

a. reproducible.

b. best revealed by repeating the same experiment several times.

c. a procedure by which you don't get the results you were hoping for.

d. usually the result of insufficient planning.

e. All of the above

4. The first people to produce and isolate large quantities of purified buckminsterfullerene were professional

a. mathematicians.

b. physicists.

c. chemists.

d. biologists.

e. None of the above

5. Technology

a. arises from the application of science.

b. grows parallel to science.

c. poses both risks and benefits.

d. is not the same thing as science.

e. All of the above

6. A scientific hypothesis is a

a. restatement of a natural phenomenon.

b. well-tested theory that has been shown to be valid.

c. testable assumption.

d. prediction of what will happen in a certain situation.

e. test designed to limit possible conclusions.

7. Which of the following statements about science is false?

a. Science deals with testable hypotheses.

b. An experiment can be used to prove that a hypothesis is correct.

c. Science deals with observations and experimentation.

d. Scientists understand that natural phenomenon have natural explanations.

e. Experiments do not always go as planned.

8. A light bulb stops working in your kitchen. You suspect that either the light bulb is burned out or the circuit breaker is tripped. Your suspicion is an example of a scientific

a. fact.

b. hypothesis.

c. law.

d. theory.

e. None of the above

9. How might the demand for reproducibility in science have the long-run effect of compelling honesty?

a. Any false claims are eventually uncovered. Scientists, therefore, stand to gain most from reporting their results truthfully.

b. A scientist who has knowingly falsified any bit of evidence runs the risk of losing credibility for all his or her life's work.

c. Science is about discovering the rules of nature. A scientist who creates his or her own rules through dishonesty, which may not be reproducible in other laboratories, is not truly doing science.

d. All of the above are true.

10. How many conversion factors can be made out of a single equality?

a. None

b. One

c. Two

d. More than two

ANSWERS TO CALCULATION CORNERS (UNIT CONVERSION)

a. 7.32 kg
b. 518 lb
c. 4.846 qt
d. 0.1 kcal
e. 400 J

Perhaps you are wondering about how many digits to include in your answers. Were you perplexed, for example, that the answer to (e) is 400 J and not 418.4 J? There are specific procedures to follow in figuring which digits from your calculator to write down. The digits you are supposed to write down are called *significant figures*. Because there are not many calculations in the chapter portions of *Conceptual Chemistry*, however, a full discussion of significant figures is left to Appendix B. It is there for those of you looking for a little more quantitative depth, which is certainly often needed when performing experiments in the laboratory.

Contextual Chemistry

Global Climate Change

There are numerous issues relating to science and society. "Contextual Chemistry" is a special feature that highlights these issues and prods you to consider implications. These features can also serve as a centerpiece for opinionated yet respectful discussions with your classmates and others.

Certain atmospheric gases trap solar heat through a process called the *greenhouse effect*, which we study in more detail in Chapter 16. These "greenhouse gases" act like blankets to keep our planet warm—if not for these gases our planet would be a chilly –18°C. The most significant greenhouse gas is water vapor. Because of the water cycle—the constant movement of water between the oceans, air, and land—the amount of water vapor worldwide remains fairly constant. Changes occur only in response to changes in the average global temperature—greater warmth means more water vapor, while cooler temperatures mean less.

Next in significance is carbon dioxide, CO_2. Direct and indirect measurements show a steady increase in atmospheric CO_2 levels since humans began large scale burning of carbon-based fuels starting with the industrial revolution in the early 1800s. Atmospheric CO_2, however, has not been rising as fast as one might expect, given the amount of CO_2 emitted by human activities. Thus, scientists estimate that about half of the CO_2 we produce is absorbed by the oceans as well as by vegetation, which uses carbon dioxide in photosynthesis.

Because CO_2 is a potent greenhouse gas, one would expect that increasing levels in the atmosphere would result in an increase in the average global temperature. This, in turn, would alter the many climate systems around our planet. Some

areas, for example, would become wetter, while others might experience a greater number of droughts. This alone would introduce significant challenges to communities. Most feared, however, is "runaway" climate change. In such a scenario, changes in the global climate would stimulate an acceleration of further changes in what is called a "positive feedback loop." For example, as temperatures get warmer, more water vapor enters the atmosphere. Because water vapor is a greenhouse gas, the global temperatures would rise further leading to even more water vapor going into the atmosphere and hence even warmer temperatures, and so on. Under such a scenario, land-locked ice caps would be expected to melt. This would raise the sea level by many meters inundating all coastlines and the billions of people who live there—a disaster of epic proportions.

These are the fears. But are these fears founded upon reliable evidence? How certain are scientists that human

CO_2 output is inducing increases in global temperatures? And are modern temperatures actually increasing? Are the chances of runaway climate change significant enough to warrant a reworking of our energy infrastructure? These are important and legitimate questions now being asked by the general public, which includes corporate executives as well as our political leaders.

There is often, however, a disconnect between what scientists are willing to say and what the general public wants to hear. The scientist may be asked for a definitive statement, such as "Human-induced global climate change is a problem." But the scientist knows he or she can do no better than to speak in terms estimated probability and that even facts are subject to interpretation.

Consider how physical data reveal a direct relationship between levels of atmospheric CO_2 and global temperatures over the past 400,000 years (see the accompanying graph). For some, this settles the direct relationship

▲ The potential effects of global warming are uncertain. Many different scenarios are possible.

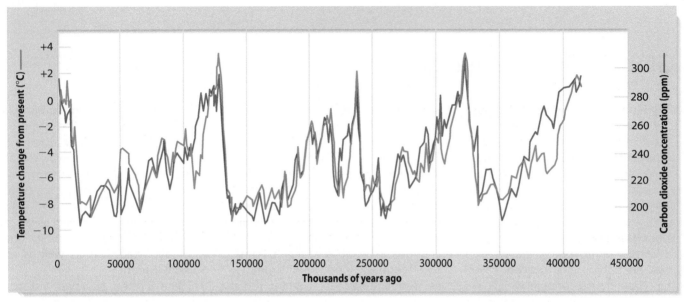

▲ Levels of atmospheric carbon dioxide and global temperatures appear to be closely related to each other.

between atmospheric CO_2 and global temperatures as a fact. Others will look deeper to note that past temperatures reliably go up some 600 years *before* CO_2 levels go up. To them it may become a fact that increasing CO_2 does *not* cause global warming. But CO_2 doesn't have to be the cause. Rather, once warming starts, the oceans begin to release more CO_2, which is a potent greenhouse gas, thus creating a significant warming positive feedback loop. Then again, it could be emphasized that increased water vapor—not the CO_2—acts as the most significant feedback propagator. However, more water vapor in the atmosphere also leads to greater cloud cover, which has a cooling effect. And so the investigation would continue from one point/counterpoint to the next. To the scientist, the world is anything but black and white. They understand the value of slow and careful deliberations as well as the need for evidence that is quantifiable, not anecdotal. Further, they understand the value of having their conclusions subject to the scrutiny of peer review.

For people seeking firm black or white conclusions, however, consider how easy it would be to stop and focus upon a single interpretation of data that best aligns with

one's personal worldview. For example, what conclusions might an oil executive be willing to accept compared to the president of a small island nation whose highest elevation is 10 meters above sea level? Objectivity is difficult to achieve, even for scientists. But what separates the scientist is his or her objectivity-driven scientific method.

Through the study of chemistry you will become equipped to understand many aspects of climate science. Water, for example, has an amazing ability to absorb heat because of the way in which atoms within each water molecule are bonded (Chapters 6 and 8). Also, carbon dioxide is more than just a greenhouse gas—it also reacts with water to form carbonic acid, which lowers the pH of our alkaline oceans (Chapter 10). Fossil fuels are valuable to us because they are so energy rich (Chapter 12). Alternative energy sources, however, offer many advantages (Chapter 17).

Climate science is not politics—it is a science. Yet climate science is now revealing potential dire consequences, such as rising sea levels. What we need, more than ever, is an open-minded dialogue between

science and society. Your efforts to learn some chemistry is a very positive step in that direction.

CONCEPTCHECK
Snow and ice reflect solar radiation, which helps to cool the Earth. Land is good at absorbing solar energy, which helps to warm the Earth. If Earth's snow and ice started melting, would this favor global warming or global cooling? How so?

CHECK YOUR ANSWER As snow and ice melt, this exposes the land, which in absorbing solar energy increases the temperature. This, in turn, would cause more snow and ice to melt, leading to even higher temperatures. This is an example of a positive feedback loop that would favor global warming.

Think and Discuss

1. What is the difference between "climate" and "weather" and which is easier to predict?

2. Some temperature gauges used for measuring global temperatures can be found in urban areas where they are affected by surrounding heat sources. Why would moving such a gauge to a more rural setting be a bad idea?

3. Two systems of regulating CO_2 emissions are the carbon tax and the emissions trading scheme (ETS), also known as cap and trade. Search the web for information on both of these systems and discuss the pros and cons of each.

4. As demonstrated at Plant Barry of Alabama Power, we have the working technology for the large-scale capture of carbon dioxide from fossil fuel burning power plants. The system works, but they don't turn it on. Why not?

5. Scientists tend to think quietly in the confines of their laboratories rather than out loud on a wide-open political platform. At what point should a scientist be compelled to do otherwise? Should all scientists be taught some political skills in training for their careers?

TABLE 1.4   Opposing Viewpoints of Human-Induced Global Climate Change

POINT
1. There has been no significant global warming over the past century. Temperatures have actually been cooling since 1998.
2. If there has been significant global warming, humans are not to blame. The most likely causes include changes in Earth's orbit and fluctuations in the Sun's energy output.
3. There is no definitive proof that humans have had a significant impact on global climate. Regulations on human activities, therefore, would be an unnecessary expense and harmful to the economy, especially for developing nations where energy needs are growing the fastest.
4. Alarm about global climate change is a movement whose underlying goal is to push forward a partisan political agenda.

COUNTER POINT
1. There has been a long-term warming trend. In 1998, a powerful El Nino created an unusual warm spell. For only a few subsequent years the temperatures were cooler.
2. To a significant extent, humans are responsible primarily due to their output of carbon dioxide, which is a potent greenhouse gas. The evidence on this is extensive and has proven highly compelling to the vast majority of climate scientists.
3. The potential for runaway climate change is risky enough to warrant action. But even if there is no climate change we would benefit from the development of green technologies in terms of new jobs, decentralized energy sources, clean energy for the developing world, and good stewardship of the environment.
4. The denial of global climate change is a movement spearheaded by the fossil fuel industry working to protect its financial interests.

▲ Climate scientists study an ice core containing ancient air captured within Greenland's ice cap. While there are many details of climate science that still puzzle us, there are also many details we understand quite well. Puzzlement in one area does not invalidate our overall successes.

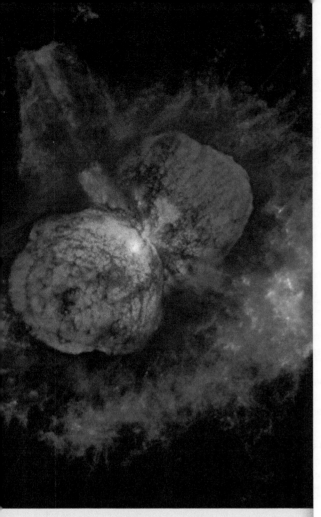

▲ Eta Carinae is a supergiant binary star system within our galaxy some 8000 light years away. Explosive turmoil is already apparent within this system, which is likely to supernova within the next million years. The atoms created within such exploding stars are ejected into the universe where they eventually recombine because of gravity to form new star systems. Amazingly, the atoms here on Earth are the remnants of stellar explosions from long ago. We are literally made of stardust.

2

Particles of Matter

THE MAIN IDEA

> Matter is made of tiny particles we call atoms.

Hydrogen atoms are the lightest of all atoms, and they make up more than 90 percent of the atoms in the known universe. Their origin goes back to the birth of the universe. Heavier atoms are produced in stars, which are massive collections of hydrogen pulled together by gravitational forces. The great pressures deep in a star's interior cause hydrogen atoms to fuse together, forming heavier atoms. With the exception of hydrogen, nearly all the atoms that occur naturally on the Earth—including those in your body—are the products of stars. A tiny fraction of these atoms came from our own star, the Sun, but most are from stars that ran their course long before our solar system came into being. You are literally made of stardust.

Atoms are so small that we can't see them directly. How then do we know they exist? How do atoms account for the mass of an object or its temperature? How can we use this idea of tiny particles called atoms to explain the nature of solids, liquids, and gases?

Chemistry

HANDS ON

The Breathing Rubber Balloon

How good is an inflated rubber balloon at holding air? As it deflates over time, how does the air escape?

PROCEDURE

1. Pour about 5 mL of water into a large rubber balloon. Inflate the balloon to full size and then tie the balloon so it remains inflated.

2. Add a few drops of a fragrant material, such as cinnamon oil or perfume, to about 5 mL of water. Pour this solution into a second large balloon. If available, use a dropper or a funnel to avoid spilling any of this fragrant solution onto the outside of the balloon. Inflate this second balloon to the same size of the first balloon and then tie it shut.

3. Switch both balloons and then present them to some one who did not see you preparing these balloons. Ask the person which balloon contains the fragrant material.

ANALYZE AND CONCLUDE

1. Explain how the fragrance gets out of the rubber balloon.

2. Silver-colored Mylar balloons stay inflated for a very long time. Can you think of a reason why this is so?

3. If the fragrance-containing balloon were heated in a microwave oven for a few moments, would the smell be more or less pronounced? Why?

4. Might water molecules also be coming out of the inflated balloons? (Hint: Fragrance molecules tend to be much larger than water molecules.)

5. Are air molecules also coming out of the inflated balloon? How do you know?

2.1 The Submicroscopic World Is Super-Small

EXPLAIN THIS

A liter of water is about how many times larger than a molecule of water?

From afar, a sand dune appears to be made of a smooth, continuous material. Up close, however, the dune reveals itself to be made of tiny grains of sand. In a similar fashion, everything around us—no matter how smooth it may appear—is made of very small fundamental units you know as **atoms.** Atoms are so small, however, that a single grain of sand contains on the order of 125 million trillion of them, which is 125,000,000,000,000,000,000. In scientific notation, this huge number is written as 1.25×10^{20} (see Appendix A). Interestingly, the number of atoms in a single grain of sand is about a quarter of a million times greater than the number of grains of sand shown in the dune of **Figure 2.1**.

As small as atoms are, there is much we have learned about them. We know, for example, that there are more than 100 different types of atoms, and we have arranged them in the widely recognized *periodic table*, as shown on the front inside cover of this textbook. Some atoms link together to form larger but still incredibly small units of matter called **molecules.** As shown in Figure 2.1, for example, two hydrogen atoms and one oxygen atom link together to form a single molecule of water, which you know as H_2O. Water molecules are so small that an 8-ounce glass of water contains about a trillion trillion of them, which in scientific notation is 1×10^{24}.

Our world can be studied at different levels of magnification, as illustrated in **Figure 2.2**. At the *macroscopic* level, matter is large enough to be seen, measured, or handled. A handful of sand and a glass of water are macroscopic

LEARNING OBJECTIVE

> Describe the particulate nature of matter.

READING CHECK

Are atoms made of molecules, or are molecules made of atoms?

▲ Figure 2.1
There are far more atoms in a single grain of sand or molecules within a glass of water than there are grains of sand within this towering sand dune.

Oxygen atom

Hydrogen atoms

Water molecule, H_2O

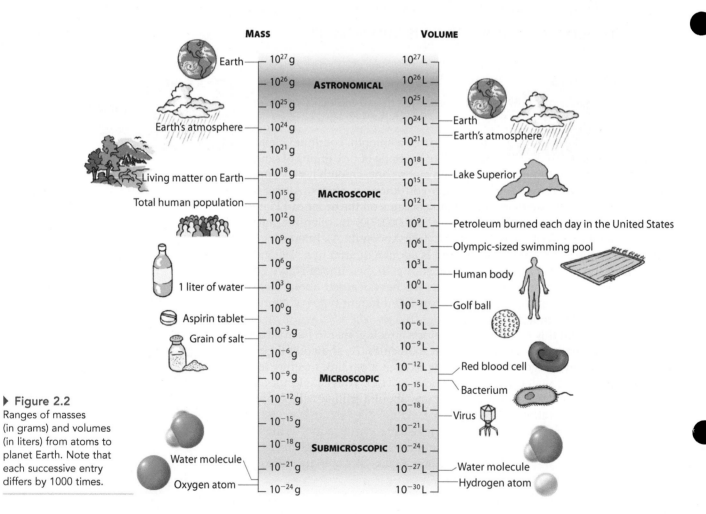

▶ Figure 2.2
Ranges of masses (in grams) and volumes (in liters) from atoms to planet Earth. Note that each successive entry differs by 1000 times.

samples of matter. At the *microscopic* level, physical structure is so fine that it can be seen only with a microscope. A biological cell is microscopic, as is the detail on a dragonfly's wing. Below the microscopic level is the **submicroscopic**—the realm of atoms and molecules, and an important focus of chemistry.

2.2 Discovering the Atom

EXPLAIN THIS

How is your nose able to smell a rose petal without actually touching the petal?

In the 4th century B.C., the influential Greek philosopher Aristotle described the composition and behavior of matter in terms of the four qualities shown in **Figure 2.3**: hot, cold, moist, and dry. Aristotle's model was a remarkable achievement for its day, and people using it in Aristotle's time found it made sense. When pottery is made, for example, wet clay changes to ceramic because the heat of the fire drives out the moist quality of the wet clay and replaces it with the dry quality of the ceramic.

Aristotle's views on the nature of matter made so much sense that less obvious views were difficult to accept. One alternative view was the forerunner of our present-day model: matter is composed of a finite number of incredibly small but discrete units we call atoms. This model was advanced by several Greek philosophers, including Democritus (460–370 B.C.), who coined the term *atom* from the Greek phrase *a tomos*, which means "not cut" or "that which is indivisible" (**Figure 2.4**). So compelling was Aristotle's reputation, however, that the atomic model would not reappear for 2000 years.

According to Aristotle, it was theoretically possible to transform any substance to another substance simply by altering the relative proportions of the four basic qualities. This meant that, under the proper conditions, a metal like lead could be transformed to gold. This concept laid the foundation of **alchemy,** a field of study concerned primarily with finding potions that would produce gold or confer immortality. Alchemists from the time of Aristotle to as late as the 1600s tried in vain to convert various metals to gold. Despite the futility of their efforts, the alchemists learned much about the behavior of many chemicals and developed many useful laboratory techniques.

With the advent of modern science, Aristotle's views on the nature of matter came into question. For example, in the late 1700s the French chemist Antoine Lavoisier discovered the law of mass conservation, as was discussed in Section 1.4. This verifiable law ran counter to Aristotle's idea that matter could lose or

LEARNING OBJECTIVE

Describe the evidence for the particulate nature of matter.

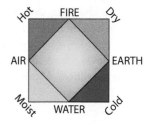

▲ Figure 2.3
Aristotle thought that all materials were made of various proportions of four fundamental qualities: hot, dry, cold, and moist. Various combinations of these qualities gave rise to the four basic elements: hot and dry gave fire, moist and cold gave water, hot and moist gave air, and dry and cold gave earth. He supposed that a hard substance like rock contained mostly the dry quality, for example, and a soft substance like clay contained more of the moist quality.

◀ Figure 2.4
In his atomic model, Democritus imagined that atoms of iron were shaped like coils—making iron rigid, strong, and malleable—and that atoms of fire were sharp, lightweight, and yellow.

READINGCHECK

How did Lavoisier define an element?

FORYOUR INFORMATION

To help finance his scientific projects, Lavoisier took part-time employment as a tax collector, in which position he introduced reforms to help ease the tax burden on peasants. But because of this employment, he was beheaded in 1794 during the French Revolution. After hearing appeals to spare Lavoisier's life, the judge determined that "the Republic needs neither scientists nor chemists; the course of justice cannot be delayed." About 18 months after Lavoisier's execution, the French government sent a formal apology to his widow, Marie-Anne.

gain mass as its hot, dry, cold, or moist qualities changed. Lavoisier further hypothesized that an *element* is any material made of a fundamental substance that cannot be broken down into anything else. Through experiments, he was able to transform water into two different substances—hydrogen and oxygen. According to Lavoisier, Aristotle was wrong to think of water as an element.

Further experimental work by Lavoisier and others led the English chemist John Dalton (1766–1844) to reintroduce the atomic ideas of Democritus (**Figure 2.5**). Dalton wrote a series of postulates—claims he assumed to be true based on experimental evidence—some of which are as follows:

1. Each element consists of indivisible, minute particles called atoms.
2. Atoms can be neither created nor destroyed in chemical reactions.
3. All atoms of a given element are identical.
4. Atoms of different elements have different masses.

It didn't matter that these tiny atoms were too small to be seen. What did matter was that Dalton's atomic model worked to explain much of what was then known about chemical reactions. Where alchemists using Aristotle's model failed, chemists using Dalton's model succeeded—not in making gold but in being able to understand and control the outcome of numerous chemical reactions.

In 1869, a Russian chemistry professor, Dmitri Mendeleev (1834–1907), produced a chart summarizing the properties of known elements for his students (**Figure 2.6**). Mendeleev's chart was unique in that it resembled a calendar. Elements were listed in horizontal rows in order of increasing mass. The first row contained the lightest elements, the second row contained the next heaviest elements, and so forth. Aligning rows of elements above and below each other (like days of a calendar) revealed that elements within the same vertical column had similar properties, such as chemical reactivity. In order to achieve this pattern, however, he had to shift some elements left or right occasionally. This left gaps—blank spaces that could not be filled by any known element (**Figure 2.7**). Instead of looking on these gaps as defects, Mendeleev boldly predicted the existence of elements that had not yet been discovered. His predictions about the properties of some of those missing elements led to their discovery.

That Mendeleev was able to predict the properties of new elements helped convince many scientists of the accuracy of Dalton's atomic hypothesis, upon which Mendeleev's periodic table was based. This in turn helped promote

▲ Figure 2.5
John Dalton was born into a very poor family. Although his formal schooling ended at age 11, he continued to learn on his own and even began teaching others when he was only 12. His primary research interest was weather, which led him to conduct many experiments with gases. Soon after publishing his conclusions on the atomic nature of matter, his reputation as a first-rate scientist increased rapidly. In 1810, he was elected into Britain's premiere scientific organization, the Royal Society.

▲ Figure 2.6
Dmitri Mendeleev was a devoted and highly effective teacher. Students adored him and would fill lecture halls to hear him speak about chemistry. Much of his work on the periodic table occurred in his spare time following his lectures. Mendeleev taught not only in the university classrooms but anywhere he traveled. During his journeys by train, he would travel third class with peasants to share his findings about agriculture.

— 70 —

но въ ней, мнѣ кажется, уже ясно выражается примѣнимость вы
ставляемаго мною начала ко всей совокупности элементовъ, пай
которыхъ извѣстенъ съ достовѣрностію. На этотъ разъ я и желалъ
преимущественно найдти общую систему элементовъ. Вотъ этотъ
опытъ:

		Ti=50	Zr=90		?=180.
		V=51	Nb=94	Ta=182.	
		Cr=52	Mo=96	W=186.	
		Mn=55	Rh=104,4	Pt=197,4	
		Fe=56	Ru=104,4	Ir=198.	
		Ni=Co=59	Pl=106,6	Os=199.	
H=1			Cu=63,4	Ag=108	Hg=200.
	Be=9,4	Mg=24	Zn=65,2	Cd=112	
	B=11	Al=27,4	?=68	Ur=116	Au=197?
	C=12	Si=28	?=70	Sn=118	
	N=14	P=31	As=75	Sb=122	Bi=210
	O=16	S=32	Se=79,4	Te=128?	
	F=19	Cl=35,5	Br=80	I=127	
Li=7	Na=23	K=39	Rb=85,4	Cs=133	Tl=204
		Ca=40	Sr=87,6	Ba=137	Pb=207.
		?=45	Ce=92		
		?Er=56	La=94		
		?Yt=60	Di=95		
		?In=75,6	Th=118?		

а потому приходится въ разныхъ рядахъ имѣть различное измѣненіе разностей,
чего нѣтъ въ главныхъ числахъ предлагаемой таблицы. Или же придется предпо-
лагать при составленіи системы очень много недостающихъ членовъ. То и
другое мало выгодно. Мнѣ кажется притомъ, наиболѣе естественнымъ составить
кубическую систему (предлагаемая есть плоскостная), но и попытки для ея образо-
ванія не повели къ надлежащимъ результатамъ. Слѣдующія двѣ попытки могутъ по-
казать то разнообразіе сопоставленій, какое возможно при допущеніи основнаго
начала, высказаннаго въ этой статьѣ.

Li	Na	K	Cu	Rb	Ag	Cs	—	Tl
7	23	39	63,4	85,4	108	133		204
Be	Mg	Ca	Zn	Sr	Cd	Ba		Pb
B	Al	—	—	—	Ur	—	—	Bi?
C	Si	Ti	—	Zr	Sn	—	—	—
N	P	V	As	Nb	Sb	—	Ta	—
O	S	—	Se	—	Te	—	W	—
F	Cl	—	Br	—	J	—	—	—
19	35,5	58	80	190	127	160	190	220.

Dalton's proposed atomic nature of matter from a hypothesis to a more widely accepted theory. Mendeleev's chart ultimately led to our modern periodic table, which we discuss more fully in Chapter 3.

Since the time of Lavoisier, Dalton, and Mendeleev, our understanding of atoms has grown substantially. Although we have not discovered the alchemist's dream of immortality, we have learned how to design medicines that cure numerous diseases. From crude oil we can make fuels, plastics, clothing, and more. From the thin air we can produce fertilizer. Virtually every aspect of modern society has been and will continue to be affected by our ability to manipulate atoms to meet our needs. Of all the discoveries made by humans, our discovery of the atom is arguably one of our greatest and most profound.

Are atoms for real? Today we have the technology to capture images of individual atoms, as shown in **Figure 2.8** and discussed further in Section 3.8. Just as important is the fact that atoms and molecules can be used to explain common observations. For example, heat transforms moist clay into ceramic by driving off water molecules. In ice, water molecules are stuck together in a fixed orientation. Warmth melts the ice by helping the water molecules break away from each other.

Consider also the Hands-On Chemistry activity at the beginning of this chapter, in which a fragrance is found to travel through the skin of a rubber balloon. We can explain how this occurs by assuming the fragrance consists of tiny molecules that can pass through the micropores of the inflated rubber balloon. Similarly, we can use the idea of molecules to explain how moisture can collect on a tabletop before disappearing, as shown in **Figure 2.9**, or how a dark-colored powdered drink mix dissolves in water with no stirring, as shown

FOR YOUR INFORMATION

Some atoms are larger than others, but they are all exceedingly small. Gold atoms, for example, are so small that about 4,000,000,000,000 (4 trillion) of them could fit within the period at the end of this sentence.

(a)

(b)

(c)

▲ Figure 2.8
(a) Scanning probe microscopes are relatively simple devices used to create submicroscopic imagery. (b) An image of gallium and arsenic atoms. (c) Each dot in the world's tiniest map consists of a few thousand gold atoms, each dot moved into its proper place by a scanning probe microscope.

▲ Figure 2.9
(a) Place your palms down on a cool, dark, and reflective table such as a slate lab benchtop. Water molecules exiting from your skin collect onto this surface. (b) Lift your hands to see this moisture, which quickly dissappears as the water molecules evaporate into the air.

in **Figure 2.10**. The explanatory powers of the atomic model are great. This, along with our hi-tech evidence, leads us to trust that matter is made of these super-small particles we call atoms.

CONCEPT CHECK

Lavoisier hypothesized that an element was a material made of a fundamental substance that cannot be broken down into anything else. According to Dalton, this fundamental substance was made of

a. water.

b. fire.

c. atoms.

d. molecules.

CHECK YOUR ANSWER The answer is (c). Dalton reintroduced the concept of atoms put forth by Democritus some 2000 years earlier. Unlike Democritus, however, Dalton assumed that the atoms of different elements differed from each other only by their mass.

(a)

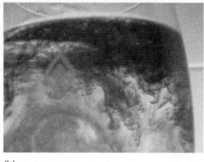

(b)

(c)

▲ Figure 2.10
(a) Kool-Aid crystals settle to the bottom of a container of water. (b) Without stirring, the crystals begin to dissolve as they are bombarded by water molecules in the liquid phase. (c) The bustling movement of the water molecules eventually causes the Kool-Aid to be uniformly mixed with the water.

2.3 Mass Is How Much, and Volume Is How Spacious

EXPLAIN THIS

For a given parcel of air, which is greatest: its mass, weight, or volume?

To describe a material object, we can quantify any number of properties, but perhaps the most fundamental of these is mass. **Mass** is the quantitative measure of how much matter a material object contains. The greater the mass of an object, the greater the amount of matter in it. A gold bar that is twice as massive as another gold bar, for example, contains twice as many gold atoms.

Mass is also a measure of an object's *inertia*, which is the resistance the object has to any change in its motion. A cement truck, for example, has a lot of mass (inertia), which is why it requires a powerful engine to get it moving and powerful brakes to cause it to come to a stop.

The standard unit of mass is the *kilogram*, and a replica of the cylinder used to determine exactly what mass "1 kilogram" describes is shown in **Figure 2.11**. An average-sized adult human male has a mass of about 70 kilograms (154 pounds). For smaller quantities, we use the *gram*. Table 1.3 in the previous chapter tells us that the prefix *kilo-* means "1000," so we see that 1000 grams is equivalent to 1 kilogram (1000 grams = 1 kilogram). For even smaller quantities, the *milligram* is used (1000 milligrams = 1 gram).

Since mass is simply a measure of the amount of matter in a sample, which is a function of how many atoms the sample contains, the mass of an object remains the same no matter where it is located. A 1-kilogram gold bar, for example, has the same mass whether it is on the Earth, on the Moon, or floating "weightless" in space. This is because it contains the same number of atoms in each location.

Weight is more complicated. By definition, **weight** is the gravitational force exerted on an object by the most massive nearby body, such as Earth. The weight of an object, therefore, depends entirely upon its location, as is shown in **Figure 2.12**. On the Moon, a gold bar weighs less than it does on Earth. This is because the Moon is much less massive than Earth; hence, the gravitational force exerted by the Moon on the bar is much less. On Jupiter, the gold bar would weigh more than it does on Earth, because of the greater gravitational force exerted on the bar by this very massive planet.

LEARNING OBJECTIVE

Distinguish between mass, weight, and volume.

READINGCHECK

What is mass?

▲ **Figure 2.11**
The standard kilogram is defined as the mass of a platinum-iridium cylinder kept at the International Bureau of Weights and Measures in Sèvres, France. The cylinder is removed from its very safe location only once a year for comparison with duplicates, such as the one shown here, which is housed at the National Institute of Standards and Technology in Washington, D.C.

▼ **Figure 2.12**
(a) A 1-kilogram gold bar resting on the Earth weighs 2.2 pounds. (b) On the Moon, this same gold bar would weigh 0.37 pound. (c) Deep in space, far from any planet, the gold bar would weigh 0 pounds, though it would still have a mass of 1 kilogram.

(a) Mass = 1 kg
Weight = 2.2 lb

(b) Mass = 1 kg
Weight = 0.37 lb

(c) Mass = 1 kg
Weight = 0 lb

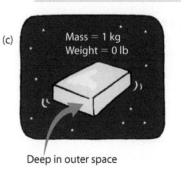

Deep in outer space

FOR YOUR INFORMATION

If you counted by ones at a rate of one number each second, counting to a million would take you 11.6 days (assuming no time off for sleeping). How about a billion? Is there a dramatic difference? Get this: counting to a billion would take you 31.8 years. Counting to a trillion, which is a thousand billions, would take you 31,800 years! Now you know why there are far more millionaires than billionaires, and yet no trillionaires.

▲ **Figure 2.13**
The volume of an object, no matter what its shape, can be measured by its displacement of water. When this rock is immersed in the water, the rise in the water level equals the volume of the rock, which in this example measures about 90 mL.

Because mass is independent of location, it is customary in science to measure matter by its mass rather than its weight. *Conceptual Chemistry* adheres to this convention by representing the mass of matter in units of kilograms, grams, and milligrams.

CONCEPT CHECK

Is there gravity on the Moon?

CHECK YOUR ANSWER Yes, absolutely! The Moon exerts a downward gravitational pull on any body near its surface, as evidenced by the fact that astronauts were able to land and walk on the Moon. This NASA photograph shows an astronaut jumping. Without gravity, this jump would have been his last, because he would never have come down.

The amount of space a material object occupies is its **volume.** The SI unit of volume is the *liter*, which is only slightly larger than the USCS unit of volume, the *quart*. A liter is the volume of space marked off by a cube measuring 10 centimeters by 10 centimeters by 10 centimeters, which is 1000 cubic centimeters. A smaller unit of volume is the *milliliter*, which is one-thousandth of a liter, or 1 cubic centimeter.

A convenient way to measure the volume of an irregular object is shown in **Figure 2.13.** The volume of water displaced is equal to the volume of the object.

2.4 Density Is the Ratio of Mass to Volume

LEARNING OBJECTIVE

Calculate the density of a material.

✔ **READING CHECK**

Density is a measure of what?

EXPLAIN THIS

Why does hot air rise?

The relationship between an object's mass and the amount of space it occupies is the object's density. Density is a measure of compactness, of how tightly mass is squeezed into a given volume. A block of lead has much more mass squeezed into its volume than does a same-sized block of aluminum. The lead is therefore more dense. We think of density as the "lightness" or "heaviness" of objects of the same size, as **Figure 2.14** shows.

Density is the amount of mass contained in a sample divided by the volume of the sample:

$$\text{density} = \frac{\text{mass}}{\text{volume}}$$

An object having a mass of 1 gram and a volume of 1 milliliter, for example, has a density of

$$\text{density} = \frac{1\text{ g}}{1\text{ mL}} = \frac{1\text{ g}}{\text{mL}}$$

An object having a mass of 2 grams and a volume of 1 milliliter is denser; its density is

$$\text{density} = \frac{2\text{ g}}{1\text{ mL}} = \frac{2\text{ g}}{\text{mL}}$$

Other units of mass and volume besides grams and milliliters may be used in calculating density. The densities of gases, for example, because they are so low, are often given in grams per liter. In all cases, however, the units are a unit of mass divided by a unit of volume.

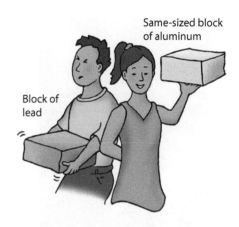

▲ **Figure 2.14**
The amount of mass in a block of lead far exceeds the amount of mass in a block of aluminum of the same size. Hence, the lead block weighs much more and is more difficult to lift.

CONCEPTCHECK

Which occupies a greater volume: 1 kilogram of lead or 1 kilogram of aluminum?

CHECK YOUR ANSWER The aluminum. Think of it this way. Because lead is so dense, you need only a little bit in order to have 1 kilogram. Aluminum, by contrast, is far less dense, so 1 kilogram of aluminum occupies much more volume than the same mass of lead.

The densities of some substances are given in Table 2.1. Which would be more difficult to pick up: a liter of water or a liter of mercury?

TABLE 2.1   Densities of Some Solids, Liquids, and Gases

SUBSTANCE	DENSITY (G/ML)	DENSITY (G/L)
Solids		
osmium	22.5	22,500
gold	19.3	19,300
lead	11.3	11,300
copper	8.92	8920
iron	7.86	7860
zinc	7.14	7140
aluminum	2.70	2700
ice	0.92	920
Liquids		
mercury	13.6	13,600
seawater	1.03	1030
fresh water at 4°C	1.00	1000
ethyl alcohol	0.81	810
Gases*		
oxygen at 0°C	0.00143	1.43
dry air		
0°C	0.00129	1.29
20°C	0.00121	1.21
helium at 0°C	0.000178	0.178

*All values at sea-level atmospheric pressure.

▲ **Figure 2.15**
The hot air inside this hot air balloon is less dense than the surrounding colder air, which is why the balloon rises.

Gas densities are much more affected by pressure and temperature than are the densities of solids and liquids. With an increase in pressure, gas molecules are squeezed closer together. This makes for less volume (but the same mass) and therefore greater density. The density of the air inside a diver's breathing tank, for example, is much greater than the density of air at normal atmospheric pressure. With an increase in temperature, gas molecules move faster and thus have a tendency to push outward, thereby occupying a greater volume. Thus, hot air is less dense than cold air, which is why hot air rises and the balloon in **Figure 2.15** can take its passengers for a breathtaking ride.

CONCEPTCHECK

1. Which has greater density: 1 gram of water or 10 grams of water?

2. Which has greater density: 1 gram of lead or 10 grams of aluminum?

CHECK YOUR ANSWERS

1. The density is the same for any amount of water. Whereas 1 gram of water occupies a volume of 1 milliliter, 10 grams occupies a volume of 10 milliliters. The ratio 1 gram/1 milliliter is the same as the ratio 10 grams/10 milliliters.

2. The lead. Density is mass per volume, and this ratio is greater for any amount of lead than for any amount of aluminum.

CALCULATION CORNER MANIPULATING AN ALGEBRAIC EQUATION

With a little algebraic manipulation, it is easy to change the equation for density around so that it can be solved for either mass or volume. The first step is to multiply both sides of the density equation by volume. Then canceling the volumes that appear in the numerator and denominator results in the equation for the mass of an object.

$$\text{density} \times \text{volume} = \frac{\text{mass}}{\text{volume}} \times \text{volume}$$

$$\text{density} \times \text{volume} = \text{mass}$$

Dividing both sides of the preceding equation by density results in an equation for the volume of on object.

$$\frac{\text{density}}{\text{density}} \times \text{volume} = \frac{\text{mass}}{\text{density}}$$

$$\text{volume} = \frac{\text{mass}}{\text{density}}$$

In summary, the three equations expressing the relationship among density, mass, and volume are

Density	Mass	Volume
$D = \dfrac{M}{V}$	$M = D \times V$	$V = \dfrac{M}{D}$

A good way to remember these relationships is to use the following diagram. Use your finger to cover the quantity you want to know, and that quantity's relationship to the other quantities is revealed. For example, covering the M shows that mass is equal to density times volume, $D \times V$.

EXAMPLE 1

A pre-1982 penny has a density of 8.92 grams per milliliter and a volume of 0.392 milliliters. What is its mass?

ANSWER 1

$$D \times V = M = 8.92\,\text{g/mL} \times 0.392\,\text{mL} = 3.50\,\text{g}$$

EXAMPLE 2

A post-1982 penny has a density of 7.40 grams per milliliter and a mass of 2.90 grams. What is its volume?

ANSWER 2

$$V = \frac{M}{D} = \frac{2.90\,\text{g}}{7.40\,\text{g/mL}} = 0.392\,\text{mL}$$

YOUR TURN

1. What is the average density of a loaf of bread that has a mass of 500 grams and a volume of 1000 milliliters?

2. The loaf of bread in the previous problem loses all its moisture after being toasted. Its volume remains at 1000 milliliters, but its density has been reduced to 0.4 grams per milliliter. What is its new mass?

3. A sack of groceries accidentally set on a 500-gram loaf of white bread increases the average density of the loaf to 5 grams per milliliter. What is its new volume?

2.5 Energy Is the Mover of Matter

EXPLAIN THIS

How does a wooden arrow lying on the ground have potential energy?

Matter is substance, and energy is that which can move substance. The concept of energy is abstract and therefore not as easy to define as the concepts of mass and volume. One definition of **energy** is the capacity to do work. If something has energy, it can do work on something else—it can exert a force and move that something else. Accordingly, energy is not something we observe directly. Rather, we only witness its effects.

An object may store energy by virtue of its position. This stored energy is called **potential energy** because it has the "potential" for doing work. As shown in **Figure 2.16**, a boulder perched on the edge of a cliff has potential energy due to the force of gravity, just as the poised arrow has potential energy due to the tension of the bow. The potential energy of an object increases as the distance over which the force is able to act increases. The higher a boulder is positioned above level ground, the more potential energy it has to do work as it falls downward under the pull of gravity. Similarly, an arrow in a fully drawn bow has more potential energy than does one in a half-drawn bow.

Kinetic energy is the energy of motion. Both a falling boulder and a flying arrow have kinetic energy. The faster a body moves, the more kinetic energy it has and therefore the more work it can do. For example, the faster an arrow flies, the more work it can do to a target, as evidenced by its deeper penetration.

Substances possess what is known as *chemical potential energy*, which is the energy that is stored within atoms and molecules. For example, any material that can burn has chemical potential energy. The firecracker in **Figure 2.17**, for instance, has chemical potential energy. This energy gets released when the

READINGCHECK

If something has energy, what can it do?

(a)

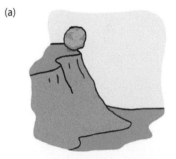

(b)

▲ Figure 2.16
(a) An elevated boulder's potential energy becomes apparent when the boulder is perched in a position where it might easily fall downhill. When it does fall, this potential energy is converted to kinetic energy. (b) Much of the potential energy in Tenny's drawn bow will be converted to the kinetic energy of the arrow upon its release.

▶ Figure 2.17
A firecracker is a mixture of solids that possess chemical potential energy. When a firecracker explodes, the solids react to form gases that fly outward and so possess a great deal of kinetic energy. Light and heat (both of which are forms of energy) are also formed.

▶ **Figure 2.18**
The energy content of this candy bar (230 Calories = 230,000 calories), when released through burning, is enough to heat up 230,000 grams (about 507 pounds) of water by 1 degree Celsius.

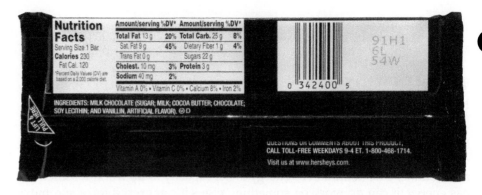

firecracker is ignited. During the explosion, some of the chemical potential energy is transformed to the kinetic energy of flying particles. Much of the chemical potential energy is also transformed to light and heat. We explore the relationship between energy and chemical reactions in Chapter 9.

The SI unit of energy is the *joule*, which is about the amount of energy released from a candle burning for only a moment. In the United States, a common unit of energy is the *calorie*. One calorie is by definition the amount of energy required to raise the temperature of 1 gram of water by 1 degree Celsius. One calorie is 4.184 times larger than 1 joule. Put differently, 4.184 joules of energy is equivalent to 1 calorie (4.184 joules = 1 calorie). So a joule is about one-fourth of a calorie.

In the United States, the energy content of food is measured by the *Calorie* (note the capital C). One Calorie equals 1 kilocalorie, which is 1000 calories (note the small c). The candy bar in **Figure 2.18** offers 230 Calories (230 kilocalories), providing a total of 230,000 calories to the consumer.

2.6 Temperature Is a Measure of How Hot—Heat It Is Not

LEARNING OBJECTIVE

Distinguish between heat and temperature.

EXPLAIN THIS

Are our bodies better at detecting heat or temperature?

The atoms and molecules that form matter are in constant motion, jiggling to and fro or bouncing from one position to another. By virtue of their motion, these particles possess kinetic energy. Their average kinetic energy is directly related to a property you can sense: how hot something is. Whenever something becomes warmer, the kinetic energy of its submicroscopic particles increases. For example, strike a penny with a hammer and the penny becomes warm because the hammer's blow causes its atoms to jostle faster, increasing their kinetic energy. (The hammer becomes warm for the same reason.) Put a flame to a liquid and the liquid becomes warmer because the energy of the flame causes the particles of the liquid to move faster, increasing their kinetic energy. For example, the molecules in the hot coffee in **Figure 2.19** are moving faster on average than those in the cold coffee.

Temperature tells us how warm or cold an object is relative to some standard. We express temperature by a number that corresponds to the degree of hotness on some chosen scale. Just touching an object certainly isn't a good way of measuring its temperature, as **Figure 2.20** illustrates. To measure temperature, therefore, we take advantage of the fact that nearly all materials expand when their temperature is raised and contract when it is lowered. With increasing temperature, the particles move faster and are on average farther apart—the material expands. With decreasing temperature, the particles move more slowly and are on average closer together—the material contracts. A **thermometer** exploits this characteristic of matter, measuring temperature by means of the expansion and contraction of a liquid, usually mercury or colored alcohol.

▲ **Figure 2.19**
The difference between hot coffee and cold coffee is the average speed of the molecules. In the hot coffee, the molecules are moving faster on average than they are in the cold coffee. (The "motion trails" on the molecules of hot coffee indicate their higher speed.)

You may have noticed telephone wires sagging on a hot day. This happens because the wires are longer in hot weather than in cold. What is happening on the atomic level to cause such changes in wire length?

CHECK YOUR ANSWER On a hot day, the atoms in the wire are moving faster, and as a result the wire expands. On a cold day, those same atoms are moving more slowly, which causes the wire to contract.

▲ **Figure 2.20**
Can we trust our sense of hot and cold? Will both fingers feel the same temperature when they are put in the warm water? Try this yourself, and you will see why we use a thermometer for an objective measurement.

The most common thermometer in the world is the Celsius thermometer, named in honor of the Swedish astronomer Anders Celsius (1701–1744), who first suggested the scale of 100 degrees between the freezing point and boiling point of fresh water. In a Celsius thermometer, the number 0 is assigned to the temperature at which pure water freezes and the number 100 is assigned to the temperature at which it boils (at standard atmospheric pressure), with 100 equal divisions called *degrees* between these two points.

In the United States, we use the Fahrenheit thermometer, named after its originator, the German scientist G. D. Fahrenheit (1686–1736), who chose to assign 0 to the temperature of a mixture containing equal weights of snow and common salt and to assign 100 to the body temperature of a human. Because these reference points are not dependable, the Fahrenheit scale has since been modified such that the freezing point of pure water is designated 32°F and the boiling point of pure water is designated 212°F. On this recalibrated scale, the average human body temperature taken orally is around 98.2°F.

A temperature scale favored by scientists is the Kelvin scale, named after the British physicist Lord Kelvin (1824–1907). This scale is calibrated not in terms of the freezing and boiling points of water but rather in terms of the motion of atoms and molecules. On the Kelvin scale, zero is the temperature at which there is no atomic or molecular motion. This is a theoretical limit called **absolute zero,** which is the temperature at which the particles of a substance have absolutely no kinetic energy to give up. Absolute zero corresponds to –459.7°F on the Fahrenheit scale and to –273.15°C on the Celsius scale. On the Kelvin scale, this temperature is simply 0 K, which is read "zero kelvin" or "zero K." Marks on the Kelvin scale are the same distance apart as those on the Celsius scale, so the temperature of freezing water is 273 kelvin. (Note that the word *degree* is not used with the Kelvin scale. To say "273 degrees kelvin" is incorrect. To say "273 kelvin" is correct.) The three scales are compared in **Figure 2.21**.

It is important to understand that temperature is a measure of the *average* amount of energy in a substance, not the *total* amount of energy, as **Figure 2.22** shows. The total energy in a swimming pool full of boiling water is much more than the total energy in a cupful of boiling water even though both are at the same temperature. Your utility bill after heating the swimming pool water to

 FOR YOUR INFORMATION

Most atoms are ancient. They have existed through imponderable ages, recycling through the universe in innumerable forms, both nonliving and living. In this sense, you don't "own" the atoms that make up your body–you are simply their present caretaker. There will be many caretakers to follow.

READING CHECK
What does temperature measure?

▶ **Figure 2.21**
Some familiar temperatures measured on the Fahrenheit, Celsius, and Kelvin scales.

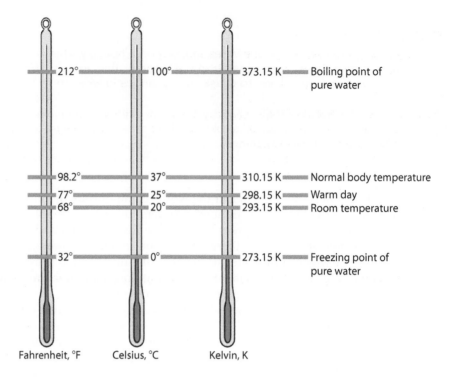

Fahrenheit, °F   Celsius, °C   Kelvin, K

100°C would show this. Whereas the total amount of energy in the pool is much more than in the cup, the *average* molecular motion is the same in both water samples. The water molecules in the swimming pool are moving on average just as fast as the water molecules in the cup. The only difference is that the swimming pool contains more water molecules and hence a greater total amount of energy.

Heat is energy that flows from a higher-temperature object to a lower-temperature object. If you touch a hot stove, heat enters your hand because the stove is at a higher temperature than your hand. When you touch a piece of ice, energy passes out of your hand and into the ice because the ice is at a lower temperature than your hand. From a human perspective, if you are absorbing heat, you experience warmth; if you are losing heat, you experience cooling. The next time you touch the hot forehead of a sick, feverish friend, ask him or her whether your hand feels hot or cold. Whereas temperature is absolute, hot and cold are relative.

In general, the greater the temperature difference between two bodies in contact with each other, the greater the rate of heat flow. This is why a hot stove can cause much more damage to your skin than a warm stove. Because heat is a form of energy, its unit is the joule.

▶ **Figure 2.22**
Bodies of water at the same temperature have the same average molecular kinetic energies. The total energy, however, depends upon how much water you have. For example, a swimming pool of water has much more total energy than a cupful of water, even when at the same temperature. Consider what your electric bill would look like after heating a swimming pool full of water to 100°C.

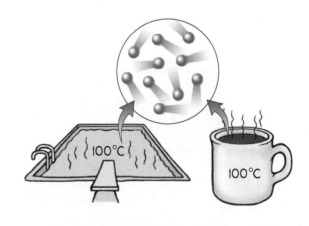

CONCEPTCHECK

When you enter a swimming pool, the water may feel quite cold. After a while, though, your body "gets used to it," and the water no longer feels so cold. Use the concept of heat to explain what is going on.

CHECK YOUR ANSWER Heat flows because of a temperature difference. When you enter the water, your skin temperature is much higher than the water temperature. The result is a significant flow of heat from your body to the water, which you experience as cold. Once you have been in the water awhile, your skin temperature is much closer to the water temperature (due to the cooling effects of the water and your body's ability to conserve heat), so the flow of heat from your body is less. With less heat flowing from your body, the water no longer feels so cold.

2.7 The Phase of a Material Depends on the Motion of Its Particles

EXPLAIN THIS

What is found between two adjacent molecules of a gas?

One of the most evident ways we can describe matter is by its physical form, which may be one of three phases (also sometimes described as physical states): *solid, liquid,* or *gas.* A **solid** material, such as a rock, occupies a constant amount of space and does not readily deform upon the application of pressure. In other words, a solid has both a definite volume and a definite shape. A **liquid** also occupies a constant amount of space (it has a definite volume), but its form changes readily (it has an indefinite shape). A liter of milk, for example, may take the shape of its carton or the shape of a puddle, but its volume is the same in both cases. A **gas** is diffuse, having neither a definite volume nor a definite shape. Any sample of gas assumes both the shape and the volume of the container it occupies. A given amount of air, for example, may assume the volume and shape of a toy balloon or be compressed into the volume and shape of a bicycle tire. Released from its container, a gas diffuses into the atmosphere, which is a collection of various gases held to our planet only by the force of gravity.

On the submicroscopic level, the solid, liquid, and gaseous phases are distinguished by the extent of interaction between the submicroscopic particles (the atoms or molecules). This is illustrated in **Figure 2.23**. In solid matter, the attractions between particles are strong enough to hold all the particles together in some fixed 3-dimensional arrangement. The particles are able to vibrate about fixed positions, but they cannot move past one another. Adding heat causes these vibrations to increase until, at a certain temperature, the vibrations are rapid enough to disrupt the fixed arrangement. The particles can then slip past one another and tumble around much like a bunch of marbles in a bag. This kind of motion is representative of the liquid phase of matter, and it is the mobility of the particles that gives rise to the liquid's fluid character—its ability to flow and take on the shape of its container.

Further heating causes the particles in the liquid to move so fast that the attractions they have for one another are unable to hold them together. They then separate from one another, forming a gas. Moving at an average speed of 500 meters per second (1100 miles per hour), the particles of a gas are widely separated from one another. Matter in the gaseous phase therefore occupies much more volume than it does in the solid or liquid phase, as **Figure 2.24** shows. Applying pressure to a gas squeezes the gas particles closer together, which reduces their volume. Enough air for an underwater diver to breathe for many minutes, for example, can be squeezed (compressed) into a tank small enough to be carried on the diver's back.

LEARNING OBJECTIVE

Describe the particulate nature of three phases of matter.

 FORYOUR
INFORMATION

Coffee and tea are decaffeinated using carbon dioxide in a fourth phase of matter known as a supercritical fluid. This phase, in which carbon dioxide has properties of both a gas and a liquid, is attained by adding lots of pressure and heat. Supercritical carbon dioxide is relatively easy to produce. To get water to form a supercritical fluid, however, requires pressures in excess of 217 atmospheres and a temperature of 374°C.

 **READING**CHECK

On a submicroscopic level, how are solids, liquids, and gases distinguished?

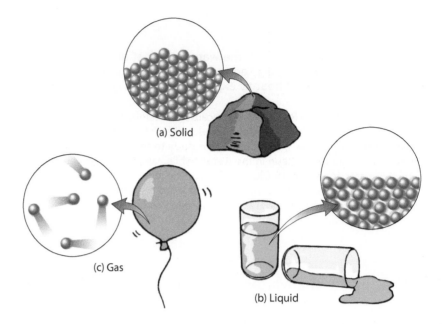

(a) Solid

(c) Gas

(b) Liquid

 CHEMICAL CONNECTIONS

How is your sense of smell connected to temperature?

(Answers to Chemical Connections questions are given at the back of Appendix C.)

Although gas particles move at high speeds, the speed at which they can travel from one side of a room to the other is relatively slow. This is because the gas particles are continually hitting one another, and the path they end up taking is circuitous. At home, you get a sense of how long it takes for gas particles to migrate each time someone opens the oven door after baking, as **Figure 2.25** shows. A shot of aromatic gas particles escapes the oven, but there is a notable delay before the aroma reaches the nose of someone sitting in the next room.

CONCEPTCHECK
Why are gases so much easier to compress into smaller volumes than solids and liquids?

CHECK YOUR ANSWER Because there is a lot of space between gas particles. The particles of a solid or liquid, on the other hand, are already close to one another, meaning there is little room left for a further decrease in volume.

Familiar Terms Are Used to Describe Phase Changes

Figure 2.26 illustrates that you must either add heat to a substance or remove heat from it if you want to change its phase. The process by which a solid transforms into a liquid is called **melting.** To visualize what happens when heat begins to

(a)

(b)

(c)

◀ Figure 2.25
In traveling from point A to point B, the typical gas particle travels an indirect path because of numerous collisions with other gas particles—about 8 billion collisions every second! The changes in direction shown here represent only a few of these collisions. Although the particle travels at very high speeds, it takes a relatively long time to cross between two distant points because of these numerous collisions.

melt a solid, imagine that you are holding hands with a group of people and each of you starts jumping around randomly. The more violently you jump, the more difficult it is to hold onto one another. If everyone jumps violently enough, keeping hold is impossible. Something like this happens to the particles of a solid when it is heated. As heat is added to the solid, the particles vibrate more and more violently. If enough heat is added, the attractive forces between the particles are no longer able to hold them together. The solid melts.

A liquid can be changed to a solid by the removal of heat. This process is called **freezing.** As heat is withdrawn from the liquid, particle motion diminishes until the particles are moving slowly enough on average for the attractive forces between them to take firm hold. The only motion the particles are capable of then is vibration about fixed positions, which means the liquid has solidified, or frozen. As we explore further in Chapter 8, freezing is merely the reverse of melting. The temperature at which a substance freezes or melts is the same. For pure water, this freezing/melting point is 0°C.

A liquid can be heated so that it becomes a gas—a process called **evaporation.** As heat is added, the particles of the liquid acquire more kinetic energy and move faster. Particles at the liquid surface eventually gain enough energy to fly out of the liquid and enter the air. In other words, they enter the gaseous phase. As more and more particles absorb the heat being added, they too acquire enough energy to escape from the liquid surface and become gas particles. Because a gas results from evaporation, this phase is also sometimes referred to as *vapor*. Water in the gaseous phase, for example, may be referred to as *water vapor*.

The rate at which a liquid evaporates increases with temperature. A puddle of water, for example, evaporates from a hot pavement more quickly than it

FOR YOUR INFORMATION

Sublimation and deposition are two less commonly mentioned phase changes. Sublimation is the transformation of a solid directly to a gas. Snow readily sublimes from mountain tops on sunny days. Dry ice, as shown in Figure 2.24, also readily sublimes. Deposition is the transformation of a gas directly to a solid. An example of deposition is the formation of frost.

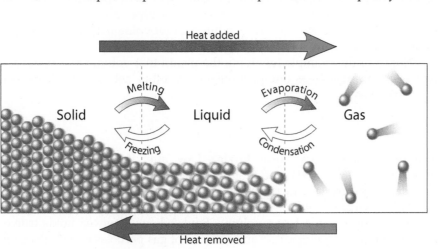

◀ Figure 2.26
Melting and evaporation involve the addition of heat; condensation and freezing involve the removal of heat.

does from your cool kitchen floor. When the temperature is hot enough, evaporation occurs beneath the surface of the liquid. As a result, bubbles form and are buoyed up to the surface. We say that the liquid is **boiling.** A substance is often characterized by its *boiling point*, which is the temperature at which it boils. At sea level, the boiling point of fresh water is 100°C.

The transformation from gas to liquid—the reverse of evaporation—is called **condensation.** This process can occur when the temperature of a gas decreases. The water vapor held in the warm daylight air, for example, may condense to form a wet dew in the cool of the night.

2.8 Gas Laws Describe the Behavior of Gases

LEARNING OBJECTIVE

Describe how the volume of a gas is affected by pressure, temperature, and number of particles.

EXPLAIN THIS

Why do bubbles of air in water get larger as they rise?

Scientists of the 17th, 18th, and 19th centuries investigated the relationships among the pressure, volume, and temperature of gaseous materials. Their observations are summed up by a set of gas laws named in their honor. These laws help us to understand the behavior of gases, including the air we breathe.

Boyle's Law: Pressure and Volume

Think of the molecules of air inside the inflated tire of an automobile. Inside the tire, the molecules behave like zillions of tiny Ping-Pong balls, perpetually moving helter-skelter and banging against the inner walls. Their impacts on the inner surface of the tire produce a jittery force that appears to our coarse senses as a steady outward push. Averaging this pushing force over a unit of area gives the *pressure* of the enclosed air.

Suppose the temperature is kept constant and twice as many molecules are pumped into the same volume, as shown in **Figure 2.27**. Then the air density—the number of molecules per given volume—is doubled. If the molecules move with the same average kinetic energy, or, equivalently, if their temperature is the same, then, to a close approximation, the number of collisions will be doubled. This means that the pressure is doubled.

We can also double the air density by compressing air to half its volume. Consider the cylinder with the movable piston in **Figure 2.28**. If the piston is pushed downward so that the volume is half the original volume, the density of molecules will double and the pressure will correspondingly double. In general, a lessening of the volume means a correspondingly greater pressure (because the gas density is greater), while a larger volume means a correspondingly smaller pressure (because the gas density is less).

Notice in the example involving the piston that volume and pressure are *inversely proportional*—as one gets bigger, the other gets smaller. This can be represented as

$$V \propto \frac{1}{P}$$

This relationship is known as **Boyle's Law,** after the 17th-century scientist Robert Boyle (1627–1691) who first described it. Boyle discovered that as volume decreases, pressure increases, and as volume increases, pressure decreases. Similarly, an increase in pressure results in a smaller volume, while a decrease in pressure results in a larger volume. Boyle's Law, however, holds true only assuming that the temperature and number of gas particles remain the same.

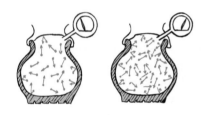

▲ Figure 2.27
When the density of gas in the tire is doubled, the pressure is doubled.

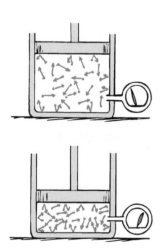

▲ Figure 2.28
When the volume of gas is decreased, the density, and therefore the pressure, are increased.

Here's directly proportional.

Here's inversely proportional.

CONCEPTCHECK

1. You bring a long snorkel with you to the bottom of a deep swimming pool. The snorkel provides a direct path between your lungs and the air above, but you still can't breathe through it. Why?

2. A scuba diver swimming underwater breathes compressed air, which counteracts the water pressure, allowing her to breathe. She then holds her breath while returning to the surface. What happens to the volume of air in her lungs?

CHECK YOUR ANSWERS The weight of the water above you squeezes you from all directions. This water is very heavy, and your lungs are not strong enough to expand against all this pressure.

As she rises to the surface, the water pressure decreases, which allows the compressed air in her lungs to expand in accordance with Boyle's Law. Rising 10 m to the surface results in half the pressure, which results in a doubling of the volume. This is enough to burst the lungs. A first lesson in scuba diving is *not* to hold your breath while ascending—to do so can be fatal.

Charles's Law: Volume and Temperature

The 18th-century French scientist and daring balloon aviator Jacques Charles (1746–1823) discovered the direct relationship between the volume of a gas and its temperature at constant pressure. Charles showed that the volume of a gas increases as its temperature increases. Likewise, the volume of a gas decreases as its temperature decreases, as is shown in **Figure 2.29**.

Remarkably, when the volume and temperature of various gases at constant pressure are plotted on a graph, it appears that at a rather cold temperature, –273.15°C, the volume of any gas dwindles down to zero, as shown in **Figure 2.30**.

Of course, the volume of a gas never reaches zero. Somewhere along the way the gas transforms into a liquid. Nonetheless, in 1848, William Thomson, also known as Lord Kelvin (1824–1907), recognized that this convergent temperature of –273.15°C would be a convenient zero point for an alternative temperature scale by which the absolute motion of particles could be measured. The number 0 is assigned to this lowest possible temperature—*absolute zero*. Accordingly, this scale became known as the Kelvin temperature scale, as was introduced in Section 2.6.

According to **Charles's Law,** the volume and absolute temperature (measured in kelvins) of a gas are *directly proportional*—as one gets bigger, so does the other. This can be represented as

$$V \propto T$$

CONCEPTCHECK

A perfectly elastic balloon holding helium is warmed. What happens to its volume?

CHECK YOUR ANSWER According to Charles's Law, as a gas is warmed, its volume expands.

▼ Figure 2.29
Evan inflates two balloons to the same size and heats one over some boiling water while the other cools down in the freezer. A quick recomparison of the two balloons shows that while the heated balloon expanded, the cooled balloon became smaller. Ask your instructor what happens when a balloon is dipped into –196°C liquid nitrogen and then pulled back out.

▶ **Figure 2.30**
A plot of experimental data showing volume versus the temperature (at constant pressure) for hydrogen, H_2, and oxygen, O_2, in their gaseous phases. Note how the gases converge to zero volume at the same temperature, −273.15°C.

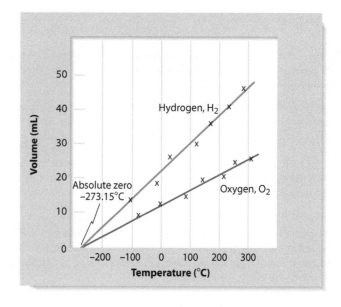

FOR YOUR INFORMATION

In 1783, Jacques Charles was the first to fly in a balloon filled with hydrogen gas, which he created by reacting scrap iron with sulfuric acid. (Two weeks earlier, the Mongolfier brothers had been responsible for the first balloon flight using hot air.) One of Charles's first flights over the French countryside lasted about 45 minutes and took him 15 miles to a small village, where the people were so terrified they tore the balloon to shreds.

READING CHECK

What four interrelated quantities can be used to describe the properties of a gas?

Avogadro's Law: Volume and Number of Particles

The 19th-century Italian scientist Amedeo Avogadro hypothesized that the volume of a gas is a function of the number of gas particles it contains. In other words, as the number of gas particles increases, so does the volume, assuming a constant pressure and temperature. This relationship is known as **Avogadro's Law.** An easy way to demonstrate this law is to grasp the opening of an empty plastic bag and blow into it. The more air molecules you blow into the bag, the bigger its volume. This can be represented as

$$V \propto n$$

CONCEPT CHECK

Assuming the same pressure and temperature, which has a greater number of gas particles: 5 liters of gaseous water, H_2O, or 5 liters of gaseous oxygen, O_2?

CHECK YOUR ANSWER According to Avogadro's Law, the same volume means the same number of gas particles. Each particle here refers to a single molecule of either water, H_2O, or oxygen, O_2. There are just as many water gas particles in 5 liters as there are oxygen gas particles. The nature of the gas particles doesn't matter. But can you answer this question: which has a greater number of atoms?

The Ideal Gas Law and Kinetic Molecular Theory

The properties of a gas can be described by four interrelated quantities: pressure, volume, temperature, and number of particles. Boyle's, Charles's, and Avogadro's gas laws each describe how one quantity varies relative to another as long as the remaining two are held constant. Mathematically, these three laws can be combined into a single law, called the **ideal gas law,** which shows the relationship of all these quantities in a single equation:

$$PV = nRT$$

where P is the pressure of the gas, V is its volume, n is the number of gas particles, and T is the temperature. The R in this equation is the *gas constant*, which is the same no matter what the identity of the gas. Its value depends only upon the chosen units of pressure, volume, and temperature.*

---

*A commonly used value for the gas constant is 0.082057 L atm/K mol. Regarding the units, L is volume in liters, atm is pressure in atmospheres, K is temperature in kelvin, and 1 mole (mol) equals 6.02×10^{23} particles. The concept of the mole is formally introduced in Section 7.3.

The ideal gas law gets its name from the fact that it accurately describes only an *ideal gas*. Such a gas is one in which individual gas particles are considered to occupy no volume in their container. Furthermore, the particles of an ideal gas experience no attractions to or repulsions from one another. Each particle of a real gas, however, does have volume, albeit very, very small. Also, these particles do interact with each other. Real gases, therefore, do not follow any of the gas laws perfectly. At normal pressures, however, the contribution of particle size to total volume is insignificant and gas laws are good predictors of gas behavior. Furthermore, when the temperature of the gas is well above its boiling point, the gas particles are moving so fast that they rebound off one another without sticking. The interactions, therefore, are negligible, so the particles closely resemble those of an ideal gas.

CONCEPTCHECK

Which do gas laws describe more accurately: a gas at high pressure and low temperature or a gas at low pressure and high temperature?

CHECK YOUR ANSWER Gas laws work best for gases at low pressures and high temperatures. The air we breathe is a good example. At atmospheric pressure, the distances between air molecules are much greater than the sizes of the air molecules. Also, air is well above its components' boiling points (N_2 boils at $-196°C$ and O_2 at $-183°C$). This behavior permitted the gas law discoveries of Boyle, Charles, Avogadro, and others.

From the gas laws a model has been developed that gives us insight as to why gases behave as they do. This model, known as the **kinetic molecular theory,** can be summarized as a set of five postulates:

1. A gas consists of tiny particles, either atoms or molecules or both.
2. Gas particles are in constant random motion, colliding with one another and with the walls of their container.
3. The impacts of gas particles on the walls of the container produce a jittery force that appears as a steady push against the inner surface. This pushing force provides the pressure of the enclosed gas.
4. Deviations from gas laws arise primarily because of the interactions occurring among gas particles and because gas particles are not infinitely small.
5. The average kinetic energy (energy due to motion) of the gas particles is directly proportional to the temperature of the gas.

We can use these postulates to rationalize the various gas laws. For Boyle's Law, pressure decreases with increasing volume because collisions of the gas particles with the walls of the container occur less frequently, and those impacts are spread out over a greater area. For Charles's Law, volume increases with increasing temperatures because faster-moving particles push against the inner sides of the container more forcefully, which causes an expansion. Similarly, for Avogadro's Law, volume increases with an increasing number of particles because each particle produces a force on the inner sides of the container, and when there are a greater number of particles, the net force on the inner surface is also greater.

The kinetic molecular theory was developed from the study of gases, but this theory also nicely sums up what we know about liquids and solids. As discussed earlier in this chapter, liquids and solids are also made of tiny particles, either atoms or molecules. These particles are in constant random motion. This motion is confined, however, because the particles are held together by electrical attractions. We'll be discussing the nature of these electrical attractions in later chapters. In the liquid phase, the particles can tumble over one another, which gives the liquid its fluid character. In the solid phase, the particles are held together so strongly that they can vibrate only about fixed positions. Kinetic molecular theory also tells us

FOR YOUR INFORMATION

With your mouth wide open, blow air from your lungs onto the palm of your hand. Now repeat the same procedure with your lips puckered so that your breath is compressed within your mouth and expands upon exiting. You'll discover that as air expands, it cools. In more technical terms, as the pressure decreases, so does the temperature. In regard to the weather, as warm air rises, it expands into less dense higher-altitude air and thus cools. This cooling causes atmospheric water vapor to condense into tiny suspended droplets, which is how clouds are formed.

▲ Figure 2.31
This gap in the roadway of a bridge is called an expansion joint; it allows the bridge to expand and contract. (Can you tell whether this photo taken on a warm day or a cold day?)

that the higher the temperature, the faster the movement of the particles, as was discussed in Section 2.6. This explains why liquids and solids, like gases, also tend to expand when heated and contract when cooled, as illustrated in **Figure 2.31**.

CONCEPTCHECK

Assuming a constant volume and number of particles, as the temperature of a gas increases, what happens to the pressure? Why?

CHECK YOUR ANSWER As the temperature increases in a gas, so does the average kinetic energy (speed) of its particles. Faster-moving particles collide more frequently with the inner surface of a container and strike it harder. More frequent and more forceful impacts mean greater pressure. So as the temperature of a gas increases, so does its pressure, provided the volume and number of gas particles remain constant. Understanding this, can you explain why the pressure inside an automobile tire is greatest just after driving?

CALCULATION CORNER SCUBA DIVING AND HOT AIR BALLOONS

We can express Boyle's Law and Charles's Law mathematically as follows:

$$P_1V_1 = P_2V_2$$

Boyle's Law

$$\frac{V_1}{T_1} = \frac{V_2}{T_2}$$

Charles's Law

Here P_1, V_1, and T_1 represent an original pressure, volume, and temperature, respectively, while P_2, V_2, and T_2 represent a new pressure, volume, and temperature, respectively. Each of the preceding equations contains four variables. For such equations, if three of the variables are known we can calculate the fourth (assuming all temperatures are expressed in kelvins).

EXAMPLE

What would be the new volume of a 1.00 liter balloon if it were brought from sea level, where the air pressure is 1.00 atmosphere, to an altitude of 2500 meters, where air pressure is about 0.743 atmospheres? Assume there is no change in temperature. (As we explore in Chapter 16, the "atmosphere" is a common unit of pressure and is equal to the average atmospheric pressure at sea level.)

$$P_1 = 1.00\,\text{atm}$$
$$P_2 = 0.743\,\text{atm}$$
$$V_1 = 1.00\,\text{liter}$$
$$V_2 = ?$$

Use algebra to rearrange the equation for Boyle's Law to solve for V_2:

$$V_2 = P_1V_1/P_2$$
$$= (1.00\,\text{atm})(1.00\,\text{liter})/(0.743\,\text{atm})$$
$$= 1.35\,\text{liters}$$

YOUR TURN

1. A scuba diver swimming underwater in the ocean breathes compressed air at a pressure of 2 atmospheres.

If she holds her breath while returning to the surface, by how much might the volume of her lungs increase?

2. A 5.00-liter rubber balloon is submerged 5.00 meters under ocean water, where its new volume is measured to be 3.38 liters. Show that the pressure at this depth is 1.48 atmospheres.

3. A perfectly elastic 419-liter balloon is heated from 25°C (298 K) to 50°C (323 K). To what new volume does it expand?

4. A hot air balloon 401,000 liters in volume is warmed from 298 K to 398 K. As the air inside the balloon expands, it is unable to stretch the fabric, which is not very elastic. Instead, the expanded air escapes out of a hole placed at the top of the balloon. Show that 135,000 liters of air escapes.

5. Air has a density of 1.18 g/L. Show that the hot air balloon in the previous question is now lighter by 159 kilograms, which helps the balloon to rise.

The answers for Calculation Corners appear at the end of each chapter.

Chapter 2 Review

LEARNING OBJECTIVES

Describe the particulate nature of matter. (2.1)	→ *Questions 1, 2, 25, 38, 44, 45, 83*
Describe the evidence for the particulate nature of matter. (2.2)	→ *Questions 3–5, 26, 46–49, 85*
Distinguish between mass, weight, and volume. (2.3)	→ *Questions 6–8, 27, 29–31, 35, 50–53*
Calculate the density of a material. (2.4)	→ *Questions 9, 10, 32, 33, 54–57*
Differentiate between potential and kinetic energy. (2.5)	→ *Questions 11–13, 34, 58–61*
Distinguish between heat and temperature. (2.6)	→ *Questions 14–16, 39–41, 62–67*
Describe the particulate nature of three phases of matter. (2.7)	→ *Questions 17–19, 42, 68–72*
Describe how the volume of a gas is affected by pressure, temperature, and number of particles. (2.8)	→ *Questions 20–24, 28, 36, 37, 43, 73–82, 84*

SUMMARY OF TERMS (KNOWLEDGE)

Absolute zero The lowest possible temperature, which is the temperature at which the atoms of a substance have no kinetic energy: 0 K = −273.15°C = −459.7°F.

Alchemy A medieval endeavor concerned with turning other metals to gold.

Atoms Extremely small fundamental units of matter.

Avogadro's Law A gas law that describes the direct relationship between the volume of a gas and the number of gas particles it contains at constant pressure and temperature. The greater the number of particles, the greater the volume.

Boiling Evaporation in which bubbles form beneath the liquid surface.

Boyle's Law A gas law that describes the indirect relationship between the pressure of a gas sample and its volume at constant temperature. The smaller the volume, the greater the pressure.

Charles's Law A gas law that describes the direct relationship between the volume of a gas sample and its temperature at constant pressure. The greater the temperature, the greater the volume.

Condensation The transformation of a gas to a liquid.

Density The amount of mass contained in a sample divided by the volume of the sample.

Energy The capacity to do work.

Evaporation The transformation of a liquid to a gas.

Freezing The transformation of a liquid to a solid.

Gas Matter that has neither a definite volume nor a definite shape, always filling any space available to it.

Heat The energy that flows from one object to another because of a temperature difference between the two.

Ideal gas law A gas law that summarizes the pressure, volume, temperature, and number of particles of a gas within a single equation often expressed as $PV = nRT$, where P is presure, V is volume, n is number of molecules, R is the gas constant, and T is temperature given in kelvin.

Kinetic energy Energy due to motion.

Kinetic molecular theory A theory that explains the properties of solids, liquids, and gases by proposing that they consist of rapidly moving tiny particles, either atoms or molecules or both.

Liquid Matter that has a definite volume but no definite shape, assuming the shape of its container.

Mass The quantitative measure of how much matter an object contains.

Melting The transformation of a solid to a liquid.

Molecule An extremely small fundamental structure built of atoms.

Potential energy Stored energy.

Solid Matter that has a definite volume and a definite shape.

Submicroscopic The realm of atoms and molecules, where objects are smaller than can be detected by optical microscopes.

Temperature How warm or cold an object is relative to some standard. Also, a measure of the average kinetic energy per molecule of a substance, measured in degrees Celsius, degrees Fahrenheit, or kelvin.

Thermometer An instrument used to measure temperature.

Volume The amount of space an object occupies.

Weight The gravitational force of attraction between two bodies (where one body is usually the Earth).

READING CHECK QUESTIONS (COMPREHENSION)

2.1 The Submicroscopic World Is Super-Small

1. It would take you 31,800 years to count to a trillion. About how many times would you have to do this to have counted all the atoms there are in a single grain of sand?

2. Is a biological cell macroscopic, microscopic, or submicroscopic?

2.2 Discovering the Atom

3. The term *atom* was derived from what Greek phrase?

4. What 18th-century chemist discovered the law of mass conservation?

5. What did Mendeleev predict based upon his newly created periodic table?

2.3 Mass Is How Much and Volume Is How Spacious

6. What is inertia, and how is it related to mass?

7. Which can change from one location to another: mass or weight?

8. What is the difference between an object's mass and its volume?

2.4 Density Is the Ratio of Mass to Volume

9. The units of density are a ratio of what two quantities?

10. What happens to the volume of a loaf of bread that is squeezed? The mass? The density?

2.5 Energy Is the Mover of Matter

11. What do we call the energy an object has because of its position?

12. What do we call the energy an object has because of its motion?

13. Which represents more energy: a joule or a calorie?

2.6 Temperature Is a Measure of How Hot—Heat It Is Not

14. In which is the average speed of the molecules less: in cold coffee or in hot coffee?

15. Which temperature scale has its zero point as the point of zero atomic and molecular motion?

16. Is it natural for heat to travel from a cold object to a warmer object?

2.7 The Phase of a Material Depends on the Motion of Its Particles

17. How does the arrangement of particles in a gas differ from the arrangement of particles in liquids and solids?

18. Which requires the removal of thermal energy: melting or freezing?

19. What is it called when evaporation takes place beneath the surface of a liquid?

2.8 Gas Laws Describe the Behavior of Gases

20. What happens to the pressure inside a tire as more air molecules are pumped into the tire?

21. What happens to the volume of a gas as its temperature is increased? (Assume constant pressure and number of particles.)

22. At what temperature do gases theoretically cease to occupy any volume?

23. What happens to the volume of a gas as more gas particles are added to it? (Assume constant pressure and temperature.)

24. Why do real gases not obey the ideal gas law perfectly?

CONFIRM THE CHEMISTRY (HANDS-ON APPLICATION)

25. A TV screen looked at from a distance appears as a smooth continuous flow of images. Up close, however, we see this is an illusion. What really exists are a series of tiny dots (pixels) that change color in a coordinated way to produce images. Use a magnifying glass to examine closely the screen of a computer monitor or television set.

26. Add a pinch of red-colored Kool-Aid crystals to a still glass of hot water. Add the same amount of crystals to a second still glass of cold water. With no stirring, which would you expect to become uniform in color first: the hot water or the cold water? Why?

27. Pennies dated 1982 or earlier are nearly pure copper, each having a mass of about 3.5 grams. Pennies dated after 1982 are made of copper-coated zinc, each having a mass of about 2.9 grams. Hold a pre-1982 penny on the tip of your index finger and a post-1982 penny on the

tip of your other index finger. Move your forearms up and down to feel the difference in inertia—the difference of 0.6 grams (600 milligrams) is subtle but not beyond a set of well-tuned senses. If one penny on each finger is below your threshold, try two pre-1982s stacked on one finger and two post-1982s stacked on the other. Share this activity with a friend.

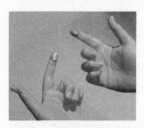

28. Air molecules stuck inside an inflated balloon are perpetually colliding with the inner surface of the balloon. Each collision provides a little push outward on the balloon. All the many collisions working together is what keeps the balloon inflated. To get a "feel" for what's happening here, add about a tablespoon of tiny beads to a large balloon. (Beans, grains of rice, etc., will also work.) Inflate the balloon to its full size and tie it shut. Hold the balloon in the palms of both hands and shake rapidly. Can you feel the collisions? As you shake the balloon wildly, the flying beads represent the gaseous phase. How should you move the balloon so that the beads represent the liquid phase? The solid phase? Absolute zero?

29. Fill a tall glass two-thirds full with water and mark the water level with masking tape. Fill a small plastic canister, such as a film canister, with pennies. Cap the canister and place it in the water. Note the new water level with a second strip of tape. Remove the canister, being careful not to splash water out of the glass. Remove half the pennies from the canister so as to decrease its mass.

Cap the canister; predict how much the water level will rise when you submerge the canister. Which of the following statements do your results support? (a) The volume of water an object displaces depends only on the dimensions of the object and not on its mass. (b) The volume of water an object displaces depends on both the dimensions and the mass of the object.

THINK AND SOLVE (MATHEMATICAL APPLICATION)

30. What is the mass in kilograms of a 130-pound human standing on Earth?

31. Gravity on the Moon is only one-sixth as strong as gravity on the Earth. What is the mass of a 10-kilogram object on the Moon, and what is its mass on the Earth?

32. Someone wants to sell you a piece of gold and says it is nearly pure. Before buying the piece, you measure its mass to be 52.3 grams and find that it displaces 4.16 mL of water. Calculate its density and consult Table 2.1 to assess its purity.

33. What volume of water would a 52.3-gram sample of pure gold displace?

34. How many joules are there in a candy bar containing 230,000 calories?

35. How many milliliters of dirt are there in a hole that has a volume of 5 liters? How many milliliters of air?

36. You measure the pressure of the four tires of your car each to be 35.0 pounds per square inch (psi). You then roll your car forward so that each tire is upon a sheet of paper. You outline the surface area of contact between each tire and the paper, which you later measure to be 32.0 square inches. What is the weight of your car?

37. A perfectly elastic balloon holding 1.0 liters of helium at 298 K is warmed to 348 K. What is the new volume of the helium-filled balloon?

THINK AND COMPARE (ANALYSIS)

38. Rank the following in order of increasing volume:
    a. Bacterium
    b. Virus
    c. Water molecule

39. Rank the following in order of increasing temperature:
    a. 100 K
    b. 100°C
    c. 100°F

40. Rank the following in order of increasing molecular kinetic energy:
    a. Cupful of boiling water at 100°C
    b. Swimming pool full of boiling water at 100°C
    c. A cup of ice at –10°C on the top floor of a skyscraper

41. Rank the following in order of increasing *average* molecular kinetic energy:
    a. Cupful of boiling water at 100°C
    b. Swimming pool full of boiling water at 100°C
    c. A cup of ice at –10°C on the top floor of a skyscraper

42. Rank the following in order of increasing force of attraction between its submicroscopic particles:
    a. Sugar   b. Water   c. Air

43. Rank the following in order of increasing temperature: A billion molecules of a gas at 1 atmosphere of pressure in a
    a. 10-liter container.
    b. 5-liter container.
    c. 1-liter container.

THINK AND EXPLAIN (SYNTHESIS)

2.1 The Submicroscopic World Is Super-Small

44. You take 50 mL of small BBs and combine them with 50 mL of large BBs and you get a total of 90 mL of BBs of mixed size. Explain.

45. You take 50 mL of water and combine it with 50 mL of purified alcohol and you get a total of 98 mL of mixture. Explain.

2.2 Discovering the Atom

46. In one of his experiments, Lavoisier used sunlight to heat a piece of tin on a floating block of wood covered by a glass jar. As the tin decomposed, the water level inside the jar rose. Explain this result using Avogadro's Law.

47. Which of Dalton's postulates accounts for Lavoisier's mass conservation principle?

48. What is wrong with the following depiction of a chemical reaction?

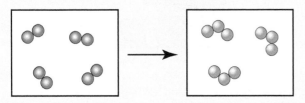

49. A friend argues that if mass were really conserved he would never need to refill his gas tank. What explanation do you offer your friend.

2.3 Mass Is How Much and Volume Is How Spacious

50. Can an object have mass without having weight? Can it have weight without having mass?

51. Does a 2-kilogram solid iron brick have twice as much mass as a 1-kilogram solid iron brick? Twice as much weight? Twice as much volume?

52. Which weighs more: a liter of water at 20°C or a liter of water at 80°C?

53. A little girl sits in a car at a traffic light holding a helium-filled balloon. The windows are closed and the car is relatively airtight. When the light turns green and the car accelerates forward, her head pitches backward but the balloon pitches forward. Explain why.

2.4 Density Is the Ratio of Mass to Volume

54. What happens to the density of a gas as the gas is compressed into a smaller volume?

55. The following three boxes represent the number of submicroscopic particles in a given volume of a particular substance at different temperatures. Which box represents the highest density? Which box represents the highest temperature? Why would this be a most unusual substance if box (a) represented the liquid phase and box (b) represented the solid phase?

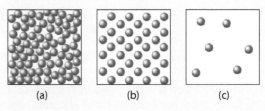

(a)   (b)   (c)

56. What happens to the density of air as it is heated?

57. What is the density of empty space?

2.5 Energy Is the Mover of Matter

58. Which is more evident: potential or kinetic energy? Explain.

59. At what point in its motion is the kinetic energy of a pendulum bob at a maximum? At what point is its potential energy at a maximum?

60. Consider a ball thrown straight up in the air. At what position is its kinetic energy at a maximum? Where is its gravitational potential energy at a maximum?

61. Does a car burn more gasoline when its lights are turned on? Defend your answer.

2.6 Temperature Is a Measure of How Hot—Heat It Is Not

62. Which has more total energy: a cup of boiling water at 100°C or a swimming pool of slightly cooler water at 90°C?

63. Under what circumstances does heat naturally travel from a cold substance to a warmer substance?

64. Distinguish between temperature and heat.

65. An old remedy for separating two nested drinking glasses stuck together is to run water at one temperature into the inner glass and then run water at a different temperature over the surface of the outer glass. Which water should be hot and which should be cold?

66. A supersonic jet heats up considerably when traveling through the air at speeds greater than the speed of sound. As a result, the jet in flight is several centimeters longer than when it is on the ground. Offer an explanation for this length change from a submicroscopic perspective.

67. Creaking noises are often heard in the attic of old houses on cold nights. Give an explanation in terms of thermal expansion.

2.7 The Phase of a Material Depends on the Motion of Its Particles

68. Which has stronger attractions among its submicroscopic particles: a solid at 25°C or a gas at 25°C? Explain.

69. The following leftmost diagram shows the moving particles of a gas within a rigid container. Which of the three boxes on the right (a, b, or c) best represents this material upon the addition of heat?

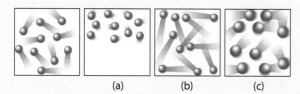

           (a)   (b)   (c)

70. The following leftmost diagram shows two phases of a single substance. In the middle box, draw what these particles would look like if heat were taken away. In the box on the right, show what they would look like if heat were added. If each particle represents a water molecule, what is the temperature of the box on the left?

71. Humidity is a measure of the amount of water vapor in the atmosphere. Why is humidity always very low inside your kitchen freezer?

72. Why is perfume typically applied behind the ear rather than on the ear?

2.8 Gas Laws Describe the Behavior of Gases

73. Would you or the gas company gain by having gas warmed before it passed through your gas meter?

74. At a depth of about 10 meters the water pressure on you is equal to the pressure of 1 atmosphere. The total pressure on you is therefore 2 atmospheres—1 atm from the water and 1 atm from the air above the water. When the following glass is pushed down to a depth of around 10 meters in depth, what will be the level of the water on the inside of the glass?

75. Why do you suppose that airplane windows are smaller than bus windows?

76. When you suck a drink through a soda straw, what causes the drink to rise into your mouth: the muscles of your lungs and cheeks or the weight of the atmosphere?

77. A soda straw fits snuggly through a cork that's wedged into a narrow-neck bottle containing a liquid beverage. You try to suck the beverage out of the bottle, but doing so is not possible for you. Why?

78. What happens to the size of the bubbles of boiling water as they rise to the surface? Why? Which gas law applies?

79. A child's lost helium-filled rubber balloon rises higher and higher into the sky. What eventually happens to the balloon? What happens to the helium?

80. Use Boyle's Law to explain why a package of chips puffs up on board a high flying airplane.

81. An airliner cruises around 30,000 feet, but the cabin is kept at more comfortable pressure that corresponds to around 8000 feet, which is about 0.743 atmospheres. For most people, the cabin would be even more comfortable if it were kept at a pressure of 1.00 atmospheres. Why don't airlines pressurize their cabins to 1.00 atmospheres? Which gas law applies?

82. Close your lips and blow air into your mouth. What happens to your cheeks? Why? Which gas law applies?

THINK AND DISCUSS (EVALUATION)

83. Your friend smells cinnamon coming from an inflated rubber balloon containing cinnamon extract. You tell him that the cinnamon molecules are passing through the micropores of the balloon. He accepts the idea that the balloon contains micropores but insists that he is simply smelling cinnamon-flavored air. You explain that scientists have discovered that gases are made of molecules, but that's not good enough for him. He needs to see the evidence for himself. How might you lead him to accept the concept of molecules?

84. You're at the top of a 200-mile tall tower that extends above the atmosphere into outer space. Because there is no air that high you are wearing a pressurized spacesuit. If a hole were poked into your suit, the air molecules in your suit would escape. Before you quickly patch the hole, what would happen to the pressure inside your suit? What would happen to the volume of your suit? Would the escaped air molecules fall up or down or remain level? Which laws apply to each of these questions?

85. You're at the top of a 200-mile tall tower that extends above the atmosphere into outer space. On the observation deck you step onto a scale, which shows you to be a couple pounds lighter than you were at the bottom of the tower. But since you're in outer space, you wonder what would happen if you stepped off the observation deck. You instead hold the scale over the edge and then let go of it. What happens to the scale? Does it float or fall? Is there gravity in outer space? Would it be possible for the space shuttle to park along side this observation deck? Why is it important for us to know these sorts of things?

86. The British diplomat, physicist, and environmentalist John Ashton in speaking to a group of scientists stated (paraphrased): "There has to be much better communication between the world of science and the world of politics. Consider the different meaning of the word 'uncertainty.' To scientists, it means uncertainty over the strength of a signal. To politicians it means 'go away and come back when you're certain.'" Pretend you are a scientist with strong but inconclusive evidence in support of impending climate change. How might you best persuade politicians to take action?

READINESS ASSURANCE TEST (RAT)

If you have a good handle on this chapter, then you should be able to score at least 7 out of 10 on this RAT. Check your answers online at www.ConceptualChemistry.com. If you score less than 7, you need to study further before moving on.

Choose the BEST answer to the following.

1. Chemistry is the study of the
   a. submicroscopic only because it deals with atoms and molecules, which can't be seen with a microscope.
   b. microscopic only because it pertains to the formation of crystals.
   c. macroscopic only because it deals with powders, liquids, and gases that fill beakers and flasks.
   d. submicroscopic, microscopic, and macroscopic because most everything is made of atoms and molecules.

2. The advent of modern chemistry began with the work of
   a. Aristotle.
   b. Democritus.
   c. Lavoisier.
   d. Dalton.
   e. Mendeleev.

3. Does a 2-kg solid gold brick have twice as much volume as a 1-kg solid block of wood?
   a. The gold brick has twice the volume.
   b. The gold brick has half the volume.
   c. The volume of the gold brick could be less than that of the block of wood.
   d. The volume of the gold brick compared to the block of wood depends on the weight of the wood.

4. A child swinging on a swing has her greatest kinetic energy when she is
   a. as far from the ground as possible.
   b. as close to the ground as possible.
   c. moving downward toward the ground.
   d. moving upward toward the sky.

5. In which are the molecules, on average, moving faster: a swimming pool of boiling water or a cup of boiling water?
   a. The swimming pool because it contains more energy.

   b. The cup because the molecules rebound off the container's inner surface more frequently.
   c. The average motion of molecules in each is the same.

6. Heat is simply another word for
   a. temperature.
   b. energy.
   c. energy that flows from hot to cold.
   d. All of the above

7. Water freezes at a temperature of
   a. 0°C.
   b. 273 K.
   c. Both of these
   d. Neither of these

8. What type of phase change does the following figure best describe?

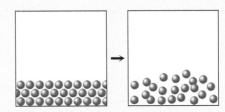

   a. Melting
   b. Condensation
   c. Evaporation
   d. Freezing

9. Gases do not follow the ideal gas law perfectly because the particles that comprise the gas
   a. are not infinitely small.
   b. can stick to each other as they collide.
   c. move in random directions.
   d. Two of the above
   e. All of the above

10. At a temperature above absolute zero, the phase in which atoms and molecules no longer move is the
   a. solid phase.
   b. liquid phase.
   c. gas phase.
   d. None of the above

ANSWERS TO CALCULATION CORNER (MANIPULATING AN ALGEBRAIC EQUATION)

1. 0.5 grams per milliliter
2. 400 grams
3. 100 milliliters

Scuba Diving and Hot Air Balloons

1. The new volume of the scuba diver will be two times that of the original volume:

$$V_2 = P_1 V_1 / P_2$$
$$= (2.00 \text{ atm})(V_1)/(1 \text{ atm})$$
$$= 2.00 V_1$$

2. Plug the following values into Boyle's Law and solve to the new pressure, P_2:

$$P_1 V_1 = P_2 V_2$$
$$(1 \text{ atm})(5.00 \text{ L}) = (P_2)(3.38 \text{ L})$$
$$P_2 = (1.00 \text{ atm})(5.00 \text{ L})/(3.38 \text{ L})$$
$$P_2 = 1.48 \text{ atm}$$

3. Plug the following values in Charles's Law and solve for the new volume, V_2. Be sure to use the absolute temperatures given in kelvin:

$$\frac{V_1}{T_1} = \frac{V_2}{T_2}$$
$$(419 \text{ L})/(298 \text{ K}) = V_2/(323 \text{ K})$$
$$V_2 = (419 \text{ L})(323 \text{ K})/(298 \text{ K})$$
$$V_2 = 454 \text{ L}$$

4. Use Charles's Law to show that the new volume at the higher temperature would be 536,000 liters. As per Appendix B, the answer needs to have three significant figures, which is why 535,564 liters rounds up to 536,000 liters. Take the difference to find the volume of gas that escapes from the balloon.

$$\frac{V_1}{T_1} = \frac{V_2}{T_2}$$
$$(401,000 \text{ L})/(298 \text{ K}) = V_2/(398 \text{ K})$$
$$V_2 = (401,000 \text{ L})(398 \text{ K})/(298 \text{ K})$$
$$V_2 = 536,000 \text{ L}$$

Volume of air that escapes = $V_2 - V_1$ = 135,000 liters

5. The density of air is 1.18 grams per liter. In other words, 1.18 grams equals 1 liter. As per the Calculation Corner at the end of Chapter 1, you should be able to recognize this as a conversion factor. Put this together with the conversion factor of 1000 grams equals 1 kilogram and we can convert from liters to grams to kilograms in a single equation:

(135,000 L of air) × (1.18 grams of air/L of air)
× (1 kilogram/1000 grams) = 159 kg

Green Chemistry

Over the past couple of decades there has been a growing effort among industries, governmental agencies, and universities to develop technologies that allow materials to be manufactured with reduced or no negative environmental impacts. This is an area of research known as *green chemistry*. To ensure implementation, green chemistry technologies are also designed to be cost competitive and profitable.

Green chemistry applied to the manufacture of materials follows a set of high standards, as outlined here:

1. The raw materials used in the manufacturing process should be sustainably renewable (such as from agriculture) rather than depleting (such as from fossil fuels).

2. Waste products from the manufacturing process should be minimized. Any waste products should be either recycled or rendered environmentally safe.

3. The manufacturing process should be energy efficient. For example, chemical reactions that run at easy-to-attain ambient temperatures and pressures are most desirable. Also, the fewer steps there are to a manufacturing process, the better. Energy can also be saved by using catalysts and by monitoring the manufacturing process carefully to ensure optimal operating conditions.

4. The manufacturing process should minimize hazards. This can be accomplished by choosing chemical reactions that involve fewer toxic chemicals and are not explosive.

5. The desired final products of the manufacturing process should have little or no toxicity and be recyclable. Furthermore, the materials of these products should degrade to innocuous substances after use and not accumulate in the environment.

To illustrate the power and potential of green chemistry, it is useful to look into a few of the growing number of green chemistry projects. To learn about many other projects, be sure to visit either www.epa.gov/greenchemistry or www.acs.org/greenchemistry.

Renewable Toner

cleaner, greener toner solution

Laser printers and copiers consume about 400 million pounds of toner each year in the United States. Most of this toner is made from nonrenewable petroleum. This toner is also not easily de-inked from the paper, which means expensive and harsh chemicals are needed for recycling. As a green chemistry alternative, the Battelle Institute (www.Battelle.org) oversaw the development of a high-quality toner made from soybeans. Much less energy is required to create this soy-based toner, which is made from a renewable resource. Furthermore, soy-based toner is easily de-inked from paper, thereby simplifying the recycling process.

Benign Pesticides

As described in Chapter 15, we rely on pesticides to protect our crops. Many modern pesticides, however, are harmful to a broad range of species, including beneficial insects, such as those that pollinate. A number of these pesticides also persist in the environment, where they are taken up into the food chain. To counter these disadvantages, scientists at Dow AgroSciences (www.dowagro.com) have been able to produce pesticides that are specific to the nervous system of certain detrimental insects. As a result, toxicity

to other species, including mammals, is many times lower than that of traditional pesticides. Furthermore, the starting materials for the synthesis of these pesticides are created from fermentation broths. In other words, the starting materials are grown much like a yogurt culture. Also, due in part to their natural origin, these pesticides decompose well, so that they do not persist in the environment.

Affordable Hydrogen Peroxide

Almost everyone is familiar with disinfecting solutions of hydrogen peroxide, H_2O_2, available from the local drugstore. What most people don't know is that the manufacture of this simple molecule on a large scale is costly, dangerous, and involves numerous steps. This is unfortunate, because H_2O_2 is an effective and environmentally friendly oxidizing agent that would be used extensively by industries if not for its high cost (see Chapter 11). Instead, companies turn to other oxidizing agents, such as chlorine. Although cheaper, chlorine tends to produce highly toxic by-products, such as dioxins. Using nanotechnology, chemists at Headwaters Technology (www.headwaters.com) have recently designed catalysts that allow the efficient and safe production of H_2O_2 in a single step directly from hydrogen, H_2, and oxygen, O_2 (see Section 3.8). Cheaper and readily available H_2O_2 is good news for the profit margins of the chemical industry

as well as for the environment, where H_2O_2 simply decomposes to water (H_2O) and oxygen.

Verstile Ionic Liquids

Salts, such as sodium chloride, NaCl, are commonly solids at room temperature. These salts can be transformed into a liquid solution by adding water. There are some salts, however, that are already liquids at room temperature—no water needed! These *ionic liquids* have attracted much attention over the past couple of decades for their many unusual properties and potential applications, especially when it comes to green chemistry.

Many ionic liquids are organic compounds, which means it is easy to modify their properties by tweaking their chemical structures (see Chapter 12). An ionic liquid made of the following compound, for example, has been found to have great potency against the polymer coats, called biofilms, that colonies of bacteria build around themselves for protection. Biofilms are a major problem in hospitals. They also foul the hulls of ships and pipes in industrial machinery. In these situations, biofilms are treated with harsh chemicals. The ionic liquids offer a potentially more effective and environmentally friendly alternative.

▲ 1-hexyl-3-methyllimidazolium chloride

Very exciting is the fact that certain ionic liquids are able to dissolve cellulose, which is, by far, the most abundant and renewable material and energy resource on this planet. As discussed in Chapter 13, Cellulose is the structural biopolymer of plants. Strands of cellulose bind very tightly to each other, which is good for the wood we use to build our houses or the cotton in our clothes. When the

▲ Switchgrass is an efficient source of cellulose in that it grows year round in many environments and never needs replanting.

cellulose is dissolved in ionic liquids, however, this tight binding relaxes, allowing the separation of individual strands. Separated strands of cellulose can then be easily broken down into sugars that can be fermented into ethanol. (Ethanol from such sources is known as cellulosic ethanol.) They can also serve as a starting material for nonpetroleum-based polymers. Both these applications would help move us away from our dependence upon nonrenewable and greenhouse gas-emitting petroleum.

CONCEPT CHECK

How can we be sure that green chemistry technology will get implemented by industries?

WAS THIS YOUR ANSWER? Green chemistry technology is good for the pocketbooks of the industries that implement this technology as well as for the environment.

Think and Discuss

1. Specify which of the principles of green chemistry are illustrated by the example of affordable hydrogen peroxide.

2. To what extent should green chemistry technologies be exported to developing nations? Why?

3. As a consumer, what can you do to support green chemistry?

4. There is now much research being dedicated to the transformation of ever-abundant and renewable cellulose into biofuels and bioplastics. Why wasn't this research started in earnest during the 1960s, which was a decade of much environmental awakening?

5. About 40 percent of all our solid waste sent to landfills consists of cellulose-containing products, such as paper, cardboard, and packaging. Might our landfills one day be considered a valuable source of cellulosic ethanol?

▲ Gold and diamond are elements in that they each consist of only one type of atom.

3

Elements of Chemistry

THE MAIN IDEA

Elements combine to form compounds, which blend together to form mixtures.

Chapter 1 introduced 9 key terms, Chapter 2 provided 28, and in this chapter you'll find another 26. Why all these new terms? In the laboratory, chemists perform experiments, make many observations, and then draw conclusions. Over time, the result is a growing body of new knowledge that inevitably exceeds the capacity of everyday language. New terms are needed as we attempt to describe the nature of matter beyond its casual appearance.

Instead of just memorizing the formal definitions of terms, you will serve yourself better by making sure that you understand the underlying concepts. Practice articulating and paraphrasing these concepts aloud to yourself or to a friend without looking at the book. When you are able to express these concepts in your own words, you will have the insight to do well in this course and beyond.

Chemistry

The Fire-Extinguishing Gas

You likely know that baking soda and vinegar combine to form a froth of bubbles. But what is the nature of the gas within these bubbles?

PROCEDURE

1. Wearing your safety goggles, add about a teaspoon of baking soda to a tall glass.
2. Place tape on the outside of the glass one-third of the way up from the bottom of the glass.
3. Slowly add about a tablespoon of white distilled vinegar to the baking soda. Allow the bubbles to subside before adding additional tablespoons of the vinegar. Do not fill the glass with vinegar beyond the taped mark.

4. Remove all flammable materials from around the glass, especially paper towels.
5. Light a wooden match and dip the flame into the mouth of the glass. At some point the flame should be extinguished. Drop the match into the glass if it does not go out.

ANALYZE AND CONCLUDE

1. At what level inside the glass is the flame extinguished? If others are doing this with you, do their matches go out at the same level? Are you able to raise and lower the flame right at the point where the flame is about to be extinguished? Your glass is not sealed on top, so why doesn't the invisible gas it contains escape?
2. Is there a limit to the number of times a flame can be

extinguished by this gas? If you were to tilt the glass part way so that no liquid poured out, would anything else pour out? How might you tell?
3. Did the gas in the glass exist before you added the vinegar to the baking soda? Is this gas heavier or lighter than air? How else is this gas different from the air we breathe?

3.1 Matter Has Physical and Chemical Properties

EXPLAIN THIS

Why are physical changes typically easier to reverse than chemical changes?

Properties that describe the look or feel of a substance, such as color, hardness, density, texture, and phase, are called **physical properties.** Every substance has its own set of characteristic physical properties that we can use to identify that substance (**Figure 3.1**).

The physical properties of a substance can change when conditions change, but that does not mean a different substance is created. Cooling liquid water to below 0°C causes the water to change to ice, but the substance is still water, no matter what the phase. The only difference is how the H_2O molecules are arranged and how rapidly they are moving. In the liquid phase, the water molecules tumble around one another, whereas in the ice phase, they vibrate about fixed positions. Water freezing is an example of what chemists call a physical change. During a **physical change,** a substance changes its phase or some other physical property but not its chemical identity. As shown in **Figure 3.2**, water in either the liquid or solid phase is still made of water

LEARNING OBJECTIVE

Describe how materials can be identified by their physical and chemical properties.

▼ Figure 3.1
Gold, diamond, and water can be identified by their physical properties.

Gold
Opacity: opaque
Color: yellowish
Phase at 25°C: solid
Density: 19.3 g/mL

Diamond
Opacity: transparent
Color: colorless
Phase at 25°C: solid
Density: 3.5 g/mL

Water
Opacity: transparent
Color: colorless
Phase at 25°C: liquid
Density: 1.0 g/mL

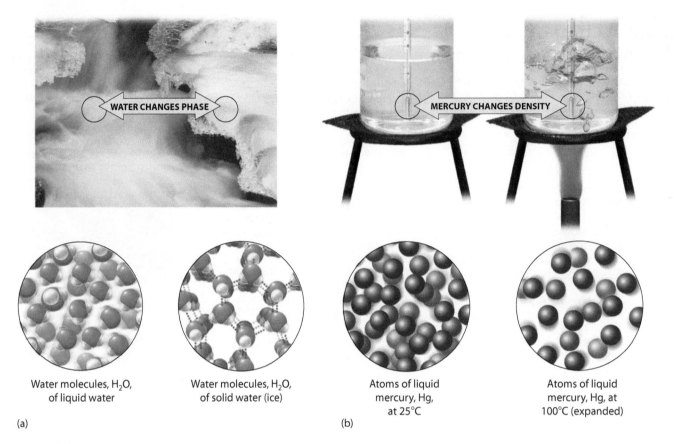

▲ Figure 3.2
Two physical changes. (a) Liquid water and ice might appear to be different substances, but at the submicroscopic level, it is evident that both consist of water molecules. (b) At 25°C, the atoms in a sample of mercury are a certain distance apart, yielding a density of 13.5 grams per milliliter. At 100°C, the atoms are farther apart, meaning that each milliliter now contains fewer atoms than at 25°C, and the density is now 13.4 grams per milliliter. The physical property we call density has changed with temperature, but the identity of the substance remains unchanged: mercury is mercury.

molecules. Likewise, the density of elemental mercury, Hg, decreases with increasing temperature because its atoms become spaced farther apart—but the mercury is still made of mercury atoms.

Chemical properties are those that characterize the ability of a substance to react with other substances or to transform from one substance to another. **Figure 3.3** shows three examples. One chemical property of methane, the main component of natural gas, is that it reacts with oxygen to produce carbon dioxide and water, along with appreciable heat energy. Similarly, it is

▶ Figure 3.3
The chemical properties of substances determine the ways in which they can change into new substances. Natural gas and baking soda, for example, can both undergo chemical reactions in which they are transformed into carbon dioxide and water. Similarly, copper can be transformed into patina.

Methane
Reacts with oxygen to form carbon dioxide and water, giving off lots of heat during the reaction.

Baking soda
Reacts with vinegar to form carbon dioxide and water, absorbing heat during the reaction.

Copper
Reacts with carbon dioxide and water to form the greenish-blue substance called patina.

a chemical property of baking soda that it can react with vinegar to produce carbon dioxide and water while absorbing a small amount of heat energy. A chemical property of copper is that it reacts with carbon dioxide and water to form a greenish-blue solid known as patina. Copper statues exposed to the carbon dioxide and water in the air become coated with patina. The patina is not copper, it is not carbon dioxide, and it is not water. It is a new substance formed by the reaction of these chemicals with one another.

CONCEPTCHECK
The melting of gold is a physical change. Why?

CHECK YOUR ANSWER During a physical change, a substance changes only one or more of its physical properties; its chemical identity does not change. Because melted gold is still gold but in a different form, its melting is only a physical change.

During a chemical change, there is a change in the way the atoms are *chemically bonded* to one another. A **chemical bond** is the force of attraction between two atoms that holds them together in a compound. A methane molecule, for example, is made of a single carbon atom bonded to four hydrogen atoms, and an oxygen molecule is made of two oxygen atoms bonded to each other. **Figure 3.4** shows the chemical change in which the atoms in a methane molecule and those in two oxygen molecules first pull apart and then form new bonds with different partners, resulting in the formation of molecules of carbon dioxide and water.

Any change in a substance that involves a rearrangement of the way atoms are bonded is called a **chemical change.** Thus, the transformation of methane and oxygen to carbon dioxide and water is a chemical change, as are the other two transformations shown in Figure 3.3.

The chemical change shown in **Figure 3.5** occurs when an electric current is passed through water. The energy of the current is sufficient to pull bonded atoms away from each other. Loose atoms then form new bonds with different atoms, which results in the formation of new molecules. Thus, water molecules are changed to molecules of hydrogen and oxygen, two substances that are very different from water. The hydrogen and oxygen are both gases at room temperature, and they can be seen as bubbles rising to the surface of the liquid.

In the language of chemistry, materials undergoing a chemical change are said to be *reacting*. Methane reacts with oxygen to form carbon dioxide and water. Water reacts when exposed to electricity to form hydrogen gas and oxygen gas. Thus, the term *chemical change* means the same thing as *chemical reaction*. During a **chemical reaction,** new materials are formed by a change in the way atoms are

READINGCHECK

What happens to the atoms within a molecule undergoing a chemical reaction?

CHEMICAL CONNECTIONS

Chemically speaking, how is your fingernail connected to the air?

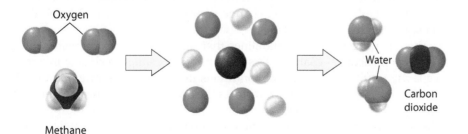

▲ Figure 3.4
The chemical change in which molecules of methane and oxygen transform to molecules of carbon dioxide and water, as atoms break old bonds and form new ones. In this sort of illustration, note that each sphere represents an atom, while a set of joined spheres represents a molecule.

▶ **Figure 3.5**
Water can be transformed to hydrogen gas and oxygen gas by applying the energy of an electric current. This is a chemical change, because new materials (the two gases) are formed as the atoms originally found in the water molecules are rearranged.

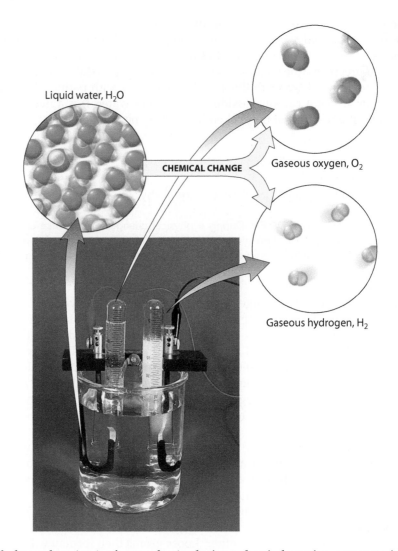

Liquid water, H_2O

CHEMICAL CHANGE

Gaseous oxygen, O_2

Gaseous hydrogen, H_2

bonded together. Again, for emphasis, *during a chemical reaction, new materials are formed by a change in the way atoms are bonded together.* We will explore chemical bonds and their role in chemical reactions when we get to Chapters 6 and 9.

Distinguishing Physical and Chemical Changes Can Be Difficult

After a physical change, the molecules are the same as the ones you started with. After a chemical change, the original molecules no longer exist and new ones are in their place. In both cases, however, there is a change in physical appearance. Frozen water and melted water, for example, look very different. Likewise, iron and rust look very different (**Figure 3.6**). So how can you quickly determine whether an observed change is physical or chemical? After all, we can't see the individual molecules.

There are two powerful guidelines that can help you distinguish between physical and chemical changes. First, in a physical change, a change in appearance is the result of a new set of conditions imposed on the original material. Restoring the original conditions restores the original appearance: frozen water melts upon warming. Second, in a chemical change, a change in appearance is the result of the formation of a new material that has its own unique set of physical properties. The more evidence you have suggesting that a different material has been formed, the greater the likelihood that the change is a chemical change. Iron is a moldable metal that can be used to build cars. Rust is a reddish solid that readily falls apart. This suggests that the rusting of iron is a chemical change.

▲ Figure 3.6
The transformation of water to ice and the transformation of iron to rust both involve changes in physical appearance. The formation of ice is a physical change, whereas the formation of rust is a chemical change. But how do we know?

CONCEPTCHECK

Each sphere in the following diagrams represents an atom. Joined spheres represent molecules. One set of diagrams shows a physical change, and the other shows a chemical change. Which is which?

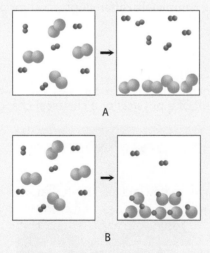

A

B

CHECK YOUR ANSWER In set A, the molecules before and after the change are the same. They differ only in their positions. Set A, therefore, represents only a physical change. In set B, new molecules, consisting of bonded red and blue spheres, appear after the change. These molecules represent a new material, so set B represents a chemical change.

Figure 3.7 shows potassium chromate, a material whose color depends on its temperature. At room temperature, potassium chromate is a bright canary yellow. At higher temperatures (above 660°C), it is a deep reddish orange. Upon cooling, the canary color returns, suggesting that the change is physical. With a chemical change, reverting to the original conditions does not restore the original appearance. Ammonium dichromate, shown in **Figure 3.8**, is an orange material that, when heated, explodes into ammonia, water vapor,

▶ Figure 3.7
When heated, potassium chromate changes from yellow to orange. After cooling, the original yellow color is restored.

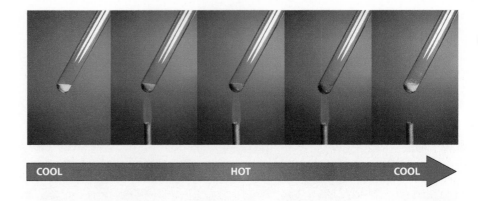

COOL   HOT   COOL

▶ Figure 3.8
When heated, orange ammonium dichromate undergoes a chemical change to ammonia, water vapor, and chromium(III) oxide. A return to the original temperature does not restore the orange color, because the ammonium dichromate is no longer there.

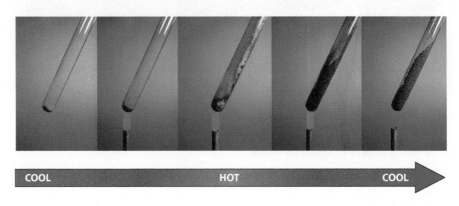

COOL   HOT   COOL

and green chromium(III) oxide. When the test tube is returned to the original temperature, there is no trace of orange ammonium dichromate. In its place are new substances having completely different physical properties.

READINGCHECK

How do we account for the great variety of substances in the world?

CONCEPTCHECK

Evan has grown an inch in height over the past year. Is this change best described as the result of a physical or a chemical change?

CHECK YOUR ANSWER Are new materials being formed as Evan grows? Absolutely—created out of the food he eats. His body is very different from, say, the spaghetti he ate yesterday. Yet, through some very advanced chemistry, his body is able to absorb the atoms of that spaghetti and rearrange them into new materials. Biological growth, therefore, is the result of chemical changes.

3.2 Elements Are Made of Atoms

EXPLAIN THIS

Why isn't water an element?

It may seem that there must be many different kinds of atoms to account for the many different type of substances—from wood to steel to chocolate ice cream. But the number of different kinds of atoms is surprisingly small. The great variety of substances results from the many ways a few kinds of atoms can be combined. Just as the three colors red, green, and blue can be combined to form any color on a computer screen or the 26 letters of the alphabet make up all the words in a dictionary, only a few kinds of atoms combine in different ways to produce all substances. To date, we know of slightly more than 100 different kinds of atoms. Of these, about 90 are found in nature. The remaining atoms have been created in the laboratory.

Any material consisting of only one type of atom is classified as an **element.** A few examples are shown in **Figure 3.9**. Pure gold, for example, is an element—it contains only gold atoms. Similarly, one of the gases in air is nitrogen, an element. Nitrogen gas is an element because it contains only nitrogen atoms. Likewise, the graphite in your pencil is an element—carbon. Graphite is made up solely of carbon atoms. All of the elements are organized in a chart called the **periodic table,** shown in **Figure 3.10**.

As you can see from the periodic table, each element is designated by its **atomic symbol,** which comes from the letters of a word associated with that element. For example, the atomic symbol for carbon is C, and that for chlorine is Cl. In many cases, the atomic symbol is derived from the element's Latin name. Gold has the atomic symbol Au after its Latin name, *aurum.* Lead has the atomic symbol Pb after its Latin name, *plumbum* (**Figure 3.11**). Elements having symbols derived from Latin names are usually those discovered earliest. Note that only the first letter of an atomic symbol is capitalized. The symbol for the element cobalt, for instance, is Co, while CO is a combination of two elements: carbon, C, and oxygen, O.

LEARNING OBJECTIVE

Recognize the elements of the periodic table as the fundamental building blocks of matter.

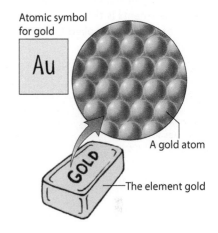

Atomic symbol for gold

A gold atom

The element gold

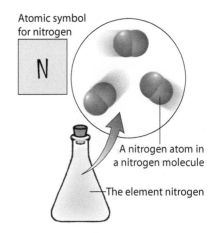

Atomic symbol for nitrogen

A nitrogen atom in a nitrogen molecule

The element nitrogen

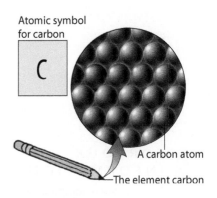

Atomic symbol for carbon

A carbon atom

The element carbon

▲ **Figure 3.9**
Any element consists of only one kind of atom. Gold consists of only gold atoms, a flask of gaseous nitrogen consists of only nitrogen atoms, and the carbon of a graphite pencil consists of only carbon atoms.

1 H																	2 He
3 Li	4 Be											5 B	6 C	7 N	8 O	9 F	10 Ne
11 Na	12 Mg											13 Al	14 Si	15 P	16 S	17 Cl	18 Ar
19 K	20 Ca	21 Sc	22 Ti	23 V	24 Cr	25 Mn	26 Fe	27 Co	28 Ni	29 Cu	30 Zn	31 Ga	32 Ge	33 As	34 Se	35 Br	36 Kr
37 Rb	38 Sr	39 Y	40 Zr	41 Nb	42 Mo	43 Tc	44 Ru	45 Rh	46 Pd	47 Ag	48 Cd	49 In	50 Sn	51 Sb	52 Te	53 I	54 Xe
55 Cs	56 Ba	57 La	72 Hf	73 Ta	74 W	75 Re	76 Os	77 Ir	78 Pt	79 Au	80 Hg	81 Tl	82 Pb	83 Bi	84 Po	85 At	86 Rn
87 Fr	88 Ra	89 Ac	104 Rf	105 Db	106 Sg	107 Bh	108 Hs	109 Mt	110 Ds	111 Rg	112 Cn	113 Uut	114 Fl	115 Uup	116 Lv	117 Uus	118 Uuo

58 Ce	59 Pr	60 Nd	61 Pm	62 Sm	63 Eu	64 Gd	65 Tb	66 Dy	67 Ho	68 Er	69 Tm	70 Yb	71 Lu
90 Th	91 Pa	92 U	93 Np	94 Pu	95 Am	96 Cm	97 Bk	98 Cf	99 Es	100 Fm	101 Md	102 No	103 Lr

▲ **Figure 3.10**
The periodic table lists all the known elements. As of this writing, scientists have yet to confirm evidence for the existence of a few of the heaviest elements shown here in gray. These ultramassive atoms are generally very unstable and difficult to create in the laboratory. When formed, they tend to exist for only fractions of a second.

FOR YOUR INFORMATION

An element whose discovery has yet to be confirmed is simply named by its number, where un = 1, bi = 2, tri = 3, and so forth. The as-yet-to-be confirmed element 118, for example, is provisionally named ununoctium, with the atomic symbol Uuo. The scientific organization in charge of naming elements is the International Union of Pure and Applied Chemistry, www.IUPAC.org.

▲ **Figure 3.11**
A plumb bob, a heavy weight attached to a string and used by carpenters and surveyors to establish a straight vertical line, gets it name from the lead (plumbum, Pb) that is still sometimes used as the weight. Plumbers got their name because they once worked with lead pipes. Because of lead's toxicity, copper or PVC pipes are now used.

The terms *element* and *atom* are often used in a similar context. You might hear, for example, that gold is an element made of gold atoms. Generally, *element* is used in reference to an entire macroscopic or microscopic sample, and *atom* is used when speaking of the submicroscopic particles in the sample. The important distinction is that elements are made of atoms and not the other way around.

The number of atoms that arrange themselves in a unit of an element is shown by the **elemental formula.** For elements in which two or more atoms are bonded into molecules, the elemental formula is the chemical symbol followed by a subscript indicating the number of atoms in each molecule. For example, elemental nitrogen, as was shown in Figure 3.9, consists of molecules containing two nitrogen atoms per molecule. Thus, N_2 is the elemental formula given for atmospheric nitrogen. Similarly, atmospheric oxygen has the elemental formula O_2, while the elemental formula for sulfur is S_8. For elements in which the basic units are individual atoms (not molecules), the elemental formula is simply the chemical symbol. This is the case for most elements. To name two examples, Au is the elemental formula for gold, and Li is the elemental formula for lithium.

FOR YOUR INFORMATION

Carbon is the only element that can form bonds with itself indefinitely. Sulfur's practical limit is S_8, and nitrogen's limit is around N_{12}. The elemental formula for a 1-carat diamond, however, is about $C_{10,000,000,000,000,000,000,000}$.

CONCEPT CHECK

The oxygen we breathe, O_2, is converted to ozone, O_3, in the presence of an electric spark. Is this a physical or chemical change?

CHECK YOUR ANSWER When atoms regroup, the result is an entirely new substance, and that is what happens here. The oxygen we breathe, O_2, is odorless and life-giving. Ozone, O_3, can be toxic and has a pungent smell commonly associated with electric motors. The conversion of O_2 to O_3 is therefore a chemical change. However, both O_2 and O_3 are elemental forms of oxygen.

3.3 The Periodic Table Helps Us to Understand the Elements

EXPLAIN THIS

How is the periodic table like a dictionary?

The periodic table is so much more than a mere listing of known elements. Most notably, the elements are organized in the table based upon their physical and chemical properties. One of the most apparent examples is the way the elements are grouped as metals, nonmetals, and metalloids.

As shown in **Figure 3.12**, most of the known elements are **metals,** which are characterized as being shiny, opaque, and good conductors of electricity and heat. Metals are *malleable*, which means they can be hammered into different shapes or bent without breaking. They are also *ductile*, which means they can be drawn into wires. All but a few metals are solid at room temperature. The exceptions include mercury, Hg; gallium, Ga; cesium, Cs; and francium, Fr, which are all liquids at a warm room temperature of 30°C (86°F). Another interesting exception is hydrogen, H, which takes on the properties of a liquid

LEARNING OBJECTIVE

Identify how elements are organized in the periodic table.

READINGCHECK

Name two physical properties of metals.

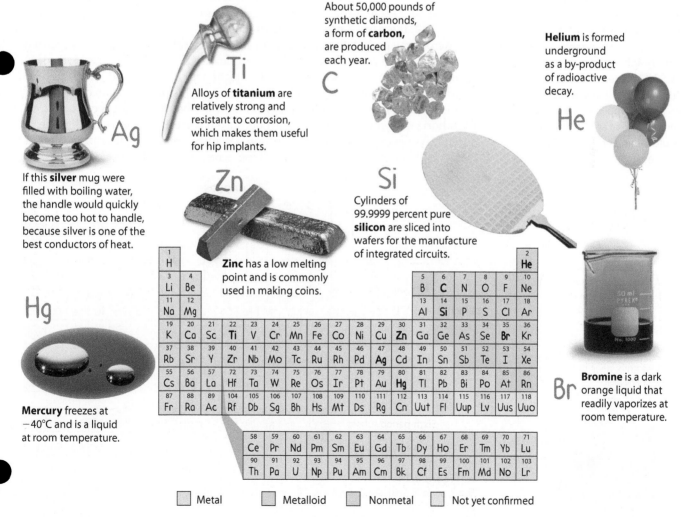

Alloys of **titanium** are relatively strong and resistant to corrosion, which makes them useful for hip implants.

About 50,000 pounds of synthetic diamonds, a form of **carbon,** are produced each year.

Helium is formed underground as a by-product of radioactive decay.

If this **silver** mug were filled with boiling water, the handle would quickly become too hot to handle, because silver is one of the best conductors of heat.

Zinc has a low melting point and is commonly used in making coins.

Cylinders of 99.9999 percent pure **silicon** are sliced into wafers for the manufacture of integrated circuits.

Mercury freezes at −40°C and is a liquid at room temperature.

Bromine is a dark orange liquid that readily vaporizes at room temperature.

Metal   Metalloid   Nonmetal   Not yet confirmed

▲ Figure 3.12
The periodic table color-coded to show metals, nonmetals, and metalloids.

▲ **Figure 3.13**
Hydrogen exists as a liquid metal deep beneath the surfaces of Jupiter (shown here) and Saturn, where pressures are exceedingly high.

 FOR YOUR INFORMATION

Most nonmetallic materials do not conduct electricity. An important exception is a material called graphene, which is a one atom thick sheet of carbon atoms. The graphite in your pencil is made of stacks of graphene. The nanotubes discussed in Section 3.8 are graphene sheets rolled into tubes. All these materials conduct electricity rather well, which leads to many useful applications. Sheets of graphene, for example, are used to make ultra-thin and flexible OLED display screens.

metal only at very high pressures (**Figure 3.13**). Under normal conditions, hydrogen behaves as a nonmetallic gas.

The nonmetallic elements, with the exception of hydrogen, are on the right hand side of the periodic table. Most **nonmetals** are poor conductors of electricity and heat. Solid nonmetals tend to be neither malleable nor ductile. Rather, they are brittle and shatter when hammered. At 30°C (86°F), some nonmetals are solid (carbon, C), others are liquid (bromine, Br), and still others are gaseous (helium, He).

Six elements are classified as **metalloids:** boron, B; silicon, Si; germanium, Ge; arsenic, As; antimony, Sb; and tellurium, Te. Situated between the metals and the nonmetals in the periodic table, the metalloids have both metallic and nonmetallic characteristics. For example, these elements are weak conductors of electricity, which makes them useful as semiconductors in the integrated circuits of computers. Note from the periodic table how germanium (number 32) is closer to more metals than nonmetals. Because of this positioning, we can deduce that germanium has more metallic properties than silicon (number 14) and is a slightly better conductor of electricity. So we find that integrated circuits fabricated with germanium operate faster than those fabricated with silicon. Because silicon is much more abundant and less expensive to obtain, however, silicon computer chips remain the industry standard.

Periods and Groups

Two other important ways in which the elements are organized in the periodic table are by horizontal rows and vertical columns. Each horizontal row is called a **period,** and each vertical column is called a **group** (or a *family*). As shown in **Figure 3.14**, there are 7 periods and 18 groups.

Across any period, the properties of elements gradually change. This gradual change is called a **periodic trend.** For example, as is shown in **Figure 3.15**, the size of atoms generally decreases from left to right across any period. Note that the trend repeats from one horizontal row to the next. This phenomenon of repeating trends is called *periodicity*, a term used to indicate that the trends recur in cycles. Each horizontal row is called a *period* because it corresponds to one full cycle of a trend. As we explore further in Section 4.9, there are many other properties of elements that change gradually in moving across a row of the periodic table.

Please put to rest any fear you may have about needing to memorize the periodic table, or even parts of it—better to focus on the many great concepts behind its organization.

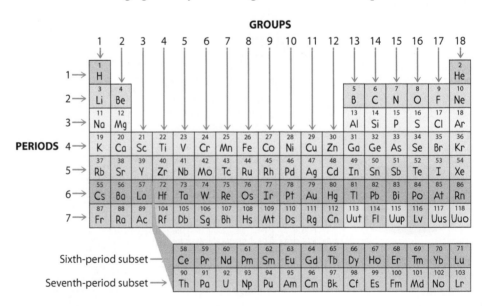

▲ **Figure 3.14**
The 7 periods (horizontal rows) and 18 groups (vertical columns) of the periodic table. Note that not all periods contain the same number of elements. Also note that, for reasons explained later, the sixth and seventh periods each include a subset of elements, which are listed apart from the main body.

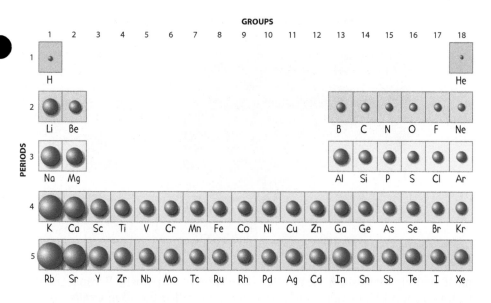

GROUPS

The size of atoms gradually decreases in moving from left to right across any period. Atomic size is a periodic (repeating) property.

CONCEPTCHECK

Which are larger: atoms of cesium, Cs (number 55), or atoms of radon, Rn (number 86)?

CHECK YOUR ANSWER Perhaps you tried looking to Figure 3.15 to answer this question and quickly became frustrated because the sixth-period elements are not shown. Well, relax. Look at the trends and you'll see that in any one period, all atoms to the left are larger than those to the right. Accordingly, cesium is positioned at the far left of period 6, and you can reasonably predict that its atoms are larger than those of radon, which is positioned at the far right of period 6. The periodic table is a road map to understanding the elements.

FOR YOUR INFORMATION

The *carat* is the common unit used to describe the mass of a gem. A 1.0 carat diamond, for example, has a mass of 0.20 grams. The *karat* is the common unit used to describe the purity of a precious metal, such as gold. A 24-karat gold ring is as pure as can be. A gold ring that is 50 percent pure is 12-karat gold.

Down any group (vertical column), the properties of elements tend to be remarkably similar, which is why these elements are said to be "grouped" or "in a family." As **Figure 3.16** shows, several groups have traditional names that describe the properties of their elements. Early in human history, people discovered that ashes mixed with water produce a slippery solution useful for removing grease. By the Middle Ages, such mixtures were described as being alkaline, a term derived from the Arabic word for ashes, al-qali. Alkaline mixtures found many uses, particularly in the preparation of soaps (**Figure 3.17**). We now know that alkaline ashes contain compounds of group 1 elements,

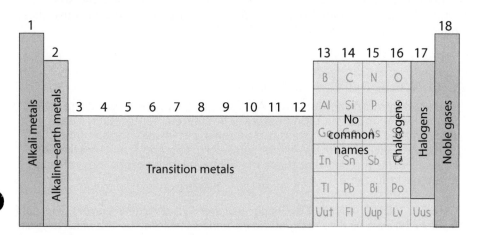

▲ Figure 3.16
The common names for various groups of elements.

▶ **Figure 3.17**
Ashes and water make a slippery alkaline solution once used to clean hands.

FOR YOUR INFORMATION

The air inside a traditional incandescent light bulb is a mixture of nitrogen and argon. As the tungsten filament is heated, minute particles of tungsten evaporate—much like steam leaving boiling water. Over time, these particles are deposited on the inner surface of the bulb, causing the bulb to blacken. Losing its tungsten, the filament eventually breaks and the bulb has "burned out."

READING CHECK

List several common transition metals.

FOR YOUR INFORMATION

A halogen incandescent light bulb contains trace amounts of a halogen gas, such as iodine or bromine. Tungsten atoms evaporating from the white hot tungsten filament combine with the halogen atoms rather than being deposited on the inner surface of the bulb, which remains clear. Furthermore, the halogen–tungsten combination splits apart when it touches the hot filament. This allows the tungsten to be deposited onto the filament, thereby restoring the filament. This is why halogen lamps have such long lifetimes.

most notably potassium carbonate, also known as potash. Because of this history, group 1 elements, which are metals, are called the *alkali metals.*

Elements of group 2 also form alkaline solutions when mixed with water. Furthermore, medieval alchemists noted that certain minerals (which we now know are made up of group 2 elements) do not melt or change when put in fire. These fire-resistant substances were known to the alchemists as "earth." As a holdover from these ancient times, group 2 elements are known as the *alkaline-earth metals.*

Over toward the right side of the periodic table, elements of group 16 are known as the *chalcogens* ("ore-forming" in Greek), because the top two elements of this group, oxygen and sulfur, are so commonly found in ores. Elements of group 17 are known as the *halogens* ("salt-forming" in Greek), because of their tendency to form various salts. Group 18 elements are all unreactive gases that tend not to combine with other elements. For this reason, they are called the inert *noble gases,* presumably because the nobility of earlier times were above interacting with common folk.

The elements of groups 3 through 12 are all metals that do not form alkaline solutions with water. These metals tend to be harder than the alkali metals and less reactive with water; hence, they are used for structural purposes. Collectively, they are known as the *transition metals,* a name that denotes their central position in the periodic table. The transition metals include some of the most familiar and important elements—iron, Fe; copper, Cu; nickel, Ni; chromium, Cr; silver, Ag; and gold, Au. They also include many lesser-known elements that are nonetheless important in modern technology. People with hip implants appreciate the transition metals titanium, Ti; molybdenum, Mo; and manganese, Mn, because these noncorrosive metals are used in implant devices.

CONCEPT CHECK

The elements copper, Cu; silver, Ag; and gold, Au, are three of the few metals that can be found naturally in their elemental state. These three metals have found great use as coins and jewelry for a number of reasons, including their resistance to corrosion and their remarkable colors. How is the fact that these metals have similar properties reflected in the periodic table?

CHECK YOUR ANSWER Copper (number 29), silver (number 47), and gold (number 79) are all in the same group in the periodic table (group 11), which suggests they should have similar—though not identical—properties.

Within the sixth period is a subset of 14 metallic elements (numbers 58 to 71) that are quite unlike any of the other transition metals. A similar subset (numbers 90 to 103) is found within the seventh period. These two subsets are the *inner transition metals.* Inserting the inner transition metals into the main body of the periodic table, as in **Figure 3.18**, produces a long and cumbersome table. So that the table can fit nicely on a standard paper size, these elements are commonly placed below the main body of the table, as shown in **Figure 3.19**.

The sixth-period inner transition metals are called the *lanthanides,* because they fall after lanthanum, La. Because of their similar physical and chemical properties, they tend to occur mixed together in the same locations in the earth. Also because of their similarities, lanthanides are unusually difficult to purify. Recently, the commercial use of lanthanides has increased. Several lanthanide elements, for example, are used in the fabrication of the light-emitting diodes (LEDs) of computer monitors and flat screen televisions.

The seventh-period inner transition metals are called the *actinides,* because they fall after actinium, Ac. They, too, all have similar properties and hence are not easily purified. The nuclear power industry faces this obstacle because it

FORYOUR INFORMATION

The Hanford nuclear facility in central Washington state, from 1943 to 1986, produced 72 tons of plutonium, nearly two-thirds the nation's supply. Creating this much plutonium generated an estimated 450 billion gallons of radioactive and hazardous liquids, which were discharged into the local environment. Today, some 53 million gallons of high-level radioactive (see Chapter 5) and chemical wastes are stored in 177 underground tanks.

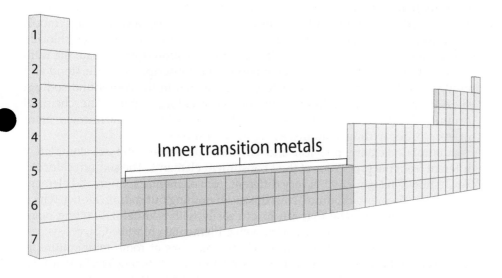

◀ **Figure 3.18**
Inserting the inner transition metals between atomic groups 3 and 4 results in a periodic table that is not easy to fit on a standard sheet of paper.

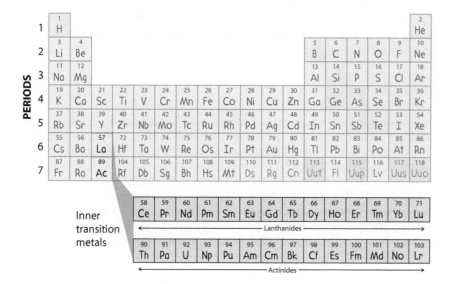

◀ **Figure 3.19**
The typical display of the inner transition metals. The count of elements in the sixth period goes from lanthanum (La, 57) to cerium (Ce, 58), on through to lutetium (Lu, 71), and then back to hafnium (Hf, 72). A similar jump is made in the seventh period.

requires purified samples of two of the most publicized actinides: uranium, U, and plutonium, Pu. Actinides heavier than uranium are not commonly found in nature but are synthesized in the laboratory.

3.4 Elements Can Combine to Form Compounds

LEARNING OBJECTIVE

Contrast compounds with the elements from which they are created.

READINGCHECK

Why is the formation of a chemical compound an example of a chemical change?

Sodium
Chlorine

Sodium chloride, NaCl

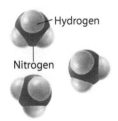

Hydrogen

Nitrogen

Ammonia, NH₃

▲ Figure 3.20
The compounds sodium chloride and ammonia are represented by their chemical formulas, NaCl and NH₃. A chemical formula shows the ratio of atoms that constitute the compound.

EXPLAIN THIS

How are compounds different from elements?

When atoms of *different* elements bond to one another, they make a **compound.** Sodium atoms and chlorine atoms, for example, bond to make the compound sodium chloride, commonly known as table salt. Nitrogen atoms and hydrogen atoms join to make the compound ammonia, which is a common household cleaner. The formation of a compound is a chemical change, because it involves the formation of a fundamentally different material.

A compound is represented by its **chemical formula,** in which the symbols for the elements are written together. The chemical formula for sodium chloride is NaCl, and that for ammonia is NH_3. Numerical subscripts indicate the ratio in which the atoms combine. By convention, the subscript 1 is understood and omitted. So the chemical formula NaCl tells us that in the compound sodium chloride there is one sodium atom for every chlorine atom. The chemical formula NH_3 tells us that in the compound ammonia there is one nitrogen atom for every three hydrogen atoms, as **Figure 3.20** shows.

Compounds have physical and chemical properties that are completely different from the properties of their constituent elements. The sodium chloride, NaCl, shown in **Figure 3.21** is very different from elemental sodium and elemental chlorine. Elemental sodium, Na, consists of nothing but sodium atoms, which form a soft, silvery metal that can be cut easily with a knife. Its melting point is 97.5°C, and it reacts violently with water. Elemental chlorine, Cl_2, consists of chlorine molecules. This material, a yellow-green gas at room temperature, is very toxic, and it was used as a chemical warfare agent during World War I. Its boiling point is –34°C. The compound sodium chloride, NaCl, is a transparent, brittle, colorless crystal with a melting point of 800°C. Sodium chloride does not react chemically with water the way sodium does. It is not toxic like chlorine—in fact, sodium chloride is an essential nutrient for all living organisms. Sodium chloride is not sodium, nor is it chlorine; it is uniquely sodium chloride, a tasty chemical when sprinkled lightly over popcorn.

CONCEPTCHECK

Hydrogen sulfide, H_2S, is an offensively smelly compound. Rotten eggs get their characteristically unpleasant smell from the hydrogen sulfide they release. Can you conclude from this information that elemental sulfur, S_8, is just as smelly?

CHECK YOUR ANSWER No, you cannot. In fact, the odor of elemental sulfur is negligible compared with that of hydrogen sulfide. Compounds are truly different from the elements from which they are formed. Hydrogen sulfide, H_2S, is as different from elemental sulfur, S_8, as water, H_2O, is from elemental oxygen, O_2.

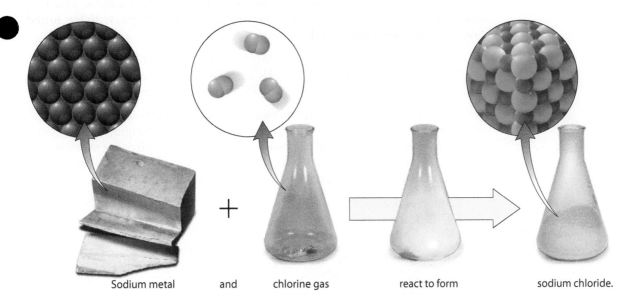

Sodium metal   and   chlorine gas   react to form   sodium chloride.

▲ **Figure 3.21**
Sodium metal and chlorine gas react together to form sodium chloride. Although the compound sodium chloride is made of sodium and chlorine, the physical and chemical properties of sodium chloride are very different from the physical and chemical properties of either sodium metal or chlorine gas.

3.5 There Is a System for Naming Compounds

EXPLAIN THIS

What information is found within the name of a compound?

A system for naming the countless number of possible compounds has been developed by the International Union for Pure and Applied Chemistry (IUPAC). This system is designed so that a compound's name reflects the elements it contains and how those elements are joined. Anyone familiar with the system can figure out the chemical identity of a compound from its systematic name.

As you might imagine, this system is very complex. However, there is no need for you to learn all its rules. Instead, learning some guidelines will prove most helpful. These guidelines alone will not enable you to name every compound. They will acquaint you with the way the system works for many simple compounds consisting of only a few elements.

As we will discuss in Chapter 6, atoms are held together in a chemical compound by their electrical charges. Whether an atom within a compound takes on a positive or negative charge can be predicted by its place in the periodic table. The atom of an element located closer to the left side of the periodic table tends to take on a positive charge, while the atom of an element placed closer to the right side of the periodic table takes on a negative charge. This applies to the naming of compounds in that the more positively charged atom is, by convention, listed first. That's why we have "sodium chloride," rather than "chloride sodium" as spelled out in guideline 1—note that sodium, Na, is on the left hand side of the periodic table and chlorine, Cl, is on the right hand side.

LEARNING OBJECTIVE

List three guidelines used to name compounds.

 **READING CHECK**

How can you tell whether an atom within a compound takes on a positive or negative charge?

GUIDELINE 1 The name of the element farther to the left in the periodic table is followed by the name of the element farther to the right, with the suffix "-ide" added to the name of the latter:

NaCl	Sodium chloride	MgO	Magnesium oxide
Li_2O	Lithium oxide	HCl	Hydrogen chloride
CaF_2	Calcium fluoride	Sr_3P_2	Strontium phosphide

GUIDELINE 2 When two or more compounds have different numbers of the same elements, prefixes are added to remove the ambiguity. This occurs primarily with compounds of nonmetals. The first four prefixes are *mono-* (one), *di-* (two), *tri-* (three), and *tetra-* (four). The prefix *mono-*, however, is commonly omitted from the beginning of the first word of the name:

Carbon and oxygen
CO   Carbon monoxide
CO_2 Carbon dioxide

Nitrogen and oxygen
NO_2 Nitrogen dioxide
N_2O_4 Dinitrogen tetroxide

Sulfur and oxygen
SO_2 Sulfur dioxide
SO_3 Sulfur trioxide

TABLE 3.1   Common Polyatomic Ions

NAME	FORMULA
Acetate ion	$CH_3CO_2^-$
Ammonium ion	NH_4^+
Bicarbonate ion	HCO_3^-
Carbonate ion	CO_3^{2-}
Cyanide ion	CN^-
Hydroxide ion	OH^-
Nitrate ion	NO_3^-
Phosphate ion	PO_4^{3-}
Sulfate ion	SO_4^{2-}

GUIDELINE 3 Atoms can clump together to form a molecular unit that acts as a single electrically charged group, called a *polyatomic ion*. For example, a carbon atom can join with three oxygen atoms to form what is known as a carbonate ion, CO_3^{2-}. Some commonly encountered polyatomic ions are shown in Table 3.1. Note that most of them are negatively charged. We'll be exploring the nature of these polyatomic ions in greater detail in Chapter 6. For now it suffices to recognize that positively charged polyatomic ions are listed first within the name (without the word ion). An example is ammonium chloride, NH_4Cl. Negatively charged polyatomic ions are placed at the end of the name. An example is lithium nitrate, $LiNO_3$.

A polyatomic ion may appear more than once within a compound. This is indicated by placing the polyatomic ion within parentheses. A subscript just outside the parentheses indicates the number of times the polyatomic ion appears. To keep it simple, the prefixes of *mono-*, *di-*, *tri-*, and *tetra-* are commonly not included for polyatomic ions.

K_2CO_3	Potassium carbonate
$AuPO_4$	Gold phosphate
$Mg(CN)_2$	Magnesium cyanide
$Al_2(SO_4)_3$	Aluminum sulfate

GUIDELINE 4 Many compounds are not usually referred to by their systematic names. Instead, they are assigned common names that are more convenient or have been used traditionally for many years. Some common names are water for H_2O, ammonia for NH_3, and methane for CH_4.

CONCEPTCHECK

What is the systematic name for $Ca(CH_3CO_2)_2$? How many oxygen atoms does it have?

CHECK YOUR ANSWER The systematic name for this compound, which consists of calcium and the polyatomic acetate ion, is calcium acetate. Each acetate ion has two oxygen atoms. This compound has two acetate ions, which means it has a total of four oxygen atoms.

3.6 Most Materials Are Mixtures

EXPLAIN THIS

Does 500 mL of sugar-sweetened water also contain 500 mL of water?

A **mixture** is a combination of two or more substances in which each substance retains its properties. Most materials we encounter are mixtures: mixtures of elements, mixtures of compounds, or mixtures of elements and compounds. Stainless steel, for example, is a mixture of the elements iron, chromium, nickel, and carbon. Seltzer water is a mixture of the liquid compound water and the gaseous compound carbon dioxide. Our atmosphere, as **Figure 3.22** illustrates, is a mixture of the elements nitrogen, oxygen, and argon plus small amounts of such compounds as carbon dioxide and water vapor.

Tap water is a mixture containing mostly water but also many other compounds. Depending on your location, your water may contain compounds of calcium, magnesium, fluorine, iron, and potassium; chlorine disinfectants; trace amounts of compounds of lead, mercury, and cadmium; organic compounds; and dissolved gases like oxygen, nitrogen, and carbon dioxide (**Figure 3.23**). While it is surely important to minimize any toxic components in your drinking water, it is unnecessary, undesirable, and impossible to remove all other substances from it. Some of the dissolved solids and gases give water its characteristic taste, and many of them promote human health: fluoride compounds protect teeth, chlorine destroys harmful bacteria, and as much as 10 percent of our daily requirement for iron, potassium, calcium, and magnesium is obtained from drinking water (**Figures 3.24**).

There is a difference between the way substances—either elements or compounds—combine to form mixtures and the way elements combine to form compounds. Each substance in a mixture retains its chemical identity. The sugar molecules in the teaspoon of sugar in **Figure 3.25**, for example, are identical to the sugar molecules already in the tea. The only difference is that

LEARNING OBJECTIVE

Recognize mixtures and show how they can be separated by physical means.

▲ **Figure 3.23**
Most of the oxygen, (O_2), in the air bubbles produced by an aquarium aerator escapes into the atmosphere. Some of the oxygen, however, mixes with the water. It is this oxygen the fish depend on to survive. Without this dissolved oxygen, which they extract with their gills, the fish would promptly drown. So fish don't "breathe" water. They breathe the oxygen, O_2, dissolved in the water.

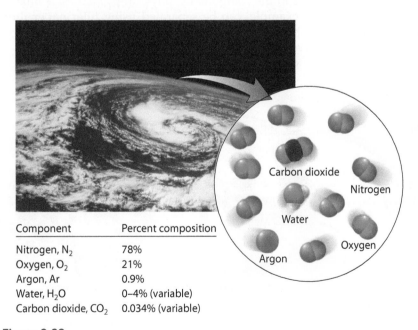

Component	Percent composition
Nitrogen, N_2	78%
Oxygen, O_2	21%
Argon, Ar	0.9%
Water, H_2O	0–4% (variable)
Carbon dioxide, CO_2	0.034% (variable)

▲ **Figure 3.22**
The Earth's atmosphere is a mixture of gaseous elements and compounds. Some of them are shown here.

▲ Figure 3.24
Tap water provides us with H_2O as well as a large number of other compounds, many of which are flavorful and help us grow. Bottoms up!

the sugar molecules in the tea are mixed with other substances, mostly water. The formation of a mixture, therefore, is a physical change.

As was discussed in Section 3.4, when elements join to form compounds, there is a change in chemical identity. Sodium chloride is not a mixture of sodium and chlorine atoms. Instead, sodium chloride is a compound, which means it is entirely different from the elements used to make it. The formation of a compound is therefore a chemical change.

CONCEPTCHECK

So far, you have learned about three kinds of matter: elements, compounds, and mixtures. Which of the following boxes contains only an element? Which contains only a compound? Which contains a mixture?

A   B   C

CHECK YOUR ANSWER The molecules in box A each contain two different types of atoms and so are representative of a compound. The molecules in box B each consist of the same atoms and so are representative of an element. Box C is a mixture of the compound and the element.

Note how the molecules of the compound and those of the element remain intact in the mixture. That is, upon the formation of the mixture, there is no exchange of atoms between the components.

Mixtures Can Be Separated by Physical Means

READINGCHECK

How do we separate the components of a mixture?

The components of mixtures can be separated from one another by taking advantage of differences in the components' physical properties. A mixture of solids and liquids, for example, can be separated using filter paper through which the liquids pass but the solids do not. This is how coffee is often made:

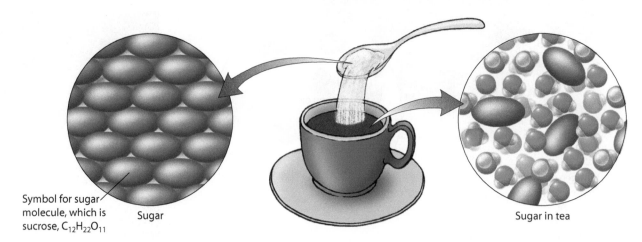

Symbol for sugar molecule, which is sucrose, $C_{12}H_{22}O_{11}$

Sugar

Sugar in tea

▲ Figure 3.25
Table sugar is a compound consisting of only sucrose molecules. Once these molecules are mixed into hot tea, they become interspersed among the water and tea molecules and form a sugar–tea–water mixture. No new compounds are formed, so this is an example of a physical change.

(a) (b)

◀ Figure 3.26
(a) The mixture is boiled in the flask on the left. The rising water vapor is channeled into a downward-slanting tube kept cool by cold water flowing across its outer surface. The water vapor inside the cool tube condenses and collects in the flask on the right. (b) A whiskey still functions on the same principle. A mixture containing alcohol is heated to the point where the alcohol, some flavoring molecules, and some water are vaporized. These vapors travel through the copper coils, where they then condense to a liquid.

the caffeine and flavor molecules after being extracted by the hot water pass through the filter and into the coffee pot while the solid coffee grounds remain behind. This method of separating a solid–liquid mixture is called *filtration* and is a common technique used by chemists.

Mixtures can also be separated by taking advantage of a difference in boiling or melting points. Seawater is a mixture of water and a variety of compounds, mostly sodium chloride. Whereas pure water boils at 100°C, sodium chloride doesn't even *melt* until 800°C. One way to separate pure water out of the mixture we call seawater, therefore, is to heat the seawater to about 100°C. At this temperature, the liquid water readily transforms to water vapor but the sodium chloride stays behind, dissolved in the remaining water. As the water vapor rises, it can be channeled into a cooler container, where it condenses to a liquid without the dissolved solids. This process of collecting a vaporized substance, called *distillation,* is illustrated in **Figure 3.26**. After all the water has been distilled from seawater, what remains consists of dry solids. These solids, also a mixture of compounds, contain a variety of valuable materials, including sodium chloride, potassium bromide, and a small amount of gold! A commercial application of this concept is shown in **Figure 3.27**.

CHEMICAL CONNECTIONS

Chemically speaking, how is an icy winter road connected to the ocean?

▲ Figure 3.27
At the southern end of San Francisco Bay, there are areas where the seawater has been partitioned off by earthen dikes. These are evaporation ponds, where the water is allowed to evaporate, leaving behind the solids that were dissolved in the seawater. These solids are further refined for commercial sale. The remarkable colors of the ponds are due to organic pigments made by salt-loving bacteria.

3.7 Matter Can Be Classified as Pure or Impure

LEARNING OBJECTIVE

Classify the states of matter under the categories of pure and impure.

READING CHECK

How does a chemist define a "pure" material?

FOR YOUR INFORMATION

White gold is a mixture of gold with smaller amounts of white metals, such as silver, palladium, or rhodium, that increase its hardness. Unlike gold, the precious metal platinum is used in jewelry in almost its pure form, about 95 percent. Platinum is a very white metal and is also very dense. A platinum ring will feel heavier than a typical yellow or white gold ring, but it will also be much more expensive.

EXPLAIN THIS

Is frozen apple juice an example of a solution, suspension, or heterogeneous mixture?

From a chemist's point of view, if a material is **pure,** it consists of only a single element or a single compound. In pure gold, for example, there is nothing but the element gold. In pure table salt, there is nothing but the compound sodium chloride. If a material is **impure,** it is a mixture and contains two or more elements or compounds. These concepts are mapped out in the classification scheme shown in **Figure 3.28.**

Because atoms and molecules are so small, there is a countless number of them in even a tiny sample. If just one atom or molecule out of many were different, then this sample could not be classified as 100 percent pure. Samples can be "purified," however, by various methods, such as distillation. When we say *pure,* it is understood to be a relative term. Comparing the purity of two samples, the purer one contains fewer impurities. A sample of water that is 99.9 percent pure has a greater proportion of impurities than does a sample of water that is 99.9999 percent pure. The 99.9999 percent pure water would be much more expensive, because this high degree of purity is rather difficult to attain (see Calculation Corner on page 80).

Sometimes naturally occurring mixtures are labeled as being pure, as in "pure orange juice." Such a statement means that nothing artificial has been added. According to a chemist's definition, however, orange juice is anything but pure, as it contains a wide variety of materials, including water, pulp, flavorings, vitamins, and sugars. Also, outside the language of chemistry, sometimes a mixture can be identified as pure. A cook, for example, might ask for pure baking powder. To a chemist, however, this doesn't make sense because baking powder is a mixture of baking soda and sodium aluminum sulfate, plus many other chemicals.

Mixtures may be heterogeneous or homogeneous. In a **heterogeneous mixture,** the different components can be seen as individual substances, such as pulp in orange juice, sand in water, or oil globules dispersed in vinegar. The different components are visible. **Homogeneous mixtures** have the same composition throughout as judged by the unaided eye. Any one region of the mixture has the same ratio of substances as does any other region. The reason for this is because the different components are mixed at a very fine level such that the components cannot be seen as individual identifiable entities. The distinction is shown in **Figure 3.29.**

▶ Figure 3.28
The chemical classification of matter.

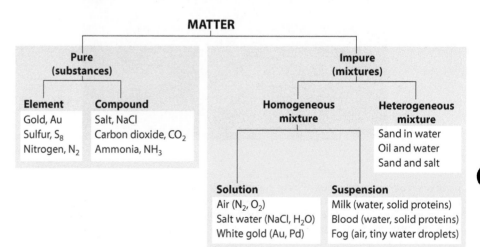

Granite   "Snow" in snow globe   Pizza

(a) Heterogeneous mixtures

Air   Clear seawater   White gold

(b) Homogeneous mixtures

◀ **Figure 3.29**
(a) In heterogeneous mixtures, the different components can be seen with the naked eye. (b) In homogeneous mixtures, the different components are mixed at a much finer level and so are not readily distinguished.

A homogeneous mixture may be either a solution or a suspension. In a **solution,** all components are in the same phase. The atmosphere we breathe is a gaseous solution consisting of the gaseous elements nitrogen and oxygen as well as minor amounts of other gaseous materials. Salt water is a liquid solution, because both the water and the dissolved sodium chloride are found in a single liquid phase. An example of a solid solution is white gold. We will discuss solutions in more detail in Chapter 7.

A **suspension** forms when the particles of a substance are finely mixed but not dissolved. The components of a suspension can be of different phases, such as solid particles suspended within a liquid or liquid droplets suspended within a gas. In a suspension, the mixing can be so thorough that the different phases are not readily distinguished. Milk is a suspension because it is a homogeneous mixture of proteins and fats finely dispersed in water. Blood is a suspension composed of finely dispersed blood cells in water. Another example of a suspension is clouds, which are homogeneous mixtures of tiny water droplets suspended in air. Shining a light through a suspension, as is shown in **Figure 3.30**, results in a visible cone as the light is reflected by the suspended components.

The easiest way to distinguish a suspension from a solution in the laboratory is to spin a sample in a centrifuge. This device, spinning at thousands of revolutions per minute, separates the components of suspensions but not those of solutions, as **Figure 3.31** shows.

▲ **Figure 3.30**
The path of light becomes visible when the light passes through a suspension.

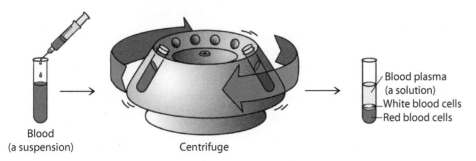

Blood
(a suspension)   Centrifuge   Blood plasma (a solution) — White blood cells — Red blood cells

◀ **Figure 3.31**
Blood, because it is a suspension, can be centrifuged into its components, which include the blood plasma (a yellowish solution) and white and red blood cells. The components of the plasma cannot be separated from one another because a centrifuge has no effect on solutions. Notice that blood plasma, white blood cells, and red blood cells can be isolated from blood. None of these components of blood, however, are in themselves pure materials.

CONCEPTCHECK

Impure water can be purified by

a. removing the impure water molecules.

b. removing everything that is not water.

c. breaking down the water to its simplest components.

d. adding some disinfectant, such as chlorine.

CHECK YOUR ANSWER The answer is b: impure water can be purified by removing everything that isn't water. H_2O is a compound made of the elements hydrogen and oxygen in a 2-to-1 ratio. Every H_2O molecule is the same as every other, and there's no such thing as an impure H_2O molecule. Just about anything, including you, beach balls, rubber ducks, dust particles, and bacteria, can be found in water. When something other than water is found in water, we say that the water is impure. It is important to see that the impurities are *in* the water and not part of the water, which means that it is possible to remove them by a variety of physical means, such as filtration or distillation.

CALCULATION CORNER   HOW PURE IS PURE?

A 100-gram sample of water contains about 3×10^{24} molecules. If this sample were ideally pure, every one of those molecules would be water. Atoms and molecules, however, are so amazingly small, and hence numerous, that the formation of a truly pure sample of macroscopic quantity is virtually impossible.

For example, consider a 100-gram sample of water that is 99.9999 percent pure. What this means is that the sample contains 99.9999 grams of water, which is still nearly 3×10^{24} water molecules. Pretty good, right? However, if the remaining 0.0001 grams were made of dissolved lead, Pb, then this would correspond to about 3×10^{17} (300,000 trillion) atoms of Pb, which is quite small compared to the number of water molecules but is still an amazingly large number.

Any material that is seemingly pure will inevitably contain impurities. Sometimes these impurities are of particular interest. For example, minor impurities in a solution might be toxic and their presence might need to be monitored. How much is present is frequently measured in units of milligrams per liter (mg/L), micrograms per liter (μg/L), or nanograms per liter (ng/L) of solution.

One liter of water contains one million milligrams of water. Because 1 L of water equals 1 million mg of water, the units of mg/L can also be expressed as 1 mg per 1 million mg. Another way of saying this is one *part per million*, or simply 1 ppm. The units of mg/L and ppm, therefore, are equivalent. Similarly, "micrograms per liter" is often expressed as *parts per billion*, ppb, and "nanograms per liter" is often expressed as *parts per trillion*, ppt. (As Table 1.3 shows, 1 milligram equals 1000 micrograms, while 1 microgram equals 1000 nanograms.)

EXAMPLE

There are about 35 grams of salts in every liter of ocean water. Express this concentration in units of ppm.

ANSWER

Convert grams of salt into milligrams of salt:

$$(35\,g\ salts)(1000\,mg/1\,g) = 35,000\,mg$$

There are about 35,000 mg of salts in a liter of ocean water, which equals 35,000 ppm.

YOUR TURN

1. Typical levels of fluoride found in fluoridated public drinking water are about 1.0 ppm. If you were to drink a liter of this water, how many milligrams of fluoride would you ingest?

2. Aquatic organisms require a dissolved oxygen concentration of about 6 ppm. At this concentration, how many grams of oxygen are present in one liter of water?

3. Chloroform, $CHCl_3$, is a common contaminant of chlorinated drinking water. A usual concentration in municipal tap water may be around 25 ppb. How many milligrams of chloroform are present in one liter of water?

4. What is your concentration in the country you live in expressed in units of ppb? (For the United States, assume a population of 310 million.) What is your concentration in the world (assume total human population of 7 billion) expressed in units of ppb? In units of ppt?

Answers to Calculation Corners appear at the end of each chapter.

3.8 The Advent of Nanotechnology

EXPLAIN THIS

Is nanotechnology the result of basic or applied research?

The age of microtechnology was ushered in some 65 years ago with the invention of the solid state transistor, a device that serves as a gateway for electronic signals. Engineers were quick to grasp the idea of integrating many transistors together to create logic boards that could perform calculations and run programs. The very transistors they could squeeze into a circuit, the more powerful the logic board. The race thus began to squeeze more and more transistors together into tinier and tinier circuits. The scales achieved were in the realm of the micron (10^{-6} meters)—thus the term *micro*technology. At the time of the transistor's invention, few people realized the impact microtechnology would have on society—from personal computers to cell phones to the Internet.

Today, we are at the beginning of a similar revolution. Technological advances have recently brought us past the realm of microns to the realm of the nanometer (10^{-9} meters), which is the realm of individual atoms and molecules—a realm where we have reached the basic building blocks of matter. Technology that works on this scale is called *nano*technology. No one knows exactly what impact nanotechnology will have on society, but people are quickly coming to realize its vast potential, which is likely to be much greater than that of microtechnology.

Nanotechnology generally concerns the manipulations of objects from 1 to 100 nanometers in scale. For perspective, a DNA molecule is about 2.0 nm wide (though about a meter long!), while a water molecule is only about 0.2 nm. Like microtechnology, nanotechnology is interdisciplinary, requiring the cooperative efforts of chemists, engineers, physicists, molecular biologists, and many others. Interestingly, there are already many products on the market that contain components developed through nanotechnology. These include sunscreens, mirrors that don't fog, dental bonding agents, automotive catalytic converters, stain-free clothing, water filtration systems, the heads of computer hard drives, and much more. Nanotechnology, however, is still in its infancy, and it will likely be decades before its potential is fully realized (**Figure 3.32**). Consider, for example, that personal computers didn't blossom until the 1990's, some 40 years after the first solid state transistor.

LEARNING OBJECTIVE

Show how nanotechnology is a novel and promising application of chemistry.

 FOR YOUR INFORMATION

Before he died in 2005, Rick Smalley, codiscoverer of the buckyball (Section 1.2), advocated that carbon nanotubes, if developed into wires, could be an ideal material for efficiently transporting electricity over vast distances. If such an infrastructure were in place, the wind energy of the Great Plains of the United States would be sufficient to supply the electrical needs of the entire country.

READING CHECK

How large are the objects that are the focus of nanotechnology?

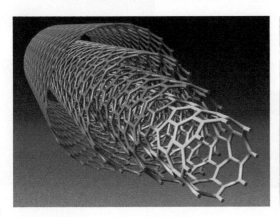 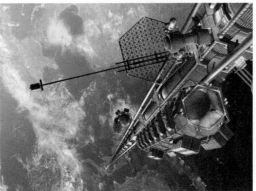

▲ Figure 3.32
Carbon nanotubes can be nested within each other to provide the strongest fiber known—a thread 1 millimeter in diameter can support a weight of about 13,000 pounds. A network of such strong fibers could be used to build the once science-fictional space elevator.

FOR YOUR INFORMATION

An interesting discovery of nanoscience is that the properties of a material at the level of its atoms can be different from its properties in bulk quantities. A bar of gold, for example, is gold in color. A thin sheet of gold atoms, by contrast, is dark red. There is much research currently being directed toward the discovery of the unique nanoproperties of materials. Many novel applications of these nanoproperties are sure to follow.

There are two main approaches to building nanoscale materials and devices: top-down and bottom-up. The top-down approach is an extension of microtechnology techniques to smaller and smaller scales. A nanosize circuit board, for example, might be carved out from a larger block of material. The bottom-up approach involves building nanosized objects atom by atom. A very important tool for either of these approaches is the **scanning probe microscope,** which detects and characterizes the surface atoms of materials by way of an ultrathin probe tip, as shown in **Figure 3.33**. The tip is mechanically dragged over the surface. Interactions between the tip and the surface atoms cause movements in the probe that are detected by way of a laser beam and translated by a computer into a topographical image. Scanning probe microscopes can also be used to move individual atoms into desired positions.

Nanotechnology will enable the continued miniaturization of integrated circuits needed for ever smaller and more powerful computers. But a computer need not rely on an integrated circuit of nanowires for processing power. A wholly new approach involves designing logic boards in which molecules (not electric circuits) read, process, and write information. One molecule that has proved most promising for such *molecular computation* is DNA, the same molecule that holds our genetic code. An advantage that molecular computing has over conventional computing is that it can run a massive number of calculations in parallel (at the same time). Because of such fundamental differences, molecular computing may one day outshine even the fastest of integrated circuits. Molecular computing, in turn, may then soon be eclipsed by other novel approaches, such as quantum or photon computing, also made possible by nanotechnology.

The ultimate expert on nanotechnology is nature. Living organisms, for example, are complex systems of interacting biomolecules all functioning on the scale of nanometers. In this sense, the living organism is nature's nanomachine. We need look no further than our own bodies to find evidence of the

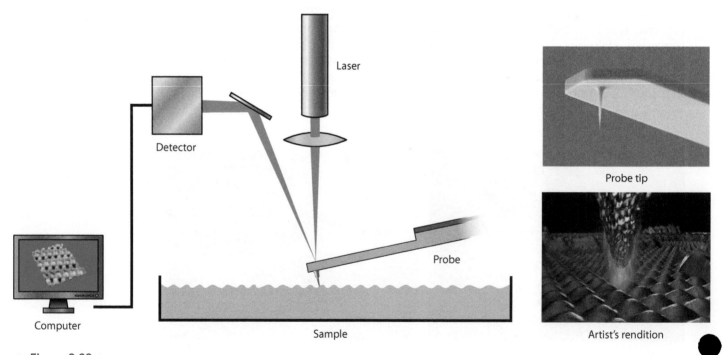

▲ Figure 3.33
A schematic of a scanning probe microscope that detects and characterizes the surface atoms of a material by way of an ultrathin probe tip attached to a miniature cantilever.

feasibility and power of nanotechnology. With nature as our teacher, we have much to learn. Such knowledge will be particularly applicable to medicine. By becoming nanotechnology experts ourselves, we would be well equipped to understand exact causes of nearly any disease or disorder (aging included) and empowered to develop innovative cures.

What are the limits of nanotechnology? As a society, how will we deal with the impending changes nanotechnology may bring? Consider the possibilities. Wall paint that can change color or be used to display video. Smart dust that the military could use to seek out and destroy an enemy. Solar cells that capture sunlight so efficiently that they render fossil fuels obsolete. Robots with so much processing power that we begin to wonder whether they experience consciousness. Nanobots that roam our circulatory systems destroying cancerous tumors or arterial plaque. Nanomachines that can "photocopy" 3-dimensional objects, including living organisms. Medicines that more than double the average human life span. Stay tuned for an exciting new revolution in human capabilities.

FOR YOUR INFORMATION

The buckyball, discussed in Chapter 1, has recently been incorporated into a new class of transparent photovoltaic cells that can be integrated into windows. This would allow a building to generate its own electricity in the sunlight.

CONCEPT CHECK

How believable would our present technology be to someone living 200 years ago? How believable might the technology of 200 years in the future be to us right now?

CHECK YOUR ANSWER Hindsight is 20/20. It's always easy to look back over time and see the progression of events that led to our present state. Much more difficult is it to think forward and project possible scenarios. Perhaps the future technology of 200 years from now will be just as unbelievable to us as our present technology is unbelievable to someone of 200 years ago.

Chapter 3 Review

LEARNING OBJECTIVES

Describe how materials can be identified by their physical and chemical properties. (3.1)	→	*Questions 1–3, 23, 24, 35–40*
Recognize the elements of the periodic table as the fundamental building blocks of matter. (3.2)	→	*Questions 4, 5, 41–44*
Identify how elements are organized in the periodic table. (3.3)	→	*Questions 6–8, 32, 45–51*
Contrast compounds with the elements from which they are created. (3.4)	→	*Questions 9–11, 31, 52–57*
List three guidelines used to name compounds. (3.5)	→	*Questions 12, 13, 58–61*
Recognize mixtures and show how they can be separated by physical means. (3.6)	→	*Questions 14–16, 25, 26, 62–67*
Classify the states of matter under the categories of pure and impure. (3.7)	→	*Questions 17–19, 27–30, 33, 34, 68–70*
Show how nanotechnology is a novel and promising application of chemistry. (3.8)	→	*Questions 20–21, 71–75*

SUMMARY OF TERMS (KNOWLEDGE)

Atomic symbol An abbreviation for an element or atom.

Chemical bond The force of attraction between two atoms that holds them together within a compound.

Chemical change The formation of new substance(s) by rearranging the atoms of the original material(s).

Chemical formula A notation that indicates the composition of a compound, consisting of the atomic symbols for the different elements of the compound and numerical subscripts indicating the ratio in which the atoms combine.

Chemical property A type of property that characterizes the ability of a substance to change into a different substance under specific conditions.

Chemical reaction A term synonymous with chemical change.

Compound A material in which atoms of different elements are bonded to one another.

Element A material consisting of only one type of atom.

Elemental formula A notation that uses the atomic symbol and (sometimes) a numerical subscript to denote how many atoms are bonded in one unit of an element.

Group A vertical column in the periodic table, also known as a family of elements.

Heterogeneous mixture A mixture in which the different components can be seen as individual substances.

Homogeneous mixture A mixture in which the components are so finely mixed that any one region of the mixture has the same ratio of substances as any other region.

Impure The state of a material that is a mixture of more than one element or compound.

Metal An element that is shiny, opaque, and able to conduct electricity and heat.

Metalloid An element that exhibits some properties of metals and some properties of nonmetals. Six elements recognized as metalloids include boron, B; silicon, Si; germanium, Ge; arsenic, As; antimony, Sb; and tellurium, Te.

Mixture A combination of two or more substances in which each substance retains its properties.

Nonmetal An element located toward the upper right of the periodic table, with the exception of hydrogen, that is neither a metal nor a metalloid.

Period A horizontal row in the periodic table.

Periodic table A chart in which all known elements are organized by physical and chemical properties.

Periodic trend The gradual change of any property in the elements across a period of the periodic table.

Physical change A change in which a substance changes its physical properties without changing its chemical identity.

Physical property Any physical attribute of a substance, such as color, density, or hardness.

Pure The state of a material that consists solely of a single element or compound.

Scanning probe microscope A tool of nanotechnology that detects and characterizes the surface atoms of materials by way of an ultrathin probe tip, which is detected by laser light as it is mechanically dragged over the surface.

Solution A homogeneous mixture in which all components are dissolved in the same phase.

Suspension A homogeneous mixture in which the various components are finely mixed, but not dissolved.

READING CHECK QUESTIONS (COMPREHENSION)

3.1 Matter Has Physical and Chemical Properties

1. What happens to the chemical identity of a substance during a physical change?
2. What changes during a chemical reaction?
3. Why is it sometimes difficult to decide whether an observed change is physical or chemical?

3.2 Elements Are Made of Atoms

4. How many types of atoms can you expect to find in a pure sample of any element?
5. Distinguish between an atom and an element.

3.3 The Periodic Table Helps Us to Understand the Elements

6. How many periods are there in the periodic table? How many groups?
7. Do properties change or remain the same for elements across any period of the periodic table?
8. Why are the lanthanides and actinides placed beneath the main body of the periodic table?

3.4 Elements Can Combine to Form Compounds

9. What is the difference between an element and a compound?
10. What does the chemical formula of a substance tell us about that substance?
11. How are the properties of a compound related to the properties of the elements used to make that compound?

3.5 There Is a System for Naming Compounds

12. What is the chemical formula for the compound titanium dioxide?
13. What is the name of the compound with the formula $NaNO_3$?

3.6 Most Materials Are Mixtures

14. What defines a material as being a mixture?
15. How can the components of a mixture be separated from one another?
16. How does distillation separate the components of a mixture?

3.7 Matter Can Be Classified as Pure or Impure

17. Why is it not practical to have a macroscopic sample that is 100 percent pure?
18. How is a solution different from a suspension?
19. How can a solution be distinguished from a suspension?

3.8 The Advent of Nanotechnology

20. How soon will nanotechnology give rise to commercial products?
21. What are the two main approaches to building nanoscale materials and devices?
22. Who is the ultimate expert at nanotechnology?

CONFIRM THE CHEMISTRY (HANDS-ON APPLICATION)

23. Place a large pot of cool water on top of a gas stove and set the burner on high. What product from the combustion of the natural gas do you see condensing on the outside of the pot? Where did it come from? Would more or less of this product form if the pot contained ice water? Where does this product go as the pot gets warmer? What physical and chemical changes can you identify?

24. When you pour a solution of hydrogen peroxide, H_2O_2, over a cut, an enzyme in your blood decomposes it to produce oxygen gas, O_2, as evidenced by the bubbling that takes place. It is this oxygen at high concentrations at the site of injury that kills off microorganisms. A similar enzyme is found in baker's yeast.

    Wear safety glasses and remove all combustibles, such as paper towels, from a clear countertop area. Pour a small packet of baker's yeast into a tall glass. Add a couple capfuls of 3 percent hydrogen peroxide and watch oxygen bubbles form. Test for the presence of oxygen by holding a lighted match with tweezers and putting the flame near the bubbles. Look for the flame to glow more brightly as the escaping oxygen passes over it. Describe oxygen's physical and chemical properties.

25. To see the gases dissolved in your water, fill a clean cooking pot with water and let it stand at room temperature for several hours. Note the tiny bubbles that adhere to the inner sides of the pot. Where did these tiny bubbles come from? What do you suppose they contain? For further experimentation, repeat this activity in two pots side by side. In one pot, use warm water from the kitchen faucet. In the second pot, use boiled water that has cooled down to the same temperature.

26. Put on your safety glasses and add several cups of tap water to a cooking pot. Boil the water to dryness. (Turn off the burner before the water is all gone. The heat from the pot will finish the evaporation. Watch out for splattering!) Examine the resulting residue by scraping it with the knife. These are the solids you ingest with every glass of water you drink.

THINK AND SOLVE (MATHEMATICAL APPLICATION)

27. The Colorado River water in Colorado has a salinity of about 50 ppm. By the time this water passes into Mexico its salinity has increased to about 1000 ppm. How many milligrams of salts have been added to each liter of water?

28. Rainwater is naturally acidic containing about 48 micrograms of acid per liter. Express this concentration in units of ppm, ppb, and ppt.

29. Dioxins are highly toxic compounds that form upon the burning of certain plastics, especially PVC. Dioxins bioaccumulate, which means that animals higher in the food chain tend to have greater concentrations within their bodies. Most of our exposure to dioxins comes from the food we eat rather than the air we breathe. How many milligrams of dioxins are there in a liter of milk containing 0.16 ppt?

30. Drinking water is routinely disinfected by adding chlorine to a concentration of about 2 ppm. How many milligrams is this per liter of water?

THINK AND COMPARE (ANALYSIS)

31. Rank the following compounds in order of an increasing number of atoms:
    a. $C_{12}H_{22}O_{12}$
    b. Buckminsterfullerene, C_{60}
    c. $Pb(C_2H_3O_2)_2$

32. Rank the following elements in order of the size of their atoms from smallest to largest:
    a. Cesium, Cs, number 55
    b. Silver, Ag, number 47
    c. Oxygen, O, number 8

33. Rank the following in order of increasing purity:
    a. Mountain spring water
    b. Distilled water
    c. Ocean water

34. The following are the concentrations of arsenic in three different water samples. Rank them in order of increasing amounts of arsenic per liter:
    a. 0.5 ppt
    b. 100 ppb
    c. 0.0007 ppm

THINK AND EXPLAIN (SYNTHESIS)

3.1 Matter Has Physical and Chemical Properties

35. A cotton ball is dipped in alcohol and wiped across a table top. Explain what happens to the alcohol molecules deposited on the table top. Is this a physical or chemical change?

36. A skillet is lined with a thin layer of cooking oil followed by a layer of unpopped popcorn kernels. Upon heating, the kernels all pop, thereby escaping the skillet. Identify any physical or chemical changes.

37. In the winter Vermonters make a tasty treat called "sugar on snow" in which they pour boiled-down maple syrup onto a scoop of clean fresh snow. As the syrup hits the snow it forms a delicious taffy. Identify the physical changes involved in the making of sugar on snow. Identify any chemical changes.

38. Each night you measure your height just before going to bed. When you arise each morning, you measure your height again and consistently find that you are 1 inch taller than you were the night before but only as tall as you were 24 hours ago! Is what happens to your body in this instance best described as a physical change or a chemical change? Be sure to try this activity if you haven't already.

39. Classify the following changes as physical or chemical.

   a. Grape juice turns to wine _____
   b. Wood burns to ashes _____
   c. Water begins to boil _____
   d. A broken leg mends itself _____
   e. Grass grows _____
   f. An infant gains 10 pounds _____
   g. A rock is crushed to powder _____

40. Is aging primarily an example of a physical or chemical change?

3.2 Elements Are Made of Atoms

41. A 24-karat piece of gold is as pure as it gets. Make an argument for why this piece of gold should not be classified as an element.

42. Hydrogen sulfide, H_2S, is a highly toxic and super stinky gas. Might you expect elemental sulfur, S_8, also to be a highly toxic and super stinky gas? Please explain.

43. Why is water not classified as an element?

44. What do diamonds, buckyballs, nanotubes, and graphite have in common?

3.3 The Periodic Table Helps Us to Understand the Elements

45. Which elements are some of the oldest known? What is your evidence?

46. Germanium, Ge (number 32), computer chips operate faster than silicon, Si (number 14), computer chips. So how might a gallium, Ga (number 31), chip compare with a germanium chip?

47. Helium, He, is a nonmetallic gas and the second element in the periodic table. Rather than being placed adjacent to hydrogen, H, however, helium is placed on the far right of the table. Why?

48. Name ten elements available to you as a modern-day consumer.

49. Strontium, Sr (number 38), is especially dangerous to humans because it tends to accumulate in calcium-dependent bone marrow tissues (calcium, Ca, number 20). How does this fact relate to what you know about the organization of the periodic table?

50. With the periodic table as your guide, describe the element selenium, Se (number 34), using as many of this chapter's key terms as you can.

51. Why not memorize the periodic table?

3.4 Elements Can Combine to Form Compounds

52. A sample of iron weighs more after it rusts. Why?

53. If you eat metallic sodium or inhale chlorine gas, you stand a strong chance of a painful death. Let these two elements react with each other, however, and you can safely sprinkle the product on your popcorn for better taste. What is going on?

54. Why can't the elements of a compound be separated from one another by physical means?

55. Water molecules contain oxygen atoms. Does this mean that oxygen, O_2, and water, H_2O, have similar properties?

56. Oxygen atoms are used to make water molecules. Why do we drown when we breathe in water despite all the oxygen atoms present in this material?

57. Which of the following boxes contains an element? A compound? A mixture? How many different types of molecules are shown altogether in all three boxes?

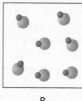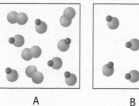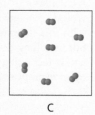

A	B	C

3.5 There Is a System for Naming Compounds

58. What is the common name for trioxygen?

59. What is the name of the compound with the formula $Sr_3(PO_4)_2$?

60. Give the name and the formula for a compound resulting from the combination of aluminum, sulfur, and oxygen.

61. Elemental copper, Cu, is copper color. Elemental sulfur, S_8, is yellow. What does this tell you about the color of the compound copper sulfide, CuS?

3.6 Most Materials Are Mixtures

62. Oxygen, O_2, has a boiling point of 90 K (–183°C), and nitrogen, N_2, has a boiling point of 77 K (–196°C). Which is a liquid and which is a gas at 80 K (–193°C)?

63. Each sphere in the following diagrams represents an atom. Joined spheres represent molecules. Which box contains a liquid phase? Why can you not assume that box B represents a lower temperature?

A   B

64. Based on the information given in the following diagrams, which substance has the lower boiling point, the compound or the element?

 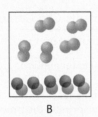

A   B

65. How might you separate a mixture of sand and salt? How about a mixture of sand and iron?

66. Mixtures can be separated into their components by taking advantage of differences in the chemical properties of the components. Why might this separation method be less convenient than taking advantage of differences in the physical properties of the components?

67. Many dry cereals are fortified with iron, which is added to the cereal in the form of small iron particles. How might these particles be separated from the cereal?

3.7 Matter Can Be Classified as Pure or Impure

68. Which of the following boxes shown best represents a suspension? A solution? A compound?

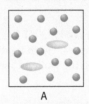 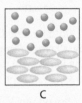

A   B   C

69. Classify the following as a(an) homogeneous mixture, heterogeneous mixture, element, or compound: table salt, blood, steel, planet Earth.

70. Classify the following as element, compound, or mixture, and justify your classifications: salt, stainless steel, tap water, sugar, vanilla extract, butter, maple syrup, aluminum, ice, milk, cherry-flavored cough drops.

3.8 The Advent of Nanotechnology

71. How does a scanning probe microscope differ from an optical microscope?

72. People often behave differently when they are in a group compared to when they are by themselves. Explain how this is similar to the behavior of atoms. Is this good news or bad news for the development of nanotechnology?

73. How is chemistry similar to nanotechnology? How is it different?

THINK AND DISCUSS (EVALUATION)

74. A calculator is useful, but certainly not exciting. Why would someone from 100 years ago vehemently disagree with this statement? We often marvel at a new technology, but how long does this marveling last? How soon before a new technology becomes taken for granted? Think of other examples. Is technology the source of happiness?

75. How might speculations about potential dangers of nanotechnology threaten public support for it? Consider Michael Crichton's science fiction novel "Prey" in which self-replicating nanobots run amok, turning everything they contact into gray goo.

READINESS ASSURANCE TEST (RAT)

If you have a good handle on this chapter, then you should be able to score at least 7 out of 10 on this RAT. Check your answers online at www.ConceptualChemistry.com. If you score less than 7, you need to study further before moving on.

Choose the BEST answer to each of the following questions.

1. What chemical change occurs when a wax candle burns?
   a. The wax near the flame melts.
   b. The molten wax is pulled upwards through the wick.
   c. The wax within the wick is heated to about 600°C.
   d. The heated wax molecules combine with oxygen molecules.

2. Is the following transformation representative of a physical change or a chemical change?

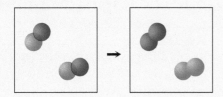

   a. Chemical because of the formation of elements

   b. Physical because a new material has been formed

   c. Chemical because the atoms have changed partners

   d. Physical because of a change in phase

3. The oldest known elements are the ones

   a. at the beginning of the periodic table.

   b. at the end of the periodic table.

   c. with atomic symbols that do NOT match their modern names.

   d. with atomic symbols that match their modern names.

4. If you have one molecule of TiO_2, how many molecules of O_2 does it contain?

   a. Three, TiO_2 contains three molecules.

   b. Two, TiO_2 is a mixture of Ti and 2 O.

   c. One, TiO_2 is a mixture of Ti and O_2.

   d. None, O_2 is a different molecule than TiO_2.

5. The largest atoms are

   a. also the heaviest atoms.

   b. the seventh period actinides.

   c. the group 1 metals.

   d. nonmetallic.

6. What is the name of the compound $CaCl_2$?

   a. Carbon chloride

   b. Dichlorocalcium

   c. Calc two

   d. Calcium chloride

7. Someone argues that he or she doesn't drink tap water because it contains thousands of molecules of some impurity in each glass. How would you respond in defense of the water's purity, if it indeed does contain thousands of molecules of some impurity per glass?

   a. Impurities aren't necessarily bad; in fact, they may be good for you.

   b. The water contains water molecules and each water molecule is pure.

   c. There's no defense. If the water contains impurities it should not be consumed.

   d. That's next to nothing compared to the amount of water.

8. What is the difference between a compound and a mixture?

   a. They both consist of atoms from different elements.

   b. The difference is the way in which their atoms are bonded together.

   c. The components of a mixture are not chemically bonded together.

   d. One is a solid and the other is a liquid.

9. The air in your house is an example of a

   a. homogeneous mixture because it is mixed very well.

   b. heterogeneous mixture because of the dust particles it contains.

   c. homogeneous mixture because it is all at the same temperature.

   d. heterogeneous mixture because it consists of different types of molecules.

10. Half-frozen fruit punch is always sweeter than the same fruit punch completely melted because

   a. the sugar sinks to the bottom.

   b. only the water freezes while the sugar remains in solution.

   c. the half-frozen fruit punch is warmer.

   d. sugar molecules solidify as crystals.

ANSWERS TO CALCULATION CORNER (HOW PURE IS PURE?)

1. One liter of 1.0 ppm fluoridated drinking water contains 1.0 mg of fluoride.

2. A dissolved oxygen concentration of 6 ppm corresponds to 6 mg/L. With one liter, you have 6 mg:

   (6 mg/L)(1 L) = 6 mg

   Use a conversion unit to express 6 mg as 0.006 grams:

   (6 mg)(1 gram/1000 mg) = 0.006 grams

3. A concentration of 25 ppb is 25 micrograms per liter. Convert from micrograms into milligrams as follows:

   (25 micrograms)(1 milligram/1000 micrograms) = 0.025 milligrams

   Thus, there are 0.025 milligrams of chloroform in each liter of this water.

4. Assuming you live in the United States, which has a population of about 310 million, then your concentration is about 1/0.31 billion = 3.1 ppb. The concentration of you in the whole world is about 1/7 billion = 0.140 ppb, which is about 140 ppt. If you can imagine how few of you there are compared to national or world populations, then you have a sense of how dilute a dissolved substance is when its concentration is measured by the ppb or ppt.

Extending the Human Life Span

Living isn't easy, especially at the level of the molecules that make us. Our cells and the molecules they contain are constantly exposed to a hostile environment of viruses, bacteria, free radicals, radiation, and random chemical reactions. We live because our bodies are able to repair themselves from perpetual molecular damage. Over time, however, our bodies lose the ability to self-heal. We age. We grow frail and eventually die.

Since the introduction of modern medicine and better health habits, the average life expectancy of humans has increased dramatically—in the United States from about 48 years old in 1900 to about 78 years old in 2000. The maximum attainable human life span of about 120 years, however, appears to have remained fairly constant. Is there truly a limit to how long we can live? Is it possible to change how long we live while still maintaining youthful vigor and resilience? After all, the quality of life matters just as much, if not more, than the quantity of life.

Scientists have long known that a healthier and longer life can be attained by reducing the intake of calories by at least one-third of the number in a normal, healthy diet while maintaining necessary nutrients. Worms fed such a calorie-restricted diet live up to 5 months longer, which for them is a life span increase of about 60 percent. Mice live about 14 months longer (50 percent increase) and dogs also about 14 months longer (10 percent increase). Would calorie restriction also help humans extend their maximum life span? The answer is likely yes, but by how much is questionable. Some scientists are optimistic that it could add 10 to 15 years. Others are more cautious in thinking 2 to 3 years may be more

reasonable. Either way, there are potential benefits to be had, and every gain counts.

Scientists are working to unravel the molecular mysteries of why calorie restriction works. This, in turn, should help them to discover compounds that mimic the effects of calorie restriction. Interestingly, one class of compounds that mimics those effects are *polyphenols*, which are abundant in highly pigmented foods, such as pomegranates, or beverages, such as red wine.

Gerontologists who study the aging process have come to recognize at least six categories of damage sustained by our cells and biomolecules (see Table 3.2). These are likely to be the underlying causes of our becoming frail and more susceptible to death as we grow older. Prevent or reverse these damages, and the result would be a rejuvenated body, which, in turn, would have a greater chance of experiencing an extended life span. The important thing to note here is that we're not talking

about keeping old people alive in their frailty. Instead, we're talking about strategies that would permit people to both look and feel better despite having lived for so many years. Humans would remain productive and active for much longer periods of time.

Let's take a brief look at senescent cell therapy, which is just one example of longevity research. In the early 1960s, Professor Leonard Hayflick, one of the founders of the biotech revolution, noted that a human cell is only able to replicate about 60 times. Once this limit is reached the cell may self-destruct or it may be attacked by the immune system. Some of these cells, however, remain in a quasi-dormant state, called *senescence*, where they produce chemicals that are generally not good for the organism. As we grow older, senescent cells in our body accumulate. We are then exposed to greater amounts of the dangerous chemicals these cells produce. This, in turn, gives rise to numerous age-related problems,

TABLE 3.2   Six Categories of Cellular Damage

PROBLEM	CURRENT AND POTENTIAL REMEDIES*
Cells die and are not replaced.	Improved health habits; growth factors; **gene doping,** stem cells; apoptosis active dephosphorylation inhibitors (**salubrinal**); **senescent cell** therapy
DNA within the cell nucleus is altered, giving rise to cancer.	Improved health habits; chemotherapy; radiation therapy; surgery; **telomere restoration therapy;** gene therapy; **nanoshells; angiogenesis inhibitors**
DNA within cellular mitochondria is altered, resulting in cell death.	Gene therapy to produce mitochondrial proteins from nuclear DNA
Unwanted cells, such as fat cells, accumulate.	Improved health habits; **calorie restriction diet;** surgery; target cell surface differences
Collagen loses elasticity because of cross-linking.	Sunscreen; advanced glycosylation end-product breaker drugs, such as **ALT-711,** to break crosslinks or inhibit their formation
Unwanted junk, such as atherosclerotic and amyloid plaques, accumulates.	**Beta-sheet breaker peptides;** inhibition of **ABAD beta amyloid complex;** genetically modified white blood cells; **lysosome replacement therapy**

*Boldface: Use as keyword phrase in your internet search engine.

such as the weakening of tissues, arthritis, and cataracts. Recently, scientists found a way to selectively kill senescent cells as they formed in mice. Remarkably, the treated mice remained healthier as they aged compared to mice who received no treatment. If a similar treatment could be applied to humans it would be a huge advance in the treatment of age-related diseases.

Might scientists be interested in finding remedies to the maladies of aging? How about the businesses or organizations that sponsor the scientists? Do you suppose more or less money will be channeled into these efforts as more promising discoveries are made? Will there be a sufficient demand for resulting products that provide for a longer and healthier life? With wrinkle-free skin? Will people of impoverished nations be able to afford this anti-aging technology? Should it be offered freely? Might we have more

or fewer remedies available to us by the time you are 40 years older? What if medical science advances faster than you age? Stay tuned. Future advances in biotech are sure to have a profound impact on your quality of living.

CONCEPTCHECK

How are people like cars?

WAS THIS YOUR ANSWER? Professor Leonard Hayflick says people are like cars because they age reliably "even though there's nothing in the blueprints that shows a process for doing it." In other words, there is no "death gene," no mechanism that kills us off after a certain time limit; aging is a consequence of living in a corrosive environment. If, however, you want to keep a car like "new," what do you do? Wait until it's ready to fall apart? Or repair as necessary

despite the expense of new body parts or skilled labor? Both people and cars need daily maintenance if their life expectancies are to be maximized. If a car can be nurtured to live for many centuries, can people too? Do we have the resources?

Think and Discuss

1. There are millions of people who don't exercise or eat right even though they know such habits will likely extend how long they live in good health. Why?

2. In the past 20 years, the average life expectancy within most nations has risen by a couple of years, but so has the "healthy life expectancy," which is a measure of how long people remain in good health. Are the two necessarily related? How so?

3. How might cures for age-related diseases also be a solution to the problem of overpopulation?

4. How would society be able to support so many people living well into their 100s?

5. The Pentagon today owes its soldiers $653 billion in future retirement benefits. What might happen to this cost if the soldiers actually lived some 40 years longer than expected? How about 100 years? 500 years? At what point should the Pentagon no longer "owe" these benefits to the soldiers? What trends do you foresee in company retirement plans?

6. How are younger people going to be able to find work when so many older people are not retiring? And how might a healthier upper age group affect the wealth distribution between generations?

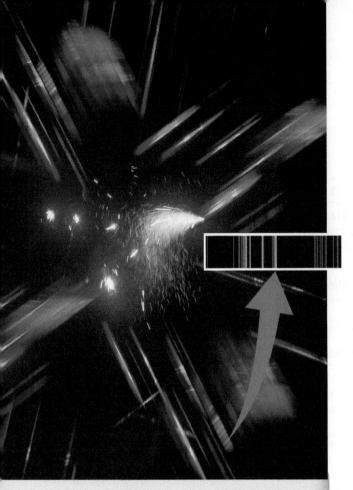

▲ Each element emits its own characteristic spectral pattern, which can be used to identify the element just as a fingerprint can be used to identify a person. The inset image shows the spectral pattern from strontium, which is a common ingredient of fireworks.

4

Subatomic Particles

THE MAIN IDEA

Atoms are made of electrons, protons, and neutrons.

You have already learned that matter is made of fundamental units we call atoms. In this chapter we'll be exploring how atoms themselves are made of even more fundamental units called *subatomic particles*, which include electrons, protons, and neutrons. We'll be looking at this from a historical perspective, beginning with the discovery of the electron in the early 20th century. You will learn that an atom is mostly empty space with nearly all of its mass concentrated in a tiny center called the *atomic nucleus*.

As electrons whiz around the atomic nucleus, they can absorb or emit energy in the form of light, which travels in distinct little packets called *photons*. Evidence of this is seen by looking at glowing elements through a glass prism, as shown in the opening photograph for this chapter. By studying this light, scientists have developed models of the atom. A few of these models are presented in this chapter. Through these models, which continue to be refined even today, we gain a powerful understanding of how atoms behave.

Chemistry

The Quantized Whistle

Matter is made of tiny particles. Each particle is a distinct *quantity* of matter, which is why we say that matter is *quantized*. When we say something is quantized, we mean that it is made of fundamental units. Light energy comes to us in extremely tiny packets called photons and is therefore also quantized. But how can energy be quantized? What does this really mean? The following activity provides an analogy.

PROCEDURE

1. Ask someone who knows how to whistle to whistle from a high pitch to a low pitch in one continuous breath. If you can whistle, then you can do the exercise yourself. Notice that the change in pitch is smooth.

2. Have the whistler repeat the same whistle into one end of a long tube, such as the cardboard tube inside wrapping paper or some piping. Notice that the whistler's smooth continuous whistle is forced into distinct steps. The whistle is quantized!

ANALYZE AND CONCLUDE

1. Try experimenting with tubes of different lengths. To hear yourself more clearly, use a flexible plastic tube and twist the outer end toward your ear.

2. Does a longer tube create fewer or more steps than a shorter tube? Why is it so difficult to whistle down a garden hose?

3. What musical instruments work by changes in the length of a tube?

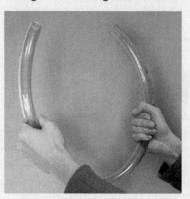

4.1 Physical and Conceptual Models

EXPLAIN THIS

How do we predict the behavior of atoms?

Atoms are so amazingly small that the number of them in a baseball is roughly equal to the number of Ping-Pong balls that could fit inside a hollow sphere as big as the Earth, as **Figure 4.1** illustrates.

Atoms are so small that we can never *see* them in the usual sense. This is because light travels in waves, and atoms are smaller than the wavelengths of visible light. As illustrated in **Figure 4.2**, a single object, at any magnification will remain invisible if it is smaller than the wavelength of light used to illuminate it. We could stack microscope on top of microscope and still not see an individual atom.

Although we cannot see atoms *directly*, we can generate images of them *indirectly*. As was discussed in Section 3.8, the scanning probe microscope

LEARNING OBJECTIVE

Distinguish between models that describe physical attributes and those that describe the behavior of a system.

Atoms in a baseball Ping-Pong balls in the Earth

◀ Figure 4.1
If the Earth were filled with nothing but Ping-Pong balls, the number of balls would be huge and roughly equal to the number of atoms in a baseball. Said differently, if a baseball were the size of the Earth, one of its atoms would be the size of a Ping-Pong ball.

▶ Figure 4.2
(a) A bacterium is visible because it is much larger than the wavelengths of visible light. We can see the bacterium through the microscope because the bacterium reflects visible light back toward the eye. (b) An atom is invisible because it is smaller than the wavelengths of visible light, which pass by the atom with no reflection—much like a blade of grass sticking up in a pond will in no way reflect any much wider water waves passing by.

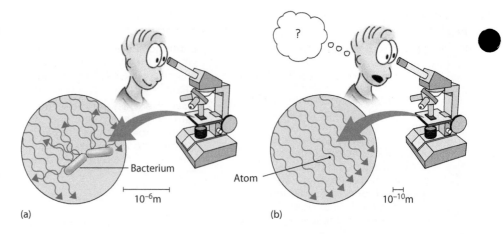

(a) Bacterium 10^{-6}m

(b) Atom 10^{-10}m

 FOR YOUR INFORMATION

Atoms are so small that there are more than 10 billion trillion of them in each breath you exhale. This number of atoms is greater than the number of breaths in the earth's atmosphere. Within a few years, the atoms of your breath are uniformly mixed throughout the atmosphere. What this means is that anyone anywhere on the earth inhaling a breath of air takes in numerous atoms that were once part of you. And, of course, the reverse is true: you inhale atoms that were once part of everyone who has ever lived. We are literally breathing one another.

 **READING CHECK**

What is an accurate conceptual model able to predict?

does this by dragging an ultrathin needle back and forth over the surface of a sample. The result is a computer-generated image of the positions of the atoms on the surface. While such an image is both remarkable and useful, it is not a photograph showing the actual appearance of the atoms. Atoms do not have solid surfaces, as implied by these images. Rather, as we'll be discussing in this chapter, atoms are made of mostly empty space. Envision what empty space looks like. Then you will have a more accurate picture of what it's really like down there at the level of atoms and molecules.

A very small or very large visible object can be represented with a **physical model,** which is a model that replicates the object at a more convenient scale. **Figure 4.3a,** for instance, shows a large-scale physical model of a microorganism that a biology student might use to study the microorganism's structure. Because atoms are invisible, however, we cannot use a physical model to represent them. In other words, we cannot simply scale up the atom to a larger size, as we might with a microorganism. So, rather than describing the atom with a physical model, chemists use a **conceptual model,** which is a representation of a system that helps us predict how the system behaves. The more accurate a conceptual model, the more accurately it predicts the behavior of the system.

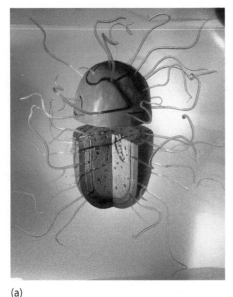

(a)

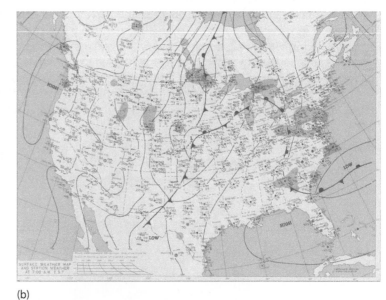

(b)

▲ Figure 4.3
(a) This large-scale model of a microorganism is a physical model. (b) Weather forecasters rely on conceptual models such as this one to predict the behavior of weather systems.

For example, the weather is best described using a conceptual model like the one shown in **Figure 4.3b**. Such a model shows how the various components of the system—humidity, atmospheric pressure, temperature, electric charge, the motion of large masses of air—interact with one another. Other systems that can be described by conceptual models are the economy, population growth, the spread of diseases, and team sports.

CONCEPT CHECK

A basketball coach describes a playing strategy to her team by way of sketches on a game card. Do the illustrations represent a physical model or a conceptual model?

CHECK YOUR ANSWER The sketches are a conceptual model the coach uses to describe a system (the players on the court), with the hope of achieving an outcome (winning the game).

Like the weather, the atom is a complex system of interacting ultra-tiny components, notably *electrons, protons,* and *neutrons.* The atom, therefore, is best described with a conceptual model. Thus, you should be careful not to interpret any visual representation of an atomic conceptual model as a recreation of an actual atom. In Section 4.6, for example, you will be introduced to the planetary model of the atom, wherein electrons are shown orbiting a dense center of protons and neutrons (the atomic nucleus) much as planets orbit the Sun. This planetary model is limited, however, in that it fails to explain many properties of atoms. Therefore, newer and more accurate (and more complicated) conceptual models of the atom have since been introduced. In these models, electrons appear as a cloud, but even these models have their limitations.

In this textbook, our focus is on conceptual atomic models that are easily represented by visual images, including the planetary model and a slightly more sophisticated model in which electrons are grouped in units called *shells.* Despite their limitations, such images are excellent guides to learning chemistry, especially for the beginning student. We will introduce these models as they were developed historically, beginning with the discovery of the subatomic particles.

4.2 The Electron Was the First Subatomic Particle Discovered

EXPLAIN THIS

Before the advent of LCD screens, why were television sets so very heavy?

LEARNING OBJECTIVE

Identify experiments leading to the discovery of the electron.

In 1752, Benjamin Franklin (1706–1790) learned from experiments with thunderstorms that lightning is a flow of electrical energy through the atmosphere. This discovery prompted 19th-century scientists to explore whether or not electrical energy could travel through gases other than the atmosphere. To find out, they passed electricity across glass tubes in which they had sealed various gases.

In every case, the result was a brightly glowing ray (**Figure 4.4a**). This meant that electrical energy was able to travel through different types of gases. To the surprise of these early investigators, a ray was also produced when the voltage was applied across a glass tube that had been evacuated and was thus empty of any gas (**Figure 4.4b**). This implied that the ray was not simply the glowing of a contained gas but rather an entity in and of itself.

(a)

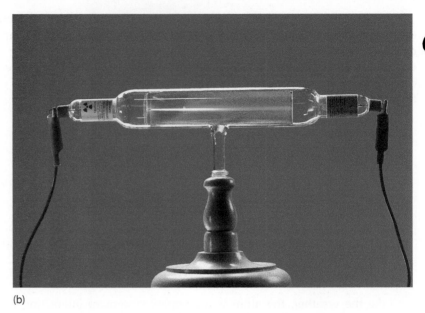

(b)

▲ **Figure 4.4**
(a) Electrical energy passing through a glass tube filled with neon gas generates a bright glowing red. (b) The ray passing through an evacuated glass tube is not usually visible. In the tube shown here, however, the ray is highlighted by a fluorescent backing that glows green as the ray passes over it.

Experiments showed that the ray emerged from the end of the tube that was negatively charged. Because this negatively charged end was called the *cathode*, the apparatus, shown in **Figure 4.5a**, was named a *cathode ray tube.* Magnetic fields deflected the ray, as did small electrically charged metal plates. When such plates were used, the ray was always deflected toward the positively charged plate and away from the negatively charged plate. Because identical charges repel each other, this meant the cathode ray was negatively charged. The speed of the ray was found to be considerably less than the speed of light. Because of these characteristics, it appeared that the ray behaved more like a beam of particles than a beam of light.

▲ A popular myth holds that Franklin discovered the electrical nature of lightning by flying a kite during a lightning storm. Franklin, however, was smart enough to know the extreme danger posed by such a foolish act. As a practical application of his discovery, Franklin invented the lightning rod, which is a sharp point of metal placed on a rooftop and connected to the ground by a long wire. Houses equipped with such rods are protected from the danger of lightning strikes.

▶ **Figure 4.5**
(a) A simple cathode ray tube. The small hole in the positively charged end of the tube, the anode, permits the passage of a narrow beam that strikes the end of the tube, producing a glowing dot as the beam interacts with the glass.
(b) The cathode ray is deflected by a magnetic field.

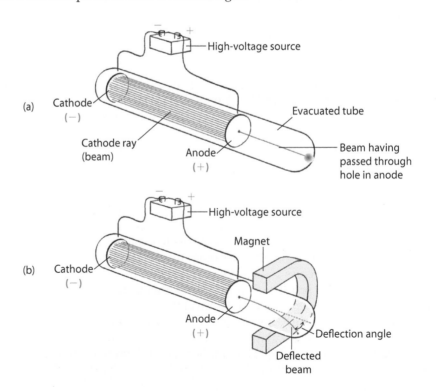

In 1897, J. J. Thomson (1856–1940) measured the deflection angles of cathode ray particles in a magnetic field, using a magnet positioned as shown in **Figure 4.5b**. He reasoned that the deflection of the particles depended on their mass and electric charge. The greater a particle's mass (inertia), the greater its resistance to a change in motion, which gives rise to a *smaller* deflection. The greater a particle's charge, the stronger the magnetic interactions, which gives rise to a *larger* deflection. The angle of deflection, he concluded, was proportional to the ratio of the particle's charge to its mass:

$$\text{angle of deflection} \propto \frac{\text{charge}}{\text{mass}}$$

Knowing only the angle of deflection, however, Thomson was unable to calculate either the charge or the mass of each particle. In order to calculate the mass, he needed to know the charge, but in order to calculate the charge, he needed to know the mass.

READING**CHECK**

The greater a particle's mass, the greater its resistance to what?

CONCEPTCHECK

For which equation is it not possible to calculate one specific value for x?

$$4 = \frac{x}{2} \qquad 3 = \frac{x}{y}$$

CHECK YOUR ANSWER In the first equation, it is possible to figure that $x = 8$ (because $8/2 = 4$). In the second equation, one specific value for x cannot be determined unless the value of y is known. Similarly, Thomson could not calculate the electron's mass without knowing its charge.

In 1909, the American physicist Robert Millikan (1868–1953) calculated the fundamental unit of electric charge on the basis of the innovative experiment shown in **Figure 4.6**. Millikan sprayed tiny oil droplets into a specially designed chamber. The droplets picked up a negative charge after passing through a hole in a charged plate. Millikan then reversed the charge on the plate so that the droplets would be pulled upward. By adjusting the strength of the plate's charge he could get droplets of various sizes to hover motionless. In such cases the upward electric force exactly balanced the downward force of gravity.

Repeated measurements showed that the electric charge on any droplet was always some multiple of a single very small value, 1.60×10^{-19} coulomb, which Millikan proposed to be the fundamental increment of all electric charge. (The *coulomb* is a unit of electric charge.) Using this value and the charge-to-mass ratio discovered by Thomson, Millikan calculated the mass of a cathode ray particle to be considerably less than that of the smallest known

▲ Joseph John Thomson, known to his colleagues as J. J., was one of the first directors of the famous Cavendish Laboratory of Cambridge University in England, where almost all the early discoveries concerning subatomic particles and their behavior were made. Seven of Thomson's students went on to receive Nobel prizes for their scientific work. Thomson himself won a Nobel prize in 1906 for his work with the cathode ray tube.

◀ Figure 4.6
Millikan determined the charge of an electron with this oil-drop experiment.

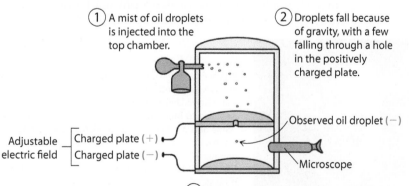

① A mist of oil droplets is injected into the top chamber.

② Droplets fall because of gravity, with a few falling through a hole in the positively charged plate.

Observed oil droplet (−)

Adjustable electric field ⎡ Charged plate (+)
⎣ Charged plate (−)

Microscope

③ The electric field is adjusted until a droplet hovers. The upward electric force exerted on the droplet by the positively charged plate is exactly balanced by the downward force of gravity exerted on the droplet.

The European science community of the 1800s viewed most American scientists as inventors—clever, but not profound in their thinking or discoveries. This attitude began to change at the turn of the 20th century, principally because of the work of American scientists such as Robert Millikan, who excelled in his experimental designs and conclusions. In addition to research, he also spent much time preparing textbooks so that his students did not have to rely so much on lectures. He won a Nobel prize in 1923 and served as the president of Caltech from 1921 to 1945.

atom, hydrogen. This was startling because at the time the atom was thought to be the smallest particle of matter. Here scientists had discovered a particle smaller than the smallest atom.

CONCEPT CHECK

What do the numbers 45, 30, 60, 75, 105, 35, 80, 55, 90, 20, and 65 have in common?

CHECK YOUR ANSWER They are all multiples of 5. In a similar fashion, Millikan noted electric charges that were multiples of a very small number, which he calculated to be 1.60×10^{-19} coulomb.

The cathode ray particle is known today as the **electron,** a name that comes from the Greek word for amber (*electrik*), which is a material the early Greeks used to study the effects of static electricity. The electron is a fundamental component of all atoms. All electrons are identical, each having a negative electric charge and an incredibly small mass of 9.1×10^{-31} kilograms. The arrangement of electrons in atoms determine many of a material's properties, including chemical reactivity and such physical attributes as taste, texture, appearance, and color.

The cathode ray—a stream of electrons—has found a great number of applications. Most notably, the original television sets (not the modern LCD screens) were cathode ray tubes with one end widened out into a phosphor-coated screen. Signals from the television station would cause electrically charged plates in the tube to control the direction of the ray in such a way that images were traced onto the screen.

4.3 The Mass of an Atom Is Concentrated in Its Nucleus

LEARNING OBJECTIVE

Defend Rutherford's conclusion that each atom contains a densely packed positively charged center.

EXPLAIN THIS

If atoms are mostly empty space, why can't we walk through walls?

It was reasoned that if atoms contained negatively charged particles, some balancing positively charged matter must also exist. From this, Thomson put forth what he called a *plum-pudding model* of the atom, shown in **Figure 4.7.** Further experimentation, however, soon proved this model to be wrong.

Around 1910, a more accurate picture of the atom came to one of Thomson's former students, the New Zealand physicist Ernest Rutherford (1871–1937). Rutherford oversaw the now-famous gold-foil experiment,

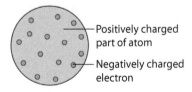

Positively charged part of atom

Negatively charged electron

Figure 4.7
Thomson's plum-pudding model of the atom. Thomson proposed that the atom might be made of thousands of tiny, negatively charged particles swarming within a cloud of positive charge, much like plums and raisins in an old-fashioned Christmas plum pudding.

◀ When Ernest Rutherford was 24, he placed second in a New Zealand scholarship competition to attend Cambridge University in England, but the scholarship was awarded to Rutherford after the winner decided to stay home and get married. In addition to discovering the atomic nucleus, Rutherford was also first to characterize and name many of the nuclear phenomena discussed in Chapter 5. He won a Nobel prize in 1908 for showing how elements such as uranium can become different elements through the process of radioactive decay. At the time, the idea of one element transforming to another was shocking and met with great skepticism, because it seemed reminiscent of alchemy.

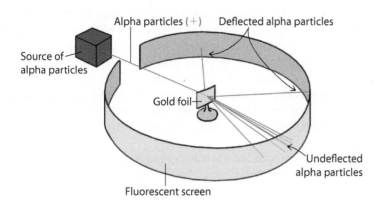

which was the first experiment to show that the atom is mostly empty space and that most of its mass is concentrated in a tiny central core called the **atomic nucleus.**

In Rutherford's experiment, shown in **Figure 4.8**, a beam of positively charged particles, called alpha particles, was directed through a very thin sheet of gold foil. Since alpha particles were known to be thousands of times more massive than electrons, it was expected that the alpha-particle stream would not be impeded as it passed through the "atomic pudding" of gold foil. This was indeed observed to be the case—for the most part. Nearly all alpha particles passed through the gold foil. However, some particles were deflected from their straight-line path as they passed through the foil. A few of them were even deflected straight back toward the source! These alpha particles must have hit something relatively massive, but what?

Rutherford reasoned that undeflected particles traveled through regions of the gold foil that were empty space, as **Figure 4.9** shows, and the deflected ones were repelled by extremely dense positively charged centers. Each atom, he concluded, must contain one of these centers, which he named the *atomic nucleus*.

Rutherford guessed that atomic nuclei must be positively charged to balance the negative charge of the electrons in the atom. He also guessed that the electrons were not part of this nucleus but still somewhere in the atom. Today we know that, as **Figure 4.10** illustrates, the electrons do indeed exist outside the nucleus, swirling around it at ultrahigh speeds.

What Figure 4.10 does not show is that an atom is mostly empty space, with the diameter of the whole atom being about 10,000 times greater than the diameter of its nucleus. If a nucleus were the size of the period at the end of this sentence, the outer edges of the atom would be located 3.3 meters (11 feet) away. Because electrons are even smaller than the nucleus, and because they are widely separated from each other (as well as from the nucleus), atoms are indeed mostly empty space—just as our solar system is mostly empty space.

✓ READINGCHECK

What did Rutherford reason about the deflected alpha particles?

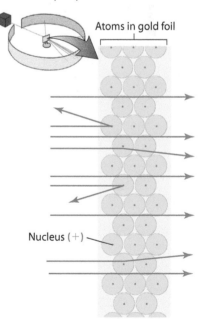

▲ Figure 4.9
Rutherford's interpretation of the results from his gold-foil experiment. Most alpha particles passed through the empty space of the gold atoms undeflected, but a few were deflected by atomic nuclei.

◀ Figure 4.10
Electrons whiz around the atomic nucleus, forming what can be best described as an ultrathin cloud. If this illustration were drawn to scale, the atomic nucleus would be too small to be seen. An atom is mostly empty space.

▲ **Figure 4.11**
As close as Tracy and Ian are in this photograph, none of their atoms meet.

So if atoms are mostly empty space, why don't they simply pass through one another? They don't because electrons repel the electrons of neighboring atoms. Therefore, two atoms can get only so close to each other before they are repelled. This explains why gravity doesn't pull you through the floor as you stand. While the force of gravity pulls you down, the electric force of repulsion between the atoms of the floor and those of your feet pushes you up. As you stand on the floor, these two forces are balanced and you find yourself neither falling nor rising—there's more to standing than most people realize!

Similarly, when the atoms of your hand push against the atoms of a wall, repulsions between electrons in your hand and electrons in the wall prevent your hand from passing through the wall. You sense this repulsion as a pressure that pushes back. Our sense of touch comes from these electrical repulsions. Interestingly, when you touch someone, your atoms and those of the other person do not meet. Instead, atoms from the two of you get close enough so that you sense an electrical repulsion. There is still a tiny, though imperceptible, gap between the two of you (**Figure 4.11**).

4.4 The Atomic Nucleus Is Made of Protons and Neutrons

LEARNING OBJECTIVE

Describe the structure of the atomic nucleus and how the atomic mass of an element is calculated.

EXPLAIN THIS

Why aren't we harmed by drinking heavy water, D2O?

The positive charge of any atomic nucleus was found to be equal in magnitude to the combined negative charge of all the electrons in the atom. It was thus reasoned, and then experimentally confirmed, that the nucleus contains positively charged subatomic particles, which we call **protons.** Protons are nearly 2000 times more massive than electrons. The number of protons in a nucleus is equal to the number of electrons whirling about it, so the charges are balanced.

Scientists have agreed to identify elements by **atomic number,** which is the number of protons in the nucleus of each atom of a given element. The modern periodic table lists the elements in order of increasing atomic number. Hydrogen, with one proton per atom, has atomic number 1; helium, with two protons per atom, has atomic number 2; and so on.

> ### CONCEPT CHECK
> How many protons are there in an iron atom, Fe (atomic number 26)?
>
> **CHECK YOUR ANSWER** The atomic number of an atom and its number of protons are the same. Thus, there are 26 protons in an iron atom. Another way to put this is that all atoms that contain 26 protons are, by definition, iron atoms.

If we compare the electric charges and masses of different atoms, we see that an atomic nucleus must be made up of more than just protons. Helium, for example, has twice the electric charge of hydrogen but *four* times the mass. The added mass is due to another subatomic particle found in the nucleus, the **neutron,** which was first detected in 1932 by the British physicist James Chadwick (1891–1974).

Neutrons have about the same mass as protons, but they have no electric charge. Any object that has no net electric charge is said to be *electrically neutral,*

TABLE 4.1   Subatomic Particles

	PARTICLE	CHARGE	RELATIVE MASS	ACTUAL MASS (KG)
	Electron	−1	1	9.11×10^{-31}*
Nucleons {	Proton	+1	1836	1.673×10^{-27}
	Neutron	0	1841	1.675×10^{-27}

*9.11×10^{-31} kg = 0.000000000000000000000000000000911 kg (see Appendix A).

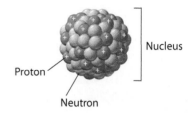

▲ Figure 4.12
An illustration of an atomic nucleus showing its protons and neutrons all clumped together. Warning! This is not what the nucleus really looks like. Interestingly, the atomic nucleus, just like the atom, is also made of mostly empty space, but is filled with a very high concentration of energy, as we discuss in Chapter 5.

and that is where the neutron got its name. We discuss the important role that neutrons play in holding the atomic nucleus together in Chapter 5.

Both protons and neutrons are called **nucleons,** a term that denotes their location in the atomic nucleus. **Figure 4.12** shows an illustration of a nucleus with many nucleons (protons and neutrons). Table 4.1 summarizes the basic facts about electrons, protons, and neutrons.

For any element, the number of neutrons in the nucleus may vary. For example, most hydrogen atoms (atomic number 1) have no neutrons. A small percentage, however, have one neutron, and a smaller percentage have two neutrons. Similarly, most iron atoms (atomic number 26) have 30 neutrons, but a small percentage have 29 neutrons. Atoms of the same element that contain different numbers of neutrons are **isotopes** of one another.

We identify isotopes by their **mass number,** which is the total number of protons plus neutrons (in other words, the number of nucleons) in the nucleus. As **Figure 4.13** shows, a hydrogen isotope with only one proton is called hydrogen-1, where 1 is the mass number. A hydrogen isotope with one proton and one neutron is therefore hydrogen-2, and a hydrogen isotope with one proton and two neutrons is hydrogen-3. Similarly, an iron isotope with 26 protons and 30 neutrons is called iron-56, and one with only 29 neutrons is iron-55.

An alternative method of indicating isotopes is to write the mass number as a superscript and the atomic number as a subscript to the left of the atomic symbol. For example, an iron isotope with a mass number of 56 and atomic number of 26 is written as

$$\text{Mass number} \diagdown \,^{56}_{26}\text{Fe—Atomic symbol}$$
$$\text{Atomic number} \diagup$$

The total number of neutrons in an isotope can be calculated by subtracting its atomic number from its mass number:

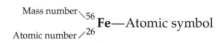

$$\begin{array}{r} \text{mass number} \\ - \ \underline{\text{atomic number}} \\ \text{number of neutrons} \end{array}$$

READING CHECK

How are isotopes identified?

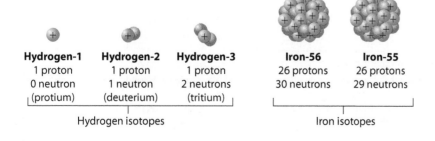

Hydrogen-1
1 proton
0 neutron
(protium)

Hydrogen-2
1 proton
1 neutron
(deuterium)

Hydrogen-3
1 proton
2 neutrons
(tritium)

Hydrogen isotopes

Iron-56
26 protons
30 neutrons

Iron-55
26 protons
29 neutrons

Iron isotopes

◀ Figure 4.13
Isotopes of an element have the same number of protons but different numbers of neutrons and hence different mass numbers. The three hydrogen isotopes have special names: protium for hydrogen-1, deuterium for hydrogen-2, and tritium for hydrogen-3. Of these three isotopes, hydrogen-1 is the most common. For most elements, such as iron, the isotopes have no special names and are indicated merely by mass number.

For example, uranium-238 has 238 nucleons. The atomic number of uranium is 92, which tells us that 92 of these 238 nucleons are protons. The remaining 146 nucleons must be neutrons:

$$238 \text{ protons and neutrons}$$
$$\underline{-92 \text{ protons}}$$
$$146 \text{ neutrons}$$

Atoms interact with one another electrically. Therefore, the way any atom behaves in the presence of other atoms is determined largely by the charged particles it contains, especially its electrons. Isotopes of an element differ only by mass, not by electric charge. For this reason, isotopes of an element share many characteristics—in fact, as chemicals they cannot be distinguished from one another. For example, a sugar molecule containing carbon atoms with seven neutrons is digested no differently from a sugar molecule containing carbon atoms with six neutrons. Interestingly, about 1 percent of the carbon we consume in food is the carbon-13 isotope, containing seven neutrons per nucleus. The remaining 99 percent of the carbon in our diet is the more common carbon-12 isotope, containing six neutrons per nucleus.

The total mass of an atom is called its **atomic mass.** This is the sum of the masses of all the atom's components (electrons and the nucleus). Because electrons are so much less massive than the nucleus, their contribution to atomic mass is negligible.

As we explore further in Section 9.2, a special unit has been developed for atomic masses. This is the *atomic mass unit*, amu; 1 atomic mass unit is equal to 1.661×10^{-24} gram, which is slightly less than the mass of a single proton. As shown in **Figure 4.14**, the atomic masses listed in the periodic table are in atomic mass units. As is explored in the Calculation Corner on page 103, the atomic mass of an element as presented in the periodic table is actually the average atomic mass of the various isotopes of that element occurring in nature.

CONCEPT**CHECK**

Distinguish between mass number and atomic mass.

CHECK YOUR ANSWER Both terms include the word *mass*, so they are easily confused. Focus your attention on the second word of each term, however, and you'll get it right every time. Mass *number* is a *count* of the number of nucleons in an isotope. An atom's mass number requires no units because it is simply a count. Atomic *mass* is a measure of the total mass of an atom, which is given in atomic mass units. If necessary, atomic mass units can be converted to grams using the relationship 1 atomic mass unit = 1.661×10^{-24} gram.

▶ Figure 4.14
Helium, He, has an atomic mass of 4.003 atomic mass units, and neon, Ne, has an atomic mass of 20.180 atomic mass units.

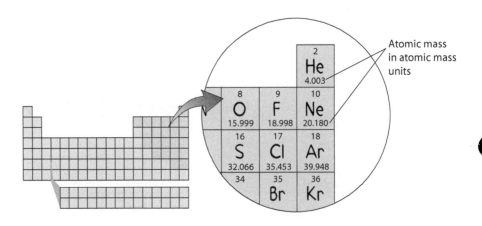

CALCULATION CORNER CALCULATING ATOMIC MASS

Roughly 99 percent of all carbon atoms, are the isotope carbon-12, and most of the remaining 1 percent are the heavier isotope carbon-13. This small amount of carbon-13 raises the average mass of carbon from 12.000 atomic mass units to the slightly greater value 12.011 atomic mass units.

To arrive at the atomic mass presented in the periodic table, you first multiply the mass of each naturally occurring isotope of an element by the fraction of its abundance and then add up all the fractions. This gives you the *weighted average* of all the isotopes.

EXAMPLE

Carbon-12 has a mass of 12.0000 atomic mass units and makes up 98.89 percent of naturally occurring carbon. Carbon-13 has a mass of 13.0034 atomic mass units and makes up 1.11 percent of naturally occurring carbon. Use this information to show that the atomic mass of carbon shown in the periodic table, 12.011 atomic mass units, is correct.

ANSWER

Recognize that 98.89 percent and 1.11 percent expressed as fractions are 0.9889 and 0.0111, respectively.

	Contributing Mass of ^{12}C	Contributing Mass of ^{13}C	
Fraction of Abundance	0.9889	0.0111	
Mass (amu)	× 12.0000	× 13.0034	step 1
	11.867	0.144	

atomic mass = 11.867 + 0.144 = 12.011   step 2

YOUR TURN

Chlorine-35 has a mass of 34.97 atomic mass units, and chlorine-37 has a mass of 36.95 atomic mass units. Determine the atomic mass of chlorine, Cl (atomic number 17), if 75.53 percent of all chlorine atoms are the chlorine-35 isotope and 24.47 percent are the chlorine-37 isotope.

4.5 Light Is a Form of Energy

EXPLAIN THIS

How are microwaves and X rays similar? How are they different?

Modern models of the atom were developed by scientists to explain how atoms emit light. To help you understand and appreciate these models, it is good to review the basic concepts regarding the nature of light.

From physics we learn that surrounding every charged particle is an *electric field*, which is a region of space through which the electric force of that charged particle may act. If an electron were to start vibrating, the electric field surrounding that electron would also start to vibrate and at the same frequency. Interestingly, such a vibrating electric field generates a complementary vibrating magnetic field, which in turn reinforces the vibrating electric field. The net result is a series of self-reinforcing electric and magnetic waves that propagate away from the vibrating electron.

By analogy, you know that if you shake the end of a stick back and forth in still water, you create waves that travel outward on the water's surface. If you similarly shake an electrically charged rod in empty space, you create electromagnetic waves in space that also travel outward. Electromagnetic waves, however, are oscillations (vibrations) of electric and magnetic fields, not oscillations of a material medium such as water. These waves are called *electromagnetic radiation*. Most of the electromagnetic radiation we encounter is generated by electrons, which can oscillate at exceedingly high rates because of their small size.

The distance between two crests of an electromagnetic wave is called the *wavelength* of the wave. **Figure 4.15** labels two wavelengths—one very long, the other very short—on a fictitious wave drawn for illustration only. Electromagnetic waves can also be characterized by their *wave frequency*, a measure of how rapidly they oscillate. The basic unit of wave frequency is the hertz (abbreviated Hz), where 1 hertz equals 1 cycle per second. The higher the frequency of a wave, the shorter its wavelength and the greater its energy.

LEARNING OBJECTIVE

Describe the nature and range of electromagnetic waves.

READINGCHECK

What are electromagnetic waves?

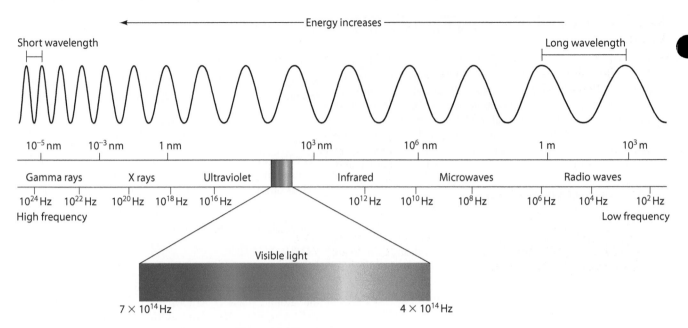

▲ **Figure 4.15**
The electromagnetic spectrum is a continuous band of wave frequencies extending from high-energy gamma rays, which have short wavelengths and high frequencies, to low-energy radio waves, which have long wavelengths and low frequencies. The descriptive names of these regions are merely a historical classification, for all waves are the same in nature, differing only in wavelength and frequency. Interestingly, the speed at which they travel is also the same. This is the speed of light, which is about 3×10^8 meters per second.

 FOR YOUR INFORMATION

Tap the end of a stick on the surface of some still water. Do this repeatedly and you will generate waves that emanate outward. The faster you tap your stick, the closer together the waves are to one another, but the speed at which they travel outward remains the same. Try it and see! Similarly, as electrons oscillate back and forth in an atom, they generate electromagnetic waves that emanate from the atom. All electromagnetic waves travel at the same speed, which is the speed of light.

Figure 4.15 shows a full range of frequencies and wavelengths of electromagnetic radiation in a display known as the **electromagnetic spectrum.** The most energetic region of the electromagnetic spectrum consists of gamma rays. Next is the region of slightly lower energy, where we find X rays, and following that is the electromagnetic radiation we call ultraviolet light. Within a narrow region from about 7×10^{14} (700 trillion) hertz to about 4×10^{14} (400 trillion) hertz are the frequencies of electromagnetic radiation known as visible light. This region includes the rainbow of colors our eyes are able to detect, from violet at 700 trillion hertz to red at 400 trillion hertz. Lower in energy than visible light are infrared waves (detected by our skin as "heat waves"), then microwaves (used to cook foods), and finally radio waves (through which radio and television signals are sent)—the waves of lowest energy.

CONCEPTCHECK
Can you see radio waves? Can you hear them?

CHECK YOUR ANSWER Your eyes are equipped to see only the narrow range of frequencies of electromagnetic radiation from about 700 trillion to 400 trillion hertz—the range of visible light. Radio waves are one type of electromagnetic radiation, but their frequency is much lower than what your eyes can detect. Thus, you can't see radio waves. Neither can you hear them. You can, however, turn on an electronic gizmo called a radio, which translates radio waves into signals that drive a speaker to produce sound waves your ears can hear.

We see white light when all frequencies of visible light reach our eye at the same time. By passing white light through a prism or through a diffraction grating, which is a glass plate or plastic sheet with microscopic lines etched into it, the color components of the light can be separated, as shown in **Figure 4.16.**

(Remember—each color of visible light corresponds to a different frequency.) A **spectroscope,** shown in **Figure 4.17,** is an instrument used to observe the color components of any light source. As we discuss in the following section, a spectroscope allows us to analyze the frequencies of light emitted by elements as they are made to glow.

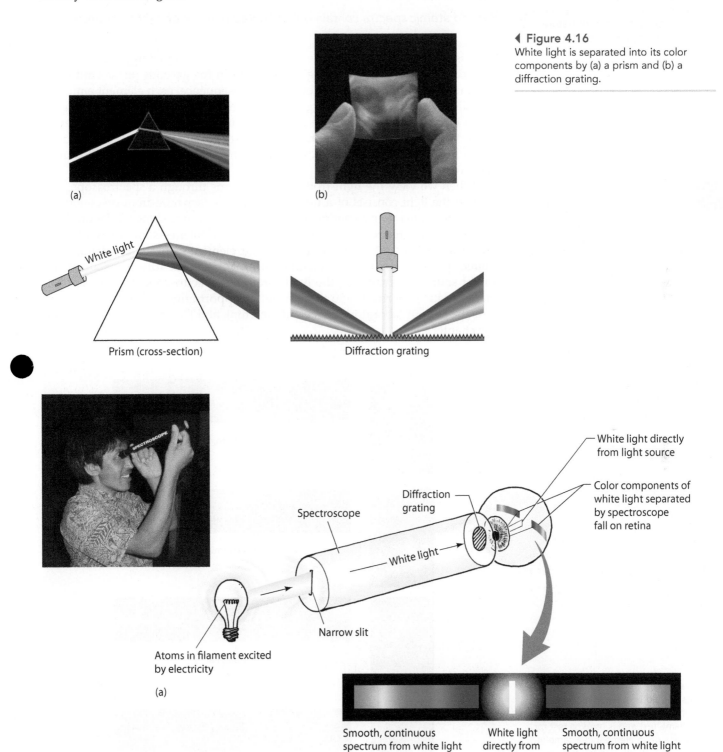

◄ **Figure 4.16**
White light is separated into its color components by (a) a prism and (b) a diffraction grating.

(a)

(b)

White light

Prism (cross-section)

Diffraction grating

White light directly from light source

Color components of white light separated by spectroscope fall on retina

Diffraction grating

Spectroscope

White light

Narrow slit

Atoms in filament excited by electricity

(a)

Smooth, continuous spectrum from white light

White light directly from light source

Smooth, continuous spectrum from white light (mirror image)

(b)

▲ Figure 4.17
(a) In a spectroscope, light emitted by atoms passes through a narrow slit before being separated into particular frequencies by a prism or (as shown here) a diffraction grating. (b) This is what the eye sees when the slit of a diffraction-grating spectroscope is pointed toward a white-light source. Spectra of colors appear to the left and right of the slit.

4.6 Atomic Spectra and the Quantum Hypothesis

LEARNING OBJECTIVE

Recount how the quantum nature of energy led to Bohr's planetary model of the atom.

FOR YOUR INFORMATION

A star's age is revealed by its elemental makeup. The first and oldest stars were composed of hydrogen and helium, because those were the only elements that existed at that time. Heavier elements were produced after many of these early stars exploded in supernovae. Later stars incorporated these heavier elements in their formation. In general, the younger a star, the greater the amounts of these heavier elements it contains.

EXPLAIN THIS

Why do atomic spectra contain only a limited number of light frequencies?

Light is given off by atoms subjected to various forms of energy, such as heat or electricity. The atoms of any given element in the gaseous phase emit only certain frequencies of light, however. As a consequence, each element emits its own distinctive glow when energized. Sodium atoms emit bright yellow light, which makes them useful as the light source in street lamps because our eyes are very sensitive to yellow light. To name just one more example, neon atoms emit a brilliant red-orange light, which makes them useful as the light source in neon signs.

When we view the light from glowing atoms through a spectroscope, we see that the light consists of a number of discrete (separate from one another) frequencies rather than a continuous spectrum like the one shown in Figure 4.17. The pattern of frequencies formed by a given element—some of which are shown in **Figure 4.18**—is referred to as that element's **atomic spectrum.** The atomic spectrum is an element's fingerprint. You can identify the elements in a light source by analyzing the light through a spectroscope and looking for characteristic patterns. If you don't have the opportunity to work with a spectroscope in your laboratory, check out the Think and Do activities at the end of this chapter.

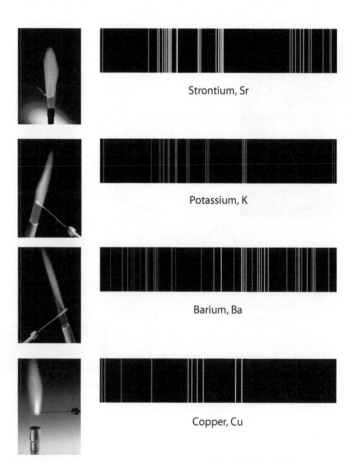

Strontium, Sr

Potassium, K

Barium, Ba

Copper, Cu

▶ **Figure 4.18**
Elements heated by a flame glow their characteristic color. This is commonly called a flame test and is used to test for the presence of an element in a sample. When viewed through a spectroscope, the color of each element is revealed to consist of a pattern of distinct frequencies known as an atomic spectrum.

CONCEPTCHECK

How might you deduce the elemental composition of a star?

CHECK YOUR ANSWER Aim a well-built spectroscope at the star, and study its spectral patterns. In the late 1800s, this was done with our own star, the Sun. Spectral patterns of hydrogen and some other known elements were observed, in addition to one pattern that could not be identified. Scientists concluded that this unidentified pattern belonged to an element not yet discovered on the Earth. They named this element helium after the Greek word for "sun," helios.

The Quantum Hypothesis

An important step toward our present-day understanding of atoms and their spectra was taken by the German physicist Max Planck (1858–1947). In 1900, Planck hypothesized that light energy is *quantized* in much the same way matter is. The mass of a gold brick, for example, equals some whole-number multiple of the mass of a single gold atom. Similarly, an electric charge is always some whole-number multiple of the charge on a single electron. Mass and electric charge are therefore said to be *quantized*, in that they consist of some number of fundamental units.

Planck identified each discrete parcel of light energy as a **quantum,** represented in **Figure 4.19**. A few years later, Einstein recognized that a quantum of light energy behaves much like a tiny particle of matter. To emphasize its particulate nature, each quantum of light was called a *photon,* a name coined because of its similarity to the word *electron.*

Using Planck's quantum hypothesis, the Danish scientist Niels Bohr (1885–1962) explained the formation of atomic spectra as follows. First, an electron has more potential energy when farther from the nucleus. This is analogous to the greater potential energy an object has when held higher above the ground. Second, Bohr recognized that when an atom absorbs a photon of light, it is absorbing energy. This energy is acquired by one of the electrons. Because this electron has gained energy, it must move away from the nucleus.

Bohr realized that the opposite is also true: when a high-potential-energy electron in an atom loses some of its energy, the electron moves closer to the nucleus and the energy lost from the electron is emitted from the atom as a photon of light. Both absorption and emission are illustrated in **Figure 4.20**.

READINGCHECK

What does an electron have when farther from the nucleus?

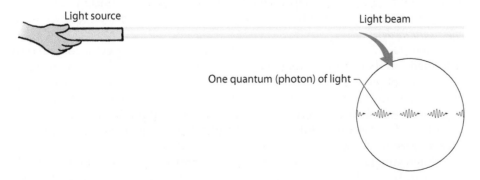

▲ Figure 4.19

Light is quantized, which means it consists of a stream of energy packets. Each packet is called a quantum, also known as a photon.

▶ **Figure 4.20**
An electron is lifted away from the nucleus as the atom it is in absorbs a photon of light and drops closer to the nucleus as the atom releases a photon of light.

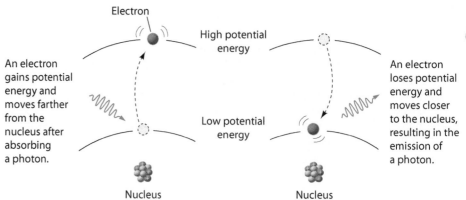

An electron gains potential energy and moves farther from the nucleus after absorbing a photon.

An electron loses potential energy and moves closer to the nucleus, resulting in the emission of a photon.

▲ Niels Bohr (1885–1962) and Albert Einstein (1879–1955) were good friends and colleagues, but they differed on their views of quantum theory and its philosophical implications. Einstein accepted quantum theory but was one of its strongest critics. His critiques were answered for the most part by Niels Bohr through a series of exchanges, called the Bohr–Einstein debates, spanning the latter parts of their lives.

Just as I can't stand between two adjacent steps, an electron can't exist between two energy levels.

Bohr reasoned that because light energy is quantized, the energy of an electron in an atom must also be quantized. In other words, an electron cannot have just any amount of potential energy. Rather, within the atom there must be a number of distinct energy levels, analogous to steps on a staircase. Where you are on a staircase is restricted to where the steps are—you cannot stand at a height that is, say, halfway between any two adjacent steps. Similarly, there are only a limited number of permitted energy levels in an atom, and an electron can never have an amount of energy between these permitted energy levels. Bohr gave each energy level a **quantum number** n, where n is always some integer. The lowest energy level has a quantum number $n = 1$. An electron for which $n = 1$ is as close to the nucleus as possible, and an electron for which $n = 2, n = 3$, and so forth is farther away, in a step-wise fashion, from the nucleus.

CONCEPT CHECK
What is released as an electron transitions from a higher to a lower energy level?

CHECK YOUR ANSWER A photon of light.

Using these ideas, Bohr developed a conceptual model in which an electron moving around the nucleus is restricted to certain distances from the nucleus, with these distances determined by the amount of energy the electron has. Bohr saw this as similar to the way the planets are held in orbit around the Sun at given distances from the Sun. The allowed energy levels for any atom, therefore, could be graphically represented as orbits around the nucleus, as shown in **Figure 4.21**. Bohr's quantized model of the atom thus became known as the *planetary model*.

Bohr used his planetary model to explain why atomic spectra contain only a limited number of light frequencies, as shown in **Figure 4.22**. According to the model, photons are emitted by atoms as electrons move from higher-energy outer orbits to lower-energy inner orbits. The energy of an emitted photon is equal to the difference in energy between the two orbits. Because an electron is restricted to discrete orbits, only particular light frequencies are emitted, as atomic spectra show.

Interestingly, any transition between two orbits is always instantaneous. In other words, the electron doesn't "jump" from a higher to a lower orbit the way a squirrel jumps from a higher branch in a tree to a lower one. Rather, it takes no time for an electron to move between two orbits. Bohr was serious when he stated that electrons could never exist between permitted energy levels!

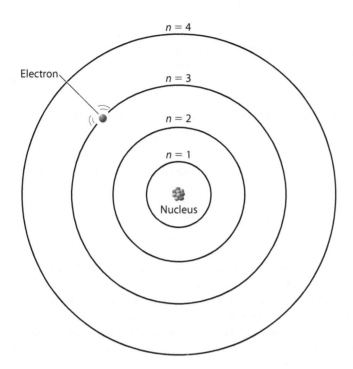

◀ **Figure 4.21**
Bohr's planetary model of the atom, in which electrons orbit the nucleus much as planets orbit the Sun, is a graphical representation that helps us understand how electrons can possess only certain quantities of energy.

CONCEPT CHECK

Is the Bohr model of the atom a physical model or a conceptual model?

CHECK YOUR ANSWER The Bohr model is a conceptual model. It is not a scaled-up version of an atom but instead is a representation that accounts for the atom's behavior.

Bohr's planetary atomic model proved to be a tremendous success. By utilizing Planck's quantum hypothesis, Bohr's model solved the mystery of atomic spectra. Despite its successes, though, Bohr's model was limited, because it did not explain why energy levels in an atom are quantized. Bohr himself was quick to point out that his model was to be interpreted only as a crude beginning and that the picture of electrons whirling about the nucleus like planets about the Sun was not to be taken literally (a warning to which popularizers of science paid no heed).

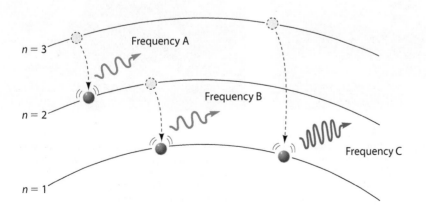

◀ **Figure 4.22**
The frequency of light emitted (or absorbed) by an atom is proportional to the energy difference between electron orbits. Because the energy differences between orbits are discrete, the frequencies of light emitted (or absorbed) are also discrete. The electron here can emit only three discrete frequencies of light—A, B, and C. The greater the transition, the higher the frequency of the photon emitted.

4.7 Electrons Exhibit Wave Properties

Summarize how electrons, when confined to an atom, behave as self-reinforcing wavelike entities represented by atomic orbitals.

READINGCHECK

What does the wave nature of an electron help to explain?

EXPLAIN THIS

How is a plucked guitar string like an electron in an atom?

If light has both wave properties and particle properties, why can't a material particle, such as an electron, also have both? This question was posed by the French physicist Louis de Broglie (1892–1987) while he was still a graduate student in 1924. His revolutionary answer was that every moving particle of matter—by virtue of its energy of motion—is endowed with the characteristics of a wave.

We now speak of waves as an essential feature of any bit of matter. An electron, or any particle, can show itself as a wave or as a particle, depending on how we examine it. This is called the wave–particle duality. Just a few years after de Broglie's suggestion, researchers in Great Britain and the United States confirmed the wave nature of electrons by observing diffraction and interference effects when electrons bounced from crystals. A practical application of the wave properties of electrons is the electron microscope, which focuses not visible-light waves but rather electron waves. Electron microscopes are able to show far greater detail than optical microscopes, as **Figure 4.23** shows.

An electron's wave nature can be used to explain why electrons in an atom are restricted to particular energy levels. Permitted energy levels are a natural consequence of the need for electron waves to form stable patterns around the atomic nucleus.

As an analogy, consider the wire loop shown in **Figure 4.24a**. This loop is affixed to a mechanical vibrator that can be adjusted to create a series of waves traveling through the wire. If the lengths of the waves are equal to the length of the wire (or some fraction thereof), then these waves form what is called a *standing wave*, as shown in **Figure 4.24b**. Within a standing wave, identically sized waves overlap one another, and they do so in a manner that creates regions of maximum intensity and other regions of minimum intensity. A region

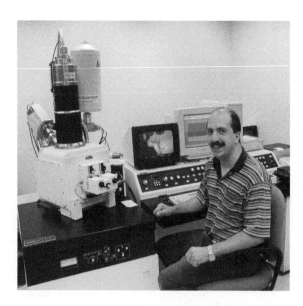

▲ Figure 4.23

(a) An electron microscope makes practical use of the wave nature of electrons. The wavelengths of electron beams are typically thousands of times shorter than the wavelengths of visible light, so the electron microscope is able to distinguish detail not visible with optical microscopes. (b) Detail of a female mosquito's head as seen with an electron microscope at a "low" magnification of 200 times. Note the remarkable resolution.

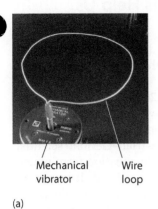

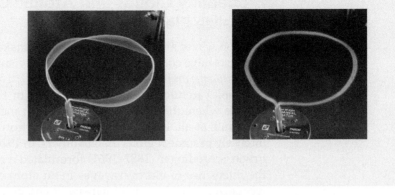

Mechanical   Wire
vibrator   loop

(a)   (b) Wavelength is self-reinforcing.   (c) Wavelength produces chaotic motion.

▲ Figure 4.24
(a) A wire loop affixed to the post of a mechanical vibrator at rest. Waves are sent through the
wire when the post vibrates. (b) Only waves of certain wavelengths are able to form standing
waves. The one shown here is 2/3 the length of the loop. (c) Most other wavelengths result in
chaotic motion with no specific regions of maximum and minimum intensities.

of maximum intensity occurs where crests are overlapping with crests and
troughs are overlapping with troughs. (Note: a "crest" is the high point of
a wave while a "trough" is the low point of a wave.) A region of minimum
intensity occurs where crests and troughs overlap so as to cancel each other
out. At the center of minimum intensity is a point of zero intensity, which is
called a *node*.

For a given size of the wire loop, only certain wavelengths will give rise
to a stable standing wave. Waves of any other wavelength are unable to align
properly and the result is chaotic motion, as suggested in **Figure 4.24c**. These
same ideas apply to electron waves within an atom. We find that only certain
electron wavelengths form standing waves. Because of their stability, these
particular electron wavelengths are favored.

But why does an electron have wavelike characteristics? Recall that according to de Broglie, the wavelength of an electron is a consequence of its energy.
Significantly, if only certain wavelengths are favored, then it follows that only
certain amounts of energy are favored. In other words, the energy of an electron
confined to an atom is *quantized*. Consider the Hands-On Chemistry activity
at the beginning of this chapter. With this activity we see that when a sound
wave is confined to a tube, the consequence is a quantization of its frequencies.
Likewise, when an electron is confined to an atom, the consequence is a quantization of the electron's energy.

Each standing wave of an electron corresponds to one of the permitted
energy levels seen in Figures 4.21 and 4.22. Only the frequencies of light that
match the difference between any two of these permitted energy levels can be
absorbed or emitted by an atom. The wave nature of electrons also explains
why they do not spiral closer and closer to the positive nucleus that attracts
them. By viewing each electron orbit as a standing wave, we see that the
circumference of the smallest orbit can be no smaller than a single wavelength.

CONCEPT CHECK
What must an electron be doing in order to have wave properties?

CHECK YOUR ANSWER According to de Broglie, particles of matter behave like
waves by virtue of their motion, which is a form of energy. An electron must therefore
be moving in order to have wave properties.

Probability Clouds and Atomic Orbitals Help Us Visualize Electron Waves

Electron waves are 3-dimensional, which makes them difficult to visualize, but scientists have come up with two ways of visualizing them: as *probability clouds* and as *atomic orbitals*.

Look carefully at the standing wave of Figure 4.24b. Note that there are regions where the wire vibrates intensely and others where it doesn't appear to vibrate at all. In a similar fashion, electron waves in an atom vibrate more intensely in some regions than in others. In 1926, the Austrian–German scientist Erwin Schrodinger (1887–1961) formulated a remarkable equation from which the intensities of electron waves in an atom could be calculated. It was soon recognized that the wave intensity at any given location determined the probability of finding the electron at that location. In other words, the electron was most likely to be found where its wave intensity was greatest and least likely to be found where its wave intensity was smallest.

If we could plot the positions of an electron of a given energy over time as a series of tiny dots, the resulting pattern would resemble what is called a **probability cloud. Figure 4.25a** shows a probability cloud for hydrogen's electron. The denser a region of the cloud, the greater the probability of finding the electron in that region. The densest regions correspond to where the electron's wave intensity is greatest. A probability cloud is therefore a close approximation of the actual shape of an electron's 3-dimensional wave. Interestingly, this cloud has no distinct surface. Rather, there is some probability that the electron could be very far away from the nucleus, but that probability quickly decreases with increasing distance.

An **atomic orbital,** like a probability cloud, specifies a volume of space where the electron is most likely to be found. By convention, atomic orbitals are drawn to delineate the volume inside which the electron is located 90 percent of the time. This gives the atomic orbital an apparent border, as shown in **Figure 4.25b**. This border is arbitrary, because the electron may exist on either side of it. Almost all of the time, though, the electron remains within the border. So, probability clouds and atomic orbitals are essentially the same thing. They differ only in that atomic orbitals specify an outer limit, which makes them easier to depict graphically.

As Table 4.2 shows, atomic orbitals come in a variety of shapes, some quite exquisite. We categorize these orbitals using the letters *s*, *p*, *d*, and *f*. The simplest is the spherical *s* orbital. The *p* orbital consists of two lobes and resembles an hourglass at the center of which there is a node. There are three kinds of *p* orbitals, which differ from one another only by their orientation in 3-dimensional space. The more complex *d* orbitals have five possible shapes, and the *f* orbitals have seven.

READINGCHECK

What does a probability cloud closely approximate?

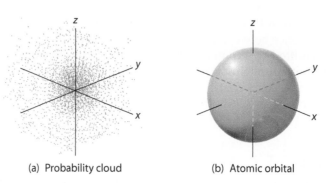

(a) Probability cloud   (b) Atomic orbital

▲ Figure 4.25
(a) The probability cloud for hydrogen's electron in an x, y, z coordinate system. The nucleus is located at the origin, which is where all three lines intersect. The more concentrated the dots, the greater the chance of finding the electron at that location. (b) The atomic orbital for hydrogen's electron. The electron is somewhere inside this spherical volume 90 percent of the time.

TABLE 4.2   Four Categories of Atomic Orbitals: *s, p, d, f*

ORBITAL TYPE	SPATTIAL ORIENTATIONS
s The *s* orbital has only one shape, which is spherical.	
p There are three *p* orbitals. They differ by orientation.	
d There are five *d* orbitals.	
f There are seven *f* orbitals.	

CONCEPTCHECK

What is the relationship between a probability cloud and an atomic orbital?

CHECK YOUR ANSWER A probability cloud indicates the most probable location of an electron of a given energy within an atom. An atomic orbital is the same as a probability cloud except that it specifies the volume in which the electron may be located 90 percent of the time.

In addition to a variety of shapes, atomic orbitals also come in a variety of sizes. In general, energetic electrons are able to spend more of their time farther away from the attracting nucleus, which means they are distributed over a greater volume of space. The higher the energy of the electron, therefore, the larger its atomic orbital. Because electron energies are quantized, however, the

1s

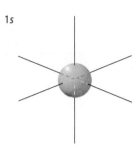

2s

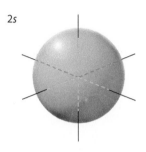

▲ **Figure 4.26**
The 2s orbital is larger than the 1s orbital because the electrons in the 2s orbital have greater energy.

▲ **Figure 4.26**
The 2s orbital is larger than the 1s orbital because the electrons in the 2s orbital have greater energy.

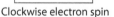

Clockwise electron spin   Counterclockwise electron spin

▲ **Figure 4.27**
Two electrons can pair together when they have opposite spins.

possible sizes of the atomic orbitals are also quantized. The size of an orbital is indicated by Bohr's quantum number $n = 1, 2, 3, 4$, and so on. The first two s orbitals are shown in **Figure 4.26**. The smallest s orbital is the 1s (pronounced "one-ess"), for which the quantum number is 1. The next largest s orbital is the 2s, and so forth.

> **CONCEPT CHECK**
> Distinguish between an *orbital* and one of Bohr's *orbits*.
>
> **CHECK YOUR ANSWER** An orbit is a *distinct path* followed by an object in its revolution around another object, like a planet around the Sun. An atomic orbital is a *volume of space* around an atomic nucleus where an electron of a given energy will most likely be found.

Each Orbital Can Hold up to Two Electrons

In the early 20th century, quantum physicists learned that electrons have a property called *spin*. Spin can have two states, analogous to a ball spinning either clockwise or counterclockwise, as shown in **Figure 4.27**. Because an electron has an electric charge, the spinning generates a tiny magnetic field. Two electrons spinning in opposite directions have oppositely aligned magnetic fields. This allows those two electrons—but no more than those two electrons—to come together as a pair within the same atomic orbital.

For atoms with multiple electrons, this idea of two electrons per orbital is of central importance. Consider the helium atom, which has two electrons. At their lowest energy states, both of these electrons will reside within the same 1s orbital because this orbital has a capacity of two electrons. What about lithium, which has three electrons? Two of its electrons will fill the 1s orbital. Lithium's third electron, however, must enter the higher energy 2s orbital.

Energy-Level Diagrams Describe How Orbitals Are Occupied

The increasing energy levels of the various atomic orbital are depicted in an **energy-level diagram,** as shown in **Figure 4.28**. Note that each orbital is represented schematically as a box and that each electron is represented by an arrow. This arrow points either up or down depending upon the spin of the electron. If two electrons are in the same orbital (box), their spins (arrows) must point in opposite directions. It is the natural tendency of electrons to occupy the lowest-energy orbitals first. These are the orbitals that bring the electron closest to the atomic nucleus, which attracts electrons because of its positive charge. So the electrons of an atom fill the orbitals of an energy-level diagram from bottom to top. Figure 4.28 shows how the orbitals of a rubidium atom are filled.

The arrangement of an atom's electrons within orbitals is called its **electron configuration.** Figure 4.28 shows the electron configuration for rubidium. An abbreviated way of presenting an electron configuration is to write the quantum number and letter of occupied orbitals and then use a superscript to indicate the number of electrons in each of those orbitals. For each of the group 1 elements, the notation is as follows:

Hydrogen, H $1s^1$
Lithium, Li $1s^2 2s^1$
Sodium, Na $1s^2 2s^2 2p^6 3s^1$
Potassium, K $1s^2 2s^2 2p^6 3s^2 3p^6 4s^1$
Rubidium, Rb $1s^2 2s^2 2p^6 3s^2 3p^6 4s^2 3d^{10} 4p^6 5s^1$
Cesium, Cs $1s^2 2s^2 2p^6 3s^2 3p^6 4s^2 3d^{10} 4p^6 5s^2 4d^{10} 5p^6 6s^1$
Francium, Fr $1s^2 2s^2 2p^6 3s^2 3p^2 4s^2 3d^{10} 4p^6 5s^2 4d^{10} 5p^6 6s^2 4f^{14} 5d^{10} 6p^6 7s^1$

Look carefully at rubidium's electron configuration as shown previously and also in Figure 4.28 and see how they correspond. Note also that all the superscripts for an atom must add up to the total number of electrons in the atom—1 for hydrogen, 3 for lithium, 11 for sodium, and so forth. Also note that the orbitals are listed in order of their energy levels as seen on the energy-level diagram. This explains why the 4s orbital is listed before the 3d orbitals even though it has a higher quantum number. Lastly, to simplify the task of writing out an electron configuration, the inner electrons can be represented using the symbol for a noble gas element in brackets. For example, the configuration for aluminum, Al, $1s^2 2s^2 2p^6 3s^2 3p^1$, can be written as [Ne] $3s^2 2p^1$, where [Ne] is shorthand for $1s^2 2s^2 2p^6$.

An atom's outermost electrons are the ones that interact most strongly with the external environment. Because of this, these outermost electrons play a key role in determining the properties of the atom—both chemical and physical. Elements that have similar electron configurations in their outermost

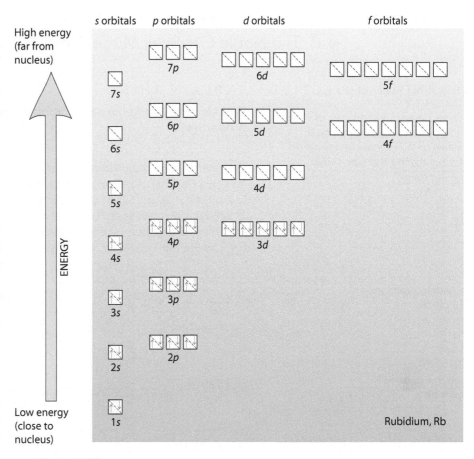

▲ **Figure 4.28**
This energy-level diagram shows the relative energy levels of atomic orbitals in a multielectron atom (in this case rubidium, Rb, atomic number 37).

orbitals, therefore, have similar properties. For example, in the alkaline metals of group 1, shown previously, the outermost occupied orbital (shown in blue) is an s orbital containing a single electron. In general, elements in the same group of the periodic table have similar electron configurations, which explains why elements in the same group have similar properties—a concept first presented in Section 3.3.

 FOR YOUR INFORMATION

In more advanced textbooks and also online you may see orbitals listed in order of quantum numbers so that the 3d is written before the 4s. This is a common convention but conceptually inconsistent with energy-level diagrams.

C O N C E P T C H E C K
What is the electron configuration for bromine, Br (atomic number 35)?

CHECK YOUR ANSWER The electron configuration for bromine is $1s^2 2s^2 2p^6 3s^2 3p^6 4s^2 3d^{10} 4p^5$. This can also be represented as [Ar] $4s^2 3d^{10} 4p^5$.

There are many interesting details to be said about energy-level diagrams. For example, why are there multiple 2p orbitals and no 1p orbitals? Why does the 4s orbital have a lower energy than the 3d orbital when it has a *higher* quantum number? And why does the 2s orbital have a slightly lower energy than any of the 2p orbitals when their quantum numbers are the same? Also, in what order do electrons fill the three 2p orbitals when each of these orbitals has the same energy level? Probing into these sorts of details under the guidance of your instructor will provide you with keen insight into the workings of chemistry. But there is also value in getting to the main point as soon as possible. For this reason, we now jump to a discussion of how the energy-level diagram is intimately connected to that all-important tool of chemistry: the periodic table.

4.8 The Noble Gas Shell Model Simplifies the Energy-Level Diagram

LEARNING OBJECTIVE

Show how atomic orbitals of similar energy can be grouped into a series of shells that can be used to explain the periodic table.

EXPLAIN THIS

Why do elements in the same group of the periodic table have similar properties?

The energy-level diagram can be organized so that orbitals of similar energy are grouped together. As shown in **Figure 4.29**, no orbital has an energy level similar to that of the 1s orbital, so this orbital is grouped by itself. The energy level of the 2s orbital, however, is close to that of the three 2p orbitals, so these four orbitals are grouped together. Likewise, the 3s and three 3p orbitals are at a similar energy level as are the 4s, five 3d, and three 4p orbitals, and so on. The result is a set of seven distinct horizontal rows of orbitals.

Interestingly, the seven rows in Figure 4.29 correspond to the seven periods in the periodic table, with the bottom row corresponding to the first period, the next row up corresponding to the second period, and so on. Furthermore, the maximum number of electrons each row can hold is equal to the number of elements in the corresponding period. The bottom row in Figure 4.29 can hold a maximum of two electrons, so there are only two elements, hydrogen and helium, in the first period of the periodic table. The second and third rows up from the bottom each have a capacity for eight electrons, and so eight elements are found in both the second and third periods. Continue analyzing Figure 4.29 in this way, and you will find 18 elements in the fourth and fifth periods, and 32 elements in the sixth and seventh periods. (As of this writing, the discovery of only 28 seventh-period elements has been confirmed.)

Recall from Section 4.7 that the higher the energy level of an orbital, the farther its electrons are from the nucleus. Electrons in the same row of orbitals in Figure 4.29, therefore, are roughly the same distance from the nucleus. Graphically, this can be represented by converging all the orbitals in a given row into a single 3-dimensional hollow shell, as shown in **Figure 4.30**. Each shell is a graphic representation of a collection of orbitals of comparable energy in a multielectron atom. To distinguish this shell from one used in more advanced chemistry courses, we call it a **noble gas shell**. We say "noble gas" because, when filled, such a shell represents the electron configuration of a noble gas element.

Seventh-row capacity: 32 electrons

Sixth-row capacity: 32 electrons

Fifth-row capacity: 18 electrons

Fourth-row capacity: 18 electrons

Third-row capacity: 8 electrons

Second-row capacity: 8 electrons

First-row capacity: 2 electrons

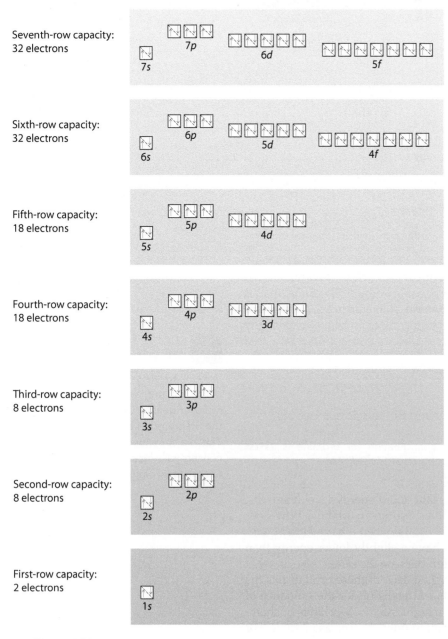

▲ **Figure 4.29**
Orbitals of comparable energy levels can be grouped together to give rise to a set of seven rows of orbitals.

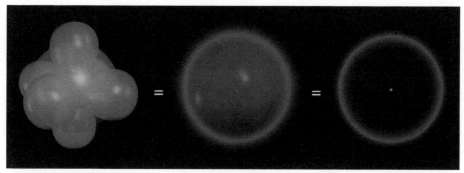

Second row of orbitals

(all 2s and 2p orbitals combined)

Second row of orbitals

(highly simplified perspective)

Second row of orbitals

(cross section of highly simplified perspective)

◀ Figure 4.30
The second row of orbitals, which consists of the 2s and three 2p orbitals, can be represented either as a single smooth, spherical shell or as a cross section of such a shell. (Note that this shell model ignores the slight difference in energy levels of closely matched orbitals.)

FOR YOUR INFORMATION

What do poets and scientists have in common? They both use metaphors to help us understand abstract concepts and relationships. The shell model of the atom is a metaphor that helps us visualize an invisible reality. Scientific models are essentially equivalent to the metaphorical language used in poetry.

The seven rows of orbitals in Figure 4.29 can thus be represented either by a series of seven concentric shells or by a series of seven cross-sectional circles of these shells, as shown in **Figure 4.31**. The number of electrons each shell can hold is equal to the number of orbitals it contains multiplied by 2 (because there can be two electrons per orbital).

You add electrons to a shell diagram just as in an energy-level diagram—electrons first fill the shells closest to the nucleus. **Figure 4.32** shows how this works for the first three periods. As with the energy-level diagram, there is one shell for each period, and the number of elements in a period is equal to the maximum number of electrons that the shell representing that period can hold.

Notice how electrons in the outermost shell begin to pair only after that shell is half filled. Carbon, for example, has four outer-shell electrons, none of which are paired. This differs from how an energy-level diagram is filled. For example, the second shell consists of the 2s and 2p orbitals, which, as shown in Figure 4.29, have different energy levels. Therefore, you might expect carbon's lower energy 2s orbital to fill with two paired electrons. In an advanced chemistry course, however, you would learn that orbitals of a similar energy level readily merge into each other through a process called *orbital hybridization*. When they do so, their energy levels become equivalent. The shell model presented here takes hybridization into account, which is why pairing doesn't occur until the shell is half filled.

This noble gas shell model is a simplification of the energy-level diagram but it retains a key feature, which is that atoms with a similar electron configuration have similar physical and chemical properties. Look carefully

READING CHECK

When do electrons in the outermost shell of an atom begin to pair with each other?

(a)

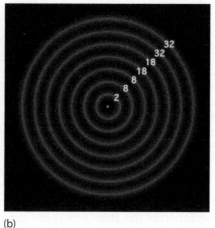

(b)

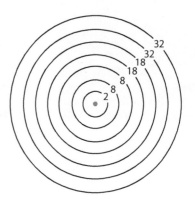

(c)

▲ **Figure 4.31**
(a) A cutaway view of the seven shells, with the number of electrons each shell can hold indicated. (b) A 2-dimensional, cross-sectional view of the shells. (c) An easy-to-draw cross-sectional view that resembles Bohr's planetary model.

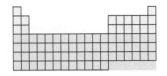

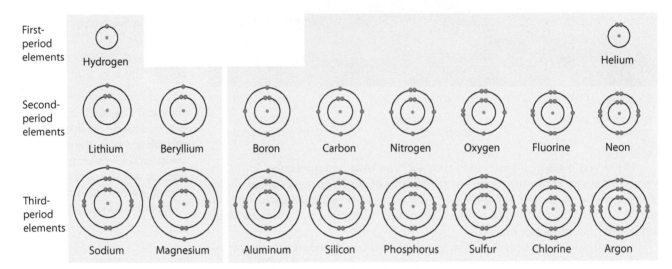

First-period elements: Hydrogen, Helium

Second-period elements: Lithium, Beryllium, Boron, Carbon, Nitrogen, Oxygen, Fluorine, Neon

Third-period elements: Sodium, Magnesium, Aluminum, Silicon, Phosphorus, Sulfur, Chlorine, Argon

▲ **Figure 4.32**
The first three periods of the periodic table according to the noble gas shell model. Elements in the same period (horizontal rows) have electrons in the same shells. Elements in the same period differ from one another by the number of electrons in the outermost shell. Elements in the same group (vertical column) have the same number of electrons in the outermost shell. These outer-shell electrons determine the character of the atom, which is why elements in the same group have similar properties.

at Figure 4.32. Can you see that the outer-shell electrons of atoms above and below one another in the periodic table—that is, within the same group—are similarly organized? For example, atoms of the first group—which includes hydrogen, lithium, and sodium—each have a single outer-shell electron. The atoms of the second group, including beryllium and magnesium, each have two outer-shell electrons. Similarly, atoms of the last group—including helium, neon, and argon—each have their outermost shells filled to capacity with electrons, two for helium and eight for both neon and argon. In general, the outer-shell electrons of atoms in the same group of the periodic table are similarly organized, which is why these atoms have similar properties.

CONCEPTCHECK

Why are there only two elements in the first period of the periodic table?

CHECK YOUR ANSWER The number of elements in each period corresponds to the number of electrons each shell can hold. The first shell has a capacity of only two electrons, which is why the first period has only two elements.

▲ Two-time Nobel laureate Linus Pauling (1901–1994) was an early proponent of teaching beginning chemistry students a shell model, from which the organization of the periodic table could be described. In 1954 Pauling won the Nobel Prize in Chemistry for his research into the nature of the chemical bond. In 1962 he was awarded the Nobel Peace Prize for his campaign against the testing of nuclear bombs, which was introducing massive amounts of radio-activity into the environment.

Remember that the shell model is not to be interpreted as an actual representation of the atom's physical structure—electrons, for example, are not really confined to the "surface" of a shell. Rather, this model serves as a tool to help us understand and predict how atoms behave. In Chapter 6 we will be using an abbreviated version of this model, called the *electron-dot structure*, to show how atoms join together to form molecules, which are tightly held groups of atoms.

4.9 The Periodic Table Helps Us Predict Properties of Elements

EXPLAIN THIS

Why does a rubber balloon inflated with helium deflate faster than one filled with argon?

Chemistry is the study of how atoms combine to form new materials. But how exactly does this happen? In order to answer this question, we first need to understand a bit about the properties of the atoms themselves. For this, we turn to the shell model, which, though limited, is able to explain a number of key atomic properties, such as atomic size and the ability of an atom to either lose or gain electrons.

Recall from Section 3.3 that properties of elements gradually change across the periodic table. The atoms of elements toward the upper right of the periodic table, for example, tend to be smaller than the atoms of elements toward the lower left (see Figure 3.15). Knowing this, you can reasonably predict that selenium atoms (atomic number 34) are smaller than calcium atoms (atomic number 20), which is indeed the case.

Recall that such a gradual change in moving across the periodic table is called a *periodic trend*. Underlying most periodic trends are two important concepts: *inner-shell shielding and effective nuclear charge*.

Consider the two electrons within the first shell of a helium atom. As shown in **Figure 4.33**, both of these electrons have an unobstructed "view" of the nucleus. Their attractions for this atomic nucleus are therefore the same.

The situation gets more complicated for atoms beyond helium, which have more than one shell occupied by electrons. In these cases, inner-shell electrons weaken the attraction between outer-shell electrons and the nucleus. Imagine, for example, you are that second-shell electron in the lithium atom shown in **Figure 4.34**. Looking toward the nucleus, what do you sense? Not just the attractive force of the positively charged nucleus. You also feel the effect of the two electrons in the first shell—which, because of their negative charge, exert a repulsive force. Thus, the two inner electrons have the effect of weakening your electrical attraction to the nucleus. This is **inner-shell shielding**—inner-shell electrons shield electrons farther out from some of the attractive pull exerted by the positively charged nucleus.

Because inner-shell electrons diminish the attraction outer-shell electrons have for the nucleus, the nuclear charge sensed by outer-shell electrons is always less than the actual charge of the nucleus. This diminished nuclear charge experienced by outer-shell electrons is called the **effective nuclear charge** and is abbreviated Z^* (pronounced zee-star), where Z stands for the nuclear charge and the asterisk indicates that this charge appears to be less than it actually is. A rough estimate of the Z^* for any electron can be calculated by subtracting the total number of inner-shell electrons from the nuclear charge. For lithium (atomic number 3), the total charge of the first-

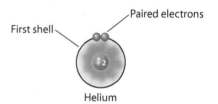

▲ Figure 4.33
The two electrons in a helium atom have equal exposure to the nucleus; hence, they experience the same degree of attraction, represented by the pink shading in the space between the nucleus and the shell boundary.

READINGCHECK

How is it possible to make a rough estimate of the effective nuclear charge experienced by an electron?

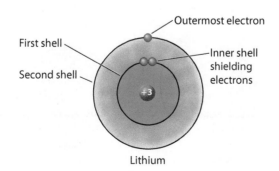

Lithium

◀ Figure 4.34
Lithium's two first-shell electrons shield the second-shell electron from the nucleus. The nuclear attraction, again represented by pink shading, is less intense in the second shell.

▶ Figure 4.35
(a) A chlorine atom has three occupied shells. The ten electrons of the inner two shells shield each of the seven electrons of the third shell from the +17 nucleus. The third-shell electrons therefore experience an effective nuclear charge of about $17 - 10 = +7$. (b) In a potassium atom, the fourth-shell electron experiences an effective nuclear charge of about $19 - 18 = +1$.

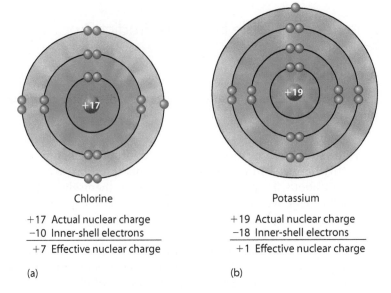

Chlorine

+17	Actual nuclear charge
−10	Inner-shell electrons
+7	Effective nuclear charge

(a)

Potassium

+19	Actual nuclear charge
−18	Inner-shell electrons
+1	Effective nuclear charge

(b)

shell electrons, –2, is subtracted from the charge of the nucleus, +3, to give a calculated effective nuclear charge of +1. The second-shell electron of lithium, therefore, senses a nuclear charge of about +1, which is much less than the actual nuclear charge of +3.

For most elements, subtracting the total number of inner-shell electrons from the nuclear charge provides a convenient estimate of the effective nuclear charge, as **Figure 4.35** illustrates.

Why Atoms toward the Upper Right Are Smaller

From left to right across any row of the periodic table, the atomic diameters get *smaller*. Let's look at this trend from the point of view of effective nuclear charge. Consider lithium's outermost electron, which experiences an effective nuclear charge of about +1. Then look across period 2 to neon, in which each outermost electron experiences an effective nuclear charge of about +8, as **Figure 4.36** shows. Because the outer-shell electrons in neon experience a greater attraction to the nucleus, they are pulled in closer to the nucleus. So neon, although nearly three times as massive as lithium, has a considerably smaller diameter.

In general, across any period from left to right, atomic diameters become smaller because of an increase in effective nuclear charge. Look back at Figure 4.32 and you will see this trend illustrated for the first three periods. In addition, **Figure 4.37** shows relative atomic diameters obtained from experimental data. Note that there are some exceptions to this trend, especially between groups 12 and 13. These exceptions can be explained by probing further into the shell model than we need to for our purposes.

Moving down a group, atomic diameters get larger because of an increasing number of occupied shells. Whereas lithium has a small diameter because it has only two occupied shells, francium (atomic number 87) has a much larger diameter because it has seven occupied shells.

▶ Figure 4.36
Lithium's outermost electron experiences an effective nuclear charge of about +1, while those of neon experience an effective nuclear charge of about +8. As a result, the outer-shell electrons in neon are closer to the nucleus and the diameter of the neon atom is smaller than the diameter of the lithium atom.

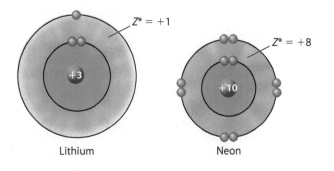

Lithium

Neon

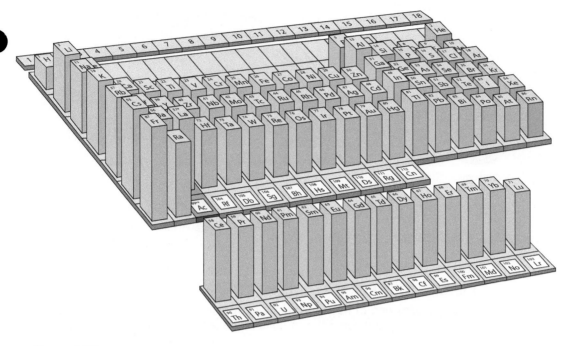

▲ Figure 4.37
This chart shows the relative atomic sizes, indicated by height. Note that atomic size generally decreases in moving to the upper right of the periodic table.

The Smallest Atoms Have the Most Strongly Held Electrons

How strongly electrons are bound to an atom is another property that changes gradually across the periodic table. In general, the trend is that the smaller the atom, the more tightly bound are its electrons.

As discussed earlier, effective nuclear charge increases in moving from left to right across any period. Thus, not only are atoms toward the right smaller, but their electrons are held more strongly. It takes about four times as much energy to remove an outer electron from a neon atom, for example, than to remove the outer electron from a lithium atom.

Moving down any group, the effective nuclear charge—as calculated by subtracting the charge of inner-shell electrons from the charge of the nucleus—stays the same. The effective nuclear charge for all group 1 elements, for example, works out to +1. Because of their greater number of shells, however, elements toward the bottom of a group (vertical column) are larger. The electrons in the outermost shell are therefore *farther* from the nucleus by an appreciable distance. From physics we learn that the electric force weakens rapidly with increasing distance. As **Figure 4.38** illustrates, an outer-shell electron in a larger atom, such as cesium, is not held as tightly as an outer-shell electron in a smaller atom, such as lithium. As a consequence, the energy needed to remove the outer electron from a cesium atom is about half the energy needed to remove the outer electron from a lithium atom. This is true even though they both have a calculated Z^* of +1. So we see that our calculated Z^* values provide only a rough estimate of the nuclear pull on electrons. The effect of distance between a shell and the nucleus also needs to be considered.

The combination of the increasing effective nuclear charge from left to right and the increasing number of shells from top to bottom creates a periodic trend in which the electrons in atoms at the upper right of the periodic table are held most strongly and the electrons in atoms at the lower left are held least strongly. This is reflected in **Figure 4.39**, which shows **ionization energy,** the amount of energy needed to pull an electron away from an atom. The greater the ionization energy, the greater the attraction between the nucleus and its outermost electrons.

CHEMICAL CONNECTIONS

How is the weather connected to an atom?

▶ **Figure 4.38**
In both lithium and cesium, the outermost electron has a calculated effective nuclear charge of +1. The outermost electron in a cesium atom, however, is not held as strongly to the nucleus because of its greater distance from the nucleus.

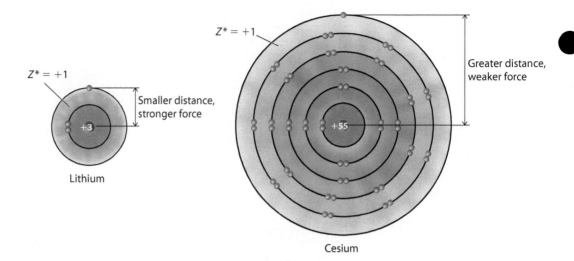

$Z^* = +1$

$Z^* = +1$

Smaller distance, stronger force

+3

Lithium

Greater distance, weaker force

+55

Cesium

CONCEPTCHECK

Which loses one of its outermost electrons more easily: a francium, Fr, atom (atomic number 87) or a helium, He, atom (atomic number 2)?

CHECK YOUR ANSWER A francium, Fr, atom loses electrons much more easily than does a helium, He, atom. Why? Because a francium atom's outer electrons are not held as tightly by its nucleus, which is buried deep beneath many layers of shielding electrons.

How strongly an atomic nucleus is able to hold on to the outermost electrons in an atom plays an important role in determining the atom's chemical behavior. What do you suppose happens when an atom that holds its outermost electrons only weakly comes into contact with an atom that has a very strong pull on its outermost electrons? As we explore in Chapter 6, the atom that pulls strongly may remove one or more electrons from the other atom. The result is that the two atoms become chemically bonded. So the shell model not only gives us insight into the workings of the periodic table, it also helps us to understand the heart of chemistry, which is the study of how new materials are created by the bonding of atoms.

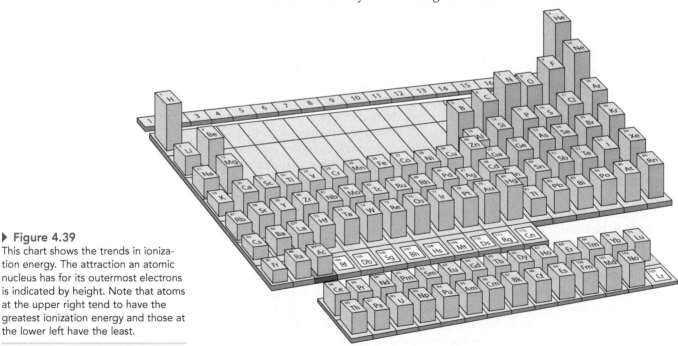

▶ **Figure 4.39**
This chart shows the trends in ionization energy. The attraction an atomic nucleus has for its outermost electrons is indicated by height. Note that atoms at the upper right tend to have the greatest ionization energy and those at the lower left have the least.

Chapter 4 Review

LEARNING OBJECTIVES

Distinguish between models that describe physical attributes and those that describe the behavior of a system. (4.1)	→	*Questions 1, 2, 44–47, 94, 95*
Identify experiments leading to the discovery of the electron. (4.2)	→	*Questions 3–5, 31, 48–50*
Defend Rutherford's conclusion that each atom contains a densely packed positively charged center. (4.3)	→	*Questions 6–8, 51–55*
Describe the structure of the atomic nucleus and how the atomic mass of an element is calculated. (4.4)	→	*Questions 9–11, 33–38, 56–62*
Describe the nature and range of electromagnetic waves. (4.5)	→	*Questions 12, 13, 30, 63–65*
Recount how the quantum nature of energy led to Bohr's planetary model of the atom. (4.6)	→	*Questions 14–17, 66–69*
Summarize how electrons, when confined to an atom, behave as self-reinforcing wavelike entities represented by atomic orbitals. (4.7)	→	*Questions 18–23, 32, 39, 70–80*
Show how atomic orbitals of similar energy can be grouped into a series of shells that can be used to explain the periodic table. (4.8)	→	*Questions 24–26, 81–84*
Use the shell model to explain periodic trends. (4.9)	→	*Questions 27–29, 40–43, 85–93*

SUMMARY OF TERMS (KNOWLEDGE)

Atomic mass The total mass of an atom. The atomic mass of each element presented in the periodic table is the *weighted average* atomic mass of the various isotopes of that element occurring in nature.

Atomic nucleus The dense, positively charged center of every atom.

Atomic number The number of protons in the atomic nucleus of each atom of a given element.

Atomic orbital A volume of space where an electron is likely to be found 90 percent of the time.

Atomic spectrum The pattern of frequencies of electromagnetic radiation emitted by the energized atoms of an element, considered to be an element's "fingerprint."

Conceptual model A representation of a system that helps us predict how the system behaves.

Effective nuclear charge The nuclear charge experienced by outer-shell electrons, diminished by the shielding effect of inner-shell electrons and also by the distance from the nucleus.

Electromagnetic spectrum The complete range of waves, from radio waves to gamma rays.

Electron An extremely small, negatively charged subatomic particle found outside the atomic nucleus.

Electron configuration The arrangement of an atom's electrons within orbitals.

Energy-level diagram A schematic drawing used to arrange atomic orbitals in order of increasing energy levels.

Inner-shell shielding The tendency of inner-shell electrons to partially shield outer-shell electrons from the attractive pull exerted by the positively charged nucleus.

Ionization energy The amount of energy needed to pull an electron away from an atom.

Isotope Any member of a set of atoms of the same element whose nuclei contain the same number of protons but different numbers of neutrons.

Mass number The number of nucleons (protons plus neutrons) in the atomic nucleus. Used primarily to identify isotopes.

Neutron An electrically neutral subatomic particle found in the atomic nucleus.

Noble gas shell A graphic representation of a collection of orbitals of comparable energy in a multielectron atom. A noble gas shell can also be viewed as a region of space about the atomic nucleus within which electrons may reside.

Nucleon Any subatomic particle found in an atomic nucleus. Another name for either proton or neutron.

Physical model A representation of an object on some convenient scale.

Probability cloud A plot of the positions of an electron of a given energy over time as a series of tiny dots.

Proton A positively charged subatomic particle found in the atomic nucleus.

Quantum A small, discrete packet of energy.

Quantum number An integer that specifies the quantized energy level within an atom.

Spectroscope A device that uses a prism or diffraction grating to separate light into its color components and measure their frequencies.

READING CHECK QUESTIONS (COMPREHENSION)

4.1 Physical and Conceptual Models

1. If a baseball were the size of the Earth, about how large would its atoms be?
2. What is the difference between a physical model and a conceptual model?

4.2 The Electron Was the First Subatomic Particle Discovered

3. Why is a cathode ray deflected by a nearby electric charge or magnet?
4. What did Thomson discover about the electron?
5. What did Millikan discover about the electron?

4.3 The Mass of an Atom Is Concentrated in Its Nucleus

6. What did Rutherford discover about the atom?
7. To Rutherford's surprise, what was the fate of a tiny fraction of alpha particles in the gold-foil experiment?
8. What kind of force prevents atoms from squishing into one another?

4.4 The Atomic Nucleus Is Made of Protons and Neutrons

9. What role does atomic number play in the periodic table?
10. Distinguish between atomic number and mass number.
11. Distinguish between mass number and atomic mass.

4.5 Light Is a Form of Energy

12. Does visible light constitute a large or small portion of the electromagnetic spectrum?
13. What does a spectroscope do to the light coming from an atom?

4.6 Atomic Spectra and the Quantum Hypothesis

14. What causes an atom to emit light?
15. Why do we say atomic spectra are like fingerprints of the elements?
16. What was Planck's quantum hypothesis?
17. Did Bohr think of his planetary model as an accurate representation of what an atom looks like?

4.7 Electrons Exhibit Wave Properties

18. Who first proposed that electrons exhibit the properties of a wave?
19. About how fast does an electron travel around the atomic nucleus?
20. How does the speed of an electron change its fundamental nature?
21. How many electrons can reside in a single atomic orbital?
22. What two elements are represented by these two energy-level diagrams?

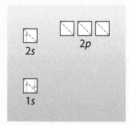 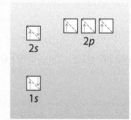

23. What element has the electron configuration $1s^2 2s^2 2p^3$?

4.8 The Noble Gas Shell Model Simplifies the Energy-Level Diagram

24. What atomic orbitals comprise the third noble gas shell?
25. Which electrons are most responsible for the properties of an atom?
26. What is the relationship between the maximum number of electrons each noble gas shell can hold and the number of elements in each period of the periodic table?

4.9 The Periodic Table Helps Us Predict Properties of Elements

27. What is effective nuclear charge?
28. How would you know from looking at the periodic table that oxygen, O (atomic number 8), molecules are smaller than nitrogen, N (atomic number 7), molecules?
29. What happens to the strength of the electric force with increasing distance?

CONFIRM THE CHEMISTRY (HANDS-ON APPLICATION)

30. Fluorescent lights contain spectal lines from the light emission of mercury atoms. Special coatings on the inner surface of the bulb help to accentuate visible frequencies, which can be seen through the diffraction grating reflection of a compact disc. Cut a narrow slit through some thick paper (or thin cardboard) and place over a bright fluorescent bulb. View this slit at an oblique angle against a CD and look for spectral lines. Place the slit over an incandescent bulb and you'll see a smooth continuous spectrum (no lines) because the incandescent filament glows at all visible frequencies. Try looking at different brands of fluorescent bulbs. You'll also be able to see spectral lines in street lights and fireworks. For those it is best to use "rainbow" glasses available from a nature, toy, or hobby store.

31. Stare at a non-LCD television set or computer monitor, and you stare down the barrel of a cathode ray tube. You can find evidence for this by holding a magnet up to the screen. Note the distortion, which results as the magnet pushes the electrons off their intented paths. *Important: use*

only a small magnet and hold it up to the screen only briefly; otherwise the distortion may become permanent.

32. Stretch a rubber band between your two thumbs and pluck one length of it. Note that no matter where along the length you pluck, the area of greatest oscillation is always at the midpoint. This is a self-reinforcing wave that occurs as overlapping waves bounce back and forth from thumb to thumb.

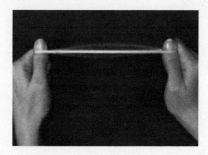

THINK AND SOLVE (MATHEMATICAL APPLICATION)

33. A class of 20 students takes an exam and every student scores 80 percent. What is the class average? Would the class average be slightly less, the same, or slightly more if one of the students instead scored 100 percent? How is this similar to how we derived the atomic masses of elements?

34. The isotope lithium-7 has a mass of 7.0160 atomic mass units, and the isotope lithium-6 has a mass of 6.0151 atomic mass units. Given the information that 92.58 percent of all lithium atoms found in nature are lithium-7 and 7.42 percent are lithium-6, show that the atomic mass of lithium, Li (atomic number 3) is 6.941 amu.

35. The element bromine, Br (atomic number 35), has two major isotopes of similar abundance, both around 50 percent. The atomic mass of bromine is reported in the periodic table as 79.904 atomic mass units. Choose the most likely set of mass numbers for these two bromine isotopes:

    a. ^{80}Br, ^{81}Br   b. ^{79}Br, ^{80}Br   c. ^{79}Br, ^{81}Br

THINK AND COMPARE (ANALYSIS)

36. Rank the three subatomic particles in order of increasing mass:
    a. The neutron
    b. The proton
    c. The electron

37. Consider these atoms: helium, He; chlorine, Cl; and argon, Ar. Rank them in terms of their atomic number, from smallest to largest.

38. Consider three 1-gram samples of matter: A, carbon-12; B, carbon-13; C, uranium-238. Rank them in terms of having the greatest number of atoms, from least to most.

39. Rank these energy-level diagrams for a fluorine atom in order of increasing energy, from lowest to highest.

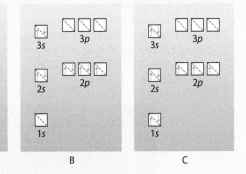

40. Consider these atoms: helium, He; aluminum, Al; and argon, Ar. Rank them, from smallest to largest: (a) in order of size and (b) in order of the number of protons.

41. Consider these atoms: potassium, K; sodium, Na; and lithium, Li. Rank them in order of the ease with which they lose a single electron, from easiest to most difficult.

42. Rank these atoms in order of the number of electrons they tend to lose, from fewest to most: sodium, Na; magnesium, Mg; and aluminum, Al.

43. Rank these atoms in order of ionization energy from lowest to highest: technetium, Tc, number 43; indium, In, number 49; aluminum, Al, number 13.

THINK AND EXPLAIN (SYNTHESIS)

4.1 Physical and Conceptual Models

44. As depicted in Figure 2.8b from Chapter 2, are gallium atoms really red and arsenic atoms green?

45. Would you use a physical model or a conceptual model to describe the following: a gold coin, a dollar bill, a car engine, air pollution, a virus, and the spread of a sexually transmitted disease?

46. What is the function of an atomic model?

47. Why is it not possible for a scanning probe microscope to make images of the inside of an atom?

4.2 The Electron Was the First Subatomic Particle Discovered

48. If the particles of a cathode ray had a greater mass, would the ray be bent more or less in a magnetic field?

49. If the particles of a cathode ray had a greater electric charge, would the ray be bent more or less in a magnetic field?

50. Thousands of magnetic marbles are thrown into a large vertically oriented wind tunnel. As they are thrown, the marbles clump together in groups of varying numbers. The wind tunnel operator is able to control the upward force of the wind so as to make different clumps of marbles hover. Notably, heavier clumps require greater upward forces. She records the various forces of wind required to maintain hovering clumps in units of ounces: 45, 30, 60, 75, 105, 35, 80, 55, 90, 20, 65. From this data, what might be the weight of a single magnetic marble? The single marble is analogous to what within Millikan's experiment? What is the force of the wind analogous to?

4.3 The Mass of an Atom Is Concentrated in Its Nucleus

51. You roll 100 marbles—one by one and in random directions—through an empty cereal box lying on the floor. They all pass through except for three, which bounce back. Is there a large or small obstruction stuck within the cereal box?

52. Why did Rutherford assume that the atomic nucleus was positively charged?

53. Which of the following diagrams best represents the size of the atomic nucleus relative to the size of the atom?

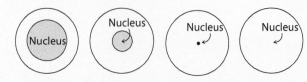

54. How does Rutherford's model of the atom explain why some of the alpha particles directed at the gold foil were deflected straight back toward the source?

55. Is the head of a politician really made of 99.99999999 percent empty space?

4.4 The Atomic Nucleus Is Made of Protons and Neutrons

56. Which contributes more to an atom's mass: electrons or protons? Which contributes more to an atom's size?

57. If two protons and two neutrons are removed from the nucleus of an oxygen-16 atom, a nucleus of which element remains?

58. Evidence for the existence of neutrons did not come until many years after the discoveries of the electron and the proton. Give a possible explanation.

59. What is the approximate mass of an oxygen atom in atomic mass units? What is the approximate mass of two oxygen atoms? How about an oxygen molecule, O_2?

60. What is the approximate mass of a hydrogen atom in atomic mass units? How about a water molecule?

61. Which is heavier, a water molecule, H_2O or a carbon dioxide molecule, CO_2?

62. Is the percentage of heavy water in rain greater than, equal to, or less than the percentage of heavy water in the oceans? Please explain.

4.5 Light Is a Form of Energy

63. What color is white light?

64. What color do you see when you close your eyes while in a dark room? Explain.

65. Do radio waves travel at the speed of light, at the speed of sound, or at some speed in between?

4.6 Atomic Spectra and the Quantum Hypothesis

66. What particle within an atom vibrates to generate electromagnetic radiation? This particle is vibrating back and forth between what?

67. How might you distinguish a sodium-vapor street light from a mercury-vapor street light?

68. How can a hydrogen atom, which has only one electron, create so many spectral lines?

69. Which color of light comes from a greater energy transition within an atom: red or blue?

4.7 Electrons Exhibit Wave Properties

70. How does the wave model of electrons orbiting the nucleus account for the fact that the electrons can have only discrete energy values?

71. Some older cars vibrate loudly when driving at particular speeds. For example, at 65 mph the car may be most quiet, but at 60 mph the car rattles uncomfortably. How is this analogous to the quantized energy levels of an electron in an atom?

72. Energy reveals itself in the form of a wave. Wherever you see a wave, there is some form of energy present. How does mass reveal itself?

73. Which of the following energy-level diagrams for carbon represents a greater amount of energy?

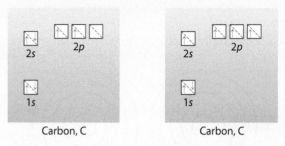

Carbon, C   Carbon, C

74. Beyond the s, p, d, and f orbitals are the g orbitals whose shapes are even more complex. Why are the g orbitals not commonly discussed by chemists?

75. Which has greater potential energy: an electron in a $3s$ orbital or an electron in a $2p$ orbital? How about an electron in the $4s$ orbital compared to the $3d$ orbital?

76. Fill in these three energy-level diagrams. Why do these three elements have such similar chemical properties? Note: electrons entering an orbital type, such as the three $2p$ orbitals, won't start pairing until each orbital has at least one electron.

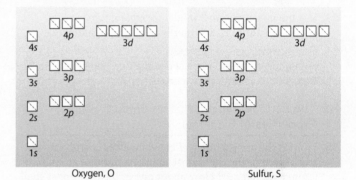

Oxygen, O   Sulfur, S

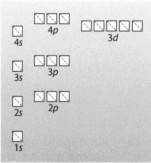

Selenium, Se

77. Which requires more energy: boosting one of lithium's $2s$ electrons to the $3s$ orbital, or boosting one of beryllium's $2s$ electrons to the $3s$ orbital?

78. Which element is represented in Figure 4.29 if all shown orbitals were filled with electrons?

79. What do the $4s$, $4p$, and $3d$ orbitals have in common?

80. Write out the electron configurations for the following atoms: phosphorus, P (atomic number 15); arsenic, As (atomic number 33); and antimony, Sb (atomic number 51). What do these configurations have in common?

4.8 The Noble Gas Shell Model Simplifies the Energy-Level Diagram

81. Does a noble gas shell have to contain electrons in order to exist?

82. Place the proper number of electrons in each shell:

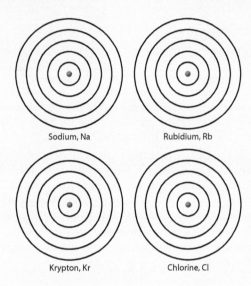

Sodium, Na

Rubidium, Rb

Krypton, Kr

Chlorine, Cl

83. Use the noble gas shell model to explain why a potassium atom, K, is larger than a sodium atom, Na.

84. Use the noble gas shell model to explain why a potassium atom, K, and a sodium atom, Na, have such similar chemical properties.

4.9 The Periodic Table Helps Us Predict Properties of Elements

85. What is the approximate effective nuclear charge for an electron in the outermost shell of a fluorine atom, F (atomic number 9)? How about one in the outermost shell of a sulfur atom, S (atomic number 16)?

86. Why is it more difficult for fluorine to lose an electron than for sulfur to do so?

87. Which of the following concepts underlies all the others: ionization energy, effective nuclear charge, or atomic size?

88. The electron configuration for sodium is $1s^2 2s^2 2p^6 3s^1$. In which of these orbitals do the electrons experience the greatest effective nuclear charge? How about the weakest effective nuclear charge?

89. Neon, Ne (atomic number 10), has a relatively large effective nuclear charge, and yet it cannot attract any additional electrons. Why not?

90. Use the noble gas shell model to explain why a lithium atom, Li, is larger than a beryllium atom, Be.

91. It is relatively easy to pull one electron away from a potassium atom but very difficult to remove a second one. Use the shell model and the idea of effective nuclear charge to explain why.

92. Why might one of cesium's electrons not be very attracted to cesium's nucleus, which has a charge of +56?

93. How is the following graphic similar to the energy-level diagram of Figure 4.29? Use it to explain why a gallium atom, Ga (atomic number 31), is larger than a zinc atom, Zn (atomic number 30).

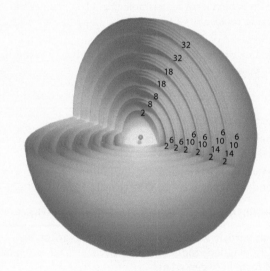

THINK AND DISCUSS (EVALUATION)

94. If matter is made of atoms and atoms are made of sub-atomic particles, what comes together to create subatomic particles? Where might you find such information?

95. Astronomical measurements reveal that most matter within the universe is invisible to us. This invisible matter, also known as dark matter, is likely to be "exotic" matter—very different from the elements that make up the periodic table. We know the dark matter is there because of its gravitational effects, but scientists can only guess as to its nature. What do you think dark matter might be made of? How soon might we know the answer?

READINESS ASSURANCE TEST (RAT)

If you have a good handle on this chapter, then you should be able to score at least 7 out of 10 on this RAT. Check your answers online at www.ConceptualChemistry.com. If you score less than 7, you need to study further before moving on.

Choose the BEST answer to the following.

1. Would you use a physical model or a conceptual model to describe the following: the brain; the mind; the solar system; the beginning of the universe?
   a. Conceptual; physical; conceptual; physical
   b. Conceptual; conceptual; conceptual; conceptual
   c. Physical; conceptual; physical; conceptual
   d. Physical; physical; physical; physical

2. The ray of light in a neon sign bends when a magnet is held up to it because
   a. neon, like iron, is attracted to magnets.
   b. the light arises from the flow of electrons within the tube.
   c. impurities within the neon plasma are attracted to the magnet.
   d. of its attraction to Earth's magnetic field.

3. Since atoms are mostly empty space, why don't objects pass through one another?
   a. The nucleus of one atom repulses the nucleus of another atom when it gets close.
   b. The electrons on the atoms repulse other electrons on other atoms when they get close.
   c. The electrons of one atom attract the nucleus of a neighboring atom to form a barrier.
   d. Atoms actually do pass through one another, but only in the gaseous phase.

4. You could swallow a capsule of germanium, Ge (atomic number 32), without significant ill effects. If a proton were added to each germanium nucleus, however, you would not want to swallow the capsule because the germanium would
   a. become arsenic.
   b. become radioactive.
   c. expand and likely lodge in your throat.
   d. have a change in flavor.

5. An element found in another galaxy exists as two isotopes. If 80.0 percent of the atoms have an atomic mass of 80.00 atomic mass units and the other 20.0 percent have an *atomic mass* of 82.00 atomic mass units, what is the approximate atomic mass of the element? (in amu)
   a. 80.4   b. 81.0   c. 81.6   d. 64.0   e. 16.4

6. How might the spectrum of an atom appear if its electrons were NOT restricted to particular energy levels?
   a. Nearly the same as it does with the energy level restrictions.
   b. There would be no frequencies within the visible portion of the electromagnetic spectrum.
   c. A broad spectrum of all colors would be observed.
   d. The frequency of the spectral lines would change with temperature.

7. What property permits two electrons to reside in the same orbital?
   a. charge   b. spin   c. mass
   d. volume   e. time

8. How many electrons are there in the third noble gas shell of a sodium atom, Na (atomic number 11)?
   a. None   b. One   c. Two   d. Three

9. An electron in the outermost shell of which group 1 element experiences the greatest effective nuclear charge?
   a. Sodium, Na   b. Potassium, K
   c. Rubidium, Rb   d. All of the above

10. List the following atoms in order of increasing atomic size: thallium, Tl; germanium, Ge; tin, Sn; phosphorus, P.
    a. Ge < P < Sn < Tl   b. Tl < Sn < P < Ge
    c. Tl < Sn < Ge < P   d. P < Ge < Sn < Tl

ANSWERS TO CALCULATION CORNERS (CALCULATING ATOMIC MASS)

	Contributing Mass of ^{35}C	Contributing Mass of ^{37}C
Fraction of Abundance	0.7553	0.2447
Mass (amu)	$\times \dfrac{34.97}{26.41}$	$\times \dfrac{36.95}{9.04}$

atomic mass = 26.41 + 0.94 = 35.48

Contextual Chemistry

Forensic Chemistry

The methods and tools of science can be used to decide questions arising from crime, such as "How did a murder victim die?" or "Who was the murderer?" The application of science to solving crimes is called *forensic* science, which can be subdivided into the various areas of science. Forensic medicine, for example, employs the methods and tools of medicine, such as autopsies, to determine a cause of death. Similarly, forensic chemistry employs the methods and tools of chemistry, such as the analysis of materials, to identify criminal suspects or, perhaps, criminal intent.

The pioneer who laid the cornerstone of modern forensic science was the early 20th-century criminologist Dr. Edmond Locard (1877–1966). Known as the Sherlock Holmes of France, Dr. Locard established the first police laboratory in Lyon, France, in 1910. His most widely recognized contribution has come to be known as the Locard Principle, which can be summarized as follows: "With contact between two items, there will

▲ Dr. Edmond Locard

be an exchange." According to this principle, a burglar cannot enter a house without leaving some trace of his presence, such as fingerprints or, perhaps, bits of hair. Knowing this, the burglar wears gloves and a cap. But the exchange of materials goes far beyond fingerprints and bits of hair. Consider the tiny grains of soil from the burglar's shoes or microfibers that shed from the burglar's clothing, including the gloves and cap! Furthermore, the burglar also carries bits of the crime scene, such as fibers from a carpet or furniture, with him as he leaves. In the words of Dr. Locard, "Wherever he steps, whatever he touches, whatever he leaves, even unconsciously, will serve as a silent witness against him."

The trained forensic chemist collects these bits of matter from the scene of the crime or from the suspect. The matter is then identified by assessing its properties, such as chemical composition or melting point. Once identified, this physical evidence may be used to support or refute the guilt of the suspect.

Samples of hair, skin, or semen left at the scene of the crime can be examined for DNA content. As we describe in Chapter 13, DNA is the biomolecule that holds a person's genetic information, which is unique for each individual. The amount of DNA recovered from a crime scene

is often quite small. The tools of genetic engineering, however, allow the forensic chemist to use a minuscule amount of collected DNA as a template for the production of much larger amounts of identical DNA. This larger quantity of DNA can then be analyzed for patterns that may be identified as belonging (or not belonging) to the criminal suspect.

Modern analytical tools allow chemists to detect and identify chemicals at ultralow concentrations. One of the most widely used and sensitive analytical instruments is the *mass spectrometer*. Within the mass spectrometer, a sample molecule is subjected to harsh energy, which breaks the molecule into fragments. These fragments are then given an electric charge and accelerated via magnets down the length of a tube. The fragments separate from each other because more massive fragments travel slower while the lighter ones

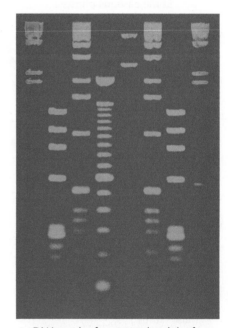

▲ DNA can be fragmented and the fragments then separated to yield a pattern characteristic of an individual.

travel faster. The fragments, sorted by mass, produce a pattern that is characteristic of the original molecule. An unknown molecule can thus be identified by comparing its fragmentation pattern to a catalog of known fragmentation patterns. The wonder of the mass spectrometer is its great sensitivity—if you can see the sample injected into a mass spectrometer, it is way too much!

The mass spectrometer is often used in tandem with other analytical tools, such as the *gas chromatograph*, which volatilizes mixtures and separates them into their individual components. Each component is then analyzed with the mass spectrometer. These are the tools of choice for testing bodily fluids for ingested compounds, such as illicit drugs or sports-enhancing steroids. Instruments at the International Olympics, for example, are standardized to detect hundreds of different agents that are prohibited for use by Olympic athletes.

A type of mass spectrometer is also employed at airports to check for compounds that may be used as explosives. The technician rubs a swab inside a piece of luggage and then places the swab within the highly sensitive spectrometer, which tests for a wide assortment of potentially dangerous compounds. For luggage, the spectrometer is used in conjunction with an X-ray machine that identifies the density of the contents. Many explosives have a density comparable to water, which is a reason why passengers are discouraged from packing water or liquid toiletries into their luggage. These X-ray machines are also able to assess the average atomic number of the atoms within the luggage. This is helpful because the chemicals of explosives tend to be made from nitrogen (atomic number 7). A region of the luggage containing an average atomic number of around 7 likely contains explosive materials.

Modern technology used by forensic scientists is not foolproof or without its limitations. Collected material evidence needs to be taken in context and weighed against other factors, such as witness accounts and the possibility that physical evidence has been tampered with—intentionally or unintentionally—prior to being collected. That said, modern technology is a very powerful tool for the forensic scientists whose primary goal is the accurate reconstruction of criminal events as they occurred in the past with the hope that these events can be deterred from occurring again in the future.

CONCEPT CHECK

Would it be a good idea for a burglar to own a type of dog that sheds?

CHECK YOUR ANSWER For the burglar this would be a bad idea, because fur from the dog could easily end up at the scene of the crime, in accordance with Locard's principle. For the greater society, however, if this dog fur led to the just conviction of the burglar's crime, then the burglar's owning the dog would be a good thing.

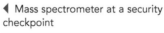
◀ Mass spectrometer at a security checkpoint

Think and Discuss

1. A dog trained to sniff out fuel is brought to the remains of a building suspected to have been burned down by an arsonist. The dog barks excitedly in evidence of residual fuel at one and only one location. Does this suggest arson? What if the dog found residual fuel in two locations? Why are cases of arson frequently difficult to solve?

2. A woman is dropped off at her apartment by her boyfriend after a night of intimate romance. Entering her bedroom, she surprises a burglar, who then attacks and strangles her to death before fleeing. DNA evidence found on the woman implicates her boyfriend as being guilty of both rape and murder. How can the boyfriend prove beyond a reasonable doubt that he is innocent? How might the outcome of this case have been different if it had occurred 100 years ago?

3. According to the Innocence Project, a group that uses DNA testing to right wrongful convictions, police lineups and similar forms of eyewitness identification are the leading cause of wrongful convictions across all DNA exonerated cases. Why might this be so? What might be done to improve the reliability of police lineups?

4. A deductive argument asserts that a conclusion necessarily follows from the truth of a premise. For example, all cats are mortal. Fluffy is a cat. Therefore, Fluffy is mortal. An inductive argument asserts that a conclusion follows, not necessarily but probably, from the truth of the premise. For example, all the cats you have ever seen are black. Therefore, all cats are black. Which of these forms of argument is used more often in a court of law? Which is used more often in science?

▲ About 20 percent of the electricity produced in the United States comes from nuclear power plants.

5

The Atomic Nucleus

THE MAIN IDEA

The atomic nucleus is the source of a tremendous amount of energy.

The atomic nucleus is the source of nuclear energy, which comes to us in different forms. Radioactivity is nuclear energy that occurs naturally in the air we breathe, in the rocks around us, and even in the food we eat. Some nuclear energy is human-controlled and is an important part of medical diagnostics and radiation therapy. Another source of nuclear energy is nuclear power plants, which transform nuclear energy into electricity. Of course, a nuclear bomb is another form of nuclear energy. Our most significant source of nuclear energy, however, is the Sun. We'll see in this chapter how solar energy comes from thermonuclear fusion and how it relates to Einstein's famous equation, $E = mc^2$. In this nuclear age it makes good sense to have a basic understanding of the atomic nucleus and how it gives rise to the various forms of nuclear energy.

Chemistry

Chain Reactions

Nuclear fission reactions, decribed in Section 5.7, can undergo what are known as explosive chain reactions. These are self-sustaining reactions in which the products of one event stimulate further events in an exponential manner. You can simulate such a chain reaction with a set of dominos.

PROCEDURE

1. Stand one domino upright so that when it topples it hits two other upright dominos, which also each hit two other upright dominos, and so forth.
2. Arrange as many upright dominos as you can in this fashion so that they fan out as

shown in the photograph to the lower right. Your challenge will be to arrange the dominos so that every one of them falls.
3. Topple the first domino and observe the chain reaction. Focus your attention on the sound.

ANALYZE AND CONCLUDE

1. What happened to the sound of the dominos as they fell?
2. Dominos set up in this fashion create a series of rows. If the first row contains two dominos, the second row contains four, and the third row contains eight, how many would there be in the seventh row? Is this more or less than the total number of dominos in all the first six rows?

3. What would happen if a Ping-Pong ball were tossed into a room in which the floor was covered with thousands of set-to-kill spring-action mousetraps? Such an explosive event can be seen by using the keywords "mousetrap chain reaction" in an Internet video search.

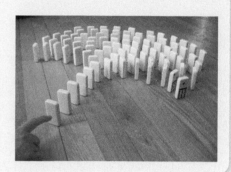

5.1 Radioactivity Results from Unstable Nuclei

EXPLAIN THIS

Why is it both impractical and impossible to prevent our exposure to radioactivity?

LEARNING OBJECTIVE

Identify three forms of radioactivity and their effects on living tissue.

As we learned in Chapter 4, atoms are made of electrons, neutrons, and protons. The neutrons and protons are at the highly dense center of each atom—its nucleus. Most atoms have stable nuclei, which means they hold together well. This stability is due to an optimal balance of the number of protons with the number of neutrons. These stable nuclei remain unchanged.

Some atoms, however, have nuclei that are unstable because they contain "off-balance" numbers of protons and neutrons. They may either contain too many neutrons and not enough protons or vice versa. Sooner or later, these nuclei spontaneously change to a more stable composition. For example, excess neutrons may transform into protons to provide a better nuclear balance. During these transformations, the nucleus ejects highly energetic particles while also emitting highly energetic electromagnetic radiation.

A material containing unstable nuclei undergoing these sorts of transformations is said to be **radioactive.** The high-energy particles and radiation emitted by a radioactive substance is called **radioactivity.** Notably, a radioactive nucleus disintegrates as it emits radioactivity. In other words, after emitting the radioactivity, the radioactive nucleus is no longer what it was. We say that the nucleus has "decayed," which is why the process of emitting radioactivity is often called *radioactive decay.*

Radioactive atoms emit three distinct types of radiation, named by the first three letters of the Greek alphabet, α, β, γ—alpha, beta, and gamma. Alpha rays carry a positive electric charge, beta rays carry a negative charge, and gamma

READINGCHECK

What types of radiation are emitted by the nuclei of radioactive elements?

▶ **Figure 5.1**
In a magnetic field, alpha rays bend one way, beta rays bend the other way, and gamma rays don't bend at all. Note that the alpha rays bend less than the beta rays. This occurs because alpha particles have more inertia (mass) than beta particles. The combined beam comes from a radioactive material placed at the bottom of a hole drilled in a block of lead.

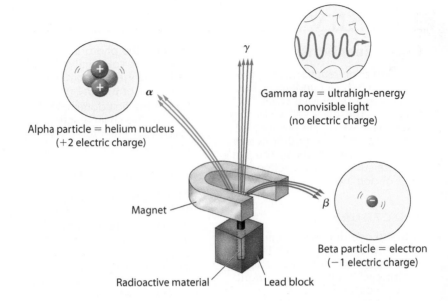

Alpha particle = helium nucleus (+2 electric charge)

Gamma ray = ultrahigh-energy nonvisible light (no electric charge)

Beta particle = electron (−1 electric charge)

Magnet

Radioactive material Lead block

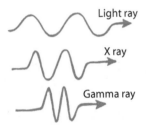

Light ray

X ray

Gamma ray

▲ **Figure 5.2**
A gamma ray is simply electromagnetic radiation, much higher in frequency and energy than light and X rays.

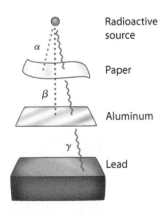

Radioactive source

Paper

Aluminum

Lead

▲ **Figure 5.3**
Alpha particles are the least penetrating and can be stopped by a few sheets of paper. Beta particles will readily pass through paper but not through a sheet of aluminum. Gamma rays penetrate several centimeters into solid lead.

rays carry no charge. The three rays can be separated by placing a magnetic field across their paths (**Figure 5.1**).

Alpha radiation is a stream of alpha particles. An **alpha particle** is the combination of two protons and two neutrons (in other words, it is the nucleus of the helium atom, atomic number 2). Alpha particles are relatively easy to shield against because of their relatively large size and their double positive charge (+2). For example, they do not normally penetrate such lightweight materials as paper or clothing. Because of their great kinetic energies, however, alpha particles can cause significant damage to the surface of a material, especially living tissue. When traveling through only a few centimeters of air, alpha particles pick up electrons and become nothing more than harmless helium. As a matter of fact, that's where the helium in a child's balloon comes from— practically all Earth's helium atoms were at one time energetic alpha particles.

Beta radiation is a stream of beta particles. A **beta particle** is merely an electron ejected from a radioactive nucleus. We will discuss in Section 5.3 how an electron arises from the nucleus. For now, just keep in mind that a beta particle is a fast-flying electron. A beta particle is normally faster than an alpha particle, and it carries only a single negative charge (–1). Beta particles are not as easily stopped as alpha particles, and they are able to penetrate light materials such as paper or clothing. They can penetrate fairly deeply into skin, where they have the potential to harm or kill living cells, but they are unable to penetrate even thin sheets of denser materials, such as aluminum. Beta particles, once stopped, simply become a part of the material they are in, like any other electron.

Gamma rays are the high-frequency electromagnetic radiation emitted by radioactive nuclei. Like photons of visible light, a gamma ray is pure energy. The amount of energy in a gamma ray, however, is much greater than that in visible light, ultraviolet light, or even X rays (**Figure 5.2**). Because they have no mass or electric charge, and because of their high energies, gamma rays are able to penetrate through most materials. However, they cannot easily penetrate very dense materials, such as lead. Lead is commonly used as a shielding material in laboratories or hospitals, where there can be much gamma radiation. Delicate molecules inside cells throughout our bodies that are zapped by gamma rays suffer structural damage. Hence, gamma rays generally cause more damage to our cells than does alpha or beta radiation.

Figure 5.3 shows the relative penetrating power of the three types of radiation. **Figure 5.4** shows an interesting practical use for gamma radiation.

◀ **Figure 5.4**
The shelf life of fresh strawberries and other perishables is markedly increased when the food is subjected to gamma rays from a radioactive source. The strawberries on the left were treated with gamma radiation, which kills the microorganisms that normally lead to spoilage. The strawberries are only a receiver of radiation and are not transformed into an emitter of radiation, as can be confirmed with a radiation detector.

CONCEPT**CHECK**

Pretend that you are given three radioactive rocks. One is an alpha emitter, one is a beta emitter, and one is a gamma emitter, and you know which is which. You can throw away one, but of the remaining two, you must hold one in your hand and place one in your pocket. What can you do to minimize your exposure to radiation?

CHECK YOUR ANSWER Throw away the gamma emitter, because it would penetrate your body from any of these locations. Hold the alpha emitter in your hand, because the skin on your hand is enough to shield you. Put the beta emitter in your pocket, because beta particles will likely be stopped by the combined thickness of your clothing and skin. (Ideally, of course, you should distance yourself as much as possible from all of the rocks.)

 FOR YOUR INFORMATION

Most helium atoms produced within the Earth find their way to the surface and then upward to outer space. Some helium, however, collects within natural gas deposits, which can contain as much as 7 percent helium. Most helium used in the world is isolated from the natural gas reserves of the Great Plains of the United States. Prior to World War II, the United States stopped supplying helium to the Germans, who then needed to use combustible hydrogen gas to fill their large zeppelin airships. One such airship, the *Hindenburg*, famously exploded in 1937.

5.2 Radioactivity Is a Natural Phenomenon

EXPLAIN THIS

Why are household smoke detectors radioactive?

It is a common misconception that radioactivity is something new in the environment. Actually, radioactivity has been around far longer than humans. It has always been in the soil we walk on and in the air that we breathe, and it warms Earth's interior. In fact, radioactive decay in Earth's interior heats the water that spurts from a geyser or wells up from a natural hot spring.

As **Figure 5.5** shows, most of the radiation we encounter is natural background radiation that originates from the Earth, from the Sun, and from other stars. At sea level, the protective blanket of the atmosphere reduces this radiation, but radiation is more intense at higher altitudes, where the air is thinner. In Denver, the "Mile-High City," a person receives more than twice as much cosmic radiation as he or she does at sea level.

LEARNING OBJECTIVE

Identify the natural sources, the units, and the applications of radioactivity.

What is a common source of radiation arising from Earth?

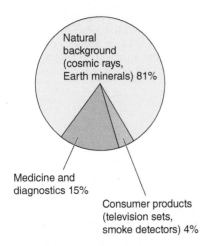

▲ **Figure 5.5**
Origins of radiation exposure for an average individual in the United States.

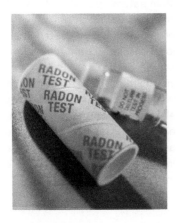

▲ **Figure 5.6**
A commercially available radon test kit for the home. The canister is unsealed in the area to be sampled. Radon seeping into the canister is adsorbed by activated carbon within the canister. After several days, the canister is resealed and sent to a laboratory that determines the radon level by measuring the amount of radiation emitted by the adsorbed radon.

▶ **Figure 5.7**
Nuclear radiation is focused on harmful tissue, such as a cancerous tumor, to selectively kill or shrink the tissue in a technique known as radiation therapy. This application of nuclear radiation has saved millions of lives—a clear-cut example of the benefits of nuclear technology. The inset shows the international symbol indicating an area where radioactive material is being handled or produced.

A common source of radiation of an Earthly origin is radon-222, an inert gas arising from uranium, which is widely found at low levels within all rock, soil, and water. Radon is a heavy gas that tends to accumulate in basements after it seeps up through cracks in the floor. Levels of radon vary from region to region, depending upon local geology. You can check the radon level in your home with a radon detector kit (**Figure 5.6**). If levels are high, corrective measures, such as sealing the basement floor and walls and maintaining adequate ventilation, should be taken. Radon gas poses a serious health risk.

About 20 percent of our annual exposure to radiation comes from sources outside of nature, primarily medical procedures. Fallout from nuclear testing and the coal and nuclear power industries are also contributors. Interestingly, the coal industry outranks the nuclear power industry as a source of radiation. The global combustion of coal annually releases about 13,000 tons of radioactive thorium and uranium into the atmosphere (in addition to other environmentally damaging molecules, including greenhouse gases). Both of these elements are found naturally in coal deposits, so their release is a natural consequence of burning coal. Nuclear power plants also produce radioactive by-products. Worldwide, the nuclear power industries generate about 10,000 tons of radioactive waste each year. Most of this waste, however, is contained and not released into the environment.

When radiation encounters our intricately structured cells, it can create chaos. Cells are able to repair most kinds of damage caused by radiation if the damage is not too severe. A cell can survive an otherwise lethal dose of radiation if the dose is spread over a long period of time to allow intervals for healing. When radiation is sufficient to kill cells, the dead cells can be replaced by new ones. Sometimes a radiated cell will survive with damaged DNA. This can alter the genetic information contained in a cell, producing one or more *mutations*. Although the effects of most mutations are inconsequential in terms of a person's health, some mutations affect the functioning of cells. Genetic mutations are the cause of most cancers, for example. In addition, mutations that occur in an individual's reproductive cells can be passed to the individual's offspring. In this case, the mutation will be present in every cell in the offspring organism's body—and may well have an effect on the functioning of the organism.

Rems Are Units of Radiation

We measure the ability of radiation to cause harm to living tissue in **rems.** Lethal doses of radiation begin at 500 rems. A person has about a 50 percent chance of surviving a dose of this magnitude received over a short period of time. During radiation therapy, a patient may receive localized doses in excess of 200 rems each day for a period of weeks (**Figure 5.7**).

Throughout our lives all the radiation we receive from natural sources and medical procedures is only a fraction of 1 rem. For convenience, the smaller unit *millirem* is used; 1 millirem (mrem) is 1/1000 of a rem.

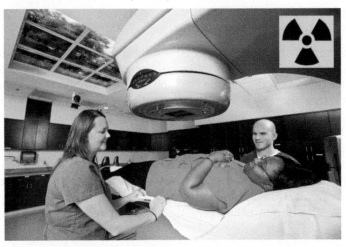

TABLE 5.1   Annual Radiation Exposure

SOURCE	TYPICAL AMOUNT RECEIVED IN 1 YEAR (MILLIREMS)
Natural Origin	
Cosmic radiation	26
Ground	33
Air (radon-222)	198
Human tissues (potassium-40; radium-226)	35
Human Origin	
Medical procedures	
Diagnostic X rays	40
Nuclear medicine	15
Television tubes, other consumer products	11
Weapons-test fallout	1

Source: U.S. Nuclear Regulatory Commission, http://www.nrc.gov/reading-rm/doc-collections/fact-sheets/bio-effects-radiation.html

The average person in the United States is exposed to about 360 millirems a year, as Table 5.1 indicates. About 80 percent of this radiation comes from natural sources, such as cosmic rays (radiation from our sun as well as other stars) and the Earth. A typical diagnostic X ray exposes a person to between 5 and 30 millirems (0.005 and 0.030 rem), less than 1/10,000 of the lethal dose.

Radioactive Tracers and Medical Imaging

Radioactive isotopes can be incorporated into molecules whose location can then be traced by the radiation they emit. When used in this way, radioactive isotopes are called *tracers*, and **Figure 5.8** shows one use for them. To check the action of a fertilizer, researchers incorporate radioactive isotopes into the molecules of the fertilizer and then apply the fertilizer to plants. The amount taken up by the plants can be measured with radiation detectors. From such measurements, scientists can tell farmers how much fertilizer to use, because fertilizer uptake is a physical and chemical process that is not affected by the radioactivity of the materials involved.

Tracers are also used in industry. Motor oil manufacturers can quantify the lubricating qualities of their products by running oil in engines containing small but measurable amounts of radioactive isotopes. As the engine runs and the pistons rub against the inner chambers, some of the metal from the engine invariably makes its way into the motor oil, and this metal carries with it the embedded radioactive isotopes. The better the lubricating qualities of a motor oil, the fewer radioactive isotopes it will contain after running in the engine for a given length of time.

◀ Figure 5.8
Tracking fertilizer uptake with a radioactive isotope.

Fertilizer with radioactive isotope applied to crop   Radioactivity detected in plant

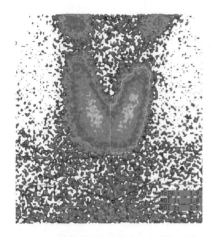

▲ **Figure 5.9**
The thyroid gland, located in the neck, absorbs much of the iodine that enters the body through food and drink. Images of the thyroid gland, such as the one shown here, can be obtained by giving a patient the radioactive isotope iodine-131. These images are useful in diagnosing metabolic disorders.

TABLE 5.2 Uses for Various Radioactive Isotopes

ISOTOPE	USAGE
Fluorine-18	Medical PET scans
Calcium-47	The study of bone formation in mammals
Californium-252	Inspection of airline luggage for explosives
Hydrogen-3 (tritium)	Drug-metabolism studies
Iodine-131	Treatment of thyroid disorders
Iridium-192	Testing the integrity of metal parts
Thallium-201	Cardiology and for tumor detection
Xenon-133	Lung-ventilation and blood-flow studies

Source: The Regulation and Use of Radioisotopes in Today's World (NUREG/BR-0217, Revision 1), U.S. Nuclear Regulatory Commission

In a technique known as *medical imaging,* isotopes are used for the diagnosis of internal disorders. Small amounts of a radioactive material such as sodium iodide, NaI, which contains the radioactive isotope iodine-131, are administered to a patient and traced through the body with a radiation detector. The result, shown in **Figure 5.9**, is an image that shows how the material is distributed in the patient's body. This technique works because the path the tracer material takes is influenced only by its physical and chemical properties, not by its radioactivity. The tracer may be introduced alone or along with some other chemical, known as a *carrier compound,* that helps target the isotope to a particular type of tissue in the body.

5.3 Radioactivity Results from an Imbalance of Forces

LEARNING OBJECTIVE

Describe how the strong nuclear force acts to hold nucleons together in the atomic nucleus.

READINGCHECK

Are protons on opposite sides of a large nucleus attracted to each other?

▶ **Figure 5.10**
(a) Two protons near each other experience both an attractive strong nuclear force and a repulsive electric force. At this tiny separation distance, the strong nuclear force overcomes the electric force, resulting in their staying together. (b) When the two protons are relatively far from each other, the electric force is more significant. The protons repel each other. This proton–proton repulsion in large atomic nuclei reduces nuclear stability.

EXPLAIN THIS

Why are larger nuclei less stable than smaller nuclei?

We know that electric charges of the same charge repel one another. So how is it possible that positively charged protons in the nucleus stay clumped together? This question led to the discovery of an attraction called the **strong nuclear force,** which acts between all nucleons. This force is very strong but only over extremely short distances. Repulsive electric forces, on the other hand, are relatively long-ranged. **Figure 5.10** suggests a comparison of the strengths of these two forces over distance.

Because the strong nuclear force decreases over distance, a large nucleus is not as stable as a small one, as shown in **Figure 5.11**. For protons that are close together, as in small nuclei, the attractive strong nuclear force easily overcomes the repulsive electric force. But for protons that are far apart, like those on opposite edges of a large nucleus, the attractive strong nuclear force may be weaker than the repulsive electric force.

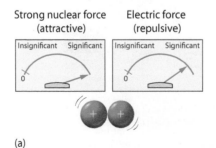

(a)

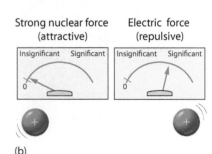

(b)

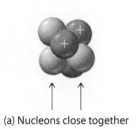

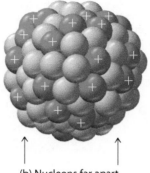

(a) All nucleons in a small atomic nucleus are close to one another; hence, they experience an attractive strong nuclear force. (b) Nucleons on opposite sides of a larger nucleus are not as close to one another, so the attractive strong nuclear forces holding them together are much weaker. The result is that the large nucleus is less stable.

(a) Nucleons close together

(b) Nucleons far apart

A large atomic nucleus is more susceptible to the repulsive forces among protons. This means there is a limit to the size of the atomic nucleus. As evidence of this, we find that all nuclei having more than 83 protons are radioactive. Furthermore, the superheavy elements, such as those above uranium, atomic number 92, are not found in nature. These superheavy elements are difficult to make in the laboratory. When they are produced, they exist for only fractions of a second.

CONCEPT CHECK

Two protons in the atomic nucleus repel each other, but they are also attracted to each other. Why?

CHECK YOUR ANSWER While two protons repel each other by the electric force, they also attract each other by the strong nuclear force. These forces act simultaneously. So long as the attractive strong nuclear force is stronger than the repulsive electric force, the protons will remain together. Under conditions in which the electric force overcomes the strong nuclear force, however, the protons fly apart from each other.

Neutrons serve as the "nuclear cement" holding an atomic nucleus together. Protons attract both protons and neutrons by the strong nuclear force. Protons also repel other protons by the electric force. Neutrons, on the other hand, have no electric charge and so attract protons and other neutrons only by the strong nuclear force. Therefore, the presence of neutrons adds to the attraction among nucleons and helps hold the nucleus together (**Figure 5.12**).

Nuclei with larger numbers of protons require larger numbers of neutrons to help balance the repulsive electric forces. For light elements, it is sufficient to have about as many neutrons as protons. For instance, the most common isotope of carbon, C-12, has equal numbers of each—six protons and six neutrons. For large nuclei, more neutrons than protons are needed. Because the strong nuclear force diminishes rapidly over distance, nucleons must be practically touching in order for the strong nuclear force to be effective. Nucleons on opposite sides

(a)

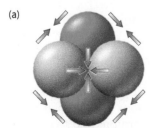

(b)

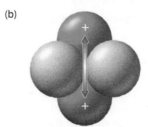

(a) The presence of neutrons helps hold the nucleus together by increasing the effect of the strong nuclear force, represented by the single-headed arrows. (b) The total strength of the attractive nuclear force, therefore, exceeds that of the repulsive electric force occurring between the two protons, represented by the double-headed arrow.

All nucleons, both protons and neutrons, attract one another by the strong nuclear force.

Only protons repel one another by the electric force.

▶ **Figure 5.13**
(a) A neutron without an adjacent proton is unstable and decays to a proton by emitting an electron. (b) Large nuclei have more neutrons than protons, which means that some of these neutrons don't have a sufficient number of adjacent protons. One of these extra neutrons may transform into a proton. Destabilized by an increase in the number of protons, the nucleus begins to shed fragments, such as alpha particles.

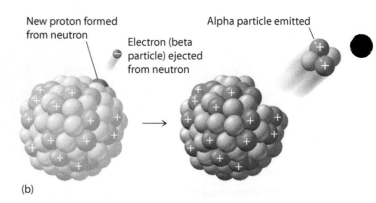

New proton formed from neutron

Electron (beta particle) ejected from neutron

Alpha particle emitted

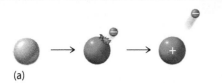

(a)

(b)

of a large atomic nucleus are not so attracted to one another. The electric force, however, does not diminish by much across the diameter of a large nucleus and so begins to win out over the strong nuclear force. To compensate for the weakening of the strong nuclear force across the diameter of the nucleus, large nuclei have more neutrons than protons. Lead nuclei, for example, have about one-and-a-half times as many neutrons as protons.

So we see that neutrons are stabilizing and large nuclei require an abundance of them. But neutrons are not always successful in keeping a nucleus intact. Interestingly, neutrons are not stable without protons. A lone neutron is radioactive and spontaneously transforms to a proton and an electron (**Figure 5.13a**). A neutron seems to need protons around to keep this from happening. After the size of a nucleus reaches a certain point, the neutrons so outnumber the protons that there are not sufficient protons in the mix to prevent the neutrons from turning into protons. As neutrons in a nucleus change into protons, the stability of the nucleus decreases because the repulsive electric force becomes increasingly significant. The result is that pieces of the nucleus fragment away in the form of radiation, as indicated in **Figure 5.13b**.

CONCEPTCHECK

What role do neutrons serve in the atomic nucleus? What is the fate of a neutron when alone or distant from one or more protons?

CHECK YOUR ANSWER Neutrons serve as a nuclear cement in nuclei and add to nuclear stability. But when alone or away from protons, a neutron becomes radioactive and spontaneously transforms to a proton and an electron.

5.4 Radioactive Elements Transmute to Different Elements

LEARNING OBJECTIVE

Name the isotope that results from a series of alpha and beta decays.

 READINGCHECK

What is the term for the changing of one element into another? Hint: It is the same word used by the alchemists.

EXPLAIN THIS

Why is the element lead often found within uranium ore?

When a radioactive nucleus emits an alpha or beta particle, the atomic number of the nucleus is changed. It becomes another element.

The changing of one element to another is called **transmutation**. Consider a uranium-238 nucleus, which contains 92 protons and 146 neutrons. When an alpha particle is ejected, the nucleus loses two protons and two neutrons. Since an element is defined by the number of protons in its nucleus, the 90 protons and 144 neutrons remaining no longer constitute a uranium atom. What we have now is the nucleus of a different element—thorium. This transmutation can be written as a nuclear equation:

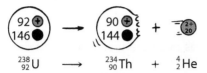

$$\ce{^{238}_{92}U} \longrightarrow \ce{^{234}_{90}Th} + \ce{^{4}_{2}He}$$

We see that $\ce{^{238}_{92}U}$ transmutes to the two elements written to the right of the arrow. When this transmutation occurs, energy is released, partly in the form of the kinetic energy of the alpha particle ($\ce{^4_2He}$), partly in the kinetic energy of the thorium atom, and partly in the form of gamma radiation. In this and all such equations, the mass numbers at the top balance ($238 = 234 + 4$) and the atomic numbers at the bottom also balance ($92 = 90 + 2$).

Thorium-234, the product of this reaction, is also radioactive. When it decays, it emits a beta particle. Because the formation of a beta particle results in an additional proton, the atomic number of the resulting nucleus is *increased* by 1. So after beta emission by thorium with 90 protons, the resulting element has 91 protons. It is no longer thorium but the element protactinium. Although the atomic number has increased by 1 in this process, the mass number (protons + neutrons) remains the same. The nuclear equation is

$$\ce{^{234}_{90}Th} \longrightarrow \ce{^{234}_{91}Pa} + \ce{^{0}_{-1}e}$$

We write an electron as $\ce{^0_{-1}e}$. The superscript 0 indicates that the electron's mass is insignificant relative to that of protons and neutrons. The subscript –1 is the electric charge of the electron.

So we see that when an element ejects an alpha particle from its nucleus, the mass number of the remaining atom is decreased by 4 and its atomic number is decreased by 2. The resulting atom is an atom of the element two spaces back in the periodic table, because this atom has two fewer protons. When an element ejects a beta particle from its nucleus, the mass of the atom is practically unaffected, meaning there is no change in its mass number but its atomic number increases by 1. The resulting atom becomes the element one space forward in the periodic table because it has one more proton.

The decay of uranium-238 to lead-206 is shown in **Figure 5.14**. Each blue arrow shows an alpha decay, and each red arrow shows a beta decay.

FOR **YOUR** INFORMATION

Beta emission is also accompanied by the emission of a neutrino, which is a neutral particle with nearly zero mass that travels at about the speed of light. Neutrinos are hard to detect because they interact very weakly with matter—a piece of lead about eight light-years thick would be needed to stop half the neutrinos produced in typical nuclear decays. Thousands of neutrinos are flying through you every second of every day, because the universe is filled with them. Only occasionally, one or two times a year, does a neutrino interact with an atom of your body.

◀ Figure 5.14
U-238 decays to Pb-206 through a series of alpha and beta decays.

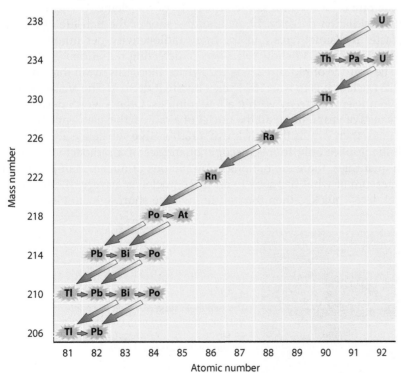

5.5 The Shorter the Half-Life, the Greater the Radioactivity

LEARNING OBJECTIVE

Recognize how a radioactive element can be identified by the rate at which it decays.

READING CHECK

How is radioactive half-life defined?

EXPLAIN THIS

How is the rate of transmutation related to half-life?

Radioactive isotopes decay at different rates. The radioactive decay rate is measured in terms of a characteristic time, the **half-life.**

The half-life of a radioactive material is the time required for half of the radioactive atoms to decay. Radium-226, for example, has a half-life of 1620 years. This means that half of any given specimen of Ra-226 decays by the end of 1620 years. In the next 1620 years, half of the remaining radium decays, leaving only one-fourth the original number of radium atoms. The other three-fourths convert, by a succession of decays, to lead. After 20 half-lives, an initial quantity of radioactive atoms is diminished to about one-millionth of the original quantity (**Figure 5.15**).

Half-lives are remarkably constant and not affected by external conditions. Some radioactive isotopes have half-lives that are less than a millionth of a second, while others have half-lives of more than a billion years. For example, uranium-238 has a half-life of 4.5 billion years. This means that in 4.5 billion years, half the uranium in the earth today will be lead-206.

It is not necessary to wait through the duration of a half-life in order to measure it. The half-life of an element can be accurately estimated by measuring the rate of decay of a known quantity of the element. This is easily done using a radiation detector. In general, the shorter the half-life of a substance, the faster it disintegrates and the more radioactivity per minute is detected. **Figures 5.16** and **5.17** show two ways of detecting radiation.

The half-life of radium-226 is 1620 years. Does this mean that an individual Ra-226 nucleus has to wait 1620 years before it decays? The answer is no. Some Ra-226 nuclei will decay within a few minutes. Others won't decay for tens of thousands of years. But if all the nuclei of a radioactive element decay at different times, then why is the half-life of a radioactive substance always the same? Half-life applies only to macroscopic quantities of a radioactive substance. In such quantities, there are billions upon billions of nuclei. Half-life is a statistical measure of what happens to all these nuclei *on average*. If, on average, the nuclei decay quickly, then the half-life will be short, which tells us that these

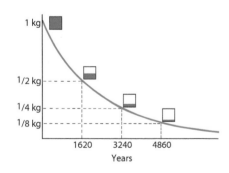

▲ **Figure 5.15**
Radium-226 has a half-life of 1620 years, meaning that every 1620 years the amount of radium decreases by half as it transmutes to other elements.

◀ **Figure 5.16**
The handheld tube of a Geiger counter contains a gas that gets ionized by incoming nuclear radiation. This completes an electric circuit, which generates audible clicks that increase with increasing radiation.

nuclei are not very stable. Conversely, if, on average, it takes a long time for the nuclei to decay, then the half-life will be long, which tells us that these nuclei are more stable.

CONCEPT CHECK

1. If a sample of radioactive isotopes has a half-life of 1 day, how much of the original sample will remain at the end of the second day? The third day?

2. Which will give a higher counting rate on a radiation detector—radioactive material that has a short half-life or a long half-life?

CHECK YOUR ANSWERS

1. In the second day, one-fourth of the original sample will be left. The three-fourths that underwent decay becomes a different element altogether. At the end of 3 days, one-eighth of the original sample will remain.

2. The material with the shorter half-life is more active and will show a higher counting rate on a radiation detector.

▲ **Figure 5.17**
The film badge worn by this technician contains audible alerts for both radiation surge and accumulated exposure. Information from the individualized badges is periodically downloaded to a database for analysis.

5.6 Isotopic Dating Measures the Ages of Materials

EXPLAIN THIS

How does radioactivity allow archeologists to measure the ages of ancient artifacts?

Cosmic rays continually bombard the Earth's atmosphere. This bombardment causes many atoms in the upper atmosphere to transmute. These transmutations result in many protons and neutrons being "sprayed out" into the environment. Most of the protons are stopped as they collide with the atoms of the upper atmosphere. These protons strip electrons from the atoms they collide with and thus become hydrogen atoms. The neutrons, however, continue for longer distances because they have no electric charge and therefore do not interact electrically with matter. Eventually, many of them collide with atomic nuclei in the lower atmosphere. A nitrogen atom that captures a neutron, for instance, becomes an isotope of carbon by emitting a proton:

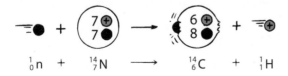

$$\,_{0}^{1}n + \,_{7}^{14}N \longrightarrow \,_{6}^{14}C + \,_{1}^{1}H$$

This carbon-14 isotope, which makes up less than one-millionth of 1 percent of the carbon in the atmosphere, is radioactive and has eight neutrons. (The most common isotope, carbon-12, has six neutrons and is not radioactive.) Because both carbon-12 and carbon-14 are forms of carbon, they have the same chemical properties. Both of these isotopes, for example, chemically react with oxygen to form carbon dioxide, which is consumed by plants through the process of photosynthesis. This means that all plants contain a tiny quantity of radioactive carbon-14.

LEARNING OBJECTIVE

Review how the age of ancient artifacts can be determined by measuring the amounts of remaining radioactivity the artifacts contain.

CHEMICAL CONNECTIONS

How is the radioactivity of your body connected to the stars?

READINGCHECK

Plants contain a little carbon-14, but why do humans also contain carbon-14?

All animals eat either plants or plant-eating animals; therefore, all animals, including humans, have a little carbon-14 in them. So we see why all living things on Earth contain some carbon-14. This radioactive isotope decays back into nitrogen as it emits a beta particle:

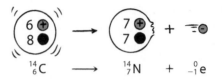

$$^{14}_{6}\text{C} \longrightarrow {^{14}_{7}\text{N}} + {^{0}_{-1}\text{e}}$$

FORYOUR **INFORMATION**

A 1 g sample of carbon from recently living matter contains about 50 trillion billion (5×10^{22}) carbon atoms. Of these carbon atoms, about 65 billion (6.5×10^{10}) are the radioactive C-14 isotope. This gives the carbon a beta disintegration rate of about 13.5 decays per minute.

Because plants absorb carbon dioxide while alive, any carbon-14 that decays is immediately replenished with fresh carbon-14 from the atmosphere. In this way, a radioactive equilibrium is reached in which there is a constant ratio of about 1 carbon-14 atom to every 100 billion carbon-12 atoms. When a plant dies, replenishment of carbon-14 ends. Then the percentage of carbon-14 decreases at a constant rate determined by its half-life. The amount of carbon-12, however, does not change, because this isotope does not undergo radioactive decay. The longer a plant or other organism is dead, the less carbon-14 it contains relative to the constant amount of carbon-12.

The half-life of carbon-14 is about 5730 years. This means that half of the carbon-14 atoms now present in a plant or animal that dies today will decay in the next 5730 years. Half of the remaining carbon-14 atoms will then decay in the following 5730 years, and so on.

With this knowledge, scientists are able to calculate the age of carbon-containing artifacts, such as wooden tools or the skeleton shown in **Figure 5.18**. Their age is measured by their current level of radioactivity. This process, known as **carbon-14 dating,** enables investigators to probe as far as 50,000 years into the past. Beyond this time span, there is too little carbon-14 remaining for accurate analysis.

Because of fluctuations in cosmic ray bombardment rates over the centuries, carbon dating has an uncertainty of about 15 percent. This means, for example, that the straw of an old adobe brick dated to be 500 years old may really be only 425 years old on the low side or 575 years old on the high side. For many purposes, this is an acceptable level of uncertainty.

▼ Figure 5.18
The amount of radioactive carbon-14 in the skeleton is reduced by one-half every 5730 years. The result is that the same skeleton today contains only a trace amount of the original carbon-14. The red arrows represent the relative amounts of carbon-14.

CONCEPTCHECK
Suppose that an archaeologist extracts 1 gram of carbon from an ancient ax handle and finds that it is one-fourth as radioactive as 1 gram of carbon extracted from a freshly cut tree branch. About how old is the ax handle?

CHECK YOUR ANSWER Assuming the ratio of C-14/C-12 was the same when the ax was made, the ax handle is as old as two half-lives of C-14, or about 11,460 years old.

22,920 years ago   17,190 years ago   11,460 years ago   5730 years ago   Present

Artifacts derived from nonliving materials can also be dated based on the radioactive minerals they contain. The naturally occurring mineral isotopes uranium-238 and uranium-235, for example, decay very slowly and ultimately become lead—but not the common isotope lead-208. Instead, as was shown in Figure 5.14, uranium-238 decays to lead-206. Uranium-235, on the other hand, decays to lead-207. Thus, the lead-206 and lead-207 that now exist in a uranium-bearing rock were at one time uranium. Older rocks contain higher percentages of these trace isotopes.

If you know the half-lives of uranium isotopes and the percentage of lead isotopes in some uranium-bearing rock, you can calculate the date of the rock's formation. Rocks dated in this manner have been found to be as much as 3.7 *billion* years old. Samples from the Moon have been dated at 4.2 billion years, which is close to the estimated age of our solar system: 4.6 billion years.

5.7 Nuclear Fission—The Splitting of Atomic Nuclei

EXPLAIN THIS

Why isn't it possible for a nuclear power plant to explode like a nuclear bomb?

Biology students know that living tissue grows by the division of cells. The splitting in half of living cells is called *fission.* In a similar way, the splitting of an atomic nucleus into two smaller halves is called **nuclear fission.**

Nuclear fission involves the delicate balance between two forces within the nucleus. One force is the *strong nuclear force,* which is a force that holds all the nucleons together. The second force is the repulsive *electric force* occurring among all the like-sign protons. In most nuclei the nuclear strong force dominates. In uranium, however, this domination is weak. If the uranium nucleus is stretched into an elongated shape (**Figure 5.20**), the electric forces may push it into an even more elongated shape. If the elongation passes a critical point, the electric forces overwhelm the strong nuclear forces, and the nucleus splits. This splitting process is called nuclear fission.

The absorption of a neutron by a uranium nucleus supplies enough energy to cause such an elongation. The resulting fission process may produce many different combinations of smaller nuclei. More significantly, the energy released by fission is enormous—about 7 million times that of a TNT molecule explosion. This energy is mainly in the form of the kinetic energy of the fission fragments, which fly apart from one another. A much smaller amount of energy is released as gamma radiation.

Here is the equation for a typical uranium fission reaction:

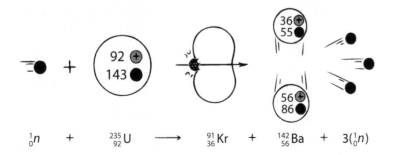

$$^1_0 n \quad + \quad ^{235}_{92}U \quad \longrightarrow \quad ^{91}_{36}Kr \quad + \quad ^{142}_{56}Ba \quad + \quad 3(^1_0 n)$$

Note that in this reaction 1 neutron starts the fission of a single uranium nucleus, which produces nuclear fragments and 3 neutrons. These 3 neutrons

LEARNING OBJECTIVE

Describe the process by which large atomic nuclei can split in half leading to the production of energy.

▲ **Figure 5.19**
Carbon-14 dating was developed by the American chemist Willard F. Libby (1908–1980) at the University of Chicago in the 1950s. For this work he received the Nobel Prize in Chemistry in 1960.

▶ **Figure 5.20**
An elongation of the nucleus may result in the repulsive electric force's overcoming the attractive strong nuclear force, in which case fission occurs.

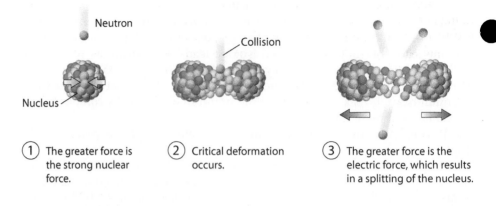

Neutron

Collision

Nucleus

(1) The greater force is the strong nuclear force.

(2) Critical deformation occurs.

(3) The greater force is the electric force, which results in a splitting of the nucleus.

READING CHECK

What exactly is a chain reaction?

can cause the fissioning of 3 more uranium atoms, releasing 9 more neutrons. If each of these 9 neutrons succeeds in splitting a uranium atom, the next step in the reaction produces 27 neutrons, and so on. Such a sequence, illustrated in **Figure 5.21**, is called a **chain reaction.** A chain reaction is a self-sustaining reaction in which the products of one reaction event initiate further reaction events.

Why do chain reactions not happen in naturally occurring uranium ore deposits? They would if all uranium atoms fissioned so easily. Fission occurs mainly in the rare isotope U-235, which makes up only 0.7 percent of the uranium in pure uranium metal, as shown in **Figure 5.22**. When the more abundant isotope U-238 absorbs neutrons created by fission of U-235, the U-238 typically does not undergo fission. So any chain reaction is snuffed out by the neutron-absorbing U-238, as well as by the rock in which the ore is imbedded.

Chain reactions are more effective in large chunks of uranium than in smaller chunks. In smaller chunks, neutrons easily find the surface and escape, as shown in **Figure 5.23**. As the neutrons escape, the chain reaction no longer builds up.

If two small chunks of uranium are suddenly pushed together, they make a larger chunk. Within this larger chunk, neutrons are no longer able to escape as easily. Instead, they continue the chain reaction, which becomes sustainable. The minimum size of a chunk needed for a sustainable chain reaction is called the **critical mass.** Any chunk at or above the critical mass produces a steady release of energy. With the correct engineering, this release of energy can be controlled, which is what happens within a nuclear power plant. A different sort of engineering, as shown in **Figure 5.24**, produces an explosion of energy, which is the basis of a nuclear fission bomb.

▼ **Figure 5.21**
A chain reaction.

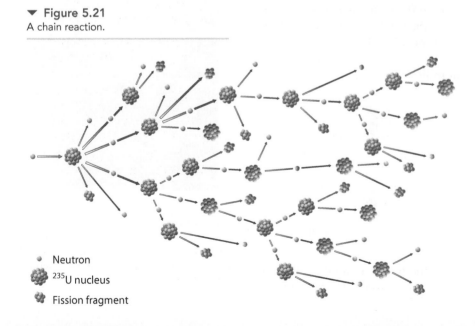

• Neutron

🟤 ^{235}U nucleus

🟤 Fission fragment

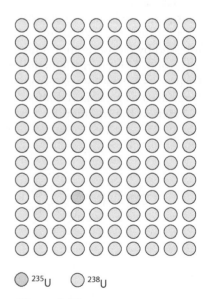

⚪ ^{235}U   ⚪ ^{238}U

▲ **Figure 5.22**
Only 1 part in 140 of naturally occurring uranium is U-235.

Constructing a fission bomb is a formidable task. The difficulty is in separating enough uranium-235 from the more abundant uranium-238. Separation is difficult because, except for their slightly different masses, the two isotopes have the same physical and chemical properties. Scientists took more than 2 years to extract enough of the 235 isotope from uranium ore to make the bomb that was detonated at Hiroshima in 1945. To this day, uranium isotope separation, also known as uranium enrichment, remains a difficult process.

Nuclear Fission Reactors

The awesome energy of nuclear fission was introduced to the world in the form of nuclear bombs, and this violent image still colors our thinking about nuclear power, making it difficult for many people to recognize its potential usefulness. Currently, about 20 percent of electric energy in the United States is generated by *nuclear fission reactors* (whereas most electric power is nuclear in some other countries—about 75 percent in France). These reactors are simply nuclear furnaces. They, like fossil fuel furnaces, do nothing more elegant than boil water to produce steam for a turbine (**Figure 5.25**). The greatest practical difference is the amount of fuel involved: a mere 1 kilogram of uranium fuel, less than the size of a baseball, yields more energy than 30 freight-car loads of coal.

A fission reactor contains four components: nuclear fuel, control rods, moderator (to slow the velocity of the neutrons, making them effective for the fission process), and a liquid (usually water) to transfer heat from the reactor to the turbine and generator. The nuclear fuel is primarily U-238, plus about 3 percent U-235. Because the U-235 isotopes are so highly diluted with U-238, an explosion like that of a nuclear bomb is not possible. The reaction rate, which depends on the number of neutrons that initiate the fission of other U-235 nuclei, is controlled by rods inserted into the reactor. The control rods are made of a neutron-absorbing material, usually metals like cadmium or boron.

Heated water around the nuclear fuel is kept under high pressure to keep it at a high temperature without boiling (see Section 8.4). It transfers heat to a second, lower-pressure water system, which operates the turbine and electric generator in a conventional fashion. In this design, two separate water systems are used so that no radioactivity reaches the turbine or the outside environment. The entire setup resides inside a building like the one shown in **Figure 5.26**.

A significant disadvantage of fission power is the generation of radioactive waste products. Light atomic nuclei are most stable when composed of equal numbers of protons and neutrons, as discussed earlier, and heavy nuclei need more neutrons than protons for stability. For example, there are 143 neutrons but only 92 protons in U-235. When uranium fissions into two medium-weight elements, the extra neutrons in their nuclei make them unstable. They are radioactive, most with very short half-lives, but some with half-lives of thousands of years. Safely disposing of these waste products, as well as materials made radioactive in the production of nuclear fuels, requires special storage casks and procedures. Although fission has been successfully producing electricity for a half century, disposing of radioactive wastes in the United States remains problematic.

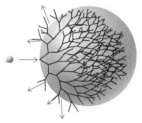

Neutrons escape small lump of uranium-235.

Neutrons trigger more reactions within large lump of uranium-235.

▲ Figure 5.23
This exaggerated view shows that a chain reaction in a small piece of pure uranium-235 runs its course before it can cause a large explosion because neutrons leak from the surface too soon. The surface area of the small piece is large relative to its mass. In a larger piece, more uranium and less surface are presented to the neutrons.

 **FOR YOUR INFORMATION**

One ton of uranium can produce more than 40 million kilowatt-hours of electricity. This is equivalent to burning 16,000 tons of coal or 80,000 barrels of oil. There are currently 104 operating U.S. nuclear power plants, which produce over 20 percent of U.S. electricity.

 FOR YOUR INFORMATION

A nuclear power plant "meltdown" occurs when the fissioning nuclear fuels are no longer submerged within a cooling fluid, such as water. The temperature rises to the point that the solid nuclear fuel and the reaction vessel itself melt into a liquid phase that has the potential of penetrating through the floor of the containment building.

◀ Figure 5.24
Simplified diagram of a uranium-fission bomb. Two smaller masses of uranium are initially apart from each other. The bomb is ignited when high explosives (gunpowder) force the two masses to come together into a single lump with a critical mass.

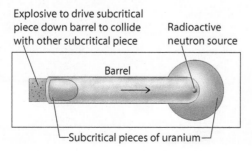

Explosive to drive subcritical piece down barrel to collide with other subcritical piece

Radioactive neutron source

Barrel

Subcritical pieces of uranium

▶ Figure 5.25

Diagram of a nuclear fission power plant. Note that the water in contact with the fuel rods is completely contained and that radioactive materials are not involved directly in the generation of electricity.

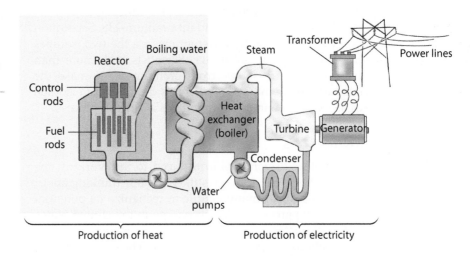

FOR YOUR INFORMATION

Notably, active safety measures failed when the Generation II Fukushima Daiichi nuclear plant of Japan was hit by the powerful tsunami of 2011. To learn more about this particular nuclear disaster and about the risks of nuclear power, look ahead to Section 17.3.

The designs for nuclear power plants have progressed over the years. The earliest designs, from the 1950s through the 1980s, are called the Generation I and II reactors. The safety systems of these reactors are "active," in that they rely on a series of active measures, such as water pumps, that come into play to keep the reactor core cool in the event of an accident. The Generation III reactors, built in the 1990s, also rely on active safety measures but are more economical to build, operate, and maintain.

While not yet operational, the latest Generation IV nuclear reactors will have fundamentally different reactor designs. For example, they will incorporate passive safety measures that will cause the reactor to shut down by itself in the event of an emergency. The fuel source may be the depleted uranium stockpiled from earlier-generation reactors. The designs will also allow the formation of hydrogen fuel from water. Furthermore, these reactors can be built as small modular units that generate between 150 and 600 megawatts of power rather than the 1500 megawatts that is the usual output of today's reactors. Smaller reactors are easier to manage and can be used in tandem to build a generating capacity suited to the community being served. The Generation IV International Forum aims to have Generation IV power plants operating within the next decade.

The benefits of fission power are plentiful electricity, conservation of the many billions of tons of fossil fuels that every year are literally turned to heat and smoke (and in the long run may be far more precious as sources of organic molecules than as sources of heat), and the elimination of the megatons of sulfur oxides and other poisons that are put into the air each year by the burning of these fossil fuels.

▶ Figure 5.26

The nuclear reactor is housed within a dome-shaped containment building that is designed to prevent the release of radioactive isotopes in the event of an accident. The Soviet built Chernobyl nuclear power plant that reached meltdown in 1986 had no such containment building, leading to the release of massive amounts of radiation into the surrounding environment.

FOR YOUR INFORMATION

Sustainable fission chain reactions are a prime source of heat found within Earth's core, which is to say that Earth itself is a planetary-sized nuclear reactor. Know nukes before you say "No nukes"!

5.8 The Mass–Energy Relationship: $E = mc^2$

EXPLAIN THIS

Why does it get easier to pull nucleons away from nuclei heavier than iron?

Clearly, a lot of energy comes from every gram of nuclear fuel when it undergoes fission. What is the source of this energy? As we will see, it comes from nucleons losing mass as they undergo nuclear reactions.

In the early 1900s, Albert Einstein discovered that mass is actually "congealed" energy. Mass and energy are two sides of the same coin, as stated in his celebrated equation, $E = mc^2$. In this equation, E stands for the energy contained by any mass when at rest, m stands for mass, and c is the speed of light. This relationship between energy and mass is the key to understanding why and how energy is released in nuclear reactions.

How easy might it be to pull a nucleon out of a nucleus? (Remember, "nucleon" is the generic name for either a proton or a neutron.) To do this, you would have to fight against the strong nuclear force, which holds the nucleon to the nucleus. So a lot of energy would be required to pull this nucleon out of the nucleus. What we learn from Einstein's equation is that the energy you put in to pull the nucleon out is not lost. Instead, this energy is absorbed by the nucleon, which thus becomes more massive as you pull it out. For example, if the mass of the nucleon were 1.00000 while in the nucleus, its mass might be a slightly greater 1.00728 after it has been pulled out of the nucleus. The energy you put into pulling the nucleon out was converted into mass. Energy and mass are two sides of the same coin. In other words, energy can become mass and mass can become energy as dictated by the equation $E = mc^2$.

So the mass of a nucleon depends upon where it is. In general, a nucleon's mass is greatest when it is free by itself outside of the nucleus. The nucleon's mass is smallest when it is tightly bound within the nucleus.

Not all nuclei, however, are the same. In one nucleus, for example, a nucleon might find itself more tightly bound than in another. The mass of the nucleon, therefore, also depends upon which nucleus it is in. As illustrated in the graph of **Figure 5.27**, a nucleon has its greatest mass when in the hydrogen nucleus and its smallest mass when in the iron nucleus. The mass of a nucleon then gradually increases as it enters heavier nuclei, such as that of uranium.

LEARNING OBJECTIVE

Show how the mass of a nucleon depends upon the identity of the nucleus within which it is contained.

READING CHECK

What do the symbols represent in the equation $E = mc^2$?

▶ Figure 5.27
This graph shows that the average mass of a nucleon depends on which nucleus it is in. A nucleon has the most mass when in the lightest (hydrogen) nucleus. The nucleon has its least mass when in an iron nucleus and an intermediate mass when in the heaviest (uranium) nucleus.

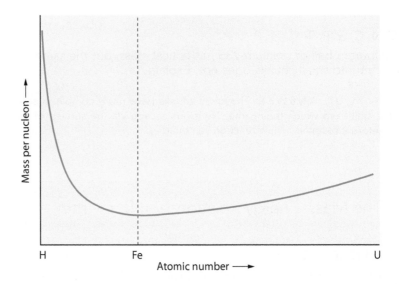

The graph of Figure 5.27 is the key to understanding the energy released in nuclear processes. Uranium, being toward the right hand side of the graph, is shown to have a relatively large amount of mass per nucleon. When the uranium nucleus splits in half, however, smaller nuclei of lower atomic numbers are formed. As shown in **Figure 5.28**, these nuclei are lower on the graph than uranium, which means that they have a smaller amount of mass per nucleon. Thus, nucleons lose mass in their transition from being in a uranium nucleus to being in one of its fragments. When this decrease in mass is multiplied by the speed of light squared (c^2 in Einstein's equation), the product is equal to the energy yielded by each uranium nucleus as it undergoes fission.

Interestingly, Einstein's mass–energy relationship applies to chemical reactions as well as to nuclear reactions. For nuclear reactions, the energies involved are so great that the change in mass is measurable, corresponding to about 1 part in 1000. In chemical reactions, the energy involved is so small that the change in mass, about 1 part in 1,000,000,000, is not detectable. This is why the law of mass conservation is best stated as follows: there is no *detectable* change in the total mass of materials as they chemically react to form new materials (see Sections 1.4, 2.2, and 9.1). In truth, there are changes in the mass of atoms during a chemical reaction. These changes, however, are too small to be of any concern to the working chemist.

▶ Figure 5.28
The mass of each nucleon in a uranium nucleus is greater than the mass of each nucleon in any one of its nuclear fission fragments. This lost mass is mass that has been transformed into energy, which is why nuclear fission is an energy-releasing process.

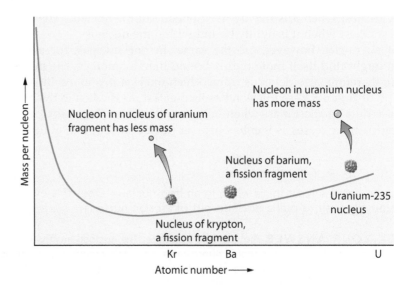

We can think of the mass-per-nucleon graph as an energy valley that starts at hydrogen (the highest point) and slopes steeply to iron (the lowest point), then slopes gradually up to uranium. Iron is at the bottom of the energy valley and has the most stable nucleus. It also has the most tightly bound nucleus; more energy per nucleon is required to separate nucleons from iron's nucleus than from any other element's nucleus.

All nuclear power today is produced by nuclear fission. A more promising long-range source of energy is to be found on the left side of the energy valley, in a process known as nuclear fusion.

5.9 Nuclear Fusion—The Combining of Atomic Nuclei

EXPLAIN THIS

How does the energy of gasoline come from nuclear fusion?

In the graphs of Figures 5.27 and 5.28, we see that the steepest part of the energy valley goes from hydrogen to iron. Energy is gained by the nucleons as light nuclei combine. This combining of nuclei is called **nuclear fusion,** and it is the opposite of nuclear fission. We can see from **Figure 5.29** that as we move along the list of elements from hydrogen to iron, the average mass per nucleon decreases. Thus, when two small nuclei fuse—for example, a pair of hydrogen isotopes—the mass of the resulting helium nucleus is less than the mass of the two small nuclei before fusion. The mass difference is released in the form of energy (**Figure 5.30**).

LEARNING OBJECTIVE

Describe the process by which small nuclei can join together, leading to the production of energy, such as occurs in the Sun.

READINGCHECK

How does the mass of a pair of hydrogen isotopes about to fuse compare with the mass of the resulting helium nucleus?

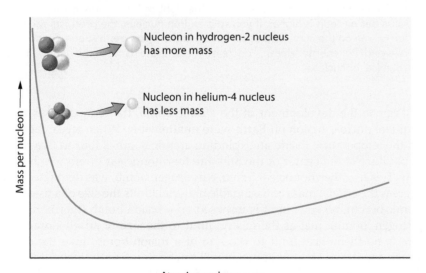

Nucleon in hydrogen-2 nucleus has more mass

Nucleon in helium-4 nucleus has less mass

Mass per nucleon ⟶

Atomic number ⟶

◀ **Figure 5.29**
The mass of each nucleon in a hydrogen-2 nucleus is greater than the mass of each nucleon in a helium-4 nucleus, which results from the fusion of two hydrogen-2 nuclei. This lost mass is mass that has been converted to energy, which is why nuclear fusion is a process that releases energy.

▶ **Figure 5.30**
The mass of a nucleus is not equal to the sum of the mass of its parts. (a) The fission fragments of a uranium nucleus are less massive than the uranium nucleus. (b) Two protons and two neutrons are more massive in their free states than when they are combined to form a helium nucleus. Can you relate this to the graphs of Figure 5.28 and 5.29?

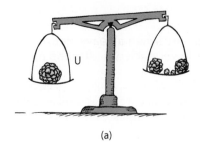
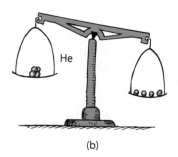

(a)   (b)

For a fusion reaction to occur, the nuclei must collide at a very high speed in order to overcome their mutual electric repulsion. The required speeds correspond to the extremely high temperatures found in the Sun and other stars. Fusion brought about by high temperatures is called **thermonuclear fusion.** In the high temperatures of the Sun, approximately 657 million tons of hydrogen is converted into 653 million tons of helium each second. The missing 4 million tons of mass is converted to energy—a tiny bit of which reaches our planet as sunshine. So thermonuclear fusion is the energy source of our sun, which is, in turn, the ultimate energy source of life on Earth (**Figure 5.31**).

Such reactions are, quite literally, nuclear burning. Thermonuclear fusion is analogous to ordinary chemical combustion. In both chemical and nuclear burning, a high temperature starts the reaction; the release of energy by the reaction maintains a high enough temperature to spread the fire. The net result of the chemical reaction is the combination of atoms into more tightly bound molecules. In nuclear fusion reactions, the net result is more tightly bound nuclei. In both cases, mass decreases as the corresponding amount of energy is released.

CHEMICAL CONNECTIONS

How are the atoms of your body connected to sunlight?

CONCEPT CHECK

a. Fission and fusion are opposite processes, yet each releases energy. Isn't this contradictory?
b. To get a release of nuclear energy from the element iron, should iron undergo fission or fusion?

CHECK YOUR ANSWERS

a. No, no, no! This is contradictory only if the same element is said to release energy by both fission and fusion. Only the fusion of light elements and the fission of heavy elements result in a decrease in nucleon mass and a release of energy.
b. Neither, because iron is at the very bottom of the "energy valley." Fusing a pair of iron nuclei produces an element to the right of iron on the curve, in which the mass per nucleon is higher. If you split an iron nucleus, the products will lie to the left of iron on the curve and again have a higher mass per nucleon. So no energy is released. For energy release, "decrease mass" is the name of the game—any game, chemical or nuclear.

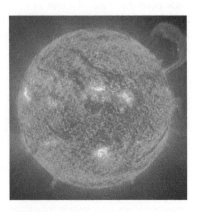

▲ **Figure 5.31**
Thermonuclear fusion takes place in stars, such as the Sun. Some day, humans may produce vast quantities of energy through thermonuclear fusion, as the stars have always done.

Prior to the development of the atomic bomb, the temperatures required to initiate nuclear fusion on Earth were unattainable. When researchers found that the temperature inside an exploding atomic bomb is four to five times the temperature at the center of the Sun, the thermonuclear bomb was but a step away. This first thermonuclear bomb, a hydrogen bomb, was detonated in 1952. Whereas the critical mass of fissionable material limits the size of a fission bomb (atomic bomb), no such limit is imposed on a fusion bomb (thermonuclear, or hydrogen, bomb). Just as there is no limit to the size of an oil-storage depot, there is no theoretical limit to the size of a fusion bomb. Like the oil in the storage depot, any amount of fusion fuel can be stored with safety until ignited. Although a mere match can ignite an oil depot, nothing less energetic than an atomic bomb can ignite a thermonuclear bomb. We can see that there is no such thing as a "baby" hydrogen bomb. A typical thermonuclear bomb stockpiled by the United States today, for example, is about 1000 times more destructive than the atomic bomb detonated over Hiroshima at the end of World War II.

The hydrogen bomb is another example of a discovery used for destructive rather than constructive purposes. The potential constructive possibility is the controlled release of vast amounts of clean energy.

Controlling Fusion

Carrying out fusion reactions under controlled conditions requires temperatures of millions of degrees. There are a variety of techniques for attaining high temperatures. No matter how the temperature is produced, a problem is that all materials melt and vaporize at the temperatures required for fusion. One solution to this problem is to confine the reaction in a nonmaterial container.

A nonmaterial container is a magnetic field, which can exist at any temperature and can exert powerful forces on charged particles in motion. "Magnetic walls" of sufficient strength provide a kind of magnetic straitjacket for hot ionized gases called *plasmas*. Magnetic compression further heats the plasma to fusion temperature.

Although there are no nuclear fusion power plants currently in operation, an international project now exists whose goal is to prove the feasibility of nuclear fusion power in the near future. This fusion power project is the International Thermonuclear Experimental Reactor (ITER). After construction at the chosen site in Cadarache, France, the first sustainable fusion reaction may begin as early as 2015 (**Figure 5.32**). The reactor will house electrically charged hydrogen gas (plasma) heated to over 100 million degrees Celsius, which is hotter than the center of the Sun. In addition to producing about 500 MW of power, the reactor could be the energy source for the creation of hydrogen, H_2, which could be used to power fuel cells, which we explore in Chapter 11.

If people are one day to dart about the universe in the same way we jet about the Earth today, their supply of fuel is assured. The fuel for fusion—hydrogen—is found in every part of the universe, not only in the stars but also in the space between them. About 90 percent of the atoms in the universe are estimated to be hydrogen. For people of the future, the supply of raw materials is also assured, because all the elements known to exist result from the fusing of more and more hydrogen nuclei. Future humans might synthesize their own elements and produce energy in the process, just as the stars have always done.

 FOR YOUR INFORMATION

Elements are created in stars as smaller nuclei fuse to form larger nuclei. This process is energy releasing only up to iron. The manufacture of elements heavier than iron cannot be sustained within a shining star. So where do heavier elements, such as gold, come from? The final stage of certain very large stars involves a mighty collapse, called a supernova. The energy of a supernova can outshine an entire galaxy, and it is this energy that makes the heavier-than-iron elements. The gold atoms in your jewelry were created using the abundant energy of a supernova that exploded very long ago, likely in a galaxy very far away.

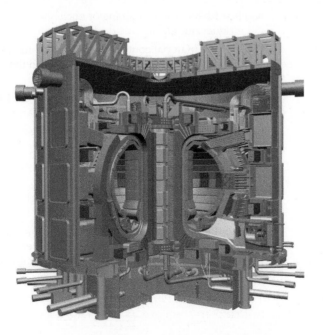

◀ Figure 5.32
A cross-sectional view of the ITER (rhymes with "fitter") planned to be built and operating in Cadarache, France, before 2020.

Chapter 5 Review

LEARNING OBJECTIVES

Identify three forms of radioactivity and their effects on living tissue. (5.1)	→	*Questions 1, 2, 33, 37, 38*
Identify the natural sources, the units, and the applications of radioactivity. (5.2)	→	*Questions 3–5, 28, 34, 39, 40, 76*
Describe how the strong nuclear force acts to hold nucleons together in the atomic nucleus. (5.3)	→	*Questions 6–8, 41–43*
Name the isotope that results from a series of alpha and beta decays. (5.4)	→	*Questions 9–11, 32, 44–49*
Recognize how a radioactive element can be identified by the rate at which it decays. (5.5)	→	*Questions 12, 13, 26, 27, 29–31, 35, 50–52, 79*
Review how the age of ancient artifacts can be determined by measuring the amounts of remaining radioactivity they contain. (5.6)	→	*Questions 14–16, 53–56*
Describe the process by which large atomic nuclei can split in half leading to the production of energy. (5.7)	→	*Questions 17–19, 57–60, 75, 78, 80*
Show how the mass of a nucleon depends upon the identity of the nucleus within which it is contained. (5.8)	→	*Questions 20–22, 61–65*
Describe the process by which small nuclei can join together leading to the production of energy, such as occurs in the Sun. (5.9)	→	*Questions 23–25, 36, 66–74, 77*

SUMMARY OF TERMS (KNOWLEDGE)

Alpha particle A subatomic particle consisting of the combination of two protons and two neutrons ejected by a radioactive nucleus. The composition of an alpha particle is the same as that of the nucleus of a helium atom.

Beta particle An electron emitted during the radioactive decay of a radioactive nucleus.

Carbon-14 dating The process of estimating the age of once-living material by measuring the amount of radioactive carbon-14 present in the material.

Chain reaction A self-sustaining reaction in which the products of one reaction event initiate further reaction events.

Critical mass The minimum mass of fissionable material needed for a sustainable chain reaction.

Gamma rays High-frequency electromagnetic radiation emitted by radioactive nuclei.

Half-life The time required for half the atoms in a sample of a radioactive isotope to decay.

Nuclear fission The splitting of the atomic nucleus into two smaller halves.

Nuclear fusion The combining of nuclei of light atoms to form heavier nuclei.

Radioactive Material containing nuclei that are unstable because of a less than optimal balance in the number of neutrons and protons.

Radioactivity The high-energy particles and electromagnetic radiation emitted by a radioactive substance.

Rem A unit for measuring the ability of radiation to harm living tissue.

Strong nuclear force The attractive force between all nucleons, effective only at very short distances.

Thermonuclear fusion Nuclear fusion brought about by high temperatures.

Transmutation The changing of an atomic nucleus of one element into an atomic nucleus of another element through a decrease or increase in the number of protons.

READING CHECK QUESTIONS (COMPREHENSION)

5.1 Radioactivity Results from Unstable Nuclei

1. Which type of radiation—alpha, beta, or gamma—results in the greatest change in mass number? The greatest change in atomic number?

2. Which of the three rays—alpha, beta, or gamma—has the greatest penetrating power?

5.2 Radioactivity Is a Natural Phenomenon

3. What is the origin of most of the natural radiation we encounter?

4. Which produces more radioactivity in the atmosphere, coal-fired power plants or nuclear power plants?

5. Is radioactivity on the Earth something relatively new? Defend your answer.

5.3 Radioactivity Results from an Imbalance of Forces

6. Why doesn't the repulsive electric force of protons in the atomic nucleus cause the protons to fly apart?

7. Which have more neutrons than protons, large nuclei or small nuclei?

8. What role do neutrons play in the atomic nucleus?

5.4 Radioactive Elements Transmute to Different Elements

9. In what form is most of the energy released by atomic transmutation?

10. What change in atomic number occurs when a nucleus emits an alpha particle? A beta particle?

11. What is the long-range fate of all the uranium that exists in the world today?

5.5 The Shorter the Half-Life, the Greater the Radioactivity

12. What is meant by the half-life of a radioactive sample?

13. What is the half-life of uranium-238?

5.6 Isotopic Dating Measures the Ages of Materials

14. What happens to a nitrogen atom in the atmosphere that captures a neutron?

15. Why is there more carbon-14 in living bones than in once-living ancient bones of the same mass?

16. Uranium-238 ultimately decays into what isotope of lead?

5.7 Nuclear Fission—The Splitting of Atomic Nuclei

17. What happens to the uranium-235 nucleus when it is stretched out?

18. Is a chain reaction more likely to occur in two separate pieces of uranium-235 or in the same pieces stuck together?

19. How is a nuclear reactor similar to a conventional fossil-fuel power plant? How is it different?

5.8 The Mass–Energy Relationship: $E = mc^2$

20. Who discovered that energy and mass are two different forms of the same thing?

21. In which atomic nucleus do nucleons have the least mass?

22. How does the mass per nucleon in uranium compare with the mass per nucleon in the fission fragments of uranium?

5.9 Nuclear Fusion—The Combining of Atomic Nuclei

23. How does the mass of a pair of atoms that have fused compare to the sum of their masses before fusion?

24. What kind of containers is used to contain plasmas at temperatures of millions of degrees?

25. What kind of nuclear power is responsible for sunshine?

CONFIRM THE CHEMISTRY (HANDS-ON APPLICATION)

26. Throw ten coins onto a flat surface. Remove all the coins that landed tails-up. Collect the remaining coins. After tossing them once again, remove all coins landing tails-up.

    Repeat this process until all the coins have been removed. Can you see how this relates to radioactive half-life? In units of "tosses," what is the average half-life of 25 coins? 50 coins? 1 million coins?

27. Repeat the preceding exercise, but use 10 dimes and 25 pennies. Let the dimes represent a radioactive isotope, such as carbon-14, and the pennies represent a nonradioactive isotope, such as carbon-12. Remove only the dimes when they land heads-up. Collect all the

pennies and add them to the dimes that were heads-up. Questions: Does the number of pennies affect the behavior of the dimes? Someone gives you two sets of coins. The first set contains 10 dimes and 25 pennies. The second set contains 2 dimes and 25 pennies. Which set of coins has gone through a greater number of tosses? Which set provides the most "radioactivity" after a single toss? Which set is analogous to a sample of once-living ancient material?

28. Calculate your estimated annual dose of radiation using the EPA's radiation dose calculator available at http://epa.gov/rpdweb00/understand/calculate.html.

THINK AND SOLVE (MATHEMATICAL APPLICATION)

29. Radiation from a point source follows an inverse-square law where the amount of radiation received is proportional to $1/d^2$, where d is distance. If a Geiger counter that is 1 meter away from a small source reads 100 counts per minute, what will be its reading 2 meters from the source? 3 meters from it?

30. Consider a radioactive sample with a half-life of one week. How much of the original sample will be left at the end of the second week? The third week? The fourth week?

31. A radioisotope is placed near a radiation detector, which registers 80 counts per second. Eight hours later, the detector registers 5 counts per second. What is the half-life of the radioactive isotope?

32. Uranium-238 absorbs a neutron and then emits a beta particle. Show that the resulting nucleus is neptunium-239.

THINK AND COMPARE (ANALYSIS)

33. Rank these three types of radiation by their ability to penetrate this page of your book, from highest to lowest:
    a. Alpha particle
    b. Beta particle
    c. Gamma ray

34. Consider the following atoms: C-12; C-14; N-14. From greatest to least, rank them by the number of
    a. protons in the nucleus.
    b. neutrons in the nucleus.
    c. nucleons in the nucleus.

35. Rank the following isotopes in order of their radioactivity starting with the most radioactive isotope to the least radioactive isotope.
    a. Nickel-59, half-life 75,000 years
    b. Uranium-238, half-life 4.5 billion years
    c. Actinium-225, half-life 10 days

36. Rank the following in order of the most energy released to the least energy released.
    a. Uranium-235 splitting into two equal fragments
    b. Uranium-235 splitting into three equal fragments
    c. Uranium-235 splitting into 92 equal fragments

THINK AND EXPLAIN (SYNTHESIS)

5.1 Radioactivity Results from Unstable Nuclei

37. Just after an alpha particle leaves the nucleus, would you expect it to speed up? Defend your answer.

38. When food is irradiated with gamma rays from a cobalt-60 source, does the food become radioactive? Defend your answer.

5.2 Radioactivity Is a Natural Phenomenon

39. People who work around radioactivity wear film badges to monitor the amount of radiation that reaches their bodies. Each badge consists of a small piece of photographic film enclosed in a lightproof wrapper. What kind of radiation do these devices monitor, and how can they determine the amount of radiation the people receive?

40. In Hawaii, on the island of Oahu, it takes about 25 years for the rain that lands on the mountain tops to seep downward and reach the water table. How is it that this information was obtained from the above-ground testing of nuclear bombs in the South Pacific?

5.3 Radioactivity Results from an Imbalance of Forces

41. A pair of protons in an atomic nucleus repels each other, but the protons are also attracted to each other. Explain.

42. In bombarding atomic nuclei with proton "bullets," why must the protons be given large amounts of kinetic energy in order to make contact with the target nuclei?

43. Without the strong nuclear force, would the periodic table be the same, less complicated, or more complicated?

5.4 Radioactive Elements Transmute to Different Elements

44. Why do different isotopes of the same element have the same chemical properties?

45. Why is lead found in all deposits of uranium ores?

46. What does the proportion of lead and uranium in rock tell us about the age of the rock?

47. What are the atomic number and atomic mass of the element formed when $_{84}^{218}Po$ emits a beta particle? What are they if the polonium emits an alpha particle?

48. Elements heavier than uranium in the periodic table do not exist in any appreciable amounts in nature because they have short half-lives. Yet there are several elements below uranium in the table that have equally short half-lives but do exist in appreciable amounts in nature. How can you account for this?

49. "Strontium-90 is a pure beta source." How could a chemist test this statement?

5.5 The Shorter the Half-Life, the Greater the Radioactivity

50. Radium-226 is a common isotope on Earth, but has a half-life of about 1620 years. Given that Earth is some 5 billions years old, why is there any radium at all?

51. The original reactor built in 1942 was just "barely" critical because the natural uranium that was used contained less than 1 percent of the fissionable isotope U-235 (half life 713 million years). What if, in 1942, the Earth had been 9 billion years old instead of 4.5 billion years old? Would this reactor have reached critical stage with natural uranium?

52. How is a smoke dectector similar to a Geiger counter?

5.6 Isotopic Dating Measures the Ages of Materials

53. Is carbon dating advisable for measuring the age of materials a few years old? How about a few thousand years old? A few million years old?

54. Why is carbon-14 dating not accurate for estimating the age of materials more than 50,000 years old?

55. The age of the Dead Sea Scrolls was determined by carbon-14 dating. Could this technique have worked if they had been carved on stone tablets? Explain.

56. If you make an account of 1000 people born in the year 2000 and find that half of them are still living in 2060, does this mean that one-quarter of them will be alive in 2120 and one-eighth of them alive in 2180? What is different about the death rates of people and the "death rates" of radioactive atoms?

5.7 Nuclear Fission—The Splitting of Atomic Nuclei

57. The uranium ores of the Athabasca Basin deposits of Saskatchewan, Canada, are unusually pure, containing up to 70 percent uranium oxides. Why doesn't this uranium ore undergo an explosive chain reaction?

58. Why will nuclear fission probably never be used directly for powering automobiles? How could it be used indirectly?

59. Why is carbon better than lead as a moderator in nuclear reactors?

60. Uranium-235 releases an average of 2.5 neutrons per fission, while plutonium-239 releases an average of 2.7 neutrons per fission. Which of these elements might you therefore expect to have the smaller critical mass?

5.8 The Mass–Energy Relationship: $E = mc^2$

61. How does the mass per nucleon in uranium compare with the mass per nucleon in the fission fragments of uranium?

62. Why does iron not yield energy if it undergoes fusion or fission?

63. Which process would release energy from gold, fission or fusion? From carbon? From iron?

64. If a uranium nucleus were to fission into three fragments of approximately equal size instead of two, would more energy or less energy be released? Defend your answer using Figure 5.28.

65. Is the mass of an atomic nucleus greater or less than the sum of the masses of the nucleons composing it? Why don't the nucleon masses add up to the total nuclear mass?

5.9 Nuclear Fusion—The Combining of Atomic Nuclei

66. Heavy nuclei can be made to fuse—for instance, by firing one gold nucleus at another one. Does such a process yield energy or cost energy? Explain.

67. Which produces more energy, the fission of a single uranium nucleus or the fusing of a pair of deuterium nuclei? The fission of a gram of uranium or the fusing of a gram of deuterium? (Why do your answers differ?)

68. If a fusion reaction produces no appreciable radioactive isotopes, why does a hydrogen bomb produce significant radioactive fallout?

69. Explain how radioactive decay has always warmed the Earth from the inside and how nuclear fusion has always warmed the Earth from the outside.

70. What percentage of nuclear power plants in operation today are based upon nuclear fusion?

71. Sustained nuclear fusion has yet to be achieved and remains a hope for abundant future energy. Yet the energy that has always sustained us has been the energy of nuclear fusion. Explain.

72. Oxygen and two hydrogen atoms combine to form a water molecule. At the nuclear level, if one oxygen atom and two hydrogen atoms were fused, what element would be produced?

73. If a pair of carbon nuclei were fused, and the product emitted a beta particle, what element would be produced?

74. Ordinary hydrogen is sometimes called a perfect fuel because of its almost unlimited supply on Earth, and when it burns, harmless water is the product of the combustion. So why don't we abandon fission energy and fusion energy, not to mention fossil-fuel energy, and just use hydrogen?

THINK AND DISCUSS (EVALUATION)

75. The 1986 accident at Chernobyl, in which dozens of people died and thousands more were exposed to cancer-causing radiation, created fear and outrage worldwide and led some people to call for the closing of all nuclear plants. Yet many people choose to smoke cigarettes in spite of the fact that 2 million people die every year from smoking-related diseases. The risks posed by nuclear power plants are involuntary risks we must all share like it or not, whereas the risks associated with smoking are voluntary because a person chooses to smoke. Why are we so unaccepting of involuntary risk but accepting of voluntary risk?

76. Your friend Paul says that the helium used to inflate balloons is a product of radioactive decay. Your mutual friend Steve says no way. Then there's your friend Alison who is fretful about living near a fission power plant. She wishes to get away from radiation by traveling to the high

mountains and sleeping out at night on granite outcroppings. Still another friend, Michele, has journeyed to the mountain foothills to escape the effects of radioactivity altogether. While bathing in the warmth of a natural hot spring she wonders aloud how the spring gets its heat. What do you tell these friends?

77. Speculate about some worldwide changes likely to follow the advent of successful fusion reactors.

78. What should be done with the nuclear wastes from nuclear power plants? Some have half-lives of many thousands of years. Discuss possible solutions.

79. Can radioactive wastes be heated to shorten the half-life?

80. As described at Hanford.gov, the world's first nuclear reactors dedicated to the full-scale production of plutonium were built at a facility in Hanford, WA. During its 40 years of operation, this facility produced about how many tons of plutonium? This allowed for the creation of about how many nuclear warheads? About how many gallons of high-level radioactive waste are presently stored at the Hanford site? What are the current plans for taking care of this radioactive waste? Were the benefits of the Hanford nuclear reactors worth the risks?

READINESS ASSURANCE TEST(RAT)

If you have a good handle on this chapter, then you should be able to score at least 7 out of 10 on this RAT. Check your answers online at www.ConceptualChemistry.com. If you score less than 7, you need to study further before moving on.

Choose the BEST answer to the following.

1. Which type of radiation from cosmic sources predominates on the inside of a high-flying commercial airplane: alpha, beta, or gamma?

    a. Alpha radiation

    b. Beta radiation

    c. Gamma radiation

    d. None of these predominates as all three are abundant.

2. Is it at all possible for a hydrogen nucleus to emit an alpha particle?

    a. Yes, because alpha particles are the simplest form of radiation.

    b. No, because it would require the nuclear fission of hydrogen, which is impossible.

    c. Yes, but it does not occur very frequently.

    d. No, because it does not contain enough nucleons.

3. A sample of radioactive material is usually a little warmer than its surroundings because

    a. it efficiently absorbs and releases energy from sunlight.

    b. its atoms are continually being struck by alpha and beta particles.

    c. it is radioactive.

    d. it emits alpha and beta particles.

4. What evidence supports the contention that the strong nuclear force is stronger than the electrical interaction at short internuclear distances?

    a. Protons are able to exist side-by-side within an atomic nucleus.

    b. Neutrons spontaneously decay into protons and electrons.

    c. Uranium deposits are always slightly warmer than their immediate surroundings.

    d. The radio interference that arises adjacent to any radioactive source.

5. When the isotope bismuth-213 emits an alpha particle, what new element results?

    a. Lead

    b. Platinum

    c. Polonium

    d. Thallium

6. A certain radioactive element has a half-life of 1 hour. If you start with a 1-g sample of the element at noon, how much of this same element will be left at 3:00 PM?

    a. 0.5 grams

    b. 0.25 grams

    c. 0.125 grams

    d. 0.0625 grams

7. The isotope Cesium-137, which has a half-life of 30 years, is a product of nuclear power plants. How long will it take for this isotope to decay to about one-half its original amount?

    a. 0 years

    b. 15 years

    c. 30 years

    d. 60 years

    e. 90 years

8. If uranium were to split into 92 pieces of equal size instead of 2, would more energy or less energy be released?

    a. Less energy would be released because of less mass per nucleon.

    b. Less energy would be released because of more mass per nucleon.

    c. More energy would be released because of less mass per nucleon.

    d. More energy would be released because of more mass per nucleon.

9. Which process would release energy from gold, fission or fusion? From carbon?

    a. Gold: fission; carbon: fusion

    b. Gold: fusion; carbon: fission

    c. Gold: fission; carbon: fission

    d. Gold: fusion; carbon: fusion

10. If an iron nucleus split in two, its fission fragments would have

    a. more mass per nucleon.

    b. less mass per nucleon.

    c. the same mass per nucleon.

    d. either more or less mass per nucleon.

Fracking for Shale Gas

We depend greatly on fossil fuels to meet our energy needs. Three major forms of fossil fuels are coal, petroleum, and natural gas. Both coal and petroleum are made of large and complicated carbon-based molecules. The main component of natural gas, by contrast, is methane, CH_4, which is a structurally simple molecule. Traditionally, coal and petroleum have been much more accessible in large quantities than natural gas. A prime reason for this is because much of the world's natural gas remains trapped miles beneath the surface within a type of rock known as shale.

Two recent technological advances are now allowing access to this otherwise difficult to reach natural gas, also called *shale gas*. The first is our ability to drill very deep and then sideways. This is important because shale deposits are laid down horizontally. As shown in the accompanying image, drilling sideways maximizes the surface area of the shale within reach of the bore hole. The second advance is the process of *hydraulic fracturing*, also known as *fracking*, in which channels are punched into the shale using a powerful explosive. A slurry of water and sand with a small amount of other chemicals, such as anticorrosion agents and lubricants, is then injected into these channels. This slurry is known as *fracking fluid*. Under high pressure, this fluid expands into natural cracks, which remain open as grains of sand become lodged within them. After the fracking fluid is removed, large volumes of natural gas, which is lighter than air, escape through the cracks and rise to the surface, where the gas is piped to a storage facility for future use.

▲ Shale deposits are thousands of feet deep and usually only about 100 feet thick, so the most efficient access is provided by horizontal drilling. Natural cracks in the shale contain large amounts of natural gas. These cracks are forced open by high-pressure fracking fluid.

Millions of gallons of water are used to frack a single well—typically 5 to 15 million gallons, and as much as 400 million gallons for the largest wells. The used fracking fluid is toxic and requires special methods of disposal. Regulations for this disposal, however, vary from state to state. Some states permit local municipal water treatment facilities to accept fracking fluid. In states with more stringent regulations, additional wells are drilled deep below the local water table. The used fracking fluid is pumped to the bottom of these wells where it is pushed into the ground.

There is currently a "gold rush" for shale gas, with thousands of new wells being drilled every year (see accompanying figure). Notably, about 30 percent of the U.S. natural gas supply now comes from fracking operations. This is up from about zero percent over the past decade and is soon expected to reach 50 percent. Hundreds of thousands of jobs have since been created either directly or indirectly from this new industry. Furthermore, land owners are profiting from fees and royalties paid to them by companies who establish wells on their property. With

Shale gas, lower 48 states

Montana Thrust Belt
Heath
Cody
Gammon
Bakken
Devonian (Ohio)
Appalachian Basin
Utica
Hilliard-Baxter-Mancos
Mowry
Michigan Basin
Antrim
Denver Basin
Forest City Basin
Monterey Temblor
Lewis
Woodford
Fayetteville
New Albany
Marcellus
Barnett
Conasauga
Barnett-Woodford
Tuscaloosa
Haynesville-Bossier

Basins

Shale plays
◼ Current plays
◻ Prospective plays

Stacked plays
— Shallowest/youngest
— Intermediate depth/age
— Deepest/oldest

▲ Fracking technology was first developed in the Barnett shale on the property of the Dallas-Fort Worth International Airport. The richest gas-bearing shale in the United States is the Marcellus formation, centered over western Pennsylvania.

fracking technology, the United States in 2009 surpassed Russia to become the world's leading producer of natural gas.

In addition to economic benefits, there are also potential environmental benefits to increased production of natural gas. As described in Chapter 17, most of the electricity produced in the United States comes from the burning of coal, which generates large amounts of pollutants, such as particulates, mercury, sulfur dioxides, and carbon dioxide. Natural gas, however, burns with much greater efficiency while generating far fewer pollutants —the output of carbon dioxide, for example, is about half that of coal. Although shale gas is not the ideal environmentally friendly fuel, its development is seen by many in the industry as a responsible way to

supply our energy needs for decades while other more sustainable energy technologies, such as solar energy, are developed.

Fracking technology, however, also involves significant environmental risks. Foremost is the issue of dealing with the large volumes of used fracking fluid. This fluid is usually stored in pools adjacent to the well before being shipped by truck to a disposal site. At each of these stages there is the potential for an accidental spill of the fluid into the environment. Furthermore, if the upper portions of the well are not properly installed, the used fracking fluid coming up the well could potentially seep into the water table, ruining local water supplies. The toxins that could be released include not only cancer-causing agents within the fluid formula but

also toxic materials coming directly from the shale, such as heavy metals and hydrocarbons, as well as unhealthy concentrations of radioactive radium-226.

An added risk is gas leakage. Consider that as natural gas moves from the well to where it is used, about 2 to 6 percent of this gas leaks directly into the atmosphere. Methane itself is about 70 times more potent a greenhouse gas than is carbon dioxide. Because methane reacts with oxygen, however, it resides in the atmosphere for only about 10 years before it is transformed into carbon dioxide. By some estimates, when well leakage is taken into account, the burning of shale gas for generating electricity actually contributes more to the global greenhouse effect than does the burning of coal.

In addition, there have been highly publicized instances in which people who live near a fracking well find methane in their water supply—so much that their water burns when lit with a match. The water supply in many areas naturally contains methane, which, though nonpoisonous, is highly combustible, so these instances may not be connected to fracking operations. In 2011, however, scientists at Duke University published a peer-reviewed study showing a strong correlation between flammable drinking water and distance from a fracking well. Although fracking occurs miles underground, this study suggests that the released methane may still reach the surface water table either by migrating up less than perfect well encasings or by migrating through geologic faults.

For people who live in areas of active drilling there are other concerns. A scenic rural area, for example, may be transformed into a zone of heavy industrial activity featuring tall drilling towers. Also, the truck traffic around an active drilling area is intense, causing damage to roads and creating much noise pollution that can at times run 24/7. Fracking for shale gas may benefit the nation as a whole, but that might not matter much to you if the fracking is happening right in your neighbor's backyard.

There are strong social, economic, and political pressures for the continued exploitation of shale gas. There is also a general agreement that "best practices" need to be developed and enforced so as to safeguard against risks. The specifics of how to regulate this nascent industry, however, have yet to be determined.

CONCEPT CHECK

What is the purpose of the sand within a fracturing fluid?

CHECK YOUR ANSWER The high pressure water opens up natural cracks within the shale. The sand gets stuck within these cracks, preventing them from reclosing. This allows gas seeping out of the shale to make its way to the bore hole.

Think and Discuss

1. Assuming a spacing of about 1 well for every 80 acres, it is estimated that the Marcellus shale of the northeastern Untied States could produce about 500 trillion cubic feet (ft^3) of natural gas over the next 50 years before being depleted. Currently, the United States consumes about 23 trillion ft^3 of natural gas each year. Assuming these rates are accurate and don't change, for how many years could the Marcellus shale meet the demand for natural gas in the United States?

2. The Safe Drinking Water Act of 1974 authorizes the EPA to regulate injection wells in order to protect underground sources of drinking water. In 2005, however, the U.S. Congress passed an amendment to this act that specifically exempts fracking operations from such regulations. Why do you suppose this amendment was approved?

Why have attempts to repeal this amendment failed?

3. The 2005 amendment exempting fracking from the Safe Water Drinking Act had the effect of shifting fracking regulations from the federal government to individual states. Why might states be more effective at regulating fracking within their own borders? Why might the federal government actually be more effective?

4. Why should corporations of the fracking industry be encouraged to get together to spell out their own regulations? Should these corporations be trusted to enforce these self-imposed regulations?

5. The northeastern United States has large methane-bearing shale deposits buried deep underground. What other important resource is also available in the northeast that permits the extraction of methane from this shale?

6

How Atoms Bond

THE MAIN IDEA

Atoms bond by exchanging or sharing electrons.

▲ These large cubic salt crystals grew many years ago from an evaporating sea in what is now Colorado.

The properties of a material can be traced to how the atoms within that material are chemically bonded to each other. Salt crystals, as shown in the opening photograph, are cubic because that's the way sodium and chlorine join together to form sodium chloride. Smash these crystals with a hammer and you'll merely get smaller cubes! Why are metals opaque to light, and why do they conduct both electricity and heat so well? Again, the answer has to do with how their atoms are bonded.

You know carbon dioxide to be a gas at room temperature and water to be a liquid. Would it surprise you to learn that, molecule for molecule, carbon dioxide is over twice as heavy as water? As we explore in this chapter, carbon dioxide molecules are nonsticky, which allows them to float away from each other into a gaseous phase. Conversely, the relatively light water molecules are sticky, which keeps them together in a liquid phase. The reason has to do with, you guessed it, how the atoms within these molecules are bonded. So if we are to understand why materials behave as they do, it's important that we have some understanding of how it is that their atoms bond.

Chemistry

Gumdrop Molecules

Create molecular models using tooth-picks and colored gumdrops. We recommend black or purple for carbon, white or yellow for hydrogen, green for chlorine, and red for oxygen.

PROCEDURE

1. To build plausible molecular structures, you need to follow these two rules:
   a. Each atom has a specific number of bonds it is able to form, as follows: carbon (4), hydrogen (1), chlorine (1), oxygen (2).
   b. When atoms are placed together within a single molecule, the atoms need to be as far apart from each other as possible while still also connected.

2. Using the preceding information, build plausible structures for the following compounds: methane, CH_4; dichloromethane, CH_2Cl_2; ethane, C_2H_6; hydrogen peroxide, H_2O_2; acetylene, C_2H_2.

ANALYZE AND CONCLUDE

1. For methane, is it possible to have all five atoms connected lying flat on the table? What is the angle between your hydrogen-carbon-hydrogen bonds? Is it possible to make all these angles greater than 90°? For hydrogen peroxide, would it be preferable to have the two hydrogens on the same side or opposite sides of the oxygen atoms?
2. Using the above rules, how many structures are possible for C_2H_6O?
3. Why do the atoms of molecules tend to be as far apart from each other as possible? Can all molecular structures be deduced from only the chemical formula for that molecule?
4. For pictures of built gumdrop molecules and for more challenges, see the second Think and Do question at the end of this chapter.

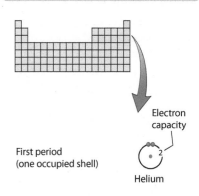

6.1 Electron-Dot Structures

EXPLAIN THIS

Why do the atoms of group 18 resist forming chemical bonds?

An atomic model is needed to help us understand how atoms bond. We begin this chapter with a brief overview of the shell model presented in Section 4.8. You may recall how electrons are arranged around an atomic nucleus. Rather than moving in neat orbits like planets around the Sun, electrons are wavelike entities that hover in various volumes of space called *shells*.

As was shown in Figure 4.31, there are seven shells available to the electrons in an atom, and the electrons fill these shells in order from innermost to outermost. Furthermore, the maximum number of electrons allowed in the first shell is 2, and for the second and third shells it is 8. The fourth and fifth shells can each hold 18 electrons, and the sixth and seventh shells can each hold 32 electrons.* These numbers match the number of elements in each period (horizontal row) of the periodic table. **Figure 6.1** shows how this model applies to the first three elements of group 18.

▶ Figure 6.1
Occupied shells in the group 18 elements helium through argon. Each of these elements has a filled outermost shell. Note that the number of electrons in each shell (2, 8, 8, and so on) corresponds to the number of elements in the periods of the periodic table. For more detail, please review Sections 4.8 and 4.9.

LEARNING OBJECTIVE

Identify paired and unpaired electrons within an electron-dot structure.

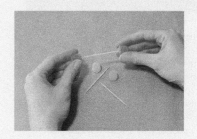

Electron capacity

First period
(one occupied shell)

Helium

Second period
(two occupied shells)

Neon

Third period
(three occupied shells)

Argon

---

*As a point of reference for physics students reading this text, these are shells of orbitals grouped by similar energy levels rather than by principal quantum number. They are the "argonian" shells developed by Linus Pauling in the 1930s to explain chemical bonding and the organization of the periodic table. This is an old atomic model, but it works well for a simple description of chemical bonding.

▲ Figure 6.2
Gilbert Newton Lewis (1875–1946) revolutionized chemistry with his theory of chemical bonding, which he published in 1916. He worked most of his life in the chemistry department of the University of California, Berkeley, where he was not only a productive researcher but also an exceptional teacher. Among his teaching innovations was the idea of providing students with problem sets as a follow-up to lectures and readings.

Electrons in the outermost occupied shell of any atom play a significant role in that atom's chemical properties, including its ability to form chemical bonds. To indicate their importance, these outermost electrons are called **valence electrons** (from the Latin *valentia*, "strength"), and the shell they occupy is called the **valence shell.** Valence electrons can be conveniently represented as a series of dots surrounding an atomic symbol. This notation is called an **electron-dot structure** or, sometimes, a *Lewis dot symbol* in honor of the American chemist Gilbert N. Lewis, who first proposed the concepts of shells and valence electrons (**Figure 6.2**).

The electron-dot structures shown in **Figure 6.3** help us to understand ionic and covalent bonds. Electron-dot structures, however, are not so useful in describing metallic bonds. The reason is because within a metallic bond, the bonding electrons readily flow from one atom to the next, as is discussed further in Section 6.4. This is why the metallic groups 3–12 are not included in Figure 6.2.

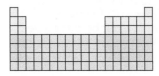

◀ Figure 6.3
The valence electrons of an atom are shown in its electron-dot structure. Note that the first three periods here parallel Figure 4.32. Also note that for larger atoms, not all the electrons in the valence shell are valence electrons. Krypton, Kr, for example, has 18 electrons in its valence shell, but only 8 of these are classified as valence electrons. You can learn the reason for this detail, which involves *suborbitals,* in a follow-up course on advanced chemistry.

READINGCHECK

Electron-dot structures are needed to help us understand what kinds of chemical bonds?

When you look at the electron-dot structure of an atom, you immediately know two important things about that element. You know how many valence electrons it has and how many of these electrons are *paired.* Chlorine, for example, has three sets of paired electrons and one unpaired electron, and carbon has four unpaired electrons:

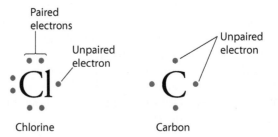

Paired valence electrons are relatively stable. In other words, they usually do not form chemical bonds with other atoms. For this reason, electron pairs in an electron-dot structure that are not participating in a chemical bond are called **nonbonding pairs.** (Do not take this term literally, however, because in Chapter 10, you'll see that, under the right conditions, even "nonbonding" pairs can form a chemical bond.)

Valence electrons that are *unpaired,* by contrast, have a strong tendency to participate in chemical bonding. By doing so, they become paired with an electron from another atom. The ionic and covalent bonds discussed in this chapter all result from either a transfer or a sharing of unpaired valence electrons.

So why do electrons like to pair up? As was discussed in Chapter 4, electrons have a property called *spin*. Spin can have two states, and these states are analogous to a ball's spinning either clockwise or counterclockwise, as was shown in Figure 4.27. Because an electron has an electric charge, the spinning generates a tiny magnetic field. Two electrons spinning in opposite directions have oppositely aligned magnetic fields, which allows them to come together as a pair.

> ### CONCEPT CHECK
> Where are valence electrons located, and why are they important?
>
> **CHECK YOUR ANSWER** Valence electrons are located in the outermost occupied shell of an atom. They are important because they play a leading role in determining the chemical properties of the atom.

FOR YOUR INFORMATION

"The rapid progress true Science now makes occasions my regretting sometimes that I was born so soon. It is impossible to imagine the heights to which may be carried, in a thousand years, the power of man over matter. O that moral Science were in as fair a way of improvement, that men would cease to be wolves to one another, and that human beings would at length learn what they now improperly call humanity."

—Benjamin Franklin, in a letter to chemist Joseph Priestley, 8 February 1780.

6.2 Atoms Can Lose or Gain Electrons to Become Ions

EXPLAIN THIS

When does a gain result in a negative?

When the number of protons and electrons in an atom are equal, the charges balance and the atom is electrically neutral. If electrons are lost or gained, as illustrated in **Figures 6.4** and **6.5**, the balance is lost and the atom takes on a net electric charge. Any atom having a net electric charge is an **ion**. When electrons are lost, protons outnumber electrons and the ion has a positive net charge. A positively charged ion is sometimes referred to as a *cation*. When electrons are gained, electrons outnumber protons and the ion has a negative net charge. A negatively charged ion is sometimes referred to as an *anion*.

Chemists use a superscript to the right of the atomic symbol to indicate the strength and sign of an ion's charge. Thus, as shown in Figures 6.4 and 6.5, the positive ion formed from the sodium atom is written Na^{1+} and the negative ion formed from the fluorine atom is written F^{1-}. Usually the numeral 1 is omitted when indicating either a 1+ or 1− charge. Hence, these two ions are most frequently written Na^+ and F^-. Two more examples: a calcium atom that loses two electrons is written Ca^{2+}, and an oxygen atom that gains two electrons is written O^{2-}.

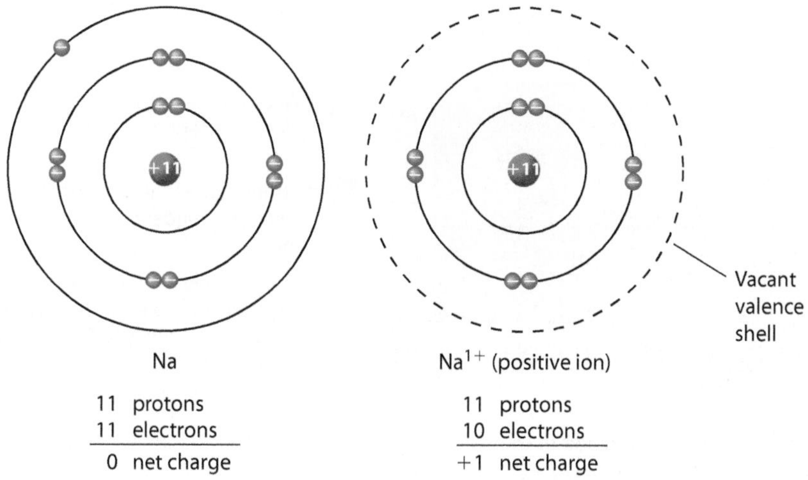

LEARNING OBJECTIVE

> Use the periodic table to predict the type of ion an atom tends to form.

◀ **Figure 6.4**
An electrically neutral sodium atom contains 11 negatively charged electrons surrounding the 11 positively charged protons of the nucleus. When this atom loses an electron, the result is a positive ion.

Na
11 protons
11 electrons
0 net charge

Na^{1+} (positive ion)
11 protons
10 electrons
+1 net charge

Vacant valence shell

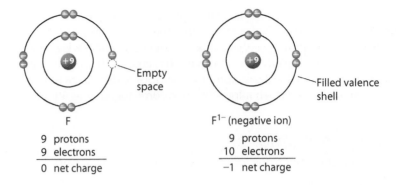

► Figure 6.5
An electrically neutral fluorine atom contains nine protons and nine electrons. When this atom gains an electron, the result is a negative ion.

F

9 protons
9 electrons
0 net charge

F^{1-} (negative ion)

9 protons
10 electrons
−1 net charge

We can use the shell model to deduce the type of ion an atom tends to form. According to this model, *atoms tend to lose or gain electrons to create a filled outermost shell.* Let's take a moment to consider this point, looking to Figures 6.4 and 6.5 as visual guides.

If an atom has only one electron or only a few electrons in its outermost occupied shell, it tends to give up (lose) these electrons so that the next shell inward, which is already filled, becomes the outermost occupied shell. The sodium atom of Figure 6.4, for example, has one electron in its third shell. In forming an ion, the sodium atom loses this electron, thereby making the second shell, which is already filled to capacity, the outermost occupied shell. Because the sodium atom has only one valence electron to lose, it tends to form only the 1+ ion. It is a *positive* ion, because the number of protons (11+) now exceeds the number of electrons (10−).

If the outermost shell of an atom is almost filled, that atom attracts electrons from another atom and so forms a negative ion. The fluorine atom of Figure 6.5, for example, has one space available for an additional electron. After this additional electron is gained, the fluorine atom achieves a filled shell. Fluorine therefore tends to form the 1− ion. Remember, electrons are negatively charged. So gaining an electron results in a *negative* ion.

The periodic table tells us the type of ion each atom tends to form. As **Figure 6.6** shows, each atom of any group 1 element, for example, has only one valence electron and so tends to form the 1+ ion. Each atom of any group 17 element has room for one additional electron in its valence shell and therefore tends to form the 1− ion. Atoms of the noble gas elements tend not to form ions of any type because their valence shells are already filled to capacity.

READINGCHECK

Why do you NOT need to memorize the type of ion each atom tends to form?

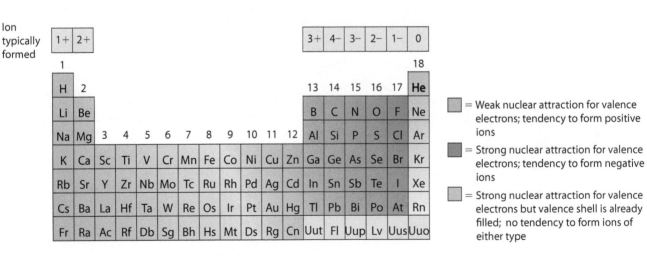

Ion typically formed

| 1+ | 2+ | | | | | | | | | | | 3+ | 4− | 3− | 2− | 1− | 0 |

▲ Figure 6.6
The periodic table is your guide to the types of ions that atoms tend to form within an ionic compound.

□ = Weak nuclear attraction for valence electrons; tendency to form positive ions

■ = Strong nuclear attraction for valence electrons; tendency to form negative ions

□ = Strong nuclear attraction for valence electrons but valence shell is already filled; no tendency to form ions of either type

What type of ion does the magnesium atom, Mg, tend to form?

CHECK YOUR ANSWER The magnesium atom (atomic number 12) is found in group 2 and has two valence electrons to lose (see Figure 6.2). Therefore, it tends to form the 2+ ion.

As is indicated in Figure 6.6, the attraction between an atom's nucleus and its valence electrons is weakest for elements on the left in the periodic table and strongest for elements on the right. From sodium's position in the table, we can see that a sodium atom's single valence electron is not held very strongly, which explains why it is so easily lost. The attraction the sodium nucleus has for its second-shell electrons, however, is much stronger, which is why the sodium atom rarely loses more than one electron.

On the other side of the periodic table, the nucleus of a fluorine atom strongly holds onto its valence electrons, which explains why the fluorine atom tends not to lose any electrons to form a positive ion. Instead, fluorine's nuclear pull on the valence electrons is strong enough to accommodate even an additional electron taken from some other atom.

The nucleus of a noble gas atom pulls so strongly on its valence electrons that they are very difficult to remove, so noble gas atoms generally do not lose electrons. However, because their outermost shells are already filled to capacity, they generally don't gain electrons either. Thus, these atoms tend not to form ions of any sort.

Why does the magnesium atom tend to form a 2+ ion?

CHECK YOUR ANSWER Magnesium is on the left in the periodic table, so atoms of this element do not hold onto their two valence electrons very strongly. Because these electrons are not held very tightly, they are easily lost, which is why the magnesium atom tends to form the 2+ ion.

Using our shell model to explain the formation of ions works well for groups 1 and 2 and groups 13 through 18. This model is too simplified to work well for the transition metals of groups 3 through 12, however, or for the inner transition metals. In general, these metal atoms tend to form positive ions, but the number of electrons lost varies. For example, depending on conditions, an iron atom may lose two electrons to form the Fe^{2+} ion or lose three electrons to form the Fe^{3+} ion.

 FOR YOUR INFORMATION

What do the ions of the following elements have in common: calcium, Ca; chromium, Cr; cobalt, Co; copper, Cu; iodine, I; iron, Fe; magnesium, Mg; manganese, Mn; molybdenum, Mo; nickel, Ni; phosphorus, P; potassium, K; selenium, Se; sodium, Na; sulfur, S; and zinc, Zn? They are all dietary minerals that are essential for good health but that can be harmful, even lethal, when consumed in excessive amounts.

Molecules Can Form Ions

We have seen that atoms form ions by losing or gaining electrons. Interestingly, molecules can also become ions. In most cases, this occurs whenever a molecule loses or gains a proton—equivalent to the hydrogen ion, H^+. (Recall that a hydrogen atom is a proton together with an electron. The hydrogen ion, H^+, therefore, is simply a proton.) For example, a water molecule, H_2O, can gain a hydrogen ion, H^+ (a proton), to form the hydronium ion, H_3O^+:

$$
\underset{\text{Water}}{H\!-\!\overset{O}{\underset{H}{}}} \quad + \quad \underset{\substack{\text{Hydrogen ion} \\ \text{(proton)}}}{H^+} \quad \longrightarrow \quad \underset{\text{Hydronium ion}}{H\!-\!\overset{H}{\underset{H}{O^+}}}
$$

TABLE 6.1   Common Polyatomic Ions

NAME	FORMULA
Hydronium ion	H_3O^+
Ammonium ion	NH_4^+
Bicarbonate ion	HCO_3^-
Acetate ion	$CH_3CO_2^-$
Nitrate ion	NO_3^-
Cyanide ion	CN^-
Hydroxide ion	OH^-
Carbonate ion	CO_3^{2-}
Sulfate ion	SO_4^{2-}
Phosphate ion	PO_4^{3-}

Similarly, the carbonic acid molecule, H_2CO_3, can lose two protons to form the carbonate ion, CO_3^{2-}:

Carbonic acid   Carbonate ion   Hydrogen ions (protons)

How these reactions occur will be explored in later chapters. For now, you should understand that the hydronium and carbonate ions are examples of **polyatomic ions,** which are molecules that carry a net electric charge. Table 6.1 lists some commonly encountered polyatomic ions.

6.3 Ionic Bonds Result from a Transfer of Electrons

LEARNING OBJECTIVE

> Describe how ions combine to form ionic compounds.

READINGCHECK

What type of force gives rise to an ionic bond?

EXPLAIN THIS

Why do ionic compounds have very high melting points?

When an atom that tends to lose electrons is placed in contact with an atom that tends to gain them, the result is an electron transfer and the formation of two oppositely charged ions. This occurs when sodium and chlorine are combined. As shown in **Figure 6.7**, the sodium atom loses one of its electrons to the chlorine atom, resulting in the formation of a positive sodium ion and a negative chloride ion. The two oppositely charged ions are attracted to each other by the electric force, which holds them close together. This electric force of attraction between two oppositely charged ions is called an **ionic bond.**

A sodium ion and a chloride ion together make the chemical compound sodium chloride, commonly known as table salt. This and all other chemical compounds containing ions are referred to as **ionic compounds.** All ionic compounds are completely different from the elements from which they are made. As discussed in Section 3.4, sodium chloride is not sodium, nor is it chlorine. Rather, it is a collection of sodium and chloride ions that form a unique material having its own physical and chemical properties.

CONCEPTCHECK

Is the transfer of an electron from a sodium atom to a chlorine atom a physical change or a chemical change?

CHECK YOUR ANSWER Recall from Chapter 3 that only a chemical change involves the formation of new material. Thus, this or any other electron transfer, because it results in the formation of a new substance, is a chemical change.

▶ **Figure 6.7**
(a) An electrically neutral sodium atom loses its valence electron to an electrically neutral chlorine atom. (b) This electron transfer results in two oppositely charged ions. (c) The ions are then held together by an ionic bond. The spheres drawn around these and subsequent illustrations of electron-dot structures indicate the relative sizes of the atoms and ions.

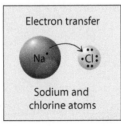

Sodium and chlorine atoms

(a)

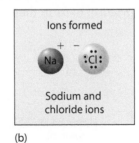

Sodium and chloride ions

(b)

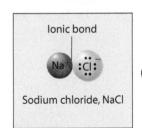

Sodium chloride, NaCl

(c)

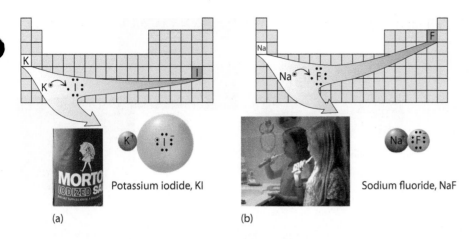

◀ **Figure 6.8**
(a) The ionic compound potassium iodide, KI, is added in minute quantities to commercial salt because the iodide ion, I⁻, that it contains is an essential dietary mineral. (b) The ionic compound sodium fluoride, NaF, is often added to municipal water supplies and toothpastes because it is a good source of the tooth-strengthening fluoride ion, F⁻.

Potassium iodide, KI

Sodium fluoride, NaF

(a)   (b)

As **Figure 6.8** shows, ionic compounds typically consist of elements that are found on opposite sides of the periodic table. Also, because of how the metals and nonmetals are organized in the periodic table, positive ions are generally derived from metallic elements and negative ions are generally derived from nonmetallic elements.

For all ionic compounds, positive and negative charges must balance. In sodium chloride, for example, there is one sodium 1+ ion for every chloride 1− ion. Charges must also balance in compounds containing ions that carry multiple charges. The calcium ion, for example, carries a charge of 2+, but the fluoride ion carries a charge of only 1−. Because two fluoride ions are needed to balance each calcium ion, the formula for calcium fluoride is CaF_2, as **Figure 6.9** illustrates. Calcium fluoride occurs naturally in the drinking water of some communities, where it is a good source of the tooth-strengthening fluoride ion, F⁻.

An aluminum ion carries a 3+ charge, and an oxide ion carries a 2− charge. Together, these ions make the ionic compound aluminum oxide, Al_2O_3, the main component of such gemstones as rubies and sapphires. **Figure 6.10** illustrates the formation of aluminum oxide. The three oxide ions in Al_2O_3 carry a total charge of 6−, which balances the total 6+ charge of the two aluminum ions. As mentioned, rubies and sapphires differ in color because of the impurities they contain. Rubies are red because of minor amounts of chromium ions, and sapphires are blue because of minor amounts of iron and titanium ions.

CONCEPT CHECK

What is the chemical formula for the ionic compound magnesium oxide?

CHECK YOUR ANSWER Because magnesium is a group 2 element, you know a magnesium atom must lose two electrons to form a Mg^{2+} ion. Because oxygen is a group 16 element, an oxygen atom gains two electrons to form an O^{2-} ion. These charges balance in a one-to-one ratio, so the formula for magnesium oxide is MgO.

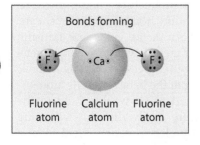

Bonds forming

Ionic bonds formed

Fluorine atom   Calcium atom   Fluorine atom

Calcium fluoride, CaF_2

Fluorite

◀ **Figure 6.9**
A calcium atom loses two electrons to a pair of fluorine atoms. In the process, the calcium atom becomes a calcium ion, Ca^{2+}, and the fluorine atoms become fluoride ions, F⁻. The oppositely charged ions join to form the ionic compound calcium fluoride, CaF_2, which occurs naturally as the mineral fluorite.

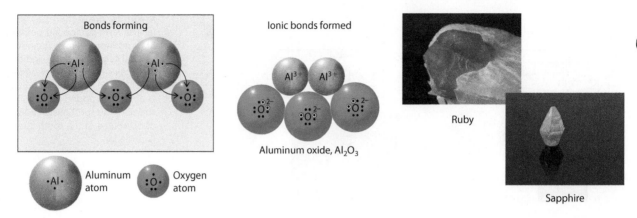

▲ **Figure 6.10**
Two aluminum atoms lose a total of six electrons to three oxygen atoms. In the process, the aluminum atoms become aluminum ions, Al^{3+}, and the oxygen atoms become oxide ions, O^{2-}. The oppositely charged ions join to form the ionic compound aluminum oxide, Al_2O_3. Certain gemstones are crystalline aluminum oxide with trace amounts of impurities, such as chromium, which makes ruby, and titanium, which makes sapphire.

▶ **Figure 6.11**
(a) Sodium chloride, as well as other ionic compounds, forms ionic crystals in which every internal ion is surrounded by ions of the opposite charge. (For simplicity, only a small portion of the ion array is shown here. A typical NaCl crystal involves millions and millions of ions.) (b) A view of crystals of table salt through a microscope shows their cubic structure. The cubic shape is a consequence of the cubic arrangement of sodium and chloride ions.

An ionic compound typically contains a multitude of ions grouped together in a highly ordered 3-dimensional array. In sodium chloride, for example, each sodium ion is surrounded by six chloride ions and each chloride ion is surrounded by six sodium ions (**Figure 6.11**). Overall, there is one sodium ion for each chloride ion, but there are no identifiable sodium–chloride pairs. Such an orderly array of ions is known as an *ionic crystal*. As mentioned at the onset of this chapter, on the atomic level, the crystalline structure of sodium chloride is cubic, which is why macroscopic crystals of table salt are also cubic. Smash a large cubic sodium chloride crystal with a hammer, and what do you get? Smaller cubic sodium chloride crystals! Similarly, the crystalline structures of other ionic compounds, such as calcium fluoride and aluminum oxide, are a consequence of how the ions pack together.

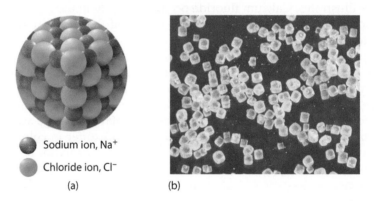

Sodium ion, Na^+

Chloride ion, Cl^-

(a) (b)

6.4 The Electrons of Metallic Bonds Are Loosely Held

LEARNING OBJECTIVE

Relate the properties of a metal to how the atoms of that metal are chemically bonded.

EXPLAIN THIS

Why aren't alloys described as metallic compounds?

In Section 3.3, you learned about the properties of metals. They conduct electricity and heat, are opaque to light, and deform—rather than fracture—under pressure. Because of these properties, metals are used to build homes, appliances, cars, bridges, airplanes, and skyscrapers. Metal wires across the landscape transmit communication signals and electric power. We wear metal jewelry, exchange metal currency, and drink from metal cans. Yet what is it that gives a metal its metallic properties? We can answer this question by looking at the behavior of its atoms.

The outer electrons of most metal atoms tend to be weakly held to the atomic nucleus. Consequently, these electrons are easily dislodged, leaving behind positively charged metal ions. The many electrons dislodged from a

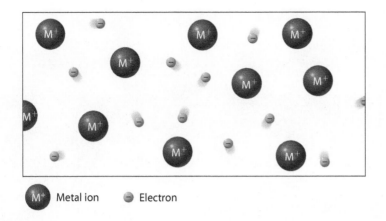

M⁺ Metal ion   ⊖ Electron

◀ **Figure 6.12**
Metal ions are held together by freely flowing electrons. These loose electrons form a kind of "electronic fluid," which flows through the lattice of positively charged ions.

large group of metal atoms flow freely through the resulting metal ions, as depicted in **Figure 6.12**. This "fluid" of electrons holds the positively charged metal ions together in the type of chemical bond known as a **metallic bond.**

The mobility of electrons in a metal accounts for the metal's significant ability to conduct electricity. Also, metals are opaque and shiny because the free electrons easily vibrate to the oscillations of any light falling on them, reflecting most of it. Furthermore, the metal ions are not rigidly held to fixed positions, as ions are in an ionic crystal. Rather, because the metal ions are held together by a "fluid" of electrons, these ions can move into various orientations relative to one another, which occurs when a metal is pounded, pulled, or molded into a different shape.

Two or more different metals can be bonded to each other by metallic bonds. This occurs, for example, when molten gold and molten palladium are blended to form the homogeneous solution known as white gold. The quality of the white gold can be modified simply by changing the proportions of gold and palladium. White gold is an example of an **alloy,** which is any mixture composed of two or more metallic elements. By playing around with proportions, metalworkers can readily modify the properties of an alloy. For example, in designing the Sacagawea dollar coin, shown in **Figure 6.13**, the U.S. Mint needed a metal having a gold color—so that it would be popular—and having the same electrical characteristics as the Susan B. Anthony dollar coin—so that the new coin could substitute for the Anthony coin in vending machines.

Only a few metals—gold and platinum are two examples—appear in nature in metallic form. Deposits of these natural metals, also known as *native metals*, are quite rare. For the most part, metals found in nature are chemical compounds. Iron, for example, is most frequently found as iron oxide, Fe_2O_3, and copper as chalcopyrite, $CuFeS_2$. Geologic deposits containing relatively high concentrations of metal-containing compounds are called **ores.** The metals industry mines these ores from the ground, as shown in **Figure 6.14**, and then processes them into metals. Although metal-containing compounds occur just about everywhere, only ores are concentrated enough to make the extraction of the metal economical.

Because our planet is chock-full of metal-containing compounds, it is difficult to imagine how we could ever incur a shortage of metals. Experts suggest, however, that if we continue with our present rate of consumption, such shortages will occur within the next two centuries. The problem is not a shortage of metal-containing compounds but rather a shortage of ores from which these compounds can be extracted *at a reasonable cost.*

Consider the recovery of gold. All the gold in the world isolated from nature so far could form a single cube 18 meters on a side, which would have a mass of about 130,000 tons. This includes all the naturally occurring elemental gold we have mined plus all the gold purified from gold-containing ores. Because the rate of gold production is steadily decreasing, one might think we have already isolated a significant portion of the Earth's total gold reserves. Our oceans, however, are laden with gold—as much as 2 milligrams per ton of seawater. Given that there is about 1.5×10^{18} tons of seawater on the planet,

READING CHECK

What holds the positively charged metal ions together within a metallic bond?

▲ **Figure 6.13**
The gold color of the Sacagawea U.S. dollar coin is achieved by an outer surface made of an alloy of 77% copper, 12% zinc, 7% manganese, and 4% nickel. The interior of the coin is pure copper.

▶ Figure 6.14
The world's biggest open-pit mine is the copper mine at Bingham Canyon, Utah.

 FOR YOUR INFORMATION

Metal ores contain ionic compounds in which the metal atoms have lost electrons to become positive ions. As we discuss in Chapter 11, converting the ores to metals requires that electrons be given back to the metal ions. This is done by heating the ore with electron-releasing materials, such as carbon, in hot furnaces that reach about 1500°C. The metal emerges in a molten state that can be cast into a variety of useful shapes.

our oceans contain 3.4 billion tons of gold! As yet, however, no method has been found for recovering gold from seawater profitably—this gold is simply too dilute (**Figure 6.15**).

Like the gold in the ocean, most of the metal-containing compounds in the Earth's crust are finely mixed with other stuff, which is to say the compounds are diluted. Ores are, by definition, parts of the Earth's crust where, for geologic reasons, the compounds have been concentrated. High-grade ores, those containing relatively large concentrations of compounds, are the first to be mined. After these are depleted, we move on to lower-grade ores, which have lower yields that translate into greater costs. Eventually, a nation's ore supplies are depleted, as are the aluminum oxide ores in the United States, as described in **Figure 6.16**. The nation is forced to import metals or their ores from other countries, which also have finite ore resources.

Interestingly, high-grade ore nodules discovered on the ocean floor are a potential new source of metals, as are metal-rich asteroids in space. Rather than investing in new ore resources, however, it is far cheaper to produce metals from recycled products. Perhaps one day in the future our recycling programs will be so strong worldwide that mining from such unusual places will not be necessary.

▲ Figure 6.15
Natural resources are unavailable to us when the energy required to collect them far exceeds the resource's inherent value. For example, most of the world's gold is found in the oceans, but this gold is too dilute for extraction to be worthwhile.

▲ Figure 6.16
An open-pit aluminum mine in Australia. Aluminum ore is no longer mined in the United States because the reserves have dwindled to the point where it is less expensive to import high-grade aluminum ore from other countries, including Australia.

6.5 Covalent Bonds Result from a Sharing of Electrons

EXPLAIN THIS

A lone proton encounters the lone pair of electrons of an ammonia molecule and forms what?

Imagine two children playing together and sharing their toys. Perhaps a force that keeps the children together is their mutual attraction to the toys they share. In a similar fashion, two atoms can be held together by their mutual attraction for electrons they share. A fluorine atom, for example, has a strong tendency to acquire an additional electron, which fills the outer shell. As shown in **Figure 6.17**, a fluorine atom can add an additional electron to its outer shell by grabbing onto the unpaired valence electron of another fluorine atom. This results in a situation in which the two fluorine atoms are mutually attracted to the same two electrons. This type of electrical attraction in which atoms are held together by their mutual attraction for shared electrons is called a **covalent bond,** where *co-* signifies sharing and *-valent* refers to the fact that it is valence electrons that are being shared.

A substance composed of atoms held together by covalent bonds is a **covalent compound.** The fundamental unit of most covalent compounds is a **molecule,** which we can now formally define as any group of atoms held together by covalent bonds. **Figure 6.18** uses the element fluorine to illustrate this principle.

When writing electron-dot structures for covalent compounds, chemists often use a straight line to represent the two electrons involved in a covalent bond. In some representations, the nonbonding electron pairs are ignored. This occurs in instances where these electrons play no significant role in the process being illustrated. Here are two frequently used ways of showing the electron-dot structure for a fluorine molecule without using spheres to represent the atoms:

$$:\ddot{F}-\ddot{F}: \qquad F-F$$

Remember—the straight line in both versions represents *two* electrons, one from each atom. Thus, we now have two types of electron pairs to keep track of. The term *nonbonding pair* refers to any pair that exists in the electron-dot structure of an individual atom, and the term *bonding pair* refers to any pair that results from formation of a covalent bond. In a nonbonding pair, both electrons originate in the same atom; in a bonding pair, one electron comes from each atom participating in the bond.

Recall from Section 6.3 that an ionic bond is formed when an atom that tends to lose electrons makes contact with an atom that tends to gain them.

LEARNING OBJECTIVE

> Describe how atoms combine to form covalent compounds.

 FOR YOUR INFORMATION

Spectroscopic studies of interstellar dust within our galaxy have revealed the presence of more than 120 kinds of molecules, such as hydrogen chloride, HCl; water, H_2O; acetylene, H_2C_2; formic acid, HCO_2H; methanol, CH_3OH; methyl amine, NH_2CH_3; acetic acid, CH_3CO_2H; and even the amino acid glycine, $NH_2CH_2CO_2H$. Notably, about half of these interstellar molecules are carbon-based organic molecules. As discussed in Chapter 5, the atoms originated from the nuclear fusion of ancient stars. How interesting that these atoms then join together to form molecules even in the deep vacuum of outer space.

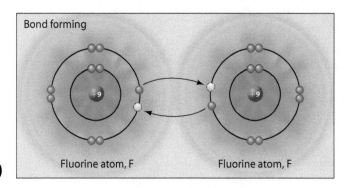

Fluorine atom, F   Fluorine atom, F

Bond forming

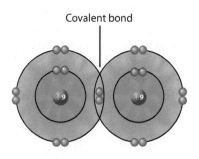

Covalent bond

Fluorine molecule, F_2

▲ **Figure 6.17**
The positive nuclear charge (represented by red shading) of a fluorine atom can cause the fluorine atom to become attracted to the unpaired valence electron of a neighboring fluorine atom. In this way, each fluorine atom achieves a filled valence shell and the two atoms are held together in a fluorine molecule by the attraction they both have for the two shared electrons. The atoms are said to be held together by a covalent bond.

▶ **Figure 6.18**
Molecules are the fundamental units of the gaseous covalent compound fluorine, F_2. Notice that in this model of a fluorine molecule, the spheres overlap, whereas the spheres shown earlier for ionic compounds do not. Now you know that this difference in representation is because of the difference in bond types.

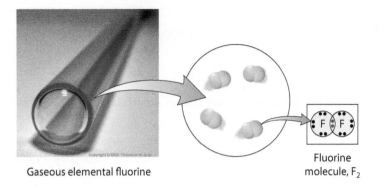

Gaseous elemental fluorine

Fluorine
molecule, F_2

A covalent bond, by contrast, is formed when two atoms that *both* tend to gain electrons are brought into contact with each other. Atoms that tend to form covalent bonds are therefore primarily atoms of the nonmetallic elements, in the upper right corner of the periodic table (with the exception of the noble gas elements, which are very stable and tend not to form bonds).

Hydrogen tends to form covalent bonds because, unlike the other group 1 elements, it has a fairly strong attraction for an additional electron. Two hydrogen atoms, for example, covalently bond to form a hydrogen molecule, H_2, as shown in **Figure 6.19**.

The number of covalent bonds an atom can form is equal to the number of additional electrons it can attract, which is the number needed to fill its valence shell. Hydrogen attracts only one additional electron, so it forms only one covalent bond. Oxygen, which attracts two additional electrons, finds them when it encounters two hydrogen atoms and reacts with them to form water, H_2O, as **Figure 6.20** shows. In water, not only does the oxygen atom have access to two additional electrons by covalently bonding to two hydrogen atoms, but each hydrogen atom has access to an additional electron by bonding to the oxygen atom. Each atom thus achieves a filled valence shell.

Nitrogen attracts three additional electrons and is thus able to form three covalent bonds, as occurs in ammonia, NH_3, shown in **Figure 6.21**. Likewise, a carbon atom can attract four additional electrons and is thus able to form four covalent bonds, as occurs in methane, CH_4, also shown in Figure 6.21. Note that the number of covalent bonds formed by these and other nonmetallic elements parallels the negative charge these ions tend to form (see Figure 6.6). This makes sense, because covalent-bond formation and negative-ion formation are both applications of the same concept: nonmetallic atoms tend to gain electrons until their valence shells are filled.

READINGCHECK

What do covalent bonds have in common with the negative ions formed by nonmetals?

▶ **Figure 6.19**
Two hydrogen atoms form a covalent bond as they share their unpaired electrons.

Before bonding

Covalent bond formed

H· ·H

H : H

Hydrogen atom   Hydrogen atom

Hydrogen molecule, H_2

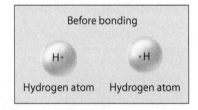

▶ **Figure 6.20**
The two unpaired valence electrons of oxygen pair with the unpaired valence electrons of two hydrogen atoms to form the covalent compound water.

Before bonding

Covalent bonds formed

Oxygen atom

O

H·

Hydrogen atom

H·

Hydrogen atom

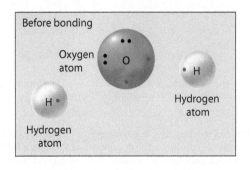

Water molecule, H_2O

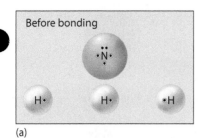

Before bonding

(a)

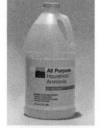

Nonbonding lone pair

H : N : H

H

Ammonia molecule, NH₃

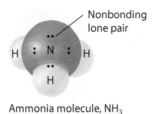

Before bonding

(b)

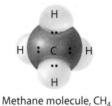

H

H : C : H

H

Methane molecule, CH₄

◀ **Figure 6.21**
(a) A nitrogen atom attracts the three electrons in three hydrogen atoms to form ammonia, NH₃, a gas that can dissolve in water to make an effective cleanser. (b) A carbon atom attracts the four electrons in four hydrogen atoms to form methane, CH₄, the primary component of natural gas. In these and most other cases of covalent-bond formation, the result is a filled valence shell for all the atoms involved.

Diamond is a most unusual covalent compound consisting of carbon atoms covalently bonded to one another in four directions. The result is a *covalent crystal*, which, as shown in **Figure 6.22**, is a highly ordered, 3-dimensional network of covalently bonded atoms. This network of carbon atoms forms a very strong and rigid structure, which is why diamonds are so hard. Also, because a diamond is a group of atoms held together only by covalent bonds, it can be characterized as a single molecule. Unlike most other molecules, a diamond molecule is large enough to be visible to the naked eye, so it is more appropriately referred to as a macromolecule.

CONCEPT CHECK

How many electrons make up a covalent bond?

CHECK YOUR ANSWER Two—one from each participating atom.

It is possible to have more than two electrons shared between two atoms, and **Figure 6.23** shows a few examples. Molecular oxygen, O₂, consists of two oxygen atoms connected by four shared electrons. This arrangement is called a

◀ **Figure 6.22**
The crystalline structure of diamond is nicely illustrated with sticks to represent the covalent bonds. The molecular nature of a diamond is responsible for its extreme hardness. Interestingly, most naturally occurring diamonds formed deep within the Earth, not millions, but *billions* of years ago.

▶ **Figure 6.23**
Double covalent bonds in molecules of oxygen, O_2, and carbon dioxide, CO_2, and a triple covalent bond in a molecule of nitrogen, N_2.

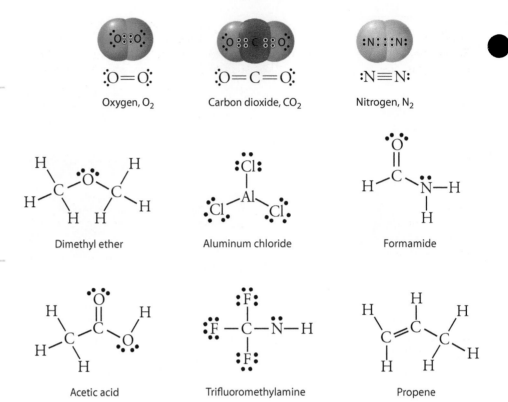

Oxygen, O_2 Carbon dioxide, CO_2 Nitrogen, N_2

▶ **Figure 6.24**
Additional electron-dot structures, also known as Lewis structures. Note that each carbon atom always forms four bonds. In formamide, two of these bonds are single bonds and the other two are part of a double bond. A double bond, therefore, counts as two bonds, and a triple bond counts as three bonds.

Dimethyl ether   Aluminum chloride   Formamide

Acetic acid   Trifluoromethylamine   Propene

Ammonium ion   Hydroxide ion

(a)   (b)

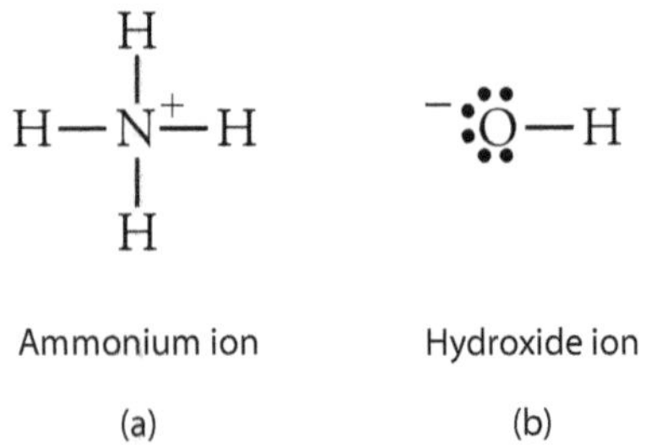

▲ **Figure 6.25**
The electron-dot structures for the ammonium and hydroxide ions.

double covalent bond or, for short, a double bond. As another example, the covalent compound carbon dioxide, CO_2, consists of two double bonds connecting two oxygen atoms to a central carbon atom.

Some atoms can form *triple covalent bonds*, in which six electrons—three from each atom—are shared. One example is molecular nitrogen, N_2. Any double or triple bond is often referred to as a multiple bond. Multiple bonds higher than these, such as the quadruple covalent bond, are not commonly observed.

Figure 6.24 shows electron-dot structures, also sometimes called *Lewis structures*, for some additional molecules. Note from these structures that each element consistently forms the same number of bonds. Hydrogen, for example, forms one bond, while carbon forms four bonds, nitrogen forms three bonds, and oxygen forms two bonds. As discussed earlier, this number of bonds is equal to the number of additional electrons the atom is able to attract as indicated by its position in the periodic table. Also note from these structures that the number of lone pairs on atoms is also consistent. Hydrogen, aluminum, and carbon show no lone pairs. Nitrogen has one lone pair, oxygen has two, and chlorine and fluorine have three. These numbers of lone pairs allow each atom to have a filled shell of electrons.

The pattern is a bit different for the charged atoms within a polyatomic ion. A positively charged atom within a polyatomic ion has one additional covalent bond and one fewer lone pair. For example, the nitrogen atom of an ammonia molecule, NH_3, has three bonds and one lone pair, as shown in Figure 6.21a. Add a hydrogen ion, H^+, to ammonia and you create the ammonium ion, NH_4^+, as shown in **Figure 6.25a**. The nitrogen of the ammonium ion has four bonds and no lone pairs. (As we discuss in Chapter 10, the reason for this is that the lone pair of ammonia reacted with a hydrogen ion to form the fourth bond.)

Similarly, a negatively charged atom within a polyatomic ion has one fewer covalent bond and one additional lone pair. For example, the oxygen of a water molecule, H_2O, has two bonds and two lone pairs, as shown in Figure 6.20. Remove a hydrogen ion, H^+, from a water molecule and you create the hydroxide ion, OH^-, as shown in **Figure 6.25b**. The oxygen of the hydroxide ion has only one bond but three lone pairs. (The additional lone pair was created as the hydrogen ion left the water molecule without an electron.)

CONCEPT CHECK

Shown below is the electron-dot structure for the polyatomic bicarbonate ion. Why is one of the oxygen atoms shown with only one covalent bond?

Bicarbonate ion

CHECK YOUR ANSWER In a neutral molecule, one would expect each oxygen atom to have two bonds. The oxygen seen to the right in this electron-dot structure is bonded only once because it is carrying a negative charge. Note that although it is bonded only once, it has three lone pairs, which with the two bonding electrons fill the valence shell.

6.6 Valence Electrons Determine Molecular Shape

EXPLAIN THIS

Why does the water molecule have a bent shape?

Molecules are 3-dimensional entities and therefore best depicted in three dimensions. We can translate the 2-dimensional electron-dot structure representing a molecule into a more accurate 3-dimensional rendering by using the model known as **valence-shell electron-pair repulsion,** also called VSEPR (pronounced ves-per). According to this model, electron pairs in a valence shell arrange themselves to get as far away as possible from all other electron pairs in the shell. This includes nonbonding pairs and any bonding pairs or groups of bonding pairs held together in a double or triple bond. This behavior is the result of simple electrostatic repulsions between the electron pairs.

The 2-dimensional electron-dot structure for methane, CH_4, is

In this structure, the bonding electron pairs (shown as straight lines representing two electrons each) are set 90° apart because that is the farthest apart they can be shown in two dimensions. When we extend the view to three dimensions, however, we can create a more accurate rendering, in which the four bonding pairs are 109.5° apart:

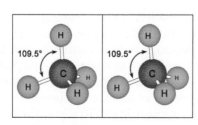

Stereo image

LEARNING OBJECTIVE

Predict the shape of a small molecule using the valence-shell electron-pair repulsion model.

 READING CHECK

What do electron pairs, either bonding or nonbonding, tend to do?

These two renderings of methane are *stereo images*—you can see them in three dimensions by looking at them cross-eyed so that they appear to overlap. To get the images to overlap, you can also try touching your nose to the page and then slowly pulling the book away from your face. If neither of these techniques works, be sure to build the gumdrop molecule as described in the Hands-On Chemistry activity at the beginning of this chapter.

You can think of this 3-dimensional structure as follows: the central carbon atom has one hydrogen atom sticking out of its top and is supported on a tripod whose legs are formed by the three lower C–H bonds.

Draw the four triangles defined by the hydrogen atoms in the above stereo image of CH_4 (one triangle being the base, the other three being the three upright faces) and you'll see that the shape of the methane molecule is a pyramid that has a triangular base supporting three other triangles that meet at the pyramid's apex. In geometry, a pyramid that has a triangular base is given the special name *tetrahedron*, and so chemists say that the methane molecule is *tetrahedral*:

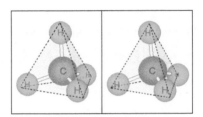

Stereo image of tetrahedral methane molecule

The VSEPR model allows us to use electron-dot structures to predict the 3-dimensional geometry of simple molecules. This geometry is determined by considering the number of atoms or nonbonding electron pairs surrounding the centrally located atom. Chemists refer to any atom or nonbonding electron pair surrounding a central atom as a *substituent*. They use this fancy term the same way people commonly use the word *appendage* to refer to either an arm or a leg. Thus, to simplify our discussion, we define a **substituent** as any atom or nonbonding pair of electrons surrounding a centrally located atom. For example, the carbon of a methane molecule has four substituents—the four hydrogen atoms. The oxygen atom of a water molecule also has four substituents—two hydrogen atoms and two nonbonding pairs of electrons:

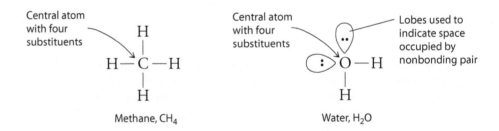

Methane, CH_4 Water, H_2O

As shown in Table 6.2, when a central atom has only two substituents, the geometry of the molecule is *linear*, meaning that a single straight line may be drawn passing through both substituents and the central atom. Three substituents arrange themselves in a triangle—the plane of which passes through the central atom—so this geometry is called *triangular planar*. Four substituents form a tetrahedron, as already discussed. Why these geometries? Simply put, these are the geometries that allow the maximum distance between substituents.

Molecular Shape Is Defined by the Substituent Atoms

When chemists talk about the shape of a molecule, they are talking about the relative positions of the atoms of the molecule. How atoms position themselves is spelled out by VSEPR. But from VSEPR, we see that the nonbonding electrons (lone pairs) also affect how atoms within a molecule get positioned.

Figuring out the shape of a molecule is a two-step process. The first step is to use VSEPR to position all substituents, both atoms and nonbonding pairs, around a central atom. The second step is to ignore all nonbonding pairs and decide what 3-dimensional shape the atoms form. Let's work through a few examples from Table 6.2 to see what all this means.

In any molecule in which there are no nonbonding pairs around the central atom, the molecular shape is the same as the VSEPR geometry. Thus, to use the examples from Table 6.2, all the molecules of the first row (BeH_2, CO_2, HCN) have a linear shape. Two molecules of the second row (BH_3 and H_2CO) have a triangular planar shape, and the first molecule of the third row (CH_4) has a tetrahedral shape.

Now let's look at molecules that have nonbonding pairs, beginning with germanium chloride, $GeCl_2$, and its one nonbonding pair. The VSEPR geometry is triangular planar, but to get the shape of the molecule, we focus only on the atoms. This reveals the germanium and two chlorine atoms held together at an angle—a shape known as *bent*. Similarly, focusing only on the atoms, a water molecule is also seen to have a bent shape. Now you know why water molecules

 FOR YOUR INFORMATION

"Dry cleaning" is the process of washing clothes without water. The most common dry cleaning solvent is perchloroethylene, C_2Cl_4. (Can you deduce its chemical structure?) The advantage of dry cleaning is that dirt, grime, and stains are typically more soluble in the dry cleaning solvent, which is also less harsh on the clothing and can do a full load in under 10 minutes. After a washing cycle, the solvent is centrifuged out of the machine, filtered, distilled, and recycled for the next load. Clothes come out of the machine already dried and ready for folding. Perchloroethylene, also known as perc, is relatively safe, but it is mildly carcinogenic and can cause dizziness in those who work with it. An up-and-coming alternative to perc is carbon dioxide, CO_2, which at super high pressures forms an unusual liquid-like and gas-like medium, called a supercritical fluid, that has remarkable cleaning properties.

TABLE 6.2   Molecular Geometries

NUMBER OF SUBSTITUENTS	3-DIMENSIONAL GEOMETRY	EXAMPLES		
2	180° Linear	H—Be—H BeH₂	O=C=O CO₂	H—C≡N HCN
3	120° Triangular planar	H—B with H, H BH₃	O=C with H, H H₂CO	Ge with Cl, Cl (lone pair) GeCl₂
4	109.5° Tetrahedral	C with H, H, H, H CH₄	N with H, H, H (lone pair) NH₃	O with H, H (two lone pairs) H₂O

▶ **Figure 6.26**
The shapes of
molecules from
Table 6.2.

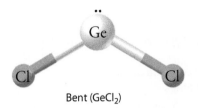

Bent (GeCl₂)

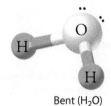

Bent (H₂O)

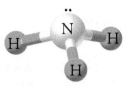

Triangular pyramidal (NH₃)

are always depicted with the two hydrogen atoms close to each other, like a set of mouse ears, rather than as far apart as possible on opposite sides of the oxygen atom—two nonbonding pairs are pushing them into this orientation.

Focusing only on the atoms of the ammonia molecule, NH₃, in Table 6.2 means the shape is not tetrahedral, because in a tetrahedron, all four corners must be equally distant from the central atom. Ammonia's shape, as shown in **Figure 6.26**, is typically described as *triangular pyramidal*.

CONCEPT CHECK

What is the shape of a chlorine trifluoride molecule, ClF₃?

$$F-Cl \quad \overset{\displaystyle F}{\underset{\displaystyle F}{|}}$$

CHECK YOUR ANSWER Describe what you see as you focus only on the atoms (ignoring the nonbonding pairs). For chlorine trifluoride, the shape has all four atoms in the same plane. They form a triangle having a fluorine atom at each corner and the chlorine atom sitting at the midpoint on one side.

Call it what you like—most chemists simply call it T-shaped. If you're curious how chlorine is able to form three covalent bonds, you should consider taking a follow-up course in general chemistry.

6.7 Polar Covalent Compounds—Uneven Sharing of Electrons

LEARNING OBJECTIVE

Differentiate between ionic, polar covalent, and nonpolar covalent chemical bonds.

EXPLAIN THIS

How is the chemical bond between sodium and chlorine mostly ionic but partially covalent?

If the two atoms in a covalent bond are identical, as shown below, their nuclei have the same positive charge. The electrons shared between these two identical atoms are shared evenly. We can represent these electrons being shared evenly using an electron-dot structure with the electrons situated exactly halfway between the two atomic symbols. Alternatively, we can draw a cloud in which the positions of the two bonding electrons over time are shown as a series of dots. Where the dots are most concentrated is where the electrons have the greatest probability of being located:

H : H

In a covalent bond between nonidentical atoms, the nuclear charges are different, and consequently, the bonding electrons may be shared *unevenly*.

This occurs in a hydrogen–fluorine bond, in which electrons are more attracted to fluorine's greater nuclear charge:

The bonding electrons spend more time around the fluorine atom. For this reason, the fluorine side of the bond is slightly negative, and because the bonding electrons have been drawn away from the hydrogen atom, the hydrogen side of the bond is slightly positive. This separation of charges is called a **dipole** (pronounced *die*-pole) and is represented either by the characters δ– and δ+ (read "slightly negative" and "slightly positive," respectively) or by a crossed arrow pointing to the negative side of the bond:

$$\overset{\delta+}{H} \overset{\delta-}{-F} \qquad \overset{\longrightarrow}{H-F}$$

So atoms forming a chemical bond engage in a tug-of-war for electrons. How strongly an atom is able to tug on bonding electrons has been measured experimentally and quantified as the atom's **electronegativity.** The range of electronegativities runs from 0.7 to 3.98, as **Figure 6.27** shows. The greater an atom's electronegativity, the greater its ability to pull electrons toward itself when bonded. Thus, in hydrogen fluoride, fluorine has a greater electronegativity, or pulling power, than hydrogen.

Electronegativity is greatest for elements at the upper right of the periodic table and lowest for elements at the lower left. Noble gases are not considered in electronegativity discussions because, as previously mentioned, they rarely participate in chemical bonding.

When the two atoms in a covalent bond have the same electronegativity, no dipole is formed (as is the case with H_2) and the bond is classified as a **nonpolar** bond. When the electronegativities of the atoms differ, a dipole may form (as with HF) and the bond is classified as a **polar** bond. Just how polar a bond is depends on the difference between the electronegativity values of the two atoms—the greater the difference, the more polar the bond. An electronegativity difference between two atoms greater than about 1.7 indicates an ionic bond. A difference of less than 1.7 but greater than about 0.4 indicates a polar covalent bond. Nonpolar covalent bonds have electronegativity differences of less than 0.4.

As can be seen in Figure 6.27, the greater the distance between two atoms in the periodic table, the greater the difference in their electronegativities, and hence the greater the polarity of the bond between them. So a chemist can predict which bonds are more polar than others without reading the electronegativities. Bond polarity can usually be inferred by looking at the relative positions of the atoms in the periodic table—the farther apart they are, especially when one is at the lower left and one is at the upper right, the greater the polarity of the bond between them.

READING CHECK

What is an atom with great electronegativity able to do?

◀ Figure 6.27
The experimentally measured electronegativities of elements.

H																	He
2.2																	—
Li	Be											B	C	N	O	F	Ne
0.98	1.57											2.04	2.55	3.04	3.44	3.98	—
Na	Mg											Al	Si	P	S	Cl	Ar
0.93	1.31											1.61	1.9	2.19	2.58	3.16	—
K	Ca	Sc	Ti	V	Cr	Mn	Fe	Co	Ni	Cu	Zn	Ga	Ge	As	Se	Br	Kr
0.82	1.0	1.36	1.54	1.63	1.66	1.55	1.83	1.88	1.91	1.90	1.65	1.81	2.01	2.18	2.55	2.96	—
Rb	Sr	Y	Zr	Nb	Mo	Tc	Ru	Rh	Pd	Ag	Cd	In	Sn	Sb	Te	I	Xe
0.82	0.95	1.22	1.33	1.6	2.16	1.9	2.2	2.28	2.20	1.93	1.69	1.78	1.96	2.05	2.1	2.66	—
Cs	Ba	La	Hf	Ta	W	Re	Os	Ir	Pt	Au	Hg	Tl	Pb	Bi	Po	At	Rn
0.79	0.89	1.10	1.3	1.5	2.36	1.9	2.2	2.20	2.8	2.54	2.00	2.04	2.33	2.02	2.0	2.2	—
Fr	Ra	Ac	Rf	Db	Sg	Bh	Hs	Mt	Ds	Rg	Cn	Uut	Fl	Uup	Lv	Uus	Uuo
0.7	0.9	1.1	—	—	—	—	—	—	—	—	—	—	—	—	—	—	—

FOR YOUR INFORMATION

The term *electronegativity* was coined by Linus Pauling in his famous 1932 publications "The Nature of the Chemical Bond, Parts I–IV." Through these articles, Pauling was the first to formally describe the periodic trends of ionic, polar covalent, and nonpolar covalent bonding. The term *electronegativity* refers to the *negative* charge that an atom in a molecule picks up due to the *electrons* it may attract. There are other electronegativity scales, but Pauling's original scale has remained the most commonly used.

CONCEPTCHECK

List these bonds in order of increasing polarity: P–F, S–F, Ga–F, Ge–F.

(least polar) _____, _____, _____, _____ (most polar)

CHECK YOUR ANSWER *If you answered the question, or attempted to, before reading this answer, hooray for you! You're doing more than reading the text—you're learning chemistry.* The greater the difference in electronegativities between two bonded atoms, the greater the polarity of the bond; so the order of increasing polarity is S–F, P–F, Ge–F, Ga–F.

Note that this answer can be obtained merely by looking at the relative positions of these elements in the periodic table rather than by calculating the differences in their electronegativities.

The magnitude of bond polarity is sometimes indicated by the size of the crossed arrow or the δ– and δ+ symbols used to depict a dipole, as shown in **Figure 6.28**. Note that the electronegativity difference between atoms in an ionic bond is relatively large. For example, the bond in NaCl has an electronegativity difference of 2.23, far greater than the difference of 1.43 shown for the C–F bond in Figure 6.28.

What is important to understand here is that there is no black-and-white distinction between ionic and covalent bonds. Rather, there is a gradual change from one to the other as the atoms that bond are located farther and farther apart in the periodic table. This continuum is illustrated in **Figure 6.29**. Atoms on opposite sides of the periodic table have great differences in electronegativity, and hence the bonds between them are highly polar—in other words, ionic. Nonmetallic atoms of the same element have the same electronegativities, so their bonds are nonpolar covalent. The polar covalent bond, with its uneven sharing of electrons and slightly charged atoms, lies between these two extremes.

▶ **Figure 6.28**
These bonds are in order of increasing polarity from left to right, a trend indicated by the larger and larger crossed arrows and δ–/δ+ symbols. Which of these pairs of elements are farthest apart in the periodic table?

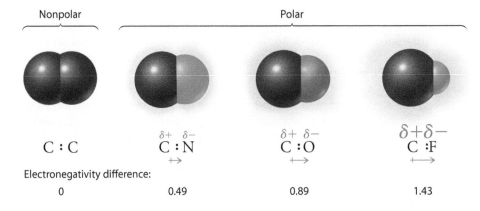

▶ **Figure 6.29**
The ionic bond and the nonpolar covalent bond represent the two extremes of chemical bonding. The ionic bond involves a transfer of one or more electrons, and the nonpolar covalent bond involves the equitable sharing of electrons. The character of a polar covalent bond falls between these two extremes.

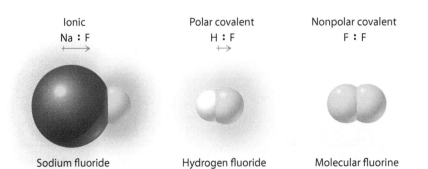

6.8 Molecular Polarity—Uneven Distribution of Electrons

EXPLAIN THIS

Which is heavier: carbon dioxide or water?

When all the bonds in a molecule are nonpolar, the molecule as a whole is also nonpolar—as is the case with H_2, O_2, and N_2. When a molecule consists of only two atoms and the bond between them is polar, the polarity of the molecule is the same as the polarity of the bond—as with HF and HCl.

Complexities arise when assessing the polarity of a molecule containing more than two atoms. Consider carbon dioxide, CO_2, shown in **Figure 6.30**. The cause of the dipole in either of the carbon–oxygen bonds is oxygen's greater pull on the bonding electrons (because oxygen is more electronegative than carbon). At the same time, however, the oxygen atom on the opposite side of the carbon pulls those electrons back to the carbon. The net result is an even distribution of bonding electrons around the entire molecule. So dipoles that are of equal strength but pull in opposite directions in a molecule effectively cancel each other, with the result that the molecule as a whole is nonpolar.

Figure 6.31 illustrates a similar situation in boron trifluoride, BF_3, where three fluorine atoms are oriented 120° from one another around a central boron atom. Because the angles are all the same and because each fluorine atom pulls on the electrons of its boron–fluorine bond with the same force, the resulting polarity of this molecule is zero.

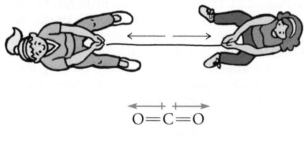

$$O=C=O$$

◀ **Figure 6.30**
There is no net dipole in a carbon dioxide molecule, so the molecule is nonpolar. This is analogous to two people in a tug-of-war. As long as they pull with equal force but in opposite directions, the rope remains stationary.

◀ **Figure 6.31**
The three dipoles of a boron trifluoride molecule oppose one another at 120° angles, which makes the overall molecule nonpolar. This is analogous to three people pulling with equal force on ropes attached to a central ring. As long as they all pull with equal force and all maintain the 120° angles, the ring will remain stationary.

▶ **Figure 6.32**
Nitrogen is a liquid at temperatures below its chilly boiling point of –196°C. Nitrogen molecules are not very attracted to one another because they are nonpolar. As a result, the small amount of heat energy available at –196°C is enough to separate them and allow them to enter the gaseous phase.

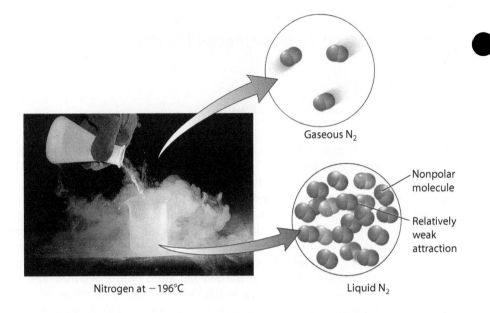

Gaseous N_2

Nonpolar molecule

Relatively weak attraction

Nitrogen at −196°C

Liquid N_2

As is discussed further in the next chapter, nonpolar molecules have only relatively weak attractions to other nonpolar molecules. This lack of attraction between nonpolar molecules explains the low boiling points of many nonpolar substances. Recall from Section 2.7 that boiling is a process wherein the molecules of a liquid separate from one another as they go into the gaseous phase. When there are only weak attractions between the molecules of a liquid, less heat energy is required to liberate the molecules from one another and allow them to enter the gaseous phase. This translates into a relatively low boiling point for the liquid, as shown for nitrogen, N_2 in **Figure 6.32**. The boiling points of hydrogen, H_2; oxygen, O_2; carbon dioxide, CO_2; and boron trifluoride, BF_3, are also quite low for the same reason.

There are many instances, however, in which the dipoles of different bonds in a molecule do *not* cancel each other. Reconsider the rope analogy of Figure 6.31. As long as everyone pulls equally, the ring stays put. Imagine, however, that one person begins to ease off on the rope. Now the pulls are no longer balanced, and the ring begins to move away from the person who is slacking off, as **Figure 6.33** shows. Likewise, if one person began to pull harder, the ring would move away from the other two people.

▶ **Figure 6.33**
The two strongly electronegative chlorine atoms in $GeCl_2$ pull germanium's lone pair inward. By analogy, if one person eases off in a three-way tug-of-war but the other two continue to pull, the ring moves in the direction of the purple arrow.

A similar situation occurs in molecules in which polar covalent bonds are not equal and opposite. Perhaps the most relevant example is water, H_2O. Each hydrogen–oxygen covalent bond has a relatively large dipole because of the great electronegativity difference. Because of the bent shape of the molecule, however, the two dipoles, shown in blue in **Figure 6.34**, do not cancel each other the way the C–O dipoles in Figure 6.30 do. Instead, the dipoles in the water molecule work together to give an overall dipole, shown in purple, for the molecule.

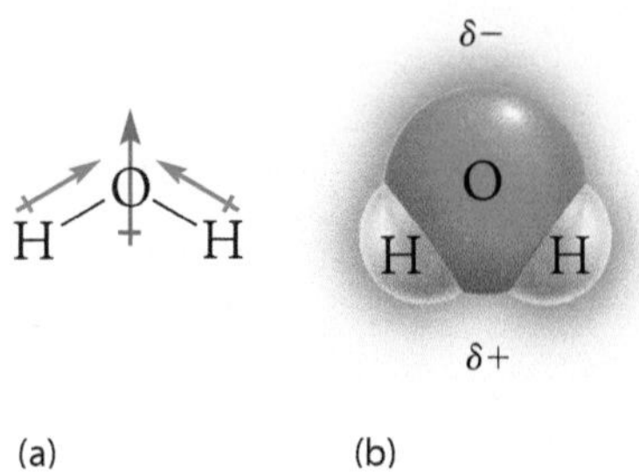

▲ **Figure 6.34**
(a) The individual dipoles in a water molecule add together to give a large overall dipole for the whole molecule, shown in purple. (b) The region around the oxygen atom is therefore slightly negative, and the region around the two hydrogen atoms is slightly positive.

C O N C E P T C H E C K

Which of these molecules is polar, and which is nonpolar?

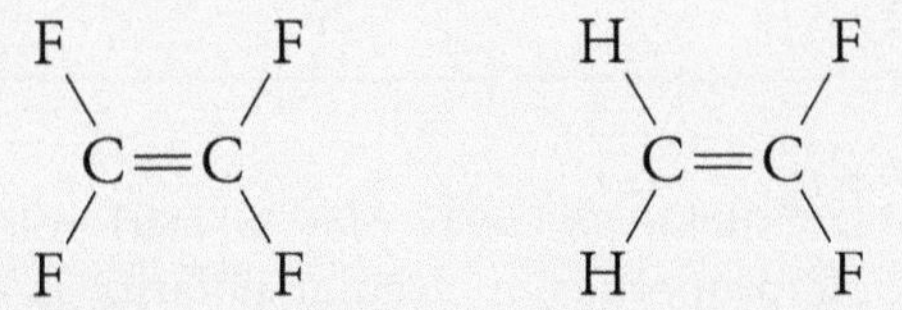

$$F \backslash \qquad /F \qquad\qquad H \backslash \qquad /F$$
$$C{=}C \qquad\qquad C{=}C$$
$$F / \qquad \backslash F \qquad\qquad H / \qquad \backslash F$$

CHECK YOUR ANSWER Symmetry is often the greatest clue for determining polarity. Because the molecule on the left is symmetrical, the dipoles on the two sides cancel each other. This molecule is therefore nonpolar.

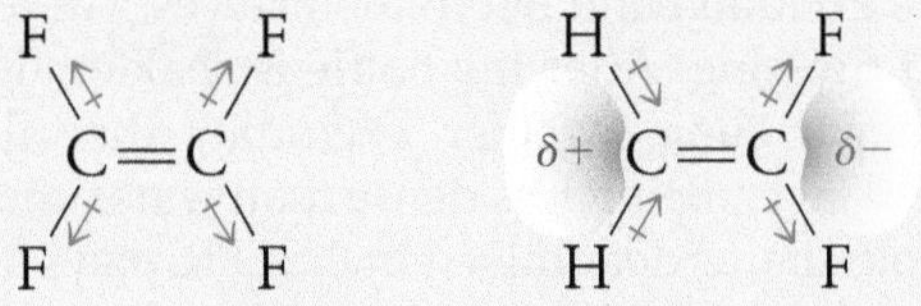

$$F \qquad F \qquad\qquad H \qquad F$$
$$C{=}C \qquad\qquad \delta+ \; C{=}C \; \delta-$$
$$F \qquad F \qquad\qquad H \qquad F$$

Because the molecule on the right is less symmetrical (more "lopsided"), it is a polar molecule.

Figure 6.35 illustrates how polar molecules electrically attract one another and, as a result, are relatively difficult to separate. In other words, polar molecules can be thought of as being "sticky," which is why it takes more energy to separate them—to change the substance's phase. For this reason, substances composed of polar molecules typically have higher boiling points than substances composed of nonpolar molecules, as Table 6.3 shows. Water, for example, boils at 100°C, whereas carbon dioxide boils at –79°C. This 179°C difference is quite dramatic when you consider that a carbon dioxide molecule is more than twice as massive as a water molecule.

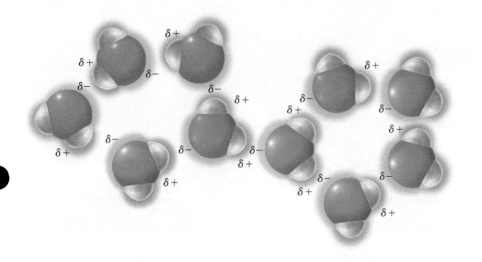

◀ **Figure 6.35**
Water molecules attract one another because each contains a slightly positive side and a slightly negative side. The molecules position themselves in such a way that the positive side of one faces the negative side of a neighbor.

TABLE 6.3   Boiling Points of Some Polar and Nonpolar Substances

SUBSTANCE	BOILING POINT (°C)
Polar	
Hydrogen fluoride, HF	20
Water, H_2O	100
Ammonia, NH_3	−33
Nonpolar	
Hydrogen, H_2	−253
Oxygen, O_2	−183
Nitrogen, N_2	−196
Boron trifluoride, BF_3	−100
Carbon dioxide, CO_2	−79

CHEMICAL CONNECTIONS

How is a glacier connected to a lone pair of electrons?

Because molecular "stickiness" can play a lead role in determining a substance's macroscopic properties, molecular polarity is a central concept of chemistry. **Figure 6.36** helps bring to mind, in a tragic way, the fact that oil (both petroleum-based oil and cooking oil) and water don't mix. It's not, however, that oil and water *repel* each other. Rather, water molecules are so attracted to themselves because of their polarity that they pull themselves together. The nonpolar oil molecules are thus excluded and left to themselves. We see this in a bottle of oil and vinegar salad dressing. After the bottle is shaken vigorously, the water molecules of the vinegar cling together, excluding the oil molecules, which separate into their own layer. Being less dense than water, the oil rises to the top.

The shape of a molecule, as determined by VSEPR, plays a major role in determining the polarity of that molecule. A molecule's polarity, in turn, has a great influence on macroscopic behavior. Consider what the world would be like if the oxygen atom in a water molecule did not have its two nonbonding pairs of electrons. Instead of being bent, each water molecule would be linear, much like carbon dioxide. The dipoles of the two hydrogen–oxygen bonds would cancel each other, which would make water a nonpolar substance and give it a relatively low boiling point. Water would not be a liquid at the ambient temperatures of our planet, and we in turn would not be here discussing these concepts. We should be thankful that the oxygen within water has two nonbonding pairs. That's good chemistry.

▶ Figure 6.36
Oil and water are difficult to mix, as is evident from this beach scene showing oil washed up from the Deepwater Horizon oil spill of 2010 in the Gulf of Mexico—the largest accidental oil spill in human history. The oil came from a sea-floor oil gusher that released an estimated 53,000 barrels every day for three months before it was finally capped. Petroleum-based oil molecules are not able to compete with the attraction water molecules have for themselves, which is why the oil and water don't mix.

Chapter 6 Review

LEARNING OBJECTIVES

Identify paired and unpaired electrons within an electron-dot structure. (6.1)	→	*Questions 1–3, 37–40*
Use the periodic table to predict the type of ion an atom tends to form. (6.2)	→	*Questions 4–6, 41–47*
Describe how ions combine to form ionic compounds. (6.3)	→	*Questions 7–9, 27, 29–32, 48–50*
Relate the properties of a metal to how the atoms of that metal are chemically bonded. (6.4)	→	*Questions 10–12, 51, 52, 74, 75*
Describe how atoms combine to form covalent compounds. (6.5)	→	*Questions 13–16, 53–58*
Predict the shape of a small molecule using the valence-shell electron-pair repulsion model. (6.6)	→	*Questions 17–19, 28, 59–62*
Differentiate between ionic, polar covalent, and nonpolar covalent chemical bonds. (6.7)	→	*Questions 20–22, 33, 35, 63–67*
Recognize the important role that molecular interactions play in determining the physical properties of a material. (6.8)	→	*Questions 23–26, 34, 36, 68–73*

SUMMARY OF TERMS (KNOWLEDGE)

Alloy A mixture of two or more metallic elements.

Covalent bond A chemical bond in which atoms are held together by their mutual attraction for one or more pairs of electrons they share.

Covalent compound A substance, such as an element or a chemical compound, in which atoms are held together by covalent bonds.

Dipole A separation of charge that occurs in a chemical bond because of differences in the electronegativities of the bonded atoms.

Electron-dot structure A shorthand notation of the shell model of the atom, in which valence electrons are shown around an atomic symbol. The electron-dot structure for an atom or ion is sometimes called a *Lewis dot symbol*, while the electron-dot structure of a molecule or polyatomic ion is sometimes called a *Lewis structure*.

Electronegativity The ability of an atom to attract a bonding pair of electrons to itself when bonded to another atom.

Ion An atom having a net electrical charge because of either a loss or gain of electrons.

Ionic bond A chemical bond in which there is an electric force of attraction between two oppositely charged ions.

Ionic compound A chemical compound containing ions.

Metallic bond A chemical bond in which positively charged metal ions are held together within a "fluid" of loosely held electrons.

Molecule The fundamental unit of a chemical compound, which is a group of atoms held tightly together by covalent bonds.

Nonbonding pairs Two paired valence electrons that are not participating in a chemical bond.

Nonpolar Said of a chemical bond or molecule that has no dipole. In a nonpolar bond or molecule, the electrons are distributed evenly.

Ore A geologic deposit containing relatively high concentrations of one or more metal-containing compounds.

Polar Said of a chemical bond or molecule that has a dipole. In a polar bond or molecule, electrons are congregated to one side. This makes that side slightly negative, while the opposite side (lacking electrons) becomes slightly positive.

Polyatomic ion An ionically charged molecule.

Substituent A term used to describe an atom or a nonbonding pair of electrons surrounding a centrally located atom.

Valence electrons The electrons in the outermost occupied shell of an atom.

Valence shell The outermost occupied shell of an atom.

Valence-shell electron-pair repulsion A model, also known as VSEPR (pronounced ves-per), that explains molecular geometries in terms of electron pairs striving to be as far apart from one another as possible.

READING CHECK QUESTIONS (COMPREHENSION)

6.1 Electron-Dot Structures

1. How many electrons can occupy the first shell? How many can occupy the second shell?

2. Which electrons are represented by an electron-dot structure?

3. How do the electron-dot structures of elements in the same group in the periodic table compare with one another?

6.2 Atoms Can Lose or Gain Electrons to Become Ions

4. How does an ion differ from an atom?

5. To become a negative ion, does an atom lose or gain electrons?

6. Why does the fluorine atom tend to gain only one electron?

6.3 Ionic Bonds Result from a Transfer of Electrons

7. Which elements tend to form ionic bonds?

8. Suppose an oxygen atom gains two electrons to become an oxygen ion. What is its electric charge?

9. What is an ionic crystal?

6.4 The Electrons of Metallic Bonds Are Loosely Held

10. Do metals more readily gain or lose electrons?

11. What is an alloy?

12. What is a native metal?

6.5 Covalent Bonds Result from a Sharing of Electrons

13. Which elements tend to form covalent bonds?

14. How many electrons are shared in a double covalent bond?

15. Within a neutral molecule, how many covalent bonds does an oxygen atom form?

16. Within a polyatomic ion, how many covalent bonds does a negatively charged oxygen form?

6.6 Valence Electrons Determine Molecular Shape

17. What does *VESPR* stand for?

18. What is meant by the term *substituent*?

19. How many substituents does the oxygen atom in a water molecule have?

6.7 Polar Covalent Bonds—Uneven Sharing of Electrons

20. What is a dipole?

21. Which element in the periodic table has the greatest electronegativity? Which has the least electronegativity?

22. Which is more polar: a carbon–oxygen bond or a carbon–nitrogen bond?

6.8 Molecular Polarity—Uneven Distribution of Electrons

23. How can a molecule be nonpolar when it consists of atoms that have different electronegativities?

24. Why do nonpolar substances boil at relatively low temperatures?

25. Which has a greater degree of symmetry: a polar molecule or a nonpolar molecule?

26. Why don't oil and water mix?

CONFIRM THE CHEMISTRY (HANDS-ON APPLICATION)

27. View crystals of table salt with a magnifying glass or, better yet, a microscope if one is available. If you do have a microscope, crush the crystals with a spoon and examine the resulting powder. Purchase sodium-free salt, which is potassium chloride, KCl, and examine these ionic crystals both intact and crushed. Sodium chloride and potassium chloride both form cubic crystals, but there are significant differences. What are they?

28. The following are the structures you may have come up with for the chemical compounds given in the Hands-On Chemistry activity at the beginning of this chapter.

Looking for more challenges? Try building the structures for carbon dioxide, CO_2; water, H_2O; and ammonia, NH_3. All of these structures were given to you in this chapter. For an ultimate challenge, try benzene, C_6H_6, or acetic acid, $C_2H_4O_2$.

Dichloromethane, CH_2Cl_2, tetrahedron

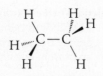

Ethane, C_2H_6, two tetrahedrons

Hydrogen peroxide, H_2O_2, two bent shapes stuck together

H—C≡C—H

Acetylene, C_2H_2, linear

THINK AND SOLVE (MATHEMATICAL APPLICATION)

29. Ores of manganese, Mn, sometimes contain the mineral rhodochrosite, $MnCO_3$, which is an ionic compound of manganese ions and carbonate ions. How many electrons has each manganese atom lost to make this compound?

30. What is the electric charge on the calcium ion in calcium chloride, $CaCl_2$?

31. Magnesium ions carry a 2+ charge, and chloride ions carry a 1– charge. What is the chemical formula for the ionic compound magnesium chloride?

32. Barium ions carry a 2+ charge, and nitrogen ions carry a 3– charge. What is the chemical formula for the ionic compound barium nitride?

THINK AND COMPARE (ANALYSIS)

33. Rank the following in order of increasing polarity
   a. C–H   b. O–H   c. N–H

34. Rank the following compounds in order of increasing boiling point:
   a. Fluorine, F_2
   b. Hydrogen fluoride, HF
   c. Hydrogen chloride, HCl

35. Rank the following in order of increasing symmetry:
   a. CH_4 b. NH_3 c. H_2O

36. Rank the following in order of increasing boiling point:
   a. CH_4 b. NH_3 c. H_2O

THINK AND EXPLAIN (SYNTHESIS)

6.1 Electron-Dot Structures

37. How do the electron-dot structures of elements in the same period in the periodic table compare with one another?

38. How many more electrons can fit within the valence shell of a fluorine atom?

39. How many more electrons can fit within the valence shell of a hydrogen atom?

40. How is the number of unpaired valence electrons in an atom related to the number of bonds that the atom can form?

6.2 Atoms Can Lose or Gain Electrons to Become Ions

41. The valence electron of a sodium atom does not sense the full 11+ of the sodium nucleus. Why?

42. Why is it so easy for a magnesium atom to lose two electrons?

43. Why doesn't the neon atom tend to lose or gain any electrons?

44. Why does an atom with few valence electrons tend to lose these electrons rather than gain more?

45. Why does an atom with many valence electrons tend to gain electrons rather than lose any?

46. Sulfuric acid, H_2SO_4, loses two protons to form what polyatomic ion? What molecule loses a proton to form the hydroxide ion, OH^-?

47. Which should be larger, the potassium ion, K^+, or the potassium atom, K? Which should be larger, the potassium ion, K^+, or the argon atom, Ar? Please explain.

6.3 Ionic Bonds Result from a Transfer of Electrons

48. Which should be more difficult to pull apart: a sodium ion from a chloride ion or a potassium ion from a chloride ion? Please explain.

Shorter distance between positive and negative charges   Longer distance between positive and negative charges

49. Which are closer together: the two nuclei within potassium fluoride, KF, or the two nuclei within molecular fluorine, F_2? Please explain.

50. Two fluorine atoms join together to form a covalent bond. Why don't two potassium atoms do the same thing?

6.4 The Electrons of Metallic Bonds Are Loosely Held

51. Given that the total number of atoms on our planet remains fairly constant, how is it ever possible to deplete a natural resource such as a metal?

52. An artist wants to create a metal sculpture using a mold so that his artwork can be readily mass-produced. He wants his sculpture to be exactly 6 inches tall. Should the mold also be 6 inches tall? Why or why not?

6.5 Covalent Bonds Result from a Sharing of Electrons

53. What happens when hydrogen's electron gets close to the valence shell of a fluorine atom?

54. What drives an atom to form a covalent bond: its nuclear charge or the need to have a filled outer shell? Please explain.

55. Atoms of nonmetallic elements form covalent bonds, but they can also form ionic bonds. How is this possible?

56. Write the electron-dot structure for the covalent compound ethane, C_2H_6.

57. Write the electron-dot structure for the covalent compound hydrogen cyanide, HCN.

58. Write the electron-dot structure for the covalent compound acetylene, C_2H_2.

6.6 Valence Electrons Determine Molecular Shape

59. Why is a germanium chloride molecule, $GeCl_2$, bent even though there are only two atoms surrounding the central germanium atom?

60. Write the electron-dot structure for the covalent compound hydrogen peroxide, H_2O_2.

61. Examine the 3-dimensional geometries of PF_5 and SF_4 shown below. Which do you think is the more polar compound?

PF₅   SF₄

62. The electron-pair geometry of water is tetrahedral, but its molecular shape is bent. Is this contradictory? Why or why not?

6.7 Polar Covalent Bonds—Uneven Sharing of Electrons

63. What is the source of an atom's electronegativity?

64. In each of the molecules shown below, which atom carries the greater positive charge?

H–Cl   Br–F   C≡O   Br–Br

65. Which is more polar, a sulfur–bromine (S–Br) bond or a selenium–chlorine (Se–Cl) bond?

66. True or False: The greater the nuclear charge of an atom, the greater its electronegativity. Please explain.

67. True or False: The more shells in an atom, the lower its electronegativity. Please explain.

6.8 Molecular Polarity—Uneven Distribution of Electrons

68. Water, H_2O, and methane, CH_4, have about the same mass and differ by only one type of atom. Why is the boiling point of water so much higher than that of methane?

69. Circle the molecule from each pair that should have a higher boiling point: (atomic numbers: $Cl = 17$; $O = 8$; $C = 6$; $H = 1$)

$$\begin{array}{cc} \text{Cl} \quad \text{Cl} & \text{H} \quad \text{Cl} \\ \diagdown \ \diagup & \diagdown \ \diagup \\ \text{C}=\text{C} & \text{C}=\text{C} \\ \diagup \ \diagdown & \diagup \ \diagdown \\ \text{H} \quad \text{H} & \text{Cl} \quad \text{H} \end{array}$$

$$\begin{array}{cc} \text{Cl} & \text{Cl} \quad \text{H} \\ \diagdown & \diagdown \ \diagup \\ \text{C}=\text{O} & \text{C}=\text{C} \\ \diagup & \diagup \ \diagdown \\ \text{Cl} & \text{Cl} \quad \text{H} \end{array}$$

70. Compared to borane, BH_3, why is ammonia, NH_3, more polar?

71. Three kids sitting equally apart around a table are sharing jelly beans. One of the kids, however, tends to take jelly beans and only rarely gives one away. If each jelly bean represents an electron, who ends up being slightly negative? Who ends up being slightly positive? Is the negative kid just as negative as one of the positive kids is positive? Would you describe this as a polar or nonpolar situation? How about if all three kids were equally greedy?

72. Which is stronger: the covalent bond that holds atoms together within a molecule or the electrical attraction between two neighboring molecules? Please explain.

THINK AND DISCUSS (EVALUATION)

73. This chapter goes into a fair amount of detail explaining WHY oil and water don't mix and WHY water has a high boiling point. The reason why, of course, can be traced all the way back to the presence of two lone pairs of electrons on the oxygen atom of each water molecule. What value is there in understanding these sorts of details?

74. What should be done with the mining pits after all the ore as been removed? Consider the open-pit copper mine of Figure 6.14.

75. What are some of the obstacles people face when trying to recycle materials? How might these obstacles be overcome in your community? Should the government require that certain materials be recycled? If so, how should this requirement be enforced?

READINESS ASSURANCE TEST (RAT)

If you have a good handle on this chapter, then you should be able to score at least 7 out of 10 on this RAT. Check your answers online at www.ConceptualChemistry.com. If you score less than 7, you need to study further before moving on.

Choose the BEST answer to the following.

1. An atom loses an electron to another atom. Is this an example of a physical or chemical change?

   a. Chemical change involving the formation of ions.

   b. Physical change involving the formation of ions.

   c. Chemical change involving the formation of covalent bonds.

   d. Physical change involving the formation of covalent bonds.

2. Aluminum ions carry a 3+ charge, and chloride ions carry a 1– charge. What would be the chemical formula for the ionic compound aluminum chloride?

   a. Al_3Cl

   b. $AlCl_3$

   c. Al_3Cl_{3+}

   d. $AlCl$

3. Which would you expect to have a higher melting point: sodium chloride, NaCl, or aluminum oxide, Al_2O_3?

   a. The aluminum oxide has a higher melting point because it is a larger molecule and has a greater number of molecular interactions.

   b. NaCl has a higher melting point because it is a solid at room temperature.

   c. The aluminum oxide has a higher melting point because of the greater charges of the ions and hence the greater force of attractions between them.

   d. The aluminum oxide has a higher melting point because of the covalent bonds within the molecule.

4. Atoms of metallic elements can form ionic bonds, but they are not very good at forming covalent bonds. Why?

   a. These atoms are too large to be able to come in close contact with other atoms.

   b. They have a great tendency to lose electrons.

   c. Their valence shells are already filled with electrons.

   d. They are on the wrong side of the periodic table.

5. Why are ores so valuable?

   a. They are sources of naturally occurring gold.

   b. Metals can be efficiently extracted from them.

   c. They tend to occur in scenic mountainous regions.

   d. They hold many clues to the Earth's natural history.

6. Distinguish between a metal and a metal-containing compound.

   a. There is no distinction between the two.

   b. Only one of these contains ionic bonds.

   c. Only one of these contains covalent bonds.

   d. Only one of these occurs naturally.

7. In terms of the periodic table, is there an abrupt or gradual change between ionic and covalent bonds?

   a. An abrupt change occurs across the metalloids.

   b. Actually, any element of the periodic table can form a covalent bond.

   c. There is a gradual change: the farther apart, the more ionic.

   d. Whether an element forms one or the other depends on nuclear charge and not on the relative positions in the periodic table.

8. A hydrogen atom does not form more than one covalent bond because it

   a. has only one shell of electrons.

   b. has only one electron to share.

   c. loses its valence electron so readily.

   d. has such a strong electronegativity.

9. When nitrogen and fluorine combine to form a molecule, the most likely chemical formula is

   a. N_3F.

   b. N_2F.

   c. NF_4.

   d. NF.

   e. NF_3.

10. A substance consisting of which molecule shown below should have a higher boiling point?

    $$S=C=O \qquad O=C=O$$

    a. The molecule on the left, SCO, because it comes later in the periodic table.

    b. The molecule on the left, SCO, because it has less symmetry.

    c. The molecule on the right, OCO, because it has more symmetry.

    d. The molecule on the right, OCO, because it has more mass.

Toxic Wastes and the Superfund Act

The story of one of the most significant environmental disasters involving chemical pollutants began in 1892. This was the year when developer William T. Love proposed the building of a canal linking the upper and lower portions of the Niagara River, which is the river feeding Niagara Falls in western New York. By 1910, however, the project had lost its financial backing and all that remained was a ditch some 60 feet wide and 3000 feet long. Located just east of the town of Niagara Falls, the ditch was purchased by the Hooker Chemical Corporation, which maintained it as a dump site for toxic chemical wastes, including poisonous heavy metals such as mercury and cancer-causing solvents such as benzene. By the early 1950s, Love Canal was completely filled and covered over with earth, creating a long vacant field.

At about this time, the population of the town of Niagara Falls was growing rapidly and more land was needed for the building of new communities. In 1953, a local school board, using the threat of eminent domain, convinced Hooker Chemical to sell the land above and around Love Canal. Hooker sold the land for $1, with a warning of the dangers of the toxic wastes and a disclaimer of all subsequent liability. An elementary school was soon built directly above the toxic wastes, along with 800 single-family homes and 240 apartments in the surrounding neighborhood.

Over the years, the buried canisters ruptured, spewing their toxic contents underground. Being close to Lake Erie, the water table of Love Canal is close to the surface. After heavy rains, the water table would rise to mix with the leaking toxic wastes, which would leach away toward the basements of neighboring houses. In some places, the wastes would leach directly to the surface. The residents complained about foul odors and health issues,

but city and county officials did not know how to respond aside from installing fans in homes or covering pools of toxic wastes with dirt.

▲ **Figure 1**
A 1978 aerial view of the south end of the Love Canal district.

Finally, in 1978, health officials began to study the situation. They found numerous toxins in the surrounding soil and were able to correlate these toxins to reported health problems such as miscarriages and birth defects. The school was closed, and 239 families living on or immediately adjacent to Love Canal were evacuated. A fence was built around the canal to keep people away from the toxic wastes. The many families living just outside the fenced area, however, grew increasingly angry. They not only continued to face unusual health problems but also were unable to move away, because their homes were no longer sellable. Their concerns captured little attention until on May 19, 1980, they held hostage two representatives of the Environmental Protection Agency (EPA), with the demand that the government buy their homes so that they could relocate. Within a few days, then-pres-

ident Jimmy Carter promised the funding that allowed all Love Canal families to leave.

Love Canal turned the nation's attention to the problem of toxic wastes. It was soon recognized that Love Canal was just one of thousands of areas where toxic wastes were improperly disposed of. To remedy this massive problem, in December 1980, Congress enacted the Comprehensive Environmental Response, Compensation, and Liability Act, which established a "Superfund" that would provide the money to clean up abandoned hazardous sites. This fund was to be supported by special taxes on industries related to the production of toxic chemicals. Furthermore, in cases where the polluter could be identified, the act granted the government the authority to enforce a "polluter pays" policy. For example, the federal government spent $101 million from the Superfund to remedy the Love Canal toxic wastes. By 1995, federal courts ruled that the Hooker Corporation, despite its original disclaimer, was liable. Occidental Petroleum, which had bought Hooker in 1969, was thus required to reimburse the $101 million to the Superfund and

▲ **Figure 2**
Clean up at Love Canal involved the removal of easily accessible toxic-waste containers, as shown here, along with polluted soil. It was deemed, however, that the bulk of the buried toxic wastes would be too dangerous to remove. Instead, the site was capped with a thick layer of water-resistant clay. Also, a trench system was dug around the site to direct groundwater to a specially designed water treatment facility.

to pay the $27 million in interest the Superfund would have earned had this money not been spent.

Over its history, however, the Superfund has received only a small fraction of its funding from the collection of fines from the polluters. The reason is that in many cases, the polluter cannot be identified or has since gone out of business or has no money to pay. Then in 1994, Congress decided that chemical industries should no longer pay the Superfund tax. As shown in Figure 3, until that year, these taxes (shown in green) were the main source of funding. Since then, the government has had to dedicate increasing amounts of public money each year to keep the Superfund program afloat. For example, the 2009 economic stimulus program of the Obama administration directed $600 million to specific Superfund sites that were in danger of no longer being supported. As of the writing of this edition, there has been a growing political will to reinstate the Superfund taxes on industries.

Since its enactment, the Superfund program has taken responsibility for about 1600 toxic-waste sites. Cleanup efforts have been completed at about 300 of these sites. Of the still-active sites, another 700 or so are near completion. Most of these sites, however, are small sites where the cleanup has been relatively easy. A greater proportion of the remaining sites are the so-called "megasites," which will require decades of cleanup efforts and environmental monitoring. To find the Superfund site nearest to where you live and to learn the story of how it came to be there, visit www.epa.gov/superfund.

CONCEPT CHECK

Why were heavy rains particularly dangerous for the former residents of Love Canal?

WAS THIS YOUR ANSWER? Heavy rains raised the water table, allowing it to mix with the hazardous wastes, which would then leach outward to the surrounding homes.

Think and Discuss

1. In a 1983 settlement, Occidental Petroleum paid $20 million to 1328 residents of Love Canal and set aside another $1 million for a medical trust fund. Do you think this was sufficient, too little, or too much? Why? Consider that the deeds of the homeowners contained clauses warning them of the health risks of the area.

2. Why did the town of Niagara Falls not have to pay into the Superfund, especially given that it pressured the Hooker Corporation into selling the land? Might the town have been liable had it taken the land against the will of Hooker by way of eminent domain?

3. In the 1940s, it was the accepted rule that if a company owned a parcel of land, the company had the right to use that land as a dump site for whatever materials it wanted to dispose of, including hazardous wastes. Accordingly, Hooker had the right to use Love Canal as a toxic-waste repository. Was it fair, then, that the parent company of the Hooker Corporation had to pay for the cleanup of Love Canal?

4. Over many years, the Homeowners Association of Wide Beach Development, New York, along the shores of Lake Erie, sprayed thousands of gallons of waste oil onto area dirt roads to control dust. Unknown to the association, this oil contained dangerous cancer-causing chemicals known as PCBs. Should the Wide Beach Development qualify as a Superfund site, or should the Homeowners Association take care of the problem themselves?

5. The American Chemical Council, which represents chemical industries, argues that Superfund taxes unfairly punish all chemical manufacturers for the mistakes of a few. Furthermore, all consumers benefit from the products (from soaps to toys to computers) of chemical manufacturers, so supporting the Superfund with public taxpayer money is most appropriate. Do you agree or disagree? Why?

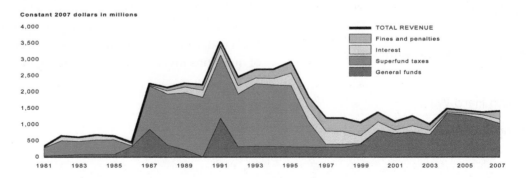

▲ Figure 3
Partial contributions from various sources to the Superfund trust fund from 1981 through 2007. Note that in 1993, Superfund taxes accounted for about 70 percent of the trust revenue. More recently, the proportion of revenue coming in from fines has been small. Thus, the "polluter pays" ethic of the original Superfund Act has been essentially lost.

7

How Molecules Mix

THE MAIN IDEA

Molecules are "sticky."

▲ Water in these hot springs is loaded with calcium carbonate, a mineral that calcifies to form these spectacular pools of Pamukkale, Turkey.

In Chapter 6, we focused on how atoms bond together to form ionic or covalent compounds. You understand from earlier chapters that the formation of these compounds is a *chemical change*. In this chapter, we look at how the ions of ionic compounds and the molecules of covalent compounds—once created—interact to form mixtures, which, as you know, is a *physical change*.

So when you stir sugar into water, the sugar crystals disappear, but where do they go? What does it means to say that one solution is more concentrated, while another is more dilute? Why does a warm soda fizzle in your mouth more than a cold one does? How does soap remove water-resistant grime? What is "hard water," and why does it hinder soap from working? How is water made safe for drinking, and why is it so difficult to create fresh water from ocean water? The answers to these sorts of questions involve an understanding of mixtures.

Chemistry

Circular Rainbows

We can separate the components of ink through a technique called paper chromatography. All you need are some felt-tip pens or water-soluble markers; some porous paper, such as a paper towel, table napkin, or coffee filter; and some water.

PROCEDURE

1. Place a concentrated dot of ink at the center of the piece of porous paper.
2. Dip your finger in some water so that a drop of water hangs from the tip of your finger. Touch this drop of water to the dot of ink and watch the water get absorbed by the paper. The

moment it is all absorbed, add another drop. Continue adding drops in this fashion, allowing the water to spread radially through the paper.
3. The different color components of the ink will separate while the water travels through the paper. Experiment using different inks and paper. Try other solvents, such as rubbing alcohol, fingernail-polish remover, or white vinegar.

ANALYZE AND CONCLUDE

1. The different components of the ink have differing affinities for the paper and the mobile solvent. Which components have a greater affinity for the

paper? Which have a greater affinity for the water?
2. Why doesn't the ink from a permanent marker work for this activity?
3. How do you suppose a chemist separates the different products she's created through a chemical reaction?

7.1 Four Different Types of Dipole Attractions

EXPLAIN THIS

Is it possible for a fish to die from drowning?

A bit of review should prove helpful for your conceptual understanding of this chapter on mixtures. First, recall from Chapter 6 that compounds can have what we call a *dipole*. What this means is that the electrons within the compound are not distributed evenly. Instead, they tend to congregate preferentially to one side of the compound. An extreme case of a dipole occurs within an ionic compound such as sodium chloride, NaCl. With sodium chloride, the bonding electrons spend almost all of their time with the chlorine. This transforms the chlorine atom into a negatively charged chloride ion, Cl^-, while the sodium becomes a positively charged sodium ion, Na^+. Sodium chloride, NaCl, has a very high boiling point, around 1413°C, because of the strong attractions among all the highly charged sodium and chloride ions.

A milder form of a dipole occurs with water, in which the oxygen atom of each water molecule pulls electrons away from the hydrogen atoms, as was shown in Figure 6.34. This makes the oxygen side of the water molecule slightly negative, while the hydrogen side is slightly positive. A molecule with such a dipole is said to be a *polar molecule*. Water has a relatively high boiling point of 100°C because of the electrical attractions among all the polar water molecules.

We begin this chapter on mixtures by describing four kinds of molecular attractions, as shown in Table 7.1. Each is electrical in nature, involving the attraction between positive and negative changes.

Ion–Dipole Attractions

What happens to polar molecules, such as water molecules, when they are near an ionic compound, such as sodium chloride? The opposite charges electrically attract one another. A positive sodium ion attracts the negative

LEARNING OBJECTIVE

Identify four different types of dipole attractions and their role in determining the physical properties of a material.

TABLE 7.1   Molecular Attractions Involving Dipoles

ATTRACTION	RELATIVE STRENGTH
Ion–dipole	Strongest
Dipole–dipole	↑
Dipole–induced dipole	
Induced dipole–induced dipole	Weakest

Figure 7.1
Electrical attractions are shown as a series of overlapping arcs. The blue arcs indicate negative charge, and the red arcs indicate positive charge.

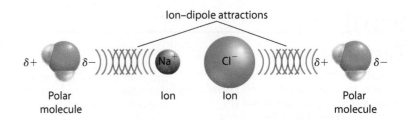

side of a water molecule, and a negative chloride ion attracts the positive side of a water molecule. This phenomenon is illustrated in **Figure 7.1**. Such an attraction between an ion and the dipole of a polar molecule is called an *ion–dipole* attraction.

Ion–dipole attractions are much weaker than ionic bonds. However, a large number of ion–dipole attractions can act collectively to disrupt ionic bonds. This is what happens to sodium chloride in water. Attractions exerted by the water molecules break the ionic bonds and pull the ions away from one another. The result, represented in **Figure 7.2**, is a solution of sodium chloride in water. (A solution in water is called an *aqueous solution*.)

Dipole–Dipole Attractions

An attraction between two polar molecules is called a *dipole–dipole* attraction. An unusually strong dipole–dipole attraction is the **hydrogen bond.** This attraction occurs between molecules that have a hydrogen atom covalently bonded to a small, highly electronegative atom, usually nitrogen, oxygen, or fluorine. Recall from Section 6.7 that the electronegativity of an atom describes how well that atom is able to pull bonding electrons toward itself. The greater the atom's electronegativity, the better it is able to attract bonding electrons and become negatively charged.

Look at **Figure 7.3** to see how hydrogen bonding works. The hydrogen side of a polar molecule (water, in this example) has a partial positive charge because the more electronegative oxygen atom pulls more strongly on the electrons of the covalent bond. The hydrogen is therefore electrically attracted to a pair of non-bonding electrons on the partially negatively charged atom of another molecule (in this case, another water molecule). This mutual attraction between hydrogen and the negatively charged atom of another molecule is a hydrogen bond.

Even though the hydrogen bond is much weaker than any covalent or ionic bond, the effects of hydrogen bonding can be very pronounced. As we explore in Chapter 8, water owes many of its properties to hydrogen bonds. The hydrogen bond is also of great importance in the chemistry of the large molecules, such as DNA and proteins, which we discuss in Chapter 13.

Figure 7.2
Sodium and chloride ions tightly bound in a crystal lattice are separated from one another by the collective attraction exerted by many water molecules to form an aqueous solution of sodium chloride.

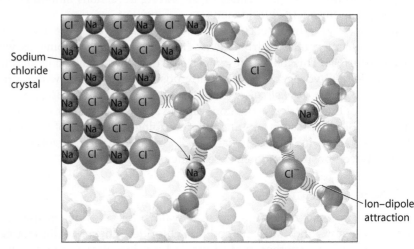

Aqueous solution of sodium chloride

Dipole–Induced Dipole Attractions

In many molecules, the electrons are distributed evenly, so there is no dipole. The oxygen molecule, O_2, is an example. Such a nonpolar molecule can be induced to become a temporary dipole, however, when it is brought close to a water molecule (or to any other polar molecule), as **Figure 7.4** illustrates. The slightly negative side of the water molecule pushes the electrons in the oxygen molecule away. Thus, the oxygen molecule's electrons are pushed to the side that is farther from the water molecule. The result is a temporarily uneven distribution of electrons called an **induced dipole.** The resulting attraction between the permanent dipole (water) and the induced dipole (oxygen) is a *dipole–induced dipole* dipole attraction.

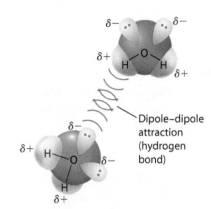

Dipole–dipole attraction (hydrogen bond)

▲ **Figure 7.3**
The dipole–dipole attraction between two water molecules is a hydrogen bond because it involves hydrogen atoms bonded to highly electronegative oxygen atoms.

CONCEPT CHECK

How does the electron distribution in an oxygen molecule change when the hydrogen side of a water molecule is nearby?

CHECK YOUR ANSWER Because the hydrogen side of the water molecule is slightly positive, the electrons in the oxygen molecule are pulled toward the water molecule, inducing in the oxygen molecule a temporary dipole in which the larger side is nearer the water molecule (rather than as far away as possible, as shown in Figure 7.4).

Remember, induced dipoles are only temporary. If the water molecule in **Figure 7.4b** were removed, the oxygen molecule would return to its normal, nonpolar state. In general, dipole–induced dipole attractions are much weaker than dipole–dipole attractions. But dipole–induced dipole attractions are strong enough to hold relatively small quantities of oxygen dissolved in water, as depicted in **Figure 7.5**. This attraction between water and molecular oxygen is vital for fish and other forms of aquatic life that rely on molecular oxygen dissolved in water.

Dipole–induced dipole attractions are also responsible for holding plastic wrap to glass, as shown in **Figure 7.6**. These wraps are made of very long nonpolar molecules that are induced to have dipoles when placed in contact with glass, which is highly polar. As we will discuss next, the molecules of a nonpolar material, such as plastic wrap, can also induce dipoles among themselves. This explains why plastic wrap sticks not only to polar materials such as glass but also to itself.

CONCEPT CHECK

Distinguish between a dipole–dipole attraction and a dipole–induced dipole attraction.

CHECK YOUR ANSWER The dipole–dipole attraction is stronger and involves two permanent dipoles. The dipole–induced dipole attraction is weaker and involves a permanent dipole and a temporary one.

Dipole–induced dipole attraction

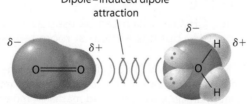

Isolated oxygen molecule (nonpolar)

(a)

Induced dipole (oxygen molecule)

Permanent dipole (water molecule)

(b)

◀ **Figure 7.4**
(a) An isolated oxygen molecule has no dipole; its electrons are distributed evenly. (b) An adjacent water molecule induces a redistribution of electrons in the oxygen molecule. (The slightly negative side of the oxygen molecule is shown as larger than the slightly positive side because the slightly negative side contains more electrons.)

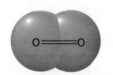

Is it possible for the electrons of an atom or a nonpolar molecule to be bunched to one side?

▲ Figure 7.5
The electrical attraction between water and oxygen molecules is relatively weak, which explains why not much oxygen is able to dissolve in water. For example, water fully aerated at room temperature contains only about 1 oxygen molecule for every 200,000 water molecules. The gills of a fish, therefore, must be highly efficient at extracting this molecular oxygen from water.

▲ Figure 7.6
Temporary dipoles induced in the normally nonpolar molecules in plastic wrap make it stick to glass.

Nonpolar   Temporary dipole
argon   in argon

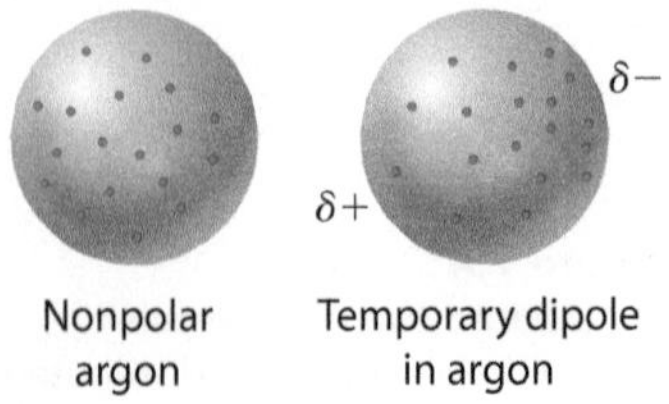

▲ Figure 7.7
The electron distribution in an atom is normally even. At any given moment, however, the electron distribution may be somewhat uneven, resulting in a temporary dipole.

Induced Dipole–Induced Dipoles Attractions

Individual atoms and nonpolar molecules, on average, have a fairly even distribution of electrons. Because of the randomness of electron distribution, however, at any given moment, the electrons in an atom or a nonpolar molecule may be bunched to one side. The result is a temporary dipole, as shown in **Figure 7.7**.

Just as the permanent dipole of a polar molecule can induce a dipole in a nonpolar molecule, a temporary dipole can do the same thing. This gives rise to the relatively weak *induced dipole–induced dipole* attraction, illustrated in **Figure 7.8**.

CONCEPTCHECK

Distinguish between a dipole–induced dipole attraction and an induced dipole–induced dipole attraction.

CHECK YOUR ANSWER The dipole–induced dipole attraction is stronger and involves a permanent dipole and a temporary one. The induced dipole–induced dipole attraction is weaker and involves two temporary dipoles.

Induced dipole–induced dipole attractions (sometimes called *dispersion forces*) help explain why natural gas is a gas at room temperature but gasoline is a liquid. The major component of natural gas is methane, CH_4, and one of the major components of gasoline is octane, C_8H_{18}. We can see in **Figure 7.9** that the number of induced dipole–induced dipole attractions between two methane molecules is appreciably less than the number between two octane molecules. You know that two small pieces of Velcro are easier to pull apart than two long pieces. Like short pieces of Velcro, methane molecules can be pulled apart with little effort. That's why methane has a low boiling point, –161°C, and is a gas at room temperature. Octane molecules, like long strips of Velcro, are relatively difficult to pull apart because of the larger number of induced dipole–induced dipole attractions. The boiling point of octane, 125°C, is therefore much higher than that of methane, and octane is a liquid at room temperature. (The greater mass of octane also plays a role in making its boiling point higher.)

Induced dipole–induced dipole attractions also explain how the gecko can race up a glass wall and support its entire body weight with only a single toe. A gecko's feet are covered with billions of microscopic hairs called *spatulae*, each of which is about 1/300 as thick as a human hair. The force of attraction between these hairs and the wall is the weak induced dipole–induced dipole attraction. But because there are so many hairs, the surface area of contact is relatively great; hence, the total force of attraction is enough to prevent the gecko from falling (**Figure 7.10**). Research is currently under way to develop a synthetic, dry glue based on gecko adhesion. Velcro, watch out!

CONCEPTCHECK

Methanol, CH_3OH, which can be used as a fuel, is not much larger than methane, CH_4, but it is a liquid at room temperature. Suggest why.

CHECK YOUR ANSWER The polar oxygen–hydrogen covalent bond in each methanol molecule leads to hydrogen bonding between molecules. These relatively strong interparticle attractions hold methanol molecules together as a liquid at room temperature.

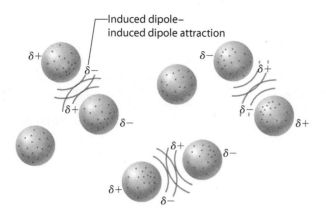

Induced dipole–
induced dipole attraction

◀ **Figure 7.8**
Because the normally even distribution of electrons in atoms can momentarily become uneven, atoms can be attracted to one another through induced dipole–induced dipole attractions.

Temporary dipoles are more significant for larger atoms. A reason for this is that their outermost electrons are relatively far from the nucleus as well as from each other. As a result, these electrons are easily "pushed around," as described in **Figure 7.11**. Thus, larger atoms are sometimes described as "soft"—more like a marshmallow than a marble. The technical term for this property is *polarizability*. We say that a larger atom is more "polarizable," which means it can form an induced dipole more easily.

Fluorine is one of the smallest atoms, which means it is a "hard" atom (like a marble) and not very polarizable. Nonpolar molecules made with fluorine atoms exhibit only very weak induced dipole–induced dipole attractions. This is the principle behind the Teflon® nonstick surface. The Teflon® molecule, part of which is shown in **Figure 7.12**, is a long chain of carbon atoms chemically bonded to fluorine atoms, and the fluorine atoms exert essentially no attractions on any material in contact with the Teflon® surface—scrambled eggs in a frying pan, for instance.

FOR YOUR INFORMATION

Dipole–induced dipole attractions are sometimes called *Debye forces*, while induced dipole–induced dipole attractions are sometimes called *London dispersion forces*, or more simply *London forces*. Each of these are the last names of the 20th-century scientists who first described them.

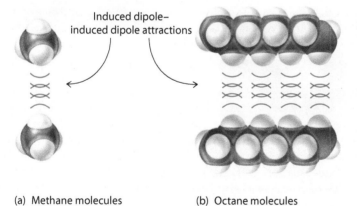

Induced dipole–
induced dipole attractions

(a) Methane molecules   (b) Octane molecules

▲ **Figure 7.9**
(a) Two nonpolar methane molecules are attracted to each other by induced dipole–induced dipole attractions, but there is only one attraction per molecule. (b) Two nonpolar octane molecules are similar to methane molecules, but they are longer. The number of induced dipole–induced dipole attractions between these two molecules is therefore greater.

▶ **Figure 7.10**
If the gecko's foot is so sticky, how does the gecko keep its feet clean? Answer: The gecko's foot is extremely nonpolar. Dirt may stick to it briefly, but after a few steps, the dirt sticks better to the surface upon which the gecko walks.

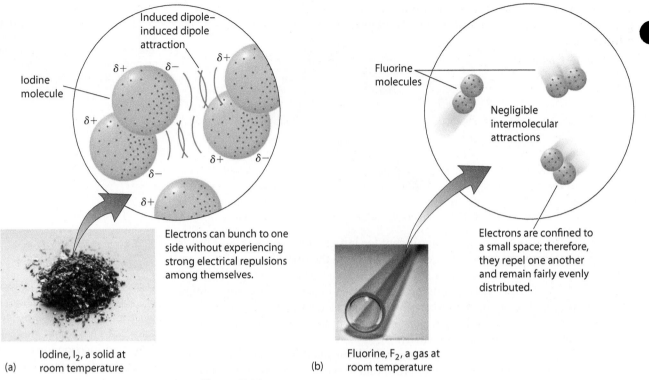

△ Figure 7.11
(a) Temporary dipoles form more readily in larger atoms, such as those in an iodine molecule, because in larger atoms, electrons are more loosely held and can bunch to one side and still be relatively far apart from one another. (b) In smaller atoms, such as those in a fluorine molecule, electrons are tightly held and cannot bunch to one side so well because the repulsive electric force increases as the electrons get closer together.

▶ **Figure 7.12**
Few things stick to Teflon® because of the high proportion of fluorine atoms that it contains. The structure depicted here is only a portion of the full length of the molecule.

$$\cdots -\overset{\overset{\displaystyle F}{|}}{\underset{\underset{\displaystyle F}{|}}{C}} -\overset{\overset{\displaystyle F}{|}}{\underset{\underset{\displaystyle F}{|}}{C}} -\overset{\overset{\displaystyle F}{|}}{\underset{\underset{\displaystyle F}{|}}{C}} -\overset{\overset{\displaystyle F}{|}}{\underset{\underset{\displaystyle F}{|}}{C}} -\overset{\overset{\displaystyle F}{|}}{\underset{\underset{\displaystyle F}{|}}{C}} -\overset{\overset{\displaystyle F}{|}}{\underset{\underset{\displaystyle F}{|}}{C}} -\cdots$$

7.2 A Solution Is a Single-Phase Homogeneous Mixture

LEARNING OBJECTIVE

Describe the formation of saturated and unsaturated solutions from a molecular point of view.

EXPLAIN THIS

In "The Wizard of Oz," was the Wicked Witch of the West melting in the water that Dorothy threw on her?

What happens when table sugar, known chemically as sucrose, is stirred into water? Is the sucrose destroyed? We know it isn't because it sweetens the water and all we have to do is evaporate the water and the sugar reappears in its solid form. When sucrose is dissolved in water, does it disappear because it somehow ceases to occupy space or because it fits within the nooks and crannies of the water? Not so, for the addition of sucrose changes the volume.

This may not be noticeable at first, but if you continue adding sucrose to a glass of water, you'll see that the water level rises, just as it would if you were adding sand.

Sucrose stirred into water loses its crystalline form. Each sucrose crystal consists of billions upon billions of sucrose molecules packed neatly together. When the crystal is exposed to water (as was first shown in Figure 3.25 and is shown again here in **Figure 7.13**), an even greater number of water molecules pull on the sucrose molecules via hydrogen bonds formed between the sucrose molecules and the water molecules. With a little stirring, the sucrose molecules soon mix throughout the water. In place of sucrose crystals and water, we have a homogeneous mixture of sucrose molecules in water. As discussed earlier, *homogeneous* means that a sample taken from any part of a mixture is the same as a sample taken from any other part. In our sucrose example, this means that the sweetness of the first sip of the solution is the same as the sweetness of the last sip.

Recall that a homogeneous mixture consisting of a single phase is called a *solution*. Sugar water is a solution in the liquid phase. Solutions aren't always liquids, however. They can also be solid or gaseous, as **Figure 7.14** shows. Gemstones are solid solutions. A ruby, for example, is a solid solution of trace quantities of red chromium compounds in transparent aluminum oxide. A blue sapphire is a solid solution of trace quantities of light green iron compounds and blue titanium compounds in aluminum oxide. Another important example of solid solutions is metal alloys, which are mixtures of different metallic elements. The alloy known as brass is a solid solution of copper and zinc, for instance, and the alloy stainless steel is a solid solution of iron, chromium, nickel, and carbon.

An example of a gaseous solution is the air we breathe. By volume, this solution is 78 percent nitrogen gas, 21 percent oxygen gas, and 1 percent other gaseous materials, including water vapor and carbon dioxide. The air we exhale is a gaseous solution of 75 percent nitrogen, 14 percent oxygen, 5 percent carbon dioxide, and around 6 percent water vapor. The difference between the air we inhale and exhale is a result of the chemical changes going on within our bodies.

In describing solutions, the component present in the largest amount is the **solvent** and any other components are **solutes.** For example, when a teaspoon of table sugar is mixed with 1 liter of water, we identify the sugar as the solute and the water as the solvent.

The process of a solute's mixing with a solvent is called **dissolving.** To make a solution, a solute must dissolve in a solvent; that is, the solute and solvent must form a homogeneous mixture. Whether one material dissolves in another is a function of the electrical attractions between their molecules. The stronger these electrical attractions are, the greater the likelihood that dissolving will occur.

FOR YOUR INFORMATION

Be it large or small, an individual sugar crystal is transparent. A teaspoon of table sugar appears white because of the way light gets scattered as it passes in and out of the numerous tiny crystals at numerous different angles. Technically speaking, when table sugar dissolves in water, it is this scattering effect that disappears, not the sugar itself.

READING CHECK

What is the difference between the solvent and the solute?

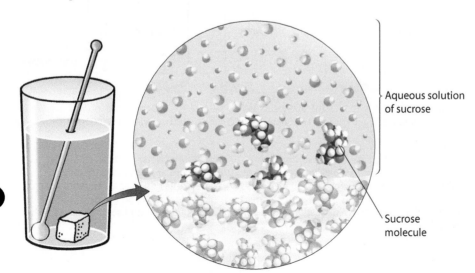

◀ Figure 7.13
Water molecules pull the sucrose molecules in a sucrose crystal away from one another. This pulling away does not, however, affect the covalent bonds within each sucrose molecule, which is why each dissolved sucrose molecule remains intact as a single molecule.

Aqueous solution of sucrose

Sucrose molecule

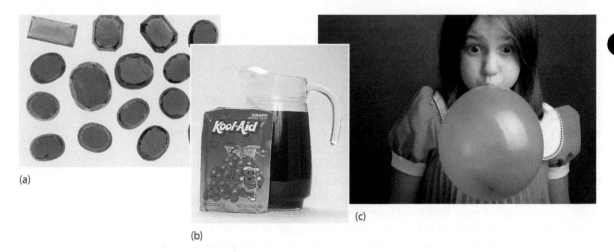

(a)

(b)

(c)

▲ Figure 7.14
Solutions may occur in (a) the solid phase, (b) the liquid phase, or (c) the gaseous phase.

CONCEPTCHECK
What is the solvent in the gaseous solution we call air?

CHECK YOUR ANSWER Nitrogen is the solvent, because it is the component present in the greatest quantity.

There is a limit to how much of a given solute can be dissolved in a given solvent, as **Figure 7.15** illustrates. We know that when you add sugar to a glass of water, for example, the sugar rapidly dissolves. As you continue to add sugar, however, there comes a point when it no longer dissolves. Instead, it collects at the bottom of the glass, even after stirring. At this point, the water is *saturated* with sugar, meaning that the water cannot accept any more sugar. When this happens, we have what is called a **saturated solution,** defined as a solution in which no more solute can be dissolved. A solution that has not reached the limit of solute that will dissolve is called an **unsaturated solution.**

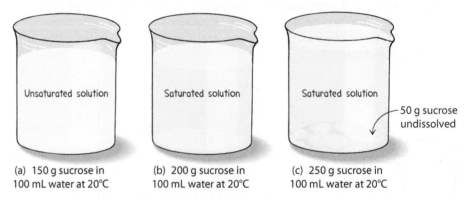

(a) 150 g sucrose in 100 mL water at 20°C

(b) 200 g sucrose in 100 mL water at 20°C

(c) 250 g sucrose in 100 mL water at 20°C

50 g sucrose undissolved

▲ Figure 7.15
A maximum of 200 grams of sucrose dissolves in 100 milliliters of water at 20°C. (a) Mixing 150 grams of sucrose in 100 milliliters of water at 20°C produces an unsaturated solution. (b) Mixing 200 grams of sucrose in 100 milliliters of water at 20°C produces a saturated solution. (c) If 250 grams of sucrose is mixed with 100 milliliters of water at 20°C, 50 grams of sucrose remains undissolved. (As we discuss later, the concentration of a saturated solution varies with temperature.)

7.3 Concentration Is Given as Moles per Liter

EXPLAIN THIS

What is the solvent in brown sugar?

The quantity of solute dissolved in a solution is described in mathematical terms by the solution's **concentration,** which is the amount of solute dissolved per amount of solution:

$$\text{concentration} = \frac{\text{amount of solute}}{\text{amount of solution}}$$

For example, a sucrose–water solution may have a concentration of 1 gram of sucrose for every liter of solution. This can be compared with concentrations of other solutions. A sucrose–water solution containing 2 grams of sucrose per liter of solution, for example, is more concentrated, and one containing only 0.5 gram of sucrose per liter of solution is less concentrated, or more dilute.

Chemists are often more interested in the number of solute particles in a solution than in the number of grams of solute. Submicroscopic particles, however, are so very small that the number of them in any observable sample is incredibly large. To avoid awkwardly large numbers, scientists use a unit called the mole. One **mole** of any type of particle is equal to 6.02×10^{23} particles. (This superlarge number is 602 billion trillion, or 602,000,000,000,000,000,000,000 particles.) (Interestingly, the term *mole* is derived from the Latin word *moles*, meaning heap, mass, or pile.)

One mole of gold atoms, for example, is 6.02×10^{23} gold atoms, and 1 mole of sucrose molecules is 6.02×10^{23} sucrose molecules.

Even if you've never heard the term *mole* before now, you are already familiar with the basic idea. Saying "one mole" is just a shorthand way of saying "six point oh two times ten to the twenty-third particles." Just as "a couple of" means two of something and "a dozen of" means 12 of something, "a mole of" means 6.02×10^{23} of some elementary unit, such as atoms, molecules, or ions. It's as simple as that:

- a couple of coconuts = 2 coconuts
- a dozen doughnuts = 12 doughnuts
- a mole of molecules = 6.02×10^{23} molecules

One mole of gold atoms, for example, is 6.02×10^{23} gold atoms, and 1 mole of sucrose molecules is 6.02×10^{23} sucrose molecules. A stack containing "1 mole" of pennies would reach a height of about 860 quadrillion kilometers, which is roughly equal to the diameter of our galaxy, the Milky Way. And "1 mole" of marbles would be enough to cover the entire land area of the 50 United States to a depth greater than 1.1 kilometers.

But sucrose molecules are so small that 6.02×10^{23} of them are in only 342 grams of sucrose, which is about a cupful. Thus, because 342 grams of sucrose contains 6.02×10^{23} molecules of sucrose, we can use our shorthand wording and say that 342 grams of sucrose contains 1 mole of sucrose. As **Figure 7.16** shows, therefore, an aqueous solution that has a concentration of 342 grams of sucrose per liter of solution also has a concentration of 6.02×10^{23} sucrose molecules per liter of solution, or, by definition, a concentration of 1 mole of sucrose per liter of solution. The number of grams tells you the mass of solute in a given solution, and the number of moles indicates the actual number of molecules.

READING CHECK

What does concentration measure?

FOR YOUR INFORMATION

It would take 11.6 days of nonstop counting to count to a million. To count to a billion would take 31.7 years. To count to a trillion would take 31,700 years! Counting to a trillion 602 billion times would take about 2 million times the estimated age of the universe. In short, 602 billion trillion, 6.02×10^{23}, is an inconceivably large number.

▶ **Figure 7.16**
An aqueous solution of sucrose that has a concentration of 1 mole of sucrose per liter of solution contains 6.02×10^{23} sucrose molecules (342 grams) in every liter of solution.

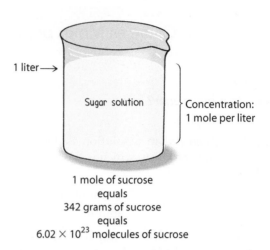

1 liter →

Sugar solution

Concentration: 1 mole per liter

1 mole of sucrose
equals
342 grams of sucrose
equals
6.02×10^{23} molecules of sucrose

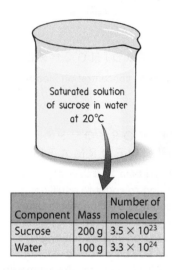

Saturated solution of sucrose in water at 20°C

Component	Mass	Number of molecules
Sucrose	200 g	3.5×10^{23}
Water	100 g	3.3×10^{24}

▲ **Figure 7.17**
Although 200 grams of sucrose is twice as massive as 100 grams of water, there are about 10 times as many water molecules in 100 grams of water as there are sucrose molecules in 200 grams of sucrose. How can this be? Each water molecule is about one-twentieth as massive (and much smaller) than each sucrose molecule.

A common unit of concentration used by chemists is **molarity,** which is the solution's concentration expressed in moles of solute per liter of solution:

$$\text{molarity} = \frac{\text{number of moles of solute}}{\text{liters of solution}}$$

A solution that contains 1 mole of solute per liter of solution is a 1-molar solution, which is often abbreviated 1 *M*. A 2-molar (2 *M*) solution contains 2 moles of solute per liter of solution.

The difference between referring to the number of molecules of solute and referring to the number of grams of solute can be illustrated by the following question. A saturated aqueous solution of sucrose contains 200 grams of sucrose and 100 grams of water. Which is the solvent: sucrose or water?

As shown in **Figure 7.17**, there are 3.5×10^{23} molecules of sucrose in 200 grams of sucrose, but there are almost 10 times as many molecules of water in 100 grams of water—3.3×10^{24} molecules. As defined earlier, the solvent is the component present in the largest amount, but what do we mean by *amount?* If *amount* means number of molecules, then water is the solvent. If amount means mass, then sucrose is the solvent. So the answer depends on how you look at it. From a chemist's point of view, *amount* typically means the number of molecules, so water is the solvent in this case.

 FORYOUR INFORMATION

A "mole" of stacked pennies would stretch across our galaxy. Make this estimation yourself. First, measure the number of stacked pennies in 1 centimeter. To find the length of 1 "mole" of stacked pennies, take the number of particles in 1 mole (6.02×10^{23}) and divide it by the number of pennies in 1 cm. Your answer will be in centimeters. To convert to kilometers, divide by the number of centimeters in a kilometer, which is 100,000. Tall stack!

CONCEPTCHECK

1. How many moles of sucrose are in 0.5 liter of a 2-molar solution? How many molecules of sucrose is this?

2. Does 1 liter of a 1-molar solution of sucrose in water contain 1 liter of water, less than 1 liter of water, or more than 1 liter of water?

CHECK YOUR ANSWER First, you need to understand that 2-molar means 2 moles of sucrose per liter of solution. To obtain the amount of solute, you should multiply solution concentration by amount of solution:

(2 moles/L)(0.5 L) = 1 mole, which is the same as 6.02×10^{23} molecules.

The definition of molarity refers to the number of liters of solution, not to the number of liters of solvent. When sucrose is added to a given volume of water, the volume of the solution increases. So if 1 mole of sucrose is added to 1 liter of water, the result is more than 1 liter of solution. Therefore, 1 liter of a 1-molar solution requires less than 1 liter of water.

CALCULATION CORNER CONCENTRATING ON SOLUTIONS

From the formula for the concentration of a solution, we can derive equations for the amount of solute and the amount of solution:

$$\text{concentration of solution} = \frac{\text{amount of solute}}{\text{volume of solution}}$$

$$\text{amount of solute} =$$

$$\text{concentration of solution} \times \text{volume of solution}$$

$$\text{volume of solution} = \frac{\text{amount of solute}}{\text{concentration of solution}}$$

In solving for any of these values, the units must always match. If concentration is given in grams per liter of solution, for example, the amount of solute must be in grams and the amount of solution must be in liters.

Note that these equations are set up for calculating the volume of solution rather than the volume of solvent. The volume of solution is greater than the volume of solvent because, in addition to containing the solvent, the solution also contains the solute. As discussed at the beginning of Section 7.2, for example, the volume of an aqueous solution of sucrose depends not only on the volume of water but also on the volume of dissolved sucrose.

EXAMPLE 1

How many grams of sucrose are in 3 liters of an aqueous solution that has a concentration of 2 grams of sucrose per liter of solution?

ANSWER 1

This question asks for amount of solute, so you should use the second of the three formulas given previously:

$$\text{amount of solute} = \frac{2\,\text{g}}{1\,\text{L}} \times 3\,\text{L} = 6\,\text{g}$$

EXAMPLE 2

A solution you are using in an experiment has a concentration of 10 grams of solute per liter of solution. If you pour enough of this solution into an empty laboratory flask so that the flask contains 5 grams of the solute, how many liters of the solution have you poured into the flask?

ANSWER 2

This question asks for amount of solution, and you will want to use the third formula:

$$\text{volume of solution} = \frac{5\,\text{g}}{10\,\text{g/L}} = 0.5\,\text{L}$$

YOUR TURN

1. At 20°C, a saturated solution of sodium chloride in water has a concentration of about 380 grams of sodium chloride per liter of solution. How many grams of sodium chloride are required to make 3 liters of a saturated solution?

2. A student is told to use 20 grams of sodium chloride to make an aqueous solution that has a concentration of 10 grams of sodium chloride per liter of solution. How many liters of solution does she end up with?

Answers to Calculation Corners appear at the end of each chapter.

7.4 Solubility Is How Well a Solute Dissolves

EXPLAIN THIS

How can oxygen be removed from water?

The **solubility** of a solute is its *ability* to dissolve in a solvent. As can be expected, solubility mainly depends on the attractions between the fundamental particles of the solute and solvent. If a solute has any appreciable solubility in a solvent, then that solute is said to be **soluble** in that solvent.

Solubility also depends on attractions of solute particles for one another and attractions of solvent particles for one another. As shown in **Figure 7.18**, for example, there are many polar hydrogen–oxygen bonds in a sucrose molecule. Sucrose molecules, therefore, can form multiple hydrogen bonds with one another. These hydrogen bonds are strong enough to make sucrose a solid at room temperature and to give it the relatively high melting point of 185°C. In order for sucrose to dissolve in water, the water molecules must first pull sucrose molecules away from

LEARNING OBJECTIVE

Discuss how solutes dissolve in solvents and how solubility changes with temperature.

READING CHECK

Solubility depends upon what?

CH_2OH

H   H
OH   H
HO

H   OH

O

H
HOCH_2
O

H   HO
H   HO

H
CH_2OH

OH   H

Sucrose

▲ Figure 7.18
A sucrose molecule contains many hydrogen–oxygen covalent bonds, in which the hydrogen atoms are slightly positive and the oxygen atoms are slightly negative. These dipoles in any given sucrose molecule result in the formation of hydrogen bonds with neighboring sucrose molecules.

 FOR YOUR INFORMATION

Water sometimes gets into the gas lines of cars, usually by condensing from moist atmosphere. In regions where the winter gets super cold, this water can clog the gas line by freezing. To prevent this from happening, the mindful driver pours in a small 12-ounce bottle of *gas line antifreeze* with each fill-up. For a fuel-injected car, isopropyl alcohol, C_3H_7OH, is the recommended gas line antifreeze. For a carbureted car, methyl alcohol, CH_3OH, is recommended. Just like ethanol, each of these alcohols is soluble in water. They are also soluble in gasoline. Their presence, therefore, helps the water to mix with the gasoline, thereby preventing a gas line freeze-up.

▶ Figure 7.19
Ethanol and water molecules are about the same size, and they both form hydrogen bonds. As a result, ethanol and water will readily mix with each other.

one another. This puts a limit on the amount of sucrose that can dissolve in water—eventually, a point is reached at which there are not enough water molecules to separate the sucrose molecules from one another. As we discussed in Section 7.2, this is the point of saturation and any additional sucrose added to the solution does not dissolve.

When the molecule-to-molecule attractions among solute molecules are comparable to the molecule-to-molecule attractions among solvent molecules, there is no practical point of saturation. As shown in **Figure 7.19**, for example, the hydrogen bonds among water molecules are about as strong as those between ethanol molecules. These two liquids therefore mix together quite well in just about any proportion. We can even add ethanol to water until the ethanol, rather than the water, can be considered the solvent.

A solute that has no practical point of saturation in a given solvent is said to be *infinitely soluble* in that solvent. Ethanol, for example, is infinitely soluble in water. Also, all gases are generally infinitely soluble in other gases, because they can be mixed together in just about any proportion.

Let's now look at the other extreme of solubility, where a solute has very little solubility in a given solvent. An example is oxygen, O_2, in water. In contrast to sucrose, which has a solubility of 200 grams per 100 milliliters of water, only 0.004 gram of oxygen can dissolve in 100 milliliters of water. We can account for oxygen's low solubility in water by noting that the only attractions that occur between oxygen molecules and water molecules are relatively weak dipole–induced dipole attractions. More important, however, is the fact that the

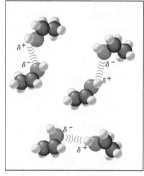

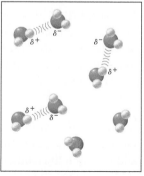

Ethanol   Ethanol and water   Water

stronger attraction of water molecules for one another—through the hydrogen bonds that the water molecules form with one another—effectively excludes oxygen molecules from intermingling with them.

A material that does not dissolve in a solvent to any appreciable extent is said to be **insoluble** in that solvent. We consider many substances to be insoluble in water, including sand and glass. Just because a material is not soluble in one solvent, however, does not mean it won't dissolve in another. Sand and glass, for example, are soluble in hydrofluoric acid, HF, which is used to give glass the decorative frosted look shown in **Figure 7.20**. Also, although Styrofoam™ is insoluble in water, it is partially soluble in acetone, a solvent used in fingernail-polish remover. Pour a little acetone into a Styrofoam™ cup, and the acetone soon causes the Styrofoam™ to deform, as demonstrated in **Figure 7.21**.

▲ Figure 7.20
Glass is frosted by dissolving its outer surface in hydrofluoric acid.

CONCEPT CHECK

Why isn't sucrose infinitely soluble in water?

CHECK YOUR ANSWER The attraction between two sucrose molecules is much stronger than the attraction between a sucrose molecule and a water molecule. Because of this, sucrose dissolves in water only as long as the number of water molecules far exceeds the number of sucrose molecules. When there are too few water molecules to dissolve any additional sucrose, the solution is saturated.

Solubility Changes with Temperature

You probably know from experience that water soluble solids usually dissolve better in hot water than in cold water. A highly concentrated solution of sucrose in water, for example, can be made by heating the solution almost to the boiling point. This is how syrup and hard candy are made.

Solubility increases with increasing temperature because hot water molecules have greater kinetic energy and therefore are able to collide with the solid solute more vigorously. The vigorous collisions facilitate the disruption of particle-to-particle attractions in the solid.

Although the solubilities of many solid solutes—sucrose, to name just one example—are greatly increased by rises in temperature, the solubilities of other solid solutes, such as sodium chloride, are only mildly affected, as **Figure 7.22** shows. This difference involves a number of factors, including the strength of the chemical bonds in the solute molecules and the way those molecules are packed together. Some chemicals, such as calcium carbonate, $CaCO_3$, actually become *less* soluble as the water temperature increases. This explains why the inner surfaces of tea kettles are often coated with calcium carbonate residues.

FOR YOUR INFORMATION

If a hot saturated solution is allowed to cool slowly without disturbance, the solute may stay in solution. The result is a *supersaturated* solution. Supersaturated aqueous solutions of sucrose are fairly easy to make. See the number 29 Think and Do activity at the end of this chapter.

◀ Figure 7.21
Is this cup melting or dissolving?

Acetone

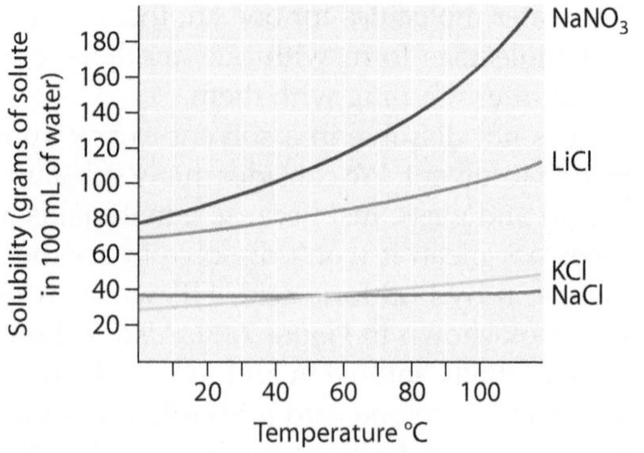

▶ Figure 7.22
The solubility of many water soluble solids increases with temperature, while the solubility of others is only very slightly affected by temperature.

When a sugar solution saturated at a high temperature is allowed to cool, some of the sugar usually comes out of solution and forms what is called a **precipitate.** When this occurs, the solute, sugar in this case, is said to have *precipitated* from the solution.

Let's put on our quantitative thinking caps and consider another example. At 100°C, the solubility of sodium nitrate, $NaNO_3$, in water is 165 grams per 100 milliliters of water. As we cool this solution, the solubility of $NaNO_3$ decreases, as shown in Figure 7.22, and this change in solubility causes some of the dissolved $NaNO_3$ to precipitate (come out of solution). At 20°C, the solubility of $NaNO_3$ is only 87 grams per 100 milliliters of water. So if we cool the 100°C solution to 20°C, 78 grams (165 grams – 87 grams) precipitates, as shown in **Figure 7.23**.

Solubility of Gases

In contrast to the solubilities of most solids, the solubilities of gases in liquids *decrease* with increasing temperature, as Table 7.2 shows. This effect occurs because with an increase in temperature, the solvent molecules have more kinetic energy. This makes it more difficult for a gaseous solute to remain in solution, because the solute molecules are literally ejected by the high-energy solvent molecules.

Perhaps you have noticed that compared to cold carbonated beverages, warm ones go flat faster. The higher temperature causes the molecules of carbon dioxide gas to leave the liquid solvent at a higher rate.

The solubility of a gas in a liquid also depends on the pressure of the gas immediately above the liquid. In general, a higher gas pressure above the liquid means more of the gas dissolves. A gas at a high pressure has many, many gas particles crammed into a given volume. The "empty" space in an unopened soft drink bottle, for example, is crammed with carbon dioxide molecules in the gaseous phase. With nowhere else to go, many of these molecules dissolve in the liquid, as shown in **Figure 7.24**. Alternatively, we might say that the great pressure forces the carbon dioxide molecules into solution. When the bottle is opened, the "head" of highly pressurized carbon dioxide gas escapes. Now the gas pressure above the liquid is lower than it was before. As a result, the solubility of the carbon dioxide drops, and the carbon dioxide molecules that were once squeezed into the solution begin to escape into the air above the liquid.

TABLE 7.2   Temperature-Dependent Solubility of Oxygen Gas in Water at a Pressure of 1 Atmosphere

TEMPERATURE (°C)	O_2 SOLUBILITY (g O_2/L H_2O)
0	0.0141
10	0.0109
20	0.0092
25	0.0083
30	0.0077
35	0.0070
40	0.0065

▶ Figure 7.23
The solubility of sodium nitrate is 165 grams per 100 milliliters of water at 100°C but only 87 grams per 100 milliliters at 20°C. Cooling a 100°C saturated solution of $NaNO_3$ to 20°C causes 78 grams of the solute to precipitate.

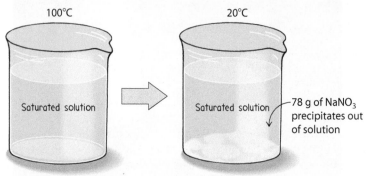

(a) 165 g $NaNO_3$ in 100 mL water   87 g $NaNO_3$ in 100 mL water

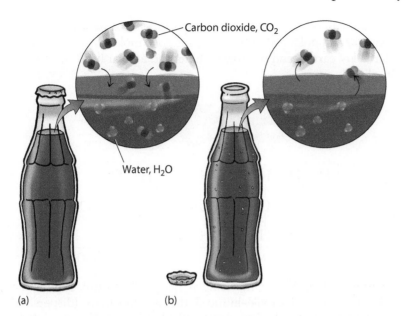

Carbon dioxide, CO_2

Water, H_2O

(a)   (b)

◀ Figure 7.24
(a) The carbon dioxide gas above the liquid in an unopened soft drink bottle consists of many tightly packed carbon dioxide molecules that are forced by pressure into solution. (b) When the bottle is opened, the pressure is released and carbon dioxide molecules originally dissolved in the liquid can escape into the air.

The rate at which carbon dioxide molecules leave an opened soft drink is relatively slow. You can increase the rate by pouring in granulated sugar, salt, or sand. The microscopic nooks and crannies on the surfaces of the grains serve as *nucleation sites* where carbon dioxide bubbles are able to form rapidly and then escape by buoyant forces. Shaking the beverage also increases the surface area of the liquid-to-gas interface, making it easier for the carbon dioxide to escape from the solution. Once the solution is shaken, the rate at which carbon dioxide escapes becomes so great that the beverage froths over. You also increase the rate at which carbon dioxide escapes when you pour the beverage into your mouth, which abounds in nucleation sites. You can feel the resulting tingly sensation.

CONCEPT CHECK

You open two cans of soft drink, one from a warm kitchen shelf and the other from the coldest depths of your refrigerator. Which provides the greater fizz in your mouth?

CHECK YOUR ANSWER The solubility of carbon dioxide in water decreases with increasing temperature. The warm drink, therefore, will fizz in your mouth more than the cold one will.

FOR YOUR INFORMATION

It is not just dipole–induced dipole attractions that keep carbon dioxide dissolved in water. As we'll discuss in Chapter 10, carbon dioxide reacts with water to form carbonic acid, which is much more soluble in water. When a can of carbonated soda is opened, much of this carbonic acid rapidly transforms back into water and carbon dioxide, which quickly bubbles out of solution because of its low solubility.

7.5 Soap Works by Being Both Polar and Nonpolar

EXPLAIN THIS

How does washing soda help to clean laundry?

Dirt and grease together make *grime*. Because grime contains many nonpolar components, it is difficult to remove from hands or clothing with water alone. To remove most grime, we can use a nonpolar solvent such as paint thinner, which dissolves the grime because of strong induced dipole–induced dipole attractions. Paint thinner is good for removing the grime left on hands after an activity such as changing a car's motor oil. But nonpolar solvents such as paint thinner are toxic, they have offensive odors, and they are harsh on the skin. Rather than washing our dirty hands and clothes with nonpolar solvents, we have a more

LEARNING OBJECTIVE

Describe the mechanism by which soaps and detergents clean and explain how this mechanism is foiled by hard water.

 READINGCHECK

What properties do soap molecules have?

pleasant alternative—soap and water. Soap works because soap molecules have both nonpolar and polar properties. A typical soap molecule has two parts: a long, nonpolar tail of carbon and hydrogen atoms and a polar head containing at least one ionic bond.

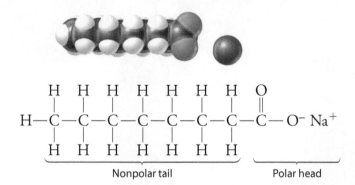

FORYOUR INFORMATION

Grease is soluble in paint thinner, which is why paint thinner can be used to clean one's hands of grease. But body oils are also soluble in paint thinner, which is why hands cleaned with paint thinner feel dry and chapped.

Because most of a soap molecule is nonpolar, it attracts nonpolar grime molecules via induced dipole–induced dipole attractions (dispersion forces) as **Figure 7.25** illustrates. In fact, grime quickly finds itself surrounded in three dimensions by the nonpolar tails of soap molecules. This attraction is usually sufficient to lift the grime away from the surface being cleaned. With the nonpolar tails facing inward toward the grime, the polar heads are all directed outward, where are attracted to water molecules by relatively strong ion–dipole attractions. If the water is flowing, the whole conglomeration of grime and soap molecules flows with it, away from your hands or clothes and then down the drain.

For the past several centuries, soaps have been prepared by treating animal fats with sodium hydroxide, NaOH, also known as caustic lye. In this reaction, which is still used today, each fat molecule is broken down into three *fatty acid* soap molecules and one glycerol molecule:

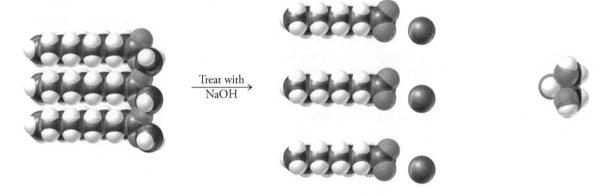

Fat molecule   Three fatty acid soap molecules   Glycerol molecule

In the 1940s, chemists began developing a class of synthetic soaplike compounds known as *detergents,* which offer several advantages over true soaps, such as stronger grease penetration and lower price.

The chemical structure of detergent molecules is similar to that of soap molecules in that both possess a polar head attached to a nonpolar tail. The polar head in a detergent molecule, however, typically consists of either a sulfate group, $-OSO_3^-$, or a sulfonate group, $-SO_3^-$, and the nonpolar tail can have an assortment of structures.

One of the most common sulfate detergents is sodium lauryl sulfate, a main ingredient of many toothpastes. A common sulfonate detergent is sodium dodecyl benzenesulfonate, also known as a linear alkylsulfonate, or LAS, often found in dishwashing liquids. Both of these detergents are biodegradable, which means that microorganisms can break down the molecules once they are released into the environment.

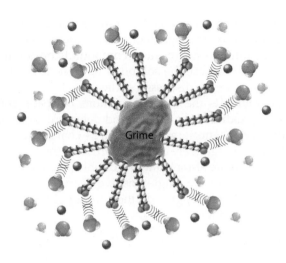

◀ Figure 7.25
Nonpolar grime attracts and is surrounded by the nonpolar tails of soap molecules, forming what is called a *micelle*. The polar heads of the soap molecules are attracted by ion–dipole attractions to water molecules, which then carry away the soap–grime combination.

CONCEPT CHECK

What type of attractions hold soap or detergent molecules to grime?

CHECK YOUR ANSWER If you haven't yet formulated an answer, why not back up and reread the question? You've got only four choices: ion–dipole, dipole–dipole, dipole–induced dipole, and induced dipole–induced dipole. The answer is induced dipole–induced dipole attractions, because the interaction is between two nonpolar entities—the grime and the nonpolar tail of a soap or detergent molecule.

$$CH_3CH_2CH_2CH_2CH_2CH_2CH_2CH_2CH_2CH_2CH_2CH_2{-}O{-}\overset{\displaystyle O}{\underset{\displaystyle O}{S}}{-}O^-\ Na^+$$

Sodium lauryl sulfate

Sodium dodecyl benzenesulfonate

7.6 Softening Hard Water

EXPLAIN THIS

Why does a scum form on the surface of boiling hard water?

LEARNING OBJECTIVE

Describe how dissolved ions can be removed from hard water.

Water containing large amounts of calcium and magnesium ions is said to be **hard water**, and it has many undesirable qualities. For example, when hard water is heated, its calcium and magnesium ions tend to bind with negatively charged ions also found in the water to form solid compounds, like those shown in **Figure 7.26**. These can clog water heaters and boilers. You'll also find coatings of these calcium and magnesium compounds on the inside surfaces of a well-used tea kettle (because the solubility of these compounds decreases with increasing temperature, as discussed earlier).

Hard water also inhibits the cleansing actions of soaps and, to a lesser extent, detergents. The sodium ions of soap and detergent molecules carry a 1+ charge, and calcium and magnesium ions carry a 2+ charge (note their positions in the periodic table). The negatively charged portion of the polar head of a soap or detergent molecule is more attracted to the double positive charge of calcium and magnesium ions than to the single positive charge of sodium ions. Soap or detergent molecules, therefore, give up their sodium ions to bind selectively with calcium or magnesium ions.

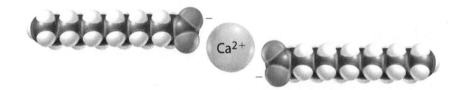

Soap or detergent molecules bound to calcium or magnesium ions tend to be insoluble in water. As they come out of solution, they form a scum, which can appear as a ring around the inside of your bathtub. Because the soap or detergent molecules are tied up with calcium and magnesium ions, more soap or detergent must be added to maintain cleaning effectiveness.

Many detergents today contain sodium carbonate, Na_2CO_3, commonly known as washing soda. The calcium and magnesium ions in hard water are more attracted to the carbonate ion with its two negative charges than they are to a soap or detergent molecule with its single negative charge. With the calcium and magnesium ions bound to the carbonate ion, as shown in **Figure 7.27**, the soap or detergent is free to do its job. Because it removes the ions that make water hard, sodium carbonate is known as a water-softening agent.

In some homes, the water is so hard that it must be passed through a *water-softening unit*. In a typical unit, illustrated in **Figure 7.28**, hard water is passed through a large tank filled with tiny beads of a water-insoluble resin known as an *ion-exchange resin*. The surface of the resin contains many negatively charged ions bound to positively charged sodium ions. As calcium and magnesium ions

 READINGCHECK

Is the polar head of a soap molecule more attracted to calcium or sodium ions?

▲ Figure 7.26
Hard water causes calcium and magnesium compounds to build up on the inner surfaces of water pipes, especially those used to carry hot water.

FOR YOUR INFORMATION

Most modern water softeners are equipped with meters that let you know the rate at which you consume water. This is a great way to keep tabs on your water-conservation efforts.

▼ Figure 7.27
(a) Sodium carbonate is added to many detergents as a water-softening agent. (b) The doubly positive calcium and magnesium ions of hard water preferentially bind with the doubly negative carbonate ion, freeing the detergent molecules to do their job.

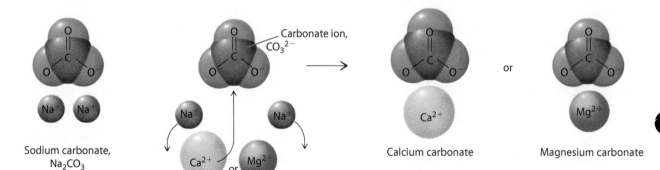

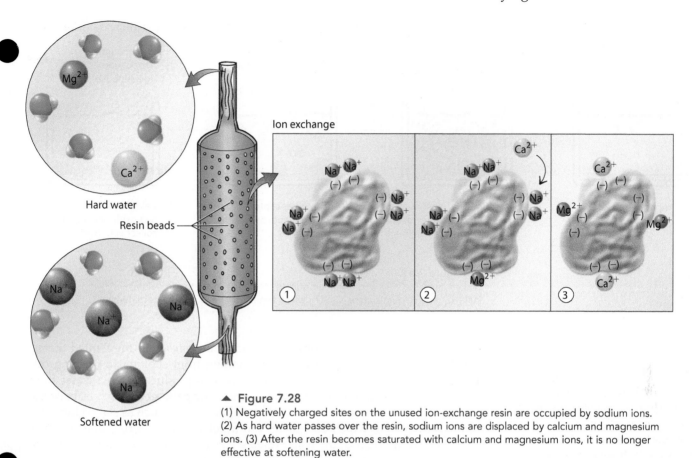

▲ **Figure 7.28**
(1) Negatively charged sites on the unused ion-exchange resin are occupied by sodium ions.
(2) As hard water passes over the resin, sodium ions are displaced by calcium and magnesium ions. (3) After the resin becomes saturated with calcium and magnesium ions, it is no longer effective at softening water.

pass over the resin, they displace the sodium ions and thereby become bound to the resin. The calcium and magnesium ions are able to do this because their positive charge (2+) is greater than that of the sodium ions (1+). The calcium and magnesium ions therefore have a greater attraction for the negative sites on the resin. The net result is that for every one calcium or magnesium ion that binds, two sodium ions are set free. In this way, the resin *exchanges* ions. The water that exits from the unit is now free of calcium and magnesium ions, but it does contain sodium ions in their place.

Eventually, all the sites for calcium and magnesium on the resin are filled, and then the resin needs to be either discarded or recharged. It is recharged by flushing it with a concentrated solution of sodium chloride, NaCl. The abundant sodium ions displace the calcium and magnesium ions (ions are *exchanged* once again), freeing up the binding sites on the resin.

7.7 Purifying the Water We Drink

EXPLAIN THIS

Why is water so difficult to purify?

As was discussed in Section 3.7, it is impossible to obtain 100 percent pure water. However, we are able to purify water to meet our needs. We do this by taking advantage of the differences in physical properties of water and the solutes or particulates it contains.

LEARNING OBJECTIVE

Identify the industrial means by which water is purified.

READING CHECK

How do we purify our drinking water?

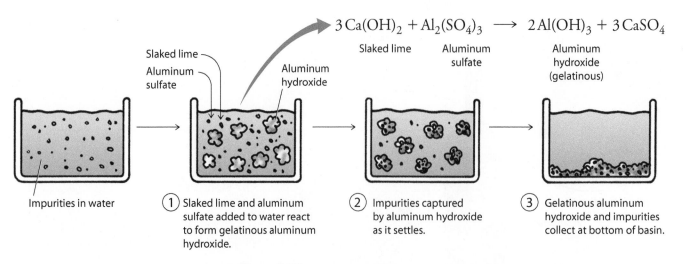

$$3\,Ca(OH)_2 + Al_2(SO_4)_3 \longrightarrow 2\,Al(OH)_3 + 3\,CaSO_4$$

Slaked lime

Aluminum sulfate

Aluminum hydroxide

Slaked lime   Aluminum sulfate   Aluminum hydroxide (gelatinous)

Impurities in water

① Slaked lime and aluminum sulfate added to water react to form gelatinous aluminum hydroxide.

② Impurities captured by aluminum hydroxide as it settles.

③ Gelatinous aluminum hydroxide and impurities collect at bottom of basin.

▲ Figure 7.29
Slaked lime, $Ca(OH)_2$, and aluminum sulfate, $Al_2(SO_4)_3$, react to form aluminum hydroxide, $Al(OH)_3$, and calcium sulfate, $CaSO_4$, which together form a gelatinous material.

 FOR YOUR INFORMATION

In the early 1990s, municipalities in Peru stopped chlorinating their drinking water. Within months, these municipalities were hit with 1.3 million new cases of cholera, resulting in 13,000 deaths.

Water that is safe for drinking is said to be *potable*. In the United States, potable water is currently used for everything from cooking to flushing our toilets. The first step most public utilities take to produce potable water from natural sources is to remove any dirt particles or pathogens, such as bacteria. This is done by mixing the water with certain minerals, such as slaked lime and aluminum sulfate, which coagulate into a gelatinous material, that becomes interspersed throughout the water (**Figure 7.29**). This is done in a large settling basin. Slow stirring causes the gelatinous material to clump together and settle to the bottom of the basin. As these clumps form and settle, they carry with them many of the dirt particles and bacteria. The water is then filtered through sand and gravel.

To improve the odor and flavor of the water, many treatment facilities also *aerate* the water by cascading it through a column of air, as shown in **Figure 7.30**. Aeration removes many unpleasant-smelling volatile chemicals, such as sulfur compounds. At the same time, air dissolves into the water, giving it a better taste—without dissolved air, the water tastes flat. As a final step, the water is treated with a disinfectant, usually chlorine gas, Cl_2, but sometimes ozone, O_3, and then stored in a holding tank that feeds into the city mains.

Developed countries have the technology and infrastructure to produce vast quantities of water suitable for drinking—with the result that many citizens take their drinking water for granted. The number of public water-treatment facilities in developing nations, however, is relatively small. In these locations, many people drink their water in the form of a hot beverage, such as tea, which is disinfected through boiling. Alternatively, disinfecting iodine tablets can be used.

Fuel for boiling and tablets for disinfecting, however, are not always available. As a result, more than 400 people in the world (mostly children) die every hour from preventable diseases or infections such as cholera, typhoid fever, dysentery, and hepatitis, which they contract by drinking contaminated water. In response, several American manufacturers have developed tabletop systems that bathe water with pathogen-killing ultraviolet light. One prototype model, shown in **Figure 7.31**, disinfects 15 gallons per minute, weighs about 15 pounds, and is powered by photovoltaic solar cells, which permit it to run unsupervised in remote locations.

Aside from pathogens, untreated water from wells or rivers may contain toxic metals that seep into the water supply from natural geologic formations. Many of the wells in Bangladesh, for example, are made very deep to avoid the pathogens that run rampant in the surface waters of the region. The water

▲ Figure 7.30
Volatile impurities are removed from drinking water by cascading it through the columns of air within these stacks.

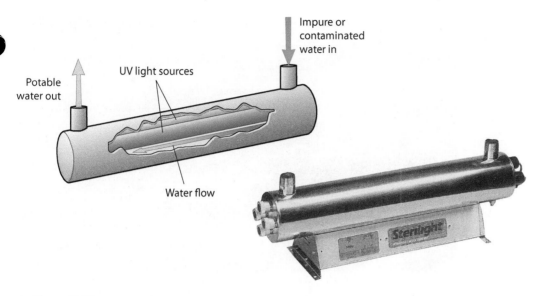

▲ Figure 7.31
Small-scale water-disinfecting units such as the one shown here hold great value in regions of the world where potable water is scarce.

obtained from these deep wells, however, is highly contaminated with arsenic—a naturally occurring element in the Earth's crust. The arsenic is in the underlying rock, which formed from river sediments carried down from the Himalayas. Because this region is so densely populated, as many as 70 million people may be subject to some level of arsenic poisoning, which manifests itself as skin lesions and a higher susceptibility to cancer. Low-cost methods for removing arsenic from well water are greatly needed, as are worldwide recognition of this problem and the political, economic, and social support to overcome it.

CONCEPTCHECK

At a water-treatment facility, how does adding slaked lime and aluminum sulfate to water purify the water?

CHECK YOUR ANSWER The water entering a water-treatment plant is usually a heterogeneous mixture containing suspended solids. Adding slaked lime and aluminum sulfate serves to capture these suspended solids, which then sink to the bottom, where they are easily removed.

Fresh Water Can Be Made from Salt Water

With the depletion of sources of natural fresh water in many regions, there has been growing interest in techniques for generating fresh water from the Earth's far larger reserves of seawater or from *brackish* (moderately salty) groundwater. Worldwide, *desalination* plants operate in about 120 countries, with a combined capacity to produce about 76 billion liters per day in 2010 and a projected 126 billion liters per day by 2016. In many areas of the Caribbean, North Africa, and the Middle East, desalinized water is the main source of municipal supply (**Figure 7.32**).

The two primary methods of removing salts from seawater or brackish water are *distillation* and *reverse osmosis*. These techniques are also highly effective in removing a host of other contaminants, such as pathogens, fertilizers, and pesticides. Distillation and reverse osmosis, therefore, are also used to purify naturally occurring fresh water. Many popular brands of bottled water, for example, contain fresh water that has been treated either by distillation or by reverse osmosis.

 FOR YOUR INFORMATION

In 1908, Jersey City, New Jersey, became the first American city to begin chlorinating its drinking water. By 1910, as disinfecting drinking water with chlorine became more widespread, the death rate from typhoid fever dropped from about 100 to 20 lives per 100,000. By 1935, the death rate fell to three lives per 100,000. In 1960, fewer than 20 persons in the entire United States died from typhoid fever.

▲ Figure 7.32
Saudi Arabia is the world's leading producer of desalinized water. Its desalination plants, such as the one shown here, have a combined generating capacity of about 4 billion liters per day.

Distillation involves vaporizing water with heat and then condensing the vapors into purified liquid water (see Section 3.6). Earth's desalinized water is produced using this technique. Because water has such a high heat of vaporization, however, this technique is energy intensive. Solar distillers avoid the burning of fuels, but they require about 1 square meter of surface area to produce 4 liters of fresh water per day, as shown in **Figure 7.33**. For a single home or a small village, this surface-area requirement may be easily accommodated.

For many regions, *reverse osmosis* is a preferable method of water desalination. In order to understand reverse osmosis, you must first understand osmosis. Osmosis involves a semipermeable membrane. A **semipermeable membrane** contains submicroscopic pores that allow the passage of water molecules but not of larger solute ions or solute molecules. When a body of fresh water is partitioned from a body of salt water by a semipermeable membrane, water molecules pass from the fresh water into the salt water at a higher rate than from the salt water into the fresh water. The reason for this is the presence of more water molecules along the fresh water face of the

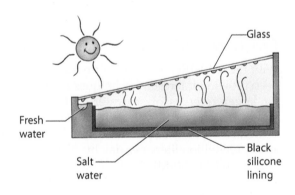

Glass

Fresh water

Salt water

Black silicone lining

▲ Figure 7.33
These solar distillers are popular in the remote communities along the Texas–Mexico border, where the waters from the Rio Grande basin are saline and tainted by the runoff of agricultural chemicals from upstream irrigation.

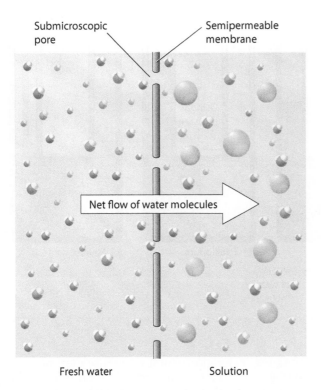

Submicroscopic pore

Semipermeable membrane

Net flow of water molecules

Fresh water   Solution

▲ **Figure 7.34**
Osmosis. The submicroscopic pores of a semipermeable membrane allow only water molecules to pass. Because there are more water molecules along the fresh water face of the membrane than along the solution face, more water molecules are available to migrate into the solution than are available to migrate into the fresh water.

membrane than along the salt water face. The result is a net movement of fresh water into the body of salt water, as illustrated in **Figure 7.34**. This net flow of water across a semipermeable membrane into a more concentrated solution is called **osmosis.**

The result of osmosis is a buildup in volume of the salt water and a decrease in volume of the fresh water. These changes in volume, in turn, allow a buildup in pressure, called *osmotic pressure*. For the system in **Figure 7.35a**, osmotic pressure is the consequence of the salt water's greater height and, therefore, weight bearing onto the semipermeable membrane. As osmotic pressure builds, the rate at which water molecules are able to pass from the salt water into the fresh water increases. The water molecules in the salt water are literally being squeezed back across the membrane by the osmotic pressure. Eventually, the rates of water molecules passing in both directions across the membrane are the same, and the system reaches equilibrium, as shown in **Figure 7.35b**. If external pressure is applied to the salt water, even more water molecules are squeezed across the membrane from the salt water into the fresh water, as shown in **Figure 7.35c**. Water forced across a semipermeable membrane into a less concentrated solution is **reverse osmosis.** So we see that reverse osmosis is a mechanism for generating fresh water from salt water.

The osmotic pressure for seawater, however, is an astounding 24.8 atmospheres (365 pounds per square inch). Generating pressures greater than this has its share of technical difficulties and is an energy-intensive process. Nonetheless, engineers have succeeded in building durable reverse osmosis units, shown in **Figure 7.36**, that can be networked together to generate fresh water from seawater at rates of millions of gallons per day. Reverse osmosis desalination facilities treating brackish water, which require much lower external pressures, are proportionately more economical.

▶ **Figure 7.35**
(a) Osmosis results in a greater volume of salt water, which causes the pressure to increase on the salt side of the membrane.
(b) When the pressure on the salt side becomes high enough, equal numbers of water molecules pass in both directions.
(c) The application of external pressure forces water molecules to pass from the salt water to the fresh water so that now the salt-to-fresh rate exceeds the fresh-to-salt rate.

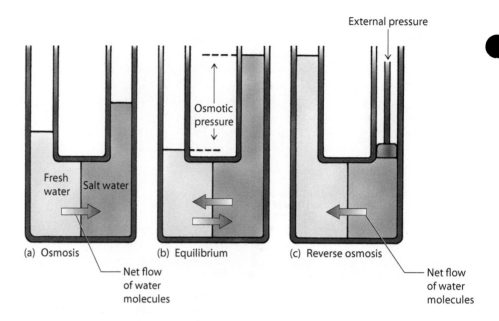

(a) Osmosis   (b) Equilibrium   (c) Reverse osmosis

Net flow of water molecules   Net flow of water molecules

CONCEPTCHECK

Biological membranes, including cucumber membranes, are semipermeable. A cucumber shrivels to a smaller size when it is left in a solution of salt water. Is this an example of osmosis or of reverse osmosis?

CHECK YOUR ANSWER No external pressure is involved, which rules out reverse osmosis. Instead, the shriveling of the cucumber tells us that the cucumber's cells are losing water to the more concentrated salt water. This is osmosis, whereby water molecules migrate across a semipermeable membrane into regions of higher salt concentrations. If you were to add a few other ingredients to the solution, such as spices and the right kinds of microorganisms, you would have a pickle.

Is Bottled Water Worth the Price?

In the United States, natural sources of fresh water are relatively plentiful, allowing utilities to sell fresh water at rates of a fraction of a penny per liter. Nonetheless, consumers are still willing to purchase bottled water at up to $2 per liter! Each year, Americans spend about $4 billion on bottled water.

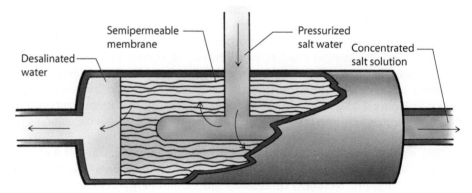

Semipermeable membrane   Pressurized salt water   Concentrated salt solution

Desalinated water

▲ **Figure 7.36**
An industrial reverse osmosis unit consists of many semipermeable membranes packed around highly pressurized salt water. As desalinated water is pushed out one side, the remaining salt water, which is now even more concentrated, exits on the other side. A network of reverse osmosis units operating parallel to one another can produce enormous volumes of fresh water from salt water.

Bottled water also comes with environmental costs. In the United States, about 29 billion water bottles are discarded each year. Of these, only about 20 percent are recycled—the rest end up primarily in landfills. Some estimates show that up to 54 million barrels of oil are burned each year to meet the market demand for bottled water in the United States alone. To put this into perspective, according to the Earth Policy Institute, the volume of oil needed to produce a single plastic bottle is about one-quarter the volume of the bottle.

In an effort to overcome the ecological negatives of bottled water and a price up to 1000 times that of tap water, many bottled-water marketers are now focusing on supposed peripheral benefits of their product. This includes the addition of dissolved oxygen, which, as was shown in Table 7.2, can be no more than 0.0083 g/L at room temperature. For comparison, a single breath of air contains about 100 times the amount of molecular oxygen found in a half liter of "oxygenated" water. Furthermore, most of the gases accumulated in your gut simply pass out the opposite end from your mouth, assuming you don't burp.

Worse still are claims of bottled water that contains "functional" water in which the structure of water has been modified using undetectable "subtle energy" to make the water more nutritious. There is also water through which an electric current has been passed. The water companies claim that this creates two forms of water: alkaline and acidic. The alkaline water, they say, is good for you because it is "ionized, restructured, micro-clustered, activated, hydrogen–saturated, and oxidation–reduced." These sorts of misleading claims take advantage of the millions of people who don't understand that modifying the structure of water, H_2O, gives you something that is no longer water. Buyer beware! Water is H_2O. If it's not H_2O, it isn't water. We'll be discussing what actually happens when an electric current passes through water in Chapter 11.

Interestingly, about 25 percent of bottled water sold in the United States is simply municipal water that has been purified via reverse osmosis. Many homeowners are now discovering that rather than purchasing purified water, it is less expensive and more ecologically sound to install a small reverse osmosis unit in their own home. For fun, carbonators can also be installed so that you can have your very own soda fountain, as shown in **Figure 7.37**.

▲ **Figure 7.37**
Carbonating your own water is not only fun but also cheaper and more ecologically sound than purchasing your soda from a store.

Chapter 7 Review

LEARNING OBJECTIVES

Identify four different types of dipole attractions and their role in determining the physical properties of a material. (7.1)	→ *Questions 1–4, 23, 33, 35–39*
Describe the formation of saturated and unsaturated solutions from a molecular point of view. (7.2)	→ *Questions 5–7, 40–42*
Describe the components of a solution and calculate a solution's concentration. (7.3)	→ *Questions 8–10, 25–32, 43–45*
Discuss how solutes dissolve in solvents and how solubility changes with temperature. (7.4)	→ *Questions 11–13, 24, 34, 46–56, 67*
Describe the mechanism by which soaps and detergents clean and explain how this mechanism is foiled by hard water. (7.5)	→ *Questions 14–16, 57–59*
Describe how dissolved ions can be removed from hard water. (7.6)	→ *Questions 17–19, 60, 61*
Identify the industrial means by which water is purified. (7.7)	→ *Questions 20–22, 62–66, 68–70*

SUMMARY OF TERMS (KNOWLEDGE)

Concentration A quantitative measure of the amount of solute per volume of solution.

Dissolving The process of mixing a solute in a solvent to produce a homogeneous mixture.

Hard water Water containing large amounts of calcium and magnesium ions.

Hydrogen bond An unusually strong dipole–dipole attraction occurring between molecules that have a hydrogen atom covalently bonded to a small highly electronegative atom, usually nitrogen, oxygen, chlorine, or fluorine.

Induced dipole A temporarily uneven distribution of electrons in an otherwise nonpolar atom or molecule.

Insoluble Said of a solute that does not dissolve to any appreciable extent in a given solvent.

Molarity A common unit of concentration equal to the number of moles of a solute per liter of solution.

Mole A very large number equal to 6.02×10^{23} and usually used in reference to the number of atoms, ions, or molecules within a macroscopic amount of a material.

Osmosis The net flow (diffusion) of water across a semipermeable membrane from a region of low solute concentration to a region of high solute concentration.

Precipitate A solute that has come out of solution.

Reverse osmosis A technique for purifying water by forcing it through a semipermeable membrane into a region of lower solute concentration.

Saturated solution A solution containing the maximum amount of solute that will dissolve in its solvent.

Semipermeable membrane A membrane containing submicroscopic pores that allow passage of water molecules but not of larger solute ions or solute molecules.

Solubility The ability of a solute to dissolve in a given solvent.

Soluble Said of a solute that is capable of dissolving to an appreciable extent in a given solvent.

Solute Any component in a solution that is not the solvent.

Solvent The component in a solution that is present in the largest amount.

Unsaturated solution A solution that is capable of dissolving additional solute.

READING CHECK QUESTIONS (COMPREHENSION)

7.1 Four Different Types of Dipole Attractions

1. What is the primary difference between a chemical bond and an attraction between two molecules?

2. Which is stronger, the ion–dipole attraction or the induced dipole–induced dipole attraction?

3. What is a hydrogen bond?

4. Are induced dipoles permanent?

7.2 A Solution Is a Single-Phase Homogeneous Mixture

5. What happens to the volume of a sugar solution as more sugar is dissolved in it?

6. Why is a ruby considered to be a solution?

7. Distinguish between a solute and a solvent.

7.3 Concentration Is Given as Moles per Liter

8. What does it mean to say that a solution is concentrated?

9. Is 1 mole of particles a very large or a very small number of particles?

10. Is concentration typically given with the volume of solvent or the volume of solution?

7.4 Solubility Is How Well a Solute Dissolves

11. Why does the solubility of a gas solute in a liquid solvent decrease with increasing temperature?

12. Why do sugar crystals dissolve faster when crushed?

13. Is sugar a polar or nonpolar substance?

7.5 Soap Works by Being Both Polar and Nonpolar

14. Which portion of a soap molecule is nonpolar?

15. Water and soap are attracted to each other by what type of molecular attraction?

16. What is the difference between a soap and a detergent?

7.6 Softening Hard Water

17. What component of hard water makes it hard?

18. Why are soap molecules so attracted to calcium and magnesium ions?

19. Calcium and magnesium ions are more attracted to sodium carbonate than to soap. Why?

7.7 Purifying the Water We Drink

20. Why is treated water sprayed into the air prior to being piped to users?

21. What are two ways people disinfect water in areas where municipal treatment facilities are not available?

22. What naturally occurring element has been contaminating the water supply of Bangladesh?

CONFIRM THE CHEMISTRY (HANDS-ON APPLICATION)

23. To see the action of the ion–dipole attraction, create a static charge on a rubber balloon by rubbing it across your hair. Hold this charged balloon close to but not touching a thin stream of water running from a faucet. Watch the charged balloon divert the path of the falling water. Your balloon is negatively charged because it picks up electrons from your hair. Why would a balloon that was positively charged also attract the stream of water?

24. Here's a quick recipe for rock candy. In a cooking pot, make a hot, saturated solution of sugar in water. Start by mixing sugar and water in a 2:1 ratio by volume. Add more sugar or water as necessary to obtain a clear, runny syrup. Let cool for 10 minutes. Roll a wet skewer stick or weight (such as metal nut) attached to a string in some granulated sugar. Pour the warm sugar syrup into a jar. Submerge the skewer or weight in the sugar syrup. Cover the top and store the sugar syrup in a cool place. The longer you wait, the larger the crystals.

25. Just because a solid dissolves in a liquid doesn't mean the solid no longer occupies space. Fill a glass to its brim with the warm water and then carefully pour all the water into the larger container. Add a couple tablespoons of sugar to the empty glass. Return half of the warm water to the glass and stir to dissolve all the sugar. Return the remaining water, and as the water gets close to the top of the glass, ask a friend to predict whether the water level will be less than, about the same as, or more than before. If your friend doesn't understand the result, ask him or her what would happen if you had added the sugar to the glass when the glass was full of water.

THINK AND SOLVE (MATHEMATICAL APPLICATION)

26. Assume the total number of molecules in a solution is about 3 million trillion. One million trillion of these are molecules of some poison, while 2 million trillion of these are water molecules. What percentage of all the molecules in the glass is water?

27. Assume the total number of molecules in a solution is about 1,000,000 million trillion. One million trillion of these are molecules of some poison, while 999,999 million trillion of these are water molecules. What percentage of all the molecules in the glass is water?

28. You drink a small glass of water that is 99.9999 percent pure water and 0.0001 percent poison. Assume the glass contains about a 1,000,000 million trillion molecules, which is about 30 mL. How many poison molecules did you just drink? Should you be concerned?

29. How much sodium chloride, in grams, is needed to make 15 L of a solution that has a concentration of 3.0 grams of sodium chloride per liter of solution?

30. If water is added to 1 mole of sodium chloride in a flask until the volume of the solution is 1 liter, what is the molarity of the solution? What is the molarity when water is added to 2 moles of sodium chloride to make 0.5 liter of solution?

31. A student is told to use 20.0 grams of sodium chloride to make an aqueous solution that has a concentration of 10.0 g/L (grams of sodium chloride per liter of solution). Assuming that 20.0 grams of sodium chloride has a volume of 7.50 milliliters, show that she will need about 1.99 liters of water to make this solution. In making this solution, should she add the solute to the solvent or the solvent to the solute?

THINK AND COMPARE (ANALYSIS)

32. Rank the following solutions in order of increasing concentration. Solution A: 0.5 moles of sucrose in 2.0 liters of solution; Solution B: 1.0 moles of sucrose in 3.0 liters of solution; Solution C: 1.5 moles of sucrose in 4.0 liters of solution.

33. List the following compounds in order of increasing boiling point: CI_4, CBr_4, CCl_4, CF_4.

34. Rank the following compounds in order of increasing solubility in water:

$CH_3CH_2—OH$ $CH_3CH_2CH_2CH_2—OH$
   Ethanol   Butanol

$CH_3CH_2CH_2CH_2CH_2CH_2—OH$
              Hexanol

THINK AND EXPLAIN (SYNTHESIS)

7.1 Four Different Types of Dipole Attractions

35. Which is stronger: the covalent bond that holds atoms together within a molecule or the electrical attraction between two neighboring molecules? Please explain.

36. The charges with sodium chloride are all balanced—for every positive sodium ion there is a corresponding negative chloride ion. Because its charges are balanced, how can sodium chloride be attracted to water, and vice versa?

37. Why are ion–dipole attractions stronger than dipole–dipole attractions?

38. Chlorine, Cl_2, is a gas at room temperature, but bromine, Br_2, is a liquid. Why?

39. Why is calcium fluoride, CaF_2, a high melting point crystalline solid while stannic chloride, $SnCl_4$, is a volatile liquid?

7.2 A Solution Is a Single-Phase Homogeneous Mixture

40. Dipole–induced dipole attractions exist between molecules of water and molecules of gasoline, and yet these two substances do not mix because water has such

a strong attraction for itself. Which compound might best help these two substances mix into a single liquid phase?

41. If table sugar is white, why is a solution of table sugar in water transparent?

42. Which is more dense: air saturated with water vapor or air unsaturated with water vapor?

7.3 Concentration Is Given as Moles per Liter

43. The volume of many liquid solvents expands with increasing temperature. What happens to the concentration of a solution made with such a solvent as the temperature of the solution is increased?

44. Which should weigh more: fresh water or the same volume of fresh sparkling seltzer water? Why?

45. How many sugar molecules are in a 2 M sugar solution?

7.4 Solubility Is How Well a Solute Dissolves

46. Explain why, for these three substances, the solubility in 20°C water (shown in g/mL) goes down as the molecules get larger but the boiling point (shown in °C) goes up.

CH$_3$—O$^{\diagup H}$
infinite
65°C

CH$_3$CH$_2$CH$_2$CH$_2$—O$^{\diagup H}$
8 g/100mL
117°C

CH$_3$CH$_2$CH$_2$CH$_2$CH$_2$—O$^{\diagup H}$
2.3 g/100mL
138°C

47. The boiling point of 1,4-butanediol is 230°C. Would you expect this compound to be soluble or insoluble in room-temperature water? Explain.

H$^{\diagup}$O—CH$_2$CH$_2$CH$_2$CH$_2$—O$^{\diagup H}$
1,4-butanediol

48. Based on atomic size, which would you expect to be more soluble in water: helium, He, or nitrogen, N$_2$?

49. If nitrogen, N$_2$, were pumped into your lungs at high pressure, what would happen to its solubility in your blood?

50. The air a scuba diver breathes is pressurized to counteract the pressure exerted by the water surrounding the diver's body. Breathing the high-pressure air causes excessive amounts of nitrogen to dissolve in body fluids, especially the blood. If a diver ascends to the surface too rapidly, the nitrogen bubbles out of the body fluids (much like the way carbon dioxide bubbles out of a soda immediately after the container is opened). This results in a painful and potentially lethal medical condition known as the *bends*. Why does breathing a mixture of helium and oxygen rather than air help divers avoid getting the bends?

51. Account for the observation that ethanol, C$_2$H$_5$OH, dissolves readily in water but dimethyl ether, CH$_3$OCH$_3$, which has the same number and kinds of atoms, does not.

H H
| |
H—C—C—O$^{\diagup H}$
| |
H H
Ethanol

H H H
| | |
H—C O C—H
| | |
H O H
Dimethyl ether

52. At 10°C, which is more concentrated: a saturated solution of sodium nitrate, NaNO$_3$, or a saturated solution of sodium chloride, NaCl? (See Figure 7.22.)

53. Why are rain and snow called precipitation?

54. Of the two structures shown next, one is a typical gasoline molecule and the other is a typical motor oil molecule. Which is which? Base your reasoning not on memorization, but on what you know about electrical attractions between molecules and the various physical properties of gasoline and motor oil.

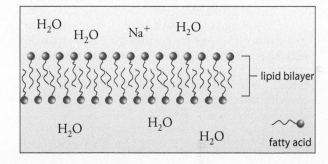

Structure A

Structure B

55. Hydrogen chloride, HCl, is a gas at room temperature. Would you expect this material to be very soluble or not very soluble in water?

56. Would you expect to find more dissolved oxygen in ocean water around the northern latitudes or in ocean water close to the equator? Why?

7.5 Soap Works by Being Both Polar and Nonpolar

Fatty acid molecules can align to form a barrier called a lipid bilayer, shown below. In this schematic, the ionic end of the fatty acid is shown as a circle and the nonpolar chain is shown as a squiggly line. Use this diagram for questions 57 through 59.

H$_2$O   H$_2$O   Na$^+$ H$_2$O

lipid bilayer

H$_2$O   H$_2$O   H$_2$O

fatty acid

57. Why do nonpolar molecules have a difficult time passing through the lipid bilayer?

58. Why do ionic compounds such as sodium chloride, NaCl, have a hard time passing through the lipid bilayer?

59. Fatty acid molecules can also align to form a lipid bilayer that extends in three dimensions. A cross section of this structure is shown next. This is similar to the micelle shown in Figure 7.25, although notably different because it contains an inner compartment of water. What is this structure called? (Hint: it forms the basis of all life.)

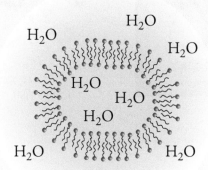

H$_2$O   H$_2$O   H$_2$O   H$_2$O   H$_2$O   H$_2$O   H$_2$O

7.6 Softening Hard Water

60. A scum forms on the surface of boiling hard water. What is this scum? Why does it form? And why do hot water heaters lose their efficiency quicker in households with hard water?

61. Phosphate ions, PO_4^{3-}, were once added to detergents to assist in cleaning. What function did they serve? These ions are no longer added to detergents because they cause excessive growth of algae in aquatic habitats receiving the wastewater. What chemical has replaced them?

7.7 Purifying the Water We Drink

62. Cells at the top of a tree have a higher concentration of sugars than cells at the bottom. How might this fact assist a tree in moving water upward from its roots?

63. What reverses with reverse osmosis?

64. Why is it significantly less costly to purify fresh water through reverse osmosis than to purify salt water through reverse osmosis?

65. Some people fear drinking distilled water because they have heard it leaches minerals from the body. Using your knowledge of chemistry, explain how these fears have no basis and how distilled water is in fact very good for drinking.

66. Many homeowners get their drinking water piped from wells dug into their property. Sometime this well water smells bad because of trace quantities of the gaseous compound hydrogen sulfide, H_2S. How might this odor be removed from water already taken from the tap?

THINK AND DISCUSS (EVALUATION)

67. Oxygen, O_2, dissolves quite well within a class of compounds known as liquid perfluorocarbons—so well that oxygenated perfluorcarbons can be inhaled in a liquid phase, as is demonstrated by the alive and liquid breathing rodent shown below the water-bound goldfish. Do you suppose perfluorocarbon molecules are polar or nonpolar? Why would the rodent drown if it were brought up to the water layer and the goldfish die if they swam down into the perfluorocarbon layer? How might perfluorocarbons be used to clean our lungs or serve as an artificial blood? When is it okay to sacrifice the lives of animals for scientific research?

68. Why are people so willing to buy bottled water when it is so expensive, both financially and environmentally?

69. It is possible to tow icebergs to coastal cities as a source of fresh water. What obstacles—technological, social, environmental, and political—do you foresee for such an endeavor?

70. In reference to human nature, Jerome Delli Priscoli, a social scientist with the U.S. Army Corps of Engineers, stated, "The thirst for water may be more persuasive than the impulse toward conflict." Do you agree or disagree with his statement? Might our universal need for water be our salvation or our demise? Consider the current water disputes between places such as Sudan and Egypt, Turkey and Iraq, or the United States and Mexico.

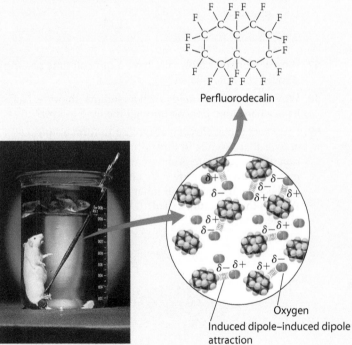

Perfluorodecalin

Oxygen

Induced dipole–induced dipole attraction

READINESS ASSURANCE TEST (RAT)

If you have a good handle on this chapter, then you should be able to score at least 7 out of 10 on this RAT. Check your answers online at www.ConceptualChemistry.com. If you score less than 7, you need to study further before moving on.

Choose the BEST answer to the following.

1. The hydrogen bond is a type of
   a. ionic bond.
   b. metallic bond.
   c. covalent bond.
   d. None of the above

2. Iodine, I_2, has a higher melting point compared to fluorine, F_2, because its
   a. atoms are larger.
   b. molecules are heavier.
   c. nonpolarity is greater.
   d. atoms have a greater electronegativity.

3. Plastic wrap is made of nonpolar molecules and is able to stick well to polar surfaces, such as glass, by way of dipole–induced dipole molecular attractions. How is it that plastic wrap also sticks to itself so well?
   a. By way of dipole–dipole molecular attractions.
   b. By way of induced dipole–induced dipole molecular attractions.
   c. By way of dipole–induced dipole molecular attractions.
   d. Ions are formed as the plastic rubs against itself.

4. Which of the following statements describe a saturated solution?
   a. A carbonated beverage with bubbles.
   b. A solution where the solvent cannot dissolve any more solute.
   c. A stirred solution of salt water with salt at the bottom.
   d. All of the above
   e. None of the above

5. Sodium chloride, NaCl, is insoluble in gasoline because
   a. its ions are so small, there is not much opportunity for the gasoline to interact with them.
   b. salt can only form dipole–induced dipole attractions.
   c. gasoline is so strongly attracted to itself, the salt, NaCl, is excluded.
   d. its ions are much more attracted to themselves.

6. Hardened water results in soap scum because it
   a. contains excessive amounts of sodium ions.
   b. is more dense than softened water.
   c. contains excessive amounts of magnesium ions.
   d. has a greater polarity than softened water.

7. Fish don't live very long in water that has just been boiled and brought back to room temperature because
   a. of a higher concentration of dissolved CO_2 in the water.
   b. the nutrients in the water have been destroyed.
   c. the salts in the water become more concentrated.
   d. the boiling process removes the air that was dissolved in the water.

8. How many moles of sugar (sucrose) are in 5 liters of sugar water that has a concentration of 0.5 M?
   a. 5.5 moles
   b. 5.0 moles
   c. 2.5 moles
   d. 1.5 moles

9. What is an advantage of using chlorine gas to disinfect drinking water supplies?
   a. It provides residual protection against pathogens.
   b. It gives the water a fresh taste.
   c. Residual chlorine in water helps to whiten teeth.
   d. Excess chlorine is absorbed in our bodies as a mineral supplement.
   e. All of the above

10. Why do red blood cells, which contain an aqueous solution of dissolved ions and minerals, burst when placed in fresh water?
    a. The dissolved ions provide a pressure that eventually bursts open the cell.
    b. More water molecules enter the cell than leave the cell.
    c. The fresh water acts to dissolve the blood cell wall.
    d. All of the above

ANSWERS TO CALCULATION CORNER (CALCULATING FOR SOLUTIONS)

1. Multiply the solution concentration by the final volume of the solution. This provides the amount of solute required: (380 g/L)(3 L) = 1140 g.

2. Divide the amount of solute by the solution concentration to obtain the amount of solution prepared: 20 g/10 g/L = 2 L.

Contextual Chemistry

Water Fluoridation

In the early 1900s, a young dentist, Frederick McKay, opened a dental practice in Colorado Springs, Colorado. He soon discovered that many local residents had brown hole-pocked teeth that, despite being unsightly, had a resistance to cavity formation. Following years of investigation, McKay and his colleagues determined that this condition was caused by unusually high levels of naturally occurring calcium fluoride in the drinking water. The condition became known as *dental fluorosis*. By the 1930s, the U.S. Public Health Service began to study the idea of adding fluoride ions to drinking water. These studies indicated that dental fluorosis could be avoided but resistance to cavity formation maintained at fluoride ion concentrations of about 1 part per million (ppm), which equals 1 mg per liter.

The first municipal fluoridation test programs began in the 1940s, soon after World War II, which was a time when the general public held great trust in the chemical industry as well as government. In Grand Rapids, Michigan, for example, sodium fluoride, once widely regarded as rat poison, was to be added to the municipal drinking water for 15 years, after which time the program would be evaluated for effectiveness. Also, the upstate New York towns of Newburgh and Kingston, which had similar demographic and water profiles, were chosen for a double study in which the water of only one town (Newburgh) would be fluoridated. After 10 years, the rates of dental decay in each city would be compared. Halfway through the Newburgh/Kingston trial, however, the U.S. Public Health Service announced a preliminary 65 percent decrease in dental decay in Newburgh. Word of this

and other trials quickly spread, and soon many municipalities were requesting water fluoridation, well before any of these initial trials could be completed.

As can be expected, there were many who advocated against the fluoridation of public drinking water. They argued that fluoride is not a natural nutrient, but a drug and industrial pollutant, and that fluoridation is unethical because individuals are not being asked for their informed consent prior to being given medication. As any physician knows, what is good for one patient is not necessarily good for another patient. A counterargument to this was that fluoridation of public water provides a benefit to the whole community, including the underprivileged, and not just those who can afford good health care. Opponents to fluoridation, however, also pointed out that the long-term side effects of fluoride were not known. Of particular concern were the risks of skeletal fluorosis, bone cancer, and hypothyroidism. Believing these risks to be minimal compared to the great dental benefits, the government, motivated in large part by professional dental associations, continued to push for fluoridation. Today, about 145 million people in the United States regularly drink fluoridated water.

As a quick Internet search reveals, the fluoridation debate continues to be most vigorous. For example, the American Dental Association, www.ada.org, publishes a well-known

booklet that clearly advocates the great benefits and minimal risks of fluoridating municipal drinking water. Each paragraph of the ADA's booklet, however, is carefully counterargued by fluoridation opponent Edward Bennett at www.fluoridedebate.com. Of note is the fact that drinking water in Europe is no longer fluoridated. Most European nations, as well as Japan, halted water fluoridation in the 1970s, after which their rates of dental caries continued to decline, as within the United States. Furthermore, the U.S. Centers for Disease Control and Prevention have noted the lack of evidence supporting the effectiveness of ingested fluoride. Rather, teeth appear to benefit by surface exposure, as occurs when the teeth are brushed with fluoridated toothpaste.

In 1987, the U.S. government–sponsored National Institute of Dental Research (NIDR) examined the teeth of about 39,000 school-children aged 5 to 17 from 84 geographical areas, with and without fluoridated water. Analysis of this data by John Yiamouyiannis, a biochemist and ardent opponent of fluoridation, showed no meaningful statistical differences between children growing up in fluoridated and nonfluoridated areas, except that 5- to 6-year-olds drinking fluoridated water tended to keep their baby teeth longer. Analysis by the NIDR of the same data indicated that children who have always lived in fluoridated areas have 18 percent fewer decayed surfaces than those who have never lived in fluoridated areas. In 2012, the ADA's position is that "water fluoridation continues to be effective in reducing tooth decay by 20–40 percent, even with widespread availability from other sources such as fluoride toothpastes."

As the debate continues, one thing remains certain: those who are familiar with the basic concepts of chemistry and methodologies of science are at a great advantage for understanding the issues and for being able to recognize well-informed arguments and decisions.

CONCEPTCHECK

Teflon® is a carbon-based molecule with lots of fluorine. Might flossing with Teflon®-coated dental floss be a good way of introducing additional fluoride to your teeth?

CHECK YOUR ANSWER It is the fluoride ion that is indicated to help prevent dental decay. Sources of these ions include the ionic compounds sodium fluoride, NaF, and stannous fluoride, SnF_2. The fluorine within Teflon® is not ionic. Rather, Teflon® as a molecule (covalent compound) contains fluorine atoms covalently bound to the carbon atoms. These

fluorine atoms cannot escape the molecule to join the structure of tooth enamel.

Think and Discuss

1. In 1962, the Kettering Laboratory of the University of Cincinnati conducted a study in which dogs were exposed to calcium fluoride dust at a rate simulating human occupational exposure. The results of the study showed significant damage to the dogs' lungs and lymph nodes. This study was funded by an industry group that was seeking evidence to help counter worker claims of crippling skeletal fluorosis. Consequently, the study remained unpublished. Should there be laws that prohibit the suppression of unfavorable research data? Try answering this question from the point of view of a company that might have contributed millions of dollars to the research.

2. What kind of pressures might a scientist face if she discovered evidence suggesting that fluoridated water had neurotoxic effects in rats? Assume she works for the government, for a government-funded university, for a private research firm not related to dental health, or for a professional dental association.

3. The most common form of fluoride added to municipal drinking water today is hydrofluosilicic acid, H_2SiF_6, which is obtained from the antipollution smokestack scrubbers of the phosphate fertilizer industry. What are the advantages and disadvantages of this system? If water fluoridation were banned, what might become of this hydrofluosilicic acid?

4. How could the Newburgh/ Kingston trials have been designed to ensure that the results were not affected by the bias of profluoridation dentists and government officials? Were government officials justified when they published the preliminary results?

5. Vitamin B12 deficiency affects about one-quarter of the U.S. population and is more common in the elderly. This deficiency is often undetected and can lead to devastating and irreversible complications. Should vitamin B12 be added to municipal drinking water? Why or why not?

6. The town of Modesto, California, claims that over the course of 10 years, it'll save at least $5 million and at most $32 million by not fluoridating its water. How might the town of Modesto use this saved money to help protect the dental health of all its citizens, including children of underemployed families? Should the saved money instead be returned to the citizens of Modesto in the form of lower taxes?

7. In many countries lacking elaborate waterworks systems, fluoride is provided to the general population by adding it to commercial table salt. What might be some of the advantages and disadvantages of this sort of fluoride delivery system?

8. People have been told that the fluoride in their drinking water helps to protect their teeth. To what extent does this prompt them to ignore good dental health habits such as eating healthfully, avoiding sweets, and brushing and flossing regularly?

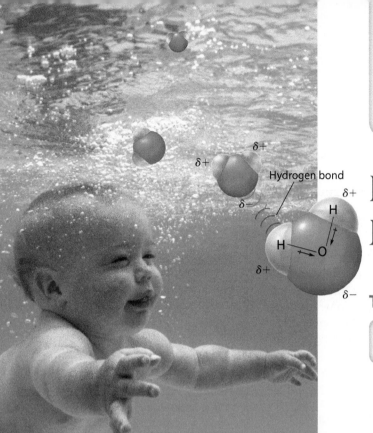

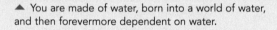

▲ You are made of water, born into a world of water, and then forevermore dependent on water.

8

How Water Behaves

THE MAIN IDEA

Water is exceptional.

Water is so common in our lives that its many unusual properties easily escape our notice. Consider, for example, that water is the only chemical substance on our planet's surface that can be found abundantly in all three phases—solid, liquid, and gas. Another unique property of water is its great resistance to any change in temperature. As a result, the water in you moderates your body temperature, just as the oceans moderate global temperatures. Unlike most other liquids, which freeze from the bottom up, liquid water freezes from the top down. To the trained eye of a chemist, water is far from a usual substance. Rather, it's downright bizarre and exotic.

Almost all of the amazing properties of water are a consequence of the ability of water molecules to cling tenaciously to one another by way of electrical attractions. In this chapter, we explore the physical behavior of water while diving into the details and consequences of the "stickiness" of water molecules. We begin by exploring first the properties of solid water (ice); then those of liquid water; and finally those of gaseous water, also known as water vapor.

Chemistry

Will the Raindrops Fall?

Water is a sticky substance, which explains why moisture in the air condenses into tiny droplets, as seen within a cloud or fog. These tiny droplets, in turn, coalesce into larger drops that fall, as seen with rain. So what happens when a glass of water covered with plastic wrap punctured with many tiny holes is turned upside down?

PROCEDURE

1. Fill a tall glass with water and cover it with plastic wrap held taut with a rubber band around the rim.
2. Use a pin to poke many tiny holes into the plastic wrap directly over the mouth of the glass.

3. Predict what will happen when you turn the glass upside down over a sink. Will there be many tiny streams of water? Will a shower of raindrops form and fall? Will the water mostly not pass through the holes?

ANALYZE AND CONCLUDE

Explain your observation in terms of the stickiness of water molecules.

1. Is there any difference when you hold the glass still and level versus when you shake the glass or hold it at an angle?
2. Swirl some liquid dishwashing detergent into the water in the glass. Do you get the same results?

3. Assume you did the same activity using a wide-mouth plastic bottle with a hole drilled into the bottom. What results would you expect then?

8.1 Water Molecules Form an Open Crystalline Structure in Ice

EXPLAIN THIS

Why Is Ice Slippery?

Experience tells us not to place a sealed glass jar of liquid water in the freezer, for we know that water expands as it freezes. Trapped in the jar, the freezing water expands outward with a force strong enough to shatter the glass into a hazardous mess or to pop the lid from the jar, as shown in **Figure 8.1**. This expansion occurs because when the water freezes, the water molecules arrange themselves in a six-sided crystalline structure that contains many open spaces. As **Figure 8.2** shows, a given number of water molecules in the liquid can get relatively close to one another and so occupy a certain volume. Water molecules in the crystalline structure of ice occupy a greater volume than they do in liquid water, however. Consequently, ice is less dense compared to liquid water, which is why ice floats in water.

LEARNING OBJECTIVE

Relate the physical properties of ice to its crystalline structure.

 READING CHECK

Why does water expand as it freezes?

▲ Figure 8.1
The expansion of the freezing water inside the jar caused the lid to rise above the mouth of the jar.

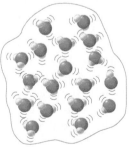

Liquid water
(dense)

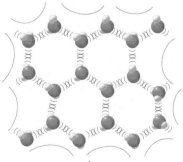

Ice
(less dense)

◀ Figure 8.2
Water molecules in the liquid phase are arranged more compactly than are water molecules in the solid phase, in which they form an open crystalline structure.

This property of expanding upon freezing is quite rare. The atoms or molecules of most frozen solids pack in such a way that the solid phase occupies a *smaller* volume than in the liquid phase (**Figure 8.3**).

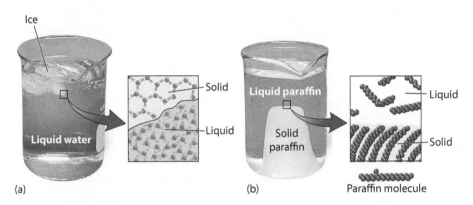

▶ Figure 8.3
(a) Because water expands as it freezes, ice is less dense compared to liquid water and so floats in the water. (b) Like most other materials, paraffin is denser in its solid phase than in its liquid phase. Solid paraffin thus sinks in liquid paraffin.

▲ Figure 8.4
The six-sided geometry of ice crystals gives rise to the six-sided structure of snowflakes.

CONCEPT CHECK

Are you interpreting correctly the illustration of water's open crystalline structure shown in Figure 8.2? If you are, you'll be able to answer this question: What's inside one of the open spaces?

a. air

b. water vapor

c. nothing

CHECK YOUR ANSWER If air were in the spaces, the illustration would have to show the molecules that make up air, such as O_2 and N_2, which are comparable in size to water molecules. Any water vapor in the spaces would have to be shown as free-roaming water molecules spaced relatively far apart. The open spaces shown here represent nothing but empty space. The answer is c.

The hexagonal crystalline structure of H_2O molecules in ice has some interesting effects. Most snowflakes, like the one in **Figure 8.4**, share a similar hexagonal shape, which is the microscopic consequence of this molecular geometry. Also, applying great pressure to ice causes the open spaces to collapse. The result is the formation of liquid water. The great weight of a glacier, for example, causes the underside of the glacier to melt. With its wet underside, the glacier slowly slips down mountains toward the ocean, where it spawns icebergs. Another application of the pressure melting of ice is given in a Think and Do activity at the end of this chapter.

Interestingly, ice is covered by a thin film of liquid water even at temperatures just below freezing. This is because ice's hexagonal structure requires support in three dimensions. At the surface, water molecules find nothing above them to cling to. The hexagonal structure at the surface is thus weakened to the point where it collapses into a thin film of liquid, which is what makes ice so slippery. An ice-skater's skate glides over this thin film of water, as shown in **Figure 8.5**.

So what happens when the flat surfaces of two ice cubes are held together? The thin liquid films are no longer on a surface. Sandwiched between the two ice cubes, this liquid film freezes, effectively gluing the two ice cubes together. Lumps of snow are similarly "glued" together in the making of a snowball. Of course, if the weather is too cold, even the surfaces of the snow crystals are solid, which prevents the formation of snowballs no matter how hard you press the snow together. Skiers tend to prefer this

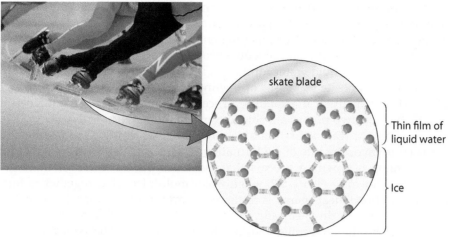

"dry snow" because it behaves more like a powder than a sticky slush. An alternative to skiing in super cold weather is skiing at high altitudes, where lower atmospheric pressure favors thinner liquid films, and hence drier, more powder-like snow.

8.2 Freezing and Melting Go On at the Same Time

EXPLAIN THIS

Why is calcium chloride, CaCl$_2$, more effective at melting ice than sodium chloride, NaCl?

As Section 2.7 discussed, *melting* occurs when a substance changes from solid to liquid, and *freezing* occurs when a substance changes from liquid to solid. When we view these processes from a molecular perspective, we see that melting and freezing occur simultaneously, as **Figure 8.6** illustrates.

The temperature 0°C is both the melting temperature and the freezing temperature of water. At this temperature, water molecules in the liquid phase are moving slowly enough that they tend to clump together to form ice crystals—they freeze. At this same temperature, however, water molecules in

LEARNING OBJECTIVE

Identify the molecular processes involved in the freezing and melting of water, explain the impact of a solute on these processes, and explain why water is most dense at 4°C.

Ice	Liquid water

Melting

Freezing

Vibrating H$_2$O molecules fixed in crystalline structure

Slowly moving H$_2$O molecules near freezing temperature

◀ Figure 8.6
At 0°C, ice crystals gain and lose water molecules simultaneously.

If left undisturbed, chilled purified bottled water can become "supercooled," which means it remains liquid at temperatures below 0°C. This happens because crystal formation usually requires a substrate, such as a microparticle, upon which to start growing. Upon opening, the bottle of supercooled water will rapidly turn into an icy slush. Similarly, during an ice storm, raindrops formed in warm upper layers of the atmosphere become supercooled as they pass through colder lower layers. Upon hitting the ground, tree branches, and power lines, the supercooled raindrops, known as freezing rain, quickly freeze. The results can be beautiful but also quite hazardous, especially for aircraft.

ice are vibrating with great commotion, much more than they vibrate at colder temperatures. Many thus break loose from the crystalline structure to form liquid water—they melt. Thus, melting and freezing occur simultaneously.

For water, 0°C is the special temperature at which the rate of ice formation equals the rate of liquid water formation. In other words, it is the temperature at which the opposite processes of melting and freezing counterbalance each other. This means that if a mixture of ice and liquid water is maintained at exactly 0°C, the two phases are able to coexist indefinitely.

Anytime we want a mixture of ice and liquid water at 0°C to freeze solid, we need to favor the rate of ice formation. This is accomplished by removing heat (cooling it down), a process that facilitates the formation of hydrogen bonds. As shown in **Figure 8.7**, when water molecules come together to form a hydrogen bond, heat energy is released. In order for the molecules to remain hydrogen-bonded, this heat energy must be removed—otherwise, it can be reabsorbed by the molecules, causing them to separate. The removal of heat energy therefore allows hydrogen bonds to remain intact after they have formed. As a result, there is the tendency for the ice crystals to grow.

Conversely, we can get a mixture of ice and liquid water at 0°C to melt completely by adding heat. This heat energy goes into breaking apart the hydrogen bonds that hold the water molecules together in the form of ice, as shown in **Figure 8.8**. Because more hydrogen bonds between water molecules are breaking, the ice crystals tend to melt.

Solutes tend to inhibit crystal formation. Anytime a solute, such as table salt or sugar, is added to water, the solute molecules take up space, as you learned in Section 7.2. When a solute is added to a mixture of ice and liquid water at 0°C, the solute molecules effectively decrease the number of liquid water molecules at the solid–liquid interface, as **Figure 8.9** illustrates. With fewer liquid water molecules available to join the ice crystals, the rate of ice formation decreases. Because ice is a relatively pure form of water, the number of molecules moving from the solid phase to the liquid phase is not affected by the presence of solute. The net result is that the rate at which water molecules leave the solid phase is greater than the rate at which they enter the solid phase. This imbalance can be compensated for by decreasing the temperature to below 0°C. At lower temperatures, water molecules in the liquid phase move more slowly and coalesce more easily. Thus, the rate of crystal formation is increased.

In general, adding anything to water lowers the freezing point. Antifreeze is a practical application of this process. The salting of icy roads is another.

▶ **Figure 8.7**
As two water molecules come together to form a hydrogen bond, attractive electric forces cause them to accelerate toward each other. This results in an increase in their kinetic energies (the energy of motion), which is perceived on the macroscopic scale as heat energy.

▶ **Figure 8.8**
Heat energy must be added in order to separate two water molecules held together by a hydrogen bond. This heat energy causes the molecules to vibrate so rapidly that the hydrogen bond breaks.

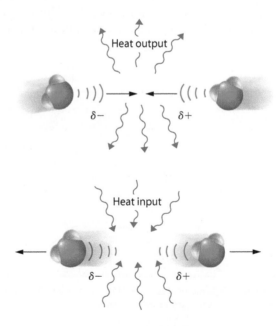

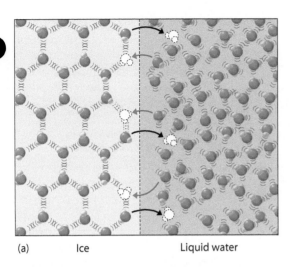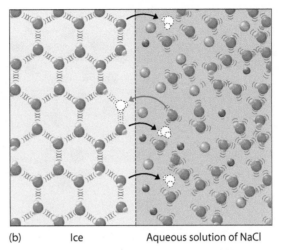

(a)   Ice   Liquid water   (b)   Ice   Aqueous solution of NaCl

▲ Figure 8.9
(a) In a mixture of ice and liquid water at 0°C, the number of molecules entering the solid phase is equal to the number of molecules entering the liquid phase. (b) Adding a solute, such as sodium chloride, decreases the number of molecules entering the solid phase, because now there are fewer liquid molecules at the interface.

CONCEPTCHECK
How does a solute decrease the rate of freezing?

CHECK YOUR ANSWER When a solute is present, there are fewer water molecules in contact with the surface of any ice crystal. This makes it more difficult for the ice crystal to grow.

Water Is Densest at 4°C

When the temperature of a substance is increased, its molecules vibrate faster and move farther apart on average. The result is that the substance expands. With few exceptions, all phases of matter—solids, liquids, and gases—expand when heated and contract when cooled. In many cases, these changes in volume are not very noticeable, but with careful observation, you can usually detect them. Telephone wires, for instance, are longer and sag more on a hot summer day than on a cold winter day. Metal lids on glass jars can often be loosened by heating them under hot water. If one part of a piece of glass is heated or cooled more rapidly than adjacent parts, the resulting expansion or contraction may break the glass.

Within any given phase, water also expands with increasing temperature and contracts with decreasing temperature. This is true of all three phases—ice, liquid water, and water vapor. Liquid water at near-freezing temperatures, however, is an exception.

Liquid water at 0°C can flow just like any other liquid, but at 0°C, the temperature is cold enough that nanoscopic crystals of ice are able to form. These crystals slightly "bloat" the liquid water's volume, as shown in **Figure 8.10**. As the temperature is increased to above 0°C, more and more of these crystals collapse, and as a result, the volume of the liquid water *decreases*.

Figure 8.11 shows that between 0°C and 4°C, liquid water *contracts* as its temperature is raised. This contraction, however, continues only up to 4°C. As near-freezing water is heated, there is a simultaneous tendency for the water to expand due to greater molecular motion. Between 0°C and 4°C, the decrease in volume caused by collapsing ice crystals is greater than the increase in volume caused by the faster-moving molecules. As a result, the water volume continues to decrease. At temperatures just above 4°C, expansion overrides contraction because most of the ice crystals have collapsed.

READINGCHECK
What forms in 0°C liquid water?

▶ **Figure 8.10**
Within a few degrees of 0°C, liquid water contains crystals of ice. The open structure of these crystals makes the volume of the water slightly greater than it would be without the crystals.

▶ **Figure 8.11**
Between 0°C and 4°C, the volume of liquid water decreases as the temperature increases. Above 4°C, water behaves the way all other substances do: its volume increases as its temperature increases. The volumes shown here are for a 1-gram sample.

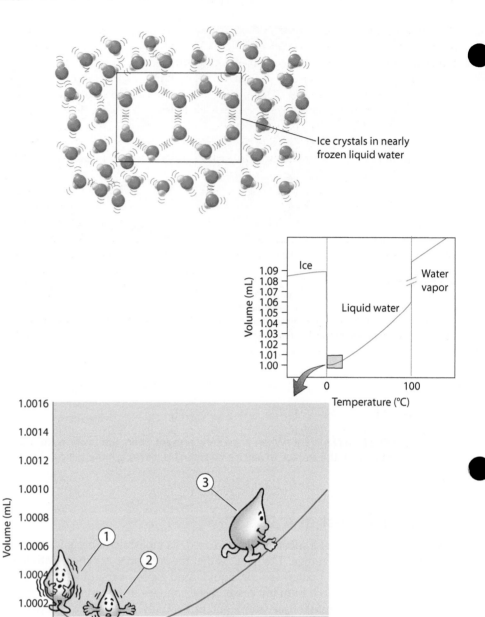

Ice crystals in nearly frozen liquid water

① Liquid water below 4°C is bloated with ice crystals.

② Upon warming, the crystals collapse, resulting in a smaller volume for the liquid water.

③ Above 4°C, liquid water expands as it is heated because of greater molecular motion.

Because of the effect of collapsing nanoscopic ice crystals, liquid water has its smallest volume and thus its greatest density at 4°C. (As discussed in Section 2.4, density is the amount of mass contained in a sample of anything divided by the volume of the sample.) By definition, 1 gram of pure water at this temperature has a volume of 1.0000 milliliter. As Figure 8.11 shows, 1 gram of liquid water at 0°C has an only slightly larger volume, 1.0002 milliliters. By comparison, 1 gram of ice at 0°C has a volume of 1.0870 milliliters. As can be seen in the small graph on the right in Figure 8.11, the volume of 1 gram of ice stays above 1.08 milliliters even below 0°C, meaning that even when ice is cooled to temperatures well below freezing, it is still less dense than liquid water.

Although liquid water at 4°C is only slightly denser than liquid water at 0°C, this small difference is of great importance in nature. Consider that if water were densest at its freezing point, as is true of most other liquids, the coldest water in a pond would settle to the bottom and the pond would freeze from the bottom up, destroying living organisms in winter months. Fortunately, this does not happen. As winter comes on and the temperature of the water drops, its density also drops. The entire volume of water in the pond does not cool all at once, however. Surface water cools first because it is in direct contact with the cold air. Being cooler than the underlying water, this surface water is denser and so sinks, with warmer water rising to replace it. That new batch of surface water then cools to the air temperature, gets denser as it does so, and sinks, only to be replaced by warmer water that cools. This process continues until the entire body of water has been cooled to 4°C. Then if the air temperature remains below 4°C, the surface water also cools to below 4°C. This surface water does not sink, however, because at this colder temperature, it is now *less* dense than the water below it. Thus, cooler, less dense water stays on the surface, where it can cool further, eventually reach 0°C, and turn to ice. While this ice forms at the surface, the organisms that require a liquid environment are happily swimming below the ice in liquid water at a "warm" 4°C, as **Figure 8.12** illustrates.

An important effect of the vertical movement of water is the creation of vertical currents that, to the benefit of organisms living in the water, transport oxygen-rich surface water to the bottom and nutrient-rich bottom water to the surface. Marine biologists refer to this vertical cycling of water and nutrients as *upwelling*.

Very deep bodies of fresh water are not ice-covered even in the coldest of winters. This is because, as noted previously, all the water must be cooled to 4°C before the temperature of the surface water can drop below 4°C. For deep water, the winter is not long enough for this to occur.

If only some of the water in a pond is 4°C, this water lies on the bottom. Because of water's ability to resist changes in temperature (Section 8.5) and its poor ability to conduct heat, the bottom of deep bodies of water in cold regions is a constant 4°C year-round.

CHEMICAL CONNECTIONS

How is a raindrop connected to a campfire?

CONCEPT CHECK

What was the precise temperature at the bottom of Lake Michigan on New Year's Eve in 1901?

CHECK YOUR ANSWER If a body of water has 4°C water, then the temperature at the bottom of that body of water is 4°C, for the same reason that rocks are at the bottom. Both 4°C water and rocks are denser than water at any other temperature. If the body of water is deep and in a region of short summers, as is the case for Lake Michigan, the water at the bottom is 4°C year-round.

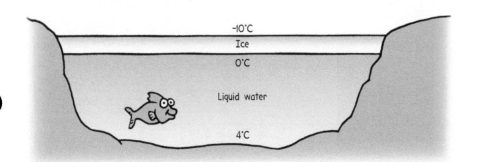

◀ **Figure 8.12**
As water cools to 4°C, it sinks. Then as water at the surface is cooled to below 4°C, it floats on top and can freeze. Only after surface ice forms can temperatures lower than 4°C extend down into the pond. This does not happen very readily, however, because surface ice insulates the liquid water from the cold air.

8.3 Liquid Water's Behavior Results from the Stickiness of Its Molecules

FOR YOUR INFORMATION

A related concept is *viscosity*, which is the resistance a fluid has to being deformed when flowing. In general, the stronger the intermolecular attractions within a fluid, the greater its viscosity. There is more hydrogen bonding in cold water than in hot water, which is why cold water is slightly more viscous than is hot water.

READING CHECK

What causes surface tension?

EXPLAIN THIS

Why does salt water have greater surface tension than fresh water does?

In this section, we explore how water molecules in the liquid phase interact with one another via **cohesive forces,** which are forces of attraction between molecules of a single substance. For water, the cohesive forces are hydrogen bonds. We also explore how water molecules interact with other polar materials, such as glass, through **adhesive forces,** forces of attraction between molecules of two *different* substances.

Cohesive and adhesive forces involving water are dynamic. It is not one set of water molecules, for example, that holds a droplet of water to the side of a glass. Rather, the billions and billions of molecules in the droplet all take turns binding with the glass surface. Keep this in mind as you read this section and examine its illustrations, which, although informative, are merely freeze-frame depictions.

The Surface of Liquid Water Behaves Like an Elastic Film

Gently lay a dry paper clip on the surface of still water. If you're careful enough, the clip will rest on the surface, as shown in **Figure 8.13**. How can this be? Don't paper clips normally sink in water?

First, you should be aware that the paper clip is not floating *in* the water the way a boat floats. Rather, the clip is resting *on* the water surface. A slight depression in the surface is caused by the weight of the clip, which pushes down on the water much like the weight of a child pushes down on a trampoline. This elastic tendency found at the surface of a liquid is known as **surface tension.**

Surface tension in water is caused by hydrogen bonds. As shown in **Figure 8.14**, beneath the surface, each water molecule is attracted in every direction by neighboring molecules, with the result that there is no tendency to be pulled in any preferred direction. A water molecule on the surface, however, is pulled only by neighbors to each side and those below; there is no pull upward. The combined effect of these molecular attractions is thus to pull the molecule from the surface into the liquid. This tendency to pull surface molecules into the liquid causes the surface to become as small as possible, and the surface behaves as if it were tightened into an elastic film. Lightweight objects that don't pierce the surface, such as a paper clip, are thus able to rest on the surface.

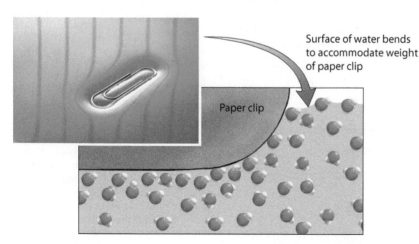

Surface of water bends to accommodate weight of paper clip

Paper clip

▲ Figure 8.13
A paper clip rests on water, pushing the surface down slightly but not sinking.

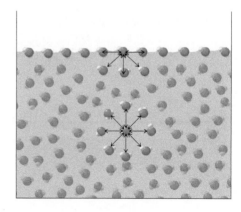

▲ Figure 8.14
A molecule at the surface is pulled only sideways and downward by neighboring molecules. A molecule beneath the surface is pulled equally in all directions.

Surface tension accounts for the spherical shape of liquid drops. Raindrops, drops of oil, and falling drops of molten metal are all spherical because their surfaces tend to contract and this contraction forces each drop into the shape having the least surface area. This is a sphere, the geometric figure that has the least surface area for a given volume. In the weightless environment of an orbiting space shuttle, a blob of water takes on a spherical shape naturally, as is shown in **Figure 8.15**. Back on the Earth, dewdrops on spider webs or drops on the downy leaves of plants are also spherical, except for the distortions caused by their weight.

(a)

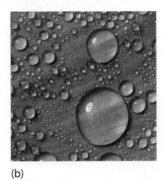

(b)

◀ **Figure 8.15**
(a) The surface tension in a freely falling weightless blob of water causes the water to take on a spherical shape. (b) Small blobs of water resting on a surface would also be spheres if it weren't for the force of gravity, which squashes them into beads.

Surface tension is greater in water than in other common liquids because the hydrogen bonds in water are relatively strong. The surface tension in water is dramatically reduced, however, by the addition of soap or detergent. **Figure 8.16** shows that soap or detergent molecules tend to aggregate at the surface of water, with their nonpolar tails sticking out away from the water. At the surface, these molecules interfere with the hydrogen bonds between neighboring water molecules, thereby reducing the surface tension. Get a metal paper clip floating on the surface of some water and then carefully touch the water a few centimeters away with the corner of a bar of wet soap or a dab of liquid detergent. You will be amazed at how quickly the surface tension is reduced.

The strong surface tension of water is what prevents it from wetting materials that have nonpolar surfaces, such as waxy leaves, umbrellas, and freshly polished automobiles. Rather than wetting (spreading out evenly), the water beads. This is good if the idea is to keep water away. If we want to clean an object, however, the idea is to get it as wet as possible. This is another way in which soaps and detergents assist in cleaning, as **Figure 8.17** illustrates. By destroying water's surface tension, soaps and detergents enhance water's ability to wet. The nonpolar grime on dirty fabrics and dishes, for example, is then penetrated by the water more rapidly, making cleaning more efficient.

 FOR YOUR INFORMATION

Hot water has less surface tension than cold water does because its faster-moving molecules are not bonded as tightly. With its lower surface tension, hot water is more effective at penetrating the fabrics of our clothing, which is why so many people choose to wash their clothes with hot water. Hot water's low surface tension also permits the grease of soups to collect in liquidy blobs on the surface of soup. When the soup cools, the surface tension of the water increases and the grease spreads over the surface of the soup, making the soup taste more "greasy." Hot soup tastes different from cold soup primarily because the surface tension of the water in the soup changes with temperature.

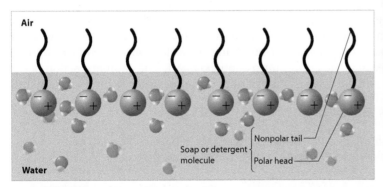

Water

Soap or detergent molecule

Nonpolar tail
Polar head

▲ **Figure 8.16**
Soap or detergent molecules align themselves at the surface of liquid water so that their nonpolar tails can escape the polarity of the water. This arrangement disrupts the water's surface tension.

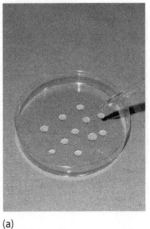

(a)

(b)

▲ **Figure 8.17**
(a) Water beads on a surface that is clean and dry. (b) On a plate smeared with a thin film of detergent, water spreads evenly, because the detergent has disrupted the water's surface tension.

C H E M I C A L
C O N N E C T I O N S

How is a paper towel
connected to a paintbrush?

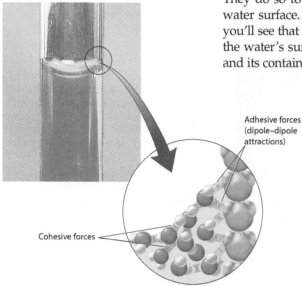

▲ Figure 8.18
Adhesive forces between water and glass cause water molecules to creep up the sides of the glass, forming a meniscus.

Capillary Action—An Interplay of Adhesive and Cohesive Forces

Because glass is a polar substance, there are adhesive forces between glass and water. These adhesive forces are relatively strong, and the many water molecules adjacent to the inner surface of a glass container compete to interact with the glass. They do so to the point of climbing up the inner surface of the glass above the water surface. Take a close look at the tube of colored water in **Figure 8.18**, and you'll see that the water is curved up the sides of the glass. We call the curving of the water's surface (or the surface of any other liquid) at the interface between it and its container a **meniscus.**

Figure 8.19 illustrates what happens when a small-diameter glass tube is placed in water. (1) Adhesive forces initially cause a relatively steep meniscus. (2) As soon as the meniscus forms, the attractive cohesive forces among water molecules respond to the steepness by acting to minimize the surface area of the meniscus. The result is that the water level in the tube rises. (3) Adhesive forces will then cause the formation of another steep meniscus. (4) This is followed by the action of cohesive forces, which cause the steep meniscus to be "filled in." This cycle is repeated until the upward adhesive force equals the weight of the raised water in the tube. This rise of the liquid due to the interplay of adhesive and cohesive forces is called **capillary action.**

In a tube that has an internal diameter of about 0.5 millimeter, the water rises slightly higher than 5 centimeters. In a tube that has a smaller diameter, there is a smaller volume and less weight for a given height, and the water rises much higher, as **Figure 8.20** illustrates.

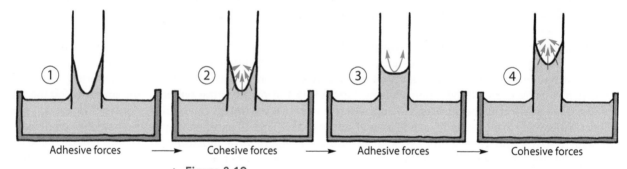

▲ Figure 8.19
Water is drawn up a narrow glass tube by an interplay of adhesive and cohesive forces.

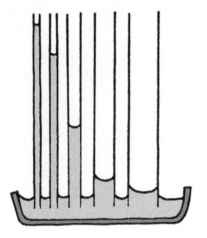

▲ Figure 8.20
Capillary tubes. The smaller the diameter
of the tube, the higher the liquid rises.

We see capillary action at work in many phenomena. If a paintbrush is dipped into water, the water rises into the narrow spaces between the bristles by capillary action. Hang your hair in the bathtub and water seeps up to your scalp the same way. This is how oil moves up a lamp wick and how water moves up a bath towel when one end hangs in water. Dip one end of a lump of sugar in coffee and the entire lump is quickly wet. The capillary action occurring between soil particles is important in bringing water to the roots of plants.

CONCEPT CHECK

An astronaut sticks a narrow glass tube into a blob of floating water while in orbit, and the tube fills with water. Why?

CHECK YOUR ANSWER Capillary action causes the water to be drawn into the tube. In the freefall environment of an orbiting spacecraft, however, there is no downward force to stop this capillary action. As a result, the water continues to creep along the inner surface of the tube until the tube is filled.

8.4 Water Molecules Move Freely between the Liquid and Gaseous Phases

EXPLAIN THIS

What is inside a bubble of boiling water?

Molecules of water in the liquid phase move about in all directions at different speeds. Some of these molecules may reach the liquid's surface moving fast enough to overcome the hydrogen bonds and escape into the gaseous phase. As presented in Section 2.6, this process of molecules converting from the liquid phase to the gaseous phase is called *evaporation* (also sometimes called *vaporization*). The opposite of evaporation is *condensation*—the changing of a gas to a liquid. At the surface of any body of water, there is a constant exchange of molecules from one phase to the other, as illustrated in **Figure 8.21**.

As evaporating molecules leave the liquid phase, they take their kinetic energy with them. This has the effect of lowering the average kinetic energy of all of the molecules remaining in the liquid, and the liquid is cooled, as **Figure 8.22** shows. Evaporation also has a cooling effect on the surrounding air, because liquid molecules that escape into the gaseous phase are moving relatively slowly compared to other molecules in the air. This makes sense when you consider that these newly arrived molecules lost a fair amount of their kinetic energy in overcoming the hydrogen bonds of the liquid phase. Adding these slower molecules to the surrounding air effectively decreases the average kinetic energy of all the molecules making up the air, and the air is cooled. So no matter how you look at it, evaporation is a cooling process. **Figure 8.23** shows a useful application of this cooling effect.

As water cools, the rate of evaporation slows down, because fewer molecules have sufficient energy to escape the hydrogen bonds of the liquid phase. A higher rate of evaporation can be maintained if the water is in contact with a relatively warm surface, such as your skin. Body heat then flows from you into the water. In this way the water maintains a higher temperature, and evaporation continues at a relatively high rate. This is why you feel cool as you dry off after getting wet—you are losing body heat to the energy-requiring process of evaporation.

When your body overheats, your sweat glands produce perspiration. The evaporation of perspiration cools you and helps you maintain a stable body

LEARNING OBJECTIVE

Show, on a molecular level, how evaporation and condensation lead to cooling and warming effects, respectively, and how they relate to the process of boiling.

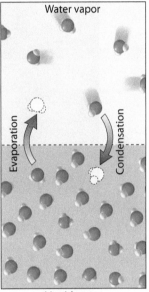

▲ Figure 8.21
The exchange of molecules at the interface between liquid and gaseous water.

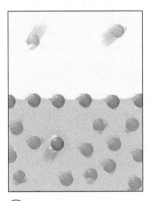

① Liquid water molecule having sufficient kinetic energy to overcome surface hydrogen bonding approaches liquid surface.

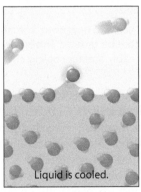

② Liquid water is cooled as it loses this high-speed water molecule.

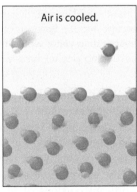

③ Molecule enters gaseous phase, having lost kinetic energy in overcoming hydrogen bonding at the liquid surface. Air is cooled as it collects these slow-moving gaseous particles.

▲ Figure 8.22
Evaporation is a cooling process.

▲ Figure 8.23
When wet, the cloth covering on this canteen promotes cooling. As the faster-moving water molecules evaporate from the wet cloth, its temperature decreases and cools the metal, which in turn cools the water in the canteen.

(b)

(a)

◀ Figure 8.24
(a) Dogs have no sweat glands (except between their toes). They cool themselves by panting. In this way, evaporation occurs in the mouth and the bronchial tract. (b) Pigs have no sweat glands and therefore cannot cool off by the evaporation of perspiration. Instead, they wallow in the mud to cool themselves.

temperature. Many animals, such as those shown in **Figure 8.24**, do not have sweat glands and must cool themselves by other means.

CONCEPT**CHECK**

If water were less "sticky," would you be cooled more or less by its evaporation?

CHECK YOUR ANSWER Water molecules leave the liquid phase only when they have enough kinetic energy to overcome hydrogen bonding. It is hydrogen bonding that makes water sticky, so to say that water is less sticky is to say that hydrogen bonds are weaker than they actually are. Then at a given temperature, more molecules in the liquid phase would have sufficient kinetic energy to overcome the weaker hydrogen bonds and escape into the gaseous phase, carrying heat away from the liquid. The cooling power of evaporating water would therefore be greater. This is why less "sticky" substances, such as rubbing alcohol, have a noticeably greater cooling effect as they evaporate.

Warm water evaporates, but so does cool water. The only difference is that cool water evaporates at a slower rate. Even frozen water "evaporates." This form of evaporation, in which molecules jump directly from the solid phase to the gaseous phase, is called **sublimation.** Because water molecules are so firmly held in the solid phase, frozen water does not release molecules into the gaseous phase as readily as liquid water does. Sublimation, however, does account for the loss of significant portions of snow and ice, especially on sunny, dry mountaintops. It's also why ice cubes left in the freezer for a long time tend to get smaller.

At the surface of any body of water, there is condensation as well as evaporation, as Figure 8.21 indicates. Condensation occurs as slow-moving water vapor molecules collide with and stick to the surface of a body of liquid water. Fast-moving water vapor molecules tend to bounce off each other or off the liquid surface, losing little of their kinetic energy. Only the slowest gas molecules condense into the liquid phase, as **Figure 8.25** illustrates. As this happens, energy is released as hydrogen bonds are formed. This energy is absorbed by the liquid and increases its temperature. Condensation involves the removal of slower-moving water vapor

CHEMICAL CONNECTIONS

How is a hurricane connected to the hydrogen bond?

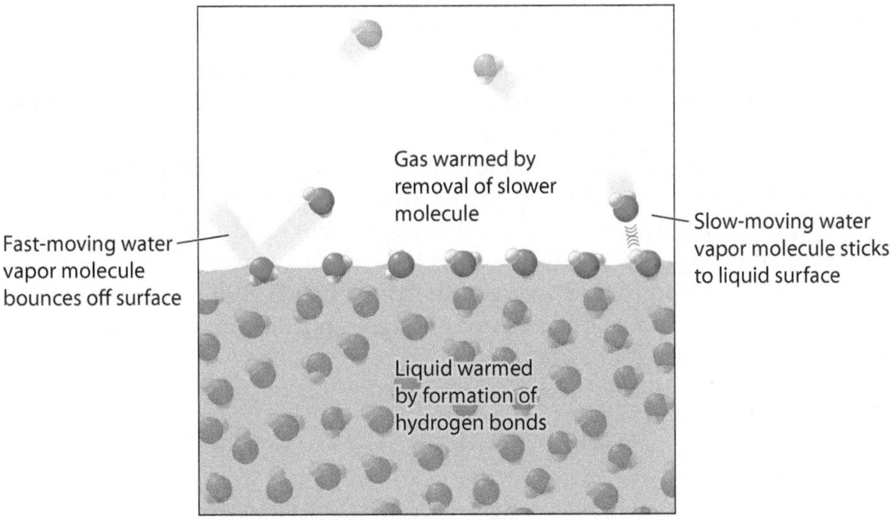

Gas warmed by
removal of slower
molecule

Fast-moving water
vapor molecule
bounces off surface

Slow-moving water
vapor molecule sticks
to liquid surface

Liquid warmed
by formation of
hydrogen bonds

◀ Figure 8.25
Condensation is a warming process.

▲ Figure 8.26
Heat is given up by water vapor when the
vapor condenses inside the radiator.

molecules from the gaseous phase. The average kinetic energy of the remaining water vapor molecules is therefore increased, which means that the water vapor is warmer. So no matter how you look at it, condensation is a warming process.

A dramatic example of the warming that results from condensation is the energy given up by water vapor when it condenses—a potentially painful experience if it condenses on you. That's why a burn from 100°C water vapor is much more damaging than a burn from 100°C liquid water; the water vapor gives up considerable energy when it condenses to a liquid and wets the skin. This energy released by condensation is utilized in heating systems, such as the household radiator shown in **Figure 8.26**.

The water vapor in our atmosphere also gives up energy as it condenses. This is the energy source for many weather systems, such as hurricanes, which derive much of their energy from the condensation of water vapor contained in humid tropical air, as **Figure 8.27** illustrates. The formation of 1 inch of rain over an area of 1 square mile yields the energy equivalent of about 32,000 tons of exploded dynamite.

After you take a shower, even a cold one, you are warmed by the heat energy released as the water vapor in the shower stall condenses. You quickly sense the difference if you step out of the stall, as the chilly guy in **Figure 8.28** is finding out. Away from the moisture, the rate of evaporation is much higher than the rate of condensation, and as a result, you feel chilly. When you remain in the shower stall, where the humidity is higher, the rate of condensation is increased, so you feel that much warmer. Now you know why you can dry yourself with a towel much more comfortably if you remain in the shower stall.

Spend a July afternoon in dry Tucson or Las Vegas and you'll soon notice that the evaporation rate is appreciably greater than the condensation rate. The result of this pronounced evaporation is a much cooler feeling than you would experience on a same-temperature July afternoon in New York City or New Orleans. In these humid locations, condensation outpaces evaporation, and you feel the warming effect as water vapor in the air condenses on your skin.

▲ Figure 8.27
As it condenses, the water vapor in humid tropical air releases ample quantities of heat. Continued condensation can sometimes lead to powerful storm systems, such as hurricanes.

CONCEPT CHECK

If the water level in a dish of water remains unchanged from one day to the next, can you conclude that no evaporation or condensation is taking place?

CHECK YOUR ANSWER Not at all, for there is much activity taking place at the molecular level. Both evaporation and condensation occur continuously and simultaneously. The fact that the water level remains constant indicates equal rates of evaporation and condensation—the number of H_2O molecules leaving the liquid surface by evaporation is equal to the number entering the liquid by condensation.

▲ Figure 8.28
If you're chilly outside the shower stall, step back inside and be warmed by the condensation of the excess water vapor there.

Boiling Is Evaporation beneath the Surface

When liquid water is heated to a sufficiently high temperature, bubbles of water vapor form beneath the surface, as we saw in Section 2.7. These bubbles are buoyed to the surface, where they escape, and we say the liquid is *boiling*. As shown in **Figure 8.29**, bubbles can form only when the pressure of the vapor inside them is equal to or greater than the combined pressure exerted by the surrounding water and the atmosphere above. At the boiling point of the liquid, the pressure inside the bubbles equals or exceeds the combined pressure of the surrounding water and the atmosphere. At lower temperatures, the pressure inside the bubbles is not enough, and the surrounding pressure collapses any bubbles that form.

 Atmospheric pressure

 Water pressure

Vapor pressure

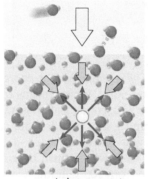

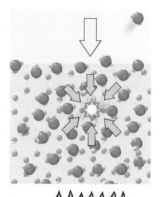

① As liquid water is heated, molecules gain enough energy to evaporate beneath the surface, forming bubbles of water vapor.

② Before the boiling point is reached, the pressure of the water vapor inside the bubbles is less than the sum of atmospheric pressure plus water pressure. As a result, the bubbles of water vapor collapse.

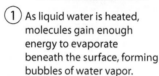

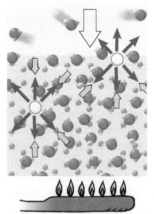

③ At the boiling point, the pressure of the water vapor inside the bubbles equals or exceeds the sum of atmospheric pressure plus water pressure. As a result, the bubbles of water vapor are buoyed to the surface and escape.

④ We see this evaporation as boiling.

At what point boiling begins depends not only on temperature but also on pressure. This includes the pressure from the weight of the surrounding liquid as well as the pressure from the weight of the atmosphere at the surface. As atmospheric pressure increases, the vapor molecules inside any bubbles that form must move faster in order to exert enough pressure from inside the bubble

READINGCHECK

What does the pressure inside a bubble of boiling water equal or exceed?

▶ **Figure 8.29**
Boiling occurs when water molecules in the liquid are moving fast enough to generate bubbles of water vapor beneath the surface of the liquid.

FORYOUR INFORMATION

Recall from Section 8.2 that adding salt lowers the freezing point of water. The reason is because the presence of a solute decreases the number of water molecules in contact with the surface of an ice crystal. Interestingly, salt water has a higher boiling point for similar reasons. In salt water, fewer water molecules at the liquid's surface are able to escape into the gaseous phase. This makes evaporation more difficult. Higher temperatures, therefore, are needed to get the salt water boiling.

to counteract the additional atmospheric pressure. So increasing the pressure exerted on the surface of a liquid raises its boiling point. A cooking application of this effect of increased pressure is shown in **Figure 8.30**.

Conversely, lower atmospheric pressure (as at high altitudes) decreases the boiling point of the liquid, as **Figure 8.31** illustrates. In Denver, Colorado, the Mile-High City, for example, water boils at 95°C instead of the 100°C boiling temperature at sea level. If you try to cook food in boiling water that is cooler than 100°C, you must wait a longer time for proper cooking. A three-minute boiled egg in Denver is runny and undercooked. If the temperature of the boiling water were very low, food would not cook at all. As the German mountaineer Heinrich Harrer noted in his book *Seven Years in Tibet*, at an altitude of 4500 meters (15,000 feet) or higher, you can sip a cup of boiling tea without any danger of burning your mouth.

Boiling, like evaporation, is a cooling process. At first thought, this may seem surprising—perhaps because we usually associate boiling with heating. But heating water is one thing; boiling it is another. As shown in **Figure 8.32**, boiling water is cooled by boiling as fast as it is heated by the energy from the heat source. So boiling water remains at a constant temperature. If cooling did not take place, continued application of heat to a pot of boiling water would raise the temperature of the water. The reason the pressure cooker in Figure 8.30 reaches higher temperatures is that boiling is forestalled by increased pressure, which in effect prevents cooling.

▲ Figure 8.30
The tight lid of a pressure cooker holds pressurized vapor above the water surface, and this inhibits boiling. In this way, the boiling temperature of the water is increased. Any food placed in this hotter water cooks more quickly than food placed in water boiling at 100°C.

CONCEPT**CHECK**

Is boiling a form of evaporation, or is evaporation a form of boiling?

CHECK YOUR ANSWER Boiling is evaporation that takes place beneath the surface of a liquid.

A simple experiment that dramatically shows the cooling effect of evaporation and boiling consists of a shallow dish of room-temperature water in a vacuum jar. When the pressure in the jar is slowly reduced by a vacuum pump, the water starts to boil. The boiling process takes heat away from the water, which consequently cools. As the pressure is further reduced, more and more of the slower-moving liquid molecules boil away. Continued boiling results in a lowering of the temperature until the freezing point of

▲ Figure 8.31
The boiling point of water (as well as other liquids) decreases with increasing altitude.

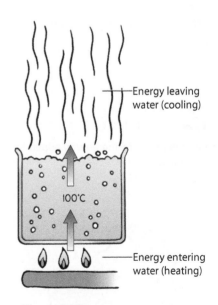

Energy leaving water (cooling)

100°C

Energy entering water (heating)

▲ Figure 8.32
Heating warms the water from below, and boiling cools it from above. The net result is a constant temperature for the water.

▲ **Figure 8.33**
In a vacuum, water can freeze and boil at
the same time.

approximately 0°C is reached. Continued cooling by boiling causes ice to form over the surface of the bubbling water. Boiling and freezing take place at the same time! The frozen bubbles of boiling water in **Figure 8.33** are a remarkable sight.

Spray some drops of coffee into a vacuum chamber and they, too, boil until they freeze. Even after they are frozen, the water molecules continue to evaporate into the vacuum until all that is left to be seen are little crystals of coffee solids. This is how freeze-dried coffee is made. The low temperature of this process tends to keep the chemical structure of the coffee solids from changing. When hot water is added, much of the original flavor of the coffee is retained.

The refrigerator also employs the cooling effect of boiling. A liquid coolant that has a low boiling point is pumped into the coils inside the refrigerator, where the liquid boils (evaporates) and draws heat from the food stored in the refrigerator. Then the coolant in its gas phase, along with its added energy, is directed outside the refrigerator to coils located in the back, appropriately called condensation coils, where heat is given off to the air as the coolant condenses back to a liquid. A motor pumps the coolant through the system as it undergoes the cyclic process of vaporization and condensation. The next time you're near a refrigerator, place your hand near the condensation coils in the back; you'll feel the heat that has been extracted from inside.

An air conditioner employs the same principle, pumping heat energy from inside a building to outside the building. Turn the air conditioner around so that inside cold is pumped outside and the air conditioner becomes a type of heater known as a heat pump.

8.5 It Takes a Lot of Energy to Change the Temperature of Liquid Water

LEARNING OBJECTIVE

Describe how the formation and breaking of hydrogen bonds are responsible for water's high specific heat and relate this to global climate.

READINGCHECK

What happens to the energy added to a material?

EXPLAIN THIS

Why does pie filling remain hot for a longer time compared to the pie crust?

Have you ever noticed that some foods stay hot much longer than others? The filling of a hot apple pie can burn your tongue while the crust does not, even when the pie was just taken out of the oven. A piece of toast may be comfortably eaten a few seconds after coming from a hot toaster, whereas you must wait several minutes before eating hot soup.

Different substances have different capacities for storing energy. This is because different materials absorb energy in different ways. The added energy may increase the jiggling motion of molecules, which raises the temperature, or it may pull apart the attractions among molecules and therefore go into potential energy, which does not raise the temperature. Generally, there is a combination of the two ways.

It takes 4.184 joules of energy to raise the temperature of 1 gram of liquid water by 1°C. As you can see in **Figure 8.34**, it takes only about one-ninth as much energy to raise the temperature of 1 gram of iron by the same amount. In other words, compared to iron, water absorbs more heat for the same change in temperature. We say water has a higher **specific heat**, defined as the quantity of heat required to change the temperature of 1 gram of a substance by 1°C.

We can think of specific heat as thermal inertia. As you learned in Section 2.3, *inertia* is a term used in physics to signify the resistance of an object to a change in its state of motion. Specific heat is like a thermal version of inertia because it signifies the resistance of a substance to a change in

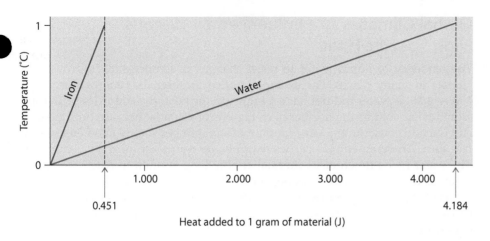

◀ Figure 8.34
It takes only 0.451 joule of heat to raise the temperature of 1 gram of iron by 1°C. A 1-gram sample of water, by contrast, requires a whopping 4.184 joules for the same temperature change.

temperature. Each substance has its own characteristic specific heat, which may be used to assist in its identification. Some typical values are given in Table 8.1.

Guess why water has such a high specific heat. Once again, the answer is hydrogen bonds. When heat is applied to water, much of the heat is consumed in breaking hydrogen bonds. Broken hydrogen bonds are a form of potential energy (just as two magnets pulled apart are a form of potential energy). Much of the heat added to water, therefore, is stored as this potential energy. Consequently, less heat is available to increase the kinetic energy of the water molecules. Because temperature is a measure of kinetic energy, we find that as water is heated, its temperature rises slowly. By the same token, when water is cooled, its temperature drops slowly—as the kinetic energy decreases, molecules slow down and more hydrogen bonds are able to re-form. This in turn releases heat that helps to maintain the temperature.

FOR YOUR INFORMATION

A material with a high specific heat has a great capacity to hold onto heat, which is why specific heat is sometimes referred to as *heat capacity*. Water has a great capacity for holding onto heat. You can think of water as a very effective "heat sponge."

TABLE 8.1   Specific Heat for Some Common Materials

MATERIAL	SPECIFIC HEAT (J/g · °C)
Ammonia, NH_3	4.70
Liquid water, H_2O	4.184
Ethylene glycol, $C_2H_6O_2$ (antifreeze)	2.42
Ice, H_2O	2.01
Water vapor, H_2Os	2.0
Aluminum, Al	0.90
Iron, Fe	0.451
Silver, Ag	0.24
Gold, Au	0.13

CONCEPT CHECK

Hydrogen bonds are not broken as heat is applied to ice (provided the ice doesn't melt) or water vapor. Would you therefore expect ice and water vapor to have specific heats that are greater or less than that of liquid water?

CHECK YOUR ANSWER As Table 8.1 shows, the specific heats of ice and water vapor are about half that of liquid water. Only liquid water has a remarkable specific heat. This is because the liquid phase is the only phase in which hydrogen bonds are continually breaking and re-forming.

FOR YOUR INFORMATION

The outer edge of a carousel platform always provides the fastest ride, because it travels a greater distance for each rotation. Similarly, because the Earth spins, those living closer to the equator travel farther, and hence faster, in a single day. When objects not fixed to the ground—such as air, water, airplanes, and ballistic missiles—move toward the equator, they are unable to keep up with the ever-faster moving ground beneath them. Because the Earth rotates to the east, these free-moving objects lag behind, which gives them an apparent motion to the west. Conversely, a free-moving object heading away from the equator deviates to the east, because it is maintaining the greater eastward-directed speed it had by being closer to the equator. This is called the *Coriolis effect.* Ocean currents heading toward the equator, for example, tend to deviate to the west, while currents heading away from the equator tend to deviate to the east. This explains why ocean currents in the northern hemisphere tend to circulate clockwise, while those in the southern hemisphere tend to circulate counterclockwise, as is evident in Figure 8.35. Why do you suppose airplane pilots need to be well acquainted with the Coriolis effect?

Global Climates Are Influenced by Water's High Specific Heat

The tendency of liquid water to resist changes in temperature improves the climate in many places. For example, notice the high latitude of Europe in **Figure 8.35**. If water did not have a high specific heat, the countries of Europe would be as cold as the northeastern regions of Canada, because both Europe and Canada get about the same amount of sunlight per square kilometer of surface area. An ocean current carries warm water northeast from the Caribbean. The water holds much of its thermal energy long enough to reach the North Atlantic off the coast of Europe, where the water then cools. The energy released, 4.184 joules per degree Celsius for each gram of water that cools, is carried by the westerly winds (winds that blow west to east) over the European continent.

The winds in the latitudes of North America are westerly. On the western coast of the continent, therefore, air moves from the Pacific Ocean to the land. Because of water's high specific heat, ocean temperatures do not vary much from summer to winter. In winter, the water warms the air, which is then blown eastward over the coastal regions. In summer, the water cools the air and the coastal regions are cooled. On the eastern coast of the continent, the temperature-moderating effects of the Atlantic Ocean are significant, but because the winds blow from the west—over land—temperature ranges in the east are much greater than in the west. San Francisco, for example, is warmer in winter and cooler in summer than is Washington, D.C., which is at about the same latitude.

Islands and peninsulas, because they are more or less surrounded by water, do not have the extremes of temperatures observed in the interior of a continent. The high summer temperatures and low winter temperatures common in Manitoba and the Dakotas, for example, are largely due to the absence of large bodies of water. Europeans, islanders, and people living near ocean air currents should be glad that water has such a high specific heat.

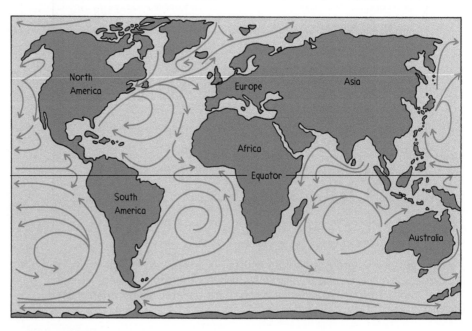

▲ Figure 8.35
Many ocean currents, shown in blue, distribute heat from the warmer equatorial regions to the colder polar regions.

CONCEPTCHECK

Which has a higher specific heat, water or sand?

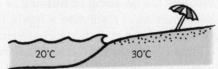

20°C   30°C

CHECK YOUR ANSWER As suggested by the illustration, the temperature of water increases less compared to the temperature of sand in the same sunlight. Water therefore has the higher specific heat. Those who visit the beach frequently know that beach sand quickly turns hot on a sunny day, while the water remains relatively cool. At night, however, the sand feels quite cool, while the water's temperature feels about the same as it did during the day.

CALCULATION CORNER   HOW HEAT CHANGES TEMPERATURE

Heat must be applied to increase the temperature of a material. Conversely, heat must be withdrawn from a material in order to decrease its temperature. The amount of heat required for a given temperature change is calculated from the equation

heat = specific heat × mass × temperature change

We can use this formula for any material, provided there is no change of phase over the course of the temperature change. The value of the temperature change is obtained by subtracting the initial temperature T_i from the final temperature T_f:

$$\text{temperature change} = T_f - T_i$$

EXAMPLE 1

How much heat is required to increase the temperature of 1.00 gram of liquid water from an initial temperature of 30.0°C to a final temperature of 40.0°C?

ANSWER 1

The temperature change is $T_f - T_i = 40.0°C - 30.0°C = +10.0°C$. To find the amount of heat needed for this temperature change, multiply this positive change by the water's specific heat and mass:

$$\text{heat} = \left(4.184 \frac{J}{g\,°C}\right)(1.00\text{ g})(+10.0°C) = 41.8\text{ J}$$

The temperature decrease that occurs when heat is removed from a material is indicated by a negative sign, as is shown in the next example.

EXAMPLE 2

A glass containing 10.0 grams of water at an initial temperature of 25.0°C is placed in a refrigerator. How much heat does the refrigerator remove from the water as the water is brought to a final temperature of 10.0°C?

ANSWER 2

The temperature change is $T_f - T_i = 10.0°C - 25.0°C = -15.0°C$. To find the heat removed, multiply this negative temperature change by the water's specific heat and mass:

$$\text{heat} = \left(4.184 \frac{J}{g\,°C}\right)(10.0\text{ g})(-15.0°C) = -628\text{ J}$$

YOUR TURN

1. A residential water heater raises the temperature of 100,000 grams of liquid water (about 26 gallons) from 25.0°C to 55.0°C. How much heat was applied?

2. How much heat must be extracted from a 10.0-gram ice cube $\left(\text{specific heat} = 2.01 \frac{J}{g\,°C}\right)$ in order to bring its temperature from a chilly –10.0°C to an even chillier –30.0°C?

The answers for Calculation Corners appear at the end of each chapter.

8.6 A Phase Change Requires the Input or Output of Energy

LEARNING OBJECTIVE

Identify the molecular processes that occur as a substance, such as water, changes phase and explain how this necessarily involves the input or output of energy.

READINGCHECK

What does the change of a substance's phase involve?

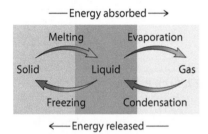

← Energy released ——

▲ Figure 8.36
Energy changes with change of phase.

EXPLAIN THIS

How does holding a wet finger in the air help to determine the direction of the wind?

Any phase change involves the breaking or forming of molecular attractions. The changing of a substance from a solid to a liquid to a gas, for example, involves the breaking of molecular attractions. Phase changes in this direction therefore require the input of energy. Conversely, the changing of a substance from a gas to a liquid to a solid involves the forming of molecular attractions. Phase changes in this direction therefore result in the release of energy. Both these directions are summarized in **Figure 8.36**.

Consider a 1-gram piece of ice at –50°C put on a stove to heat. A thermometer in the container reveals a slow increase in temperature up to 0°C, as shown in **Figure 8.37**. At 0°C, the temperature stops rising, even though heat is still being added, and for now, all the added heat goes into melting the 0°C ice, as indicated in **Figure 8.38**. This process of melting 1 gram of ice requires 335 joules. Only when all of the ice has melted does the temperature begin to rise again. Then for every 4.184 joules absorbed, the water increases its temperature by 1°C until the boiling temperature, 100°C, is reached. At 100°C, the temperature again stops rising even though heat is still being added, for now all the added heat is going into evaporating the liquid water to water vapor. The water must absorb a stunning 2259 joules of heat to evaporate all the liquid water. Finally, when all the liquid water has become vapor at 100°C, the temperature begins to rise once again and continues to rise as long as heat is added.

When the phase changes are in the opposite direction, the amounts of heat energy shown in Figure 8.37 are the amounts *released*—2259 joules per gram when water vapor condenses to liquid water and 335 joules per gram when liquid water turns to ice. The processes are reversible.

CONCEPTCHECK

From Figure 8.37, deduce how much energy (in joules) is transferred when 1 gram of

a. water vapor at 100°C condenses to liquid water at 100°C.

b. liquid water at 100°C cools to liquid water at 0°C.

c. liquid water at 0°C freezes to ice at 0°C.

d. water vapor at 100°C turns to ice at 0°C.

CHECK YOUR ANSWERS

a. 2259 joules

b. 418 joules

c. 335 joules

d. 3012 joules (2259 joules + 418 joules + 335 joules)

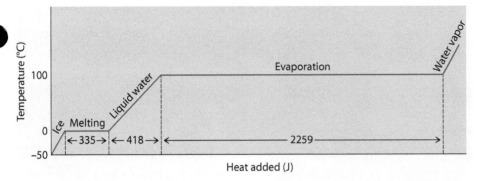

A graph showing the heat energy involved in converting 1 gram of ice initially at 250°C to water vapor. The horizontal portions of the graph represent regions of constant temperature.

The amount of heat energy required to change a solid to a liquid is called the **heat of melting,** and the amount of heat energy released when a liquid freezes is called the **heat of freezing***. Water has a heat of melting of +335 joules per gram. The positive sign indicates that this is the amount of heat energy that must be *added to* ice to melt it. Water's heat of freezing is –335 joules per gram. The negative sign indicates that this is the amount of heat energy that is *released* from liquid water as it freezes—the same amount that was required to melt it.

The amount of heat energy required to change a liquid to a gas is called the **heat of vaporization.** For water, this is +2259 joules per gram. The amount of heat energy released when a gas condenses is called the **heat of condensation.** For water, this is –2259 joules per gram. These values are high relative to the heats of vaporization and condensation for most other substances. This is due to the relatively strong hydrogen bonds between water molecules that must be broken or formed during these processes.

▲ Figure 8.38
Add heat to melting ice and there is no change in temperature. The heat is consumed in breaking hydrogen bonds.

CONCEPT CHECK
Can you add heat to ice without melting it?

CHECK YOUR ANSWER A common misconception is that ice cannot have a temperature lower than 0°C. In fact, ice can have any temperature below 0°C, down to absolute zero, –273°C. Adding heat to ice below 0°C raises its temperature, say, from –200°C to –100°C. As long as its temperature stays below 0°C, the ice does not melt.

Although water vapor at 100°C and liquid water at 100°C have the same temperature, each gram of the vapor contains an additional 2259 joules of potential energy. As the molecules bind together to a liquid phase, this 2259 joules per gram is released to the surroundings in the form of heat. In other words, the potential energy of the far-apart water vapor molecules transforms to heat as the molecules get closer together. This is like the potential energy of two attracting magnets separated from each other; when released, their potential energy is converted first to kinetic energy and then to heat as the magnets strike each other.

Water's high heat of vaporization allows you to *briefly* touch your *wetted* finger to a hot skillet or hot stove without harm. This is because energy that ordinarily would go into burning your finger goes instead into changing the phase of the moisture on your finger from liquid to vapor. You can judge the hotness of a clothes iron in the same way—with a *wet* finger.

 CHEMICAL CONNECTIONS

How is the time it takes to warm a pot of water connected to the consistently warm climate of Hawaii?

---
*The term *heat of fusion* is commonly used to indicate either heat of melting or heat of freezing.

▲ Figure 8.39
Water extinguishes a flame by absorbing much of the heat the fire needs to sustain itself, as well as by wetting, which blocks oxygen from reaching the burning material.

The firefighters in **Figure 8.39** know that certain types of flames are best extinguished with a fine mist of water rather than a steady stream. The fine mist readily turns to water vapor and in doing so quickly absorbs heat energy and cools the burning material.

Water's high heat of vaporization makes walking barefooted on red-hot coals more comfortable, as shown in **Figure 8.40**. When your feet are wet, either because they are perspiring or because you are stepping off wet grass, much of the heat from the coals is absorbed by the water and not by your skin. (Firewalking also relies on the fact that wood is a poor conductor of heat, even when it is in the form of red-hot coals.)

▶ Figure 8.40
Tracy Suchocki walks with wetted bare feet across red-hot wood coals without harm.

Many of the properties of water we have explored in this chapter at the molecular level are nicely summarized in the scene shown in **Figure 8.41**. First, notice that the massive bear is jumping from floating ice. The ice floats because hydrogen bonds hold together the H_2O molecules in the ice in an open crystalline structure that makes the ice less dense than the liquid water. Because so much energy is required both to melt ice and to evaporate liquid water, most of the Arctic ice remains in the solid phase throughout the year. Where the seawater is in contact with the ice, the temperature is lower than 0°C because salts dissolved in the seawater inhibit the formation of ice crystals and thereby lower the freezing point of the seawater. The high specific heat of the Arctic Ocean beneath the bear moderates the Arctic climate. It gets cold in the Arctic in winter, yes, but not as cold as it gets in Antarctica, where the specific heat of the mile-thick ice is only half that of the liquid water predominating in the Arctic. Covered by ice with its much lower specific heat, Antarctica experiences much greater extremes in temperature than does the Arctic.

In the previous chapter, we focused on the physical behavior of molecules and ions in general. In this chapter, the main focus has been on the extraordinary physical behavior of water molecules in particular. For the next several chapters, we shift our attention to the *chemical* behavior of molecules and ions, which change their fundamental identities as they chemically react with one another.

▲ Figure 8.41
What remarkable properties of water can you find in this photograph?

Chapter 8 Review

LEARNING OBJECTIVES

Relate the physical properties of ice to its crystalline structure. (8.1)	→	*Questions 1–3, 22, 36, 40–43, 77*
Identify the molecular processes involved in the freezing and melting of water, explain the impact of a solute on these processes, and explain why water is most dense at 4°C. (8.2)	→	*Questions 4–6, 34–35, 44–51, 75*
Describe how cohesive and adhesive forces give rise to surface tension and capillary action. (8.3)	→	*Questions 7–9, 23, 52–56*
Show, on a molecular level, how evaporation and condensation lead to cooling and warming effects, respectively, and how they relate to the process of boiling. (8.4)	→	*Questions 10–13, 37, 57–62*
Describe how the formation and breaking of hydrogen bonds are responsible for water's high specific heat and relate this to global climate. (8.5)	→	*Questions 14–17, 24–28, 33, 38, 63–68, 76*
Identify the molecular processes that occur as a substance, such as water, changes phase and explain how this necessarily involves the input or output of energy. (8.6)	→	*Questions 18–21, 29–32, 39, 69–74*

SUMMARY OF TERMS (KNOWLEDGE)

Adhesive force An attractive force between molecules of two different substances.

Capillary action The rising of liquid into a small vertical space due to the interplay of cohesive and adhesive forces.

Cohesive force An attractive force between molecules of the same substance.

Heat of condensation The energy released by a substance as it transforms from gas to liquid.

Heat of freezing The heat energy released by a substance as it transforms from liquid to solid.

Heat of melting The heat energy absorbed by a substance as it transforms from solid to liquid.

Heat of vaporization The heat energy absorbed by a substance as it transforms from liquid to gas.

Meniscus The curving of the surface of a liquid at the interface between the liquid surface and its container.

Specific heat The quantity of heat required to change the temperature of 1 gram of a substance by 1 Celsius degree.

Sublimation The process of a material transforming from a solid directly to a gas without passing through the liquid phase.

Surface tension The elastic tendency found at the surface of a liquid.

READING CHECK QUESTIONS (COMPREHENSION)

8.1 Water Molecules Form an Open Crystalline Structure in Ice

1. What accounts for the fact that ice is less dense as compared to water?

2. What is inside one of the open spaces of an ice crystal?

3. What happens to ice when great pressure is applied to it?

8.2 Freezing and Melting Go On at the Same Time

4. What is released when a hydrogen bond forms between two water molecules?

5. When the temperature of 0°C liquid water is increased slightly, does the water undergo a net expansion or a net contraction?

6. At what temperature do the competing effects of contraction and expansion produce the smallest volume for liquid water?

8.3 Liquid Water's Behavior Results from the Stickiness of Its Molecules

7. What is the difference between cohesive forces and adhesive forces?

8. In what direction is a water molecule on the surface pulled?

9. Does liquid water rise higher in a narrow tube or a wide tube?

8.4 Water Molecules Move Freely between the Liquid and Gaseous Phases

10. Do all the molecules in a liquid have about the same speed?

11. Why do we feel uncomfortably warm on a hot, humid day?

12. Is it the pressure or the higher temperature that cooks food faster in a pressure cooker?

13. What condition permits liquid water to boil at a temperature below 100°C?

8.5 It Takes a Lot of Energy to Change the Temperature of Liquid Water

14. Is it easy or difficult to change the temperature of a substance that has a low specific heat?

15. Does a substance that heats up quickly have a high or a low specific heat?

16. How does the specific heat of liquid water compare with the specific heats of other common materials?

17. What within liquid water is most responsible for its unusual specific heat?

8.6 A Phase Change Requires the Input or Output of Energy

18. When liquid water freezes, is heat released to the surroundings or absorbed from the surroundings?

19. Why doesn't the temperature of melting ice rise as the ice is heated?

20. How much heat is needed to melt 1 gram of ice? Give your answer in joules.

21. Why does it take so much more energy to boil 10 grams of liquid water than it takes to melt 10 grams of ice?

CONFIRM THE CHEMISTRY (HANDS-ON APPLICATION)

22. The principle behind ice-skating can be used to pass a metal wire through a block of ice. Create a large bubble-free ice block by adding recently boiled water to a plastic tub and freezing it in your freezer. Drape a thin metal wire across the ice block as shown in the drawing. Dumbbells or water-filled jugs can be used as the weights. Once the wire has passed through, knock the ice with a hammer and see where it breaks. In the days before refrigerators, this was the way large ice blocks were cut to size for the kitchen icebox. What do you suppose would happen if string were used instead of wire? Why is thin wire preferable to thick wire?

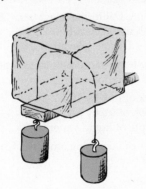

23. Glue-gun a plastic mesh screen in the mouth of an opaque plastic bottle just beneath the rim. Impress your friends by pouring water back and forth between this bottle and another one just like it but without the secret screen. With the water in your specially prepared bottle, place a card on top of the bottle. Hold the card and invert the bottle. Ask your friends what will happen when you take your hand off the card. They'll be amazed to see that the card stays. Be sure to explain the role of atmospheric pressure. Next, ask them what will happen when you knock off the card. They'll be even more amazed to see that the water doesn't come out. Finally, take a thin skewer stick and poke it up into the screen and ask what will happen when you let go of the stick. Let go of the stick and tell your friends that they're witnessing a "stickup." Of course, follow up with an explanation of the role that water's surface tension plays.

24. Place a cup of salt and a cup of rice side by side on a piece of aluminum foil on a baking sheet. Heat the salt and rice for 10 minutes in an oven preheated to 250°C. Pour the salt and rice into two separate mugs. Which cools down faster? If you don't have a thermometer, judge their cooling rates by cautious touch. Which has the lower specific heat? Why does the heated rice adhere to the side of the mug?

25. Put the high specific heat of water to practical use by keeping warm on cold evenings or by soothing painful cramps. Fill a clean sock three-quarters full with rice, which by its nature absorbs a lot of moisture from the air. Tie the open end closed with a string (don't use metal wire!) and cook in a microwave for a couple of minutes. (Don't use a conventional oven!) The moisture in the grains becomes apparent when you take the sock out of the oven—the released moisture has made the sock slightly damp. Wrap the sock around your neck for instant gratification. Need a neck cooler? Store the rice-filled sock in the freezer. The moisture in the rice stays cold for a long time. These devices make great homemade gifts when a pretty fabric is used in place of the sock and a mild fragrance is added.

THINK AND SOLVE (MATHEMATICAL APPLICATION)

26. A walnut stuck to a pin is burned beneath a can containing 100.0 grams of water at 21°C. After the walnut has completely burned, the water's final temperature is 28°C. How much heat energy came from the burning walnut?

27. How much heat is required to raise the temperature of 100,000 grams of iron by 30°C?

28. By how much will the temperature of 5.0 grams of liquid water increase upon the addition of 230 joules of heat?

29. How much heat is required to raise the temperature of 1.00 gram of water from −5.00°C to +5.00°C?

30. How much heat is required to raise the temperature of 1.00 gram of water from absolute zero, −273°C, to 100°C?

31. How much energy is required to transform 1.00 gram of water from 100°C liquid to 100°C water vapor?

32. Why does transforming water from 100°C liquid to 100°C gas require much more energy than is needed to raise the temperature of the same amount of water from absolute zero all the way to 100°C liquid?

33. To increase the temperature of 575 grams of ocean water by 5.0°C requires 2674 J. Use this information to calculate ocean water's specific heat.

THINK AND COMPARE (ANALYSIS)

34. Rank the following in order of decreasing melting point: a 0.1 M aqueous solution of
    a. sodium bromide, NaBr.
    b. magnesium chloride, $MgCl_2$.
    c. scandium iodide, ScI_3.

35. Rank the following in order of increasing temperature: water at the bottom of lake Michigan on
    a. February 1, 2011.
    b. June 15, 2011.
    c. October 15, 2011.

36. Rank the following in order of driest snow conditions during ski season: the top of
    a. Jupiter Peak (10,000 ft, −10°C), Park City, Utah.
    b. Lincoln Peak (4000 ft, −8°C), Mad River Valley, Vermont.
    c. Wintergreen Peak (2000 ft, −5°C), Wintergreen, Virginia.

37. Rank the following in order of increasing boiling point: water boiling at
    a. Death Valley, California (elevation -282 ft).
    b. Winfield, Kansas (elevation 1129 ft).
    c. Tampa Bay, Florida (elevation 2 ft).

38. Rank the following in order of increasing amount of energy required to heat
    a. a 5-g block of iron by 10°C.
    b. a 10-g block of ice by 5°C.
    c. a 1-g sample of liquid water by 2°C.

39. Rank in order of increasing amount of energy required to change 1 g of water from
    a. solid to liquid.
    b. liquid to gas.
    c. solid to gas.

THINK AND EXPLAIN (SYNTHESIS)

8.1 Water Molecules Form an Open Crystalline Structure in Ice

40. Why is it important to protect water pipes so that they don't freeze?

41. What happens to a soda can left in the freezer? Why?

42. How does the combined volume of the billions and billions of hexagonal open spaces in the crystals in a piece of ice compare with the portion of the ice that floats above the water line?

43. Review Figure 6.22. What do ice and diamond have in common? How are they different?

8.2 Freezing and Melting Go On at the Same Time

44. Ice floats in room-temperature water, but does it float in boiling water? Why or why not?

45. How is it that pure water at 4°C is not purely liquid?

46. What happens to the freezing point of a solution of salt in water as the solution becomes more concentrated with salt?

47. Which should have a lower freezing point: a 1 M solution of sodium chloride, NaCl, or a 1 M solution of calcium chloride, $CaCl_2$? Why?

48. Which should have a lower freezing point: a 1 M solution of sodium chloride, NaCl, or a 1 M solution of sucrose, $C_{12}H_{22}O_{11}$? Why?

49. Ice cubes are not usually perfectly clear. Instead, they contain clouds, which form as air comes out of solution, forming tiny opaque bubbles. Why are ice cubes typically cloudy in the center but not around the edges?

50. Suppose liquid water instead of mercury is used in a thermometer. If the temperature is initially 4°C and then changes, why can't the thermometer indicate whether the temperature is rising or falling?

51. If cooling occurred at the bottom of a pond instead of at the surface, would a lake freeze from the bottom up? Explain.

8.3 Liquid Water's Behavior Results from the Stickiness of Its Molecules

52. Capillary action causes water to climb up the internal walls of narrow glass tubes. Why does the water not climb as high when the glass tube is wider—even in the absence of gravity?

53. Mercury forms a convex meniscus with glass rather than the concave meniscus shown in Figure 8.18. What does this tell you about the cohesive forces between mercury atoms versus the adhesive forces between mercury atoms and glass? Which forces are stronger? Do you suppose the surface tension of mercury is stronger or weaker than that of water?

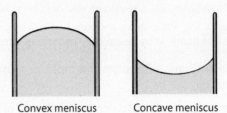

Convex meniscus   Concave meniscus

54. Would you expect the surface tension of water to increase or decrease with temperature? Defend your answer.

55. Dip a paper clip into water and then slowly pull it upward to the point where it is nearly free from the surface. You'll find that for a short distance, the water is brought up with the metal. Are these adhesive or cohesive forces at work?

56. A thin film of fresh water can be made to stretch across a small wire loop with a diameter of about 1 centimeter. At larger diameters, however, the film collapses into a water drop. If water's surface tension were greater, might greater diameters of a thin film be achieved?

8.4 Water Molecules Move Freely between the Liquid and Gaseous Phases

57. Why do bubbles in boiling water get larger as they rise to the surface?

58. Where is the boiling point of water the greatest: at the bottom of a pot of water or just below the surface of the pot of water? Why?

59. Why does wetting a cloth in a bucket of cold water and then wrapping the wet cloth around a bottle produce a cooler bottle than placing the bottle directly in the bucket of cold water?

60. A lid on a cooking pot filled with water shortens both the time it takes the water to come to a boil and the time the food takes to cook in the boiling water. Explain what is going on in each case.

61. Because boiling is a cooling process, would it be a good idea to cool your hot, sticky hands by dipping them into boiling water?

62. What would happen to the oceans if the atmosphere suddenly disappeared?

8.5 It Takes a Lot of Energy to Change the Temperature of Liquid Water

63. Describe the motion of two magnetic marbles as they roll closer to each other and then come into contact. What happens to two water molecules after they come together to form a hydrogen bond—do they vibrate more rapidly or less rapidly?

64. Why does liquid water have such a high specific heat?

65. Which has a higher specific heat: fresh water or ocean water?

66. Bermuda is close to North Carolina, but unlike North Carolina, it has a tropical climate year-round. Why?

67. If the winds at the latitude of San Francisco and Washington, D.C., were from the east rather than from the west, why might San Francisco be able to grow cherry trees and Washington, D.C., would be able to grow palm trees?

68. You can bring water in a paper cup to a boil by placing it over a hot flame. Why doesn't the paper cup burn?

8.6 A Phase Change Requires the Input or Output of Energy

69. Maple syrup is made by concentrating the sap of the sugar maple tree. Why does it take so much energy to make maple syrup?

70. Why in cold winters does a large tub of liquid water placed in a farmer's canning cellar help prevent canned food from freezing?

The following diagram describes the phase of water relative to temperature and pressure. Use this diagram for questions 71–74.

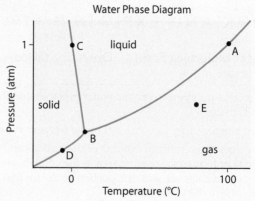

Water Phase Diagram

71. At which point is water boiling?

72. At which point does water exist in all three phases?

73. What happens to the melting temperature of water as pressure is increased? What happens to water's boiling temperature as pressure is increased?

74. Is it possible for ice to transform to water vapor without ever becoming liquid?

THINK AND DISCUSS (EVALUATION)

75. Explain why for about 50 meters beneath the floating Arctic ice cap the ocean water is relatively fresh.

76. About 250 meters beneath the surface of the Arctic Ocean, the water is warm enough to melt the Arctic polar ice cap. Deep ocean currents bring this warm water from the adjacent Atlantic Ocean. Why doesn't this deep warm water in the Arctic Ocean float to the surface to melt the ice cap? What would be some of the negative consequences if the Arctic ice cap completely melted? What positive consequences might there be?

77. The ice cap over the Arctic Ocean grows to about 15 million cubic kilometers (km^3) each winter. The ice cap over the subcontinent of Greenland, also in the northern hemisphere, has a volume of about 3 million km^3. Which would be more devastating: the melting of the Arctic ice cap or the melting of the Greenland ice cap? Discuss why.

READINESS ASSURANCE TEST (RAT)

If you have a good handle on this chapter, then you should be able to score at least 7 out of 10 on this RAT. Check your answers online at www.ConceptualChemistry.com. If you score less than 7, you need to study further before moving on.

Choose the BEST answer to the following.

1. An ice skater skates on
   a. blade wax.
   b. liquid water.
   c. ice.
   d. a thin layer of air.

2. As an ice cube floating in a glass of water melts, water level in the glass
   a. rises to the point where the ice floats above the water.
   b. does not change.
   c. gets lower.
   d. rises because of air captured within the ice cube.

3. Unlike fresh water, ocean water contracts as it is cooled all the way down to its freezing point, which is about −18°C. Why?
   a. As the water cools, the salt ions in the ocean water pull the molecules closer together.

   b. Macrocrystals instead of nanocrystals are formed during the freezing of ocean water.

   c. The presence of a solute disrupts the rate of ice nanocrystal formation.

   d. As the water cools, it becomes denser and then sinks. The pressure from the weight of the ocean exerted on the ice below the surface of the water causes it to be denser.

4. Which graph most appropriately shows the density of water plotted against temperature?

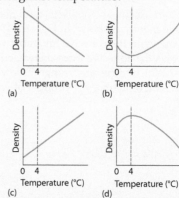

5. A glass can be filled above its brim with water without any spilling over the edge due to

   a. cohesive forces within water.

   b. adhesive forces within water.

   c. cohesive forces and gravity.

   d. adhesive forces and gravity.

6. What is the predominant gas found within a bubble of boiling water?

   a. Air

   b. Chlorine

   c. Water vapor

   d. Hydrogen

7. Water coming out of volcanic sea vents at the bottom of the ocean can reach temperatures in excess of 300°C without boiling because

   a. the heat is immediately transferred to the surrounding very cold water.

   b. the heat is immediately transferred to the surrounding volcanic rock.

   c. the water and water vapor reach equilibrium under the extreme pressure of the ocean.

   d. of the extreme pressure exerted by the miles of ocean water.

8. The specific heat of your automobile's radiator fluid should be as

   a. high as possible to better distribute engine heat.

   b. low as possible so that it requires more energy to boil.

   c. high as possible to better absorb engine heat.

   d. low as possible so that the fluid does not turn viscous.

9. To impart a hickory flavor to a roasted turkey, a cook places a pot of water containing hickory chips in an oven with the turkey. The turkey takes longer than expected to cook because

   a. much of the oven heat is consumed in changing the phase of the water.

   b. the turkey is being steamed instead of baked.

   c. to get a hickory flavor, you need to roast the turkey at a low temperature.

   d. the hickory chips do not conduct heat very well.

10. Describe how a boiling liquid can be used to cool something (i.e., a refrigerator).

    a. A low-boiling liquid absorbs heat energy, dropping the temperature.

    b. Heat is used to convert vapor into a liquid, which is cooler than a gas.

    c. As a liquid condenses, heat energy is required and the temperature drops.

    d. Heat energy is only used with solids such as ice.

    e. None of the above.

ANSWERS TO CALCULATION CORNER (HOW HEAT CHANGES TEMPERATURE)

1. The temperature change is final temperature minus initial temperature: 50.0°C−25.0°C = +30.0°C. Multiply this positive temperature change by the water's specific heat and mass:

$$\text{heat} = \left(4.184\,\frac{J}{g°C}\right)(100{,}000\text{ g})(+30.0°C)$$

$$= 12{,}552{,}000\,J$$

This large number helps to explain how an electric water heater consumes about 25 percent of all household electricity. With the proper number of significant figures (see Appendix B), this answer should be expressed as 10,000,000 joules.

2. The temperature change is −30.0°C−(10.0°C) = −20.0°C. Multiply this negative temperature change by the ice's specific heat and mass:

$$\text{heat} = \left(2.01\,\frac{J}{g°C}\right)(10.0\text{ g})(-20.0°C)$$

$$= -402\,J$$

The next time you're near a refrigerator/freezer, place your hand near the back; you'll feel the heat that has been extracted from the food inside.

Making Paper

No material has been more vital to the development of modern civilization than paper. With paper, and the subsequent invention of the printing press, knowledge became accessible to a broad range of people. Technological, political, and social revolutions followed, and from these movements arose our complex modern societies.

Paper is reported to have been made in China as early as 100 A.D., when cellulose fibers from mulberry bark were pounded into thin sheets. Finer paper was eventually produced by lifting a silk screen up through a suspension of cellulose fibers in water so that entangled fibers collected on the screen. After drying, the fibers remained intertwined, forming a sheet of paper, as shown in Figure 1.

Only handmade single sheets of paper were fabricated until, in 1798, a machine that could make continuous rolls of paper was invented in France. This machine consisted of a conveyor belt submerged at one end in a vat of suspended cellulose fibers. The conveyor belt was a screen so that water would drain from it as fibers were pulled out of the suspension. The entangled fibers were then squeezed through a series of rollers to create a long, continuous sheet of paper. Modern papermaking machines work using the same principle.

▲ **Figure 1**
Take a close look at paper and you will see that it is made of numerous tiny cellulose fibers.

▲ **Figure 2**
About 14 million tons of paper are produced in the United States each year.

Paper was originally made from fibrous plant materials, such as bark, shrubs, and various grasses. Wood was not used because its fibers are embedded in a highly adhesive material called *lignin* that does not dissolve in water. In 1867, researchers in the United States discovered that wood fibers could be isolated by soaking wood chips in sulfurous acid, which dissolves the lignin. The ability to make paper from wood was a great boon to entrepreneurs in North America because of that continent's once vast supply of trees. At about the same time, additives, such as alum, were added to strengthen paper and to help it absorb ink. Today, about 75 percent of the paper made in the United States comes from wood pulp, with the remainder coming from recycled wastepaper.

Paper made with alum or similar additives tends to turn yellow and brittle within a matter of decades, as shown in Figure 3. This is because of the acidic nature of these additives, which catalyze the decomposition of the cellulose fibers. Noncorrosive alkaline (see Chapter 10) alternatives were developed in the 1950s. Paper made with alkaline binders are designated *archival*, which means they last a minimum of 300 years. Paper companies were prompted to make the switch to acid-free paper in the late 1980s when the cost of wood pulp skyrocketed. Paper made with acidic alum binders must be at least 90 percent fiber in order to maintain strength, whereas paper made with alkaline binders maintains strength with as little as 75 percent fiber. Less fiber translates to greater profits.

A secondary benefit of alkaline binders is that in a nonacidic environment, cheap and abundant calcium carbonate can be used as a whitener in place of relatively expensive titanium dioxide. Acid-free paper has the added advantage that peroxide bleaches can be used on it instead of chlorine, which was widely used to bleach paper until it was discovered that bleaching paper with chlorine produces toxic chemicals called dioxins. Furthermore, consumers have come to recognize the value of archival-quality paper and are now demanding its production.

We use trees to meet our paper needs because they have been abundant. They are also renewable—for

▲ **Figure 3**
By 1984, the Library of Congress estimated that about 25 percent, or 3 million volumes, of its collection had become too brittle for circulation.

every tree cut down, a new one can be planted. However, growing and harvesting trees has some significant drawbacks. Timber takes decades to mature. This time frame is not fast enough to keep up with our increasing demand. As a consequence, the amount of forested land is dwindling. Also, replanting timber does not recreate a forest. Timber is typically replanted as a monoculture suited primarily for industrial needs. A true forest is a more evolved system of many species that thrive in one another's presence.

There are a number of possible alternatives to trees for the mass production of paper, including switch grass, kenaf, and industrial hemp; the last of these is shown being harvested in Figure 4. Each of these plants produces more than three times as much fiber per acre than trees do. They are also fast-growing plants. Whereas it takes 20 years before trees can be harvested, these plants, when grown in favorable climates, can be harvested three times in a single year. Also, the lignin content of these types of plants is quite low, which means their cellulose fibers are relatively easy to separate. If paper mills switched from wood to one of these alternatives, they could avoid the use of countless gallons of sulfurous acid, which is not only hazardous to the environment but also largely responsible for the stink associated with paper production.

▲ **Figure 4**
Industrial hemp being harvested for its fiber-rich content.

Today, we use an enormous amount of paper—about 70 million tons a year in the United States alone. On average, that's about 230 kilograms, or six full-sized trees, for each U.S. citizen. The good news is that our recycling efforts have improved markedly over the past several decades. In 1996, we recycled about 40 percent of paper waste. Today, we recycle just over 60 percent of paper waste. This is good, because recycling paper does more than just save trees. A lot of energy is required to make paper from a tree. Less than half as much energy is needed to process old paper into new.

CONCEPT CHECK
Was paper made before the mid-1800s acid-free?

CHECK YOUR ANSWER Yes, because acidic alum binders were not used in papermaking until the mid-1800s. Interestingly, books from the early 1800s can be found in pristine condition.

Think and Discuss

1. According to researchers at the Laboratory for Sustainable Technology, Australia, pulp made from agricultural wastes, such as straw, could meet the global demand for paper five times over. Today, however, less than 10 percent of the world's paper is made from agricultural wastes, which are instead burned or plowed into the soil. Why do you suppose pulp from agricultural wastes is not commonly used to make paper?

2. Many people thought that much less paper would be used after the invention of the personal computer. Paper use, however, remained strong, because, as it turned out, people like to print from their computers. Now we have the advent of e-readers. What impact, if any, might these devices have on the amount of paper we use?

3. Agricultural wastes such as straw and fast-growing crops such as switchgrass can be used for the making of paper. These materials, however, can also be used to create cellulosic ethanol, also known as grassoline, which is a liquid fuel. At current estimates, the United States could produce enough grassoline to replace about half its current annual consumption of gasoline. Will a move toward grassoline encourage or discourage the use of material such as straw for the making of paper?

4. The pulp from sawdust dust does not make good paper. Can you figure out why? (Hint: look at Figure 1).

5. Industrial hemp is the original strain of the marijuana plant that has been cultivated for millennia for its fiber content, especially for the production of rope and canvas. Industrial hemp contains negligible quantities of the psychoactive drug THC. If planted side by side with marijuana, it would pollinate the marijuana, thereby depleting the marijuana's psychotropic properties. Growing industrial hemp without a permit from the Drug Enforcement Administration (DEA), however, is a federal crime with a maximum penalty of life in prison. For what reasons do you suppose the DEA refuses to grant these permits?

▲ The atmosphere is awash with chemicals such as water, nitrogen, and oxygen, that are forced to undergo chemical changes during a high-energy lightning bolt.

9

How Chemicals React

THE MAIN IDEA

Atoms change partners during a chemical reaction.

Scientists have learned how to control chemical reactions to produce many useful materials—nitrogen-based fertilizers from the atmosphere, metals from rocks, plastics and pharmaceuticals from petroleum. These materials and the thousands of others produced by chemical reactions have dramatically improved our standard of living, as has the abundant energy released when fossil fuels take part in the chemical reaction called *combustion*. What happens in a chemical reaction? Why are new materials produced? Why must chemical equations always be balanced? If two molecules meet, will they always react to form new molecules? What is a catalyst, and what role do catalysts play in our environment? Why do some chemical reactions, such as the burning of wood, produce energy, while other reactions, such as those occurring in the cooking of food, require the input of energy? What is the ultimate driving force for all chemical reactions? The goal of this chapter is to provide you with a stronger handle on the basics of chemical reactions, which were introduced in Chapter 3.

Chemistry

Warming and Cooling Mixtures

The focus of this chapter is chemical changes. For practical reasons, however, we begin with a hands-on activity involving physical changes that aptly illustrate the role that energy plays when different materials are brought together.

PROCEDURE

1. Hold some room-temperature water in the cupped palm of your hand over a sink. Predict whether this water will become cool or warm as a small amount of rubbing alcohol is added to it. (Warning: rubbing alcohol causes severe and painful stomach problems if ingested. Never ingest rubbing alcohol.)

2. Add lukewarm water to two clear plastic cups. Transfer the water back and forth to ensure equal temperatures. Predict whether this water will become cool or warm as salt is added to it.

3. Stir several tablespoons of table salt into only one of the cups. Compare the temperatures of the two cups by holding them up to your temperature-sensitive cheeks.

ANALYZE AND CONCLUDE

1. If two round magnets were to roll toward each other, would they accelerate to faster speeds just before they collided? If two attracting molecules were to float toward each other, would they accelerate to faster speeds just before they collided? What happens to the temperature of a material when its molecules are accelerated to faster speeds?

2. Do you get energy from pulling two magnets apart from each other, or do you lose energy? Is energy released or absorbed when sodium and chloride ions are pulled away from their crystalline structure? Why do many athletic "cold packs" contain salts?

9.1   Chemical Reactions Are Represented by Chemical Equations

EXPLAIN THIS

How can 50 g of wood burn to produce more than 50 g of products?

As was discussed in Chapter 3, during a chemical reaction, atoms are rearranged to create one or more new compounds. This process is neatly summed up in written form as a **chemical equation.** A chemical equation shows the reacting substances, called **reactants,** before an arrow that points to the newly formed substances, called **products:**

$$\text{reactants} \rightarrow \text{products}$$

Typically, reactants and products are represented by their elemental or chemical formulas. Sometimes molecular models or, simply, names may be used instead. Phases are also often shown: (*s*) for solid, (*l*) for liquid, and (*g*) for gas. Compounds dissolved in water are designated (*aq*) for aqueous solution. Finally, numbers are placed in front of the reactants or products to show the ratio in which they either combine or form. These numbers are called *coefficients*, and they represent numbers of individual atoms and molecules. For instance, to represent the chemical reaction in which coal burns in the presence of oxygen to form gaseous carbon dioxide, we write the chemical equation using coefficients of 1:

LEARNING OBJECTIVE

Identify whether a chemical equation is balanced or not balanced.

$$1\ C\ (s) + 1\ O_2\ (g) \longrightarrow 1\ CO_2\ (g)$$

Reactants   Products

One of the most important principles of chemistry is the **law of mass conservation,** which we discussed in Sections 1.4 and 2.2. The law of mass conservation states that matter is neither created nor destroyed during a chemical reaction. The atoms present at the beginning of a reaction are merely rearranged. This means that no atoms are lost or gained during any reaction. The chemical equation must therefore be *balanced.* In a balanced equation, each atom must appear on both sides of the arrow the same number of times. The equation for the formation of carbon dioxide is balanced because each side shows one carbon atom and two oxygen atoms. You can count the number of each type of atom in the models to see this for yourself.

In another chemical reaction, two hydrogen gas molecules, H_2, react with one oxygen gas molecule, O_2, to produce two molecules of water, H_2O, in the gaseous phase:

$$2\,H_2\,(g) + 1\,O_2\,(g) \longrightarrow 2\,H_2O\,(g)$$

This equation for the formation of water is also balanced—there are four hydrogen and two oxygen atoms before and after the arrow.

A coefficient in front of a chemical formula tells us the number of times that element or compound must be counted. For example, $2\,H_2O$ indicates two water molecules, which contain a total of four hydrogen atoms and two oxygen atoms.

By convention, the coefficient 1 is omitted; so the preceding chemical equations are typically written

$$C\,(s) + O_2\,(g) \rightarrow CO_2\,(g) \qquad \text{(balanced)}$$
$$2\,H_2\,(g) + O_2\,(g) \rightarrow 2\,H_2O\,(g) \qquad \text{(balanced)}$$

CONCEPT CHECK

How many oxygen atoms are indicated by the following balanced equation?
$3\,O_2\,(g) \rightarrow 2\,O_3\,(g)$

CHECK YOUR ANSWER There are six oxygen atoms. Before the reaction, these six oxygen atoms are found in three O_2 molecules. After the reaction, these same six atoms are found in two O_3 molecules.

An unbalanced chemical equation shows the reactants and products without the correct coefficients. For example, the equation

$$NO\,(g) \rightarrow N_2O\,(g) + NO_2\,(g) \quad \text{(unbalanced)}$$

is not balanced because there are one nitrogen atom and one oxygen atom before the arrow but three nitrogen atoms and three oxygen atoms after the arrow.

You can balance unbalanced equations by adding or changing coefficients to produce correct ratios. (It's important not to change subscripts, however, because to do so changes the compound's identity—H_2O is water, but H_2O_2 is hydrogen peroxide.) For example, to balance the preceding equation, place a 3 before the NO:

$$3\,NO\,(g) \rightarrow N_2O\,(g) + NO_2\,(g) \quad \text{(balanced)}$$

Now there are three nitrogen atoms and three oxygen atoms on each side of the arrow, and the law of mass conservation is not violated.

CONCEPT CHECK

Write a balanced equation for the reaction showing hydrogen gas, H_2, and nitrogen gas, N_2, forming ammonia gas, NH_3:

$$\_\_\_H_2\,(g) + \_\_\_N_2\,(g) \rightarrow \_\_\_NH_3\,(g)$$

CHECK YOUR ANSWER Initially, we see two hydrogen atoms before the reaction arrow and three on the right. This can be remedied by placing a coefficient of 3 by the hydrogen, H_2, and a coefficient of 2 by the ammonia, NH_3. This makes for six hydrogen atoms both before and after the reaction arrow. Meanwhile, the coefficient of 2 by the ammonia also makes for two nitrogen atoms after the arrow, which balances out the two nitrogen atoms appearing before the arrow. The full balanced equation, therefore, is

$$3\,H_2\,(g) + N_2\,(g) \longrightarrow 2\,NH_3\,(g)$$

Practicing chemists develop a skill for balancing equations. This skill involves creative energy and, like other skills, improves with experience. More important than being an expert at balancing equations is knowing why they need to be balanced. And the reason is the law of mass conservation, which tells us that atoms are neither created nor destroyed in a chemical reaction—they are simply rearranged. So every atom present before the reaction must be present after the reaction, even though the groupings of atoms are different.

Many students enjoy learning how to balance chemical equations. Perhaps this is because balancing an equation is akin to solving a crossword or Sudoku puzzle. There are many different approaches one can take to balancing a chemical equation. There are also some crafty tricks one can perform to solve some difficult equations. Your instructor may share with you some of his or her favorite approaches and/or tricks to balancing equations. To get you started, we offer you the following guidelines, which you can follow as you work to balance the chemical equations appearing at the end of this chapter.

READING CHECK

Why must the chemical equation be balanced?

A Quick Guide to Balancing Chemical Equations

Unbalanced equations are balanced by changing the coefficients. Subscripts should never be changed, because this changes the chemical's identity—H_2O is water, but H_2O_2 is hydrogen peroxide. Also, a coefficient is a multiplier for all the atoms within a molecule. Writing $2\,Fe_2O_3$, for example, means four Fe atoms and six O atoms.

1. Focus on balancing only one element at a time. Start with the leftmost element and modify the coefficients so that this element appears on both sides of the arrow the same number of times.
2. Move to the next element and modify the coefficients so as to balance this element. Do not worry if you incidentally unbalance the previous element. You will come back to it in subsequent steps.
3. Continue from left to right focusing on only one element at a time. Change coefficients as needed for that one element and don't worry about messing up the other elements by doing so.
4. Repeat steps 1–3 until all the elements are balanced.

9.2 Counting Atoms and Molecules by Mass

Correlate the formula mass of a substance with the number of molecules or atoms that substance contains.

EXPLAIN THIS

Why does the combination of 4 grams of hydrogen, H_2, with 16 grams of oxygen, O_2, produce only 18 grams of water, H_2O?

In any chemical reaction, a specific number of reactants react to form a specific number of products. For example, when carbon and oxygen combine to form carbon dioxide, they always combine in the ratio of one carbon atom to one oxygen molecule. A chemist who wants to carry out this reaction in the laboratory would be wasting chemicals and money if she were to try to combine, say, four carbon atoms for every one oxygen molecule. The excess carbon atoms would have no oxygen molecules to react with and would remain unchanged.

But how is it possible to measure out a specific number of atoms or molecules? Rather than counting these particles individually, chemists can use a scale that measures the mass of bulk quantities. Because different atoms and molecules have different masses, however, a chemist can't simply measure out equal masses of each. Say, for example, he needs the same number of carbon atoms as oxygen molecules. Measuring equal masses of the two materials would not provide equal numbers.

Consider this analogy: You know that Ping-Pong balls are much lighter than golf balls. So if you were to measure out 1 kilogram of each, you understand that you would need many more Ping-Pong balls to create a full kilogram, as shown in **Figure 9.1**. You understand that the *number* of something and the *mass* of something are two different things and that they should never be confused with each other.

Because different atoms and molecules have different masses, there are different numbers of them in a sample of the same mass. Carbon atoms are less massive than oxygen molecules. That's why there are more carbon atoms in a 1-gram sample of carbon than there are oxygen molecules in a 1-gram sample of oxygen. So clearly, equal masses of these two substances do not yield equal numbers of carbon atoms and oxygen molecules.

READINGCHECK

Samples of two different materials have the same mass. Do they necessarily have the same number of atoms?

Equal masses

▲ Figure 9.1

For the same mass, the number of Ping-Pong balls is greater than the number of golf balls because a single Ping-Pong ball is lighter than a single golf ball.

The mass of one Ping-Pong ball is 2 grams.

The mass of one golf ball is 40 grams.

A Ping-Pong ball is 2/40, or 1/20, as massive as a golf ball.

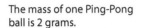

Number of Ping-Pong balls = Number of golf balls

◀ Figure 9.2
The number of golf balls in 200 grams of golf balls equals the number of Ping-Pong balls in 10 grams of Ping-Pong balls.

If we know the *relative masses* of different materials, we can measure equal numbers. Golf balls, for example, are about 20 times more massive than Ping-Pong balls are, which is to say the relative mass of Ping-Pong balls to golf balls is 20 to 1. Measuring out 20 times as much mass of golf balls as of Ping-Pong balls, therefore, gives equal numbers of each, as is shown in **Figure 9.2**.

CONCEPT CHECK

A customer wants to buy a 1:1 mixture of blue and red jelly beans. Each blue bean is twice as massive as each red bean. If the clerk measures out 5 pounds of red beans, how many pounds of blue beans must she measure out?

CHECK YOUR ANSWER Because each blue jelly bean has twice the mass of each red one, the clerk needs to measure out twice as much mass of blues in order to have the same count, which means 10 pounds of the blues. If the clerk did not know that the blue beans were twice as massive as the red ones, she would not know what mass of blues was needed for the 1:1 ratio.

FOR YOUR INFORMATION

The number of atoms in your breath is about 100 times the number of breaths in the atmosphere.

So where do we turn to find the relative masses of atoms and molecules? You guessed it! The masses of elements shown in the periodic table are relative masses. Using these masses, we can measure out equal numbers of atoms or molecules.

For example, as illustrated in **Figure 9.3**, the mass of carbon is 12.011 amu. (As discussed in Section 4.4, one *atomic mass unit* (amu) equals 1.661×10^{-24} gram.) The **formula mass** of a substance is the sum of the atomic masses of the elements in its chemical formula. So the formula mass of an oxygen molecule, O_2, is about 32 amu (15.999 amu + 15.999 amu). A carbon atom, therefore, is about 12/32 as massive as an oxygen molecule. To measure out equal numbers of carbon atoms and oxygen molecules, we could measure out 12 grams of carbon and 32 grams of molecular oxygen. Any proportion equal to 12/32, such as 6/16 or 3/8, would do. For example, 3 grams of carbon would have the same number of particles as 8 grams of molecular oxygen.

CHEMICAL CONNECTIONS

How are the atoms in your lungs connected to the atoms that were once in the lungs of Leonardo da Vinci?

▶ **Figure 9.3**
Measuring carbon, C, and oxygen, O_2, in a 12:32 ratio, which is the same as a 3:8 ratio, provides the same number of particles. (Note: the volumes shown are exaggerated. Three grams of solid carbon would occupy only about 4 mL, while eight grams of gaseous oxygen would occupy more than 6 L.)

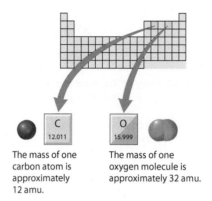

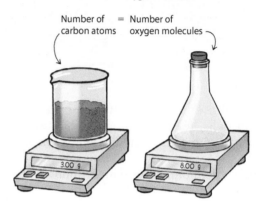

CONCEPTCHECK

1. Reacting 3 grams of carbon, C, with 8 grams of molecular oxygen, O_2, results in 11 grams of carbon dioxide, CO_2. Does it follow that 1.5 grams of carbon will react with 4 grams of oxygen to form 5.5 grams of carbon dioxide?

2. Would reacting 5 grams of carbon with 8 grams of oxygen also result in 11 grams of carbon dioxide?

CHECK YOUR ANSWERS

1. Yes, it does. The quantities are only half as much, but their ratio is the same as when 11 grams of carbon dioxide is formed: 1.5/4/5.5 = 3/8/11.

2. A common error of many students is to think that no reaction will occur if the proper ratios of reactants are not provided. You should realize, however, that in this reaction, only 3 of the 5 grams of carbon can react with 8 grams of oxygen to form 11 grams of carbon dioxide. Two grams of carbon will be unreacted after the reaction. Reacting this remaining 2 grams of carbon would require more oxygen.

9.3 Converting between Grams and Moles

LEARNING OBJECTIVE

Use the concept of moles to calculate the mass of reactants needed to produce a given mass of products.

EXPLAIN THIS

How many molecules of carbon dioxide are in 44 g of carbon dioxide?

Atoms and molecules react in specific ratios by number. In the laboratory, however, chemists work with bulk quantities of materials, which are measured by mass. Chemists therefore need to know the relationship between the mass of a given sample and the number of atoms or molecules contained in that mass. The key to this relationship is the *mole*. Recall from Section 7.3 that we use the word *mole* to indicate a number, much as we use the word *dozen* to indicate a number. While a dozen is equal to 12, a mole is equal to a number that is incredibly large: 6.02×10^{23}. This number is known as **Avogadro's number,** in honor of the 18th-century scientist Amedeo Avogadro. Again, for emphasis, one mole is equal to 6.02×10^{23}.

As **Figure 9.4** illustrates, if you express the numeric value of the atomic mass of any element in *grams*, the number of atoms in a sample of the element having this mass is always 6.02×10^{23}, which is 1 mole. For example, a 22.990-gram sample of sodium metal, Na (atomic mass = 22.990 amu), contains 6.02×10^{23} sodium atoms, and a 207.2-gram sample of lead, Pb (atomic mass = 207.2 amu), contains 6.02×10^{23} lead atoms.

The same concept holds for molecules (or any compound). Express the numeric value of the formula mass in *grams* and a sample having that mass contains 6.02×10^{23} molecules. For example, there are 6.02×10^{23} O_2 molecules in 31.998 grams of molecular oxygen, O_2 (formula mass = 31.998 amu), and 6.02×10^{23} CO_2 molecules in 44.009 grams of carbon dioxide, CO_2 (formula mass = 44.009 amu).

An Avogadro's number of grains of sand would cover the United States to a depth of about 2 meters. A stack of an Avogadro's number of pennies would be about 800,000 trillion kilometers tall, which is about the diameter of our galaxy. Placed side by side, these pennies would reach to the Andromeda galaxy, which is about 2.5 million light-years away. Avogadro's number is beyond huge, yet that's the number of water molecules found in a mere 18 grams of water. Water molecules are that small!

CONCEPT CHECK

1. How many atoms are in a 6.941-gram sample of lithium, Li (atomic mass = 6.941 amu)?

2. How many molecules are in an 18.015-gram sample of water, H_2O (formula mass = 18.015 amu)?

CHECK YOUR ANSWERS

1. Because this number of grams of lithium is numerically equal to the atomic mass, there are 6.02×10^{23} atoms in the sample, which is 1 mole of lithium atoms.

2. Because this number of grams of water is numerically equal to the formula mass, there are 6.02×10^{23} water molecules in the sample, which is 1 mole of water molecules.

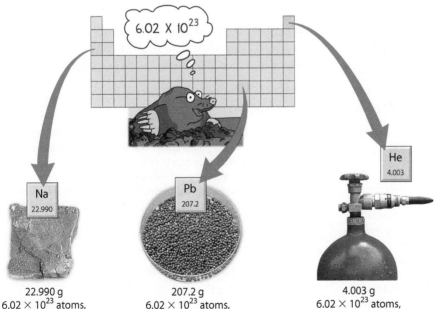

22.990 g
6.02×10^{23} atoms,
which is 1 mole

207.2 g
6.02×10^{23} atoms,
which is 1 mole

4.003 g
6.02×10^{23} atoms,
which is 1 mole

◀ Figure 9.4
Express the numeric value of the atomic mass of any element in grams, and that many grams contains 6.02×10^{23} atoms.

CALCULATION CORNER MASSES OF REACTANTS AND PRODUCTS

The coefficients of a chemical equation tell us the ratio by which reactants react and products form. The following equation, for example, tells us that every 1 mole of methane, CH_4, reacts to produce 2 moles of water, H_2O.

$$CH_4 + 2O_2 \rightarrow CO_2 + 2H_2O$$

So if you were given 16 g of methane, CH_4, how many grams of water, H_2O, would form? From the text, you should know that 16 g of methane, CH_4, is 1 mole (formula mass 16 amu). So 16 g of methane would yield 2 moles of water. But how many grams are 2 moles of water? Well, if 1 mole of water, H_2O, equals 18 g (formula mass 18 amu), then 2 moles equals 36 g. Thus, 16 g of methane, CH_4, would yield 36 g of water, H_2O. Let's look at this process from a step-by-step mathematical point of view.

SAMPLE PROBLEM 1

What mass of water is produced when 16 g of methane, CH_4 (formula mass 16 amu), reacts with oxygen, O_2, in the following reaction?

$$CH_4 + 2O_2 \rightarrow CO_2 + 2H_2O$$

SOLUTION:

Step 1. Convert the given mass to moles:

Conversion factor

$$(16 \text{ g } CH_4)\left(\frac{1 \text{ mole } CH_4}{16 \text{ g } CH_4}\right) = 1 \text{ mole } CH_4$$

Step 2. Use the coefficients of the balanced equation to find out how many moles of H_2O are produced from this many moles of CH_4:

Conversion factor

$$(1 \text{ mole } CH_4)\left(\frac{2 \text{ moles } H_2O}{1 \text{ mole } CH_4}\right) = 2 \text{ moles } H_2O$$

Step 3. Now that you know how many moles of H_2O are produced, convert this value to grams of H_2O:

Conversion factor

$$(2 \text{ moles } H_2O)\left(\frac{18 \text{ g } H_2O}{1 \text{ mole } H_2O}\right) = 36 \text{ g } H_2O$$

Steps 1 through 3 can be combined into a single equation that starts with what you are given (16 g of CH_4) and calculates that which is asked for:

$$(16 \text{ g } CH_4)\underbrace{\left(\frac{1 \text{ mole } CH_4}{16 \text{ g } CH_4}\right)}_{\text{step 1}}\underbrace{\left(\frac{2 \text{ moles } H_2O}{1 \text{ mole } CH_4}\right)}_{\text{step 2}}\underbrace{\left(\frac{18 \text{ g} H_2O}{1 \text{ mole } H_2O}\right)}_{\text{step 3}}$$

$$= 36 \text{ g } H_2O$$

Note that this technique is an application of what you learned in the unit conversion Calculation Corner of Chapter 1. Also, within each conversion factor, the unit that you want to change goes in the denominator and that which you want to obtain goes in the numerator.

The method of converting grams of a substance to moles (step 1), then from moles of this substance to moles of that substance (step 2), followed by moles of that substance to grams (step 3) is called *stoichiometry*. Using stoichiometry, a scientist can calculate the amounts of reactants or products

in any chemical reaction. The methods of stoichiometry are developed much further in general chemistry courses. For this course, all you need to do is be familiar with what stoichiometry is all about, which is keeping tabs on atoms and molecules as they react to form products. Nonetheless, for a special assignment, you might try your analytical thinking skills on the following problems. First, try to deduce the answer based on what you know about the law of mass conservation, and then follow the steps given here to check your answers.

SAMPLE PROBLEM 2

Show that 44 g of carbon dioxide, CO_2, is produced when 16 g of methane, CH_4, reacts with oxygen, O_2. How many grams of oxygen, O_2, are needed for this reaction?

SOLUTION:

The 16 g of methane, CH_4, is one mole, which reacts with oxygen to produce one mole of carbon dioxide, CO_2, which you know because of the coefficients in the balanced chemical equation shown previously. One mole of carbon dioxide (formula mass 44 amu) is 44 g. So the 16 g of methane reacts with oxygen to produce 44 g of carbon dioxide plus 36 g of water. The mass of the products (44 g + 36 g = 80 g) must be equal to the mass of the reactants (16 g + ? = 80 g). So we can calculate that the methane reacted with 64 g of oxygen, which, interestingly enough, is 2 moles, as we see from the balanced equation.

The mass of carbon dioxide produced can also be calculated as follows:

$$(16 \text{ g } CH_4)\left(\frac{1 \text{ mole } CH_4}{16 \text{ g } CH_4}\right)\left(\frac{1 \text{ mole } CO_2}{1 \text{ mole } CH_4}\right)\left(\frac{44 \text{ g } CO_2}{1 \text{ mole } CO_2}\right)$$

$$= 44 \text{ g } CO_2$$

Similarly, the mass of oxygen can be calculated as follows:

$$(16 \text{ g } CH_4)\left(\frac{1 \text{ mole } CH_4}{16 \text{ g } CH_4}\right)\left(\frac{2 \text{ moles } O_2}{1 \text{ mole } CH_4}\right)\left(\frac{32 \text{ g } O_2}{1 \text{ mole } O_2}\right)$$

$$= 64 \text{ g } O_2$$

SAMPLE PROBLEM 3

How many grams of ozone (O_3, 48 amu) can be produced from 64 g of oxygen (O_2, 32 amu) in the reaction shown?

$$3O_2 \rightarrow 2O_3$$

SOLUTION:

According to the law of mass conservation, the amount of mass in the products must equal the amount of mass in the reactants. Given that this reaction involves only one reactant and one product, you should not be surprised to learn that 64 g of reactant produces 64 g of product. Here are the stepwise calculations, which start with *grams* of oxygen and then finds the *moles* of oxygen. The moles of oxygen are then converted to moles of ozone using the coefficients of the equation. This is followed by a conversion of *moles* of ozone to *grams* of ozone:

$$(64 \text{ g } O_2)\left(\frac{1 \text{ mole } O_2}{32 \text{ g } O_2}\right)\left(\frac{2 \text{ moles } O_3}{3 \text{ moles } O_2}\right)\left(\frac{48 \text{ g } O_3}{1 \text{ mole } O_3}\right)$$

$$= 64 \text{ g } O_3$$

SAMPLE PROBLEM 4

What mass of nitrogen monoxide (NO, 30 amu) is formed when 28 g of nitrogen (N_2, 28 amu) reacts with 32 g of oxygen (O_2, 32 amu) in the reaction shown?

$$N_2 + O_2 \rightarrow 2\,NO$$

SOLUTION:

There are several ways to answer this question. One way would be to recognize that 28 g of N_2 is 1 mole of N_2 and 32 g of O_2 is 1 mole of O_2. According to the balanced equation, combining 1 mole of N_2 with 1 mole of O_2 yields 2 moles of NO. Because the mass of 1 mole of NO is 30 g, the mass of 2 moles of NO must be 30 g × 2 = 60 g, which is simply the sum of all the reactants (28 g + 32 g = 60 g). To calculate stepwise, you can start with either the 28 g of N_2 or the 32 g of O_2.

$$(28 \text{ g } N_2)\left(\frac{1 \text{ mole } N_2}{28 \text{ g } N_2}\right)\left(\frac{2 \text{ moles NO}}{1 \text{ mole } N_2}\right)\left(\frac{30 \text{ g NO}}{1 \text{ mole NO}}\right)$$

$$= 60 \text{ g NO}$$

In this reaction, N_2 and O_2 were combined in exactly the right amounts. Do you recognize that there would be too much oxygen if 28 g of N_2 were reacted with 40 g of O_2? If that were the case, then only 32 g of the 40 g of O_2

would react. There would be 8 g of unreacted O_2 left over. Stoichiometry is an area of chemistry rich in opportunities for analytical thinking.

YOUR TURN

1. How many grams of iron oxide, Fe_2O_3, can be produced from the reaction of 25.0 g of iron, Fe, with an unlimited amount of oxygen, O_2? (Not sure where to start? Here's a helpful rule: always start your calculation by writing down that which you are given, which in this case is 25.0 g of Fe. Then proceed with the conversions.)

$$4\,Fe + 3\,O_2 \rightarrow 2\,Fe_2O_3$$

2. How many grams of iron oxide, Fe_2O_3, can be produced from the reaction of 25.0 g of oxygen, O_2, with an unlimited amount of iron, Fe?

3. How many grams of oxygen, O_2, are required to react with 25.0 g of iron, Fe, to form iron oxide, Fe_2O_3?

4. A 25.0-g sample of iron, Fe, powder is mixed with 25.0 g of oxygen, O_2, gas and heated until all of the Fe has reacted. How many grams of O_2 remain after the reaction is complete?

Answers to Calculation Corners appear at the end of each chapter.

The **molar mass** of any substance, be it element or compound, is defined as the mass of 1 mole of the substance. Thus, the units of molar mass are *grams per mole*. For instance, the atomic mass of carbon is 12.011 amu, which means that 1 mole of carbon has a mass of 12.011 grams, and we say that the molar mass of carbon is 12.011 grams per mole. The molar mass of molecular oxygen, O_2 (formula mass 31.998 amu), is 31.998 grams per mole. For convenience, values such as these are often rounded to the nearest whole number. The molar mass of carbon, therefore, might also be presented as 12 grams per mole and that of molecular oxygen as 32 grams per mole.

CONCEPT CHECK

What is the molar mass of water (formula mass = 18 amu)?

CHECK YOUR ANSWER From the formula mass, you know that 1 mole of water has a mass of 18 grams. Therefore, the molar mass is 18 grams per mole.

Because 1 mole of any substance always contains 6.02×10^{23} particles, the mole is an ideal unit for chemical reactions. For example, 1 mole of carbon (12 grams) reacts with 1 mole of molecular oxygen (32 grams) to give 1 mole of carbon dioxide (44 grams).

In many instances, the ratio in which chemicals react is not 1:1. As shown in **Figure 9.5**, for example, 2 moles (4 grams) of molecular hydrogen react with 1 mole (32 grams) of molecular oxygen to give 2 moles (36 grams) of water. Note how the coefficients of the balanced chemical equation can be conveniently interpreted as the number of moles of reactants or products. A chemist therefore need only convert these numbers of moles to grams in order to know how much mass of each reactant he should measure out to have the proper proportions.

Cooking and chemistry are similar in that both require measuring ingredients. Just as a cook looks to a recipe to find the necessary quantities measured by the cup or the tablespoon, a chemist looks to the periodic table to find the necessary quantities measured by the number of grams per mole for each element or compound.

▶ **Figure 9.5**
Two moles of H_2 reacts with 1 mole of O_2 to give 2 moles of H_2O. This is the same as saying 4 grams of H_2 reacts with 32 grams of O_2 to give 36 grams of H_2O or, equivalently, that 12.04×10^{23} H_2 molecules react with 6.02×10^{23} O_2 molecules to give 12.04×10^{23} H_2O molecules.

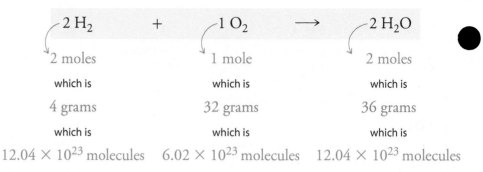

$2\,H_2$	$+$	$1\,O_2$	$\longrightarrow$	$2\,H_2O$
2 moles		1 mole		2 moles
which is		which is		which is
4 grams		32 grams		36 grams
which is		which is		which is
12.04×10^{23} molecules		6.02×10^{23} molecules		12.04×10^{23} molecules

9.4 Chemical Reactions Can Be Exothermic or Endothermic

LEARNING OBJECTIVE

Calculate the amount of energy released or absorbed by a chemical reaction using the bond energies of reactants and products.

READINGCHECK

What do the breaking and forming of chemical bonds involve?

EXPLAIN THIS

What changes during a chemical reaction?

Once a reaction is complete, there may be either a net release or a net absorption of energy. Reactions in which there is a net release of energy are called **exothermic.** Rocket ships lift into space and campfires glow red hot as a result of exothermic reactions. Reactions in which there is a net absorption of energy are called **endothermic.** Photosynthesis, for example, involves a series of endothermic reactions that are driven by the energy of sunlight. Both exothermic and endothermic reactions, illustrated in **Figure 9.6**, can be understood through the concept of bond energy.

During a chemical reaction, chemical bonds are broken and atoms rearranged to form new chemical bonds. Such breaking and forming of chemical bonds involves changes in energy. As an analogy, consider a pair of magnets. To separate them requires an input of "muscle energy." Conversely, when the two separated magnets collide, they become slightly warmer than they had been, and this warmth is evidence of energy released. Energy must be absorbed by the magnets if they are to move apart, and energy is released as they come together. The same principle applies to atoms. To pull bonded atoms apart requires an energy input. When atoms combine, there is an energy output, usually in the form of faster-moving atoms and molecules, electromagnetic radiation, or both.

▶ **Figure 9.6**
For the chemical reactions that occur when wood is burning, there is a net release of energy. For the chemical reactions that occur in a photosynthetic plant, there is a net absorption of energy.

TABLE 9.1   Selected Bond Energies

BOND	BOND ENERGY (KJ/MOLE)	BOND	BOND ENERGY (KJ/MOLE)
H–H	436	N–N	159
H–C	414	O–O	138
H–N	389	Cl–Cl	243
H–O	464	C=O	803
H–F	569	N=O	631
H–Cl	431	O=O	498
C–O	351	C≡C	837
C–C	347	N≡N	946

FOR YOUR INFORMATION

The bonds that form in the product molecules are not the same bonds that broke in the reactant molecules. The amount of energy released upon product formation, therefore, will not be the same as the amount of energy consumed in reactant deformation.

The amount of energy required to pull two bonded atoms apart is the same as the amount released when they are brought together. This energy is called **bond energy.** Each chemical bond has its own characteristic bond energy. The hydrogen–hydrogen bond energy, for example, is 436 kilojoules per mole. This means that 436 kilojoules of energy is absorbed as 1 mole of hydrogen–hydrogen bonds break apart and 436 kilojoules of energy is released upon the formation of 1 mole of hydrogen–hydrogen bonds. Different bonds involving different elements have different bond energies, as Table 9.1 shows. You can refer to the table as you study this section, but please do not memorize these bond energies. Instead, focus on understanding what they mean.

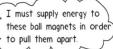

I must supply energy to these ball magnets in order to pull them apart.

By convention, a positive bond energy represents the amount of energy absorbed as a bond breaks and a negative bond energy represents the amount of energy released as a bond forms. Thus, when you are calculating the net energy released or absorbed during a reaction, you need to be careful about plus and minus signs. It is standard practice when doing such calculations to assign a plus sign to energy absorbed and a minus sign to energy released. For instance, when dealing with a reaction in which 1 mole of H–H bonds are broken, you write +436 kilojoules to indicate energy absorbed. The positive sign indicates that the molecules gained energy, which was used to break the bonds. When dealing with the formation of 1 mole of H–H bonds, you write –436 kilojoules to indicate energy released. The negative sign indicates that the molecules *lost* energy, which was released to the environment. We'll do some sample calculations in a moment.

Energy is released when they come together!

CONCEPT CHECK

Do all covalent single bonds have the same bond energy?

CHECK YOUR ANSWER No. Bond energy depends on the types of atoms bonding. The H–H single bond, for example, has a bond energy of 436 kilojoules per mole, but the H–O single bond has a bond energy of 464 kilojoules per mole. Not all covalent single bonds have the same bond energy.

An Exothermic Reaction Involves a Net Release of Energy

For most chemical reactions, the total amount of energy absorbed in breaking bonds in reactants is different from the total amount of the energy released as bonds form in the products. Consider the reaction in which hydrogen and oxygen react to form water:

$$H—H + H—H + O=O \longrightarrow \overset{H}{\underset{O}{\diagdown}}\overset{H}{\diagup} + \overset{H}{\underset{O}{\diagdown}}\overset{H}{\diagup}$$

In the reactants, hydrogen atoms are bonded to hydrogen atoms and oxygen atoms are double-bonded to oxygen atoms. The total amount of energy absorbed as these bonds break is +1370 kilojoules. Note that we use the plus sign to indicate the amount of energy *absorbed* to break bonds.

Type of bond	Number of bonds	Bond energy	Total energy
H—H	2 moles	+436 kJ/mole	+872 kJ
O=O	1 mole	+498 kJ/mole	+498 kJ
		Total energy absorbed:	+1370 kJ

In the products, there are four moles of hydrogen–oxygen bonds. The total amount of energy released as these bonds form is −1856 kilojoules. Note that we use the minus sign to indicate the amount of energy *released* as bonds are formed.

Type of bond	Number of bonds	Bond energy	Total energy
H—O	4 moles	−464 kJ/mole	−1856 kJ
		Total energy released:	−1856 kJ

The amount of energy released in this reaction exceeds the amount of energy absorbed. The net energy of the reaction is found by adding the two quantities:

$$\text{net energy of reaction} = \text{energy absorbed} + \text{energy released}$$
$$= +1370 \text{ kJ} + (-1856 \text{ kJ})$$
$$= -486 \text{ kJ}$$

The negative sign on the net energy indicates that there is a net release of energy, so the reaction is exothermic. For any exothermic reaction, energy can be considered a product and is thus sometimes included after the arrow of the chemical equation:

$$2 H_2 + O_2 \rightarrow 2 H_2O + \text{energy}$$

In an exothermic reaction, the potential energy of atoms in the product molecules is lower than their potential energy in the reactant molecules. This is illustrated in the reaction profile shown in **Figure 9.7**. The lowered potential energy of the atoms in the product molecules is due to their being more tightly held together. This is analogous to two attracting magnets, whose potential energy decreases as they come closer together. The loss of potential energy is balanced by a gain in kinetic energy. Like two free-floating magnets coming together and accelerating to higher speeds, the potential energy of the reactants is converted to faster-moving atoms and molecules, electromagnetic radiation, or both. This kinetic energy released by the reaction is equal to the difference between the potential energy of the reactants and the potential energy of the products, as is indicated in Figure 9.7.

▶ Figure 9.7
In an exothermic reaction, the product molecules are at a lower potential energy than the reactant molecules. The net amount of energy released by the reaction is equal to the difference in the potential energies of the reactants and the products. In Section 9.6, we explore why graphs such as this have a bump in the middle.

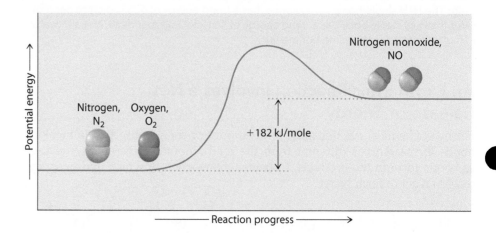

The total amount of energy released in an exothermic reaction depends on the amounts of the reactants. The reaction of large amounts of hydrogen and oxygen, for example, provided the energy to lift the Space Shuttle shown in **Figure 9.8** into orbit. There were two compartments in the large central tank, to which the orbiter was attached—one filled with liquid hydrogen and the other filled with liquid oxygen. Upon ignition, these two liquids mixed and reacted chemically to form water vapor, which produced the needed thrust as it is expelled the rocket cones. Additional thrust was obtained from a pair of solid-fuel rocket boosters containing a mixture of ammonium perchlorate, NH_4ClO_4, and powdered aluminum, Al. Upon ignition, these chemicals reacted to form products that were expelled at the rear of the rocket. The balanced equation representing this aluminum reaction is

$$3\,NH_4ClO_4 + 3\,Al \rightarrow Al_2O_3 + AlCl_3 + 3\,NO + 6\,H_2O + energy$$

▲ Figure 9.8
The Space Shuttle used exothermic chemical reactions to lift off from the earth's surface.

CONCEPT CHECK

Where does the net energy released in an exothermic reaction go?

CHECK YOUR ANSWER This energy goes into increasing the speeds of reactant atoms and molecules and often into electromagnetic radiation such as light.

An Endothermic Reaction Involves a Net Absorption of Energy

When the amount of energy released in product formation is *less* than the amount of energy absorbed when reactant bonds break, the reaction is endothermic. An example is the reaction of atmospheric nitrogen and oxygen to form nitrogen monoxide:

$$N \equiv N + O = O \rightarrow N = O + N = O$$

The amount of energy absorbed as the chemical bonds in the reactants break is

Type of bond	Number of bonds	Bond energy	Total energy
N ≡ N	1 mole	+946 kJ/mole	+946 kJ
O = O	1 mole	+498 kJ/mole	+498 kJ
		Total energy absorbed:	+1444 kJ

The amount of energy released upon the formation of bonds in the products is

Type of bond	Number of bonds	Bond energy	Total energy
N = O	2 moles	−631 kJ/mole	−1262 kJ
		Total energy released:	−1262 kJ

As before, the net energy of the reaction is found by adding the two quantities:

$$\begin{aligned} \text{net energy of reaction} &= \text{energy absorbed} + \text{energy released} \\ &= +1444\ kJ + (-1262\ kJ) \\ &= +182\ kJ \end{aligned}$$

The positive sign indicates that there is a net *absorption* of energy, meaning the reaction is endothermic. For any endothermic reaction, energy can be considered a reactant and is thus sometimes included before the arrow of the chemical equation:

$$energy + N_2 + O_2 \rightarrow 2\,NO$$

FOR YOUR INFORMATION

NASA scientists routinely test various materials for their durability against atomic oxygen, O, which is abundant in the low orbit of the Space Shuttle. They discovered that atomic oxygen effectively transforms surface organic materials into gaseous carbon dioxide. The scientists realized atomic oxygen's usefulness for restoring paintings damaged by smoke or other organic contaminants. Together with art conservationists, they used atomic oxygen to restore certain damaged paintings, and it worked spectacularly.

▶ Figure 9.9
In an endothermic reaction, the product molecules are at a higher potential energy than the reactant molecules. The net amount of energy absorbed by the reaction is equal to the difference in the potential energies of the reactants and the products.

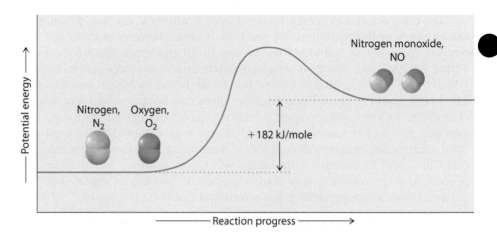

In an endothermic reaction, the potential energy of atoms in the product molecules is higher than their potential energy in the reactant molecules. This is illustrated in the reaction profile shown in **Figure 9.9**. Raising the potential energy of the atoms in the product molecules requires a net input of energy, which must come from some external source, such as electromagnetic radiation, electricity, or heat. Nitrogen and oxygen react to form nitrogen monoxide with the application of large amounts of heat, as occurs adjacent to a lightning bolt or in an internal-combustion engine.

CONCEPT CHECK

Should the following reaction be endothermic or exothermic?

$$O_2NNO_2 \rightarrow O_2N + NO_2$$

CHECK YOUR ANSWER No calculations are necessary. This reaction is endothermic because it involves only the breaking of a chemical bond. For practice with examples requiring calculations, see the questions at the end of this chapter.

Energy Is Conserved in a Chemical Reaction

We have been focusing on the key role energy plays in chemical reactions. This is an area of science known as **thermodynamics,** which stems from Greek words meaning "movement of heat." The concepts of exothermic and endothermic reactions fit neatly within what is known as the *first law of thermodynamics,* which can be paraphrased as follows:

Energy is conserved. It may be converted from one form to another, say, from potential to kinetic energy, but the total amount of energy remains the same. The energy that an exothermic reaction releases always goes somewhere into the environment, often in the form of thermal energy (heat), but other forms, such as light energy or electric energy, are also possible. Likewise, the energy that an endothermic reaction absorbs always comes from somewhere in the environment, again, in various forms.

This is a common sense type of law, from which we derive the expression "You can't get something for nothing." Energy doesn't just appear or disappear. It either comes from somewhere or goes to somewhere. In the case of an exothermic reaction, energy comes from the formation of bonds in which atoms are held more tightly but vibrate faster. This is a downhill transformation of potential energy into kinetic energy. In the case of an endothermic reaction, energy is absorbed in the formation of bonds in which atoms are held more loosely and vibrate slower. This is an uphill transformation of kinetic energy into potential energy.

The first law of thermodynamics tells us that energy can be neither created nor destroyed. This is an important law to understand, but it does not tell us the whole picture. Why, for example, do reactants react to form products? By doing nothing instead, they are following the first law of thermodynamics quite perfectly. What then compels them to move forward in the formation of products? The answer to this question is aptly provided by what is known as the *second law of thermodynamics*, which we explore next.

9.5 Chemical Reactions Are Driven by the Spreading of Energy

EXPLAIN THIS

Why are exothermic reactions self-sustaining?

Energy tends to disperse. It flows from where it is concentrated to where it is spread out. The energy of a hot pan, for example, does not stay concentrated in the pan once the pan is taken off the stove. Instead, the energy spreads away from the pan into the cooler surroundings. Similarly, the concentrated chemical energy in gasoline disperses, when ignited, in the formation of many very hot smaller molecules that scatter explosively. Some of this released thermal energy is used by the engine to get the car moving. The rest spreads into the engine block and radiator fluid and out the exhaust pipe.

Scientists consider this tendency of energy to disperse to be one of the central reasons for both physical and chemical changes. In other words, changes that result in energy spreading out tend to occur on their own—they are favored. This includes the cooling down of a hot pan or the burning of ignited gasoline. In both cases, there is a dispersal of energy to the environment.

The opposite holds true, too. Changes that result in the concentration of energy do *not* tend to occur—they are not favored. Heat from the room, for example, will never spontaneously move into a pan to heat it up. Likewise, low-energy exhaust molecules coming out of a car's tailpipe will not spontaneously come back together to form higher-energy gasoline molecules. The natural flow of energy is always a one-way trip from where it is concentrated to where it is less concentrated, or "spread out."

That energy tends to disperse is spelled out by the *second law of thermodynamics*, which can be paraphrased as follows:

> Any process that happens by itself results in the net dispersal of energy. For example, heat naturally flows from a higher-temperature object to a lower-temperature object because in doing so, energy is dispersed from where it is concentrated (a hot pan) to where it is spread out (the cooler kitchen).

Entropy is a measure of this natural spreading of energy. Wherever there is a spreading of energy, there is a corresponding *increase* in entropy. Applied to chemistry, entropy helps us to answer a most fundamental question: if you take two materials and put them together, will they react to form new materials? If the reaction results in the dispersal of energy (an overall increase in entropy), then the answer is yes. Conversely, if the reaction results in the concentration of energy (an overall decrease in entropy), then the reaction will *not* occur by itself. Instead, such a reaction will occur only if an external source of energy is supplied to it.

Using this concept of entropy, you are now in a position to understand why exothermic reactions are self-sustaining—occurring on their own without need of external help—while most endothermic reactions need a continual prodding. Exothermic reactions spread energy out to the surroundings, much like a

A quick way to determine whether a reaction might be favorable is to assess whether the reaction leads to an overall dispersal of energy, which is the same thing as an increase in entropy.

Because energy naturally tends to disperse, a reaction that leads to an increase in entropy will likely occur, while a reaction that leads to a decrease in entropy will _not_ likely occur.

cooling hot pan. This is an increase in entropy; hence, exothermic reactions are favored to occur. An endothermic reaction, by contrast, requires that energy from the surroundings be absorbed by the reactants. This is a concentration of energy, which is counter to energy's natural tendency to disperse. Endothermic reactions, therefore, can progress from reactants to products only with the continual input of energy. But from where might this energy come? The answer is "from some self-sustaining exothermic reaction occurring elsewhere."

The classic example is photosynthesis, which is an endothermic reaction by which plants use solar energy to create carbohydrates and oxygen from carbon dioxide and water, as represented by the following equation:

$$\text{sunlight} \ + \ 6\,CO_2\,(g) \ + \ 6\,H_2O\,(g) \ \rightarrow \ C_6H_{12}O_6\,(s) \ + \ 6\,O_2\,(g)$$

carbon   water   carbohydrate   oxygen
dioxide

In photosynthesis, energy dispersed from the Sun becomes contained within the carbohydrate and oxygen products, which, of course, are the primary fuels of living organisms (**Figure 9.10**). Likewise, most modern materials, such as plastics, synthetic fibers, pharmaceuticals, fertilizers, and metals such as iron and aluminum, are made or purified using endothermic reactions. Our ability to produce these new and useful materials has been the hallmark of modern chemistry. Creating these products, however, necessarily requires the input of energy, which we must obtain from some external source, such as electricity from a power plant that burns fossil or nuclear fuels.

▲ Figure 9.10
The Sun is truly a "hothouse"—dispersing enormous amounts of energy from exothermic nuclear reactions. A tiny fraction of the Sun's energy is used to drive photosynthesis, which is vital for plants and plant-eating creatures like us.

CONCEPT CHECK

Sugar crystals form naturally within a supersaturated solution of sugar water. Does the formation of these crystals result in an increase or a decrease in entropy?

CHECK YOUR ANSWER The formation of these sugar crystals results in an _increase_ in entropy. Your clue to an increase in entropy here is that the crystals form "on their own," a spontaneous process and thus one that must result in an entropy increase. Recall from Section 8.6 that energy is released when molecules come together to form a solid (the heat of freezing). This release of heat involves the spreading out of energy, which is, by definition, an increase in entropy.

FOR YOUR INFORMATION

There are examples of endothermic reactions that proceed spontaneously absorbing heat from the environment. A classic example is the mixing of ions (salt) in water, as presented in the Hands-On Chemistry activity at the beginning of this chapter. In such cases, the environment disperses energy to the dissolving ions, which then spread this energy more widely into the volume of the solvent.

While on the subject of the second law of thermodynamics, we would be remiss not to discuss the close relationship between entropy and our psychological sense of time. That energy tends to spread out is part of our human experience. We _expect_ a hot pan to cool, just as we _expect_ hot gases to come out of an exhaust pipe. But what if we saw the reverse? For example, what would we think if we saw smoke moving _into_ a smokestack? Or what if we saw a diver fly out of the water and rise upward to the diving platform? If we were watching these energy-concentrating events on video, we would quickly conclude that the video was running backward. However, if in real life we actually saw such things—if we could survive the shock—we would immediately sense that time itself was running backward. Thus, when we watch smoke shoot out of a smokestack and a diver diving into water, as shown in **Figure 9.11**, we have the sense that time is moving forward. The second law of thermodynamics, therefore, gives us our psychological sense of time; it is the "arrow of time."

▲ **Figure 9.11**
As the diver dives, his potential energy is converted into kinetic energy. As he splashes into the water, this kinetic energy is spread to make the water molecules move faster and to heat up just a little. Will this dissipated energy reconcentrate itself to push him back through the air to the diving platform? The second law of thermodynamics says no.

CONCEPT CHECK

The energy of a diver diving into a pool is dispersed as lots of moving water and a little heat after the diver hits the water. How, then, can the diver get back up to the platform?

CHECK YOUR ANSWER As a living organism, the diver has a supply of biochemical energy, obtained ultimately from photosynthesis, that he can tap to climb upward against gravity to get back to the diving platform.

Students often state the laws of thermodynamics this way: you can't win because you can't get any more energy out of a system than you put into it. You can't break even because no matter what you do, some of your energy will be dispersed as ambient heat. Finally, you can't get out of the game because you depend on entropy-increasing processes, such as solar thermonuclear fusion, to remain alive.

9.6 Chemical Reactions Can Be Slow or Fast

EXPLAIN THIS

Why does blowing into a campfire make the fire burn brighter?

How exactly do reactants react to form products? This is the focus of the remaining sections of this chapter. We begin by showing how the speed of a reaction depends on the concentration and temperature of reacting molecules.

Some chemical reactions, such as the rusting of iron, are slow, while others, such as the burning of gasoline, are fast. The speed of any reaction is indicated by its *reaction rate*, which is an indicator of how quickly the reactants transform to products. As shown in **Figure 9.12**, initially, a flask may contain only reactant

LEARNING OBJECTIVE

Describe the requirements that must be met in order for a chemical reaction to occur.

◀ **Figure 9.12**
Over time, the reactants in this reaction flask may transform to products. If this happens quickly, the reaction rate is fast. If it happens slowly, the reaction rate is slow.

Time

Low concentration of products, high concentration of reactants

Reactants ⟹ Products

High concentration of products, low concentration of reactants

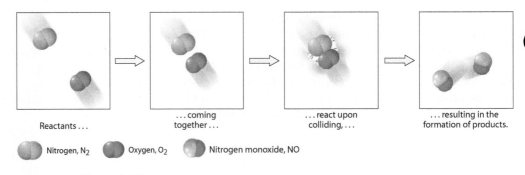

Reactants . . .   . . . coming   . . . react upon   . . . resulting in the
                         together . . .   colliding, . . .   formation of products.

Nitrogen, N_2 Oxygen, O_2 Nitrogen monoxide, NO

▲ Figure 9.13
During a reaction, reactant molecules collide with one another.

READING CHECK
What must two molecules do in order to react with each other?

molecules. Over time, these reactants form product molecules, and as a result, the concentration of product molecules increases. The **reaction rate,** therefore, can be defined either as how quickly the concentration of products increases or as how quickly the concentration of reactants decreases.

What determines the rate of a chemical reaction? The answer is complex, but one important factor is that reactant molecules must physically come together. Because molecules move rapidly, this physical contact is appropriately described as a collision. We can illustrate the relationship between molecular collisions and reaction rate by considering the reaction of gaseous nitrogen and gaseous oxygen to form gaseous nitrogen monoxide, as shown in **Figure 9.13.**

Because reactant molecules must collide in order for a reaction to occur, the rate of a reaction can be increased by increasing the number of collisions. An effective way to increase the number of collisions is to increase the concentration of the reactants. **Figure 9.14** shows that with higher concentrations, there are more molecules in a given volume, which makes collisions between molecules more probable. As an analogy, consider a group of people on a dance floor—as the number of people increases, so does the rate at which they bump into one another. An increase in the concentration of nitrogen and oxygen molecules, therefore, leads to a greater number of collisions between these molecules—hence, a greater number of nitrogen monoxide molecules will form in a given period of time.

Not all collisions between reactant molecules lead to products, however, because the molecules must collide in a certain orientation in order to react. Nitrogen and oxygen, for example, are much more likely to form nitrogen monoxide when the molecules collide in the parallel orientation shown in Figure 9.13. When they collide in the perpendicular orientation shown in **Figure 9.15**, nitrogen monoxide does not form. For larger molecules, which can have numerous orientations, this orientation requirement is even more restrictive.

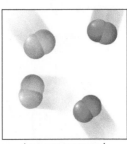

Less concentrated

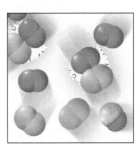

More concentrated

▲ Figure 9.14
The more concentrated a sample of nitrogen and oxygen, the greater the probability that N_2 and O_2 molecules will collide and form nitrogen monoxide.

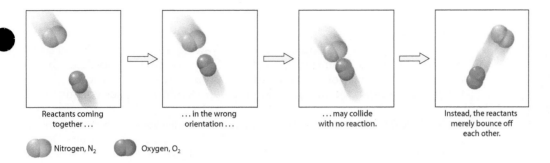

Nitrogen, N_2 Oxygen, O_2

▲ **Figure 9.15**
The orientation of reactant molecules in a collision can determine whether a reaction occurs. A perpendicular collision between N_2 and O_2 tends not to result in formation of a product molecule.

A second reason that not all collisions lead to product formation is that the reactant molecules must also collide with enough kinetic energy to break their bonds. Only then is it possible for the atoms in the reactant molecules to change bonding partners and form product molecules. The bonds in N_2 and O_2 molecules, for example, are quite strong. In order for these bonds to be broken, collisions between the molecules must contain enough energy to break the bonds. As a result, collisions between slow-moving N_2 and O_2 molecules, even those that collide in the proper orientation, may not form NO, as shown in **Figure 9.16**.

**FOR YOUR
INFORMATION**

The life sciences involve fantastic applications of chemistry, nitrogen fixation being just one example. Others include photosynthesis, cellular respiration, and molecular genetics. So there are distinct advantages to learning about chemistry *before* advancing to the life sciences.

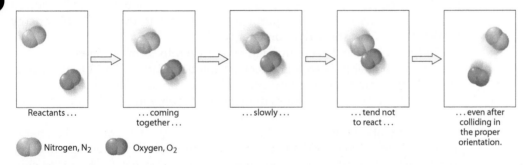

Nitrogen, N_2 Oxygen, O_2

▲ **Figure 9.16**
Slow-moving molecules may collide with insufficient force to break their bonds. As a result, they cannot react to form product molecules.

The higher the temperature of a material, the faster its molecules move and the more forceful and frequent the collisions between them. Higher temperatures, therefore, increase reaction rates. The nitrogen and oxygen molecules that make up our atmosphere, for example, are continually colliding with one another. At the normal temperatures of our atmosphere, however, these molecules do not generally have sufficient kinetic energy for the formation of nitrogen monoxide. The heat of a lightning bolt, as shown in the opening photo of this chapter, dramatically increases the kinetic energy of these molecules, to the point that a large portion of the collisions in the vicinity of the bolt result in the formation of nitrogen monoxide. The nitrogen monoxide formed in this manner undergoes further atmospheric reactions to form chemicals known as nitrates, which plants depend on for survival. This is an example of *nitrogen fixation,* which you may have learned about in a biology class.

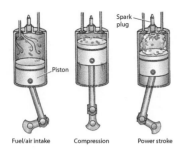

Spark plug

Piston

Fuel/air intake Compression Power stroke

The energy required to break bonds can also come from the absorption of electromagnetic radiation. As the radiation is absorbed by reactant molecules, the atoms in the molecules may start to vibrate so rapidly that the bonds between them are easily broken. In many instances, the direct absorption of electromagnetic radiation is sufficient to break chemical bonds and to initiate a chemical reaction. The common atmospheric pollutant nitrogen dioxide, NO_2, for example, may transform to nitrogen monoxide and atomic oxygen merely upon exposure to sunlight:

$$NO_2 + sunlight \rightarrow NO + O$$

Activation Energy Is Needed for Reactants to React

Whether they result from collisions, from the absorption of electromagnetic radiation, or both, broken bonds are a necessary first step in most chemical reactions. The energy required for this initial breaking of bonds can be viewed as an *energy barrier*. The minimum energy required to overcome this energy barrier is known as the **activation energy** (E_a).

In the reaction between nitrogen and oxygen to form nitrogen monoxide, the activation energy is so high (because the bonds in N_2 and O_2 are strong) that only the fastest-moving nitrogen and oxygen molecules possess sufficient energy to react. **Figure 9.17** shows the activation energy in this chemical reaction as a vertical hump.

The activation energy of a chemical reaction is analogous to the energy a car needs to drive over the top of a hill. Without sufficient energy to climb to the top of the hill, it isn't possible for the car to get to the other side. Likewise, reactant molecules can transform to product molecules only if the reactant molecules possess an amount of energy equal to or greater than the activation energy.

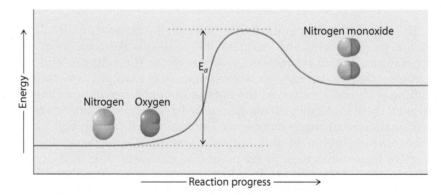

▲ **Figure 9.17**
Reactant molecules must gain a minimum amount of energy, called the activation energy, E_a, before they can transform to product molecules.

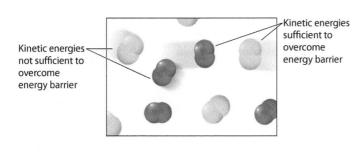

Kinetic energies not sufficient to overcome energy barrier

Kinetic energies sufficient to overcome energy barrier

◀ **Figure 9.18**
Because fast-moving reactant molecules possess sufficient energy to pass over the energy barrier, they are the ones to transform to product molecules.

At any given temperature, there is a wide distribution of kinetic energies in reactant molecules. Some are moving slowly, and others are moving quickly. As we discussed in Section 2.6, the temperature of a material is related to the average of all these kinetic energies. The few fast-moving reactant molecules in **Figure 9.18** have enough energy to pass over the energy barrier and are the ones to transform to product molecules.

When the temperature of the reactants is increased, the number of reactant molecules possessing sufficient energy to pass over the barrier also increases, which is why reactions are generally faster at higher temperatures. Conversely, at lower temperatures, fewer molecules have sufficient energy to pass over the barrier. Hence, reactions are generally slower at lower temperatures.

Most chemical reactions are influenced by temperature in this manner, including those reactions occurring in living bodies. The body temperature of animals such as humans, that regulate their internal temperature, is fairly constant. However, the body temperature of some animals, such as the alligator shown in **Figure 9.19**, rises and falls with the temperature of the environment. On a warm day, the chemical reactions occurring in an alligator are "up to speed" and the animal is more active. On a chilly day, however, the chemical reactions proceed at a lower rate; as a consequence, the alligator's movements are unavoidably sluggish.

CONCEPT CHECK

What kitchen device is used to lower the rate at which microorganisms grow on food?

CHECK YOUR ANSWER The refrigerator. Microorganisms, such as bread mold, are everywhere and difficult to avoid. By lowering the temperature of microorganism-contaminated food, the refrigerator decreases the rate of the chemical reactions that these microorganisms depend on for growth, thereby increasing the food's shelf life.

◀ **Figure 9.19**
Drivers beware! Alligators can become immobilized on pavement after being caught in the cold night air. By mid-morning, shown here, the temperature becomes warm enough to allow the alligator to get up and walk away.

9.7 Catalysts Speed Up the Destruction of Stratospheric Ozone

LEARNING OBJECTIVE

Discuss how a catalyst can speed up a chemical reaction using the destruction of stratospheric ozone as an example.

 **READING**CHECK

A catalyst speeds up a chemical reaction by lowering what?

FORYOUR INFORMATION

The stratospheric ozone layer is about 10 kilometers thick, but the concentration of ozone, O_3, within the layer is quite low. If this ozone layer were brought down to Earth's surface, atmospheric pressure would squeeze it to a thickness of only 3 millimeters. Not much ozone is up there protecting us. But thankfully, even small amounts of ozone are very good at shading us from solar ultraviolet light.

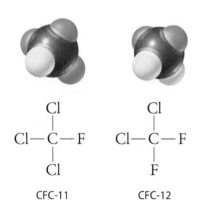

CFC-11   CFC-12

▲ **Figure 9.20**
Two of the most common CFCs, also known as freons, were CFC-11, trichlorofluoromethane, and CFC-12, dichlorodifluoromethane. At the height of CFC production in 1988, some 1.13 million tons was produced worldwide. Because of their inertness, CFCs were once thought to pose little threat to the environment.

EXPLAIN THIS

Chew a salt-free soda cracker for a few minutes and the cracker begins to taste sweet. Why?

As discussed in the previous section, a chemical reaction can be made to go faster by increasing the concentration of the reactants or by increasing the temperature. A third way to increase the speed of a reaction is to add a **catalyst,** which is any substance that increases the rate of a chemical reaction without itself being consumed.

To learn how catalysts work, let's consider some chemical reactions that take place in our atmosphere. One of the most important reactions occurs at altitudes higher than 20 km, within a region known as the *stratosphere*. Here the Sun's ultraviolet (UV) rays cause oxygen molecules to transform into ozone molecules, as follows:

$$3\,O_2(g) \; + \; UV \rightarrow 2\,O_3(g)$$
molecular oxygen   ozone

Furthermore, the ozone itself, once formed, is able to transform ultraviolet light into heat:

$$O_3(g) \; + \; UV \; \rightarrow \; O_3(g) \; + \; heat$$
ozone   ozone

In all, these ozone reactions prevent about 95 percent of the ultraviolet rays that come to our planet from the Sun from reaching its surface. This is of great benefit to life on the Earth because ultraviolet rays are most harmful to living tissue. Stratospheric ozone is our planet's safety shield.

Starting in the 1970s, scientists began to recognize that human-made chlorofluorocarbons, also known as CFCs, posed a significant threat to stratospheric ozone. Because CFCs are inert gases, they were once commonly used in air conditioners and aerosol propellants. Two of the most frequently used CFCs are shown in **Figure 9.20**.

The CFCs, it was discovered, drift their way to the stratosphere, where the intense ultraviolet rays break apart these molecules, producing chlorine atoms. The chlorine atoms, in turn, speed up the rate at which ozone molecules are destroyed. They do this by offering a reaction pathway that has a lower activation energy, as shown in **Figure 9.21**.

Chlorine atoms provide an alternate pathway involving intermediate reactions, each having a lower activation energy than the uncatalyzed reaction. This alternate pathway involves two steps. Initially, the chlorine reacts with the ozone to form chlorine monoxide and oxygen:

$$Cl \; + \; O_3 \; \longrightarrow \; ClO \; + \; O_2$$
Chlorine   Ozone   Chlorine   Oxygen
                          monoxide

The chlorine monoxide then reacts with another ozone molecule to re-form the chlorine atom, as well as to produce two additional oxygen molecules:

$$ClO \; + \; O_3 \; \longrightarrow \; Cl \; + \; 2\,O_2$$
Chlorine   Ozone   Chlorine   Oxygen
monoxide

A catalyst speeds up a chemical reaction, but interestingly, it is not destroyed by the reaction it catalyzes. Notice that although chlorine is depleted in the first

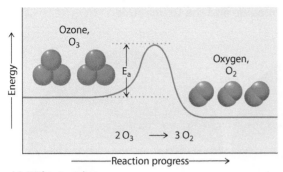

(a) Without catalyst

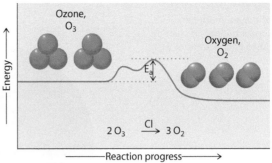

(b) With chlorine catalyst

▲ **Figure 9.21**
(a) The high activation energy indicates that only the most energetic ozone molecules can react to form oxygen molecules. (b) The presence of chlorine atoms allows conversion of ozone to oxygen with a lower activation energy, which means more ozone molecules have sufficient energy to form the product. (Note that the convention is to write the catalyst above the reaction arrow.)

reaction, it is regenerated in the second reaction. As a result, there is no net consumption of chlorine. Instead, it survives; so it can speed up the reaction repeatedly. One chlorine atom in the ozone layer, for example, can catalyze the transformation of 100,000 ozone molecules to oxygen molecules in the one or two years before the chlorine atom is removed by natural processes. In short, chlorine is bad news for stratospheric ozone, which is bad news for life on the Earth.

The fragility of stratospheric ozone came to the world's attention in 1985, with the discovery of a seasonal depletion of stratospheric ozone over the Antarctic continent, a phenomenon known as the *ozone hole*. The fact that chlorine atoms play an active role in the destruction of Antarctic ozone is revealed by measuring chlorine monoxide concentrations over this region, as shown in **Figure 9.22**. Furthermore, the satellite images show that the shape of the ozone hole typically matches the shape of a map showing chlorine monoxide distribution (**Figure 9.23**).

There has been an unprecedented level of international cooperation toward banning ozone-destroying substances such as CFCs. As a result, stratospheric concentrations of chlorine have been declining. Even with these treaties in place, however, the ozone-destroying actions of CFCs will be with us for some time. Atmospheric CFC levels are not expected to drop back to the levels found before the ozone hole was formed until sometime in the 22nd century.

FOR YOUR INFORMATION

Numerous oil-drilling sites in Siberia were once allowed to vent natural gas freely into the atmosphere. After the fall of the Soviet Union, these wells were capped to prevent this venting. Within weeks, instruments at the Mauna Loa weather observatory on the other side of the planet noted a significant drop in atmospheric levels of methane and its by-product, carbon dioxide. The effect that we humans have on global atmospheric conditions is very measurable.

◀ **Figure 9.22**
Concentrations of stratospheric ozone and chlorine monoxide in southern latitudes. As chlorine monoxide levels increase, ozone levels decrease. The yellow highlighting shows where small fluctuations in ClO concentrations result in large fluctuations in O_3 concentrations. This is consistent with catalytic behavior.

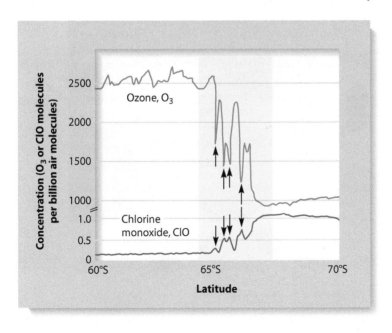

▶ **Figure 9.23**
Satellite images of the southern hemisphere showing concentrations of chlorine monoxide adjacent to concentrations of stratospheric ozone.

Chlorine Monoxide and the Ozone Hole

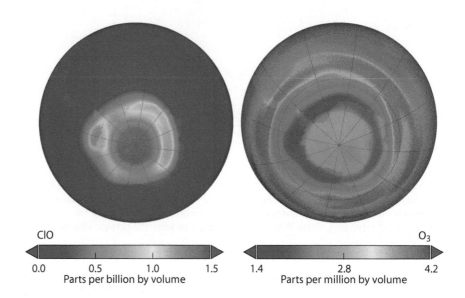

ClO
0.0   0.5   1.0   1.5
Parts per billion by volume

O_3
1.4   2.8   4.2
Parts per million by volume

FOR YOUR INFORMATION

Chlorofluorocarbons are distributed fairly evenly across the Earth's stratosphere. Over the poles, however, stratospheric clouds are able to form during the dark and extremely cold winters. CFC molecules bind to ice crystals within these clouds. When hit by sunlight in the spring, the crystal-bound CFC molecules break apart, producing a large dose of ozone-destroying chlorine atoms. In Antarctica, this process reaches a peak in September, which is when the largest ozone holes are typically observed. The ozone hole over the South Pole is typically more intense than the one over the North Pole. The reason is because the geography of the southern ocean favors a stable vortex around the South Pole, which, in turn, favors the formation of the stratospheric clouds.

Catalysts Are Also Most Beneficial

Chemists have been able to harness the power of catalysts for numerous beneficial purposes. The exhaust that comes from an automobile engine, for example, contains a wide assortment of pollutants, such as nitrogen monoxide, carbon monoxide, and uncombusted fuel vapors (hydrocarbons). To reduce the amount of these pollutants entering the atmosphere, most automobiles are equipped with *catalytic converters*, as shown in **Figure 9.24**. Metal catalysts in a converter speed up reactions that convert exhaust pollutants to less toxic substances. Nitrogen monoxide is transformed to nitrogen and oxygen, carbon monoxide is transformed to carbon dioxide, and unburned fuel is converted to carbon dioxide and water vapor. Because catalysts are not consumed by the reactions they speed up, a single catalytic converter may continue to operate effectively for the lifetime of the car.

Catalytic converters, along with microchip-controlled fuel–air ratios, have led to a significant drop in the per-vehicle emission of pollutants. This improvement, however, has been offset by an increase in the number of cars being driven, as exemplified by the traffic jam shown in **Figure 9.25**.

▶ **Figure 9.24**
A catalytic converter reduces the pollution caused by automobile exhaust by converting such harmful combustion products as NO, CO, and hydrocarbons to harmless N_2, O_2, CO_2, and H_2O. The catalyst is typically platinum, Pt; palladium, Pd; or rhodium, Rd.

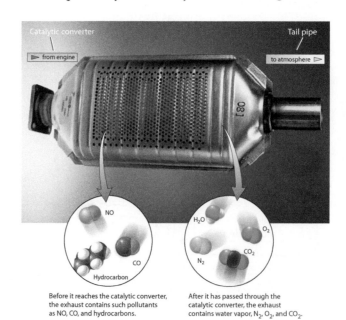

Before it reaches the catalytic converter, the exhaust contains such pollutants as NO, CO, and hydrocarbons.

After it has passed through the catalytic converter, the exhaust contains water vapor, N_2, O_2, and CO_2.

◀ Figure 9.25
The exhaust from automobiles today is much cleaner than before the advent of the catalytic converter, but there are many more cars on the road. In 1960, there were about 74 million registered motor vehicles in the United States. In 2009, there were more than 254 million.

Catalysts lower manufacturing costs by lowering required temperatures and by providing greater product yields without being consumed. Indeed, more than 90 percent of all manufactured goods are produced with the assistance of catalysts. The production of polymer plastics, as described in Chapter 12, is particularly dependent upon catalysts. Also, without catalysts, the price of gasoline would be much higher, as would be the price of such consumer goods as rubber, pharmaceuticals, automobile parts, clothing, and food grown with chemical fertilizers. Living organisms rely on special types of catalysts known as *enzymes,* which allow exceedingly complex biochemical reactions to occur with ease. You will learn more about the nature and behavior of enzymes in Chapter 13.

CHEMICAL CONNECTIONS

How is the invention of air conditioning connected to the quality of sunshine in Antarctica?

CONCEPT CHECK

How is a catalyst different from a chemical reactant?

CHECK YOUR ANSWER A catalyst is not used up during a chemical reaction. Instead, it is released as its original self. The catalyst can then serve to catalyze more reactions. A chemical reactant, by contrast, is consumed during a reaction as it is transformed into a chemical product.

Chapter 9 Review

LEARNING OBJECTIVES

Identify whether a chemical equation is balanced or not balanced. (9.1)

→ *Questions 1–4, 44–53*

Correlate the formula mass of a substance with the number of molecules or atoms that substance contains. (9.2)

→ *Questions 5–7, 35, 40, 54–60*

Use the concept of moles to calculate the mass of reactants needed to produce a given mass of products. (9.3)

→ *Questions 8–12, 30–34, 36–37, 61–63*

Calculate the amount of energy released or absorbed by a chemical reaction using the bond energies of reactants and products. (9.4)

→ *Questions 13–15, 27, 38–39, 42, 64–68*

Recognize that all chemical reactions are driven by the tendency of energy to disperse. (9.5)

→ *Questions 16–18, 43, 69–74, 83*

Describe the requirements that must be met in order for a chemical reaction to occur. (9.6)

→ *Questions 19–22, 28, 41, 75–78*

Discuss how a catalyst can speed up a chemical reaction using the destruction of stratospheric ozone as an example. (9.7)

→ *Questions 23–26, 29, 79–82, 84–85*

SUMMARY OF TERMS (KNOWLEDGE)

Activation energy The minimum energy required for a chemical reaction to proceed.

Avogadro's number The number of particles—6.02×10^{23}—contained in 1 mole of anything.

Bond energy The amount of energy required to pull two bonded atoms apart, which is the same as the amount of energy released when the two atoms are brought together into a bond.

Catalyst Any substance that increases the rate of a chemical reaction without itself being consumed by the reaction.

Chemical equation A representation of a chemical reaction in which reactants are drawn before an arrow that points to the products.

Endothermic A term that describes a chemical reaction in which there is a net absorption of energy.

Entropy A measure of an amount of energy that has been dispersed. Wherever there is a spreading of energy, there is a corresponding *increase* in entropy.

Exothermic A term that describes a chemical reaction in which there is a net release of energy.

Formula mass The sum of the atomic masses of the elements in a chemical formula.

Law of mass conservation The principle that states that matter is neither created nor destroyed during a chemical reaction—atoms merely rearrange, without any apparent loss or gain of mass, to form new molecules.

Molar mass The mass of 1 mole of a substance.

Products The new materials formed in a chemical reaction.

Reactants The reacting substances in a chemical reaction.

Reaction rate A measure of how quickly the concentration of products in a chemical reaction increases or the concentration of reactants decreases.

Thermodynamics An area of science concerned with the role energy plays in chemical reactions and other energy-dependent processes.

READING CHECK QUESTIONS (COMPREHENSION)

9.1 Chemical Reactions Are Represented by Chemical Equations

1. What is the purpose of coefficients in a chemical equation?

2. How many oxygen atoms are indicated on the right side of this balanced chemical equation?

$$4\,Cr\,(s)\, +\, 3\,O_2\,(g)\, \rightarrow\, 2\,Cr_2O_3\,(g)$$

3. Why is it important that a chemical equation be balanced?

4. Why is it important never to change a subscript in a chemical formula when balancing a chemical equation?

9.2 Counting Atoms and Molecules by Mass

5. Why don't equal masses of golf balls and Ping-Pong balls contain the same number of balls?

6. Why don't equal masses of carbon atoms and oxygen molecules contain the same number of particles?

7. What is the formula mass of nitrogen monoxide, NO, in atomic mass units?

9.3 Converting between Grams and Moles

8. If you had 1 mole of marbles, how many marbles would you have? What if you had 2 moles of marbles?

9. How many moles of water are in 18 grams of water?

10. How many molecules of water are in 18 grams of water?

11. Why is saying you have 1 mole of water molecules the same as saying you have 6.02×10^{23} water molecules?

12. What is the mass of an oxygen atom in atomic mass units?

9.4 Chemical Reactions Can Be Exothermic or Endothermic

13. If it takes 436 kilojoules to break a bond, how many kilojoules are released when the same bond is formed?

14. What is released by an exothermic reaction?

15. What is absorbed by an endothermic reaction?

9.5 Chemical Reactions Are Driven by the Spreading of Energy

16. As energy disperses, where does it go?

17. What in the universe is always increasing?

18. Why are exothermic reactions self-sustaining?

9.6 Chemical Reactions Can Be Slow or Fast

19. Why don't all collisions between reactant molecules lead to product formation?

20. What generally happens to the rate of a chemical reaction with increasing temperature?

21. Which reactant molecules are the first to pass over the energy barrier?

22. What term is used to describe the minimum amount of energy required for a reaction to proceed?

9.7 Catalysts Speed Up the Destruction of Stratospheric Ozone

23. What catalyst is effective in the destruction of atmospheric ozone, O_3?

24. What does a catalyst do to the energy of activation for a reaction?

25. What net effect does a chemical reaction have on a catalyst?

26. Why are catalysts so important to our economy?

CONFIRM THE CHEMISTRY (HANDS-ON APPLICATION)

27. If you ever have the opportunity to play with an electric train set, smell the engine car after it has been operating. You will note a slight "electric smell." This is the smell of ozone gas, which is created as the oxygen in the air is zapped with electric sparks. Why is this smell also sometimes apparent when a lightning storm strikes or when you pull your fuzzy sweater off in the dry air of winter? Is the formation of ozone from oxygen an endothermic or exothermic reaction?

28. An Alka-Seltzer antacid tablet reacts vigorously with water. But how does this tablet react to a solution of half water and half corn syrup? Propose an explanation involving the relationship between reaction speed and the frequency of molecular collisions.

29. Baker's yeast contains a biological catalyst known as *catalase*, which catalyzes the transformation of hydrogen peroxide, H_2O_2, into oxygen, O_2, and water, H_2O. Write a balanced equation for this reaction. Add a couple milliliters of 3 percent hydrogen peroxide to a glass containing a small amount of baker's yeast. What happens? Why?

THINK AND SOLVE (MATHEMATICAL APPLICATION)

30. Show that there are 1.0×10^{22} carbon atoms in a 1-carat pure diamond, which has a mass of 0.20 grams.

31. How many gold atoms are in a 5.00-gram sample of pure gold, Au (197 amu)?

32. Show that 1 mole of $KClO_3$ contains 122.55 grams.

33. Small samples of oxygen gas needed in the laboratory can be generated by a number of simple chemical reactions, such as

$$2 \, KClO_3 \, (s) \rightarrow 2 \, KCl \, (s) + 3 \, O_2 \, (g)$$

According to this balanced chemical equation, how many moles of oxygen gas are produced from the reaction of 2 moles of $KClO_3$ solid?

34. Small samples of oxygen gas needed in the laboratory can be generated by a number of simple chemical reactions, such as

$$2 \, KClO_3 \, (s) \rightarrow 2 \, KCl \, (s) + 3 \, O_2 \, (g)$$

What mass of oxygen (in grams) is produced when 122.55 grams of $KClO_3$ (formula mass 122.55 atomic mass units) takes part in this reaction?

35. Show that the mass of 2-propanol, C_3H_8O, is 60 amu and that the formula mass of propene, C_3H_6, is 42 amu. Also show that the formula mass of water, H_2O, is 18 amu.

36. How many grams of water, H_2O, and propene, C_3H_6, can be formed from the reaction of 6.0 grams of 2-propanol, C_3H_8O?

$$C_3H_8O \rightarrow C_3H_6 + H_2O$$

37. A 16-g sample of methane, CH_4, is combined with a 16-g sample of molecular oxygen, O_2, in a sealed container.

Upon ignition, what is the maximum amount of carbon dioxide, CO_2, that can be formed?

$$CH_4 + 2 \, O_2 \rightarrow CO_2 + 2 \, H_2O$$

38. Use the bond energies in Table 9.1 and the accounting format shown in Section 9.4 to determine whether these reactions are exothermic or endothermic:

$$H_2 + Cl_2 \rightarrow 2 \, HCl$$
$$2 \, HC \equiv CH + 5 \, O_2 \rightarrow 4 \, CO_2 + 2 \, H_2O$$

39. Use the bond energies in Table 9.1 and the accounting format shown in Section 9.4 to determine whether these reactions are exothermic or endothermic:

(a)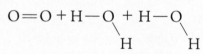

(b)

THINK AND COMPARE (ANALYSIS)

40. Rank the following in order of increasing number of atoms:

   a. 52 g of vanadium, V
   b. 52 g of chromium, Cr
   c. 52 g of manganese, Mn

41. Rank the following reaction profiles in order of increasing reaction speed.

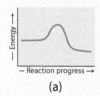 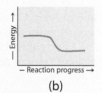 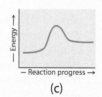

   (a)   (b)   (c)

42. Rank the following covalent bonds in order of increasing bond strength.

   a. $C \equiv C$
   b. $C = C$
   c. $C - C$

43. Rank the following in order of increasing entropy. A deck of playing cards

   a. at 45°C, new and unshuffled sitting in a room at 25°C.
   b. at 233°C, new and unshuffled sitting in a room at 25°C.
   c. at 25°C, used and shuffled sitting in a room at 25°C.

THINK AND EXPLAIN (SYNTHESIS)

9.1 Chemical Reactions Are Represented by Chemical Equations

44. Balance these equations:
   a. __Fe (s) + __O_2 (g) → __Fe_2O_3 (s)
   b. __H_2 (g) + __N_2 (g) → __NH_3 (g)

   c. __Cl_2 (g) + __KBr (aq) → __Br_2 (l) + __KCl (aq)
   d. __CH_4 (g) + __O_2 (g) → __CO_2 (g) + __H_2O (l)

45. Balance these equations:
   a. __Fe (s) + __S (s) → __Fe_2S_3 (s)
   b. __P_4 (s) + __H_2 (g) → __PH_3 (g)

c. __NO (g) + __Cl$_2$ (g) → __NOCl (g)

d. __SiCl$_4$ (l) + __Mg (s) → __Si (s) + __MgCl$_2$ (s)

46. Which of the following chemical equations is balanced?

$$2\ C_4H_{10}\,(g) + 13\ O_2\,(g) \rightarrow 8\ CO_2\,(g) + 10\ H_2O\,(l)$$

$$4\ C_6H_7N_5O_{16}\,(s) + 19\ O_2\,(g) \rightarrow 24\ CO_2\,(g) + 20\ NO_2\,(g) + 14\ H_2O\,(g)$$

Use the following illustrations to answer exercises 47–49.

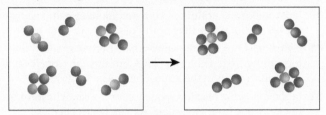

47. Assume the preceding illustrations are two frames of a movie—one before the reaction and the other after the reaction. How many diatomic molecules are represented in this movie?

48. There is an excess of at least one of the reactant molecules. Which one?

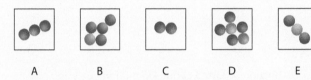

A   B   C   D   E

49. Which equation best describes this reaction?

a. 2 WX$_2$ + 2 ZYX$_3$ + X$_2$ → 2 ZXW$_4$ + 2 YW$_2$

b. 2 WX$_2$ + 2 YZW$_3$ + X$_2$ → 2 Y$_2$W$_4$ + 2 ZXW

c. 2 WX$_2$ + 2 YZW$_3$ + W$_2$ → 2 ZXW$_4$ + 2 YW$_2$

d. 2 XW$_2$ + 2 ZYW$_3$ + W$_2$ → 2 ZXW$_4$ + 2 YW$_2$

50. The reactants shown schematically on the left represent methane, CH$_4$, and water, H$_2$O. Write out the full balanced chemical equation that is depicted.

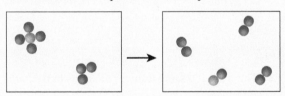

51. The reactants shown schematically on the left represent iron oxide, Fe$_2$O$_3$, and carbon monoxide, CO. Write out the full balanced chemical equation that is depicted.

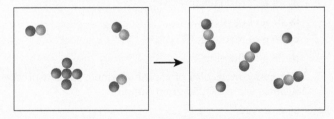

52. The following is a tough equation to balance. Hint: treat the PO$_4$ polyatomic ion as a single entity.

$$Mg(OH)_2 + H_3PO_4 \rightarrow Mg_3(PO_4)_2 + H_2O$$

53. The following is a tough equation to balance. Hint: temporarily use a fraction for a coefficient.

$$FeS_2 + O_2 \rightarrow Fe_2O_3 + SO_2$$

9.2 Counting Atoms and Molecules by Mass

54. Which has more atoms: 17.031 grams of ammonia, NH$_3$, or 72.922 grams of hydrogen chloride, HCl?

55. How many moles of molecules are in

a. 28 grams of nitrogen, N$_2$?

b. 32 grams of oxygen, O$_2$?

c. 32 grams of methane, CH$_4$?

d. 38 grams of fluorine, F$_2$?

56. How many moles of atoms are in

a. 28 grams of nitrogen, N$_2$?

b. 32 grams of oxygen, O$_2$?

c. 16 grams of methane, CH$_4$?

d. 38 grams of fluorine, F$_2$?

57. What is the mass of a water molecule in atomic mass units?

58. What is the mass of a water molecule in grams?

59. Is it possible to have a sample of oxygen that has a mass of 14 atomic mass units? Explain.

60. Which has the greater mass, 1.204×10^{24} molecules of molecular hydrogen or 1.204×10^{24} molecules of water?

9.3 Converting between Grams and Moles

61. How many grams of gallium are in a 145-gram sample of gallium arenside, GaAs?

62. How many atoms of arsenic are in a 145-gram sample of gallium arsenide, GaAs?

63. How is it possible for a jet airplane carrying 110 tons of jet fuel to emit 340 tons of carbon dioxide?

9.4 Chemical Reactions Can Be Exothermic or Endothermic

64. During a chemical reaction, bonds break and then reform. Why isn't the energy required to break these bonds equal to the energy released when the bonds are reformed?

65. Are the chemical reactions that take place in a disposable battery exothermic or endothermic? What evidence supports your answer?

66. Is the reaction going on in a rechargeable battery while it is recharging exothermic or endothermic?

67. Exothermic reactions are favored because they release heat to the environment. Would an exothermic reaction be more favored or less favored if it were carried out within a superheated chamber?

68. Why do exothermic reactions typically favor the formation of products?

9.5 Chemical Reactions Are Driven by the Spreading of Energy

69. What role does entropy play in chemical reactions?

70. Under what conditions will a hot pie not lose heat to its surroundings?

71. As the Sun shines on a snowcapped mountain, much of the snow sublimes instead of melts. How is this favored by entropy?

72. Estimate whether entropy increases or decreases with the following reaction. Use data from Table 9.1 to confirm your estimation.

$$2\ C\,(s) + 3\ H_2\,(g) \rightarrow C_2H_6\,(g)$$

73. According to the second law of thermodynamics, exothermic reactions, such as the burning of wood, are favored because they result in the dispersal of energy. Wood, however, will not spontaneously burn even when exposed to pure oxygen, O_2. Why?

74. Wild plants readily grow "all by themselves," yet the molecules of the growing plant have *less* entropy than the materials used to make the plant. How is it possible for there to be this *decrease* in entropy for a process that occurs all by itself?

9.6 Chemical Reactions Can Be Slow or Fast

75. In the laboratory, endothermic reactions are usually performed at elevated temperatures, while exothermic reactions are usually performed at lower temperatures. What are some possible reasons for this?

76. Does a refrigerator prevent or delay the spoilage of food? Explain.

77. Why does a glowing splint of wood only burn slowly in air but burst into flames when placed in pure oxygen?

78. Give two reasons why heat is often added to chemical reactions performed in the laboratory.

9.7 Catalysts Speed Up the Destruction of Stratospheric Ozone

79. Explain the connection between photosynthetic life on the Earth and the ozone layer.

80. Does the ozone pollution from automobiles help alleviate the ozone hole over the South Pole? Defend your answer.

81. Chlorine is put into the atmosphere by volcanoes in the form of hydrogen chloride, HCl, but this form of chlorine does not remain in the atmosphere very long. Why?

82. In the following reaction sequence for the catalytic formation of ozone from molecular oxygen, which chemical compound is the catalyst: nitrogen monoxide or nitrogen dioxide?

$$O_2 + 2\ NO \rightarrow 2\ NO_2$$
$$2\ NO_2 \rightarrow 2\ NO + 2\ O$$
$$2\ O + 2\ O_2 \rightarrow 2\ O_3$$

THINK AND DISCUSS (EVALUATION)

83. Discuss how any device that keeps track of time is dependent upon an increase in entropy.

84. Many people hear about atmospheric ozone depletion and wonder why we don't simply replace what has been destroyed. Knowing about chlorofluorocarbons and knowing how catalysts work, explain why this would not be a lasting solution.

85. Throughout the history of life on the Earth, there have been at least six major mass extinctions. The largest mass

extinction occurred about 450 million years ago and may have been initiated by an intense burst of ozone-depleting gamma rays produced by the explosion of a nearby star. Scientists point to the most recent sixth mass extinction as occurring right now. Discuss possible causes of this mass extinction. What creatures might survive? Should humans do anything to minimize this mass extinction, or should they just accept it as a natural course of the Earth's history?

READINESS ASSURANCE TEST (RAT)

If you have a good handle on this chapter, then you should be able to score at least 7 out of 10 on this RAT. Check your answers online at www.ConceptualChemistry.com. If you score less than 7, you need to study further before moving on.

Choose the BEST answer to the following.

1. What coefficients balance the following equation?

    __P_4(s) + __H_2(g) $\rightarrow$ __PH_3(g)

    a. 4, 2, 3
    b. 1, 6, 4
    c. 1, 4, 4
    d. 2, 10, 8

2. For the following generic balanced chemical equation where each letter represents a reactant or product, which has the greatest number of atoms?

    $$4\,B \rightarrow 2\,A + 3\,C$$

    a. The reactants, 4 B
    b. The products, 2A + 3C

    c. There are the same number of atoms in reactants and products.
    d. Not enough information is provided.

3. Which has the greatest number of atoms?

    a. 28 g of nitrogen, N_2
    b. 32 g of oxygen, O_2
    c. 16 g of methane, CH_4
    d. 38 g of fluorine, F_2

4. How many molecules of aspirin (formula mass aspirin = 180 amu) are in a 0.250-gram sample?

    a. 6.02×10^{23}
    b. 8.36×10^{20}
    c. 1.51×10^{23}
    d. More information is needed.

5. Is the synthesis of ozone, O_3, from oxygen, O_2, an example of an exothermic or endothermic reaction?

    a. Exothermic because ultraviolet light is emitted during its formation.

b. Endothermic because ultraviolet light is emitted during its formation.

c. Exothermic because ultraviolet light is absorbed during its formation.

d. Endothermic because ultraviolet light is absorbed during its formation.

6. How much energy, in kilojoules, are released or absorbed from the reaction of 1 mole of nitrogen, N_2, with 3 moles of molecular hydrogen, H_2, to form 2 moles of ammonia, NH_3? Consult Table 9.1 for bond energies.

   a. +899 kJ/mol

   b. −993 kJ/mol

   c. +80 kJ/mol

   d. −80 kJ/mol

7. How is it possible to cause an endothermic reaction to proceed when the reaction causes energy to become less dispersed?

   a. The reaction should be placed in a vacuum.

   b. The reaction should be cooled down.

   c. The concentration of the reactants should be increased.

   d. The reaction should be heated.

8. The yeast in bread dough feeds on sugar to produce carbon dioxide. Why does the dough rise faster in a warmer area?

   a. There is a greater number of effective collisions among reacting molecules.

   b. Atmospheric pressure decreases with increasing temperature.

   c. The yeast tends to "wake up" with warmer temperatures, which is why baker's yeast is best stored in the refrigerator.

   d. The rate of evaporation increases with increasing temperature.

9. What can you deduce about the activation energy of a reaction that takes billions of years to go to completion? How about a reaction that takes only fractions of a second?

   a. The activation energy of both these reactions must be very low.

   b. The activation energy of both these reactions must be very high.

   c. The slow reaction must have a high activation energy, while the fast reaction must have a low activation energy.

   d. The slow reaction must have a low activation energy, while the fast reaction must have a high activation energy.

10. What role do CFCs play in the catalytic destruction of ozone?

    a. Ozone is destroyed upon binding to a CFC molecule.

    b. CFC molecules are not likely to play a significant role in the catalytic destruction of ozone.

    c. CFC molecules activate chlorine atoms into their catalytic action.

    d. CFC molecules migrate to the upper stratosphere where they generate chlorine atoms.

ANSWERS TO CALCULATION CORNER
MASSES OF REACTANTS AND PRODUCTS

1. With an unlimited amount of O_2 available, 35.8 g of Fe_2O_3 can be produced:

$$(25.0 \text{ g Fe})\left(\frac{1 \text{ mole Fe}}{55.8 \text{ g Fe}}\right)\left(\frac{2 \text{ moles Fe}_2O_3}{4 \text{ moles Fe}}\right)\left(\frac{159.7 \text{ g Fe}_2O_3}{1 \text{ mole Fe}_2O_3}\right)$$

$$= 35.8 \text{ g Fe}_2O_3$$

2. With an unlimited amount of Fe available, 83.2 g of Fe_2O_3 can be produced:

$$(25.0 \text{ g O}_2)\left(\frac{1 \text{ mole O}_2}{32.0 \text{ g O}_2}\right)\left(\frac{2 \text{ moles Fe}_2O_3}{3 \text{ moles O}_2}\right)\left(\frac{159.7 \text{ g Fe}_2O_3}{1 \text{ mole Fe}_2O_3}\right)$$

$$= 83.2 \text{ g Fe}_2O_3$$

3. The 25.0 g of Fe will require 10.8 g of O_2:

$$(25.0 \text{ g Fe})\left(\frac{1 \text{ mole Fe}}{55.8 \text{ g Fe}}\right)\left(\frac{3 \text{ moles O}_2}{4 \text{ moles Fe}}\right)\left(\frac{32 \text{ g O}_2}{1 \text{ mole O}_2}\right)$$

$$= 10.8 \text{ g O}_2$$

4. From the previous question we know that 25.0 g of Fe requires 10.8 g of O_2. For this question, however, we were supplied with 25.0 g of O_2, which is much more than is needed. After the reaction is complete, there should be 25.0 g − 10.8 g = 14.2 g of O_2 remaining after the reaction is complete.

Mercury Emissions

Of all the metals in the periodic table, mercury, Hg (atomic number 80), is the only one to exist as a liquid at ambient temperatures. Mercury is also volatile, which means that uncontained mercury atoms evaporate into the atmosphere. Today, the atmosphere carries a load of about 5000 tons of mercury. Of this amount, about 2900 tons are from current human activities, such as the burning of coal, and 2100 tons appear to be from natural sources, such as outgassing from the Earth's crust and oceans. Since the mid-19th century, however, humans have emitted an estimated 200,000 tons of mercury into the atmosphere, most of which has since subsided onto the land and into sea. It is probable, therefore, that a large portion of the mercury emitted from "natural" sources is actually the re-emission of mercury originally put there by humans over the last 150 years.

Mercury is a poison to the nervous system. Its most dangerous form is that of the methyl mercury ion, CH_3Hg^+, which forms from elemental mercury within aquatic habitats. This form of mercury tends to bioaccumulate, so that organisms higher up in the marine food chain, such as pike, tuna, and swordfish, tend to have the highest levels. People who eat these fish regularly may be exposing themselves to high levels of mercury.

Typical consequences of mercury poisoning include a loss of mental focus and personality changes. In babies and children, the damaging effects are more severe, because the mercury disrupts brain development. Pregnant or nursing mothers are advised to avoid mercury-tainted fish, because methyl mercury passes through the placenta and into breast milk.

Because of a growing awareness of the dangers of mercury and because of imposed governmental regulations, over the past several decades, there has been a gradual phasing out of the use of mercury. This includes use in various commercial products, such as thermometers, and manufacturing processes, most notably in the manufacture of another element, chlorine. As shown in Table 9.2 however, the most significant source of human produced atmospheric mercury is the combustion of coal, which has remained largely unregulated.

One facet of the 1970 Clean Air Act was the exemption of existing coal-burning power plants. It wasn't until the Clean Air Act Amendments of 1990 that Congress gave the EPA the power to create regulations that would force all coal-burning power plants to be fitted with mercury-reducing mechanisms. The EPA was to provide these regulations by 1994, but for a number of political reasons, this never happened. A committee of state regulators and pollution control experts

TABLE 9.2 EPA Inventory of National Mercury Emission Rates*

SOURCE OF MERCURY	TONS/YEAR
Nonpoint Sources	
Lamp Breakage	1.5
General Laboratory Use	1.1
Dental Preparations	0.7
Combustion Sources	
Coal (electric utilities)	51.6
Coal (commercial use)	20.7
Oil	11.1
Municipal Waste	29.6
Medical Waste	16.0
Manufacturing	
Chlorine	7.1
Portland Cement	4.8
Pulp and Paper	1.9

*Source: Keating, Martha, et.al. Mercury Study Report to Congress, Volume I: Executive Summary, Environmental Protection Agency December 1997

was therefore convened to propose regulations. Their recommendations were put forth at the end of 2000 in a report that declared that U.S. mercury emissions could be reduced by up to 90 percent by 2007 using currently available technology. Coal-burning utilities disagreed, as did the newly elected Bush administration, which disbanded the committee.

Going back to the drawing board, the EPA took another two years to develop alternative regulations that called for a 34-ton cap (a 30 percent reduction) on the annual mercury output of coal utilities by 2007. This cap was set because it was achievable via regulations designed to curb other pollutants—no mercury-specific technologies would be needed. The cap would then be gradually lowered to 15 tons (a 70 percent reduction) by 2018. State regulators and environmental groups, as well as the EPA's oversight agency, the Office of the Inspector General, cited these regulations as too lenient and a violation of the 1990 Clean Air Act Amendment.

To override the Clean Air Act and to legitimize the EPA's proposed regulations, the Bush administration

and like-minded members of Congress introduced the Clear Skies Act of 2003, which adopted the EPA's proposed mercury regulations. By March 2005, however, Congress remained deadlocked on approving this act. In response, the Bush administration took the essential provisions of the Clear Skies Act and ordered the EPA to implement them through regulations regardless of conflicts with the Clean Air Act. This resulted in the EPA-mandated Clean Air Rules, one of which was the Clean Air Mercury Rules (CAMR). Litigation culminated in the U.S. Supreme Court's decision in February 2009 upholding a lawsuit claiming these rules to be fundamentally flawed. The EPA was once again ordered back to the drawing board to come up with regulations more in line with the original Clean Air Act Amendment of 1990.

This led to the Mercury and Air Toxics Standards, which the EPA released in December 2011 (www. epa.gov/mats). These regulations target primarily coal-fired power plants. Interestingly, by the time these regulations were released, about half of these power plants had already implemented the necessary air pollution control measures. But under these standards, less efficient older power plants are pressured to shut down and all the power plants are required to come into compliance by 2016. The EPA estimates that these regulations will prevent about 11,000 premature deaths and 130,000 cases of childhood asthma each year.

Meanwhile, a 2009 study by the U.S. Geological Survey showed that total mercury levels in North Pacific Ocean water have risen about 30 percent over the last 20 years. The study forecasts another 50 percent increase by 2050. According to the study, this increase in mercury is attributable to increases in global mercury atmospheric emission rates, particularly from Asia.

Besides the United States, other major contributors to mercury pollution include China, India, and Russia, but all nations contribute

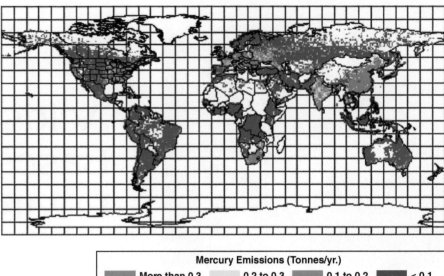

Mercury Emissions (Tonnes/yr.)

| More than 0.3 | 0.2 to 0.3 | 0.1 to 0.2 | < 0.1 |

to some extent. Atmospheric mercury remains airborne for about a year, which allows it to reach all regions of the planet. There are definitely "hot spots" that occur within 100 kilometers downwind of a coal-burning power plant, but on average, much of the mercury North Americans are exposed to originated in Asia. Similarly, much of the mercury Europeans are exposed to originated in North America. Asians get it from Europe. All of us are exposed to mercury arising from our past. Atmospheric mercury is a global problem, as is human inertia when it comes to finding solutions.

CONCEPTCHECK

There are about 4.8 billion mercury atoms in each breath of air that enters your lungs. What is the source of most of these mercury atoms?

CHECK YOUR ANSWER Almost all of these mercury atoms were put into the environment by recent or past human activities. Note that this number of mercury atoms is actually quite small compared to the number of other atoms in the air. The greatest threat arises when mercury in aquatic environments is transformed into the methyl mercury ions that bioaccumulate in the food chain.

Think and Discuss

1. A key argument for the Clear Skies Act of 2003 was that it was practical and took into consideration economic factors. A key argument against this act was that it was less than ideal. Ultimately, because it was less than ideal, the act was defeated, which furthered the delay of needed mercury emission regulations. Would you have voted for or against this act? Why?

2. Is it reasonable and practical to regulate the sale of mercury? Consider that small-scale gold miners around the world use mercury to help them extract gold. Tens of thousands of remote gold-mining sites release an estimated 1000 tons of mercury each year.

3. The cost of equipping coal-burning utilities with mercury-specific anti-pollution technologies gets passed along to the consumer. The exemption for coal-burning utilities given by the 1970 Clean Air Act Amendment has therefore saved consumers billions of dollars, which, in turn, has been of benefit to the economy. Is this a good or bad thing?

▲ The carbon dioxide from ancient atmospheres is captured within limestone sediments, such as those of the White Cliffs of Dover.

CaCO₃

10

Acids and Bases in Our Environment

THE MAIN IDEA

Acids donate protons and bases accept them.

As rainwater falls, it absorbs atmospheric carbon dioxide. Once in the rainwater, the carbon dioxide reacts to form carbonic acid, H_2CO_3, which, as we discuss in this chapter, makes rainwater naturally acidic. The acidic rain falls into the oceans, which happen to be basic because of the many alkaline minerals they contain. The carbonic acid of the rain reacts with the bases of the oceans to form salts, such as calcium carbonate, $CaCO_3$. The consquence is the removal of massive amounts of atmospheric carbon dioxide, which is a notable greenhouse gas. Furthermore, the calcium carbonate in the oceans is taken up by countless marine organisms that use this compound to form their protective shells. Upon death, the bulk of these creatures sink to the ocean floor, where over millions of years, their shells transform into a hardened rock known as limestone. Through geologic processes, ocean floors can be lifted to become parts of continents. The rock beneath the Great Plains of the United States, for example, is made of an ancient limestone sea floor, as are the impressive White Cliffs of Dover, shown on this page. These rocks are full of the calcium carbonate fossils of the ancient sea creatures, who, quite literally, are made from the atmospheres of the Earth's ancient past.

Chemistry

Rainbow Cabbage

The pH of a solution can be approximated with a *pH indicator*, which is any chemical whose color changes with pH. As Kai and Maile from Conceptual Chemistry Alive! demonstrate, many pH indicators are found in plants; the pigment of red cabbage is a good example. This pigment is red at low pH values (acidic), light purple at slightly acidic pH values, blue at neutral pH values, light green at moderately alkaline pH values, and dark green at very alkaline pH values.

PROCEDURE

1. Boil shredded red cabbage in water for about 5 minutes. Strain the broth from the cabbage and allow to cool.
2. Add the cooled blue broth to at least three clear cups so that each cup is less than half-filled or to three white porcelain bowls.
3. Add a teaspoon of white vinegar to one cup and a teaspoon of baking soda to a second cup. Watch for color changes. Add nothing to the third cup so that it remains blue.

ANALYZE AND CONCLUDE

1. What color changes do you see? What color is the red cabbage before being boiled? Are the juices within red cabbage more or less acidic than vinegar?
2. What would happen to the baking soda solution if you were to slowly add vinegar to it? What color would you get if a teaspoon of concentrated broth were added to a glass of water?

10.1 Acids Donate Protons and Bases Accept Them

EXPLAIN THIS

Why are many pharmaceuticals treated with hydrogen chloride?

The term *acid* comes from the Latin *acidus*, which means "sour." The sour taste of vinegar and citrus fruits is due to the presence of acids. Acids are essential in the chemical industry. For example, more than 40 million tons of sulfuric acid is produced annually in the United States, making this the number-one manufactured chemical. Sulfuric acid is used to make fertilizers, detergents, paint dyes, plastics, pharmaceuticals, and storage batteries, as well as to produce iron and steel. It is so important in the manufacturing of goods that its production is considered a standard measure of a nation's industrial strength. **Figure 10.1** shows only a few of the acids we commonly encounter.

Bases are characterized by their bitter taste and slippery feel. Interestingly, bases themselves are not slippery. Rather, they cause skin oils to transform into slippery solutions of soap. Most commercial preparations for unclogging drains contain sodium hydroxide, NaOH (also known as lye), which is extremely basic and hazardous when concentrated. Bases are also heavily used in industry. Each year in the United States, about 10 million tons of sodium hydroxide is manufactured for use in the production of various chemicals and in the pulp and paper industry. Solutions containing bases are often called *alkaline*, a term derived from the Arabic *al-qali* ("the ashes"). Ashes are slippery when wet because of the presence of bases such as potassium carbonate, K_2CO_3. **Figure 10.2** shows some familiar bases.

The Brønsted–Lowry Definition Focuses on Protons

Acids and bases may be defined in several ways. For our purposes, an appropriate definition is the one suggested in 1923 by the Danish chemist Johannes Brønsted (1879–1947) and the English chemist Thomas Lowry (1874–1936). In the Brønsted–Lowry definition, an **acid** is any chemical that donates hydrogen ions, H^+, and a **base** is any chemical that accepts hydrogen ions. Recall that

LEARNING OBJECTIVE

> Identify when a chemical behaves as an acid or a base.

READING CHECK

What is the Brønsted–Lowry definition of an acid and a base?

(a)

(b)

(c)

(d)

▲ **Figure 10.1**
Examples of acids. (a) Citrus fruits contain many types of acids, including ascorbic acid, $C_6H_8O_6$, which is vitamin C. (b) Vinegar contains acetic acid, $C_2H_4O_2$, and can be used to preserve foods. (c) Many toilet bowl cleaners are formulated with hydrochloric acid, HCl. (d) All carbonated beverages contain carbonic acid, H_2CO_3, while many also contain phosphoric acid, H_3PO_4.

a hydrogen atom consists of one electron surrounding a one-proton nucleus. A hydrogen ion, H^+, formed by the loss of an electron, therefore, is nothing more than a lone proton. Thus, it is also sometimes said that an acid is a chemical that donates a proton and a base is a chemical that accepts a proton.

▼ **Figure 10.2**
Examples of bases. (a) Reactions involving sodium bicarbonate, $NaHCO_3$, cause baked goods to rise. (b) Ashes contain potassium carbonate, K_2CO_3. (c) Soap is made by reacting bases with animal or vegetable oils. The soap itself, then, is slightly alkaline. (d) Powerful bases, such as sodium hydroxide, NaOH, are used in drain cleaners.

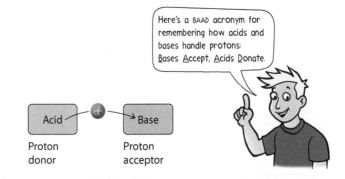

Here's a BAAD acronym for remembering how acids and bases handle protons: Bases Accept, Acids Donate.

Acid — Proton donor

Base — Proton acceptor

(a)

(b)

(c)

(d)

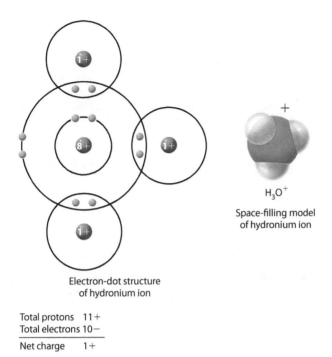

◀ **Figure 10.3**
The hydronium ion's positive charge is a consequence of the extra proton this molecule has acquired. Hydronium ions, which play a role in many acid-base reactions, are polyatomic ions, which, as mentioned in Section 6.2, are molecules that carry a net electric charge.

H_3O^+

Space-filling model
of hydronium ion

Electron-dot structure
of hydronium ion

Total protons   11+
Total electrons 10−

Net charge   1+

Consider what happens when hydrogen chloride is mixed into water:

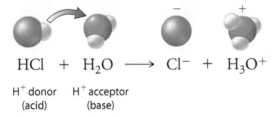

$$HCl \ + \ H_2O \ \longrightarrow \ Cl^- \ + \ H_3O^+$$

H⁺ donor   H⁺ acceptor
(acid)   (base)

Hydrogen chloride donates a hydrogen ion to one of the nonbonding electron pairs on a water molecule, resulting in a third hydrogen bonded to the oxygen. In this case, hydrogen chloride behaves as an acid (proton donor) and water behaves as a base (proton acceptor). The products of this reaction are a chloride ion and a **hydronium ion,** H_3O^+, which, as **Figure 10.3** shows, is made by adding a proton (hydrogen ion) to a water molecule.

When added to water, ammonia behaves as a base, as its nonbonding electrons (see Section 15.1) accept a hydrogen ion from water, which, in this case, behaves as an acid:

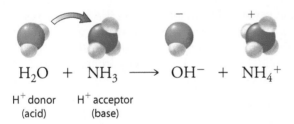

$$H_2O \ + \ NH_3 \ \longrightarrow \ OH^- \ + \ NH_4{}^+$$

H⁺ donor   H⁺ acceptor
(acid)   (base)

This reaction results in the formation of an ammonium ion and a **hydroxide ion,** which, as shown in **Figure 10.4**, is made by removing a proton (hydrogen ion) from a water molecule.

An important aspect of the Brønsted–Lowry definition is that it uses a *behavior* to define a substance as an acid or a base. We say, for example, that hydrogen chloride *behaves* as an acid when mixed with water, which *behaves* as a base. Similarly, ammonia *behaves* as a base when mixed with water, which, under this circumstance, *behaves* as an acid. Because labeling as an acid or a base depends on behavior, there is really no contradiction when a chemical such as water behaves as a base in one instance but as an acid in another

Recall that a hydrogen ion with a positive charge is simply a lone proton. These models I'm holding, of course, are not to scale because an atom's nucleus is many times smaller than the size of the atom.

Hydrogen
atom

Positive hydrogen
ion (lone proton)

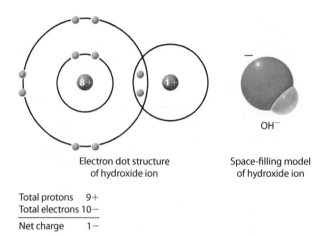

Electron dot structure of hydroxide ion

Space-filling model of hydroxide ion

OH^-

Total protons	9+
Total electrons	10−
Net charge	1−

instance. By analogy, consider yourself. You are who you are, but your behavior changes depending on whom you are with. Likewise, it is a chemical property of water to behave as a base (to accept H^+) when mixed with hydrogen chloride and as an acid (to donate H^+) when mixed with ammonia.

The products of an acid-base reaction can also behave as acids or as bases. An ammonium ion, for example, may donate a hydrogen ion back to a hydroxide ion to re-form ammonia and water:

$$H_2O + NH_3 \longleftarrow OH^- + NH_4^+$$

<table>
<tr><td></td><td></td><td></td><td>H^+ acceptor
(base)</td><td>H^+ donor
(acid)</td></tr>
</table>

Forward and reverse acid-base reactions proceed simultaneously and can therefore be represented as occurring at the same time by using two oppositely facing arrows:

$$H_2O + NH_3 \rightleftarrows OH^- + NH_4^+$$

H^+ donor (acid)	H^+ acceptor (base)	H^+ acceptor (base)	H^+ donor (acid)

When the equation is viewed from left to right, the ammonia behaves as a base, because it accepts a hydrogen ion from the water, which therefore acts as an acid. Viewed in the reverse direction, the equation shows that the ammonium ion behaves as an acid, because it donates a hydrogen ion to the hydroxide ion, which therefore behaves as a base.

CONCEPTCHECK

Identify the behavior as an acid or a base of each participant in the reaction

$$H_2PO_4^- + H_3O^+ \rightleftarrows H_3PO_4 + H_2O$$

CHECK YOUR ANSWER In the forward reaction (left to right), $H_2PO_4^-$ gains a hydrogen ion to become H_3PO_4. In accepting the hydrogen ion, $H_2PO_4^-$ is behaving as a base. It gets the hydrogen ion from the H_3O^+, which is behaving as an acid. In the reverse direction, H_3PO_4 loses a hydrogen ion to become $H_2PO_4^-$ and is thus behaving as an acid. The recipient of the hydrogen ion is the H_2O, which is behaving as a base as it transforms to H_3O^+.

The Lewis Definition Focuses on Lone Pairs

The Brønsted–Lowry definition of acids and bases is restricted to molecules that can donate or accept protons. A more general definition of acids and bases was proposed in the 1930s by Gilbert Lewis, the same chemist who introduced the idea of shells to explain the organization of the periodic table as well as chemical bonding (Chapters 4 and 6). According to the Lewis definition, a molecule behaves as a base when it donates a lone pair of electrons. Conversely, a molecule behaves as an acid when it accepts a lone pair. So while the Brønsted–Lowry definition focuses on the action of a proton, the Lewis definition focuses on the action of a lone pair.

Looking back to the previous examples of bases, we see that a lone pair of each of these bases is being donated to a positively charged proton, as illustrated here. Note how the curved arrows indicate the movement of electrons:

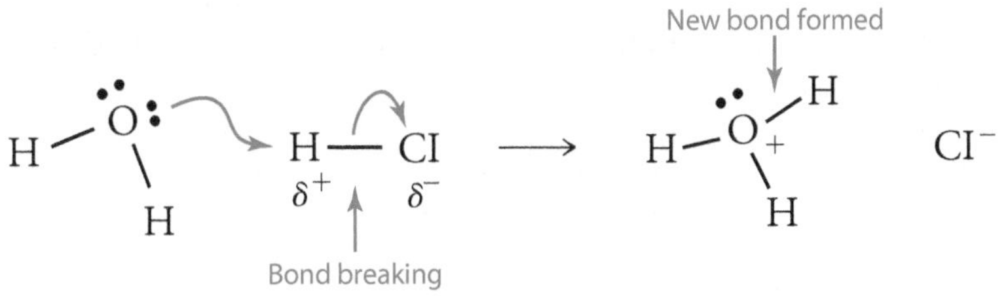

In this case, the water is behaving as a base, because its lone pair seeks out a positive charge (the proton), while the hydrochloric acid, HCl, behaves as an acid, because it accepts the lone pair.

Consider the formation of carbonic acid from water and carbon dioxide, as shown here:

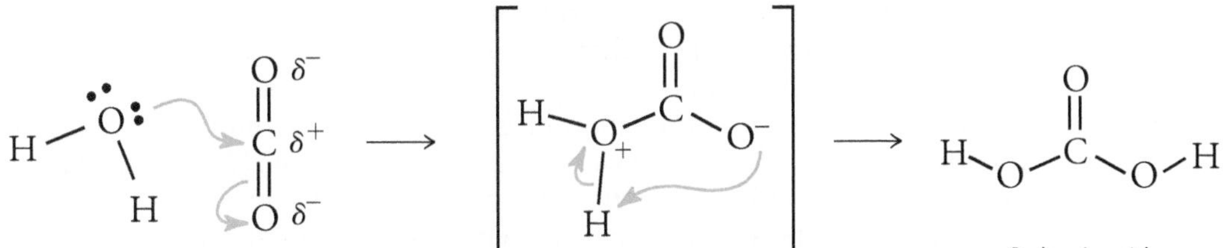

Carbonic acid

This reaction starts as a typical dipole–induced dipole attraction between the oxygen of the water and the carbon of the carbon dioxide. As the carbon accepts electrons from the lone pair of the water, however, it begins to lose electrons to one of its two adjacent oxygen atoms. This permits the formation of a covalent bond between the water and the carbon dioxide, which then become a larger molecule (shown in brackets) with both a positive and negative charge. This molecule exists only briefly before transforming into the more stable non-charged product, carbonic acid, which, as we described at the beginning of this chapter, is responsible for the natural acidity of rainwater.

CONCEPT CHECK

How is it possible for carbon dioxide, CO_2, to behave as an acid when it has no hydrogen ions to donate?

CHECK YOUR ANSWER Carbon dioxide behaves as an acid when it accepts the lone pair on the oxygen atom of a water molecule.

▲ Figure 10.5
"Lite" table salt substitutes contain potassium chloride in place of sodium chloride. Caution is advised in using these products, however, because excessive quantities of potassium salts can lead to serious illness. Furthermore, sodium ions are a vital component of our diet and should never be totally excluded. For a good balance of these two important ions, you might inquire about commercially available half-and-half mixtures of sodium chloride and potassium chloride, such as the one shown here.

A Salt Is the Ionic Product of an Acid-Base Reaction

In everyday language, the word *salt* implies sodium chloride, NaCl, table salt. In the language of chemistry, however, a **salt** is an ionic compound commonly formed from the reaction between an acid and a base. Hydrogen chloride and sodium hydroxide, for example, react to produce the salt sodium chloride and water:

HCl	+	NaOH	→	NaCl	+	H$_2$O
Hydrogen		Sodium		Sodium		Water
chloride		hydroxide		chloride		
(acid)		(base)		(salt)		

Similarly, the reaction between hydrogen chloride and potassium hydroxide yields the salt potassium chloride and water:

HCl	+	KOH	→	KCl	+	H$_2$O
Hydrogen		Potassium		Potassium		Water
chloride		hydroxide		chloride		
(acid)		(base)		(salt)		

Potassium chloride is the main ingredient in "lite" table salt, as noted in **Figure 10.5**.

Salts are generally far less corrosive than the acids and bases from which they are formed. A corrosive chemical has the power to disintegrate a material or wear away its surface. Hydrogen chloride is a remarkably corrosive acid, which makes it useful for cleaning toilet bowls and etching metal surfaces. Sodium hydroxide is a very corrosive base used for unclogging drains. Mixing hydrogen chloride and sodium hydroxide together in equal portions, however, produces an aqueous solution of sodium chloride—saltwater, which is not nearly as destructive as either starting material.

There are as many salts as there are acids and bases. Sodium cyanide, NaCN, is a deadly poison. "Saltpeter," which is potassium nitrate, KNO$_3$, is useful as a fertilizer and in the formulation of gunpowder. Calcium chloride, CaCl$_2$, is commonly used to de-ice walkways, and sodium fluoride, NaF, helps to prevent tooth decay. The acid-base reactions forming these salts are shown in Table 10.1.

The reaction between an acid and a base is called **neutralization.** As can be seen in the color-coding of the neutralization reactions in Table 10.1, the positive ion of a salt comes from the base and the negative ion comes from the acid. The remaining hydrogen and hydroxide ions join to form water.

Not all neutralization reactions result in the formation of water. In the presence of hydrogen chloride, for example, the drug pseudoephedrine behaves as a base by accepting H$^+$ from the hydrogen chloride. The negative Cl$^-$ ions then join the pseudoephedrine–H$^+$ ions to form the salt pseudoephedrine hydrochloride, which is a common nasal decongestant, shown in **Figure 10.6**.

TABLE 10.1   Acid-Base Reactions and the Salts Formed

ACID		BASE		SALT		WATER
HCN	+	NaOH	→	NaCN	+	H$_2$O
Hydrogen cyanide		Sodium hydroxide		Sodium cyanide		
HNO$_3$	+	KOH	→	KNO$_3$	+	H$_2$O
Nitric acid		Potassium hydroxide		Potassium nitrate		
2 HCl	+	Ca(OH)$_2$	→	CaCl$_2$	+	2 H$_2$O
Hydrogen chloride		Calcium hydroxide		Calcium chloride		
HF	+	NaOH	→	NaF	+	H$_2$O
Hydrogen fluoride		Sodium hydroxide		Sodium fluoride		

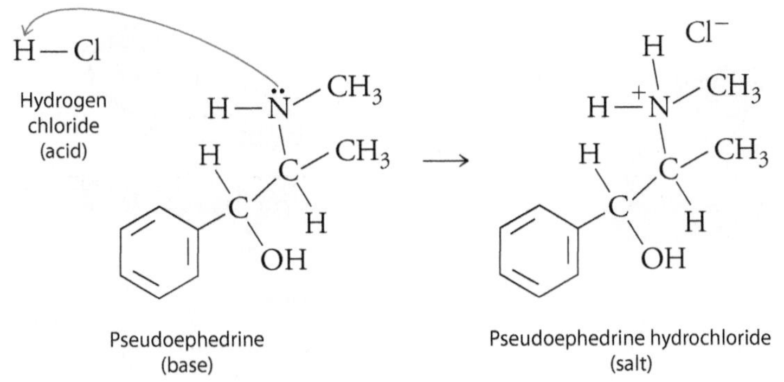

H—Cl

Hydrogen
chloride
(acid)

H—N̈—CH₃

H

H—C—CH₃

H

C—H

OH

Pseudoephedrine
(base)

→

H   Cl⁻

H—⁺N—CH₃

H

H—C—CH₃

H

C—H

OH

Pseudoephedrine hydrochloride
(salt)

◀ **Figure 10.6**
Hydrogen chloride and pseudoephedrine react to form the salt *pseudoephedrine hydrochloride*, which, because of its solubility in water, is readily absorbed into the body. Most pharmaceuticals that can be taken orally are bases that have been converted to a salt form.

Because it is a salt, psedoephedrine hydrochloride is much more water soluble than is the pseudoephedrine base, sometimes called the *free base* because it is "free" from ions. Because pseudoephedrine hydrochloride is water soluble, it is readily absorbed through the digestive system, which means it can be taken orally.

CONCEPTCHECK

Is a neutralization reaction best described as a physical change or a chemical change?

CHECK YOUR ANSWER New chemicals are formed during a neutralization reaction, meaning the reaction is a chemical change.

10.2 Some Acids and Bases Are Stronger than Others

EXPLAIN THIS

Why is hydrogen chloride, HCl, such a strong acid?

In general, the stronger an acid, the more readily it donates hydrogen ions. Likewise, the stronger a base, the more readily it accepts hydrogen ions. An example of a strong acid is hydrogen chloride, HCl, and an example of a strong base is sodium hydroxide, NaOH. The corrosiveness of these materials is a result of their strength.

One way to assess the strength of an acid or a base is to measure how much of it remains after it has been added to water. If little remains, the acid or base is strong. If a lot remains, the acid or base is weak. To illustrate this concept, consider what happens when the strong acid hydrogen chloride, HCl, is added to water and what happens when the weak acid acetic acid, $C_2H_4O_2$ (the active ingredient of vinegar), is added to water.

Being an acid, hydrogen chloride donates hydrogen ions to water, forming chloride ions and hydronium ions. Because HCl is such a strong acid, all of it is converted to these ions, as shown in **Figure 10.7**.

Because acetic acid is a weak acid, it has much less of a tendency to donate hydrogen ions to water. When this acid is dissolved in water, only a small portion of the acetic acid molecules are converted to ions, a process that occurs as the polar O—H bonds are broken. (The C—H bonds of acetic acid are unaffected by the water because of their nonpolarity.) The majority of acetic acid molecules remain intact in their original un-ionized form, as shown in **Figure 10.8**.

Figures 10.7 and 10.8 show the submicroscopic behavior of strong and weak acids in water. Because molecules and ions are too small to see, how, then, does a chemist measure the strength of an acid? One way is by measuring a

LEARNING OBJECTIVE

Describe how the strength of an acid or a base affects the number of ions in solution.

READINGCHECK

What is one way to assess the strength of an acid?

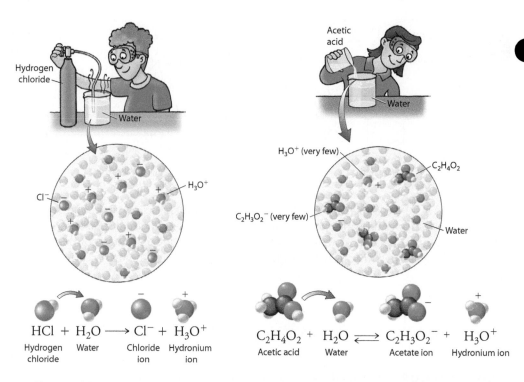

$$HCl + H_2O \longrightarrow Cl^- + H_3O^+$$

Hydrogen   Water   Chloride   Hydronium
chloride   ion   ion

$$C_2H_4O_2 + H_2O \rightleftharpoons C_2H_3O_2^- + H_3O^+$$

Acetic acid   Water   Acetate ion   Hydronium ion

▲ **Figure 10.7**
Immediately after gaseous hydrogen chloride is added to water, it reacts with the water to form hydronium ions and chloride ions. That very little HCl remains (none shown here) lets us know that HCl acts as a strong acid.

▲ **Figure 10.8**
When liquid acetic acid is added to water, only a few acetic acid molecules react with water to form ions. The majority of the acetic acid molecules remain in their un-ionized form, which implies that acetic acid is a weak acid.

 FOR YOUR INFORMATION

Aspirin is an acidic molecule, but not nearly as acidic as the hydrochloric acid, HCl, found in your stomach and used to digest food. So how is it that aspirin can cause damage to your stomach? Stomach acid is so strong that within this environment, aspirin is unable to donate its hydrogen ion, which means that it remains "un-ionized." This un-ionized aspirin is nonpolar and water insoluble, which helps it to penetrate through the largely nonpolar mucous membrane that lines the stomach. After passing through this membrane, the aspirin finds itself in a less acidic environment, where it can finally donate its hydrogen ion. This lowers the pH of the submucous inner wall of the stomach, which can damage the tissues and even cause bleeding. The remedy, of course, is to swallow specially coated aspirin tablets, which delay the release of the aspirin until after it passes through the stomach.

solution's ability to conduct an electric current, as **Figure 10.9** illustrates. In pure water, there are practically no ions to conduct electricity. When a strong acid is dissolved in water, many ions are generated, as indicated within the circle of Figure 10.7. The presence of these ions allows the flow of a strong electric current. A weak acid dissolved in water generates only a few ions, as indicated within the circle of Figure 10.8. The presence of fewer ions means there can be only a weak electric current.

This same trend is seen with strong and weak bases. Strong bases, for example, tend to accept hydrogen ions more readily than weak bases. In solution, a strong base allows the flow of a strong electric current and a weak base allows the flow of a weak electric current.

CONCEPT CHECK

According to the aqueous solutions illustrated here, which is the stronger base, NH_3 or $NaOH$?

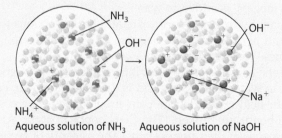

Aqueous solution of NH_3 Aqueous solution of $NaOH$

CHECK YOUR ANSWER The solution on the right contains the greater number of ions, meaning sodium hydroxide, $NaOH$, is the stronger base. Ammonia, NH_3, is the weaker base, indicated by the relatively few ions in the solution on the left.

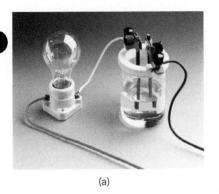
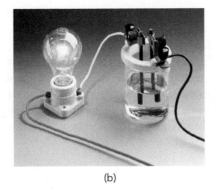

(a) (b) (c)

▲ Figure 10.9
(a) The pure water in this circuit does not allow the bulb to light because the water contains practically no ions. The lightbulb in the circuit therefore remains unlit. (b) Because HCl is a strong acid, nearly all of its molecules break apart in water, yielding a high concentration of ions, which are able to conduct an electric current that lights the bulb. (c) Acetic acid, $C_2H_4O_2$, is a weak acid; in water, only a small portion of its molecules break up into ions. Because fewer ions are generated, only a weak current exists, and the bulb is therefore dimmer.

Just because an acid or a base is strong doesn't mean a solution of that acid or base is corrosive. A *very* dilute solution of a strong acid or a strong base may have little corrosive action because in such solutions, there are only a few hydronium or hydroxide ions. Almost all the molecules of the strong acid or base break up into ions, but because the solution is dilute, there are only a few acid or base molecules to begin with. As a result, there are only a few hydronium or hydroxide ions. You shouldn't be too alarmed, therefore, when you discover that some toothpastes are formulated with small amounts of sodium hydroxide, one of the strongest bases known. The concentration of sodium hydroxide in these products is quite small.

On the other hand, a concentrated solution of a weak acid, such as the acetic acid in vinegar, may be just as corrosive as or even more corrosive than a dilute solution of a strong acid, such as hydrogen chloride. The relative strengths of two acids in solution or two bases in solution, therefore, can be compared only when the two solutions have the same concentration.

 **FOR YOUR INFORMATION**

What makes one acid strong and another weak? One factor involves the stability of the negative ion that remains after the proton has been donated. Hydrogen chloride is a strong acid because the chloride ion is able to accommodate the negative charge rather well. Acetic acid, however, is a weaker acid because the resulting oxygen ion is less able to accommodate the negative charge.

10.3 Solutions Can Be Acidic, Basic, or Neutral

EXPLAIN THIS

Why can't water be absolutely pure?

A substance whose ability to behave as an acid is about the same as its ability to behave as a base is said to be **amphoteric.** Water is a good example. Because it is amphoteric, water has the ability to react with itself. In behaving as an acid, a water molecule donates a hydrogen ion to a neighboring water molecule, which, in accepting the hydrogen ion, is behaving as a base. This reaction produces a hydroxide ion and a hydronium ion, which react together to reform the water molecule:

LEARNING OBJECTIVE

Calculate the pH of a solution given the hydronium-ion concentration.

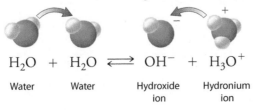

$$H_2O \ + \ H_2O \ \rightleftarrows \ OH^- \ + \ H_3O^+$$

Water   Water   Hydroxide   Hydronium
ion   ion

CHEMICAL CONNECTIONS

How is a piece of paper connected to a rock at the bottom of the ocean?

When a water molecule gains a hydrogen ion, a second water molecule must lose a hydrogen ion. So for every one hydronium ion formed, one hydroxide ion also forms. In pure water, therefore, the total number of hydronium ions must be the same as the total number of hydroxide ions. Experiments reveal that the concentration of hydronium and hydroxide ions in pure water is extremely low—about 0.00000010 M for each, where M stands for molarity, or moles per liter (Section 7.3). Water by itself, therefore, is a very weak acid as well as a very weak base, as evidenced by the unlit lightbulb in Figure 10.9a.

CONCEPTCHECK

Do water molecules react with one another?

CHECK YOUR ANSWER Yes, but not to any large extent. When they do react, they form hydronium and hydroxide ions. (Note: make sure you understand this point, because it serves as a basis for most of the rest of the chapter.)

Further experiments reveal an interesting rule pertaining to the concentrations of hydronium and hydroxide ions in any solution that contains water. The concentration of hydronium ions in any aqueous solution multiplied by the concentration of the hydroxide ions in the solution always equals the constant K_w, which is a very, very small number:

$$\text{concentration } H_3O^+ \times \text{concentration } OH^- = K_w = 0.000000000000010$$

Concentration is usually given as molarity, which is indicated by abbreviating this equation using brackets:

$$[H_3O^+] \times [OH^-] = K_w = 0.000000000000010$$

The brackets mean this equation is read "the molarity of H_3O^+ times the molarity of OH^- equals K_w." Writing in scientific notation, we have

$$[H_3O^+][OH^-] = K_w = 1.0 \times 10^{-14}$$

For pure water, the value of K_w is the concentration of hydronium ions, 0.00000010 M, multiplied by the concentration of hydroxide ions, 0.00000010 M, which can be written in scientific notation as

$$[1.0 \times 10^{-7}][1.0 \times 10^{-7}] = K_w = 1.0 \times 10^{-14}$$

The constant value of K_w is quite significant because it means that *no matter what is dissolved in the water,* the product of the hydronium-ion and hydroxide-ion concentrations always equals 1.0×10^{-14}. So if the concentration of H_3O^+ goes up, the concentration of OH^- must go down, and the product of the two remains 1.0×10^{-14}.

FORYOUR INFORMATION

The outer surface of hair is made of microscopic scale-like structures called cuticles that, like window shutters, are able to open and close. Alkaline solutions cause the cuticles to open up, which makes the hair "porous." Acidic solutions cause the cuticles to close down, which makes the hair "resistant." A beautician can control how long hair retains artificial coloring by modifying the pH of the hair-coloring solution. With an acidic solution, the cuticles close shut so that the dye binds only to the outside of each shaft of hair. This results in a temporary hair coloring, which may come off with the next hair washing. Using an alkaline solution permits the dye to penetrate through the cuticles into the hair for a more permanent effect.

CONCEPTCHECK

1. In pure water, the hydroxide-ion concentration is 1.0×10^{-7} M. What is the hydronium-ion concentration?

2. What is the concentration of hydronium ions in a solution if the concentration of hydroxide ions is 1.0×10^{-3} M?

CHECK YOUR ANSWERS

1. 1.0×10^{-7} M, because in pure water, $[H_3O^+] = [OH^-]$.

2. 1.0×10^{-11} M, because $[H_3O^+][OH^-]$ must equal $1.0 \times 10^{-14} = K_w$.

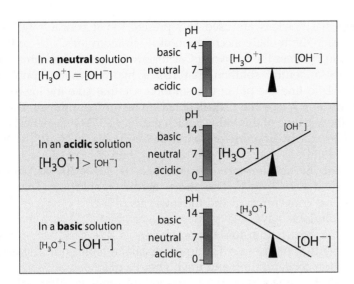

pH
14
basic
neutral 7
acidic 0

◀ **Figure 10.10**
The relative concentrations of hydronium and hydroxide ions determine whether a solution is neutral acidic, or basic. When these concentrations are equal, the solution is neutral. When the hydronium ion concentration increases, the hydroxide ion concentration necessarily decreases and the solution is acidic. Conversely, when the hydroxide ion concentration increases, the hydronium ion concentration necessarily decreases and the solution is basic.

Any solution containing an equal number of hydronium and hydroxide ions is said to be **neutral**. Pure water is an example of a neutral solution—not because it contains so few hydronium or hydroxide ions, but because it contains equal numbers of these ions. A neutral solution is also obtained when equal quantities of acid and base are combined, which explains why acids and bases are said to *neutralize* each other.

The balance of hydronium and hydroxide ions in a neutral solution is upset by adding either an acid or a base. Add an acid and the water will react with that acid to produce more hydronium ions. Many of these additional hydronium ions neutralize hydroxide ions, which then become fewer. The final result is that the hydronium-ion concentration is greater than the hydroxide-ion concentration. Such a solution is said to be **acidic**.

Add a base to water and the reverse happens. The water will react with that base to produce more hydroxide ions. Many of these additional hydroxide ions will neutralize hydronium ions, which then become fewer. The final result is that the hydronium-ion concentration is less than the hydroxide-ion concentration. Such a solution is said to be **basic**, or sometimes *alkaline*. This is all summarized in **Figure 10.10**.

READINGCHECK

Why is pure water a neutral solution?

CONCEPTCHECK

How does adding ammonia, NH_3, to water make a basic solution when there are no hydroxide ions in the formula for ammonia?

CHECK YOUR ANSWER Ammonia indirectly increases the hydroxide-ion concentration by reacting with water:

$$NH_3 + H_2O \rightleftharpoons NH_4^+ + OH^-$$

This reaction raises the hydroxide-ion concentration, which has the effect of lowering the hydronium-ion concentration. With the hydroxide-ion concentration now higher than the hydronium-ion concentration, the solution is basic.

FORYOUR
INFORMATION

Above temperatures of 374°C and pressures of 218 atmospheres, water transforms into a state of matter known as a supercritical fluid, which resembles both a liquid and a gas. For a neutral solution of supercritical water, the pH equals about 2, which means that it has high concentrations of both hydronium and hydroxide ions and is highly corrosive. Research is under way to learn how supercritical water might be used to destroy toxic chemicals, such as chemical warfare agents.

The pH Scale Is Used to Describe Acidity

The *pH scale* is a numeric scale used to express the acidity of a solution. Mathematically, **pH** is equal to the negative logarithm of the hydronium-ion concentration:

$$pH = -\log[H_3O^+]$$

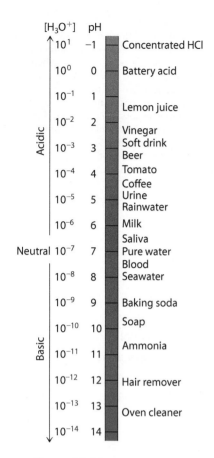

$[H_3O^+]$	pH	
10^1	−1	Concentrated HCl
10^0	0	Battery acid
10^{-1}	1	
		Lemon juice
10^{-2}	2	
		Vinegar
10^{-3}	3	Soft drink
		Beer
10^{-4}	4	Tomato
		Coffee
10^{-5}	5	Urine
		Rainwater
10^{-6}	6	Milk
		Saliva
10^{-7}	7	Pure water
		Blood
10^{-8}	8	Seawater
10^{-9}	9	Baking soda
10^{-10}	10	Soap
10^{-11}	11	Ammonia
10^{-12}	12	Hair remover
10^{-13}	13	Oven cleaner
10^{-14}	14	

Acidic — Neutral — Basic

▲ **Figure 10.11**
The pH values of some common solutions.

Note again that brackets are used to represent molar concentrations, meaning $[H_3O^+]$ is read "the molar concentration of hydronium ions." To understand the logarithm function, see the following Calculation Corner: Logarithms and pH.

Consider a neutral solution that has a hydronium-ion concentration of 1.0×10^{-7} M. To find the pH of this solution, we first take the logarithm of this value, which is −7 (see the Calculation Corner on logarithms). The pH is by definition the negative of this value, which means $-(-7) = 7$. Hence, in a neutral solution, where the hydronium-ion concentration equals 1.0×10^{-7} M, the pH is 7.

Acidic solutions have pH values less than 7. For an acidic solution in which the hydronium-ion concentration is 1.0×10^{-4} M, for example, $pH = -\log(1.0 \times 10^{-4}) = 4$. The more acidic a solution, the greater its hydronium-ion concentration and the lower its pH.

Basic solutions have pH values greater than 7. For a basic solution in which the hydronium-ion concentration is 1.0×10^{-8} M, for example, $pH = -\log(1.0 \times 10^{-8}) = 8$. The more basic a solution, the smaller its hydronium-ion concentration and the higher its pH.

Figure 10.11 shows typical pH values of some familiar solutions, and **Figure 10.12** shows two common ways of determining pH values.

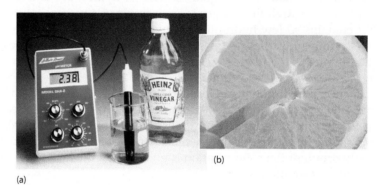

(a)

(b)

▲ **Figure 10.12**
(a) The pH of a solution can be measured electronically using a pH meter. (b) A rough estimate of the pH of a solution can be obtained with pH paper, which is coated with a dye that changes color with pH.

CALCULATION CORNER LOGARITHMS AND PH

Logarithms are super easy. When you ask for the *logarithm*, you are simply asking "to what power is 10 raised?" For example, the logarithm of 10^2 is 2, because that is the power to which 10 is raised. So what is the logarithm of 10^3? Just look at to what power 10 is raised and there is your answer: 3. If you know that 10^2 is equal to 100, then you'll understand that the logarithm of 100 is also 2. Check this out on your [log] button–equipped calculator. Similarly, the logarithm of 1000 is 3, because 10 raised to the third power, 10^3, equals 1000.

Any positive number, including a very small one, has a logarithm. The logarithm of 0.0001, which equals 10^{-4}, is −4 (the power to which 10 is raised to equal this number).

As a review of exponents, study the pattern of the numbers shown at the top of the next column. Please keep in mind that no negative numbers are shown—all numbers are greater than zero! The negative sign in the exponent merely means that the number is less than 1 (but greater than zero).

EXAMPLE 1

What is the logarithm of 0.01?

$10^3 = 1000$	(three zeros)
$10^2 = 100$	(two zeros)
$10^1 = 10$	(one zero)
$10^0 = 1$	(no zeros)
$10^{-1} = 0.1$	(one zero in the opposite direction)
$10^{-2} = 0.01$	(two zeros in the opposite direction)
$10^{-3} = 0.001$	(three zeros in the opposite direction)

ANSWER 1

The number 0.01 is 10^{-2}, the logarithm of which is −2 (the power to which 10 is raised).

The concentration of hydronium ions in most solutions is much less than 1 M. Recall, for example, that in neutral water, the hydronium-ion concentration is 0.0000001 M (10^{-7}M). The logarithm of any number smaller than 1 (but greater than zero) is a negative number. The definition of pH includes the minus sign so as to transform the logarithm of the hydronium-ion concentration to a positive number.

Note that when a solution has a hydronium-ion concentration of 1 M, the pH is 0 because 1 M = 10^0 M. A 10 M solution has a pH of −1 because 10 M = 10^1 M.

10.4 Buffer Solutions Resist Changes in pH

EXPLAIN THIS

At what point does a buffer solution lose its ability to resist changes in pH?

A **buffer solution** is any solution that resists large changes in pH. Buffer solutions work by containing two components. One component neutralizes any added acid, and the other neutralizes any added base. Buffer solutions can be prepared by mixing a weak acid with the salt of that weak acid, which, interestingly, is generally a weak base. An example is a solution of acetic acid, $C_2H_4O_2$ (a weak acid), and sodium acetate, $NaC_2H_3O_2$ (a weak base).

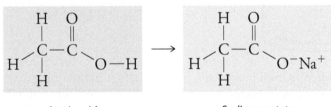

Acetic acid
(weak acid)

Sodium acetate
(salt of weak acid)

Any strong acid added to this particular buffer solution is neutralized by the sodium acetate, as shown in **Figure 10.13**. Similarly, any strong base added is neutralized by the acetic acid, as shown in **Figure 10.14**. So a buffer works by containing both a weak acid and a weak base, each of which will neutralize any incoming material that might drastically alter the pH of the solution.

So both strong acids and strong bases are neutralized by a buffer solution. This does not mean, however, that the solution's pH remains unchanged. When NaOH is added to the buffer system we are using as our example, sodium acetate is produced. Because sodium acetate behaves as a weak base (it accepts hydrogen ions but not very well), there is a slight increase in pH. When HCl is added, acetic acid is produced. Because acetic acid behaves as a weak acid, there is a slight decrease in pH. What makes a buffer solution special is its ability to resist *large* changes in pH.

CONCEPT CHECK

Why do most buffer solutions contain two dissolved components?

CHECK YOUR ANSWER The weak base component neutralizes any incoming acid, and the weak acid component neutralizes any incoming base.

▶ **Figure 10.13**
Hydrochloric acid added to a solution containing acetic acid and sodium acetate is neutralized by the sodium acetate to form additional acetic acid and sodium chloride.

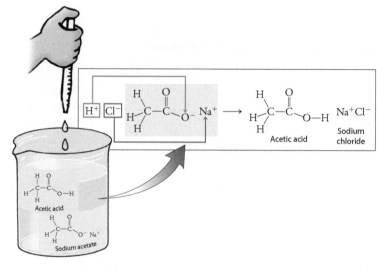

▶ **Figure 10.14**
Sodium hydroxide added to a solution containing acetic acid and sodium acetate is neutralized by the acetic acid to form additional sodium acetate and water.

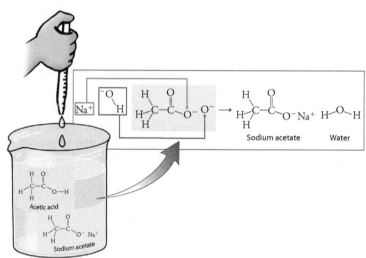

Many different buffer systems are useful for maintaining particular pH values. The acetic acid–sodium acetate system is good for maintaining a pH around 4.8. Buffer solutions containing equal mixtures of a weak base and a salt of that weak base maintain alkaline pH values. For example, a buffer solution of the weak base ammonia, NH_3, and its salt ammonium chloride, NH_4Cl, is useful for maintaining a pH of about 9.3.

Blood has several buffer systems that work together to maintain a narrow pH range between 7.35 and 7.45. A pH value above or below these levels can be lethal, primarily because it causes cellular proteins to become *denatured*, which is what happens to milk when vinegar is added to it.

The primary buffer system of our blood is a combination of carbonic acid and its salt sodium bicarbonate, shown in **Figure 10.15**. Any acid that builds up in the bloodstream is neutralized by the basic action of sodium bicarbonate, and any base that builds up is neutralized by the acidic action of carbonic acid.

The carbonic acid in your blood is formed as the carbon dioxide produced by your cells enters the bloodstream and reacts with water—this is the same

▶ **Figure 10.15**
Carbonic acid and sodium bicarbonate.

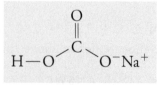

Carbonic acid
(weak acid)

Sodium bicarbonate
(salt)

(a)   (b)

(a) Hold your breath and CO_2 builds up in your bloodstream. This increases the amount of carbonic acid, which lowers your blood pH. (b) Hyperventilate and the amount of CO_2 in your bloodstream decreases. This decreases the amount of carbonic acid, which raises your blood pH.

reaction that occurs in a raindrop, as we discussed earlier. You fine-tune the levels of blood carbonic acid, and hence your blood pH, by your breathing rate, as **Figure 10.16** illustrates. Breathe too slowly or hold your breath and the amount of carbon dioxide (and hence carbonic acid) builds up, causing a slight but significant drop in pH. Hyperventilate and the carbonic acid level decreases by exhaling too much CO_2, causing a slight but significant increase in pH. Your body uses this mechanism to protect itself from changes in blood pH. One of the symptoms of a severe overdose of aspirin, for example, is hyperventilation. Aspirin, also known as acetylsalicylic acid, is an acidic chemical that, when taken in large amounts, can overwhelm the blood buffering system, causing a dangerous drop in blood pH. As you hyperventilate, however, your body loses carbonic acid, which helps to maintain the proper blood pH despite the overabundance of the acidic aspirin.

To summarize the acid/base concepts presented so far, consider the tractor shown in **Figure 10.17**. Using a pH measuring kit, the gardener has found that the soil pH is unacceptably low, perhaps because of local atmospheric pollutants, which may arise from natural or human-made sources. At this low pH, the soil contains an overabundance of hydronium ions. These hydronium ions react with alkaline nutrients of the soil, such as ammonia, to form water-soluble salts. Because of their water solubility, these nutrients in their salt form are readily washed away with the rainwater and as a result, the soil becomes nutrient-poor. The mechanism by which plants absorb whatever nutrients do remain in the soil is also disturbed by the soil's low pH.

As a result of all this, most plants do not grow well in acidic soil. To remedy this, the gardener spreads powdered limestone, a form of calcium carbonate, $CaCO_3$, which neutralizes the hydronium ions, thus raising the pH toward neutral.

Interestingly, the calcium carbonate reacts with the acidic soil to form carbon dioxide gas, which in the atmosphere helps to keep rainwater slightly acidic, which we describe next. This is the same gas that is generated by the cells of our bodies and that tends to acidify our blood. The blood pH, however, is kept fairly constant at around 7.4 because it is buffered.

▲ Figure 10.17
Raising the pH of soil by the addition of an alkaline mineral is known as liming.

10.5 Rainwater Is Acidic

EXPLAIN THIS

What happens to the pH of soda water as it loses its carbonation?

As previously mentioned, rainwater is naturally acidic. A main source of this acidity is carbon dioxide, the same gas that gives fizz to soda drinks. There are

LEARNING OBJECTIVE

Identify sources of acidity in rainwater and explain how this acidity can impact the environment.

FOR YOUR INFORMATION

Acid rain remains a serious problem in many regions of the world. Significant progress, however, has been made toward fixing the problem. In the United States, for example, sulfur dioxide and nitrogen oxide emissions have been reduced by nearly half since 1980. Also, in 2009, federal courts approved the 2005 Clean Air Interstate Rule (CAIR), which is designed to reduce levels of these pollutants even further, especially for areas downwind of heavily industrialized regions.

about 829 billion tons of CO_2 in the atmosphere, most of it from such natural sources as volcanoes and decaying organic matter but a growing amount (about 230 billion tons) from human activities.

Water in the atmosphere reacts with carbon dioxide to form *carbonic acid:*

$$CO_2(g) \quad + \quad H_2O(l) \quad \leftrightarrows \quad H_2CO_3(aq)$$

Carbon   Water   Carbonic

dioxide   acid

Carbonic acid, as its name implies, behaves as an acid and lowers the pH of water. The CO_2 in the atmosphere brings the pH of rainwater to about 5.6—noticeably below the neutral pH value of 7. Because of local fluctuations, the normal pH of rainwater varies between 5 and 7. This natural acidity of rainwater may accelerate the erosion of land and, under certain circumstances, can lead to the formation of underground caves.

By convention, *acid rain* is a term used for rain having a pH lower than 5. Acid rain is created when airborne pollutants, such as sulfur dioxide, are absorbed by atmospheric moisture. Sulfur dioxide is readily converted to sulfur trioxide, which reacts with water to form *sulfuric acid:*

$$2SO_2(g) \quad + \quad O_2(g) \quad \rightarrow \quad 2SO_3(g)$$

Sulfur   Oxygen   Sulfur

dioxide   trioxide

$$SO_3(g) \quad + \quad H_2O(l) \quad \rightarrow \quad H_2SO_4(aq)$$

Sulfur   Water   Sulfuric

trioxide   acid

Each year, about 20 million tons of SO_2 is released into the atmosphere by the combustion of sulfur-containing coal and oil. Sulfuric acid is much stronger than carbonic acid, and as a result, rain laced with sulfuric acid and therefore higher H_3O^+ concentrations eventually corrodes metal, paint, and other exposed substances. Each year, the damage costs billions of dollars. The cost to the environment is also high (**Figure 10.18**). Many rivers and lakes receiving acid rain become less capable of sustaining life. Much vegetation that receives acid rain doesn't survive. This is particularly evident in heavily industrialized regions.

(a)

(b)

(c)

▲ **Figure 10.18**

(a), (b) These two photographs show the same obelisk in New York City's Central Park before and after the effects of acid rain. (c) Many forests downwind from heavily industrialized areas, such as in the northeastern United States and in Europe, have been noticeably hard-hit by acid rain.

◀ Figure 10.19
(a) The damaging effects of acid rain do
not appear in bodies of fresh water lined
with calcium carbonate, which neutralizes
any acidity. (b) Lakes and rivers lined with
inert materials are not protected.

①　Rain is acidified as it
falls through the air.

②　Acid enters lake from rain.

③　Hydronium ions are neutralized
by calcium carbonate released
from limestone.

$CaCO_3$　H_3O^+

Limestone

(a)　$2\,H_3O^+ + CaCO_3 \longrightarrow 3\,H_2O + CO_2 + Ca^{2+}$

①　Rain is acidified as it
falls through the air.

②　Acid enters lake from rain.

③　Hydronium-ion concentration
increases, with potential
harm to the ecosystem.

H_3O^+

H_3O^+　H_3O^+

(b)　Granite rock

CONCEPTCHECK

When sulfuric acid, H_2SO_4, is added to water, what makes the resulting
aqueous solution corrosive?

CHECK YOUR ANSWER Because H_2SO_4 is a strong acid, it readily forms
hydronium ions when dissolved in water. Hydronium ions are responsible for the
corrosive action.

The environmental impact of acid rain depends on local geology, as
Figure 10.19 illustrates. In certain regions, such as the midwestern United
States, the ground contains significant quantities of the alkaline compound
calcium carbonate (limestone), deposited when these lands were submerged
under oceans, as has occurred several times over the past 500 million years. Acid
rain pouring into these regions is often neutralized by the calcium carbonate
before any damage is done. (**Figure 10.20** shows calcium carbonate neutralizing
an acid.) In the northeastern United States and many other regions, however,
the ground contains very little calcium carbonate and is composed primarily of
chemically less reactive materials, such as granite. In these regions, the effect of
acid rain on lakes and rivers accumulates.

One demonstrated solution to this problem is to raise the pH of acidified
lakes and rivers by adding calcium carbonate—a process known as liming. The
cost of transporting the calcium carbonate, coupled with the need to monitor
treated water systems closely, limits liming to only a small fraction of the vast
number of water systems already affected. Furthermore, as acid rain continues
to pour into these regions, the need to lime continues.

▶ Figure 10.20
Most chalks are made from calcium carbonate, which is the same chemical found in limestone.
The addition of even a weak acid, such as the acetic acid of vinegar, produces hydronium ions
that react with the calcium carbonate to form several products, the most notable being carbon
dioxide, which rapidly bubbles out of solution. Try this for yourself! If the bubbling is not as
vigorous as shown here, then the chalk is made of other mineral components.

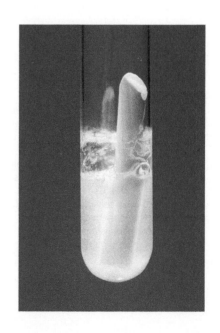

READINGCHECK

What is a longer-term solution to the problem of acid rain?

A longer-term solution to acid rain is to prevent most of the generated sulfur dioxide and other pollutants from entering the atmosphere in the first place. Toward this end, smokestacks have been designed or retrofitted to minimize the quantities of pollutants released. Although costly, the positive effects of these adjustments have been demonstrated. An ultimate long-term solution, however, would be a shift from fossil fuels to cleaner energy sources, such as nuclear and solar energy.

CONCEPTCHECK

What kind of lakes are most protected against acid rain?

CHECK YOUR ANSWER Lakes that have a floor consisting of basic minerals, such as limestone, are more resistant to acid rain, because the chemicals of the limestone (mostly calcium carbonate, $CaCO_3$) neutralize any incoming acid.

10.6 Carbon Dioxide Acidifies the Oceans

LEARNING OBJECTIVE

Describe the impact atmospheric carbon dioxide has on the ocean's pH and mineral composition.

EXPLAIN THIS

Why are colder ocean waters more vulnerable to the effect of carbon dioxide–mediated ocean acidification?

READINGCHECK

What chemical forms when carbon dioxide dissolves in water?

The amount of carbon dioxide put into the atmosphere by human activities is growing. Surprisingly, however, the atmospheric concentration of CO_2 is not increasing proportionately. One explanation has to do with the oceans, as described in **Figure 10.21**. When atmospheric CO_2 dissolves in any body of water—a raindrop, a lake, or an ocean—it forms carbonic acid, H_2CO_3. In fresh water, this carbonic acid transforms back to water and carbon dioxide, which is released back into the atmosphere. In the ocean, however, the carbonic acid is quickly neutralized by dissolved alkaline minerals. (The ocean is alkaline, pH ≈ 8.1.) As discussed in the introduction to this chapter, the products of this neutralization eventually end up on the ocean floor as insoluble solids. Thus, carbonic acid neutralization in the ocean prevents CO_2 from being released back into the atmosphere. The ocean, therefore, is a carbon dioxide *sink*—most of the CO_2 that goes in doesn't come out. Pushing more CO_2 into our atmosphere means pushing more of it into our vast oceans. So far, the oceans have been absorbing about one-third of our CO_2 emissions.

▶ **Figure 10.21**
Carbon dioxide forms carbonic acid upon entering any body of water. In fresh water, this reaction is reversible, and the carbon dioxide is released back into the atmosphere. In the alkaline ocean, the carbonic acid is neutralized to such compounds as calcium bicarbonate, $Ca(HCO_3)_2$, which precipitate to the ocean floor. As a result, most of the atmospheric carbon dioxide that enters our oceans remains there.

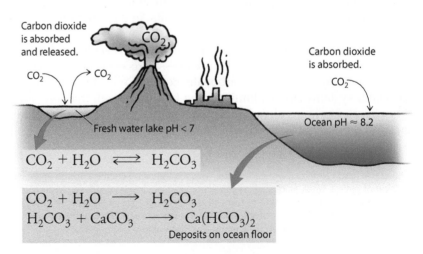

Carbon dioxide is absorbed and released.

Carbon dioxide is absorbed.

Fresh water lake pH < 7

Ocean pH ≈ 8.2

$$CO_2 + H_2O \rightleftharpoons H_2CO_3$$

$$CO_2 + H_2O \longrightarrow H_2CO_3$$
$$H_2CO_3 + CaCO_3 \longrightarrow Ca(HCO_3)_2$$

Deposits on ocean floor

The movement of our CO_2 into the oceans, however, comes at a cost. Notice from Figure 10.21 that the CO_2-derived carbonic acid, H_2CO_3, reacts with calcium carbonate, $CaCO_3$. The addition of carbon dioxide into our oceans, therefore, has the effect of decreasing the amount of calcium carbonate in the ocean water, as well as other carbonates, such as magnesium carbonate, Mg_2CO_3. Coral, shelled organisms, and many other marine species, however, use these carbonates to build and maintain their bodily structures. Importantly, they can only do this when the ocean water is saturated with these minerals, which is currently the case in most surface waters. But the addition of CO_2 leads to ocean water unsaturated with carbonates. In regions where this occurs, the carbonate-based creatures begin to dissolve, which means they perish. This, in turn, can have a significant impact throughout the marine ecosystem, which would affect us as well. Consider this example: pink salmon off the southern coast of Alaska live on a diet of sea snails, also known as pteropods, which are dependent on carbonate minerals. The destruction of this pteropod population by ocean acidification would also mean an end to the Alaskan pink salmon fishing industry.

CHEMICAL CONNECTIONS

How is petroleum connected to the pH of the oceans?

CONCEPTCHECK

How does adding CO_2 to the oceans cause a decrease in the concentration of carbonate ions, $CO_3{}^{2-}$?

CHECK YOUR ANSWER The CO_2 reacts with water, H_2O, to form carbonic acid, H_2CO_3, which effectively removes the carbonate ion, $CO_3{}^{2-}$, by reacting with it to form the bicarbonate ion, $HCO_3{}^{-}$.

Is the effect of human-produced CO_2 on ocean pH measurable? Absolutely, and it is significant. Over the past 100 years, increases in atmospheric carbon dioxide—primarily due to the burning of fossil fuels—has lowered the average pH of the ocean by about 0.1 pH unit. On a geologic time scale, this is lightning fast. The last comparable decrease in ocean pH occurred about 56 million years ago. At that time, major increases in atmospheric carbon dioxide caused the pH of the ocean to decrease by about 0.45 unit. These changes, however, took place over 5000 years for an average decrease in ocean pH of about 0.01 unit per century. The result was a huge die-off of marine organisms, which can be seen as a distinct layer of brown sediment within core samplings of the ocean floor. We are on track for surpassing this event at a rate that is about ten times as fast.

The media give much attention to the role atmospheric carbon dioxide plays in global climate (**Figure 10.22**). We discuss many of the details of this important issue in Chapter 16. You should understand, however, that our production of copious amounts of carbon dioxide is a twofold problem. One is the potential for a not-so-predictable change in global climate. The other is a very predictable change in ocean chemistry. Both need to be considered carefully.

READINGCHECK

When was the last decrease in ocean pH that would have been comparable to what is taking place today?

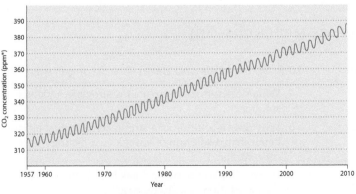

◀ Figure 10.22
Researchers at the Mauna Loa Weather Observatory in Hawaii have recorded increasing concentrations of atmospheric carbon dioxide since they began collecting data in the 1950s. This famous graph is known as the Keeling curve, after the scientist Charles Keeling, who initiated this project and first noted the trends. The oscillations within the Keeling curve reflect seasonal changes in CO_2 levels.

*ppm = parts per million, which tells us the number of carbon dioxide molecules for every million molecules of air.

Chapter 10 Review

LEARNING OBJECTIVES

Identify when a chemical behaves as an acid or a base. (10.1)	→	*Questions 1–3, 32–38*
Describe how the strength of an acid or a base affects the number of ions in solution. (10.2)	→	*Questions 4–6, 29–30, 39–47*
Calculate the pH of a solution given the hydronium-ion concentration. (10.3)	→	*Questions 7–10, 20–21, 23–28, 48–55*
Describe the chemical nature of a buffer solution and how it resists changes in pH. (10.4)	→	*Questions 11–13, 56–59*
Identify sources of acidity in rainwater and explain how this acidity can impact the environment. (10.5)	→	*Questions 14–16, 22, 31, 60–63*
Describe the impact atmospheric carbon dioxide has on the ocean's pH and mineral composition. (10.6)	→	*Questions 17–19, 64–67, 68–70*

SUMMARY OF TERMS (KNOWLEDGE)

Acid A substance that donates hydrogen ions or accepts electron pairs.

Acidic Said of a solution in which the hydronium-ion concentration is higher than the hydroxide-ion concentration.

Amphoteric A description of a substance that can behave as either an acid or a base.

Base A substance that accepts hydrogen ions or donates electron pairs.

Basic Said of a solution in which the hydroxide-ion concentration is higher than the hydronium-ion concentration. Also sometimes called *alkaline*.

Buffer solution A solution that resists large changes in pH, made from either a weak acid and one of its salts or a weak base and one of its salts.

Hydronium ion A polyatomic ion made by adding a proton (hydrogen ion) to a water molecule.

Hydroxide ion A polyatomic ion made by removing a proton (hydrogen ion) from a water molecule.

Neutral Said of a solution in which the hydronium-ion concentration is equal to the hydroxide-ion concentration.

Neutralization A reaction between an acid and a base.

pH A measure of the acidity of a solution, equal to the negative logarithm of the hydronium-ion concentration.

Salt An ionic compound commonly formed from the reaction between an acid and a base.

READING CHECK QUESTIONS (COMPREHENSION)

10.1 Acids Donate Protons and Bases Accept Them

1. What are the Brønsted–Lowry definitions of acid and base?

2. When an acid is dissolved in water, what ion does the water form?

3. When a chemical loses a hydrogen ion, is it behaving as an acid or a base?

10.2 Some Acids and Bases Are Stronger than Others

4. What does this mean: an acid is strong in aqueous solution?

5. Why does a solution of a strong acid conduct electricity better than a solution of a weak acid having the same concentration?

6. When can a solution of a weak base be more corrosive than a solution of a strong base?

10.3 Solutions Can Be Acidic, Basic, or Neutral

7. Is water a strong acid or a weak acid?

8. What is true about the relative concentrations of hydronium and hydroxide ions in an acidic solution? What about a basic solution? A neutral solution?

9. What does the pH of a solution indicate?

10. As the hydronium-ion concentration of a solution increases, does the pH of the solution increase or decrease?

10.4 Buffer Solutions Resist Changes in pH

11. What is a buffer solution?

12. A strong acid quickly drops the pH when it is added to water. This is not the case when a strong acid is added to a buffer solution. Why?

13. Why is it so important that the pH of our blood be maintained within a narrow range of values?

10.5 Rainwater Is Acidic

14. What is the product of the reaction between carbon dioxide and water?

15. What does sulfur dioxide have to do with acid rain?

16. How do humans generate the air pollutant sulfur dioxide?

10.6 Carbon Dioxide Acidifies the Oceans

17. Why are atmospheric levels of carbon dioxide not rising as rapidly as might be expected based on the increased output of carbon dioxide resulting from human activities?

18. As carbon dioxide is absorbed by the ocean, it reacts with water and then carbonate ions to form what?

19. By about how many pH units has the ocean become acidified over the past century?

CONFIRM THE CHEMISTRY (HANDS-ON APPLICATION)

20. Make a concentrated solution of red cabbage extract by boiling a cup of shredded red cabbage in a cup of water for about 5 minutes. For smaller quantities, combine 3 tablespoons of shredded red cabbage with 3 tablespoons of 70 percent isopropyl alcohol and heat in a microwave for 30 seconds. Allow to cool. How might you use red cabbage extract to measure the pH of soil? Of ocean water? What about water from a swimming pool? What about seltzer water? Red cabbage extract can measure pH, but what does pH measure?

21. Make a concentrated solution of red cabbage extract by boiling a cup of shredded red cabbage in a cup of water for about 5 minutes. Pour this extract into a large, transparent container, such as a vase or a 2-liter plastic bottle with the top cut off. Add a tablespoon of white vinegar to acidify the solution. Ask yourself or a classmate what would happen to the pH of this solution if you were to fill the container with plain water. Would the pH go up

or down or stay the same? Test your prediction. As you dilute the solution, the color grows lighter, but what happens to its hue? How can adding plain water change the pH of a solution? Might plain water be used to make this solution alkaline?

22. Add about an inch of water to a large test tube; then add a couple drops of phenolphthalein pH indicator, which you will likely need to obtain from your classroom. Add a small pinch of washing soda, which contains sodium carbonate, Na_2CO_3. Upon mixing, the washing soda turns the solution basic as evidenced by the pink color that forms. Neutralize this base by adding an acid, but not just any acid—use the acid of your breath. Bubble your breath into the solution through a straw until the pink color disappears. What acid are you adding? How does this activity relate to the acidity of rain? Why do you want to add only a small pinch of washing soda and not a tablespoon?

THINK AND SOLVE (MATHEMATICAL APPLICATION)

23. Show that the hydroxide-ion concentration in an aqueous solution is 1×10^{-4} M when the hydronium-ion concentration is 1×10^{-10} M. Recall that $10^a \times 10^b = 10^{(a + b)}$.

24. When the hydronium-ion concentration of a solution is 1×10^{-10} M, what is the pH of the solution? Is the solution acidic or basic?

25. When the hydronium-ion concentration of a solution is 1×10^{-4} M, what is the pH of the solution? Is the solution acidic or basic?

26. Show that an aqueous solution having a pH of 5 has a hydroxide-ion concentration of 1×10^{-9} M.

27. When the pH of a solution is 1, the concentration of hydronium ions is 10^{-1} M = 0.1 M. Assume that the volume of this solution is 500 mL. What is the pH after 500 mL of water is added? You will need a calculator with a logarithm function to answer this question.

28. Show that the pH of a solution is −0.301 when its hydronium-ion concentration equals 2 moles per liter. Is the solution acidic or basic?

THINK AND COMPARE (ANALYSIS)

29. Rank the following solutions in order of increasing concentration of hydronium ions, H_3O^+.

    a. Hydrogen chloride, HCl (concentration = 2 M)
    b. Acetic acid, $C_2H_4O_2$ (concentration = 2 M)
    c. Ammonia, NH_3 (concentration = 2 M)

30. The three chemicals listed next are very weak acids because they have a difficult time losing a hydrogen ion, H^+. Upon losing this hydrogen ion, the central atom of each of these molecules takes on a negative charge. Holding onto this negative charge isn't easy, especially when there are positively charged hydrogen ions, H^+, floating about ready to combine with the negative charge. Review the concept of electronegativity in Section 6.7 and rank the acidity of these molecules in order from strongest to weakest.

    a. Ammonia, NH_3
    b. Water, H_2O
    c. Methane, CH_4

31. Rank in order of decreasing pH: the rain that fell on the Hawaiian island of Kauai on the morning of

    a. January 18, 1778
    b. December 7, 1941
    c. May 8, 2010

    Assume there were no active volcanoes on or around Kauai on these dates.

THINK AND EXPLAIN (SYNTHESIS)

10.1 Acids Donate Protons and Bases Accept Them

32. An acid and a base react to form a salt, which consists of positive and negative ions. Which forms the positive ions: the acid or the base? Which forms the negative ions?

33. Water is formed from the reaction between an acid and a base. Why is water not classified as a salt?

34. Identify the acid or base behavior of each substance in these reactions.

    a. $H_3O^+ + Cl^- \rightleftharpoons H_2O + HCl$

       ____   ____   ____   ____

    b. $H_2PO_4^- + H_2O \rightleftharpoons H_3O^+ + HPO_4^{2-}$

       ____   ____   ____   ____

    c. $HSO_4^- + H_2O \rightleftharpoons OH^- + H_2SO_4$

       ____   ____   ____   ____

    d. $O^{2-} + H_2O \rightleftharpoons OH^- + OH^-$

       ____   ____   ____   ____

35. Does the phosphate ion, a common additive to automatic dishwasher detergent, tend to behave as an acid or a base? Explain.

$$^-O-\overset{\overset{\displaystyle O}{\|}}{\underset{\underset{\displaystyle O^-}{|}}{P}}-O^-$$

Phosphate ion

36. Which reactant behaves as an acid and which behaves as a base in the following reaction?

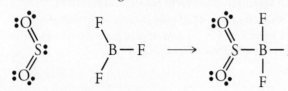

Sulfur dioxide   Boron trifluoride

37. A water molecule, H_2O, becomes strongly held to a dissolved aluminum ion, Al^{3+}. According to the Lewis definition of acids and bases, which is behaving as an acid: the water or the aluminum ion? Which is behaving as a base?

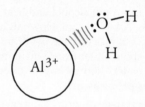

38. Does the hydride ion, H^-, tend to behave as an acid or a base?

10.2 Some Acids and Bases Are Stronger than Others

39. The main component of bleach is sodium hypochlorite, NaOCl, which consists of sodium ions, Na^+, and hypochlorite ions, ^-OCl. What products are formed when this compound is reacted with the hydrochloric acid, HCl, of toilet bowl cleaner?

40. Acetic acid, shown here, has four hydrogen atoms—one bonded to an oxygen and three bonded to a carbon. When this molecule behaves as an acid, it donates only the hydrogen bonded to the oxygen. The hydrogens bonded to the carbon remain in tact. Why?

$$H-\overset{\overset{\displaystyle H}{|}}{\underset{\underset{\displaystyle H}{|}}{C}}-\overset{\overset{\displaystyle O}{\|}}{C}-O-H \longrightarrow H-\overset{\overset{\displaystyle H}{|}}{\underset{\underset{\displaystyle H}{|}}{C}}-\overset{\overset{\displaystyle O}{\|}}{C}-O^-\ H^+$$

bond breaking

Acetic acid

41. How readily an acid donates a hydrogen ion is a function of how well the acid is able to accommodate the resulting negative charge it gains after donating. Of the two structures shown at the top of the next page, which should be the stronger acid: water or hypochlorous acid? Explain.

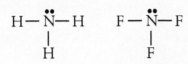

Water   Hypochlorous acid

42. Which should be a stronger base: ammonia, NH_3, or nitrogen trifluoride, NF_3?

Ammonia   Nitrogen trifluoride

43. Why is the H—F bond so much stronger than the H—I bond? (Hint: think atomic size.)

44. Which bond is easier to break: H—F or H—I?

45. Which is the stronger acid: H—F or H—I?

46. Some molecules are able to stabilize a negative charge by passing it from one atom to the next by a flip-flopping of double bonds. This occurs when the negative charge is one atom away from an oxygen double bond as follows. Note that the curved arrows indicate the movement of electrons.

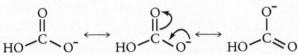

Why then is sulfuric acid so much stronger an acid compared to carbonic acid?

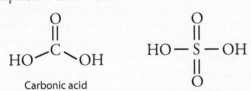

Carbonic acid   Sulfuric acid

47. Suggest a reason why the carbonate ion, CO_3^{2-}, is a stronger base than the bicarbonate ion, HCO_3^-.

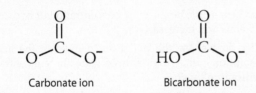

Carbonate ion   Bicarbonate ion

10.3 Solutions Can Be Acidic, Basic, or Neutral

48. Why do we use the pH scale to indicate the acidity of a solution rather than simply state the concentration of hydronium ions?

49. The amphoteric reaction between two water molecules is endothermic, which means the reaction requires the input of heat energy in order to proceed.

$$Energy + H_2O + H_2O \rightleftharpoons H_3O^+ + OH^-$$

The warmer the water, the more heat energy that is available for this reaction and the more hydronium and hydroxide ions that are formed.

a. Which has a lower pH: pure water that is hot or pure water that is cold?

b. Is it possible for water to be neutral but have a pH less than or greater than 7.0?

50. Within a neutral solution of supercritical water (374°C, 218 atm), the pH equals about 2. What is the concentration of hydronium ions within this neutral solution? What is the concentration of hydroxide ions? Why is supercritical water so corrosive?

51. The pOH scale indicates the "basicity" of a solution, where $pOH = -\log[OH^-]$. For any solution, what does the sum pH + pOH always equal?

52. What is the concentration of hydronium ions in a solution that has a pH of −3? Why is such a solution impossible to prepare?

53. Can an acidic solution be made less acidic by adding an acidic solution?

54. Bubbling carbon dioxide into water causes the pH of the water to go down (become more acidic) because of the formation of carbonic acid. Will the pH also drop when carbon dioxide is bubbled into a solution of 1M hydrochloric acid, HCl?

55. What happens to the pH of water as you blow bubbles into it through a drinking straw?

10.4 Buffer Solutions Resist Changes in pH

56. Sodium bicarbonate, $NaHCO_3$, shown here, is the active ingredient of baking soda. Compare this structure with those of the weak acids and weak bases presented in this chapter. Explain how this compound by itself in solution moderates changes in pH.

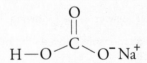

Sodium bicarbonate (salt)

57. Hydrogen chloride is added to a buffer solution of ammonia, NH_3, and ammonium chloride, NH_4Cl. What is the effect on the concentration of ammonia? On the concentration of ammonium chloride?

58. Sodium hydroxide is added to a buffer solution of ammonia, NH_3, and ammonium chloride, NH_4Cl. What is the effect on the concentration of ammonia? On the concentration of ammonium chloride?

59. Why does holding your breath cause the pH of your blood to decrease?

10.5 Rainwater Is Acidic

60. Why is vinegar so good at removing water spots from your bathroom faucet?

61. Pour vinegar onto beach sand from the Caribbean and the result is a lot of froth and bubbles. Pour vinegar onto beach sand from California, however, and nothing happens. Why?

62. How might you tell whether your toothpaste contained calcium carbonate, $CaCO_3$, or perhaps baking soda, $NaHCO_3$, without looking at the ingredients label?

63. Lakes lying in silicon dioxide, SiO_2, containing granite basins tend to become acidified by acid rain more readily than lakes lying in calcium carbonate, $CaCO_3$, limestone basins. Explain why.

10.6 Carbon Dioxide Acidifies the Oceans

64. How might warmer oceans have the effect of slowing down ocean acidification?

65. How does burning fossil fuels lower the pH of the ocean?

66. Why is it that acid raid is not so much of a cause of ocean acidification?

67. Which is the greater problem: acid rain or ocean acidification?

THINK AND DISCUSS (EVALUATION)

68. Scientists have experimented with ways of enhancing the ocean's ability to absorb atmospheric carbon dioxide. They found that adding powdered iron to a small plot of the ocean has the effect of fostering the growth of microorganisms that enhance the rate at which carbon dioxide is absorbed. Might this be a solution to the problem of global climate change? Why or why not?

69. Can industries be trusted to self-regulate the amount of pollution they produce? Is government really necessary to enforce these regulations? Shouldn't the mind of the consumer and the economic advantages of sustainable practices be sufficient to motivate industries to protect the environment?

70. Air pollution regulations can be implemented and enforced by local, state, federal, and international levels of government. Which one of these four levels should be granted the greatest authority? For each level, list the reasons it should be granted the greatest authority.

READINESS ASSURANCE TEST (RAT)

If you have a good handle on this chapter, then you should be able to score at least 7 out of 10 on this RAT. Check your answers online at www.ConceptualChemistry.com. If you score less than 7, you need to study further before moving on.

Choose the BEST answer to the following.

1. What is the relationship between the hydroxide ion and a water molecule?
   a. A hydroxide ion is a water molecule plus a proton.
   b. A hydroxide ion and a water molecule are the same thing.
   c. A hydroxide ion is a water molecule minus a hydrogen nucleus.
   d. A hydroxide ion is a water molecule plus two extra electrons.

2. What happens to the corrosive properties of an acid and a base after they neutralize each other? Why does this happen?
   a. The corrosive properties are neutralized because the acid and base no longer exist.
   b. The corrosive properties are embedded into the non-corrosive salt.
   c. The corrosive properties are doubled because the acid and base are combined in the salt.
   d. The corrosive properties are neutralized by the salt.

3. Sodium hydroxide, NaOH, is a strong base, which means that it readily accepts hydrogen ions. What products are formed when sodium hydroxide accepts a hydrogen ion from a water molecule?
   a. Sodium ions, hydroxide ions, and water
   b. Sodium ions, hydroxide ions, and hydronium ions
   c. Sodium ions and hydronium ions
   d. Sodium ions and water

4. A weak acid is added to a concentrated solution of hydrochloric acid. Does the solution become more or less acidic?
   a. More acidic because more hydronium ions are being added to the solution.
   b. Less acidic because the solution becomes more dilute with a less concentrated solution of hydronium ions being added to the solution.
   c. No change in acidity because the concentration of the hydrochloric acid is too high to be changed by the weak solution.
   d. Less acidic because the concentration of hydroxide ions will increase.

5. Why do we use the pH scale to indicate the acidity of a solution rather than simply state the concentration of hydronium ions?
   a. It includes the concentration of hydronium and hydroxide ions.
   b. It is used because the general public understands it.
   c. It is more accurate to use the pH scale.
   d. It is more convenient because the concentration of hydronium ions is usually so small.

6. When the hydronium-ion concentration equals 1 mole per liter, what is the pH of the solution? Is the solution acidic or basic?
   a. pH = 0; this is an acidic solution.
   b. pH = 1; this is an acidic solution.
   c. pH = 10; this is a basic solution.
   d. pH = 7; this is a neutral solution.

7. Sometimes an individual going through a traumatic experience cannot stop hyperventilating. In this circumstance, it is recommended that the individual breathe into a paper bag or cupped hands as a useful way to avoid an increase in blood pH, which can cause the person to pass out. How does this work?

   a. It helps to slow the person's breathing to prevent excess oxygen from entering the blood.

   b. The paper absorbs CO_2 to prevent the accumulation of excess CO_2 in the blood.

   c. CO_2 initially exhaled is reinhaled to help maintain adequate levels of carbonic acid in the blood.

   d. The paper bag helps to retain the water vapor leaving the mouth to prevent an increase in the concentration of lactic acid in the body.

8. Cutting back on the pollutants that cause acid rain is one solution to the problem of acidified lakes. What is another solution?

   a. Stop using NaCl to salt roads in the winter.

   b. Add a neutralizing substance such as limestone.

   c. Add ammonium ions to the lakes.

   d. Filter the water to remove any acidity in the lakes.

9. Why might a small piece of chalk be useful for alleviating acid indigestion?

   a. Calcium carbonate accepts protons, which will neutralize any excess acids.

   b. Calcium carbonate is able to absorb excess acid and carry it out of the system.

   c. Calcium carbonate is able to add an extra lining to the stomach to protect it from the excess acid.

   d. Calcium carbonate is able to turn the acid into a gas that can relieve the buildup of excess acid.

10. The pH of the ocean is currently dropping at a rapid rate. The last time this happened on a comparable level was about

    a. 600 years ago.

    b. 10,000 years ago.

    c. 56 million years ago.

    d. 540 million years ago.

ANSWERS TO CALCULATION CORNER LOGARITHMS AND PH

1. "What is the logarithm of 10^5?" can be rephrased as "To what power is 10 raised to give the number 10^5?" The answer is 5.

2. You should know that 100,000 is the same as 10^5. Thus, the logarithm of 100,000 is 5.

3. The pH is 9, which means this is a basic solution:

$$\begin{aligned} pH &= -\log[H_3O^+] \\ &= -\log(10^{-9}) \\ &= -(-9) \\ &= 9 \end{aligned}$$

Making Glass

We humans have long tinkered with materials around us and used them to our advantage. Moldable wet clay, for example, was found to harden to ceramic when heated by fire. Around 5000 B.C., pottery fire pits gave way to furnaces hot enough to convert copper ores to metallic copper and tin ores to metallic tin. The mixing of molten copper and tin created bronze, which is a durable alloy useful for the making of swords, plows, and other artifacts of the so-called Bronze Age. By 1200 B.C., hotter furnaces were converting iron ores to the even more durable iron, ushering in the Iron Age.

The metal ores used by early metalworkers contained many impurities, such as sand, which would also melt within the hot furnaces. Upon cooling, these impurities would solidify to a material known as *slag*, which was similar to the highly prized volcanic black glassy mineral known as *obsidian*. By tinkering with the ingredients, slags of many colors were produced, including some that were transparent. When set to a useful purpose, such as in the making of beads, containers, or windows, the hardened slag became known as *glass*. Over many centuries, glass-making gradually improved. Clearer and stronger glass made possible many important inventions, such as those shown in Figure 1.

Eyeglasses were first made in northern Italy in the 13th century. Their use quickly spread, and the effect on society was significant, especially in that eyeglasses allowed people to read and remain intellectually active throughout their lives. Another significant achievement made possible by the development of glass lenses was the telescope, which in 1609 was pointed skyward by Galileo, who observed moons orbiting the planet Jupiter and opened the door to a

▲ **Figure 1**
Eyeglasses, the telescope, and the distillation apparatus—pivotal inventions made from glass.

golden age of astronomy. Improved glass also made possible the design of distillation equipment used to isolate alcohol from fermented broths. Distilled spirits, as they were called, became noted not only for their intoxicating effects but also for their ability to disinfect and promote the healing of wounds.

Glass is an *amorphous material*. In such a substance, submicroscopic units are randomly oriented relative to one another. This is different from a *crystal*, in which submicroscopic units are arranged in a periodic and orderly fashion (Section 6.3). Glass can have a variety of chemical compositions. This allows the glass maker to customize the chemical and physical

properties of the glass and then process it into all kinds of products. Common glass is a mixture of sodium silicate, NaSiO, and calcium silicate, CaSiO. It is formed by heating a blend of sodium carbonate, calcium carbonate, and silica:

$$Na_2CO_3 \;+\; SiO_2 \;\rightarrow$$
Sodium carbonate   Silica

$$Na_2SiO_3 \;+\; CO_2$$
Sodium silicate   Carbon dioxide

$$CaCO_3 \;+\; SiO_2 \;\rightarrow$$
Calcium carbonate   Silica

$$CaSiO_3 \;+\; CO_2$$
Calcium silicate   Carbon dioxide

Various additives provide glass with special properties. Adding potassium oxide, K_2O, for example, makes a very hard glass commonly used for optical applications. When lead oxides are added, glass becomes dense and has a high refractive index, which means it readily bends white light into a rainbow of colors. Such glass is called *crystal glass*, because these properties resemble those of true crystals, such as quartz, which is a crystalline form of silica, SiO_2. Adding boron oxide, B_2O_3, markedly lowers the rate at which glass expands when it is warmed and contracts when it is cooled. This enables the glass to withstand sudden changes in temperature without breaking. Such glass is known as *Pyrex® glass* and is used in laboratory glassware and cooking utensils.

Glass can be colored by adding various chemicals. Cobalt oxide yields a blue glass used in specialty dinnerware. Adding the element selenium gives glass a red color, and the element iridium makes glass black. Glassmaking artists have experimented with chemical additives to produce a variety of colored glasses, such as those shown in Figure 2.

▲ Figure 2
Glass as an art form.

The physical properties of glass are wholly remarkable. It is easily melted, can be poured or blown into almost any shape, and retains that shape when it cools. Moreover, pieces of glass can be melted together to become sealed without glue or cement. This allows the manufacture of many intricate glassware designs. Glass is transparent and resistant to even the most corrosive chemicals, which is why chemical reactions in the laboratory are usually carried out in glass containers. Long, thin strands of glass are flexible enough to be encased in cables that can stretch for hundreds of kilometers. These are the *fiber-optic cables* shown in Figure 3, through which pulses of light can travel, carrying information data with remarkable efficiency.

Glass science and technology remains a very active area of research. For example, what sort of glass should be used to encase nuclear waste? Archeologists are looking into this question by studying the integrity of glass obtained from ships that sunk thousands of years ago. In medicine, borate glass fibers have been found to promote the healing of deep skin wounds, including ulcers. Also, radioactive glass micro-beads that selectively kill cells within a cancerous tumor have been developed. These beads are exactly the right size to become lodged within the tumor's blood vessels, which tend to be unusually narrow.

Fortunately, the starting materials for glass are abundant, so unlike the situation with paper, plastics, and metals, we are not in imminent danger of a glass shortage. Resources can still be saved, however, by recycling glass. The energy needed to produce glass from recycled products is only 70 percent of the energy needed to produce it from raw materials. Glass by any standard has been one of humankind's best bargains.

▼ Figure 3
Information-bearing light pulses travel through the glass of fiber-optic cables, which provided a revolution in long-distance communication.

Think and Discuss

1. Commercial beverages are transported in both plastic and glass containers. How might it be that the glass containers are responsible for the release of more atmospheric pollution than the plastic containers?

2. Is transparent glass a homogeneous or heterogeneous mixture? Please explain.

3. Wet clay is a mixture of microcapsules of aluminum oxides and silicon oxides surrounded by water. It can be shaped easily because the water serves as a lubricant that allows the microcapsules to slip over one another. When dry, the microcapsules become locked in position and the clay holds its shape. Dried clay, however, easily breaks. Why does the clay become much more durable after being heated to very high temperatures within a kiln?

4. What forms when the ingredients for a glass are coated onto clay pottery and heated to high temperatures within a kiln?

5. Pretend you have just earned a college degree in chemistry. You have job offers from a paper mill ($40K), a company that produces plastics ($35K), a steel mill ($60K), and a start-up glassblowing company that produces specialty laboratory glassware ($32K). You also have the option of going to graduate school where you will receive a tuition waiver along with an annual stipend of $15K. Which do you choose? What factors besides salary and type of employment contribute to your choice?

▲ The chemical reactions going on in your body are quite similar to those going on within burning wood. In both cases, the products are carbon dioxide, water, and energy.

11

Oxidations and Reductions Charge the World

THE MAIN IDEA

Oxidation is the loss of electrons, and reduction is the gain of electrons.

What do our bodies have in common with the burning of a campfire or the rusting of old farm equipment? Why does silver tarnish? How can aluminum restore tarnished silver? Why is it unwise for people with metal fillings in their teeth to bite down on aluminum foil? How are metals produced from minerals? How do batteries work, and what is their source of energy? What are fuel cells, and how do they generate electricity so efficiently? How do photovoltaic cells convert sunlight into electricity? Why is hydrogen such an environmentally friendly fuel but not itself a source of energy? The answers to all these questions involve a class of reactions called *oxidation-reduction* reaction. These reactions are similar to the acid-base reactions you learned about in the previous chapter in that they both involve the transfer of subatomic particles between reactants. For acid-base reactions, these subatomic particles are typically protons. Oxidation-reduction reactions, by contrast, involve the transfer of electrons.

Chemistry

The Penny Copper Nail

Copper metal slowly reacts with the oxygen in air to form reddish copper (I) oxide, Cu_2O, which is a compound that coats the surface of older pennies, making them look tarnished. When such a penny is placed in a solution of salt in vinegar, the copper (I) oxide acts as a base and reacts with the vinegar to form copper salts, which dissolve in the vinegar. This effectively cleans the penny. The copper salts can then be transformed back into copper metal when exposed to an iron nail.

PROCEDURE

1. Stir about half a teaspoon of salt into about half a cup of white distilled vinegar. Use a nonmetal container such as a ceramic or plastic bowl.
2. Dip a tarnished penny halfway into the solution and notice the rapid cleaning effect.
3. Add at least a dozen tarnished pennies to the solution. As they are cleaned, the concentration of copper ions in solution will increase.
4. Sandpaper an iron nail to give it a clean surface and then rest the nail in the vinegar solution for about 10 minutes. Watch for the formation of copper metal on the nail.

ANALYZE AND CONCLUDE

1. Are copper ions positively or negatively charged?
2. What is the charge on the copper atoms within Cu_2O? What is the charge on the oxygen?
3. What must copper ions gain in order to transform back into a metallic form?
4. What was the iron of the nail able to do for the copper ions?

11.1 Losing and Gaining Electrons

EXPLAIN THIS

Why is the chlorine atom such a strong oxidizing agent?

Oxidation is the process whereby a reactant loses one or more electrons. **Reduction** is the opposite process, whereby a reactant gains one or more electrons. Oxidation and reduction are complementary processes that occur at the same time. They always occur together; you cannot have one without the other. The electrons lost by one chemical in an oxidation reaction don't simply disappear; they are gained by another chemical in a reduction reaction.

An oxidation-reduction reaction occurs when sodium and chlorine react to form sodium chloride, as shown in **Figure 11.1**. The equation for this reaction is

$$2Na(s) + Cl_2(g) \rightarrow 2NaCl(s) + heat$$

To see how electrons are transferred in this reaction, we can look at each reactant individually. Each electrically neutral sodium atom changes to a positively charged ion. At the same time, we can say that each atom loses an electron and is therefore oxidized:

$$2Na(s) \rightarrow 2Na^+ + 2e^- \qquad \text{Oxidation}$$

Each electrically neutral chlorine molecule changes to two negatively charged ions. Each of these atoms gains an electron and is therefore reduced:

$$Cl_2(g) + 2e^- \rightarrow 2Cl^- \qquad \text{Reduction}$$

LEARNING OBJECTIVE

Identify when a chemical undergoes oxidation or reduction.

READING CHECK

Is it possible for oxidation to occur without reduction?

▲ **Figure 11.1**
In the exothermic formation of sodium chloride, sodium metal is oxidized by chlorine gas and chlorine gas is reduced by sodium metal.

The net result is that the two electrons lost by the sodium atoms are transferred to the chlorine atoms. Therefore, each of the two equations shown actually represents one-half of an entire process, which is why they are each a **half-reaction.** In other words, an electron won't be lost from a sodium atom without the presence of a chlorine atom available to pick up that electron. Both half-reactions are required to represent the *whole* oxidation-reduction process. Half-reactions are useful for showing which reactant loses electrons and which reactant gains them, which is why half-reactions are used throughout this chapter.

Because the sodium causes reduction of the chlorine, the sodium is acting as a *reducing agent.* A reducing agent is any reactant that causes another reactant to be reduced. Note that sodium is oxidized when it behaves as a reducing agent—it loses electrons. Conversely, the chlorine causes oxidation of the sodium and so is acting as an *oxidizing agent.* Because it gains electrons in the process, an oxidizing agent is reduced. Just remember that loss of **e**lectrons is **o**xidation and **g**ain of **e**lectrons is **r**eduction. Here is a helpful mnemonic adapted from a once-popular children's story: **Leo** the lion went "**ger.**" Another widely used mnemonic is OIL RIG—**o**xidation **i**s **l**oss and **r**eduction **i**s **g**ain.

Different elements have different oxidation and reduction tendencies. For example, metals lose electrons readily, while most nonmetals tend to gain electrons, as **Figure 11.2** illustrates.

CONCEPT CHECK

True or false:

1. Reducing agents are oxidized in oxidation-reduction reactions.

2. Oxidizing agents are reduced in oxidation-reduction reactions.

CHECK YOUR ANSWERS   Both statements are true.

Whether a reaction is classified as an oxidation-reduction reaction is not always immediately apparent. The chemical equation, however, can provide some important clues. First, look for changes in the ionic states of elements. Sodium metal, for example, consists of neutral sodium atoms. In the formation of sodium chloride, these atoms transform into positively charged sodium ions; this occurs as sodium atoms lose electrons (oxidation). A second way to identify a reaction as an oxidation-reduction reaction is to determine whether an element is gaining or losing oxygen atoms. As the element gains the oxygen atom, it is losing electrons to that oxygen, because of the oxygen's high electronegativity. The gain of oxygen, therefore, is oxidation (loss of electrons), while the loss of oxygen is reduction (gain of electrons). For example, hydrogen, H_2, reacts with oxygen, O_2, to form water, H_2O, as follows:

$$H–H + H–H + O=O \rightarrow H–O–H + H–O–H$$

Note that the element hydrogen becomes attached to an oxygen atom through this reaction. The hydrogen, therefore, is oxidized.

A third way to identify a reaction as an oxidation-reduction reaction is to see whether an element is gaining or losing hydrogen atoms. The gain of hydrogen is reduction, while the loss of hydrogen is oxidation. For the formation of water shown previously, we see that the element oxygen is gaining hydrogen atoms, which means that the oxygen is being reduced—that is, the oxygen is gaining electrons from the hydrogen, which is why the oxygen atom within water is slightly negative, as discussed in Section 6.8. The three ways of identifying a reaction as an oxidation-reduction type of reaction are summarized in **Figure 11.3**.

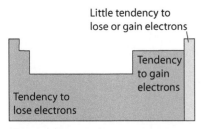

Little tendency to lose or gain electrons

Tendency to gain electrons

Tendency to lose electrons

☐ More likely to behave as oxidizing agent (be reduced)

☐ More likely to behave as reducing agent (be oxidized)

☐ Noble gas elements resist both oxidation and reduction

▲ **Figure 11.2**
The ability of an atom to gain or lose electrons is indicated by its position in the periodic table. Those at the upper right (nonmetals) tend to gain electrons, and those at the lower left (metals) tend to lose them.

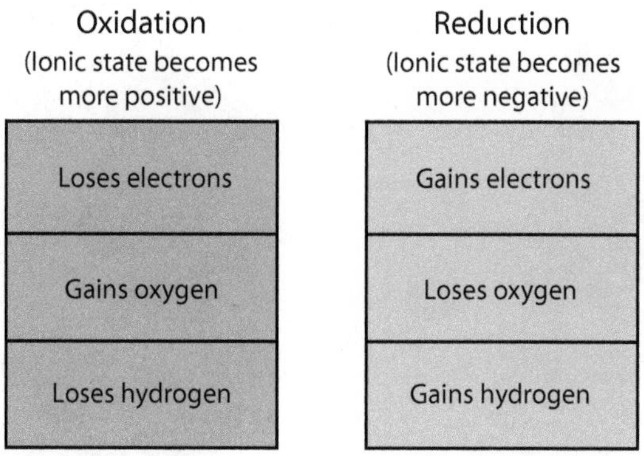

Oxidation
(Ionic state becomes
more positive)

| Loses electrons |
| Gains oxygen |
| Loses hydrogen |

Reduction
(Ionic state becomes
more negative)

| Gains electrons |
| Loses oxygen |
| Gains hydrogen |

◀ **Figure 11.3**
Oxidation results in a greater positive charge, which can be achieved by losing electrons, gaining oxygen atoms, or losing hydrogen atoms. Reduction results in a greater negative charge, which can be achieved by gaining electrons, losing oxygen atoms, or gaining hydrogen atoms.

CONCEPTCHECK

In the following equation, is carbon oxidized or reduced?

$$CH_4 \;+\; 2\,O_2 \;\rightarrow\; CO_2 \;+\; 2\,H_2O$$

CHECK YOUR ANSWER As the carbon of methane, CH_4, forms carbon dioxide, CO_2, it is losing hydrogen and gaining oxygen, which tells us that the carbon is being oxidized.

11.2 Harnessing the Energy of Flowing Electrons

EXPLAIN THIS

Why must a salt bridge also contain water?

Electrochemistry is the study of the relationship between electrical energy and chemical change. It involves either the use of an oxidation-reduction reaction to produce an electric current or the use of an electric current to produce an oxidation-reduction reaction.

To understand how an oxidation-reduction reaction can generate an electric current, consider what happens when a reducing agent is placed in direct contact with an oxidizing agent: electrons flow from the reducing agent to the oxidizing agent. This flow of electrons is an electric current, which is a form of kinetic energy that can be harnessed for useful purposes.

Iron atoms, Fe, for example, are better reducing agents than are copper ions, Cu^{2+}. So when a piece of iron metal and a solution containing copper ions are placed in contact with each other, electrons flow from the iron to the copper ions, as **Figure 11.4** illustrates. The result is the oxidation of iron atoms and the reduction of copper ions.

The elemental iron and copper ions need not be in physical contact for electrons to flow between them. If they are in separate containers but bridged by a conducting wire, the electrons can flow from the iron through the wire to the copper ions. The resulting electric current in the wire can be attached to some useful device, such as a lightbulb. But alas, an electric current is not sustained by this arrangement.

The reason the electric current is not sustained is shown in **Figure 11.5**. An initial flow of electrons through the wire immediately results in a buildup of electric charge in both containers. The container on the left builds up positive charge as it accumulates Fe^{2+} ions from the nail. The container on the right builds up negative charge as electrons accumulate on this side. This situation prevents

LEARNING OBJECTIVE

Show how electricity can be generated using materials that tend to lose or gain electrons.

READINGCHECK

Why won't electrons flow continuously through a wire connecting iron and copper in two containers?

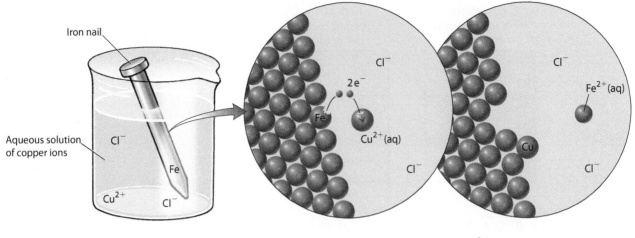

Iron nail

Aqueous solution of copper ions

Cl^-

$2e^-$

Fe

Cu^{2+}(aq)

Cl^-

Cl^-

Fe^{2+}(aq)

Cl^-

Cu

Cl^-

Oxidation   Fe $\longrightarrow$ $Fe^{2+} + 2\,e^-$

Reduction $Cu^{2+} + 2\,e^- \longrightarrow$ Cu

▲ **Figure 11.4**
A nail made of iron placed in a solution of Cu^{2+} ions oxidizes to Fe^{2+} ions, which dissolve in the water. At the same time, copper ions are reduced to metallic copper, which coats the nail. (Negatively charged ions, such as chloride ions, Cl^-, must also be present to balance these positively charged ions in solution.)

any further migration of electrons through the wire. Recall that electrons are negative, so they are repelled by the negative charge in the right container and attracted to the positive charge in the left container. The net result is that the electrons do not flow through the wire, and the bulb remains unlit. Put it this way: if you were an electron in a positively charged container, would YOU want to leave to go to another chamber that was negatively charged? Absolutely not. Remember, opposite signs attract and like signs repel.

The solution to this problem is to allow ions to migrate between containers so that neither builds up any positive or negative charge. This is accomplished with a *salt bridge*, which may be as simple as a paper towel soaked in salt water

▶ **Figure 11.5**
An iron nail is placed in water and connected by a conducting wire to a solution of copper ions. Not much happens, because this arrangement results in a buildup of charge that prevents the further flow of electrons.

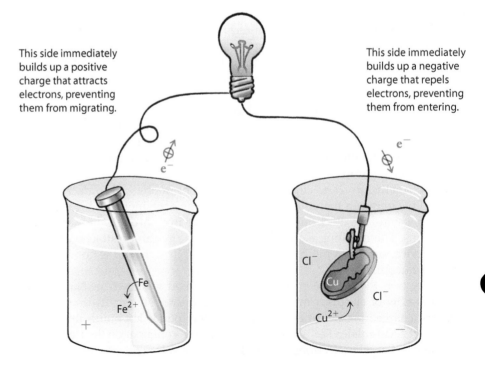

This side immediately builds up a positive charge that attracts electrons, preventing them from migrating.

This side immediately builds up a negative charge that repels electrons, preventing them from entering.

e^-

e^-

Cl^-

Fe

Fe^{2+}

+

Cu

Cl^-

Cu^{2+}

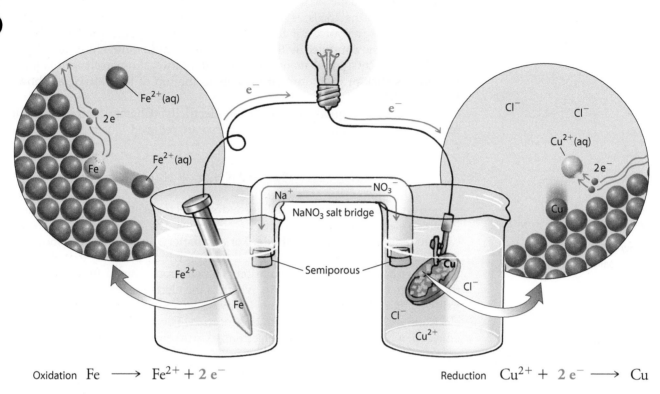

$$\text{Oxidation} \quad Fe \longrightarrow Fe^{2+} + 2\,e^-$$

$$\text{Reduction} \quad Cu^{2+} + 2\,e^- \longrightarrow Cu$$

▲ **Figure 11.6**
The salt bridge completes the electric circuit. Electrons freed as the iron is oxidized pass through the wire to the container on the right. Nitrate ions, NO_3^-, from the salt bridge flow into the left container to balance the positive charges of the Fe^{2+} ions that form, thereby preventing any buildup of positive charge. Meanwhile, Na^+ ions from the salt bridge enter the right container to balance the Cl^- ions "abandoned" by the Cu^{2+} ions as the Cu^{2+} ions pick up electrons to become metallic copper.

stretching between both containers. Commonly, a salt bridge is made of a U-shaped tube filled with a paste of salts, such as sodium nitrate, $NaNO_3$, and closed with semiporous plugs. **Figure 11.6** shows how such a salt bridge allows ions to enter either container as necessary. This creates a complete electric circuit, which permits the flow of electrons through the conducting wire.

11.3 Batteries Consume Chemicals to Generate Electricity

EXPLAIN THIS

Why is lithium a preferred metal for making batteries?

So we can see that, with the proper setup, it is possible to harness electrical energy from an oxidation-reduction reaction. The apparatus shown in Figure 11.6 is one example. Such devices are called *voltaic cells*. Instead of two containers, a voltaic cell can be an all-in-one self-contained unit, in which case it is called a *battery.* Batteries are either disposable or rechargeable, and here we explore some examples of each. Although the two types differ in design and composition, they function by the same principle: two materials that oxidize and reduce each other are connected by a medium through which ions travel to balance an external flow of electrons.

LEARNING OBJECTIVE

Describe how oxidation and reduction occur within a device that generates electricity.

READINGCHECK

What is the principle by which any battery works?

Let's look at disposable batteries first. The common *dry-cell battery*, which was invented in the 1860s, is still used today, and it is probably the cheapest disposable energy source for flashlights, toys, and the like. The basic design consists of a zinc cup filled with a thick paste of ammonium chloride, NH_4Cl; zinc chloride, $ZnCl_2$; and manganese dioxide, MnO_2. Immersed in this paste is a porous stick of graphite that projects to the top of the battery, as shown in **Figure 11.7**.

Graphite is a good conductor of electricity. Chemicals in the paste receive electrons at the graphite stick and so are reduced. The reaction for the ammonium ions is

$$2\,NH_4{}^+(aq) + 2\,e^- \rightarrow 2\,NH_3(g) + H_2(g) \quad \text{Reduction}$$

CHEMICAL CONNECTIONS

How is a flashlight connected to a coal-fired power plant?

An **electrode** is any material that conducts electrons into or out of a medium in which electrochemical reactions are occurring. The electrode where chemicals are reduced is called a **cathode.** For any battery, such as the one shown in Figure 11.7, the cathode is always positive (+), which indicates that electrons are naturally attracted to this location. The electrons gained by chemicals at the cathode originate at the **anode,** which is the electrode where chemicals are oxidized. For any battery, the anode is always negative (–), which indicates that electrons are streaming away from this location. The anode in Figure 11.7 is the zinc cup, where zinc atoms lose electrons to form zinc ions:

$$Zn(s) \rightarrow Zn^{2+}(aq) + 2\,e^- \quad \text{Oxidation}$$

The reduction of ammonium ions in a dry-cell battery produces two gases—ammonia, NH_3, and hydrogen, H_2—that need to be removed to avoid a pressure buildup and a potential explosion. Removal is accomplished by having the ammonia and hydrogen react with the zinc chloride and manganese dioxide:

$$ZnCl_2(aq) + 2\,NH_3(g) \rightarrow Zn(NH_3)_2Cl_2(s)$$

$$2\,MnO_2(s) + H_2(g) \rightarrow Mn_2O_3(s) + H_2O(l)$$

The life of a dry-cell battery is relatively short. Oxidation causes the zinc cup to deteriorate, and eventually, the contents leak out. Even while the battery

▶ **Figure 11.7**
A common dry-cell battery with a graphite rod immersed in a paste of ammonium chloride, manganese dioxide, and zinc chloride in a zinc cup.

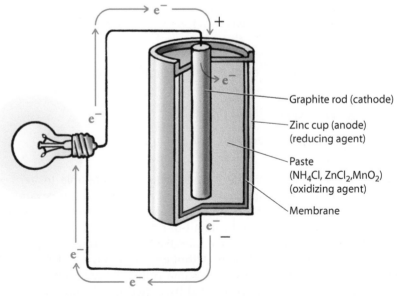

Reduction $2\,NH_4{}^+ + 2e^- \longrightarrow 2\,NH_3 + H_2$

Graphite rod (cathode)

Zinc cup (anode) (reducing agent)

Paste (NH_4Cl, $ZnCl_2$, MnO_2) (oxidizing agent)

Membrane

Oxidation $Zn \longrightarrow Zn^{2+} + 2e^-$

is not operating, the zinc corrodes as it reacts with ammonium ions. This zinc corrosion can be inhibited by storing the battery in a refrigerator. As discussed in Section 9.6, chemical reactions slow down with decreasing temperature. Chilling a battery, therefore, slows down the rate at which the zinc corrodes, which increases the life of the battery.

Another type of disposable battery, the more expensive *alkaline battery*, shown in **Figure 11.8**, avoids many of the problems of dry-cell batteries by operating in a strongly alkaline paste. In the presence of hydroxide ions, the zinc is oxidized to insoluble zinc oxide:

$$\text{Zn(s)} + 2\,\text{OH}^-\text{(aq)} \rightarrow \text{ZnO(s)} + \text{H}_2\text{O(l)} + 2\,\text{e}^- \quad \text{Oxidation}$$

At the same time, manganese dioxide is reduced:

$$2\,\text{MnO}_2\text{(s)} + \text{H}_2\text{O(l)} + 2\,\text{e}^- \rightarrow \text{Mn}_2\text{O}_3\text{(s)} + 2\,\text{H}_2\text{O(l)} \quad \text{Reduction}$$

Note how these two reactions avoid the use of the zinc-corroding ammonium ion (which means alkaline batteries last a lot longer than dry-cell batteries) and prevent formation of any gaseous products. Furthermore, these reactions are better suited to maintaining a given voltage during longer periods of operation.

The small mercury and lithium disposable batteries used for calculators and cameras are variations of the alkaline battery. In the mercury battery, mercuric oxide, HgO, rather than manganese dioxide, is reduced. Manufacturers are phasing out these batteries because of the environmental hazard posed by mercury, which is poisonous. In the lithium battery, lithium metal, rather than zinc, is used as the source of electrons. Lithium not only is able to maintain a higher voltage than zinc but also is about one-thirteenth as dense, which makes for a lighter battery.

Disposable batteries have relatively short lives, because electron-producing chemicals are consumed in their use. The main feature of *rechargeable* batteries is the reversibility of the oxidation and reduction reactions. A noteworthy example is the nickel metal hydride, NiMH, battery. Charging this battery causes the nickel metal to extract hydrogen from water to form the negatively charged hydride ion, shown next as H:, where the two dots represent two electrons.[*]

H_2O	+	Ni	+	e^-	$\rightarrow$	H: Ni	+	HO^-
water		nickel metal				nickel hydride		hydroxide ion

The role of the nickel is to stabilize the two electrons on the hydrogen, which because it contains an added electron is called the *hydride* ion, much the way chlorine with an extra electron is called the *chloride* ion, using the *-ide* suffix. A fully charged battery thus contains an abundance of nickel hydride. As the battery provides electricity, the hydride ion releases electrons, which allows it to join with the hydroxide ion to re-form water:

H: Ni	+	HO^-	$\rightarrow$	H_2O	+	Ni	+	e^-
nickel hydride		hydroxide ion		water		nickel metal		

So recharging a rechargeable battery simply means regenerating the chemicals that can release electrons on demand. For the NiMH battery, this chemical is nickel hydride, H:Ni. For a traditional car battery, this chemical is simply lead, Pb, which transforms into lead sulfate, PbSO_4, as it releases electrons. As the car battery is recharged, the PbSO_4 is transformed back into lead, Pb. This is an endothermic reaction, which requires the input of energy that comes from an on-board generator, also known as the alternator. The alternator, in turn, gets its energy from the car's internal-combustion engine, which is driven by the energy of the fuel in the gas tank.

[*]The nickel in this reaction is actually an intermetallic compound of nickel and various rare-earth elements, such as lanthanum, La.

▲ Figure 11.8
Alkaline batteries last a lot longer than dry-cell batteries and give a steadier voltage, but they are more expensive.

 **FOR YOUR INFORMATION**

A new generation of batteries are now being developed in which the anode is made of a 3-dimensional network of nanowires that cover a surface area 10,000 times greater than a traditional battery. These batteries will hold more energy per unit mass and can be re-charged in a matter of minutes, rather than hours. For the latest information, visit PrietoBattery.com.

▲ **Figure 11.9**
As of 2012, over 2.5 million Prius hybrids have been sold worldwide. Look now for hybrid vehicles that can be plugged into your home electrical outlet, charged at night, and driven the next day using no gasoline for up to 60 miles.

 FOR YOUR INFORMATION

As a battery provides electricity, electrons move from the negative anode to the positive cathode. When a rechargeable battery is being recharged, however, what was once the negative anode now becomes a negative cathode, which is where electrons are needed for reduction. Because electrons won't travel to a negative cathode on their own, they must be forced to do so. Hence, recharging is an energy-consuming process.

Rechargeable lithium-ion batteries have found a wide range of applications, from computer laptops to cell phones. Safer lithium phosphate ion batteries are used for hybrid cars, such as the popular Toyota Prius shown in **Figure 11.9**. Hybrids have improved gas mileage because as the car slows down, its kinetic energy is transformed into the electric potential energy of the battery rather than being wasted as heat from the car's brake pads. The captured electrical energy of the battery is subsequently used to assist the gas-powered engine to get the car moving. Also, the hybrid's battery system allows the engine to shut off when the car is merely idling or moving slowly, as occurs in heavy traffic.

Continued improvements in battery technology are permitting the production of next-generation hybrids, known as plug-in hybrids, which have much larger batteries and smaller fuel tanks. These hybrids can be plugged into electrical outlets, charged overnight, and driven the next day for up to 60 miles without using any gasoline. This is significant because the typical driver in the United States drives less than 40 miles a day. Furthermore, utility companies can sell electricity at cheaper rates at night, because their massive generators are underutilized at that time. Alternatively, the plug-in hybrid can be charged via residential photovoltaic panels or a small wind turbine. Cars that remained plugged in during the day would be available to contribute energy back into the grid during peak demand—owners of such cars could even be rebated for this energy. Furthermore, during an electricity blackout, a family's plug-in hybrid would store enough energy to serve as emergency backup for household needs. Plug-in hybrids, with their large and highly efficient batteries, offer much to help move individuals and the nation as a whole toward energy conservation and independence.

CONCEPT CHECK

What chemicals are produced as a nickel metal hydride battery is recharged?

CHECK YOUR ANSWER Nickel hydride, H:Ni, and hydroxide ions, HO⁻.

11.4 Fuel Cells Consume Fuel to Generate Electricity

LEARNING OBJECTIVE

Identify how a fuel cell generates electricity.

 **READING CHECK**

What is a fuel cell?

EXPLAIN THIS

What is the net balanced equation for the fuel cell depicted in Figure 11.12?

A *fuel cell* is a device that converts the energy of a fuel to electrical energy. Fuel cells are an efficient means of generating electricity. A hydrogen–oxygen fuel cell is shown in **Figure 11.10**. It has two compartments, one for entering hydrogen fuel and the other for entering oxygen fuel, separated by a set of porous electrodes. Hydrogen is oxidized upon contact with hydroxide ions at the hydrogen-facing electrode (the anode). The electrons from this oxidation flow through an external circuit and provide electric power before meeting up with oxygen at the oxygen-facing electrode (the cathode). The oxygen readily picks up the electrons (in other words, the oxygen is reduced) and reacts with water to form hydroxide ions. To complete the circuit, these hydroxide ions migrate across the porous electrodes and through an ionic paste of potassium hydroxide, KOH, to join with hydrogen at the hydrogen-facing electrode.

As the oxidation equation shown at the top of Figure 11.10 demonstrates, the hydrogen and hydroxide ions react to produce energetic water molecules that arise in the form of steam. This steam may be used for heating or for generating

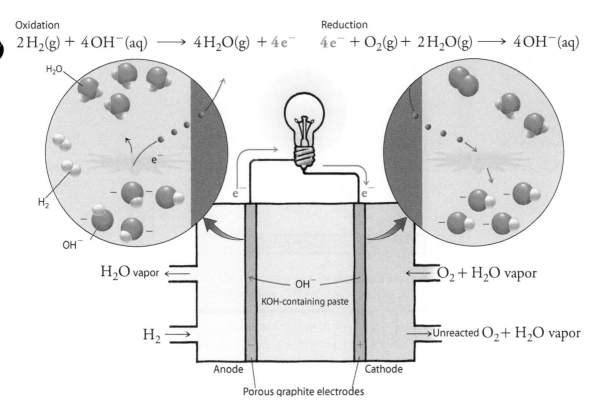

Oxidation

$$2\,H_2(g) + 4\,OH^-(aq) \longrightarrow 4\,H_2O(g) + 4\,e^-$$

Reduction

$$4\,e^- + O_2(g) + 2\,H_2O(g) \longrightarrow 4\,OH^-(aq)$$

H₂O

e^-

H₂

OH⁻

e^- e^-

H₂O vapor ←   ← O₂ + H₂O vapor

OH⁻
KOH-containing paste

H₂ →   → Unreacted O₂ + H₂O vapor

Anode   Cathode
Porous graphite electrodes

▲ **Figure 11.10**
In the hydrogen–oxygen fuel cell, hydrogen, H₂, combines with hydroxide ions, OH⁻, at the anode to form water in addition to electrons that power an external circuit. The electrons then return to the cathode where oxygen, O₂, combines with water to form hydroxide ions that migrate back to the anode.

electricity in a steam turbine. Furthermore, the water that condenses from the steam is pure water, suitable for drinking.

Although fuel cells are similar to batteries, they will run continuously as long as fuel is supplied. The International Space Station uses hydrogen–oxygen fuel cells to meet its electricity needs. The cells also produce drinking water for the astronauts. Back on the Earth, researchers are developing fuel cells for buses and automobiles. As shown in **Figure 11.11**, experimental fuel-cell buses are already operating in several cities, including Vancouver, British Columbia, and Chicago, Illinois. These vehicles produce very few pollutants and can run much more efficiently than vehicles that burn fossil fuels.

A number of problems are associated with using hydrogen as a fuel for fuel cells. Perhaps foremost is that hydrogen gas, H₂, is not a naturally abundant material. The hydrogen, therefore, must be manufactured, which is an energy-requiring process that defeats much of the utility of the fuel cell. For example, a most direct way of producing hydrogen gas is from the electrolysis of water. As described in Section 11.6, electrolysis requires a source of electrical energy. If this energy comes from a pollution-producing power plant, the environmental benefits of the fuel cell are lost. Alternatively, as discussed in Section 11.5, photovoltaic cells can produce the electricity needed for electrolysis. But today's photovoltaic cells are relatively expensive to produce, and they only work when the Sun is shining.

Hydrogen gas, however, can be produced from fossil fuels. The most efficient and environmentally friendly fossil fuel is methane, CH₄, also known as natural gas, which is now in great supply because of the advent of fracking, as was discussed on page 159. When heated with steam in the presence of a catalyst, this methane can be converted to hydrogen and carbon dioxide:

$$CH_4(g) \quad + \quad 2\,H_2O(g) \quad \rightarrow \quad 4\,H_2 \quad + \quad CO_2$$
methane   steam   hydrogen   carbon dioxide

▲ **Figure 11.11**
Because this bus is powered by a fuel cell, its tail pipe emits mostly water vapor.

▶ **Figure 11.12**
In the molten carbonate fuel cell (MCFC), carbonate ions facilitate the oxidation of hydrogen, H_2, at the anode and the reduction of oxygen, O_2, at the cathode. While the oxygen comes from the air, the hydrogen can come from the transformation of a gaseous hydrocarbon, such as methane. Even compost gas, which is mostly methane, can be used as the fuel source.

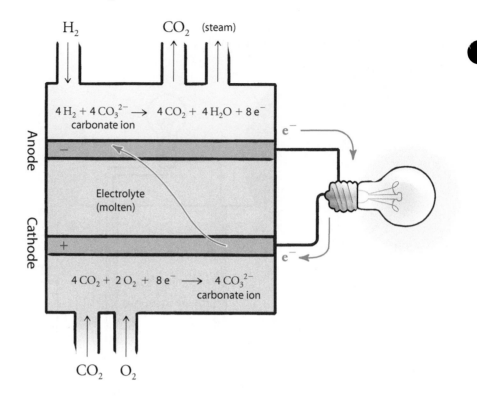

A fuel cell now used on an industrial scale is the molten carbonate fuel cell (MCFC). These fuel cells have been engineered to work in tandem with the on-site production of hydrogen from methane. As shown in **Figure 11.12**, the resulting hydrogen is oxidized at the anode as it reacts with carbonate ions to produce carbon dioxide, water, and electrons. The electrons pass through an external circuit providing electric power. Meanwhile, the carbon dioxide is piped to the cathode where electrons are received by oxygen, which reacts with the carbon dioxide to form carbonate ions. The carbonate ions migrate through the electrolyte back to the anode for subsequent reactions.

The MCFC can be scaled to create a power plant from a few kilowatts to as many as 50 MW. For reference, a typical coal-fired power plant is about 500 MW. The coal-fired power plant, however, is centralized, producing electricity for an entire region. The fuel cell power plant, by contrast, is suitable for individual buildings, such as a hospital, factory, or skyscraper (**Figure 11.13**). It runs quietly, producing few exhausts from the basement or just outside the building it serves. Aside from electricity, the MCFC also produces much thermal energy, which can be used to heat water and ventilation systems, which boosts the overall efficiency.

The electricity from a fuel cell can be used to charge the batteries of an electric vehicle. But might the fuel cell also be placed on board to power the electric vehicle directly? While only 3 kilograms of hydrogen are needed to drive a car some 500 kilometers, the volume of this much hydrogen is about 36,000 liters. Compressing the gas into smaller volumes (as occurs in the bus shown in Figure 11.11) or chilling it to the liquid phase is an energy-intensive process, and this lowers the efficiencies. Researchers, however, are looking to porous materials, such as the carbon nanofibers shown in **Figure 11.14**, that can hold large amounts of hydrogen to their surfaces, behaving in effect like hydrogen "sponges." The hydrogen can be "squeezed" out of these porous materials on demand by controlling the temperature—the warmer the material, the more hydrogen is released.

Alternatively, liquid hydrocarbon such as methanol, CH_3OH, can also be used for fuel cells. With a liquid fuel, miniature cells can be built to power portable electronic devices such as smartphones and tablets. Such devices

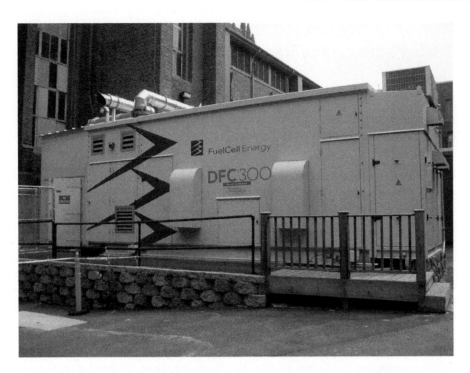

◀ Figure 11.13
This energy-efficient 250-kW fuel cell power plant built by the Fuel Cell Energy Company services the electrical and heating needs of the Environmental Science Center at Yale University.

could operate for long periods of time on a single "ampoule" of liquid methanol available at a local supermarket. The fuel cell industry holds much promise. Stay tuned for further developments.

CONCEPT**CHECK**

As long as fuel is available to it, a given fuel cell can supply electrical energy indefinitely. Why can't dry-cell or alkaline batteries do the same?

CHECK YOUR ANSWER Dry-cell and alkaline batteries generate electricity as the chemical reactants they contain are reduced and oxidized. Once these reactants are consumed, the battery can no longer generate electricity.

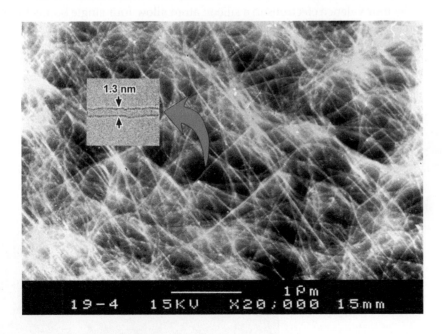

◀ Figure 11.14
Carbon nanofibers consist of near-submicroscopic tubes of carbon atoms. They outclass almost all other known materials in their ability to absorb hydrogen molecules.

11.5 Photovoltaics Transform Light into Electricity

LEARNING OBJECTIVE

Describe the nature of *n*-type and *p*-type silicon and explain how they can be used to create a photovoltaic cell.

EXPLAIN THIS

Does the charge buildup at the n-type/p-type junction increase, stay the same, or decrease in the presence of sunlight?

Photovoltaic cells are the most direct way of converting sunlight to electrical energy. Since their invention in the 1950s, photovoltaics have made remarkable strides. The first major application was in the 1960s with the U.S. space program—space satellites carried photovoltaic cells to power radios and other small electronic devices. Photovoltaics gained further momentum during the energy crisis of the mid-1970s.

The cost of receiving electricity through photovoltaics is primarily a function of the cost of purchasing the equipment and having it installed. As of 2010, a residential 2-kilowatt photovoltaic system is about $15,000. This system would pay for itself in savings from utility bills in roughly 12 years. In sunny climates, however, the payback time would be quicker. Also, payback time will become quicker as the cost of purchasing electricity from fossil fuel-based utility companies continues to rise. Furthermore, keep in mind that the price of photovoltaic systems continues to drop as photovoltaic cell technology improves. Beyond economic reasons, many people are switching to photovoltaic systems simply because it is a clean source of energy and good for the environment. Today's average residential photovoltaic system has an estimated life span of 25 years.

Worldwide, sales of photovoltaics have grown exponentially over the last several decades. Photovoltaics are now a multi-billion-dollar industry with a strong promise of continued growth. Because they need minimal maintenance and no water, they are well suited to remote or arid regions. Photovoltaics can operate on any scale and are on the verge of being cost-competitive with electricity from fossil and nuclear fuels. More than a billion handheld calculators, several million watches, several million portable lights and battery chargers, and thousands of remote communications facilities are powered by photovoltaic cells. As **Figure 11.15** illustrates, the range of their usefulness is huge.

Conventional photovoltaic cells are made from thin slabs of ultrapure silicon. The four valence electrons in a silicon atom allow four single bonds to four silicon atoms, as shown in **Figure 11.16a**. This configuration can be changed by incorporating trace amounts of other elements that have either more or fewer than four valence electrons. Arsenic atoms, for example, have five valence electrons each, also shown in Figure 11.16a. Within the silicon lattice, four of these

▶ **Figure 11.15**
Photovoltaic cells come in many sizes, from those in a handheld calculator to those in a rooftop unit that provides electricity to a house.

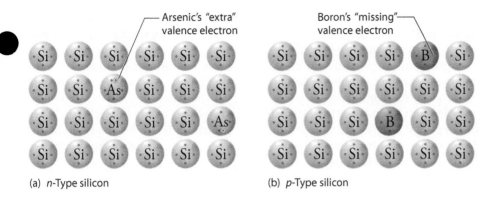

Arsenic's "extra" valence electron

Boron's "missing" valence electron

(a) *n*-Type silicon

(b) *p*-Type silicon

�C **Figure 11.16**
(a) The four valence electrons in a silicon atom can form four bonds. The fifth valence electron of arsenic is unable to participate in bonding in the silicon lattice and so remains free. Silicon that contains trace amounts of arsenic (or any other element whose atoms have five valence electrons) is called *n*-type silicon. (b) Boron has only three valence electrons for bonding with four silicon atoms. One boron–silicon pair therefore lacks an electron for covalent bonding. Silicon containing trace amounts of boron (or any other element whose atoms have three valence electrons) is called *p*-type silicon.

arsenic electrons participate in bonding with four silicon atoms, but the fifth electron remains free. This is *n-type silicon*, so called because of the presence of free negative charges (electrons) brought in by the arsenic. Boron atoms, on the other hand, have only three valence electrons, represented in **Figure 11.16b**. Incorporating boron into a silicon crystal lattice creates "electron holes," which are bonding sites where electrons should be but are not. This is *p-type silicon*, so called because any passing electron is attracted to this hole as though the hole were a positive charge.

What happens when a slice of *n*-type silicon is pressed against a slice of *p*-type silicon? Remember that the *n*-type contains free electrons and the *p*-type contains electron holes just waiting to attract any available electrons. Sure enough, electrons migrate across the junction from the *n*-type to the *p*-type, as shown in **Figure 11.17a**. This occurs only up to a certain point, however, because losing or gaining electrons upsets the electron-to-proton balance. As the *n*-type silicon loses electrons, it develops a positive charge on its side of the junction. As the *p*-type silicon gains electrons, it develops a negative charge. This charge buildup at the junction serves as a barrier to continued electron migration, as **Figure 11.17b** illustrates.

READINGCHECK

What happens to *n*-type silicon as it loses electrons to *p*-type silicon?

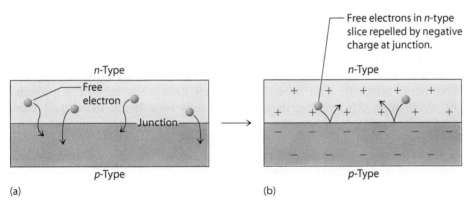

n-Type

Free electron

Junction

p-Type

(a)

Free electrons in *n*-type slice repelled by negative charge at junction.

n-Type

p-Type

(b)

◀ **Figure 11.17**
(a) Initially, free electrons migrate from the *n*-type silicon to the *p*-type silicon. (b) Quickly, however, charge buildup at the junction prevents the continued flow of electrons.

Photovoltaic cells rely on the **photoelectric effect,** which is the ability of light to knock electrons away from the atoms in an object. In most materials, the electrons either are completely ejected from the object or simply fall back to their original positions. In a select number of materials, however, including silicon, dislodged electrons roam about randomly in and among neighboring atoms without being locked into any one place, as represented in **Figure 11.18**. However, *random electron motion is not an electric current*, because for every one electron moving to the left, there is another canceling that motion by moving to the right. Greater random motion simply means higher temperatures. This is why solar energy shining onto a piece of silicon produces nothing more than heat.

The junction barrier between adjacent slices of *n*-type and *p*-type silicon acts like a one-way valve. Electrons knocked loose in the *n*-type slice are inhibited from migrating across the junction to the *p*-type slice. Electrons knocked

Sunlight

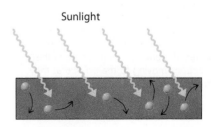

▲ **Figure 11.18**
The photoelectric effect in silicon. Light rays knock out bonding electrons, which are free to travel about the crystal lattice.

loose in the *p*-type slice, however, are readily drawn into the *n*-type slice, as shown in **Figure 11.19**. So the flow of electrons across the barrier is unidirectional, always from the *p*-type to the *n*-type.

As Figure 11.19 shows, a complete electric circuit can be made by connecting a wire to the outside faces of the two slices of silicon. When light hits either slice, dislodged electrons are forced by the junction barrier to move in one direction: from the *n*-type slice through the wire. This, in turn, creates holes in the *p*-type slice which accepts electrons returning from the external circuit. Thus, rather than being transformed completely to heat, energy from sunlight is transformed to electricity.

C O N C E P T C H E C K

Why is the unidirectional nature of the *n*-type/*p*-type silicon junction so important?

CHECK YOUR ANSWER The unidirectional nature of this junction forces electrons knocked loose by the sunlight to flow in a specific direction, which is an electric current. Otherwise, the silicon would merely get warmer.

The goal of the photovoltaic industry is to create a cell that is highly efficient, inexpensive to manufacture, and easily mass-produced. Traditional photovoltaic cells manufactured from ultrapure crystalline silicon yield efficiencies as high as 15 percent, which is good. However, the high cost of producing ultrapure crystalline silicon prevents these cells from being cost-competitive. Although great strides have been made in the past 20 years, electricity from these cells still remains three to four times more expensive than electricity from conventional sources.

A promising area of photovoltaic research involves the so-called second- and third-generation photovoltaic cells. These cells are not produced from expensive crystalline silicon. Rather, they are formed as vaporized silicon or some other photovoltaic material is deposited on a glass or plastic substrate. The resulting films are about 400 times thinner than traditional silicon wafers, which saves in material costs. Furthermore, the films are easy to mass-produce. The silicon "first-generation" cells, however, are well entrenched in the market, which means that it will still be a few years until these updated cells become commonly used. By that time, however, all these cells may be supplanted by even more efficient fourth-generation cells now being developed using nanotechnology.

Photovoltaic cells are just one type of sustainable energy technology. Many other technologies, such as wind energy, biomass conversion, and hydroelectric power, are discussed in Chapter 17.

▶ **Figure 11.19**
Free electrons in the *n*-type wafer are repelled by the negative charge of the *p*-type wafer. Free electrons in the *p*-type wafer, however, are attracted to the positive charge of the *n*-type wafer. The energy of sunlight gets these electrons moving to the point that they are pushed into an external circuit in a unidirectional fashion.

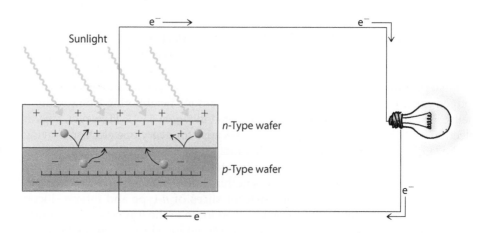

11.6 Electrolysis Produces Chemical Change

EXPLAIN THIS

How might electrolysis be used to raise a sunken ship?

Electrolysis is the use of electrical energy to produce chemical change. The recharging of a car battery is an example of electrolysis. Another, shown in **Figure 11.20**, is passing an electric current through water, a process that breaks down the water into its elemental components:

$$\text{electrical energy} \ + \ 2\,H_2O(l) \rightarrow 2\,H_2(g) + O_2(g)$$

Electrolysis is used to purify metals from metal ores. An example is aluminum, the third most abundant element in the Earth's crust. Aluminum occurs naturally bonded to oxygen in an ore called bauxite. Aluminum metal wasn't known until about 1827, when it was prepared by reacting bauxite with hydrochloric acid. This reaction gave the aluminum ion, Al^{3+}, which was reduced to aluminum metal, with sodium metal acting as the reducing agent:

$$Al^{3+} + 3\,Na \rightarrow Al + 3\,Na^{+}$$

This chemical process was expensive, which is why the initial price of aluminum was well beyond the means of the average person. Aluminum was thus considered a rare and precious metal. In 1855, aluminum dinnerware and other items were exhibited in Paris with the crown jewels of France. Then, in 1886, two men working independently, Charles Hall (1863–1914) in the United States and Paul Heroult (1863–1914) in France, almost simultaneously discovered a process whereby aluminum could be produced from aluminum oxide, Al_2O_3, a main component of bauxite. In what is now known as the Hall–Heroult process, shown in **Figure 11.21**, a strong electric current is passed through a molten mixture of aluminum oxide and cryolite, Na_3AlF_6, a naturally occurring mineral. The fluoride ions of the cryolite react with the aluminum oxide to form various aluminum fluoride ions, such as $AlOF_3^{2-}$, which are then oxidized

$$\text{Oxidation} \quad 2\,AlOF_3^{2-} + 6\,F^{-} + C \longrightarrow 2\,AlF_6^{3-} + CO_2 + 4\,e^{-}$$

READING CHECK

What is electrolysis?

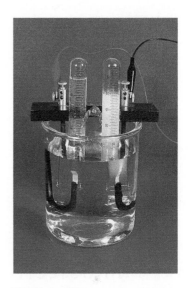

▲ **Figure 11.20**
The electrolysis of water produces hydrogen gas and oxygen gas in a 2:1 ratio by volume, in accord with the chemical formula for water: H_2O. In order for this process to work, ions must be dissolved in the water so that electric charge can be conducted between the electrodes.

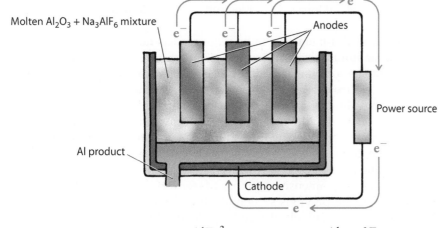

Molten Al_2O_3 + Na_3AlF_6 mixture — Anodes — e^- — Power source — Al product — Cathode

$$\text{Reduction} \quad AlF_6^{3-} + 3\,e^{-} \longrightarrow Al + 6\,F^{-}$$

◀ **Figure 11.21**
The melting point of aluminum oxide (2030°C) is too high for efficiently electrolyzing it to aluminum metal. Aluminum oxide, however, dissolves in molten cryolite at a more reasonable 980°C. A strong electric current passed through the molten aluminum oxide–cryolite mixture generates aluminum metal at the cathode, where aluminum ions pick up electrons and are thus reduced to elemental aluminum.

to the aluminum hexafluoride ion, AlF_6^{3-}. The Al^{3+} in this ion is then reduced to elemental aluminum, which collects at the bottom of the reaction chamber. This process, which is still in use by manufacturers today, greatly facilitated mass production of aluminum metal, and by 1890, the price of aluminum had dropped to about $2 per pound.

Today, worldwide production of aluminum is about 16 million metric tons annually. For each ton produced from ore, about 16,000 kilowatt-hours of electrical energy is required, as much as a typical American household consumes in 18 months. Processing recycled aluminum, on the other hand, consumes only about 700 kilowatt-hours for every ton. Thus, recycling aluminum not only reduces litter but also helps to reduce the load on power companies, which in turn reduces air pollution. Furthermore, reserves of high-quality aluminum oxide ores are already depleted in the United States. Recycling aluminum, therefore, also helps to minimize the need to develop new bauxite mines in foreign countries.

CONCEPT CHECK

Is the exothermic reaction in a hydrogen–oxygen fuel cell an example of electrolysis?

CHECK YOUR ANSWER No. During electrolysis, electrical energy is used to produce chemical change. In the hydrogen–oxygen fuel cell, chemical change is used to produce electrical energy.

11.7 Metal Compounds Can Be Converted to Metals

LEARNING OBJECTIVE

Review the oxidation-reduction chemistry involved in the creation of common metals.

READING CHECK

Which metals are most difficult to reduce?

EXPLAIN THIS

What type of chemical bond might you expect between two sodium atoms?

To convert a metal-containing compound to a metal requires an oxidation-reduction reaction. In the metal-containing compound, the metal exists as a positively charged ion, because it has lost one or more of its electrons to its bonding partner. To convert metal ions to neutral metal atoms requires that they gain electrons; that is, they must be reduced:

$$M^+ \quad + \quad e^- \quad \rightarrow \quad M^0$$

Metal ion · · · · · Electron · · · · · Metal atom

The tendency of a metal ion to be reduced depends on its location in the periodic table, as summarized in **Figure 11.22**. As discussed in Chapter 6, metals on the left in the periodic table readily lose electrons. This means it is relatively difficult to give electrons back to these metal ions—in other words, they are difficult to reduce. A sodium atom, for example, being on the left in the periodic table, loses electrons easily. Any ionic compound it forms, such as sodium chloride, tends to be very stable. Reducing the sodium ion, Na^+, to sodium metal, Na^0, is difficult because doing so requires giving electrons to the sodium ion.

Metals on the left and especially the lower left of the periodic table therefore require the most energy-intensive methods of recovery, which include *electrolysis*. As was shown in the previous section, during electrolysis, an electric current supplies electrons to positively charged metal ions, thus reducing them. Metals commonly recovered by electrolysis include the metals of groups 1–3, which occur most frequently as halides, carbonates, and phosphates. In addition, aluminum is commonly recovered by electrolysis, and other metals are obtained using electrolysis when very high purity is needed. The reactions involved when copper is produced this way are shown in **Figure 11.23**.

CONCEPTCHECK

Why is it so difficult to obtain a group 1 metal from a compound containing ions of that metal?

CHECK YOUR ANSWER The metal ions do not readily accept electrons to form metal atoms.

Some Metals Are Most Commonly Obtained from Metal Oxides

Ores containing metal oxides can be converted to metals fairly efficiently in a *blast furnace*. First, the ore is mixed with limestone and coke, which is a concentrated form of carbon obtained from coal. Then the mixture is dropped into the furnace, where the coke is ignited and used as a fuel. At high temperatures, the coke also behaves as a reducing agent, yielding electrons to the positively charged metal ions in the oxide and reducing them to metal atoms. **Figure 11.24** shows this method being used to produce iron, which is found in nature primarily as iron oxide.

In the furnace, the limestone reacts with ore impurities—predominantly silicon compounds—to form *slag*, which is primarily calcium silicate:

$$SiO_2(s) \ + \ CaCO_3(s) \ \rightarrow \ CaSiO_3(l) \ + \ CO_2(g)$$

Silica sand   Limestone   Molten slag   Carbon
(ore impurity)   (calcium silicate) dioxide

Because of the high temperatures, both the metal and the slag are molten. They drain to the bottom of the blast furnace, where they collect in two layers, the less dense slag on top. The metal layer is then tapped off through an opening at the bottom of the blast furnace.

▲ **Figure 11.22**
The ions of metallic elements at the lower left of the periodic table are most difficult to reduce. For this reason, obtaining these elements from the metal-containing compounds they form is energy intensive. Metallic elements at the upper right of the periodic table tend to form compounds that require less energy to convert to metals.

☐ Transforming the metal-containing compound to a metal is less energy intensive

■ Transforming the metal-containing compound to a metal is more energy intensive

◀ **Figure 11.23**
High-purity copper is recovered by electrolysis. Pure copper metal deposits on the negative electrode as copper ions in solution gain electrons. The source of these copper ions is a positively charged electrode made of impure copper.

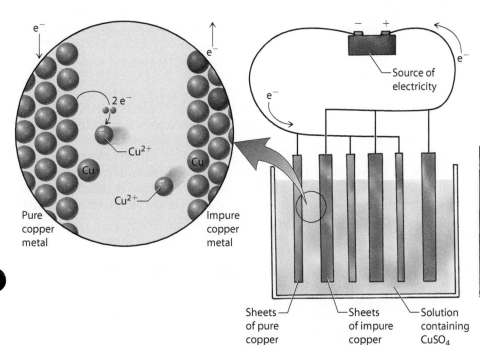

Pure copper metal

Impure copper metal

$2\,e^-$

Cu^{2+}

Cu

Cu

Cu^{2+}

e^-

e^-

e^-

e^-

Source of electricity

Sheets of pure copper

Sheets of impure copper

Solution containing $CuSO_4$

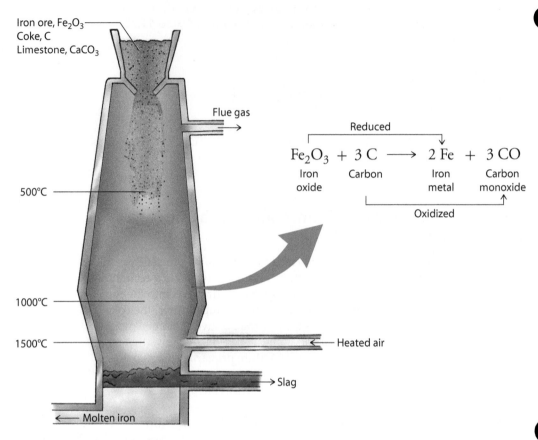

Iron ore, Fe_2O_3
Coke, C
Limestone, $CaCO_3$

Flue gas

$$Fe_2O_3 + 3\,C \longrightarrow 2\,Fe + 3\,CO$$

Iron   Carbon   Iron   Carbon
oxide   metal   monoxide

Reduced

Oxidized

500°C

1000°C

1500°C

Heated air

Slag

Molten iron

▲ Figure 11.24

A mixture of iron oxide ore, coke, and limestone is dropped into a blast furnace, where the iron ions in the oxide are reduced to metal atoms. Note how the carbon is oxidized as it gains an oxygen atom to form carbon monoxide.

FOR YOUR INFORMATION

World crude steel production has grown steadily, from about 200 million metric tons in 1950 to about 1.5 billion metric tons in 2011. The world's leading producer is China, which in 2011 produced about 680 million metric tons (mmt). This was followed by Japan (107 mmt) and then the United States (86 mmt). So what is the difference between a ton and a metric ton? A ton is the United States Customary Standard unit for 2000 pounds. A metric ton is 1000 kilograms, which is equal to 2205 pounds. So which would you rather have: a ton of gold or a metric ton of gold?

Once cooled, the metal from a blast furnace is known as a *cast metal*. (When the ore is an iron ore, the cast metal is known as *pig iron*.) A cast metal is brittle and relatively soft for a metal because it still contains impurities, such as phosphorus, sulfur, and carbon. To remove these impurities, oxygen is blown through the molten cast metal in a *basic oxygen furnace*, shown in **Figure 11.25**. The oxygen oxidizes the impurities to form additional slag, which floats to the surface and is skimmed off.

Most phosphorus and sulfur impurities are removed in a basic oxygen furnace, but the purified metal still contains about 3 percent carbon. For the production of iron, this carbon is desirable. Iron atoms are relatively large, and when they pack together, small voids are created between atoms, as shown in **Figure 11.26**. These voids tend to weaken the iron. Carbon atoms are small enough to fill the voids, and having the voids filled strengthens the iron substantially. Iron strengthened by small percentages of carbon is called **steel**. The tendency of steel to rust can be inhibited by alloying the steel with noncorroding metals, such as chromium or nickel. This yields the *stainless steel* used to manufacture eating utensils and countless other items.

Other Metals Are Most Commonly Obtained from Metal Sulfides

Metal sulfides can be purified by *flotation*, a technique that takes advantage of the fact that metal sulfides are relatively nonpolar and therefore attracted to oil. An ore containing a metal sulfide is first ground to a fine powder and then

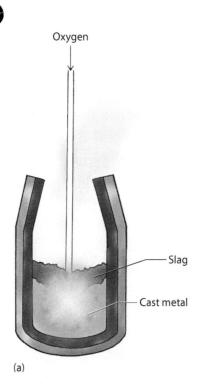

Oxygen

Slag

Cast metal

(a)

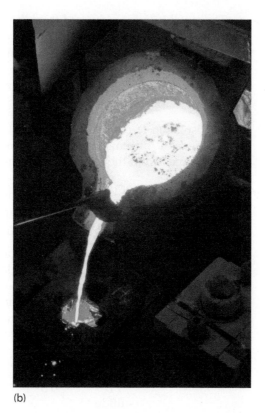

(b)

◀ **Figure 11.25**
(a) A flow of oxygen through a basic oxygen furnace oxidizes most of the impurities in a cast metal to form slag that may be skimmed away as it floats to the surface. (b) The basic oxygen furnace is hoisted and its purified contents poured into a reservoir used for molding iron pieces.

mixed vigorously with a lightweight oil and water. Compressed air is then forced up through the mixture. As air bubbles rise, they become coated with oil and metal sulfide particles. At the surface of the liquid, the coated bubbles form a floating froth, as shown in **Figure 11.27**. This froth, which is very rich in the metal sulfide, is then skimmed off.

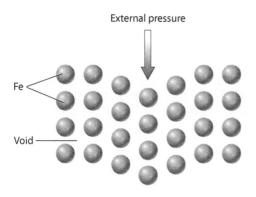

External pressure

Fe

Void

Pure iron is fairly soft and malleable because of voids between atoms.

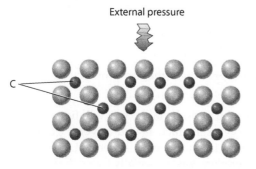

External pressure

C

When the voids are filled with carbon atoms, the carbon helps hold the iron atoms in their lattice. This strengthened metal is called steel.

▲ **Figure 11.26**
Steel is stronger than iron because of the small amounts of carbon it contains.

▲ **Figure 11.27**
Air bubbles rising through a flotation container transport metal sulfide particles to the surface.

The metal sulfides recovered from the froth are then *roasted* in the presence of oxygen. The net reaction is the oxidation of S^{2-} in the sulfide to S^{4+} in sulfur dioxide and the reduction of the metal ion to its elemental state:

$$MS(s) \; + \; O_2(g) \; \rightarrow \; M(l) \; + \; SO_2(g)$$

Metal sulfide   Oxygen   Metal   Sulfur dioxide

The isolation of copper from its most common ore, chalcopyrite, $CuFeS_2$, requires several additional steps because of the presence of iron. First, the chalcopyrite is roasted in the presence of oxygen:

$$2\,CuFeS_2 \; + \; O_2(g) \; \rightarrow \; CuS(s) \; + \; 2\,FeO(s) \; + \; 2\,SO_2(g)$$

Chalcopyrite   Oxygen   Copper   Iron   Sulfur
                    sulfide   oxide   dioxide

The copper sulfide and iron oxide from this reaction are then mixed with limestone, $CaCO_3$, and sand, SiO_2, in a blast furnace, where CuS is converted to Cu_2S. The limestone and sand form molten slag, $CaSiO_3$, in which the iron oxide dissolves. The copper sulfide melts and sinks to the bottom of the furnace. The less dense iron-containing slag floats above the molten copper sulfide and is drained off. The isolated copper sulfide is then roasted to copper metal:

$$Cu_2S(s) \; + \; O_2(g) \; \rightarrow \; 2\,Cu(l) \; \rightarrow \; SO_2(g)$$

Copper   Oxygen   Copper   Sulfur
sulfide   metal   dioxide

Roasting metal sulfides requires a fair amount of energy. Furthermore, sulfur dioxide is a toxic gas that contributes to acid rain, so its emission must be minimized. Most companies comply with EPA emissions standards by converting the sulfur dioxide to marketable sulfuric acid, H_2SO_4.

CONCEPT CHECK

Of metal chlorides, metal oxides, and metal sulfides, which are most polar? Which are least polar?

CHECK YOUR ANSWER Metal chlorides, such as sodium chloride, tend to be most polar, while the metal sulfides, such as copper sulfide, tend to be least polar. The oxides, such as iron oxide, tend to be in between.

11.8 Oxygen Is Responsible for Corrosion and Combustion

LEARNING OBJECTIVE

Compare and contrast the processes of corrosion and combustion.

READINGCHECK

What is oxygen so good at doing?

EXPLAIN THIS

Do our bodies gradually oxidize or reduce the food molecules we eat?

If you look to the upper right of the periodic table, you will find one of the most common oxidizing agents—oxygen. In fact, if you haven't guessed already, the term *oxidation* is derived from the name of this element. Oxygen is able to pluck electrons from many other elements, especially those that lie at the lower left of the periodic table. Two common oxidation-reduction reactions involving oxygen as the oxidizing agent are *corrosion* and *combustion*.

◀ Figure 11.28
Rust itself is not harmful to the iron structures on which it forms. It is the loss of metallic iron that ruins the structural integrity of these objects.

CONCEPT CHECK

Oxygen is a good oxidizing agent, but so is chlorine. What does this indicate about their relative positions in the periodic table?

CHECK YOUR ANSWER Chlorine and oxygen must lie in the same area of the periodic table. Both have strong effective nuclear charges and are strong oxidizing agents.

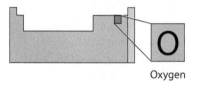

Oxygen

Corrosion is the process whereby a metal deteriorates. Corrosion caused by atmospheric oxygen is a widespread and costly problem. About one-quarter of the steel produced in the United States, for example, goes into replacing corroded iron at a cost of billions of dollars annually. Iron corrodes when it reacts with atmospheric oxygen and water to form hydrated iron oxide, which is the naturally occurring reddish-brown substance you know as rust, shown in **Figure 11.28**.

$$4\,Fe + 3\,O_2 + 3\,H_2O \rightarrow 2\,Fe_2O_3 \cdot 3\,H_2O$$
Iron   Oxygen Water   Rust

Another common metal oxidized by oxygen is aluminum. The product of aluminum oxidation is aluminum oxide, Al_2O_3, which does not flake away from the aluminum. Rather it forms a protective coat that shields the metal from further oxidation. This coat is so thin that it's transparent, which is why aluminum maintains its metallic shine.

A protective, oxidized coat is the principle underlying a process called *galvanization*. Zinc has a slightly greater tendency to oxidize than does iron. For this reason, many iron objects, such as the nails pictured in **Figure 11.29**,

FOR YOUR INFORMATION

The metals used for cathodic protection are "sacrificing" themselves to be anodes (to lose electrons) so that the desired metal, such as the copper pipe, is spared from oxidation. These sacrificing metals, therefore, are sometimes called *sacrificial anodes*.

◀ Figure 11.29
The galvanized nail (*bottom*) is protected from rusting by the sacrificial oxidation of zinc.

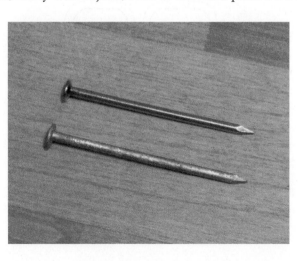

▶ **Figure 11.30**
Zinc strips help to protect the iron hull of an oil tanker from oxidizing. The zinc strip shown here is attached to the hull's exterior surface.

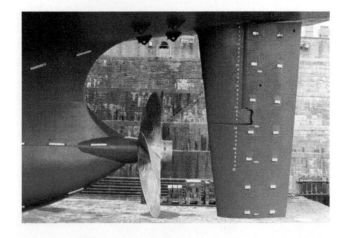

are *galvanized* by coating them with a thin layer of zinc. The zinc oxidizes to zinc oxide, an inert, insoluble substance that protects the iron underneath it from rusting.

In a technique called *cathodic protection,* iron structures can be protected from oxidation by placing them in contact with certain metals, such as zinc or magnesium, that have a greater tendency to oxidize. This forces the iron to accept electrons, which means that it is behaving as a cathode. (Rusting occurs only where iron behaves as an anode.) Ocean tankers, for example, are protected from corrosion by strips of zinc affixed to their hulls, as shown in **Figure 11.30**. Similarly, outdoor steel pipes are protected by being connected to magnesium rods inserted into the ground.

Yet another way to protect iron and other metals from oxidation is to coat them with a corrosion-resistant metal, such as chromium, platinum, or gold. *Electroplating* is the operation of coating one metal with another by electrolysis, and it is illustrated in **Figure 11.31**. The object to be electroplated is connected to a negative battery terminal and then submerged in a solution containing ions of the metal to be used as the coating. The positive terminal of the battery is connected to an electrode made of the coating metal. The circuit is completed when this electrode is submerged in the solution. Dissolved metal ions are attracted to the negatively charged object, where they pick up electrons and are deposited as metal atoms. The ions in solution are replenished by the forced oxidation of the coating metal at the positive electrode.

Combustion is a rapid oxidation-reduction reaction between a material and molecular oxygen. The burning of a campfire is a good example. Combustion reactions are characteristically exothermic (energy-releasing). A violent combustion reaction is the formation of water from hydrogen and oxygen.

▶ **Figure 11.31**
As electrons flow into the hubcap and give it a negative charge, positively charged chromium ions move from the solution to the hubcap and are reduced to chromium metal, which is deposited as a coating on the hubcap. The solution is supplied with ions as chromium atoms in the cathode are oxidized to Cr^{3+} ions.

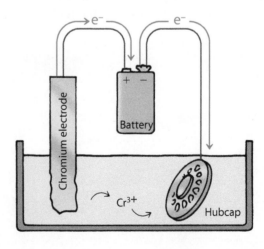

As discussed in Section 9.4, the energy from this reaction is used to power rockets into space. More common examples of combustion include the burning of wood and fossil fuels. The combustion of these and other carbon-based chemicals forms carbon dioxide and water. Consider, for example, the combustion of methane, the major component of natural gas:

$$CH_4 \;+\; 2\,O_2 \;\rightarrow\; CO_2 \;+\; 2\,H_2O \;+\; energy$$

Methane   Oxygen   Carbon   Water
                   dioxide

In combustion, electrons are transferred as polar covalent bonds are formed in place of nonpolar covalent bonds or vice versa. (This is in contrast with the other examples of oxidation-reduction reactions presented in this chapter, which involve the formation of ions from atoms or, conversely, atoms from ions.) This concept is illustrated in **Figure 11.32**, which compares the electronic structures of the combustion starting material, molecular oxygen, and the combustion product, water. Molecular oxygen is a nonpolar covalent compound. Although each oxygen atom in the molecule has fairly strong electronegativity, the four bonding electrons are pulled equally by both atoms and thus are unable to congregate on one side or the other. After combustion, however, the electrons are shared between the oxygen and hydrogen atoms in a water molecule and are pulled to the oxygen. This gives the oxygen a slight negative charge, which is another way of saying it has gained electrons and has thus been reduced. At the same time, the hydrogen atoms in the water molecule develop a slight positive charge, which is another way of saying they have lost electrons and have thus been oxidized. This gain of electrons by oxygen and loss of electrons by hydrogen is an energy-releasing process. Typically, the energy is released either as molecular kinetic energy (heat) or as light (the flame).

Interestingly, oxidation-reduction reactions chemically related to combustion occur throughout your body. You can visualize a simplified model of your metabolism by reviewing Figure 11.32 and substituting a food molecule for the methane. Food molecules relinquish their electrons to the oxygen molecules you inhale. The products are carbon dioxide, water vapor, and energy. You exhale the carbon dioxide and water vapor, but much of the energy from the reaction is used to keep your body warm, and the rest is used to drive the many other biochemical reactions necessary for life.

CHEMICAL CONNECTIONS

How is your breath connected to a campfire?

◀ Figure 11.32
(a) Neither atom in an oxygen molecule is able to preferentially attract the bonding electrons. (b) The oxygen atom of a water molecule pulls the bonding electrons away from the hydrogen atoms on the water molecule, making the oxygen slightly negative and the two hydrogens slightly positive.

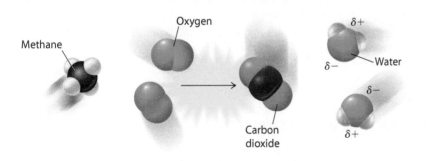

Methane

Oxygen

Carbon dioxide

Water

$\delta+$ $\delta-$ $\delta-$ $\delta+$

$O :: O$

$$\delta+H \overset{\delta-}{:}\overset{\delta-}{O}$$
$$\delta+H$$

(a) Reactant oxygen atoms share electrons equally in O_2 molecules.

(b) Product oxygen atoms pull electrons away from H atoms in H_2O molecules and are reduced.

Chapter 11 Review

LEARNING OBJECTIVES

Identify when a chemical undergoes oxidation or reduction. (11.1)	→	*Questions 1–4, 24–25, 32–33, 35–45*
Show how electricity can be generated using materials that tend to lose or gain electrons. (11.2)	→	*Questions 5–6, 46–49*
Describe how oxidation and reduction occur within a device that generates electricity. (11.3)	→	*Questions 7–9, 26, 50–52*
Identify how a fuel cell generates electricity. (11.4)	→	*Questions 10–11, 53–55*
Describe the nature of *n*-type and *p*-type silicon and explain how they can be used to create a photovoltaic cell. (11.5)	→	*Questions 12–14, 56–58, 78–80*
Describe examples of electrolysis as an application of oxidation/reduction reactions. (11.6)	→	*Questions 15–16, 27–31, 59–63*
Review the oxidation-reduction chemistry involved in the creation of common metals. (11.7)	→	*Questions 17–19, 64–66*
Compare and contrast the processes of corrosion and combustion. (11.8)	→	*Questions 20–23, 34, 67–77*

SUMMARY OF TERMS (KNOWLEDGE)

Anode The electrode where chemicals are oxidized.

Cathode The electrode where chemicals are reduced.

Combustion The rapid exothermic oxidation-reduction reaction between a material and molecular oxygen.

Corrosion The deterioration of a metal, typically caused by atmospheric oxygen.

Electrochemistry The study of the relationship between electrical energy and chemical change.

Electrode Any material that conducts electrons into or out of a medium in which electrochemical reactions are occurring.

Electrolysis The use of electrical energy to produce chemical change.

Half-reaction One-half of an oxidation-reduction reaction, represented by an equation showing electrons as either reactants or products.

Oxidation The process whereby a reactant loses one or more electrons.

Photoelectric effect The ability of light to knock electrons away from atoms.

Reduction The process whereby a reactant gains one or more electrons.

Steel Iron strengthened by small percentages of carbon.

READING CHECK QUESTIONS (COMPREHENSION)

11.1 Losing and Gaining Electrons

1. Which elements have the greatest tendency to behave as oxidizing agents?

2. What elements have the greatest tendency to behave as reducing agents?

3. Write an equation for the half-reaction in which a potassium atom, K, is oxidized.

4. What happens to a reducing agent as it reduces?

11.2 Harnessing the Energy of Flowing Electrons

5. What is electrochemistry?

6. What is the purpose of the salt bridge in a voltaic cell?

11.3 Batteries Consume Chemicals to Generate Electricity

7. What type of reaction occurs at the cathode?

8. What type of reaction occurs at the anode?

9. Does lithium metal accept or donate electrons in a lithium battery?

11.4 Fuel Cells Consume Fuel to Generate Electricity

10. What is the prime difference between a battery and a fuel cell?

11. What else do fuel cells produce besides electricity?

11.5 Photovoltaics Transform Light into Electricity

12. How is *p*-type silicon made?

13. What happens when a slice of *n*-type silicon is joined to a slice of *p*-type silicon?

14. Light shining on a slab of pure silicon dislodges electrons from the silicon atoms. What happens to these electrons?

11.6 Electrolysis Produces Chemical Change

15. What is electrolysis? How does it differ from what goes on inside a battery?

16. In what year was the efficient electrolysis of aluminum discovered?

11.7 Metal Compounds Can Be Converted to Metals

17. Which group of metals requires the most energy to be recovered from metal-containing compounds?

18. When iron ions are reduced to neutral iron atoms in a blast furnace, where do the electrons come from?

19. How are copper (II) sulfide, CuS, and iron (III) oxide, Fe_2O_3, separated from each other in a blast furnace in the preparation of copper metal?

11.8 Oxygen Is Responsible for Corrosion and Combustion

20. What metal coats a galvanized nail?

21. What do the oxidation of zinc and the oxidation of aluminum have in common?

22. What are some differences between corrosion and combustion?

23. What happens to the polarity of oxygen atoms as they transform from molecular oxygen, O_2, into water molecules, H_2O?

CONFIRM THE CHEMISTRY (HANDS-ON APPLICATION)

24. Silver tarnishes because it reacts with the small amounts of smelly hydrogen sulfide, H_2S, we put into the air upon digesting our food. In this reaction, the silver loses electrons to the sulfur. You can reverse this reaction by allowing the silver to get its electrons back from aluminum.

    Flatten a piece of aluminum foil on the bottom of a cooking pot. Fill the pot halfway with water and bring the water to a boil. Once the water is boiling, remove the pot from its heat source. Add 2 tablespoons of baking soda to the hot water. Slowly immerse a tarnished piece of silver into the water and allow the silver to touch the aluminum foil. You should see an immediate effect once the silver and aluminum make contact. (Add more baking soda if you don't.) If your silver piece is very tarnished, you may notice the unpleasant odor of hydrogen sulfide as it is released back into the air.

    As silver tarnishes, is it oxidized or reduced? Is baking soda, $NaHCO_3$, an ionic compound? Why is it needed in this activity? Does aluminum behave as an oxidizing agent or a reducing agent as it restores the silver to its untarnished state?

25. Tie, tape, or super-glue an old penny to the end of a piece of string. Slowly dip the penny in and out of a solution of salt (1/2 teaspoon) and white vinegar (1/4 cup). This will clean the penny and dissolve surface copper ions, which are necessary for this activity to work. Without touching or drying the penny, lower it into a bottle of fresh seltzer water so that the penny remains suspended above the

water and within the "head" of carbon dioxide gas. Cap the bottle so that the penny remains suspended. Wait 24 hours and look for signs of blue-green copper carbonate. Why do old copper rooftops turn blue-green, while old pennies simply become dull?

26. A battery is made by connecting a metal that tends to lose electrons with another metal that tends to gain electrons. Two metals that make for a good battery are copper and zinc. For copper, use a penny. For zinc, find a galvanized nail and smooth out the tip with some sand paper. Clean both the penny and the nail and then hold the two metals simultaneously to the tip of your tongue. Then tilt the penny and nail toward each other so that they touch each other while also touching your tongue. The moment the metals make contact you should sense a small tingling sensation. That's the electric current running between the two metals. Notice this happens only when the zinc and copper are touching, which completes the circuit. You have made a battery out of a penny and a nail! Why is the effect more pronounced when your tongue is wet with salt water?

27. Sharpen *both* ends of two small pencils. Hold a tip of each pencil to the two terminals of a 9-V battery. Submerge the other two tips into some salt water. After a few moments, bubbles will start to rise from the submerged tips. Which tip is releasing hydrogen? Which is releasing oxygen? Add some phenolphthalein solution to the salt water and then submerge the top of the battery without using the pencils. Why does the water around the positive terminal turn pink? Why does this activity quickly ruin the battery (which therefore should not be used again)?

THINK AND SOLVE (MATHEMATICAL APPLICATION)

28. Each year about 1.6×10^7 (16 million) metric tons (mt) of aluminum are produced. How many grams is this? (Recall that 1 mt is 1000 kg.)

29. Use the following balanced chemical equation to show that through electrolysis each year, the production of 1.6×10^7 metric tons of aluminum, Al, produces about 2.0×10^7 metric tons of carbon dioxide, CO_2. Interestingly, the total mass of our atmosphere is only about 5×10^{15} metric tons.

$$4\,Al_2O_3 + 3\,C \rightarrow 4\,Al + 3\,CO_2$$

30. Use the ideal gas law as described in Section 2.8 to show that the volume 1.9×10^{13} g of CO_2 occupies at room temperature (298 K) and 1 atmosphere of pressure is 1.1×10^{13} liters. Interestingly, this equals about 11 km^3, while the total volume of our lower atmosphere (the troposphere) is about 8 million km^3.

31. Converting aluminum oxide, Al_2O_3, into aluminum, Al, creates significant amounts of carbon dioxide, CO_2, through the direct chemical reaction. How else does the conversion of Al_2O_3 into Al create large amounts of CO_2?

THINK AND COMPARE (ANALYSIS)

32. Review the concept of electronegativity in Section 6.7 and rank the following elements from the weakest to strongest oxidizing agent:
    a. Chlorine, Cl
    b. Sulfur, S
    c. Sodium, Na

33. Review the concept of electronegativity in Section 6.7 and rank the following elements from the weakest to strongest reducing agent:
    a. Chlorine, Cl
    b. Sulfur, S
    c. Sodium, Na

34. Rank the following molecules from least oxidized to most oxidized:

Ethane

Ethanol

Acetaldehyde

Acetic aid

THINK AND EXPLAIN (SYNTHESIS)

11.1 Losing and Gaining Electrons

35. What element is oxidized in the following equation? What element is reduced?

$$I_2 + 2\,Br^- \rightarrow 2\,I^- + Br_2$$

36. What element behaves as the oxidizing agent in the following equation? What element behaves as the reducing agent?

$$Sn^{2+} + 2\,Ag \rightarrow Sn + 2\,Ag^+$$

37. Hydrogen sulfide, H_2S, burns in the presence of oxygen, O_2, to produce water, H_2O, and sulfur dioxide, SO_2. Through this reaction, is sulfur oxidized or reduced?

$$2\,H_2S + 3\,O_2 \rightarrow 2\,H_2O + 2\,SO_2$$

38. Unsaturated fatty acids, such as $C_{12}H_{22}O_2$, react with hydrogen gas, H_2, to form saturated fatty acids, such as $C_{12}H_{24}O_2$. Are the unsaturated fatty acids being oxidized or reduced through this process?

39. Which atom is oxidized, ⬤ or ⬤ ?

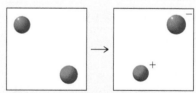

40. In the previous exercise, which atom behaves as the oxidizing agent, ⬤ or ⬤ ?

41. Which element is closer to the upper right corner of the periodic table, ⬤ or ⬤ ?

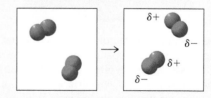

42. A chemical equation for the combustion of propane, C_3H_8, is shown here. Through this reaction, is the carbon oxidized or reduced?

$$C_3H_8 + 5\,O_2 \rightarrow 3\,CO_2 + 4\,H_2O$$

43. Upon ingestion, grain alcohol, C_2H_6O, is metabolized into acetaldehyde, C_2H_4O, which is a toxic substance that causes headaches as well as joint pain typical of a hangover. Is the grain alcohol oxidized or reduced as it transforms into acetaldehyde?

44. Your body creates chemical energy from the food you eat, which contains carbohydrates, lipids, and proteins. These are large molecules that first need to be broken down into simpler molecules, such as monosaccharides, $C_6H_{12}O_6$, fatty acids, $C_{14}H_{28}O_2$, and amino acids, $C_2H_5NO_2$. Of these molecules, which has the greatest supply of hydrogen atoms per molecule? Which is most reduced in terms of the fewest number of oxygen atoms per carbon atom? Which can react with the most oxygen molecules to produce the most energy? Which provides the most dietary Calories per gram?

45. A single hair is made of many intertwined strands of keratin, which is a sulfur-rich fibrous protein. As discussed in Chapter 13, adjacent strands of keratin are held together by sulfur–sulfur bonds. For a permanent wave, these bonds are temporarily broken so that the hair can be reshaped. A chemical agent is then added so that these sulfur–sulfur bonds can be reformed as shown here. Should this agent be an oxidizing agent or a reducing agent?

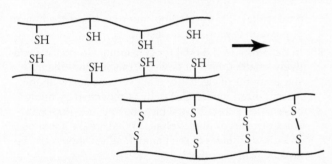

11.2 Harnessing the Energy of Flowing Electrons

46. Why does a salt bridge only last so long?

47. Chemical equations need to be balanced not only in terms of the number of atoms but also by the charge. In other words, just as there should be the same number of atoms before and after the arrow of an equation, there should be the same charge. Take this into account to balance the following chemical equation:

$$Sn^{2+} + Ag \rightarrow Sn + Ag^+$$

48. Study question 47 before attempting to balance both the atoms and charges of the following chemical equation:

$$Fe^{3+} + I^- \rightarrow Fe^{2+} + I_2$$

49. Study question 47 before attempting to balance both the atoms and charges of the following chemical equation:

$$Ce^{4+} + Cl^- \rightarrow Ce^{3+} + Cl_2$$

11.3 Batteries Consume Chemicals to Generate Electricity

50. Some car batteries require the periodic addition of water. Does adding the water increase or decrease the battery's ability to provide electric power to start the car? Explain.

51. Why is the anode of a battery indicated with a minus sign?

52. How does turning on the radio while you are driving affect the mileage of your gasoline-consuming car?

11.4 Fuel Cells Consume Fuel to Generate Electricity

53. What are some key advantages that a fuel cell power plant has over a coal-fired power plant?

54. Is the transformation of carbon dioxide, CO_2, into the carbonate ion, CO_3^{2-}, an example of oxidation or reduction? Please explain.

55. Why would a miniaturized fuel cell require a liquid fuel?

11.5 Photovoltaics Transform Light into Electricity

56. How does blending arsenic atoms into a sample of silicon increase the silicon's ability to conduct electricity?

57. When you have p-type silicon and n-type silicon joined together, why don't electrons dislodged by light in the n-type cross the junction barrier and enter the p-type?

58. In an operating photovoltaic cell, electrons move through the external circuit to the negatively charged p-type silicon wafer. How can the electrons move to the negatively charged silicon when they themselves are negatively charged? Don't like signs repel each other?

11.6 Electrolysis Produces Chemical Change

59. A major source of chlorine gas, Cl_2, is from the electrolysis of brine, which is concentrated salt water, NaCl (aq). What other two products result from this electrolysis reaction? Write the balanced chemical equation.

60. Jewelry is often manufactured by electroplating an expensive metal such as gold over a cheaper metal. Sketch a setup for this process.

61. Clorox^R is a laundry bleaching agent used to remove stains from white clothes. Suggest why the name begins with *Clor-* and ends with *-ox*.

62. Iron atoms have a greater tendency to oxidize than do copper atoms. Is this good news or bad news for a home in which much of the plumbing consists of connected iron and copper pipes? Please explain.

63. Copper atoms have a greater tendency to be reduced than do iron atoms. Was this good news or bad news for the Statue of Liberty, whose copper exterior was originally held together by steel rivets?

11.7 Metal Compounds Can Be Converted to Metals

64. Metal ores are isolated from rock by taking advantage of differences in both physical and chemical properties. Cite examples given in the text where differences in physical properties are used. Cite examples given in the text where differences in chemical properties are used.

65. How does the basic oxygen furnace remove impurities from cast metal?

66. Why isn't iron metal commonly obtained by roasting iron ores?

11.8 Oxygen Is Responsible for Corrosion and Combustion

67. Water is 88.88 percent oxygen by mass. Oxygen is what a fire needs to grow brighter and stronger. So why doesn't a fire grow brighter and stronger when water is added to it?

68. Pennies manufactured after 1982 are made of zinc metal, Zn, within a coat of copper metal, Cu. Zinc is more easily oxidized than copper. Why, then, don't these pennies quickly corrode?

69. The type of iron that the human body needs for good health is the Fe^{2+} ion. Cereals fortified with iron, however, usually contain small grains of elemental iron, Fe. What must the body do to this elemental iron to make good of it—oxidation or reduction?

70. The general chemical equation for photosynthesis is shown here. Through this reaction, are the oxygens of the water molecules, H_2O, oxidized or reduced?

$$6 CO_2 + 6 H_2O \rightarrow C_6H_{12}O_6 + 6 O_2$$

71. Why is the air over an open flame always moist?

72. The active ingredient in most dental teeth whitening formulas is hydrogen peroxide, H_2O_2, which readily decomposes into water and what other molecule? Write the balanced chemical equation for this decomposition.

73. The digestion and subsequent metabolism of foods and drugs is an oxidation process. Does this process tend to make the molecules of the foods and drugs more or less polar?

74. Why is it easier for the body to excrete a polar molecule than it is to excrete a nonpolar molecule?

75. Steel wool wetted with vinegar is sealed inside a balloon inflated with air. After several hours, the steel is turning to rust. What happens to the volume of the balloon?

76. Steel wool wetted with vinegar is stuffed into a narrow-mouth round glass bottle. A rubber balloon is then sealed over the mouth of the bottle. After several hours, the balloon is inflated in the bottle in an inverted manner. Explain.

77. Does breathing cause you to lose or gain weight?

THINK AND DISCUSS (EVALUATION)

78. What if Mars were nicely inhabitable by humans and we had developed the technology to transport and settle half the world's population there. At present birth rates, how long would it be before we were faced with not one but two world populations of 7 billion? (Hint: the world population hit 3.5 billion in 1968.) (Also see the Contextual Chemistry essay appearing at the end of Chapter 17.)

79. Use your thoughts from the previous question to explain why it is a good idea for developed countries to export sustainable energy technology, such as fuel cells and photovoltaics, to developing countries where the human population is growing the fastest.

80. In the centralized model for generating electricity, a relatively small number of power plants produce the massive amounts of electricity that everyone needs. In the decentralized model, electricity is generated by numerous smaller substations, which may include personal wind turbines or photovoltaics. What are the advantages and disadvantages of each model? When should one be favored over the other?

READINESS ASSURANCE TEST (RAT)

If you have a good handle on this chapter, then you should be able to score at least 7 out of 10 on this RAT. Check your answers online at www.ConceptualChemistry.com. If you score less than 7, you need to study further before moving on.

Choose the BEST answer to the following.

1. When lightning strikes, nitrogen molecules, N_2, and oxygen molecules, O_2, in the air react to form nitrates, NO_3^-, which come down in the rain to help fertilize the soil. Is this an example of oxidation or reduction?

    a. Oxidation

    b. Reduction

    c. Both

    d. Neither

2. What element is oxidized in the following equation? What element is reduced?

$$Sn^{2+} + 2Ag \rightarrow Sn + 2Ag^+$$

    a. The tin ion, Sn^{2+}, is oxidized, while the silver, Ag, is reduced.

    b. The tin ion, Sn^{2+}, is reduced, while the silver, Ag, is oxidized.

    c. Both the tin ion, Sn^{2+}, and the silver, Ag, are reduced.

    d. Both the tin ion, Sn^{2+}, and the silver, Ag, are oxidized.

3. What is the purpose of the salt bridge in Figure 11.6?

    a. To prevent any further migration of electrons through the wire.

b. To allow for the buildup of positively charged ions in one container and negatively charged ions in the other container.

c. To allow the Fe^{2+} and the Cu^{2+} to flow freely between the two containers.

d. To prevent a buildup of charge in either of the two chambers.

4. How does an atom's electronegativity relate to its ability to act as an oxidizing agent?

a. The greater the electronegativity of an atom, the lower its ability to act as an oxidizing agent.

b. The lower the electronegativity of an atom, the greater its ability to act as an oxidizing agent.

c. The greater the electronegativity of an atom, the greater its ability to act as an oxidizing agent.

d. Electronegativity does not affect the atom's ability to act as an oxidizing agent.

5. Sodium metal is

a. oxidized in the production of aluminum.

b. reduced in the production of aluminum.

c. both oxidized and reduced in the production of aluminum.

d. neither oxidized nor reduced in the production of aluminum.

6. Why does a battery that has thick zinc walls last longer than a battery that has thin zinc walls?

a. Thick zinc walls prevent the battery from overheating.

b. Thicker zinc walls prevent electrons from being lost into the surrounding environment.

c. Thicker zinc walls are chemically more resistant to battery acid.

d. The zinc walls are transformed into zinc ions as the battery provides electricity.

7. How is the junction barrier between n-type and p-type silicon much like a one-way valve?

a. Only one side of the barrier is exposed to light, which causes the electrons to flow in the opposite direction.

b. The orientation of the charge around the barrier allows the electrons to flow in only one direction.

c. The electrons can only be lost from the n-type silicon wafer.

d. The electrons can only be lost from the p-type silicon wafer.

8. Iron is useful for reducing copper ions to copper metal. Might it also be used to reduce sodium ions to sodium metal?

a. No, because sodium ions are too difficult to reduce.

b. Yes, especially in that sodium has such a low ionization energy.

c. Yes, especially in that sodium has such a high ionization energy.

d. No, because the sodium ions in turn oxidize the iron.

9. The general chemical equation for photosynthesis is shown here. Through this reaction, are the carbons of the carbon dioxide molecules oxidized or reduced?

$$6\,CO_2 + 6\,H_2O \rightarrow C_6H_{12}O_6 + 6\,O_2$$

a. These carbon atoms are oxidized.

b. These carbon atoms are reduced.

c. Half of these carbon atoms are reduced, and half are oxidized.

d. These carbon atoms are neither oxidized nor reduced.

10. Most of the carbon of the carbon dioxide you exhale ultimately comes from

a. the carbon dioxide you inhale.

b. the carbonic acid dissolved in the water you drink.

c. stomach gases.

d. bacteria in your gut.

e. the food you eat.

Contextual Chemistry

The Wonder Chemical, but...

What chemical holds the record for saving the most lives? Penicillin? Aspirin? Guess again. When it comes to saving lives, perhaps the most successful chemical is chlorine, which has saved many millions of us by disinfecting our drinking water. For this reason alone, chlorine ranks as a "wonder chemical," but wait, there's more! About 85 percent of all pharmaceuticals and about 95 percent of all crop protection chemicals are synthesized using chlorine. Chlorine is also an important component of many plastics, and it is used in the manufacture of a countless number of products. In all, about 45 percent of the U.S. gross domestic product is in some way rooted in chlorine chemistry. In this Spotlight, we focus on some of the details of the wonders of chlorine, beginning with its role in disinfecting water and ending with a mindful look at some of chlorine's downsides.

Municipalities chlorinate both drinking water and wastewater by bubbling chlorine gas, Cl_2, through the water. The chlorine reacts with the water to produce hypochlorous acid, $HOCl$, and hydrogen and chloride ions. The hypochlorous acid is a weak acid, so it stays in a molecular form, as shown next. The hypochlorous acid molecule is able to penetrate the nonpolar microbial cell walls, which makes it an effective disinfectant.

$$Cl_2(g) \; + \; H_2O(l) \; \rightarrow \; HOCl(aq)$$

Chlorine   Water   Hypochlorous
gas   acid

$$+ \; H^+(aq) \; + \; Cl^-(aq)$$

Hydrogen   Chloride
ion   ion

Chlorine and hypochlorous acid, as well as water-disinfecting bleach solutions, can react with organic components of drinking and waste

waters, creating chlorinated hydrocarbons, many of which are cancer-causing. Some municipalities have therefore switched to chlorine dioxide, ClO_2, which doesn't form chlorinated hydrocarbons so readily.

Most chlorine is manufactured from the electrolysis of salt water, as shown next. Notably, this reaction also produces two other valuable chemicals, sodium hydroxide, $NaOH$, and hydrogen, H_2.

$$2\,NaCl(aq) + 2\,H_2O(l) + \text{electricity} \rightarrow$$
$$Cl_2(g) \; + \; 2\,NaOH(aq) \; + \; H_2(g)$$

Chlorine is useful for the manufacture of many chemical products, even though these products themselves contain no chlorine. To create titanium metal, for example, the mineral titanium dioxide is reacted with chlorine to form titanium tetrachloride, $TiCl_4$ (sometimes humorously called "tickle"), which is then reduced by magnesium metal, Mg, as shown in the following equations:

$$TiO_2 + 2\,Cl_2 + \text{carbon} \rightarrow TiCl_4 + CO_2$$
$$TiCl_4 + 2\,Mg \rightarrow 2\,MgCl_2 + Ti$$

Interestingly, titanium tetrachloride is the "ink" used by skywriters. When the pilot is ready to maneuver the airplane to spell out words, she releases

a spray of $TiCl_4$, which reacts with atmospheric moisture to form visibly white titanium dioxide particles.

Chemists designing pharmaceuticals and crop protection chemicals often add chlorine atoms to the structure in order to modify their potency. The anti-anxiety agent Valium® is an example.

About one-third of the 40 million tons of chlorine produced annually goes to the manufacture of polyvinyl chloride, PVC, which is one of the most versatile of all plastics. PVC is ubiquitous, being used for pipes, flooring, electrical insulation, wallpaper, school supplies, swimming pools, and many other common products.

Valium

So chlorine has become an integral part of modern life. There is an important aspect of chlorine chemistry, however, that everyone should be aware of. Specifically, chlorine reacts with organic molecules to form a class of toxic molecules known as persistent organic pollutants, also known as POPs. A well-known group of POPs are the dioxins, which are represented by the compound 2, 3, 7, 8-tetrachloro-benzo-p-dioxin, also known as TCDD. These agents cause cancer and disrupt many bodily systems, especially those related to reproduction, immune responses, and hormones. Children are particularly susceptible because their bodies are still developing.

TCDD

All persistent organic pollutants share at least three qualities: (1) They are chemically stable, so they don't decompose in the environment—that is, they can "persist" for many years. (2) They are nonpolar toxins that tend to bioaccumulate within fatty tissues, especially in organisms higher in the food chain, such as humans. (3) They are semivolatile, which enables them to evaporate into the atmosphere and be carried long distances by the wind, so they are found globally.

Some POPs, such as the insecticide DDT (Chapter 15), are created on purpose. Many persistent organic pollutants, however, are created inadvertently, such as POPs that form upon the chlorine bleaching of paper. When possible, these POPs are sequestered as wastes. In the past, these wastes were not known to be harmful and they were often improperly buried, only to create later environmental hazards, as occurred at Love Canal, New York, in the late 1970s. (See the Contextual Chemistry essay at the end of Chapter 6.)

Wherever chlorine organic compounds are burned, persistent organic pollutants are formed, especially when there is a less-than-optimal quantity of oxygen available and the burning is incomplete. Since the 1970s, however, industrial emissions of POPs have declined over 90 percent, largely due to enforced regulations and technology that allows cleaner burning. Today, the most significant source of POPs, especially dioxins, is backyard trash burning, typically done in large metal barrels. Many municipalities now outlaw this hazardous practice.

Just as we are surrounded by the beneficial products of chlorine, so are we surrounded by chlorine's toxic by-products. The average American diet, for example, provides about 0.10 nanograms of dioxins per day. Even remote regions such as Arctic habitats have measurable quantities of POPs. In response, nations organized the Stockholm Convention on Persistent Organic Pollutants in 2001, during which a global protocol for containing POPs was created (http://chm.pops.int).

You are encouraged to do your own literature research. Many websites, such as www.ejnet.org/dioxin, will tell you POPs are a problem of epic proportions. Others, such as http://chlorine.americanchemistry.com, will point to evidence showing that the concentration of POPs in humans has been reduced dramatically over the past 30 years, while the production of PVC and other chlorine compounds has more than tripled. Chlorine is indeed a wonder chemical, but as with any technology, we need to make sure that the benefits are well worth the risks. Becoming a well-informed citizen is a good place to start.

CONCEPT CHECK

Why should you never discard plastic wrap in a hot barbeque grill?

CHECK YOUR ANSWER Many plastic wraps contain chlorine. The combustion of this plastic puts forth a large dose of dioxin and other POPs directly onto your food and into the air you breathe.

Think and Discuss

1. We live in a time when 1.1 billion people around the world lack access to safe drinking water. Chlorine powders, such as sodium hypochlorite, $NaClO$, however, can provide a family with safe drinking water for as little as 10 cents a day. What agencies or institutions should be involved in getting such water disinfectants to people who need them? What obstacles might be encountered? How might they be overcome?

2. What would you do if you discovered that your neighborhood was built upon a leaking toxic waste dump? Might you be able to sell your house? What would you do if governmental agencies told you that there was no proof that the toxic wastes were causing you harm?

3. The compound tetraethyl lead, $Pb(CH_2CH_3)_4$, was first formulated with gasoline in the 1920s to help car engines run more smoothly. It was known to be poisonous to humans, yet a ban on leaded gasoline in the United States didn't occur until the 1970s. In many countries, leaded gasoline is still in use. Explain how this does or doesn't relate to the chlorine industry.

4. Why are we not in danger of ever running out of chlorine? Does this mean that the cost of producing chlorine will remain relatively stable?

5. How much effort should really be put into controlling POPs when average life spans of humans continue to increase so remarkably?

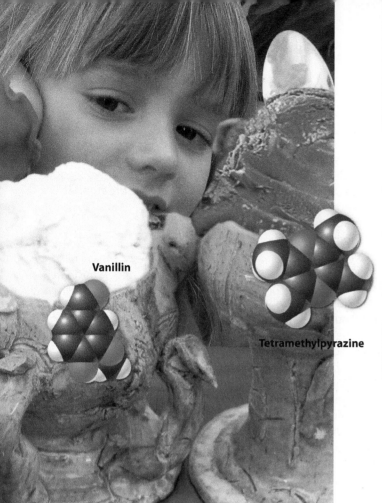

Vanillin

Tetramethylpyrazine

▲ Organic molecules can be quite tasty, especially those that provide the flavor of our favorite ice cream.

12

Organic Compounds

THE MAIN IDEA

Carbon can form a limitless number of chemical structures.

Carbon atoms have the rare ability to bond with themselves repeatedly. This allows the building of large molecular stuctures. Add to this the fact that carbon atoms can also bond with atoms of other elements and you see the possibility of an endless number of different carbon-based molecules. Each molecule has its own unique set of physical, chemical, and biological properties. The flavor of vanilla, for example, is perceived when the compound *vanillin* is absorbed by the sensory organs in the nose. Vanillin is the essential ingredient in anything having the flavor of vanilla—without vanillin, there is no vanilla flavor. The flavor of chocolate, on the other hand, is generated when not just one, but a wide assortment of carbon-based molecules are absorbed in the nose. One of the more significant of these molecules is *tetramethylpyrazine*. Life is based on carbon's ability to form diverse structures. Reflecting this fact, the branch of chemistry that is the study of carbon-containing chemical compounds has come to be known as **organic chemistry.** Because organic compounds are so closely tied to living organisms and because they have many applications— from flavorings to fuels, polymers, medicines, agriculture, and more—it is important to have a basic understanding of them.

Chemistry

Rubbing the Wrong Way

See firsthand the destructive action of isopropyl alcohol on proteins.

PROCEDURE

1. Crack open an egg and place the egg white and the yolk into two separate bowls.
2. Pour a capful of isopropyl alcohol into the egg white and observe what happens.
3. In the second bowl, scramble the yolk with a fork. Add a capful of isopropyl alcohol to the stirred yolk and observe what happens.

ANALYZE AND CONCLUDE

1. What happens to the color of egg whites when they are cooked? What happened to the color of the egg whites when they were treated with isopropyl alcohol? How is this similar to or different from what happened with the egg yolk?

2. How might the proteins within microbes respond to isopropyl alcohol? How might the tissues of your mouth and digestive system respond to isopropyl alcohol? (Hint: isopropyl alcohol is a serious poison when ingested. Do NOT ingest isopropyl alcohol.) Are all proteins destroyed by isopropyl alcohol? Might the proteins of your skin be different from the type of proteins found in your digestive system?

3. Why is isopropyl alcohol useful for cleaning your skin prior to a shot? Might skin oils, grime, and dirt be easily wiped away with a cotton swab soaked in isopropyl alcohol? Why or why not?

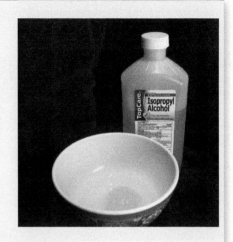

12.1 Hydrocarbons Contain Only Carbon and Hydrogen

EXPLAIN THIS

How is a road like an oil spill?

We begin with the simplest organic compounds—those consisting of only carbon and hydrogen. Organic compounds that contain only carbon and hydrogen are called **hydrocarbons.** The simplest hydrocarbon is methane, CH_4, with only one carbon per molecule as shown in **Figure 12.1** on the next page. Methane is the main component of natural gas. The hydrocarbon octane, C_8H_{18}, has eight carbons per molecule and is a component of gasoline. The hydrocarbon polyethylene contains hundreds of carbon and hydrogen atoms per molecule. Polyethylene is a plastic used to make many familiar items, such as milk containers and plastic bags.

Hydrocarbons may differ in the way the carbon atoms connect to one another. **Figure 12.2** shows the three hydrocarbons pentane, isopentane, and neopentane. These hydrocarbons have the same molecular formula, C_5H_{12}, but they are structurally different from one another. The carbon framework of pentane is a chain of five carbon atoms. In isopentane, the carbon chain branches, so that the framework is a *four*-carbon chain branched at the second carbon. In neopentane, a central carbon atom is bonded to four surrounding carbon atoms.

LEARNING OBJECTIVE

Identify the structures of hydrocarbons.

READING CHECK

What types of atoms are found in hydrocarbons?

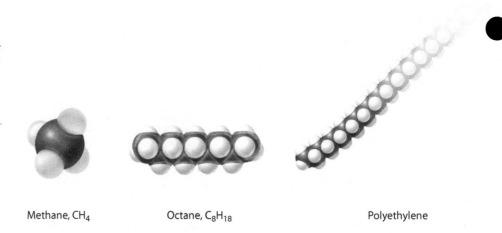

Methane, CH_4 Octane, C_8H_{18} Polyethylene

We can see the different structural features of pentanes, isopentane, and neopentane more clearly by drawing the molecules in two dimensions, as shown in the middle row of Figure 12.2. Alternatively, we can represent them by the *stick structures* (sometimes called line-angle drawings or skeletal structures) shown in the bottom row. A stick structure is a commonly used shorthand notation for representing an organic molecule. Each line (stick) represents a covalent bond, and carbon atoms are understood to exist at the end of any line wherever two or more straight lines meet (unless another type of atom is drawn at the end of the line). Any hydrogen atoms bonded to the carbons also typically are not shown. Instead, their presence is assumed so that the focus can remain on the structure formed by the carbon atoms.

Molecules such as pentane, isopentane, and neopentane have the same molecular formula, which means they have the same number of the same kinds of atoms. The way these atoms are put together, however, is different. We say that each has its own *configuration*, where the term **configuration** refers to how the atoms are connected. Different configurations result in different chemical structures. Molecules with the same chemical formula but different configurations (and hence, different structures) are known as **structural isomers.** Structural isomers are different from each other, having different physical and chemical properties. For example, pentane has a boiling point of 36°C, isopentane's boiling point is 30°C, and neopentane's is 10°C.

The number of possible structural isomers for a chemical formula increases rapidly as the number of carbon atoms increases. There are three structural isomers for compounds having the formula C_5H_{12}, 18 for C_8H_{18}, 75 for $C_{10}H_{22}$, and a whopping 366,319 for $C_{20}H_{42}$!

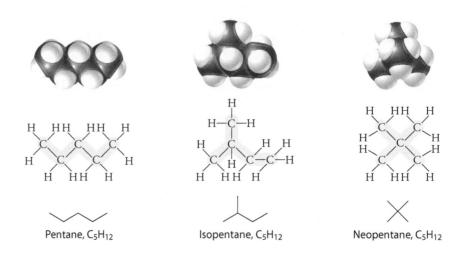

Pentane, C_5H_{12} Isopentane, C_5H_{12} Neopentane, C_5H_{12}

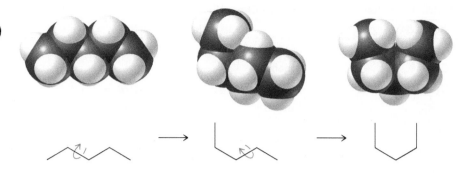

▲ **Figure 12.3**
Three conformations for a molecule of pentane. The conformation changes as bonds rotate as indicated by the blue arrows. The molecule looks different in each conformation, but the five-carbon framework is the same in all three conformations. In a sample of liquid pentane, the molecules are found in all conformations—not unlike a bucket of worms.

A carbon-based molecule can have different spatial orientations, called **conformations.** Flex your wrist, elbow, and shoulder joints and you'll find your arm passing through a range of conformations. Likewise, organic molecules can twist and turn about their carbon–carbon single bonds and thus have a range of conformations. The structures in **Figure 12.3**, for example, are different conformations of pentane. In the language of organic chemistry, we say that the *configuration* of a molecule such as pentane has a broad range of *conformations*. Change the configuration of pentane, however, and you no longer have pentane. Rather, you have a different structural isomer, such as isopentane, which has its own range of different conformations.

CONCEPTCHECK

Which carbon–carbon bond was rotated to go from the "before" conformation of isopentane to the "after" conformation?

Before   After

CHECK YOUR ANSWER   Bond "c." This rotation is similar to that of the arm of an arm wrestler who, with the arm just above the table while on the brink of losing, suddenly gets a surge of strength and swings the opponent's arm through a half-circle arc and wins. The best way to understand the geometrical shapes of organic molecules is to get your hands on a set of molecular models.

Before   After

The number of carbon atoms within a hydrocarbon is indicated by the hydrocarbon's name, as shown in Table 12.1.

When the hydrocarbon is branched, the name of the hydrocarbon is based upon the longest carbon chain. Smaller branches off this longest chain are written as a prefix ending in *-yl*. As shown in Table 12.1, a one-carbon branch would be indicated by the prefix methyl, where *meth-* means a single carbon. Also,

TABLE 12.1 Names of Simple Straight-Chain Hydrocarbons

FORMULA	HYDROCARBON NAME	–YL PREFIX
CH_4	methane	methyl
C_2H_6	ethane	ethyl
C_3H_8	propane	propyl
C_4H_{10}	butane	butyl
C_5H_{12}	pentane	pentyl
C_6H_{14}	hexane	hexyl
C_7H_{16}	heptane	heptyl
C_8H_{18}	octane	octyl
C_9H_{20}	nonane	nonyl
$C_{10}H_{22}$	decane	decyl

the longest chain is numbered to indicate where the branching takes place. For example, the following compound is 3-methylhexane because it has a methyl group branching off the third carbon of hexane:

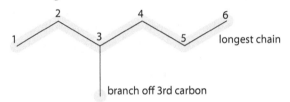

longest chain

branch off 3rd carbon

CHEMICAL CONNECTIONS

How is aspirin connected to petroleum?

Below is the structure for 2,3-dimethylhexane, which tells us that there are two methyl groups—one located at the second carbon and another located at the third carbon. Number the longest chain backwards and you would have 4,5-dimethylhexane, which is the identical structure. The convention, however, is to always use the lowest numerals possible when naming an organic compound.

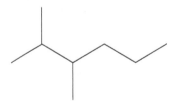

CONCEPTCHECK

Neopentane, shown in Figure 12.2, has an alternate name based upon the above naming methodology. What is this alternate name for neopentane?

CHECK YOUR ANSWER Neopentane also goes by the name 2,2-dimethylpropane. The "di" is a prefix that means two, which in this case means two methyls. The numerals tell us on what carbon these two methyl groups are attached. Note that in this case, the numeral 2 is used twice because each group needs its own number. For more practice at naming structures, see the questions at the back of this chapter.

Hydrocarbons are obtained primarily from coal and petroleum. Most of the coal and petroleum that exist today were formed between 290 million and 354 million years ago, when plant and animal matter decayed in the absence of oxygen. At that time, the Earth was covered with extensive swamps that, because they were close to sea level, periodically became submerged. The organic matter of the swamps was buried beneath layers of marine sediments and was eventually transformed to either coal or petroleum.

Coal is a solid material containing many large, complex hydrocarbon molecules. Most of the coal mined today is used for the production of steel and for generating electricity at coal-burning power plants.

Petroleum, also called crude oil, is a liquid readily separated into its hydrocarbon components through a process known as *fractional distillation,* shown in **Figure 12.4.** The crude oil is heated to a temperature high enough to vaporize most of its components. The hot vapor flows into the bottom of a fractionating tower, which is warmer at the bottom than at the top. As the vapor rises in the tower and cools, the various components begin to condense. Hydrocarbons that have high boiling points, such as tar, condense first at warmer temperatures. Hydrocarbons that have low boiling points, such as gasoline, travel to the cooler regions at the top of the tower before condensing. Pipes drain the various liquid hydrocarbon fractions from the tower. Natural gas, which is primarily methane, does not condense. It remains a gas and is collected at the top of the tower.

CONCEPT CHECK

Do the heavier or the lighter molecules of crude oil rise to the top of the fractionation tower?

CHECK YOUR ANSWER The lighter molecules found in crude oil are the ones that rise highest within the fractionation tower. These are the molecules that have the lower boiling points. So the lower the boiling point is, the higher the molecules rise.

Differences in the strength of molecular attractions explain why different hydrocarbons condense at different temperatures. As discussed in Section 7.1, in our comparison of induced dipole–induced dipole attractions in methane and octane, larger hydrocarbons experience many more of these attractions than smaller hydrocarbons do. For this reason, the larger hydrocarbons condense readily at high temperatures and so are found at the bottom of the tower. Smaller molecules, because they experience fewer attractions to neighbors, condense only at the cooler temperatures found at the top of the tower.

The gasoline obtained from the fractional distillation of petroleum consists of a wide variety of hydrocarbons having similar boiling points. Some of these components burn more efficiently than others in a car engine. The straight-chain hydrocarbons, such as hexane, burn too quickly, causing what is called *engine knock,* as illustrated in **Figure 12.5.** Gasoline hydrocarbons that have more branching, such as isooctane, burn slowly and result in the engine's running more smoothly. These two compounds, heptane and isooctane, are used as standards in assigning *octane ratings* to gasoline. An octane number of 100 is arbitrarily assigned to isooctane, and heptane is assigned an octane number of 0. The anti-knock performance of a particular gasoline is compared with those of various mixtures of isooctane and heptane, and an octane number is assigned. **Figure 12.6** shows the octane information appearing on a typical gasoline pump.

FOR YOUR INFORMATION

As discussed in Section 10.6, we are placing unusually large amounts of carbon dioxide into the atmosphere by burning fossil fuels. The atmosphere, however, is not the only possible repository for the carbon dioxide we produce. The smokestacks of power plants, for example, can be modified to capture CO_2, which is then liquefied and pumped kilometers deep into the ground. Underground storage of carbon dioxide is already being employed at the Salah natural gas refinery in Algeria. Such a system, however, has its costs. The price of electricity from a CO_2-capturing coal-fired power plant would rise by about 20 percent. The long-term costs of not implementing such systems, however, may be even greater.

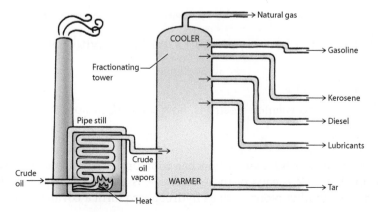

◀ Figure 12.4
A schematic for the fractional distillation of petroleum into its useful hydrocarbon components.

▶ **Figure 12.5**
(a) A straight-chain hydrocarbon, such as hexane, can be ignited from the heat generated as gasoline is compressed by the piston—before the spark plug fires. This upsets the timing of the engine cycle, giving rise to a knocking sound.
(b) Branched hydrocarbons, such as isooctane, burn less readily and are ignited not by compression alone, but only when the spark plug fires.

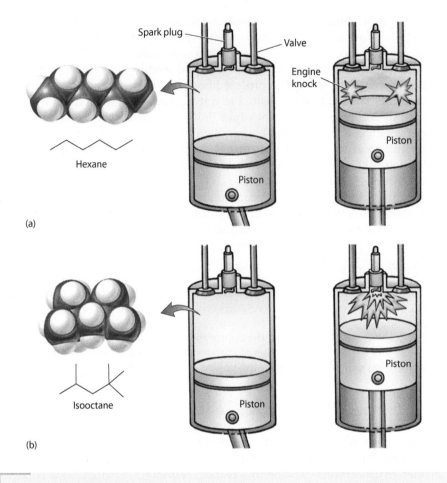

▲ **Figure 12.6**
Octane ratings are posted on gasoline pumps. Most modern car engines are designed to run optimally on regular octane gasoline. For such cars, higher-octane gasoline may provide poorer performance at greater cost.

CONCEPTCHECK

Which structural isomer shown in Figure 12.2 should have the highest octane rating?

CHECK YOUR ANSWER The structural isomer with the greatest amount of branching in the carbon framework will likely have the highest octane rating, making neopentane the clear winner. For your information, the ratings are as follows:

Compound	Octane Rating
Pentane	61.7
Isopentane	92.3
Neopentane	116.0

12.2 Unsaturated Hydrocarbons Have Multiple Bonds

LEARNING OBJECTIVE

Identify the structures of unsaturated hydrocarbons.

EXPLAIN THIS

With four unpaired valence electrons, how can carbon bond to only three adjacent atoms?

Recall from Section 6.1 that carbon has four unpaired valence electrons. As shown in **Figure 12.7**, each of these electrons is available for pairing with an electron from another atom, such as hydrogen, to form a covalent bond.

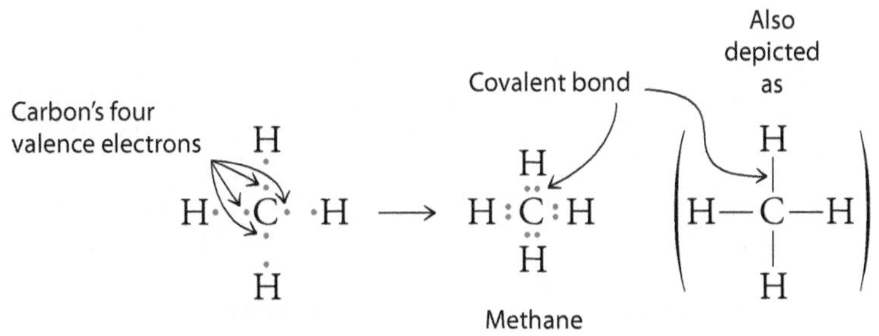

Carbon's four
valence electrons

Covalent bond

Also
depicted
as

$$H\cdot\overset{\cdot\cdot}{\underset{\cdot}{C}}\cdot H \longrightarrow H:\overset{\cdot\cdot}{\underset{H}{C}}:H \quad \left(H-\overset{\mid}{\underset{\mid}{C}}-H \right)$$

Methane

◀ Figure 12.7
Carbon has four valence electrons. Each electron pairs up with an electron from a hydrogen atom in the four covalent bonds of methane.

In all the hydrocarbons discussed so far, including the methane shown in Figure 12.7, each carbon atom is bonded to four neighboring atoms by four single covalent bonds. Such hydrocarbons are known as **saturated hydrocarbons.** The term *saturated* means that each carbon has as many atoms bonded to it as possible. We now explore cases where one or more carbon atoms in a hydrocarbon are bonded to fewer than four neighboring atoms. This occurs when at least one of the bonds between a carbon and a neighboring atom is a multiple bond. (See Section 6.5 for a review of multiple bonds.) A hydrocarbon that has a multiple bond—either double or triple—is known as an **unsaturated hydrocarbon.** Because of the multiple bond, two of the carbons are bonded to fewer than four other atoms. These carbons are thus said to be *unsaturated*.

Figure 12.8 compares the saturated hydrocarbon butane with the unsaturated hydrocarbon *cis*-2-butene. The number of atoms that are bonded to each of the two middle carbons of butane is four, whereas each of the two middle carbons of *cis*-2-butene is bonded to only three other atoms—a hydrogen and two carbons.

A generic word for a saturated hydrocarbon is *alkane*. All of the molecules discussed in Section 12.1 are examples of **alkanes.** This is indicated by the suffix *-ane*, as in *butane*. Unsaturated hydrocarbons containing one or more double bonds are called **alkenes.** This is indicated by the suffix *-ene*, as in *butene*.

An important aspect of the double bond of alkenes is that it cannot rotate more than a few degrees. By analogy, connect two gumdrops with a single toothpick and you'll be able to hold one gumdrop while rotating the other. Connect the two gumdrops with two toothpicks, however, and this rotational motion doesn't work because the two toothpicks can twist only so far. The significance of this for alkenes is what we call *cis/trans isomerism.* We use the prefix *cis* to indicate a structure in which the bulk of the carbons are on the same side of the length of the double bond, as shown in **Figure 12.9.** We use the prefix *trans* to indicate a structure in which the bulk of the carbons are on opposite sides of the length of the double bond.

An important unsaturated hydrocarbon (alkene) is benzene, C_6H_6, which may be drawn as three double bonds contained within a flat hexagonal ring, as is shown in **Figure 12.10a.** Unlike the double-bond electrons in most other unsaturated hydrocarbons, the electrons of the double bonds in benzene are not fixed between any two carbon atoms. Instead, these electrons move freely around the ring. This is commonly represented by drawing a circle within the ring, as shown in **Figure 12.10b,** rather than showing individual double bonds.

READING CHECK

What type of hydrocarbon has double or triple bonds?

Saturated hydrocarbon

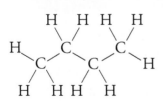

Butane, C_4H_{10}

Unsaturated hydrocarbon

$$\underset{H}{\overset{H}{\diagdown}}\overset{H}{\underset{}{C}}\overset{H}{\diagup}\quad\underset{H}{\overset{H}{\diagdown}}\overset{H}{\underset{}{C}}\overset{H}{\diagup}$$

cis-2-Butene, C_4H_8

◀ Figure 12.8
The carbons of the hydrocarbon butane are *saturated*, each being bonded to four other atoms. Because of the double bond, two of the carbons of the unsaturated hydrocarbon *cis*-2-butene are bonded to only three other atoms, which makes the molecule an unsaturated hydrocarbon.

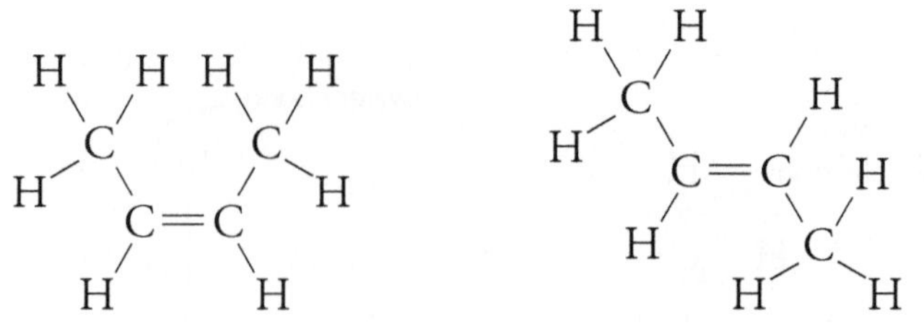

▶ **Figure 12.9**
The *cis*-isomer has the bulk of carbon atoms on the same side of the length of the double bond. The *trans*-isomer has these carbons on opposite sides. As a mnemonic, think "*cis*-same-side."

cis-2-butene   trans-2-butene

Many organic compounds contain one or more benzene rings in their structure. Because many of these compounds are fragrant, any organic molecule containing a benzene ring is classified as an **aromatic** compound (even if it is not particularly fragrant). **Figure 12.11** shows a few examples. Toluene, a common solvent used as a paint thinner, is toxic and gives airplane glue its distinctive odor. Some aromatic compounds, such as naphthalene, contain two or more benzene rings fused together. At one time, mothballs were made of naphthalene. Most mothballs sold today, however, are made of the less toxic 1,4-dichlorobenzene.

Unsaturated hydrocarbons containing a triple bond are called **alkynes.** An example of an unsaturated hydrocarbon containing a triple bond is ethyne, which traditionally is more widely known as acetylene, C_2H_2. A confined flame of acetylene burning in oxygen is hot enough to melt iron, which makes acetylene a choice fuel for welding (**Figure 12.12**). Alkynes are generally not as common as alkanes or alkenes.

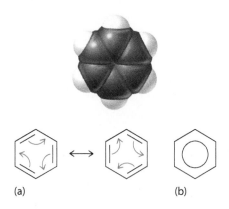

(a)   (b)

▲ **Figure 12.10**
(a) The double bonds of benzene, C_6H_6, are able to migrate around the ring.
(b) For this reason, they are often represented by a circle within the ring.

CONCEPT CHECK

Prolonged exposure to benzene increases the risk of developing certain cancers. The structure of aspirin contains a benzene ring. Does this indicate that prolonged exposure to aspirin will increase a person's risk of developing cancer?

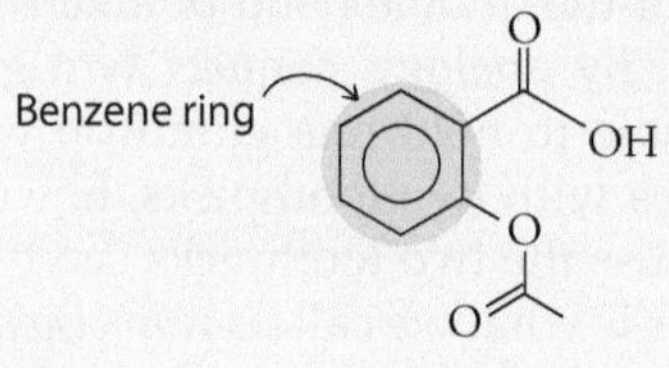

Benzene ring

Aspirin

CHECK YOUR ANSWER No. Although benzene and aspirin both contain a benzene ring, these two molecules have different overall structures and quite different chemical properties. Each carbon-containing organic compound has its own set of unique physical, chemical, and biological properties. While benzene may cause cancer, aspirin works as a safe remedy for headaches.

▶ **Figure 12.11**
The structures for three odoriferous organic compounds containing one or more benzene rings: toluene, naphthalene, and 1,4-dichlorobenzene.

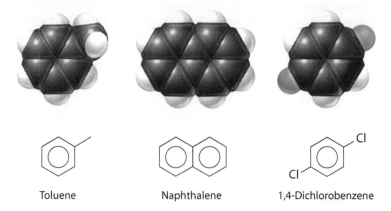

Toluene   Naphthalene   1,4-Dichlorobenzene

◀ **Figure 12.12**
The unsaturated alkyne acetylene, C_2H_2 (formally known as ethyne), when burned in this torch, produces a flame hot enough to melt iron.

$$H—C≡C—H$$

Acetylene

12.3 Functional Groups Give Organic Compounds Character

EXPLAIN THIS

Why are there so many different organic compounds?

Carbon atoms can bond to one another and to hydrogen atoms in many ways, which results in an incredibly large number of hydrocarbons. But carbon atoms can bond to atoms of other elements as well, further increasing the number of possible organic molecules. In organic chemistry, any atom other than carbon or hydrogen in an organic molecule is called a **heteroatom,** where *hetero-* means "different from either carbon or hydrogen."

A hydrocarbon structure can serve as a framework for the attachment of various heteroatoms. This is analogous to a Christmas tree's serving as the scaffolding on which ornaments are hung. Just as the ornaments give character to the tree, heteroatoms give character to an organic molecule. Heteroatoms have profound effects on the properties of an organic molecule.

Consider ethane, C_2H_6, and ethanol, C_2H_6O, which differ from each other by only a single oxygen atom. Ethane has a boiling point of $-88°C$, making it a gas at room temperature, and it does not dissolve in water very well. Ethanol, by contrast, has a boiling point of $+78°C$, making it a liquid at room temperature. It is infinitely soluble in water, and it is the active ingredient of alcoholic beverages. Consider further ethylamine, C_2H_7N, which has a nitrogen atom on the same basic two-carbon framework. This compound is a corrosive, pungent, highly toxic gas—most unlike either ethane or ethanol.

Organic molecules are classified according to the functional groups they contain. We define a **functional group** as a combination of atoms that behave as a unit. Most functional groups are distinguished by the heteroatoms they contain, and some common groups are listed in Table 12.2. Notice that the structures shown in Table 12.2 are not the structures of complete molecules. The atoms highlighted in blue, however, are all the atoms that go together to make a particular functional group.

The remainder of this chapter introduces the classes of organic molecules shown in Table 12.2. The role heteroatoms play in determining the properties of each class is the underlying theme. As you study this material, focus on understanding the chemical and physical properties of the various classes of compounds, for doing so will give you a greater appreciation of the remarkable diversity of organic molecules and their many applications.

LEARNING OBJECTIVE

Discuss the significance of heteroatoms in organic compounds.

READINGCHECK

What is a heteroatom?

 FOR YOUR INFORMATION

It is common for organic chemists to use the letter *R* to indicate an unspecified hydrocarbon coming off a functional group. For example, an alcohol can be indicated as R—OH, while an ether can be indicated as R—O—R. The *R* could equal a methyl group, $—CH_3$, an ethyl group, $—CH_2CH_3$, or any other group of carbon atoms as shown in Table 12.1.

 FOR YOUR INFORMATION

The chemistry of hydrocarbons is interesting, but start adding heteroatoms to these organic molecules and the chemistry becomes extraordinarily interesting. The organic chemicals of living organisms, for example, all contain heteroatoms.

H H
| |
H—C—C—H   Ethane
| |
H H

H H H
| | |
H—C—C—O   Ethanol
| |
H H

H H H
| | |
H—C—C—N   Ethylamine
| | \
H H H

TABLE 12.2   Functional Groups in Organic Molecules

GENERAL STRUCTURE	CLASS	GENERAL STRUCTURE	CLASS
—C—OH	Alcohols	O=C—H	Aldehydes
C=C ... C—OH	Phenols	O=C—N	Amides
—C—O—C—	Ethers	O=C—OH	Carboxylic acid
—C—N	Amines	O=C—O—C—	Esters
O=C (ketone)	Ketones		

CONCEPT CHECK

What is the significance of heteroatoms in an organic molecule?

CHECK YOUR ANSWER Heteroatoms largely determine an organic molecule's physical and chemical properties.

12.4 Alcohols, Phenols, and Ethers Contain Oxygen

LEARNING OBJECTIVE

Review the general properties of alcohols, phenols, and ethers.

EXPLAIN THIS

What do alcohols, phenols, and ethers have in common?

—C—OH

Alcohols are organic molecules in which a *hydroxyl group* is bonded to a saturated carbon. The hydroxyl group consists of an oxygen bonded to a hydrogen. Because of the polarity of the oxygen–hydrogen bond, low-formula-mass alcohols are often soluble in water, which is itself very polar. Some common alcohols and their melting and boiling points are listed in Table 12.3.

More than 11 billion pounds of methanol, CH_3OH, is produced annually in the United States. Most of it is used for making formaldehyde and acetic acid, important starting materials in the production of plastics. In addition, methanol is used as a solvent, an octane booster, and an anti-icing agent in gasoline. Sometimes called wood alcohol because it can be obtained from wood, methanol should never be ingested because in the body, it is metabolized to formaldehyde, H_2CO, and formic acid, HCO_2H. Formaldehyde is harmful to the eyes, can lead to blindness, and was once used to preserve dead biological

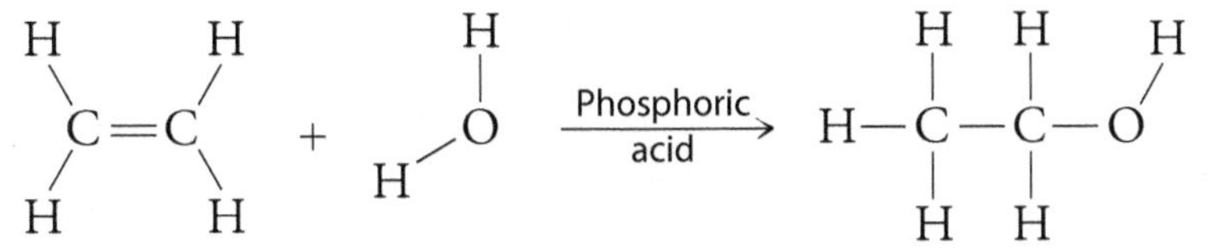

Ethene   Water   Ethanol

◀ **Figure 12.13**
Ethanol can be synthesized from the unsaturated hydrocarbon ethene, with phosphoric acid as a catalyst.

specimens. Formic acid, the active ingredient in an ant bite, can lower the pH of the blood to dangerous levels. Ingesting only about 15 milliliters (about 3 tablespoons) of methanol may lead to blindness, and about 30 milliliters can cause death.

Ethanol, C_2H_5OH, on the other hand, is the "alcohol" of alcoholic beverages, and it is one of the oldest chemicals manufactured by humans. Ethanol is prepared by feeding the sugars of various plants to certain yeasts, which produce ethanol through a biological process known as *fermentation*. Ethanol is also widely used as an industrial solvent. For many years, ethanol intended for this purpose was made by fermentation, but today industrial-grade ethanol is more cheaply manufactured from petroleum by-products such as ethene, as **Figure 12.13** illustrates.

The liquid produced by fermentation has an ethanol concentration no greater than about 12 percent, because at this concentration, the yeast cells begin to die. This is why most wines have an alcohol content of about 12 percent—they are produced solely by fermentation. To attain the higher ethanol concentrations found in such "hard" alcoholic beverages as gin and vodka, the fermented liquid must be distilled. In the United States, the ethanol content of distilled alcoholic beverages is measured as *proof,* which is twice the percentage of ethanol. An 86-proof whiskey, for example, is 43 percent ethanol by volume. The term *proof* evolved from a crude method once employed to test alcohol content. Gunpowder was wetted with a beverage of suspect alcohol content. If the beverage was primarily water, the powder would not ignite. If the beverage contained a significant amount of ethanol, the powder would burn, thus providing "proof" of the beverage's worth.

A third well-known alcohol is isopropyl alcohol, also called 2-propanol. This is the rubbing alcohol you buy at the drugstore. Although 2-propanol has a relatively high boiling point, it evaporates readily, leading to a pronounced cooling effect when it is applied to skin—an effect once used to reduce fevers. (Isopropyl alcohol is very toxic if ingested. See the Hands-On

FOR YOUR INFORMATION

Just as the body metabolizes methanol into formaldehyde, $HCOH$, it metabolizes ethanol into acetaldehyde, CH_3COH. Acetaldehyde won't cause blindness, but it does lead to some painful side effects, which people who drink too much experience as part of their hangover.

FOR YOUR INFORMATION

Alcohols can be classified as being primary, secondary, or tertiary. In a *primary* alcohol, the hydroxyl group is attached to the terminal carbon of a carbon chain. An example is ethanol, shown in Table 12.3. In a *secondary* alcohol, the hydroxyl group is attached to a central carbon, as seen with 2-propanol. For a *tertiary* alcohol, the hydroxyl group is attached to a carbon that is bonded to three other carbon atoms, as occurs in 2-methyl-2-propanol. Can you draw this structure?

TABLE 12.3 Some Simple Alcohols

STRUCTURE	SCIENTIFIC NAME	COMMON NAME	MELTING POINT (°C)	BOILING POINT (°C)
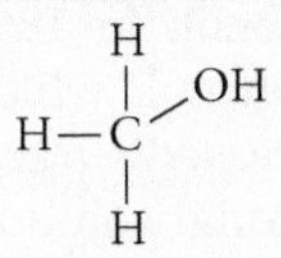	Methanol	Methyl alcohol	−97	65
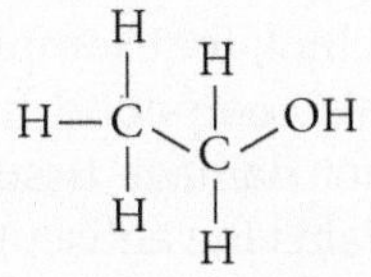	Ethanol	Ethyl alcohol	−115	78
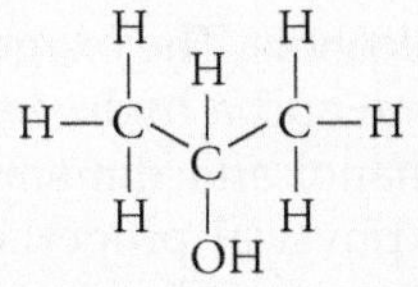	2-Propanol	Isopropyl alcohol	89	82

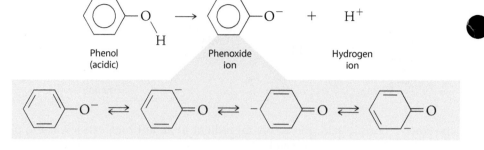

Phenol (acidic) → Phenoxide ion + Hydrogen ion

▶ **Figure 12.14**
The negative charge of the phenoxide ion is able to migrate on certain positions on the benzene ring. This mobility helps to accommodate the negative charge, which is why the phenolic group readily donates a hydrogen ion.

Chemistry activity at the beginning of this chapter to understand why. In place of isopropyl alcohol, washcloths wetted with cold water are nearly as effective in reducing fever, and they are far safer.) You are probably most familiar with the use of isopropyl alcohol as a topical disinfectant.

Phenols contain a phenolic group, which consists of a hydroxyl group bonded to a benzene ring. Because of the presence of the benzene ring, the hydrogen of the hydroxyl group is readily lost in an acid-base reaction, which makes the phenolic group mildly acidic.

The reason for this acidity is illustrated in **Figure 12.14**. How readily an acid donates a hydrogen ion is a function of how well the acid is able to accommodate the resulting negative charge it gains after donating the hydrogen ion. After a phenol donates the hydrogen ion, it becomes a negatively charged phenoxide ion. The negative charge of the phenoxide ion, however, is not restricted to the oxygen atom. Recall that the electrons of the benzene ring are able to migrate around the ring. In a similar manner, the electrons responsible for the negative charge of the phenoxide ion are also able to migrate around the ring, as shown in Figure 12.14. Just as it is easy for several people to hold a hot potato by quickly passing it around, it is easy for the phenoxide ion to hold the negative charge because the charge gets passed around. Because the negative charge of the ion is so nicely accommodated, the phenolic group is more acidic than it would be otherwise.

✔ R E A D I N G C H E C K

Is the phenol group slightly acidic or slightly basic?

Phenolic group

**F O R Y O U R
I N F O R M A T I O N**

We're classifying organic molecules based upon the functional groups they contain. As you will see shortly, however, organic molecules may contain more than one type of functional group. A single organic molecule, therefore, might be classified as both a phenol and an ether.

C O N C E P T C H E C K
Why are alcohols less acidic than phenols?

CHECK YOUR ANSWER An alcohol does not contain a benzene ring adjacent to the hydroxyl group. If the alcohol were to donate the hydroxyl hydrogen, the result would be a negative charge on the oxygen. Without an adjacent benzene ring, this negative charge has nowhere to go. As a result, an alcohol behaves only as a very weak acid, much the way water does.

The simplest phenol, shown in **Figure 12.15**, is called phenol. In 1867, Joseph Lister (1827–1912) discovered the antiseptic value of phenol, which, when applied to surgical instruments and incisions, greatly increased surgery survival rates. Phenol was the first purposefully used antibacterial solution, or *antiseptic*. Phenol damages healthy tissue, however, so a number of milder phenols have since been introduced. The phenol 4-hexylresorcinol, for example, is commonly used in throat lozenges and mouthwashes. This compound has even greater antiseptic properties than phenol, yet it does not damage tissue. Listerine® brand mouthwash (named after Joseph Lister) contains the antiseptic phenols thymol and methyl salicylate.

Ethers are organic compounds structurally related to alcohols. The oxygen atom in an ether group, however, is bonded not to a carbon and a hydrogen, but rather to two carbons. As we see in **Figure 12.16**, ethanol and dimethyl ether have the same chemical formula, C_2H_6O, but their physical properties are vastly different. Whereas ethanol is a liquid at room temperature (boiling

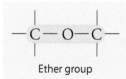

Ether group

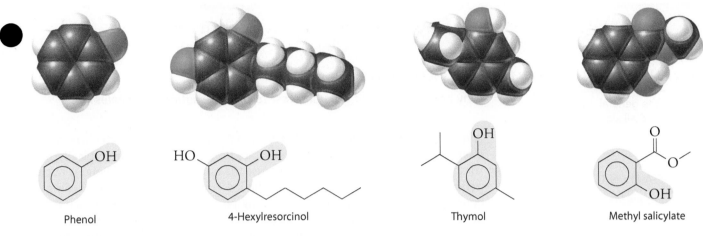

Phenol   **4-Hexylresorcinol**   **Thymol**   **Methyl salicylate**

▲ **Figure 12.15**
Every phenol contains a phenolic group (highlighted in blue).

point 78°C) and mixes quite well with water, dimethyl ether is a gas at room temperature (boiling point –25°C) and is much less soluble in water.

Ethers are not very soluble in water because, without the hydroxyl group, they are unable to form strong hydrogen bonds with water (Section 7.1). Furthermore, without the polar hydroxyl group, the molecular attractions among ether molecules are relatively weak. As a result, little energy is required to separate ether molecules from one another. This is why low-formula-mass ethers have relatively low boiling points and evaporate so readily.

Diethyl ether, shown in **Figure 12.17**, was one of the first general anesthetics. The anesthetic properties of this compound were discovered in the early 1800s, and its use revolutionized the practice of surgery. Because of its high volatility at room temperature, inhaled diethyl ether rapidly enters the bloodstream. Because this ether has low solubility in water and high volatility, it quickly leaves the bloodstream once introduced. Because of these physical properties, a surgical patient can be brought into and out of *general anesthesia* (a state of unconsciousness) simply by regulating the gases breathed. Modern-day gaseous anesthetics have fewer side effects than diethyl ether, but they operate on the same principle.

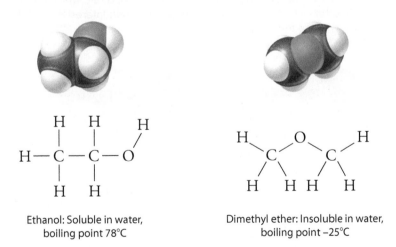

Ethanol: Soluble in water,
boiling point 78°C

Dimethyl ether: Insoluble in water,
boiling point –25°C

▲ **Figure 12.16**
The oxygen in an alcohol, such as ethanol, is bonded to one carbon atom and one hydrogen atom. The oxygen in an ether, such as dimethyl ether, is bonded to two carbon atoms. Because of this difference, alcohols and ethers of similar molecular mass have vastly different physical properties.

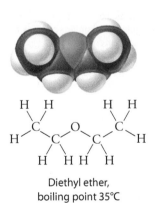

Diethyl ether,
boiling point 35°C

▲ **Figure 12.17**
Diethyl ether is the systematic name for the "ether" historically used as an anesthetic.

12.5 Amines and Alkaloids Contain Nitrogen

LEARNING OBJECTIVE

Review the general properties of amines and alkaloids.

READINGCHECK

What atoms are in an amine group?

Amine group

FOR YOUR INFORMATION

As we discuss in Chapter 14, nearly all pharmaceuticals that can be administered orally contain the nitrogen heteroatom in the water-soluble salt form.

EXPLAIN THIS

Why are rainforests of great interest to pharmaceutical companies?

Amines are organic compounds that contain the amine group, which is a nitrogen atom bonded to one, two, or three saturated carbons. Amines are typically less soluble in water than are alcohols, because the nitrogen–hydrogen bond is not as polar as the oxygen–hydrogen bond. The lower polarity and weaker hydrogen bonding of amines also means their boiling points are typically lower than those of alcohols of similar formula mass. Table 12.4 lists three simple amines.

One of the most notable physical properties of many low-formula-mass amines is their offensive odor. **Figure 12.18** shows two appropriately named amines, putrescine and cadaverine, which are responsible, in part, for the odor of decaying flesh.

Amines are typically alkaline, because the nitrogen atom readily accepts a hydrogen ion, as **Figure 12.19** illustrates. A class of alkaline amines found in nature are the *alkaloids*. Because many alkaloids have medicinal or other biological effects, there is great interest in isolating these compounds from the plants or marine organisms that contain them. As shown in **Figure 12.20**, an alkaloid, such as caffeine, reacts with an acid to form a salt that is usually quite soluble in water. This is in contrast to the unionized form of the alkaloid, known as a *free base*, which is typically insoluble in water.

Most alkaloids exist in nature not in their free-base form, but rather as the salts of naturally occurring acids known as *tannins*. The alkaloid salts of these acids are usually much more soluble in hot water than in cold water. The caffeine in coffee and tea exists in the form of the tannin salt, which is why coffee and tea are more effectively brewed in hot water. As **Figure 12.21** relates, tannins are also responsible for the stains caused by these beverages.

CONCEPTCHECK

Why do most caffeinated soft drinks also contain phosphoric acid?

CHECK YOUR ANSWER The phosphoric acid, as shown in Figure 12.20, reacts with the caffeine to form the caffeine–phosphoric acid salt, which is much more soluble in cold water than the naturally occurring tannin salt. Interestingly, it also adds a desirable tingle to the tongue.

▶ **Figure 12.18**
Low-formula-mass amines such as these tend to have offensive odors.

▶ **Figure 12.19**
Ethylamine acts as a weak base and accepts a hydrogen ion from water to become the ethyl ammonium ion. This reaction generates a small amount of hydroxide ions, which slightly increases the pH of the solution.

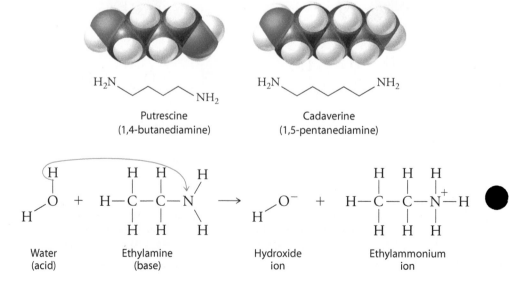

H_2N — Putrescine (1,4-butanediamine)

H_2N — NH_2 Cadaverine (1,5-pentanediamine)

Water (acid) + Ethylamine (base) → Hydroxide ion + Ethylammonium ion

TABLE 12.4 Three Simple Amines

STRUCTURE	NAME	MELTING POINT (°C)	BOILING POINT (°C)
 NH_2	Ethylamine	−81	17
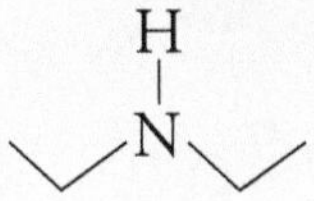 H N	Diethylamine	−50	55
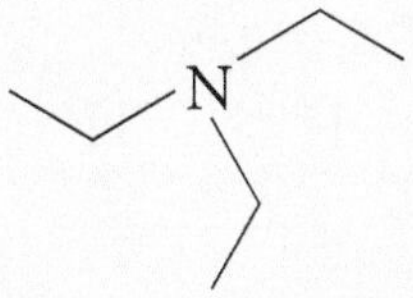 N	Triethylamine	−7	89

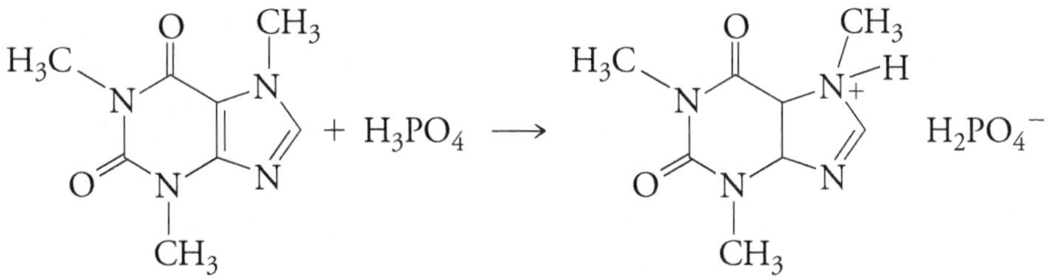

▲ Figure 12.21
Tannins are responsible for the brown stains in coffee mugs or on a coffee drinker's teeth. Because tannins are acidic, they can be readily removed with an alkaline cleanser such as baking soda.

$$\text{Caffeine, free-base form} + H_3PO_4 \longrightarrow \text{Caffeine–phosphoric acid salt} \quad H_2PO_4^-$$

Caffeine, free-base form (water insoluble)   Phosphoric acid   Caffeine–phosphoric acid salt (water soluble)

▲ Figure 12.20
All alkaloids are bases that react with acids to form salts. An example is the alkaloid caffeine, shown here reacting with phosphoric acid; both are common ingredients in soda beverages.

12.6 Carbonyl-Containing Compounds

EXPLAIN THIS

Why does the carbon of the carbonyl usually have a slightly positive charge?

The **carbonyl group** consists of a carbon atom double-bonded to an oxygen atom. It occurs in the organic compounds known as ketones, aldehydes, amides, carboxylic acids, and esters.

A **ketone** is a carbonyl-containing organic molecule in which the carbonyl carbon is bonded to two carbon atoms. A familiar example of a ketone is *acetone*, which is often used in fingernail-polish remover and is shown in **Figure 12.22a**. In an **aldehyde**, the carbonyl carbon is bonded either to one carbon atom and one hydrogen atom, as in **Figure 12.22b**, or, in the case of formaldehyde, the simplest aldehyde, to two hydrogen atoms.

Many aldehydes are particularly fragrant. A number of flowers, for example, owe their pleasant odor to the presence of simple aldehydes. The smells of lemons, cinnamon, and almonds are due to the aldehydes citral, cinnamaldehyde, and benzaldehyde, respectively. The structures of these three aldehydes are shown in **Figure 12.23**. Another aldehyde, vanillin, which was introduced at the beginning of this chapter, is the key flavoring molecule derived from seed pods of the vanilla orchid. You may have noticed that vanilla seed pods and vanilla extract are fairly expensive. Imitation vanilla flavoring is less expensive because it

LEARNING OBJECTIVE

Review the general properties of carbonyl compounds.

READING CHECK

What type of bond is found between the carbon and the oxygen of a carbonyl group?

Aldehyde group

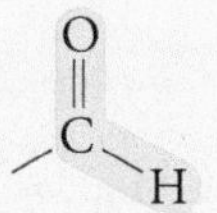

Ketone group

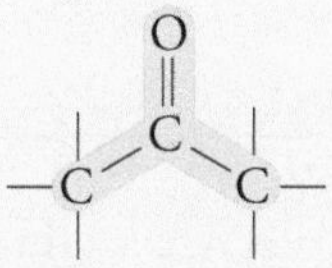

▶ Figure 12.22
(a) When the carbon of a carbonyl group is bonded to two carbon atoms, the result is a ketone. An example is acetone. (b) When the carbon of a carbonyl group is bonded to at least one hydrogen atom, the result is an aldehyde. An example is propanal. (c) The simplest aldehyde is formaldehyde, which has two hydrogens bonded to the carbon of the carbonyl.

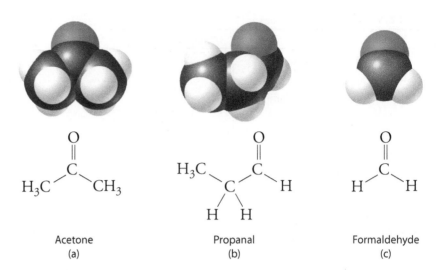

Acetone
(a)

Propanal
(b)

Formaldehyde
(c)

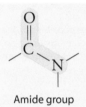

Amide group

▼ Figure 12.23
Aldehydes are responsible for many familiar fragrances.

is merely a solution of the compound vanillin, which is economically synthesized from waste chemicals from the wood pulp industry. Imitation vanilla does not taste the same as natural vanilla extract, however, because in addition to vanillin, many other flavorful molecules contribute to the complex taste of natural vanilla. Many books manufactured in the days before acid-free paper smell of vanilla because of the vanillin formed and released as the paper ages, a process that is accelerated by the acids the paper contains.

An **amide** is a carbonyl-containing organic molecule in which the carbonyl carbon is bonded to a nitrogen atom. The active ingredient of most mosquito repellents is an amide whose chemical name is *N,N*-diethyl-*m*-toluamide but is commercially known as DEET, shown in **Figure 12.24**. This compound is actually not an insecticide. Rather, it causes certain insects, especially mosquitoes, to lose their sense of direction, which effectively protects DEET wearers from being bitten.

A **carboxylic acid** is a carbonyl-containing organic molecule in which the carbonyl carbon is bonded to a hydroxyl group. As its name implies, this functional group is able to donate hydrogen ions. Organic molecules that contain it are therefore weakly acidic. An example is acetic acid, $C_2H_4O_2$, which, after water, is the main ingredient of vinegar. You may recall that this organic compound was used as an example of a weak acid back in Chapter 10.

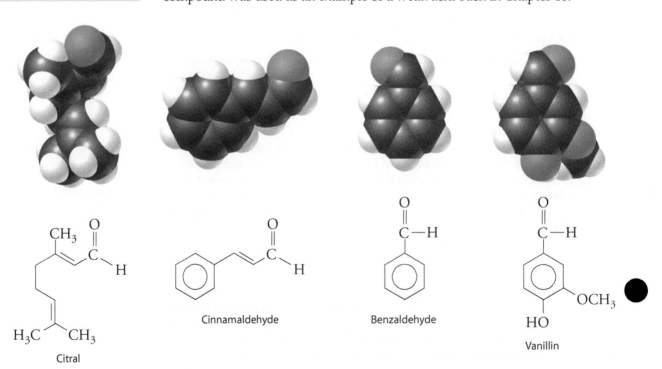

Citral

Cinnamaldehyde

Benzaldehyde

Vanillin

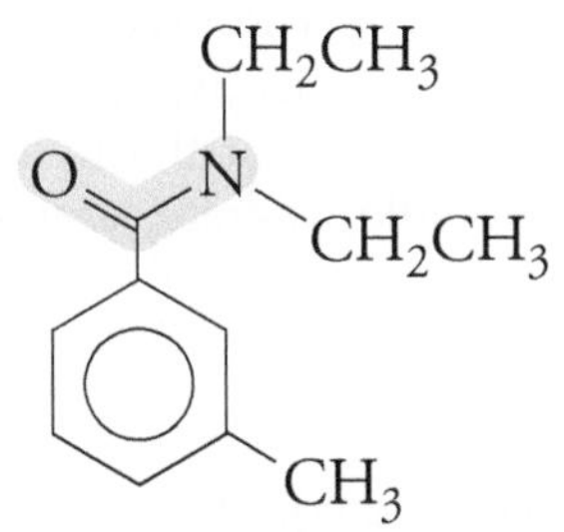

CH$_2$CH$_3$

N,N-Diethyl-*m*-toluamide
(DEET)

▲ **Figure 12.24**
N,N-diethyl-*m*-toluamide is an example of an amide. Amides contain the amide group, shown highlighted in blue.

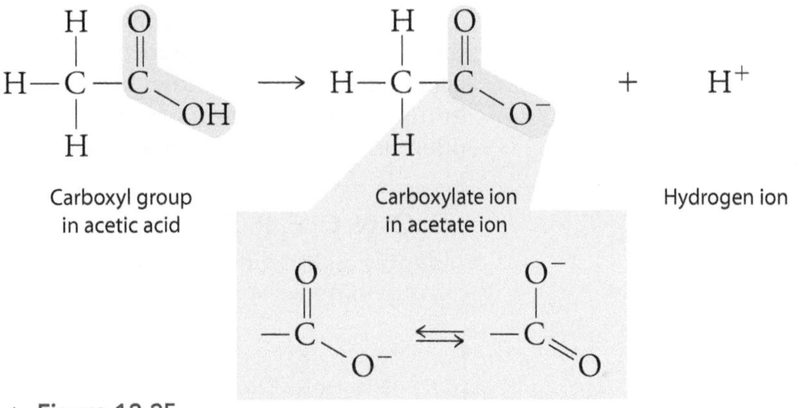

Carboxyl group
in acetic acid

Carboxylate ion
in acetate ion

Hydrogen ion

▲ **Figure 12.25**
The negative charge of the carboxylate ion is spread across the two oxygen atoms of the carboxyl group.

As with phenols, the acidity of a carboxylic acid results in part from the ability of the functional group to accommodate the negative charge of the ion that forms after the hydrogen ion has been donated. As shown in **Figure 12.25**, a carboxylic acid transforms to a carboxylate ion as it loses the hydrogen ion. The negative charge of the carboxylate ion is then distributed between the two oxygens. This spreading out helps to accommodate the negative charge.

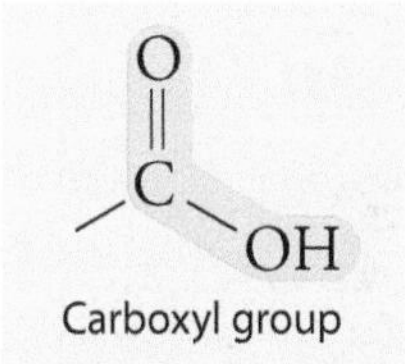

An interesting example of an organic compound that contains both a carboxylic acid and a phenol is salicylic acid, found in the bark of willow trees and illustrated in **Figure 12.26a**. At one time brewed for its antipyretic (fever-reducing) effect, salicylic acid is an important analgesic (painkiller), but it causes nausea and stomach upset due to its relatively high acidity, a result of the presence of two acidic functional groups. In 1899, Friedrich Bayer and Company, in Germany, introduced a chemically modified version of this compound in which the acidic phenolic group was transformed into an ester functional group. The result was the less acidic and more tolerable drug called acetylsalicylic acid, the chemical name for aspirin, shown in **Figure 12.26b**.

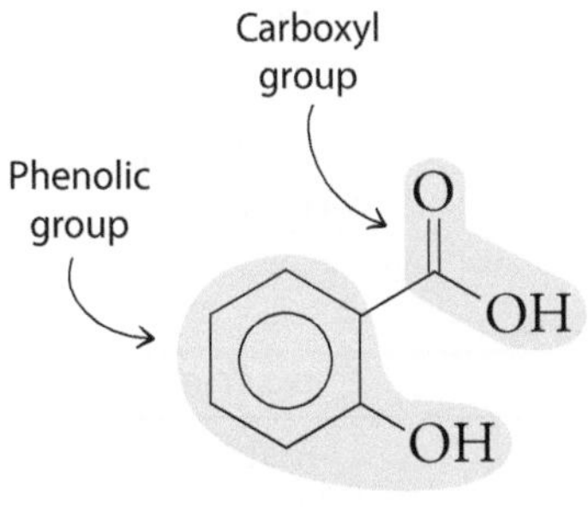

Carboxyl group

Phenolic group

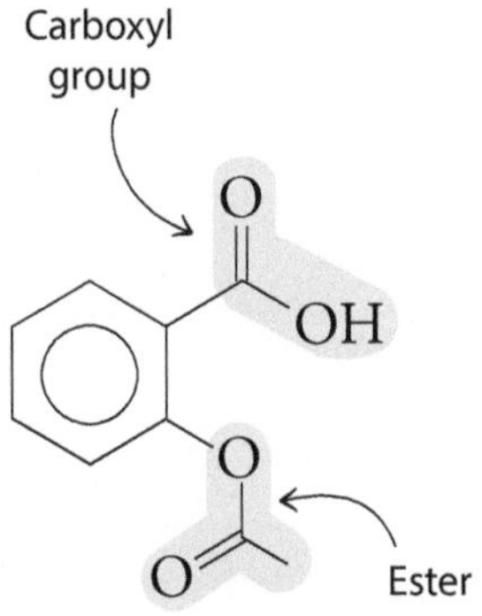

Carboxyl group

(a) Salicylic acid

◀ **Figure 12.26**
(a) Salicylic acid, which is found in the bark of willow trees, is an example of a molecule containing both a carboxyl group and a phenolic group. (b) Aspirin, acetylsalicylic acid, is less acidic than salicylic acid because it no longer contains the acidic phenolic group, which has been converted to an ester.

Carboxyl group

Ester

(b) Aspirin
(acetylsalicylic acid)

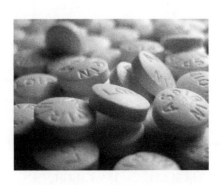

Ester group

An **ester** is an organic molecule similar to a carboxylic acid except that in the ester, the hydroxyl hydrogen is replaced by a carbon. Unlike carboxylic acids, esters are not acidic, because they lack the hydrogen of the hydroxyl group. Like aldehydes, many simple esters have notable fragrances and are often used as flavorings. Some familiar ones are listed in Table 12.5.

CONCEPTCHECK

Identify all the functional groups in these two molecules (ignore the sulfur group in penicillin G).

Testosterone

Penicillin G

CHECK YOUR ANSWERS Testosterone: alcohol and ketone. Penicillin G: amide (two amide groups) and carboxylic acid.

Esters are fairly easy to synthesize by dissolving a carboxylic acid in an alcohol and then bringing the mixture to a boil in the presence of a strong acid, such as sulfuric acid, H_2SO_4. Shown below is the synthesis of methyl salicylate from salicylic acid and methanol. Methyl salicylate is responsible for the smell of wintergreen and is a common ingredient of hard candies.

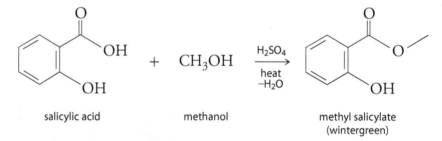

salicylic acid + methanol → methyl salicylate (wintergreen)

FOR YOUR INFORMATION

In the 1800s, most salicylic acid used by people was produced not from willow bark, but from coal tar. Tar residues within the salicylic acid had a nasty taste. This, combined with salicylic acid's stomach irritation, led many to view the salicylic acid cure as being worse than the disease. Felix Hoffmann was the chemist working at Bayer & Co, who in 1897, added the acetyl group to the phenol group of salicylic acid. According to Bayer, Hoffmann was inspired by his father, who had been complaining about the side effects of salicylic acid. To market the new drug, Bayer invented the name aspirin, where a- is for acetyl, -spir- is for the spirea flower, another natural source of salicylic acid, and in is a common suffix for medications. After World War I, Bayer, a German company, lost the rights to use the name aspirin. Bayer didn't regain these rights until 1994, for a steep price of $1 billion.

TABLE 12.5 Some Esters and Their Flavors and Odors

STRUCTURE	NAME	FLAVOR/ODOR
	Ethyl formate	Rum
	Isopentyl acetate	Banana
	Octyl acetate	Orange
	Ethyl butyrate	Pineapple

12.7 An Example of Organic Synthesis

EXPLAIN THIS

Is a pharmaceutical obtained from nature more effective than the same pharmaceutical obtained from the laboratory?

There is much more to organic chemistry than just learning functional groups and their general properties. Many, if not most, practicing organic chemists dedicate much of their time to the synthesis of organic molecules that have practical applications, such as for agriculture or pharmaceuticals. Often these target molecules are organic compounds that have been isolated from nature, where they can be found only in small quantities. To create large amounts of these chemicals, the organic chemist devises a pathway through which the compound can be synthesized in the laboratory from readily available smaller compounds. Once synthesized, the compound produced in the laboratory is chemically identical to that found in nature. In other words, it will have the same physical and chemical properties and will have the same biological effects, if any.

How is it that an organic chemist synthesizes a complex organic molecule? One common approach is to look at the structure of the desired compound and to imagine cutting certain bonds that the chemist knows would be easy to re-form. The chemist, for example, might imagine cutting the structure into two halves. Each half is then divided into even simpler portions. In essence, the chemist is working backward, starting with the desired chemical product and arriving at a list of smaller reactant molecules that would be needed to build this product. This is called a *retrosynthesis analysis.* The chemist then goes to the laboratory to attempt the actual synthesis. Things rarely go exactly as planned, so the synthetic scheme is modified as necessary. Overall, the process requires a strong working knowledge of how to build various chemical bonds, along with a healthy dose of creativity, luck, and perseverance.

To illustrate how this is done, we present a retrosynthesis analysis of the compound multistriatin, shown in **Figure 12.27**. This compound is one of the pheromones of the elm bark beetle. It is a fragrant compound released by a virgin female beetle when she has found a good source of food, such as an elm tree. Male beetles follow this scent to the female beetle at the elm tree. The male beetle carries the fungus for Dutch elm disease, so the tree upon which it lands gets infected. Larger quantities of this pheromone, which is produced only in small quantities by the beetle, are useful for traps that capture the male beetles, thereby preventing the spread of the disease. This approach of targeting a specific pest with traps is more desirable than the application of a pesticide, many of which kill beneficial insects and have the potential to contaminate foodstuffs.

The chemical structure for mulitstriatin, **1,** as shown in **Figure 12.28**, may look complicated. The knowledgeable organic chemist, however, recognizes that two oxygen atoms bonded to a single carbon atom can be easily made by reacting a carbonyl group with two adjacent hydroxyl groups, as shown in the first retrosynthetic step. So if the chemist could create compound **2,** as shown in Figure 12.28, she would be one step away from creating the desired product. The chemist then recognizes that compound **2** could be made from compound **3,** which, in turn, could be made from compound **4.** Studying compound **4** carefully, the experienced chemist recognizes that a bond two atoms away from a carbonyl group is easy to form. This leads her to break compound **4** evenly into compounds **5** and **6.** Compound **5** can be bought from a chemical supply company, while compound **6** requires only a few more steps to come to compound **8,** which is also commercially available.

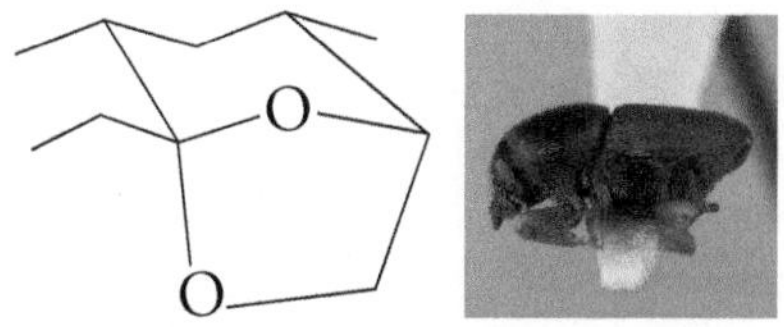

LEARNING OBJECTIVE

Summarize the retrosynthetic strategy used to plan the synthesis of a complex organic molecule.

READINGCHECK

How might an organic chemist plan for the synthesis of a complex organic molecule?

Multistriatin
(Elm bark beetle pheromone)

▲ Figure 12.27
Multistriatin is a pheromone produced only in small quantities by the female elm bark beetle.

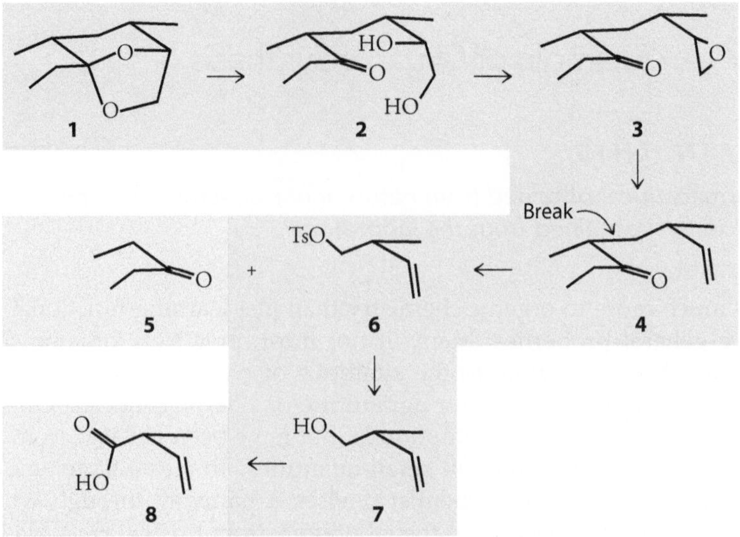

After developing this retrosynthetic plan and seeking helpful input from her colleagues, the chemist heads to the laboratory to try to make some variation of the forward synthesis work.

Naturally, some organic molecules are more difficult to synthesize than others. Look ahead to Chapter 14 for the structure of the anticancer agent Taxol®, shown in Figure 14.4. This compound is very useful for the treatment of breast cancer, but unfortunately, it is produced in nature by the yew tree in only very small quantities. This prompted teams of organic chemists to synthesize this molecule from easily obtained starting materials. The first total synthesis of Taxol® by K. C. Nicolaou of the Scripps Research Institute in 1994 was a major accomplishment. Other researchers have since refined the synthesis of Taxol®, which has been of great benefit in the fight against cancer. Online, use the keywords *taxol total synthesis* to find the stunning synthetic pathways used by these practicing world-class chemists.

12.8 Organic Molecules Can Link to Form Polymers

LEARNING OBJECTIVE

Describe how polymers are
synthesized from monomers.

EXPLAIN THIS

Why are plastics generally so inexpensive?

Polymers are exceedingly long molecules that consist of repeating molecular units called **monomers**, as **Figure 12.29** illustrates. Monomers have relatively simple structures, consisting of anywhere from 4 to 100 atoms per molecule. When monomers are chained together, they can form polymers consisting of hundreds of thousands of atoms per molecule. These large molecules are still too small to be seen with the unaided eye. They are, however, giants in the submicroscopic world—if a typical polymer molecule were as thick as a kite string, it would be 1 kilometer long.

▶ **Figure 12.29**
A polymer is a long molecule consisting of
many smaller monomer molecules linked
together.

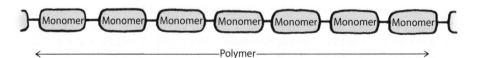

Polymer

TABLE 12.6   Addition and Condensation Polymers

POLYMERS	REPEATING UNIT	COMMON USES	RECYCLING CODE
Addition			
Polyethylene (PE)	$\cdots C - C \cdots$ with H, H above and H, CH_3 below	Plastic bags, bottles	HDPE 2 / LDPE 4
Polypropylene (PP)	$\cdots C - C \cdots$ with H, H above and H, CH_3 below	Indoor-outdoor carpets	5 PP
Polystyrene (PS)	$\cdots C - C \cdots$ with H, H above and H, phenyl ring below	Plastic utensils, insulation	6 PS
Polyvinyl chloride (PVC)	$\cdots C - C \cdots$ with H, H above and H, Cl below	Shower curtains, tubing	3 V
Condensation			
Polyethylene terephthalate (PETE)	$\cdots C{-}\text{(ring)}{-}C{-}O-CH_2CH_2-O\cdots$ (with O double bonds)	Clothing, plastic bottles	1 PET

Human-made polymers, also known as synthetic polymers, make up the class of materials commonly known as plastics. Two major types of synthetic polymers are used today—*addition polymers* and *condensation polymers.*

As shown in Table 12.6, addition and condensation polymers have a wide variety of uses. Solely the product of human design, these polymers pervade modern living. In the United States, for example, synthetic polymers have surpassed steel as the most widely used material.

Addition Polymers

Addition polymers form simply by the joining together of monomer units. For this to happen, each monomer must contain at least one double bond. As shown in **Figure 12.30**, polymerization occurs when two of the electrons from each double bond split away from each other to form new covalent bonds with neighboring monomer molecules. During this process, no atoms are lost; so the total mass of an addition polymer is equal to the sum of the masses of all its monomers.

Nearly 12 million tons of polyethylene is produced annually in the United States; that's about 90 pounds per U.S. citizen. The monomer from which it is synthesized, ethylene, is an unsaturated hydrocarbon produced in large quantities from petroleum.

High-density polyethylene (HDPE), shown schematically in **Figure 12.31a**, consists of long strands of straight-chain molecules packed closely together. The tight alignment of neighboring strands makes HDPE a relatively rigid, tough plastic useful for such things as bottles and milk jugs. Low-density polyethylene (LDPE), shown in **Figure 12.31b**, is made of strands with many branched chains, a pattern that prevents the strands from packing closely together. This makes LDPE more bendable than HDPE and gives it a lower melting point. While HDPE holds its shape in boiling water, LDPE deforms. It is most useful for such items as plastic bags, photographic film, and insulation for electric wire.

READINGCHECK

Which has more mass: an addition polymer or the monomers that combined to make that addition polymer?

▶ **Figure 12.30**
The addition polymer polyethylene is formed as electrons from the double bonds of ethylene monomer molecules split away and become unpaired valence electrons. Each unpaired electron then joins with an unpaired electron of a neighboring carbon atom to form a new covalent bond that links two monomer units together.

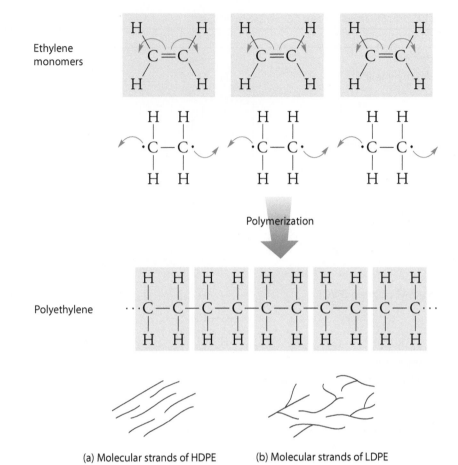

Ethylene monomers

Polymerization

Polyethylene

▶ **Figure 12.31**
(a) The polyethylene strands of HDPE are able to pack closely together, much like strands of uncooked spaghetti.
(b) The polyethylene strands of LDPE are branched, which prevents the strands from packing well.

(a) Molecular strands of HDPE   (b) Molecular strands of LDPE

Addition polymers are also created by using other monomers. The only requirement is that the monomer must contain a double (or triple) bond. The monomer propylene, for example, yields polypropylene, as shown in **Figure 12.32**. Polypropylene is a tough plastic material useful for pipes, hard-shell suitcases, and appliance parts. Fibers of polypropylene are used for upholstery, indoor-outdoor carpet, and even thermal underwear.

Figure 12.33 shows that using styrene as the monomer yields polystyrene. Transparent plastic cups are made of polystyrene, as are thousands of other household items. Blowing air into liquid polystyrene generates Styrofoam™, which is widely used for coffee cups, packing material, and insulation.

▶ **Figure 12.32**
Propylene monomers polymerize to form polypropylene.

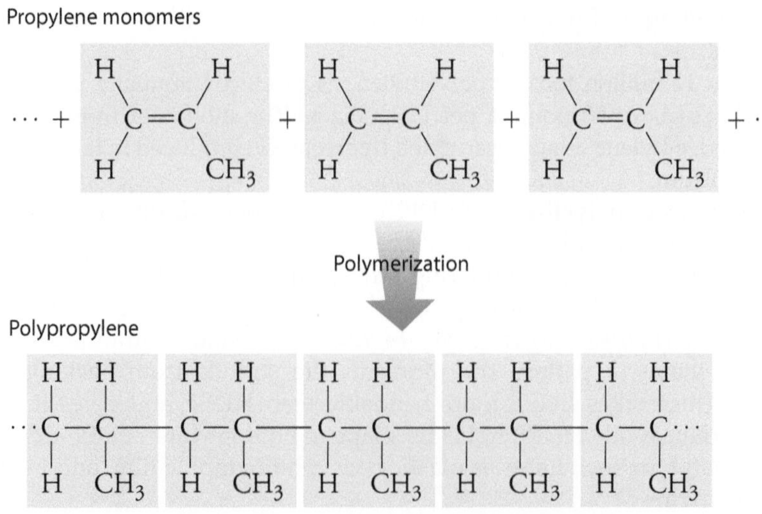

Propylene monomers

Polymerization

Polypropylene

Styrene monomers

Polymerization

Polystyrene

Another important addition polymer is polyvinyl chloride (PVC), which is tough and easily molded. Floor tiles, shower curtains, and pipes are most often made of PVC, shown in **Figure 12.34**.

Rigid polymers such as PVC can be made soft by incorporating small molecules called plasticizers. Pure PVC, for example, is a tough material great for making pipes. Mixed with a plasticizer, the PVC becomes soft and flexible and thus useful for making shower curtains, toys, and many other products now found in most households. One of the more commonly used types of plasticizers is the phthalates, some of which have been shown to disrupt the development of reproductive organs, especially in fetuses and in growing children. Governments and manufacturers are now working to phase out these plasticizers. But some phthalates, such as DINP, have been shown to be much less dangerous. Should all phthalates be banned or just the ones proven to be harmful? This is a social and political question that has yet to be resolved.

◀ **Figure 12.34**
PVC is tough and easily molded, which is why it often is used to fabricate many household items.

Polyvinyl chloride (PVC)

▶ **Figure 12.35**
The many large chlorine atoms in polyvinylidene chloride make this addition polymer sticky.

Polyvinylidene chloride (Saran)

The addition polymer polyvinylidene chloride (trade name Saran™), shown in **Figure 12.35**, was once used to make plastic wrap for food. The large chlorine atoms in this polymer help it stick to such surfaces as glass by dipole-induced dipole attractions, as discussed in Section 7.1. Because of environmental concerns, however, Saran™ Wrap brand food wrap is now made from polyethylene. The original polyvinylidene chloride polymer contains twice the number of chlorine atoms as PVC and thus readily transforms into harmful dioxins upon burning. (See the Contextual Chemistry essay at the end of Chapter 11.)

The addition polymer polytetrafluoroethylene, shown in **Figure 12.36**, is what you know as Teflon®. In contrast to the chlorine-containing Saran, fluorine-containing Teflon has a nonstick surface, because the fluorine atoms tend not to experience any molecular attractions. In addition, because carbon–fluorine bonds are unusually strong, Teflon can be heated to high temperatures before decomposing. These properties make Teflon an ideal coating for cooking surfaces. It is also relatively inert, which is why many corrosive chemicals are shipped or stored in Teflon containers.

CONCEPTCHECK

What do all monomers that are used to make addition polymers have in common?

CHECK YOUR ANSWER A multiple covalent bond between two carbon atoms.

▶ **Figure 12.36**
The fluorine atoms in polytetrafluoroethylene tend not to experience molecular attractions, which is why this addition polymer is used as a nonstick coating and lubricant.

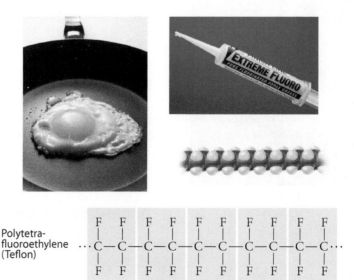

Polytetra-fluoroethylene (Teflon)

Condensation Polymers

A **condensation polymer** is formed when the joining of monomer units is accompanied by the loss of a small molecule, such as water or hydrochloric acid. Any monomer capable of becoming part of a condensation polymer must have a functional group on each end. When two such monomers come together to form a condensation polymer, one functional group of the first monomer links up with one functional group of the other monomer. The result is a two-monomer unit that has two terminal functional groups, one from each of the two original monomers. Each of these terminal functional groups in the two-monomer unit is now free to link with one of the functional groups of a third monomer, then a fourth, and so on. In this way, a polymer chain is built.

Figure 12.37 shows this process for the condensation polymer called *nylon*. This polymer is composed of two different monomers, as shown in Figure 12.37. One monomer is adipic acid, which contains two reactive end groups, both carboxyl groups. The second monomer is hexamethylenediamine, in which two amine groups are the reactive end groups. One end of an adipic acid molecule and one end of a hexamethylamine molecule can be made to react with each other, splitting off a water molecule in the process. After two monomers have joined, reactive ends still remain for further reactions, which leads to a growing polymer chain. Aside from its use in hosiery, nylon also finds important uses in the manufacture of ropes, parachutes, clothing, and carpets.

CONCEPT CHECK

The structure of 6-aminohexanoic acid is the following:

Is this compound a suitable monomer for forming a condensation polymer?

CHECK YOUR ANSWER Yes, because the molecule has two reactive ends. You know both ends are reactive because they are the ends shown in Figure 12.37. The only difference here is that both types of reactive ends are on the same molecule. Monomers of 6-aminohexanoic acid combine by splitting off water molecules to form the polymer known as nylon-6.

Another widely used condensation polymer is polyethylene terephthalate (PET or PETE), which is formed from the polymerization of ethylene glycol and terephthalic acid, as shown in **Figure 12.38**. Plastic soda bottles are made from this polymer. Also, PETE fibers are sold as Dacron™ polyester, a product used in clothing and stuffing for pillows and sleeping bags. Thin films of PETE, which are called Mylar®, can be coated with metal particles to make magnetic recording tape or those metallic-looking balloons you see for sale at grocery store checkout counters.

Monomers that contain three reactive functional groups can also form polymer chains. These chains become interlocked in a rigid 3-dimensional network that lends considerable strength and durability to the polymer. Once

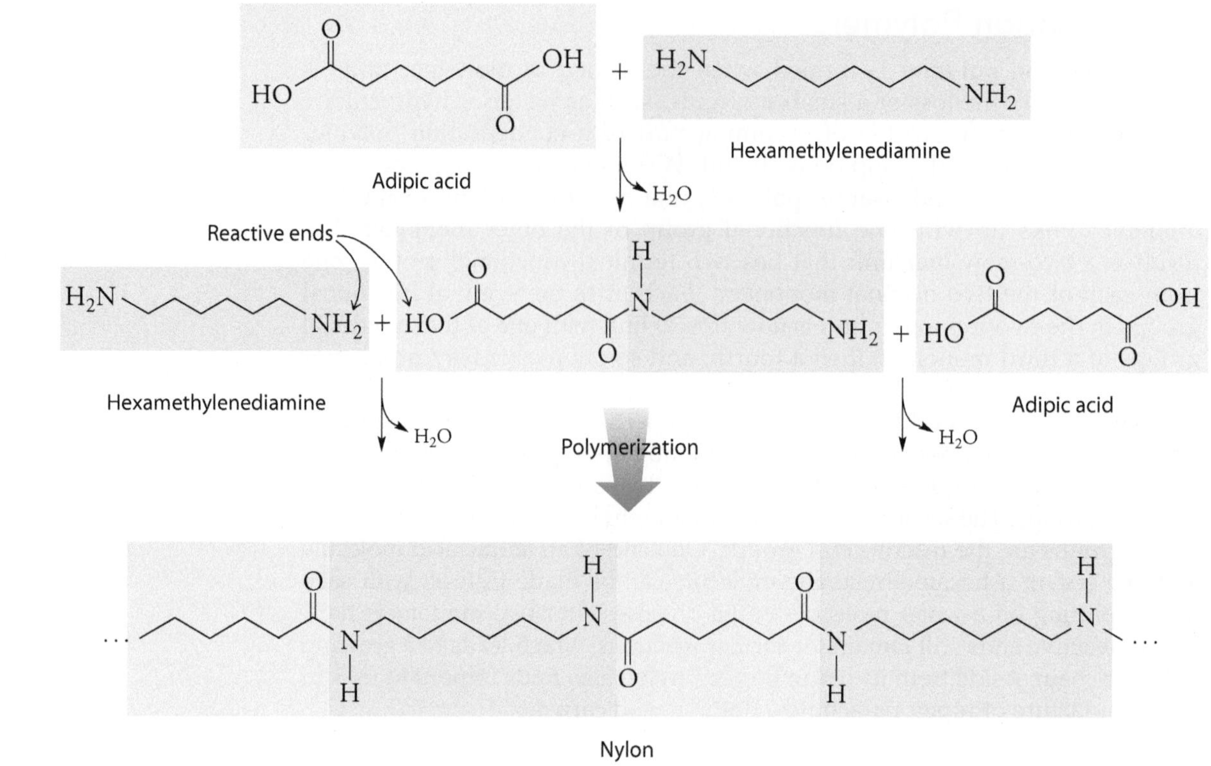

▲ Figure 12.37
Adipic acid and hexamethylenediamine polymerize to form the condensation polymer nylon.

formed, these condensation polymers cannot be remelted or reshaped, which makes them hard-set, or *thermoset*, polymers. Hard plastic dishes (Melmac™) and countertops (Formica™) are made of this material. A similar polymer, Bakelite™, made from formaldehyde and phenols containing multiple oxygen atoms, is used to bind plywood and particleboard. Bakelite was synthesized in the early 1900s, and it was the first widely used polymer.

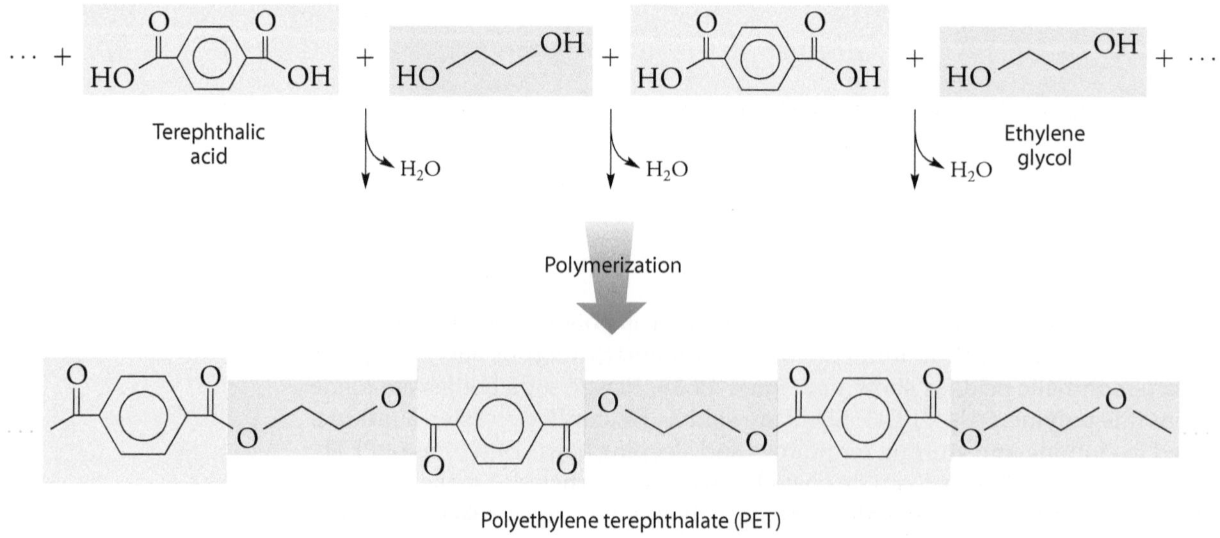

▲ Figure 12.38
Terephthalic acid and ethylene glycol polymerize to form the condensation polymer polyethylene terephthalate.

▲ Figure 12.39
Flexible and flat video displays known as OLEDs (organic light-emitting diodes) are now fabricated from polymers.

The synthetic polymer industry has grown remarkably over the past half century. Annual production of polymers in the United States alone has grown from 3 billion pounds in 1950 to more than 100 billion pounds in 2010. Today, it is a challenge to find any consumer item that does *not* contain a plastic of one sort or another.

In the near future, watch for new kinds of polymers having a wide range of remarkable properties. One interesting application is shown in **Figure 12.39**. We already have polymers that conduct electricity, others that emit light, others that replace body parts, and still others that are stronger but much lighter than steel. Imagine synthetic polymers that mimic photosynthesis by transforming solar energy to chemical energy or that efficiently separate fresh water from ocean water. These are not dreams. They are realities that chemists have already been demonstrating in the laboratory. Polymers hold a clear promise for the future.

12.9 A Brief History of Plastics

EXPLAIN THIS

Besides chance discovery, what else allowed for the development of plastics?

The search for a lightweight, nonbreakable, moldable material began with the invention of vulcanized rubber. This material is derived from natural rubber, which is a semisolid, elastic, natural polymer. The fundamental chemical unit of natural rubber is polyisoprene, which plants produce from isoprene molecules, as shown in **Figure 12.40**. In the 1700s, natural rubber was noted for its ability to rub off pencil marks, which is the origin of the term *rubber*. Natural rubber has few other uses, however, because it turns gooey at warm temperatures and brittle at cold temperatures.

In 1839, an American inventor, Charles Goodyear, discovered *rubber vulcanization*, a process in which natural rubber and sulfur are heated together. The product, vulcanized rubber, is harder than natural rubber and retains

LEARNING OBJECTIVE

Recount the history of the development of plastics starting from the early 1900s.

CHEMICAL CONNECTION

How is celluloid connected to the many exits now in movie theaters?

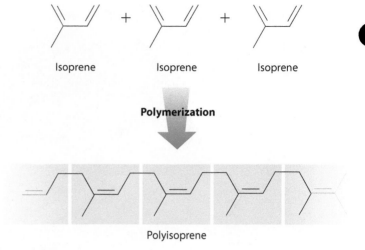

Isoprene + Isoprene + Isoprene

Polymerization

Polyisoprene

▲ **Figure 12.40**
Isoprene molecules react with one another to form polyisoprene, which is abundant within the sap of rubber trees. Isoprene is the fundamental chemical unit, or monomer, of natural rubber.

its elastic properties over a wide range of temperatures. This is the result of *disulfide cross-linking* between polymer chains, as illustrated in **Figure 12.41**. To help quench our ever-growing thirst for vulcanized rubber, the polymer of rubber (polyisoprene) is now also produced from petroleum.

Charles Goodyear was the classic eccentric inventor. He lived most of his life in poverty, obsessed with transforming rubber into a useful material. Goodyear was a man of ill health who died in debt, yet he remained stubbornly optimistic. The present-day Goodyear Corporation was founded not by Goodyear, but by others who sought to pay tribute to his name 15 years after he died.

In 1845, as vulcanized rubber was becoming popular, the Swiss chemistry professor Christian Schönbein wiped up a spilled mixture of nitric and sulfuric acids with a cotton rag that he then hung up to dry. Within a few minutes, the rag burst into flames and then vanished, leaving only a tiny bit of ash. Schönbein had discovered nitrocellulose, in which most of the hydroxyl groups in cellulose are bonded to nitrate groups, as **Figure 12.42** illustrates. Schönbein's attempts to market nitrocellulose as a smokeless gunpowder (*guncotton*) were unsuccessful, mainly because of a number of lethal explosions at plants producing the material.

Researchers in France discovered that solvents such as diethyl ether and alcohol transformed nitrocellulose to a gel that could be molded into various shapes. This workable nitrocellulose material was dubbed *collodion*, and its first application was as a medical dressing for cuts.

▶ **Figure 12.41**
(a) When stretched, the individual polyisoprene strands in natural rubber slip past one another and the rubber stays stretched. (b) When vulcanized rubber is stretched, the sulfur crosslinks hold the strands together, allowing the rubber to return to its original shape.

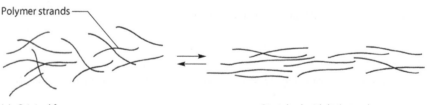

Polymer strands

(a) Original form

Stretched with little tendency to snap back to original form

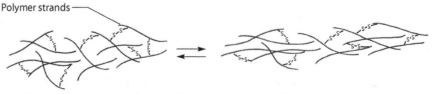

Polymer strands

(b) Vulcanized form with disulfide crosslinks

Stretched with great tendency to snap back because of crosslinks

Nitrate group

O—NO$_2$

···

O—NO$_2$

NO$_2$

O$_2$N   NO$_2$

O$_2$N

···

Nitrocellulose (cellulose nitrate)

◀ **Figure 12.42**
Nitrocellulose, also known as cellulose nitrate, is highly combustible because of its many nitrate groups, which facilitate oxidation.

In 1870, John Hyatt, a young inventor from Albany, New York, discovered that collodion's moldable properties were vastly improved by using camphor as a solvent. This camphor-based nitrocellulose material was named *celluloid,* and it became the plastic of choice for the manufacture of many household items, such as combs and hair fasteners. In addition, thin transparent films of celluloid made excellent supports for photosensitive compounds, a boon to the photography industry and a first step in the development of motion pictures.

As wonderful as celluloid was, it still had the major drawback of being highly flammable. Today, one of the few commercially available products made of celluloid is Ping-Pong balls, shown in **Figure 12.43**.

Bakelite Was the First Widely Used Plastic

About 1899, Leo Baekeland, a chemist who had immigrated to the United States from Belgium, developed an emulsion for photographic paper that was exceptional in its sensitivity to light. He sold his invention to George Eastman, who had made a fortune selling celluloid-based photographic film along with his portable Kodak camera. Expecting no more than $50,000 for his invention, Baekeland was shocked at Eastman's initial offer of $750,000 (in today's dollars, that would be about $25 million). Suddenly a very wealthy man, Baekeland was free to pursue his chemical interests.

Baekeland explored a tar-like solid once produced in the laboratories of Alfred von Baeyer, the German chemist who played a role in the development of aspirin. Whereas Baeyer had dismissed the solid as worthless, Baekeland saw it as a virtual gold mine. After several years, he produced a resin that, when poured into a mold and then heated under pressure, solidified into a transparent positive of the mold. Baekeland's resin was a mixture of formaldehyde and a phenol that polymerized into the complex network shown in **Figure 12.44**.

The solidified material, which he called *Bakelite,* was impervious to harsh acids or bases, wide temperature extremes, and just about any solvent. Bakelite quickly replaced celluloid as a molding medium, finding a wide variety of uses for several decades. It wasn't until the 1930s that alternative thermoset polymers (Section 12.8) began to challenge Bakelite's dominance in the evolving plastics industry.

The First Plastic Wrap Was Cellophane

Cellophane had its beginnings in 1892, when Charles Cross and Edward Bevan of England found that treating cellulose with concentrated sodium hydroxide followed by carbon disulfide created a thick, syrupy yellow liquid they called *viscose.* Extruding the viscose into an acidic solution generated a tough cellulose filament that could be used to make a synthetic silky cloth today called *rayon* (**Figure 12.45**).

In 1904, Jacques Brandenberger, a Swiss textile chemist, observed restaurant workers discarding fine tablecloths that had only slight stains on them. Working with viscose at the time, he had the idea of extruding it not as a

✔ R E A D I N G C H E C K

What was a major drawback of celluloid?

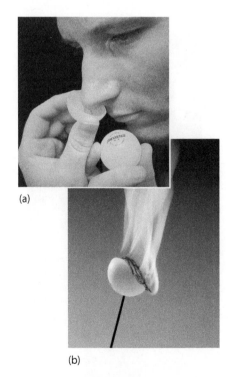
(a)

(b)

▲ **Figure 12.43**
(a) Smell a freshly cut Ping-Pong ball, and you will note the distinct odor of camphor, which is the same smell that arises from heat cream for sore muscles. This camphor comes from the celluloid from which the ball is made. (b) Ping-Pong balls burn rapidly because they are made of nitrocellulose.

O
‖
C
H⁄ ⸜H
Formaldehyde

Polymerization →

OH

Phenol

Phenol–formaldehyde resin
(Bakelite)

▲ **Figure 12.44**
(a) The molecular network of Bakelite shown in two dimensions. (The actual structure projects in all three dimensions.) (b) The first handset telephones were made of Bakelite.

◀ **Figure 12.45**
Viscose is still used today in the manufacture of fibers used to make the synthetic fabric called rayon. The fibers are formed as viscose and extruded through holes in metal dies, such as shown here.

▲ **Figure 12.46**
Cellophane transformed the way foods and other items were marketed.

fiber, but as a thin, transparent sheet that might be adhered to tablecloths and provide an easy-to-clean surface. By 1913, Brandenberger had perfected the manufacture of a viscose-derived, thin, transparent sheet of cellulose, which he named *cellophane.*

Within several years, the Du Pont Corporation bought the rights to cellophane. Hermetically sealed by cellophane, a product could be kept free of dust and germs. And unlike paper or tin foil—the alternatives of the day—cellophane was transparent and thus allowed the consumer to view the packaged contents, as shown in **Figure 12.46**. With properties such as these, cellophane played a great role in the success of supermarkets, which first appeared in the 1930s. Perhaps cellophane's greatest appeal to the consumer, however, was its shine. As marketing people soon discovered, nearly any product—soaps, canned goods, and golf balls—would sell faster when wrapped in cellophane.

Polymers Win in World War II

In the 1930s, more than 90 percent of the natural rubber used in the United States came from Malaysia. In the days after Pearl Harbor was attacked in December 1941 and the United States entered World War II, however, Japan captured Malaysia. As a result, the United States—the land with plenty of everything, except rubber—faced its first natural resource crisis. The military implications were devastating, because without rubber for tires, military airplanes and jeeps were useless. Petroleum-based synthetic rubber had been developed in 1930 by Du Pont chemist Wallace Carothers but was not widely

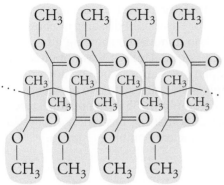

Poly(methyl methacrylate)

◀ **Figure 12.47**
The bulky side groups in poly(methyl methacrylate) prevent the polymer chains from aligning with one another. This makes it easy for light to pass through the material, which is tough, transparent, lightweight, and moldable. (Plexiglas® is a registered trademark belonging to ATOFINA.)

used because it was much more expensive than natural rubber. With Malaysian rubber impossible to get and a war on, however, cost was no longer an issue. Synthetic rubber factories were constructed across the nation, and within a few years, the annual production of synthetic rubber rose from 2000 tons to about 800,000 tons.

Also in the 1930s, British scientists developed radar as a way to track thunderstorms. With war approaching, these scientists turned their attention to the idea that military planes could be equipped with radar. To make the radar equipment light enough to fly on an airplane, however, they needed the wires of the device coated with some sort of electrical insulator. Fortunately, the polymer polyethylene, recently developed in Britain, turned out to be an ideal electrical insulator. The radar-equipped planes were slow, but flying at night or in poor weather, they could detect, intercept, and destroy enemy aircraft.

Other polymers that had a significant impact on the outcome of World War II included Nylon®, Plexiglas®, PVC, Saran™, and Teflon®. Nylon was invented in 1937 and, as discussed in the previous section, found great use in the manufacture of ropes, pharachutes, and clothing. Interestingly, tear-resistant Nylon stockings, which had just been introduced, were extremely popular during the war and often used as a commodity for favors or to repay debts among soldiers. Plexiglas, shown in **Figure 12.47**, is a polymer known to chemists as poly(methyl methacrylate). This glass-like but moldable and lightweight material made excellent domes for the gunner's nests on fighter planes and bombers.

Polyvinyl chloride (PVC) had been developed by a number of chemical companies in the 1920s. The problem with this material, however, was that it lost resiliency when heated. In 1929, Waldo Semon, a chemist at BFGoodrich, found that PVC could be made into a workable material by the addition of a plasticizer. Semon got the idea of using plasticized PVC as a shower curtain when he observed his wife sewing together a shower curtain made of rubberized cotton. Other uses for PVC were slow to appear, however, and it wasn't until World War II that this material became recognized as an ideal waterproof material for tents and rain gear. After the war, PVC replaced Bakelite as the medium for making phonograph records.

Originally designed as a covering to protect theater seats from chewing gum, Saran found great use in World War II as a protective wrapping for artillery equipment during sea voyages. (Before Saran, the standard operating procedure had been to disassemble and grease the artillery to avoid corrosion.) After the war, this polymer soon pushed cellophane aside to become the most popular food wrap of all time (**Figure 12.48**). Because of environmental issues associated with Saran, polyethylene now holds the title of most widely used food wrap.

In the late 1930s, the discoverers of Teflon were impressed by the long list of things this new material would *not* do. It would not burn, and it would not completely melt. Instead, at 620°F it congealed into a gel that could be

▲ **Figure 12.48**
The now-familiar plastic food wrap carton with a cutting edge was introduced in 1953 by Dow Chemical for its Saran™ Wrap brand.

conveniently molded. It would not conduct electricity, and it was impervious to attack by mold or fungus. No solvent, acid, or base could dissolve or corrode it. And most remarkably, nothing would stick to it, not even chewing gum.

Because of all the things Teflon would not do, DuPont was not quite sure what to do with it. Then in 1944, the company was approached by governmental researchers in desperate need of a highly inert material to line the valves and ducts of an apparatus being built to isolate uranium-235 in the manufacture of the first nuclear bomb. Thus, Teflon found its first application, and one year later, World War II came to a close with the nuclear bombing of Japan.

CONCEPTCHECK

Name at least four polymers that had a significant impact on the effectiveness of Allied forces in World War II.

CHECK YOUR ANSWER Here are six: synthetic rubber, polyethylene, Nylon, Plexiglas, PVC, Saran, and Teflon.

Attitudes about Plastics Have Changed

With a record of wartime successes, plastics were readily embraced in the postwar years. In the 1950s, Dacron polyester was introduced as a substitute for wool. Also, the 1950s were the decade during which the entrepreneur Earl Tupper created a line of polyethylene food containers, known as Tupperware®.

By the 1960s, a decade of environmental awakening, many people began to recognize the negative attributes of plastics. Being inexpensive, disposable, and non-biodegradable, plastic readily accumulated as litter and as landfill. With petroleum so readily available and inexpensive, however, and with a growing population of plastic-dependent baby boomers, little stood in the way of an ever-expanding array of plastic consumer products. By 1977, plastics surpassed steel as the number-one material produced in the United States. Environmental concerns also continued to grow, and in the 1980s, plastics recycling programs began to appear. Although the efficiency of plastics recycling still holds room for improvement, we now live in a time when sports jackets made of recycled plastic bottles are a valued commodity.

Chapter 12 Review

LEARNING OBJECTIVES

Identify the structures of hydrocarbons. (12.1)	→	*Questions 1–4, 34–44*
Identify the structures of unsaturated hydrocarbons. (12.2)	→	*Questions 5–7, 26, 28–30, 45–49*
Discuss the significance of heteroatoms in organic compounds. (12.3)	→	*Questions 8, 9, 50–53*
Review the general properties of alcohols, phenols, and ethers. (12.4)	→	*Questions 10–12, 31, 54–57*
Review the general properties of amines and alkaloids. (12.5)	→	*Questions 13–15, 32, 58–63, 82, 83*
Review the general properties of carbonyl compounds. (12.6)	→	*Questions 16–18, 33, 64–70*
Summarize the retrosynthetic strategy used to plan the synthesis of a complex organic molecule. (12.7)	→	*Questions 19, 20, 71–74, 84*
Describe how polymers are synthesized from monomers. (12.8)	→	*Questions 21–23, 27, 75–79*
Recount the history of the development of plastics starting from the early 1900s. (12.9).	→	*Questions 24, 25, 80, 81, 85*

SUMMARY OF TERMS (KNOWLEDGE)

Addition polymer A polymer formed by the joining together of monomer units with no atoms being lost as the polymer forms.

Alcohol (AL-co-hol) An organic molecule that contains a hydroxyl group bonded to a saturated carbon.

Aldehyde (AL-de-hyde) An organic molecule containing a carbonyl group, the carbon of which is bonded either to one carbon atom and one hydrogen atom or to two hydrogen atoms.

Alkane (AL-kane) A generic word for a saturated hydrocarbon.

Alkene (AL-keen) An unsaturated hydrocarbon containing one or more double bonds.

Alkyne (AL-kine) An unsaturated hydrocarbon containing a triple bond.

Amide (AM-id) An organic molecule containing a carbonyl group, the carbon of which is bonded to a nitrogen atom.

Amine (Uh-MEEN) An organic molecule containing a nitrogen atom bonded to one or more saturated carbon atoms.

Aromatic Said of any organic molecule containing a benzene ring.

Carbonyl group (CAR-bon-EEL) A carbon atom double-bonded to an oxygen atom; found in ketones, aldehydes, amides, carboxylic acids, and esters.

Carboxylic acid (CAR-box-IL-ic) An organic molecule containing a carbonyl group, the carbon of which is bonded to a hydroxyl group.

Condensation polymer A polymer formed by the joining together of monomer units accompanied by the loss of small molecules, such as water.

Configuration A term used to describe how the atoms within a molecule are connected. For example, two structural isomers will consist of the same number and same kinds of atoms, but in different configurations.

Conformation One of a wide range of possible spatial orientations of a particular configuration.

Ester (ESS-ter) An organic molecule containing a carbonyl group, the carbon of which is bonded to one carbon atom and one oxygen atom bonded to another carbon atom.

Ether (EETH-er) An organic molecule containing an oxygen atom bonded to two carbon atoms.

Functional group A specific combination of atoms that behaves as a unit in an organic molecule.

Heteroatom Any atom other than carbon or hydrogen in an organic molecule.

Hydrocarbon An organic compound containing only carbon and hydrogen atoms.

Ketone (KEY-tone) An organic molecule containing a carbonyl group, the carbon of which is bonded to two carbon atoms.

Monomers The small molecular units from which a polymer is formed.

Organic chemistry The study of carbon-containing compounds.

Phenol (fen-ALL) An organic molecule in which a hydroxyl group is bonded to a benzene ring.

Polymer A long organic molecule made of many repeating units.

Saturated hydrocarbon A hydrocarbon containing no multiple covalent bonds, with each carbon atom bonded to four other atoms.

Structural isomers Molecules that have the same chemical formula but different chemical structures.

Unsaturated hydrocarbon A hydrocarbon containing at least one multiple covalent bond.

READING CHECK QUESTIONS (COMPREHENSION)

12.1 Hydrocarbons Contain Only Carbon and Hydrogen

1. How do two structural isomers differ from each other?

2. How are two structural isomers similar to each other?

3. What physical property of hydrocarbons is used in fractional distillation?

4. What types of hydrocarbons are more abundant in higher-octane gasoline?

12.2 Unsaturated Hydrocarbons Have Multiple Bonds

5. What is the difference between a saturated hydrocarbon and an unsaturated hydrocarbon?

6. How many multiple bonds must a hydrocarbon have in order to be classified as unsaturated?

7. Aromatic compounds contain what kind of ring?

12.3 Functional Groups Give Organic Compounds Character

8. What is a heteroatom?

9. Why do heteroatoms make such a difference in the physical and chemical properties of an organic molecule?

12.4 Alcohols, Phenols, and Ethers Contain Oxygen

10. Why are low-formula-mass alcohols soluble in water?

11. What distinguishes an alcohol from a phenol?

12. What distinguishes an alcohol from an ether?

12.5 Amines and Alkaloids Contain Nitrogen

13. Which heteroatom is characteristic of an amine?

14. Do amines tend to be acidic, neutral, or basic?

15. Are alkaloids found in nature?

12.6 Carbonyl-Containing Compounds

16. Which elements make up the carbonyl group?

17. How are ketones and aldehydes related to each other? How are they different from each other?

18. How are amides and carboxylic acids related to each other? How are they different from each other?

12.7 An Example of Organic Synthesis

19. Why might an organic chemist want to synthesize a molecule that already occurs in nature?

20. What is a retrosynthesis analysis?

12.8 Organic Molecules Can Link to Form Polymers

21. What happens to the double bond of a monomer participating in the formation of an addition polymer?

22. What is released in the formation of a condensation polymer?

23. Why is plastic wrap made of polyvinylidene chloride stickier than plastic wrap made of polyethylene?

12.9 A Brief History of Plastics

24. What is one of the major drawbacks of celluloid?

25. What role did Teflon play in World War II?

CONFIRM THE CHEMISTRY (HANDS-ON APPLICATION)

26. Two carbon atoms connected by a single bond can rotate relative to each other. This ability to rotate can give rise to numerous conformations (spatial orientations) of an organic molecule. Is it also possible for two carbon atoms connected by a double bond to rotate relative to each other?

Hold two toothpicks side by side and attach one jellybean to each end such that each jellybean has both toothpicks poked into it. Hold one jellybean while rotating the other. What kind of rotations are possible? Relate what you observe to the carbon–carbon double bond. Which structure of Figure 12.8 do you suppose has more possible conformations: butane or *cis*-2-butene? What do you suppose is true about the ability of atoms connected by a carbon–carbon triple bond to twist relative to each other?

27. A property of polymers is their glass transition temperature, T_g, which is the approximate temperature below which the polymer is hard and rigid, but above which the polymer is soft and flexible. The T_g of polyethylene is a chilly −125°C, which is why polyethylene food wrap is flexible at ambient temperatures. Consider the two polymers polyethylene terephthalate (PETE) and polystyrene (PS). Which do you suppose has the higher T_g? Dip some plastics of these two polymers in boiling water to find out. A common polymer used to make chewing gum is polyvinyl acetate, which has a T_g of about 28°C, which is below body temperature but above room temperature. That's why most chewing gums are hard until they soften up in your warm mouth. Drink ice water while chewing and note how the gum quickly hardens.

THINK AND COMPARE (ANALYSIS)

28. Rank the following hydrocarbons in order of increasing number of hydrogen atoms:

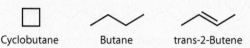

Cyclobutane   Butane   trans-2-Butene

29. Rank the following hydrocarbons in order of increasing number of hydrogen atoms:

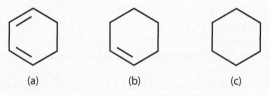

(a)   (b)   (c)

30. Rank the following hydrocarbons in order of smallest to largest carbon chain:

(a)   (b)   (c)

31. Rank the following organic molecules in order of increasing solubility in water:

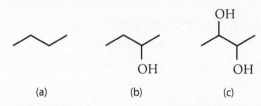

(a)   (b)   (c)

32. Rank the following organic molecules in order of increasing solubility in water:

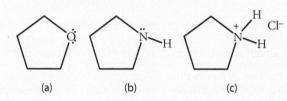

(a)   (b)   (c)

33. Rank the following compounds in order of least oxidized to most oxidized:

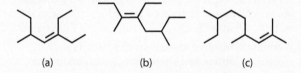

(a)   (b)   (c)   (d)

THINK AND EXPLAIN (SYNTHESIS)

12.1 Hydrocarbons Contain Only Carbon and Hydrogen

34. What property of carbon allows for the formation of so many different organic molecules?

35. Why does the melting point of hydrocarbons increase as the number of carbon atoms per molecule increases?

36. Draw all the structural isomers for hydrocarbons having the molecular formula C_6H_{14}.

37. How many structural isomers are shown here?

38. What do the compounds cyclopropane and propene have in common?

Cyclopropane   Propene

39. According to Figure 12.4, which has a higher boiling point: gasoline or kerosene?

40. The temperatures in a fractionating tower at an oil refinery are important, but so are the pressures. Where might the pressure in a fractionating tower be greatest, at the bottom or at the top? Defend your answer.

41. There are five atoms in the methane molecule, CH_4. One out of these five is a carbon atom, which is $1/5 \times 100 = 20\%$ carbon. What is the percent of carbon in ethane, C_2H_6? Propane, C_3H_8? Butane, C_4H_{10}?

42. Do heavier hydrocarbons tend to produce more or less carbon dioxide upon combustion compared to lighter hydrocarbons? Why?

43. Shown below is the structure of 2-methylpentane. What is the structure of 3-methylpentane?

2-methylpentane

44. Name the following structures:

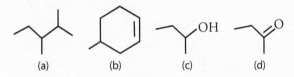

(a)   (b)

12.2 Unsaturated Hydrocarbons Have Multiple Bonds

45. Question 29 shows three molecules, each with a 6-carbon ring. Match the following names to each of these structures: cyclohexane, cyclohexene, 1,3-cyclohexadiene.

46. Draw the structure for 1,4-cyclohexadiene.

47. What are the chemical formulas for the following structures?

(a)   (b)   (c)   (d)

48. Remember that carbon–carbon single bonds can rotate, while carbon–carbon double bonds cannot. How many different structures are shown below?

49. Which of the structures shown in the previous question need not be designated as either a *cis* or *trans* isomer?

12.3 Functional Groups Give Organic Compounds Character

50. What must be added to a double bond to transform it into an alcohol? Would this be an example of oxidation or reduction?

51. What do phenols and carboxylic acids have in common?

52. What is the difference between a ketone and an aldehyde?

53. An ether is to an alcohol as an ester is to

   a. a phenol.
   b. a carboxylic acid.
   c. an amide.
   d. an ether.

12.4 Alcohols, Phenols, and Ethers Contain Oxygen

54. Why do ethers typically have lower boiling points compared to alcohols?

55. What is the percent volume of water in 80-proof vodka?

56. One of the skin-irritating components of poison oak is tetrahydrourushiol:

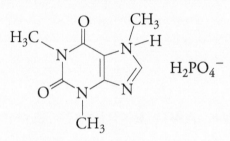

The long nonpolar hydrocarbon tail embeds itself in a person's oily skin, where the molecule initiates an allergic response. Scratching the itch spreads tetrahydrourushiol molecules over a greater surface area, causing the zone of irritation to grow. Is this compound an alcohol or a phenol? Defend your answer.

57. Cetyl alcohol, $C_{16}H_{34}O$, is a common ingredient of soaps and shampoos. It was once commonly obtained from whale oil, which is where it gets its name (*cetyl* is derived from *cetacean*). Review the discussion of soaps in Section 7.5 and draw the likely chemical structure for this primary alcohol.

12.5 Amines and Alkaloids Contain Nitrogen

58. A common inactive ingredient in products such as sunscreen lotions and shampoos is triethylamine, also known as TEA. What is the chemical structure for this compound?

59. A common inactive ingredient in products such as sunscreen lotions and shampoos is triethanolamine. What is the chemical structure for this tri-alcohol?

60. The phosphoric acid salt of caffeine has the following structure:

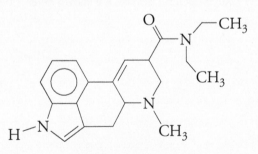

Caffeine–phosphoric acid salt

This molecule behaves as an acid in that it can donate a hydrogen ion, which is created from the hydrogen atom bonded to the positively charged nitrogen atom. What are all the products formed when 1 mole of this salt reacts with 1 mole of sodium hydroxide, NaOH, a strong base?

61. Draw all the structural isomers for amines having the molecular formula C_3H_9N.

62. In water, does the following molecule act as an acid, a base, neither, or both?

Lysergic acid diethylamide

63. If you saw the label "phenylephrine HCl" on a decongestant, would you worry that consuming it would expose you to the strong acid hydrochloric acid? Explain.

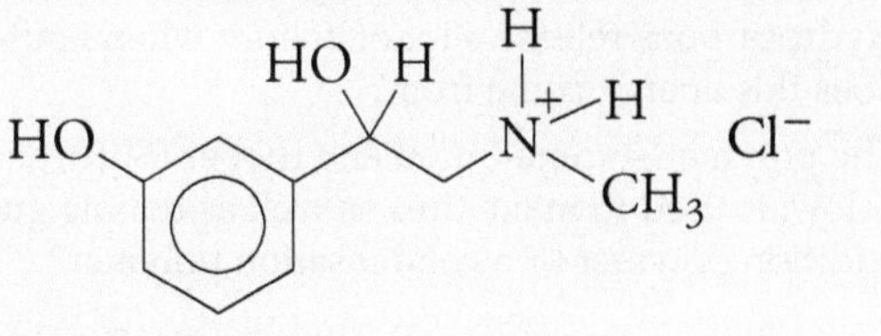

Phenylephrine–hydrochloric acid salt

12.6 Carbonyl-Containing Compounds

64. An amino acid is an organic molecule that contains both an amine group and a carboxyl group. At an acidic pH, which structure is more likely? Explain your answer.

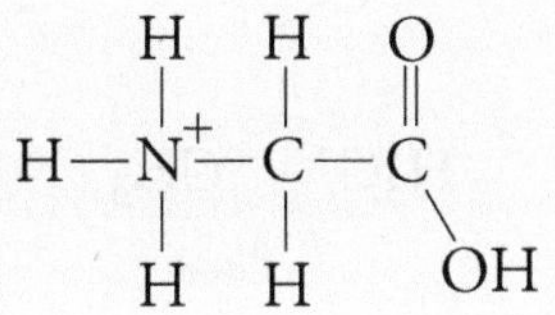

65. Identify the following functional groups in this organic molecule—amide, ester, ketone, ether, alcohol, aldehyde, amine.

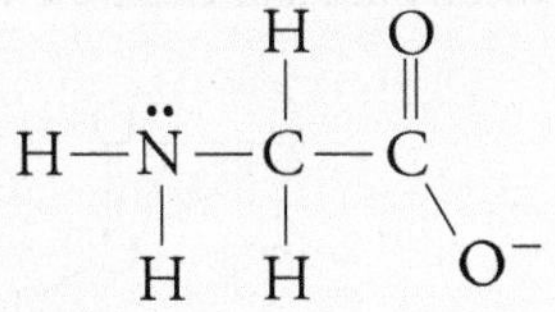

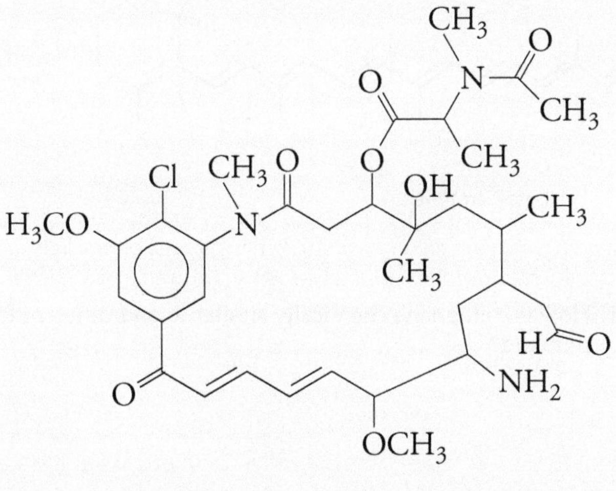

66. Suggest an explanation for why aspirin has a sour taste.

67. Benzaldehyde is a fragrant oil. If stored in an uncapped bottle, this compound will slowly transform into benzoic acid along the surface. Is this an oxidation or a reduction?

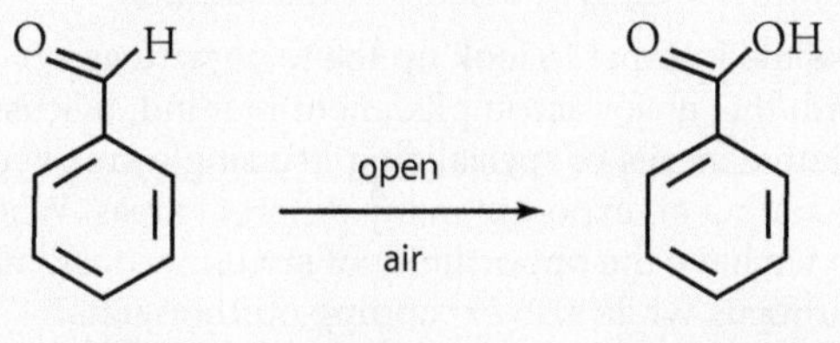

Benzaldehyde   Benzoic acid

68. What products are formed upon the reaction of benzoic acid with sodium hydroxide, NaOH? One of these products is a common food preservative. Can you name it?

69. The disodium salt of ethylenediaminetetraacetic acid, also known as EDTA, has a great affinity for lead ions, Pb^{2+}. Why? What are some useful applications of this chemistry?

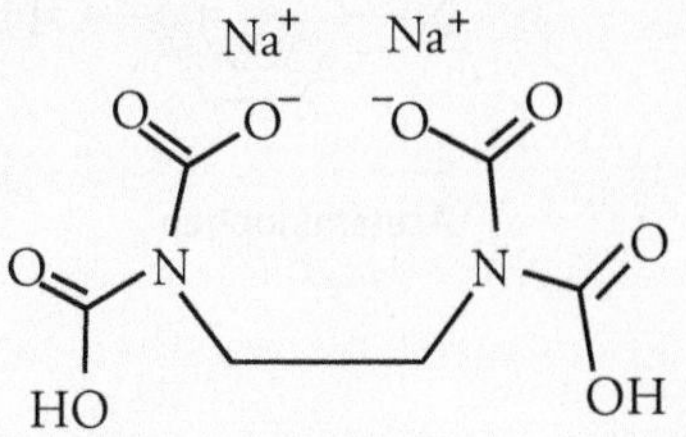

EDTA

70. The amino acid lysine is shown below. What functional group must be removed in order to produce cadaverine, as shown in Figure 12.18?

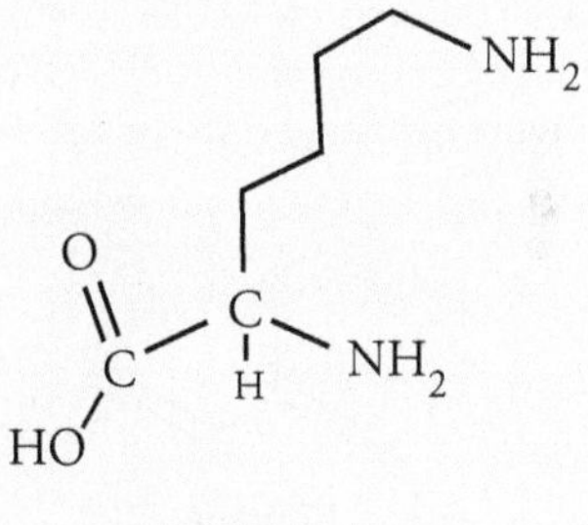

Lysine

12.7 An Example of Organic Synthesis

71. How does the total synthesis of multistriatin help to protect the elm tree from disease?

72. What two bonds within the structure of menthol can be simultaneously cut to produce two molecules, each with the same number of carbon atoms in the same configuration?

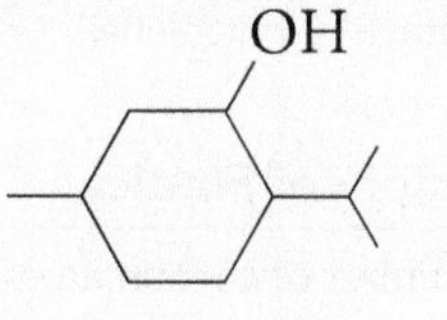

Methanol

73. Which bond would be best to break in a retrosynthetic analysis of the following molecule?

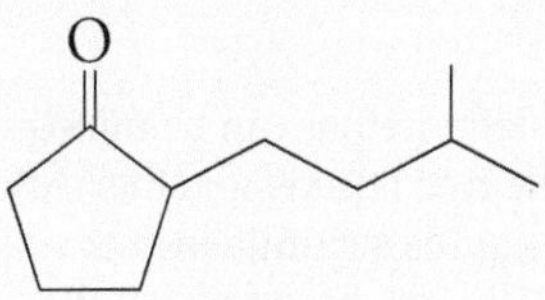

74. Under the right conditions, a carboxylic acid and an amine can join to form an amide. Knowing this, draw the structures of two molecules that can be combined to form acetaminophen.

Acetaminophen

12.8 Organic Molecules Can Link to Form Polymers

75. Would you expect polypropylene to be more or less dense than low-density polyethylene? Why?

76. Hydrocarbons release a lot of energy when ignited. Where does this energy come from?

77. The polymer styrene-butadiene rubber (SBR), shown below, is used to make tires as well as bubble gum. Is it an addition polymer or a condensation polymer?

SBR

78. Citral and camphor are both 10-carbon odoriferous natural products made from the joining of two isoprene units plus the addition of a carbonyl functional group. Shown below are their chemical structures. Find and circle the two isoprene units in each of these molecules.

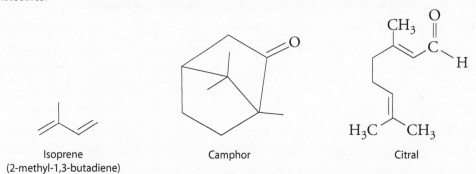

Isoprene
(2-methyl-1,3-butadiene)   Camphor   Citral

79. Many of the natural product molecules synthesized by photosynthetic plants are formed by the joining together of isoprene monomers via an addition polymerization. A good example is the nutrient beta-carotene. How many isoprene units are needed to make one beta-carotene molecule? Find and circle these units in the beta-carotene structure shown below.

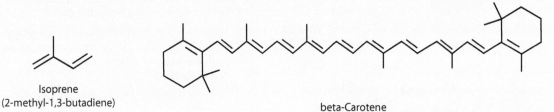

Isoprene
(2-methyl-1,3-butadiene)

beta-Carotene

12.9 A Brief History of Plastics

80. Why is the number of carbon atoms in a sample of vulcanized rubber always some multiple of five?

81. How is cellophane chemically similar to and different from celluloid?

THINK AND DISCUSS (EVALUATION)

82. The solvent diethyl ether can be mixed with water, but only by shaking the two liquids together. After the shaking is stopped, the liquids separate into two layers, much like oil and vinegar. The free-base form of the alkaloid caffeine is readily soluble in diethyl ether but not in water. Suggest what might happen to the caffeine of a caffeinated beverage if the beverage is first made alkaline with sodium hydroxide and then shaken with some diethyl ether.

83. Alkaloid salts are not very soluble in the organic solvent diethyl ether. What might happen to the free-base form of caffeine dissolved in diethyl ether if gaseous hydrogen chloride, HCl, is bubbled into the solution?

84. Use the Internet to look up the total synthesis of Taxol®. With this major accomplishment in mind, discuss the relative merits of specializing in a single area versus becoming an expert in many different areas. When in life do we have the opportunity of simultaneously narrowing our focus while also expanding our horizons?

85. Handheld calculators were first made available to the general public in the 1970s. They were all the rage, but soon faded into the background. The desktop computer was the technical wonder of the 1980s. During the 1990s, the World Wide Web came to be. What were the technical wonders of the first decade after Y2K? What might be the wonder 10 years from now? 20 years from now? 100 years from now? How grateful should we be for technological advances? Have they really made our lives better? If so, how? Which brings more happiness: getting what you need or getting what you want?

READINESS ASSURANCE TEST (RAT)

If you have a good handle on this chapter, then you should be able to score at least 7 out of 10 on this RAT. Check your answers online at www.ConceptualChemistry.com. If you score less than 7, you need to study further before moving on.

Choose the BEST answer to the following.

1. Why does the melting point of hydrocarbons increase as the number of carbon atoms per molecule increases?

    a. An increase in the number of carbon atoms per molecule also means an increase in the density of the hydrocarbon.

    b. There are more induced dipole–induced dipole molecular attractions.

    c. Larger hydrocarbon chains tend to be branched.

    d. The molecular mass also increases.

2. What is the number of structural isomers for hydrocarbons having the molecular formula C_4H_{10}?

    a. None

    b. One

    c. Two

    d. Three

3. What is the name for the structure shown below?

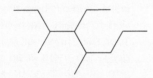

    a. 3-methyl-4-ethyl-6-methyl-octane

    b. 4-ethyl-3,5-dimethyloctane

    c. 2-ethyl-4-ethyl-5-methyl-heptane

    d. 2,4-diethyl-5-methyl-heptane

4. Why do heteroatoms make a difference in the physical and chemical properties of an organic molecule?

    a. They add extra mass to the hydrocarbon structure.

    b. Each heteroatom has its own characteristic chemistry.

    c. They can enhance the polarity of the organic molecule.

    d. All of the above are true.

5. Why might a high-formula-mass alcohol be insoluble in water?

    a. A high-formula-mass alcohol should be very soluble in water.

    b. The bulk of a high-formula-mass alcohol would likely consist of nonpolar hydrocarbons.

    c. Such an alcohol would likely be in a solid phase.

    d. For two substances to be soluble in each other, their molecules need to be of comparable mass.

6. How many possible chemical structures (configurations) can be made from the formula C_2H_2BrF?

    a. One

    b. Two

    c. Three

    d. Four

7. Explain why caprylic acid, $CH_3(CH_2)_6COOH$, dissolves in a 5 percent aqueous solution of sodium hydroxide but caprylaldehyde, $CH_3(CH_2)_6CHO$, does not.

    a. With two oxygen atoms, the caprylic acid is about twice as polar as the caprylaldehyde.

    b. The caprylaldehyde is a gas at room temperature.

    c. The caprylaldehyde behaves as a reducing agent, which neutralizes the sodium hydroxide.

    d. The caprylic acid reacts to form the water-soluble salt.

8. How many oxygen atoms are bonded to the carbon of the carbonyl of an ester functional group?

    a. None

    b. One

    c. Two

    d. Three

9. One solution to the problem of overflowing landfills is to burn plastic objects instead of burying them. What would be some of the advantages and disadvantages of this practice?

    a. Disadvantage: toxic air pollutants. Advantage: reduced landfill volume.

    b. Disadvantage: loss of vital petroleum-based resources. Advantage: generation of electricity.

    c. Disadvantage: discourages recycling. Advantage: provides new jobs.

    d. All of the above.

10. Which would you expect to be more viscous, a polymer made of long molecular strands or a polymer made of short molecular strands? Why?

    a. Long molecular strands because they tend to tangle among themselves.

    b. Short molecular strands because of a greater density.

    c. Long molecular strands because of a greater molecular mass.

    d. Short molecular strands because their ends are typically polar.

Contextual Chemistry

Hair and Skin Care

Through our modern lifestyles, we have become accustomed to chemical-based products that clean and protect our hair and skin. These include soaps, shampoos, conditioners, moisturizers, sunscreens, and many other products formulated to help us maintain a healthy look.

The essential ingredients of all these hair and skin products are part of a broad class of compounds called *surfactants*, which is short for "surface-active agents." All surfactant molecules are polar on one end and nonpolar on the other. Because of their nonpolar ends, they resist dissolving in water and instead cling to water's surface, where their nonpolar ends stick out of the water. Pour some oil onto the surfactant-containing water and the nonpolar tails will cling to the oil's surface. In this way, surfactants bind together the surfaces of water and a nonpolar material.

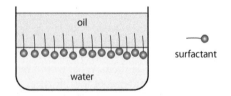

With a little agitation, the surfactant's surface-seeking behavior causes the formation of high-surface-area suds. With such an increased surface area, significant amounts of the oil are able to mix with the water, forming a single phase known as an *emulsion*. So surfactants don't help nonpolar materials to "dissolve in water." Instead, they just help these two materials to become mixed.

The prototypical shampoo surfactant is the detergent sodium lauryl sulfate. A big advantage of detergents is that in hard water, they don't form scum (see Section 7.5). For a shampoo, this is particularly

important because, otherwise, the scum would adhere to the hair. Rinsing hair first allows the removal of polar materials, such as salts, dirt, or any loose debris. Wetted hair also swells, and this opens the hair cuticles, which are scale-like structures lining the outside of the hair. The main action of the detergent is to remove sebum, a skin oil that migrates up the hair and into the cuticles. Shampoo formulas vary in their cleansing strength. Those formulated for "oily" hair are generally stronger than those formulated for "normal" hair.

$$\text{\Huge /\!\!\backslash\!\!/\!\!\backslash\!\!/\!\!\backslash} OSO_3^- \ Na^+$$

Sodium lauryl sulfate

Following the shampoo, many people apply a hair conditioner, which helps to make the hair feel soft and helps to minimize frizz due to the buildup of static charge. Notably, the polar end of a shampoo's surfactant is a negatively charged ion. Such a surfactant is called an *anionic* surfactant. The polar end of a hair conditioner's surfactant, however, is a positively charged ion. Such a surfactant is called a *cationic* surfactant. Shampoos and conditioners are generally not combined into a single formula because the attraction these oppositely charged surfactants have for each other would defeat their functions. Products that are "all-in-one" have the cationic surfactants microencapsulated.

The prototypical hair conditioning surfactant is cetyl trimethyl ammonium chloride, also known as cetrimonium chloride. Its positive charge helps it to bind with the hair so that much of it remains on the hair even after rinsing. The result is hair that feels smoother and is easier to manage. Laundry fabric softeners work by the same principle.

Cetyl trimethyl ammonium chloride

The trick to having comfortably moist skin is to maintain the moisture already found within the skin. The deeper parts of your skin are up to 80 percent water. Because skin has so much more water compared to the surrounding air, the general movement of moisture is from the skin outward. Skin oils help the outer layers of skin to retain much of this moisture. Your skin, however, will begin to feel dry when there is an increase in the rate at which your skin's moisture evaporates. This happens when the air is very dry or when your skin oils are removed by strong soaps or detergents. Rather than splashing water on your dry skin, a more practical solution is to coat your skin with a moisture-protecting nonpolar material that can enhance or mimic the action of skin oils. Commonly used materials include compounds such as propylene glycol, glycerin, and dimethicone. The best way to apply these compounds to the skin is with an emulsion made using nonionic surfactants, such as polyethylene glycol stearate, also known as PEG-100. Nonionic surfactants are used for moisturizers because they tend to be less irritating and less likely to remove skin oils.

Propylene glycol

Dimethicone

Polyethylene glycol stearate

Add UV-absorbing compounds to a skin moisturizer and you have sunblock lotion. Making a sunblock lotion water resistant, however, is problematic. Upon exposure to water, the surfactants within the lotion help to remove the lotion. Manufacturers generally recommend that you reapply the lotion after swimming. Lotions that minimize the use of surfactants have increased water resistance, but they tend to be oily or greasy. For these reasons, many water-resistant formulas include compounds that polymerize into a nonpolar, water-resistant film that helps to hold the sunscreen agent on the skin when wet.

A special consideration for sunscreens is the portion of the UV spectrum they are able to block. Nearly all sunscreens effectively block the region from 290 to 320 nm, known as UV-B. The traditional Sun Protection Factor (SPF) rating system indicates the effectiveness of a sunscreen lotion in blocking UV-B, which causes sunburn. Scientists, however, are finding that the range from 320 to 400 nm, known as UV-A, is more responsible for skin wrinkling as well as skin cancer. Sunscreens blocking both UV-A and UV-B are labeled "Broad Spectrum."

As of 2011, the U.S. Food and Drug Administration (FDA) has approved only 17 compounds for use as sunscreens. Of these, only four are UV-A blockers. The most widely used of these four, avobenzone, is easily formulated into a lotion, but it quickly loses potency if not properly formulated with stabilizers. The more stable UV-A blocker mexoryl SX was approved by the FDA in 2006 after its use for decades in other countries. The other two UV-A blockers are titanium dioxide and zinc oxide, both opaque minerals that are impractical to apply over large body-surface areas. The lists of approved UV-B and UV-A sunscreens in other countries, such as Australia and Canada, and in European nations are much longer. These countries treat sunscreens as cosmetics, while the United States treats them as drugs, which are subject to more rigorous regulations. Since 1978, only three additional sunscreen agents—avobenzone, mexoryl SX, and zinc oxide—have been approved by the FDA.

Avobenzone

Another way to help skin maintain a youthful look is to apply alpha-hydroxy acids, such as glycolic acid. These compounds serve as exfoliants, which means they help in the removal of dead outer skin. This stimulates the skin to generate new cells that help to hide thin-line wrinkles.

Discussed here are only the main chemicals used to formulate hair- and skin-care products. As any casual look at an ingredients list tells you, many other chemicals serve other purposes.

Looking to the immediate future, there are some areas where skin care may soon make some significant advances. As we grow older, our skin wrinkles, because it has lost much of its elasticity. One of the reasons for this loss of elasticity is the formation of chemical bonds, called crosslinks, between adjacent strands of collagen, which is the fundamental fiber skin cells create to form skin. Numerous agents that effectively inhibit and even reverse these collagen crosslinks have been discovered and are currently undergoing clinical trials. One such agent is 2-amino-3,4-dimethylthiazole hydrogen bromide. Search the Internet for recent developments.

CONCEPT CHECK

What do shampoos, hair conditioners, moisturizers, and sunscreen lotions all have in common?

CHECK YOUR ANSWER They all contain surfactants, which are molecules that have both polar and nonpolar ends and permit the formation of emulsions.

Think and Discuss

1. In an effort to speed up the drug approval process, the U.S. FDA created a special "fast-track" system for agents that show unusual promise or for agents that have been shown to be generally safe after years of use in other countries. Why are there so few sunscreen agents currently within this fast-track system? Should sunscreen agents continue to be classified as drugs?

2. If a collagen cross-linking agent is determined to be safe by the FDA, should this agent be available only by prescription, or should it be made available to consumers over the counter? If this agent noticeably helped to reduce skin wrinkling, how popular might it be? Assuming you were in your 50s, how much would you be willing to pay for a year's supply?

3. What criteria do you use in deciding what shampoo to buy? List the following in order of importance: the price, the brand name, the ingredients named on the front label, those listed on the back label, the fragrance, commercial advertisements, recommendation from a friend, and your experience with the shampoo.

Nutrients of Life

THE MAIN IDEA

There are four main classes of biomolecules.

▲ We are what we eat.

We absorb molecules from the food we eat and either use them for energy or incorporate them into the various structures that give our bodies both form and function. Interestingly, no molecule in a living organism is a permanent resident; rather, there is perpetual change as ingested food molecules are transformed and utilized to replace older molecules. Within a decade, most of the molecules in a human body have been replaced by new ones—the body you have today is literally not the same as the body you had seven years ago.

Although your molecules are continually replaced, the genetic code that guides the assembly of those molecules is still pretty much the same. This is similar to what we see with identical twins who are made of different molecules but share pretty much the same genetic code. No one would claim two identical twins to be the same person. Similarly, perhaps you are as different from your past self as two identical twins are from each other. Perhaps an individual's identity is not static, but is continually re-established every moment.

Although this chapter cannot promise any insights into the intriguing questions of existence, it will give you a basic understanding of *biomolecules*—the molecules that make up living organisms—and the remarkable roles they play in your body.

Chemistry

Sweet Enzymes

Look at a soda cracker and contemplate eating it. If you're hungry, this should help to stimulate your salivary glands to produce extra enzyme-containing saliva. Use salt-free or low-salt crackers with no additive flavorings, if available.

PROCEDURE

1. Place the cracker in your mouth and start chewing for as long as possible without swallowing.
2. Note any apparent change in sweetness.

ANALYZE AND CONCLUDE

1. An enzyme is a large biomolecule that helps to speed up chemical reactions. In other words,

it is a *catalyst* for biomolecular reactions, such as the breaking down of food molecules. Enzymes in your saliva work to break down complex starch molecules, which have little flavor, into simple carbohydrate molecules, which are sweet. Did you notice that the cracker tasted sweeter the longer it was in your mouth?

2. Rice candies are popular in Asian cultures. What do you suppose is done to the rice to convert it to a "candy"?
3. Why is it important to chew your food thoroughly before swallowing it?

13.1 Biomolecules Are Produced and Utilized in Cells

EXPLAIN THIS

Why aren't nucleic acids commonly listed on nutrition labels?

LEARNING OBJECTIVE

Identify the basic components of a cell and the four major classes of biomolecules.

The fundamental unit of almost all organisms is the *cell*. Typically, cells are so small that you need a microscope to see them individually. About ten average-sized human cells, for example, could fit within the period at the end of this sentence. **Figure 13.1** shows a typical animal cell and a typical plant cell.

Every cell has a *plasma membrane*. More than just a boundary, the plasma membrane controls the passage of molecules into and out of the cell and provides sites where important chemical reactions occur. In animal cells, the plasma membrane is the outermost part of the cell, but the plasma membranes of plant cells are bounded by a rigid *cell wall* that protects the cells and gives them structure.

Housed within each of these cells is a *cell nucleus*, which contains the genetic code. Everything between the plasma membrane and the cell nucleus is the *cytoplasm*, which consists of a variety of microstructures suspended in a viscous liquid. These microstructures, called *organelles*, work together in the synthesis, storage, and export of important biomolecules and in the production of energy from food and oxygen.

The great majority of the biomolecules used by cells are carbohydrates; lipids; proteins; and, in lesser quantities, the nucleic acids. In addition, most cellular reactions require small amounts of vitamins and minerals in order to function properly. We now discuss all these categories of biomolecules.

 READING CHECK

What does every cell have?

▶ **Figure 13.1**
Macroscopic, microscopic, and submicroscopic views of an animal and a plant. The atoms of the biomolecules are indicated by color: white = hydrogen; black = carbon; red = oxygen; blue = nitrogen; and orange = phosphorus.

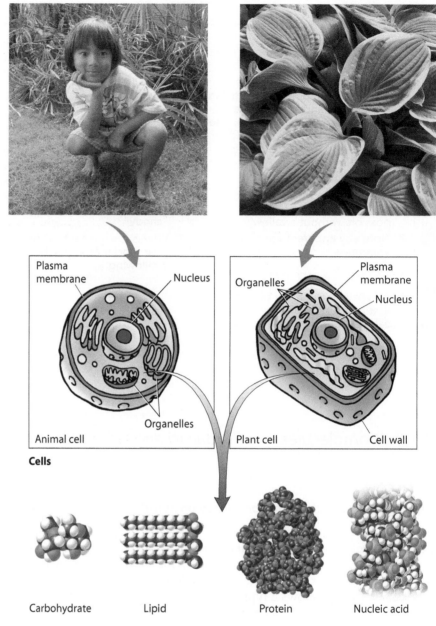

Cells

Carbohydrate   Lipid   Protein   Nucleic acid

Biomolecules

13.2 Carbohydrates Give Structure and Energy

LEARNING OBJECTIVE

Recognize the molecular structures of simple and complex carbohydrates.

EXPLAIN THIS

How many water molecules are needed to make a single molecule of glucose?

Carbohydrates are organic molecules produced by photosynthetic plants, containing carbon, hydrogen, and oxygen. The term *carbohydrate* is derived from the fact that plants make these molecules from carbon (from atmospheric carbon dioxide) and water. The term **saccharide** is a synonym for *carbohydrate*, and a *monosaccharide* ("one saccharide") is the fundamental carbohydrate unit. In most

Glucose

Fructose

monosaccharides, each carbon atom is bonded to at least one oxygen atom, most often within a hydroxyl group. There are many kinds of monosaccharides. The structures of the two most common monosaccharides, glucose and fructose, are shown in **Figure 13.2**.

Monosaccharides are the building blocks of *disaccharides*, which are carbohydrate molecules containing two monosaccharide units linked together. **Figure 13.3** shows table sugar—sucrose—the best-known disaccharide. In the digestive tract, sucrose is readily cleaved, or separated, into its monosaccharide units, glucose and fructose.

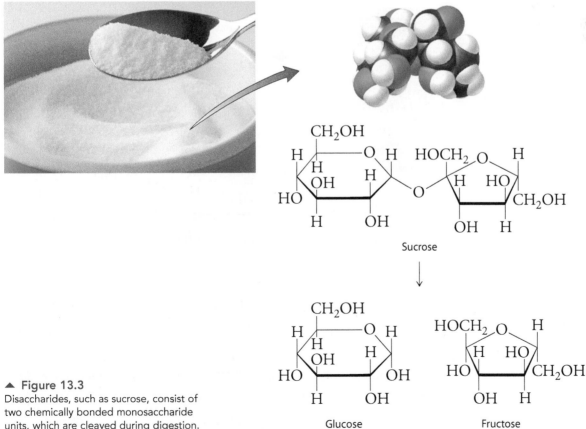

Sucrose

Glucose

Fructose

▲ Figure 13.3
Disaccharides, such as sucrose, consist of two chemically bonded monosaccharide units, which are cleaved during digestion.

Lactose, shown in **Figure 13.4**, is another important disaccharide. Lactose is the main carbohydrate in milk. In the digestive tract, it is cleaved into the monosaccharides galactose and glucose by the enzyme lactase, which most children produce in abundance up to about the age of 6. Thereafter, the production of this enzyme decreases, with the result that some adults produce little or none. This leads to *lactose intolerance*, a condition in which ingestion of milk or milk products leads to bloating, flatulence, and painful cramps. These symptoms result as certain intestinal bacteria vigorously digest the lactose. In doing so, they generate large amounts of gases, such as hydrogen, H_2. To relieve these symptoms, some milk products are treated with *Acidophilus bifidus*, a strain of bacteria that destroy gas-causing bacteria in the digestive tract. Some lactose-intolerant individuals add commercially available lactase enzyme to milk or milk products before consuming them.

Monosaccharides and disaccharides are classified as *simple carbohydrates*. The word *simple* is used because these food molecules consist of only one or two monosaccharide units. Most simple carbohydrates have some degree of sweetness and are also known as *sugars*.

Polysaccharides Are Complex Carbohydrates

Recall from Chapter 12 that polymers are large molecules made of repeating monomer units. Monosaccharides are the monomers that link to form the biomolecular polymers called *polysaccharides*, which contain hundreds to thousands of monosaccharide units. Polysaccharides can be built from any type of monosaccharide units. Our bone joints are lubricated by the polysaccharide hyaluronic acid, for example, which consists of alternating glucuronic acid and N-acetylglucosamine, as shown in **Figure 13.5a**.

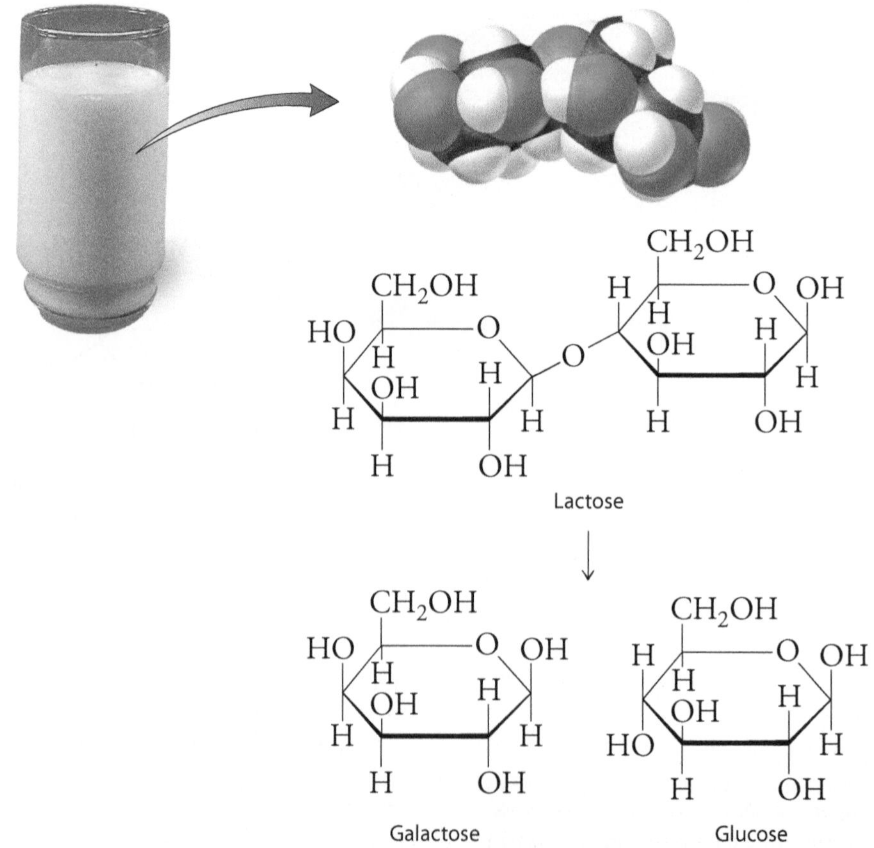

▶ **Figure 13.4**
Milk and milk products contain the disaccharide lactose, which is digested to galactose and glucose.

Lactose

Galactose   Glucose

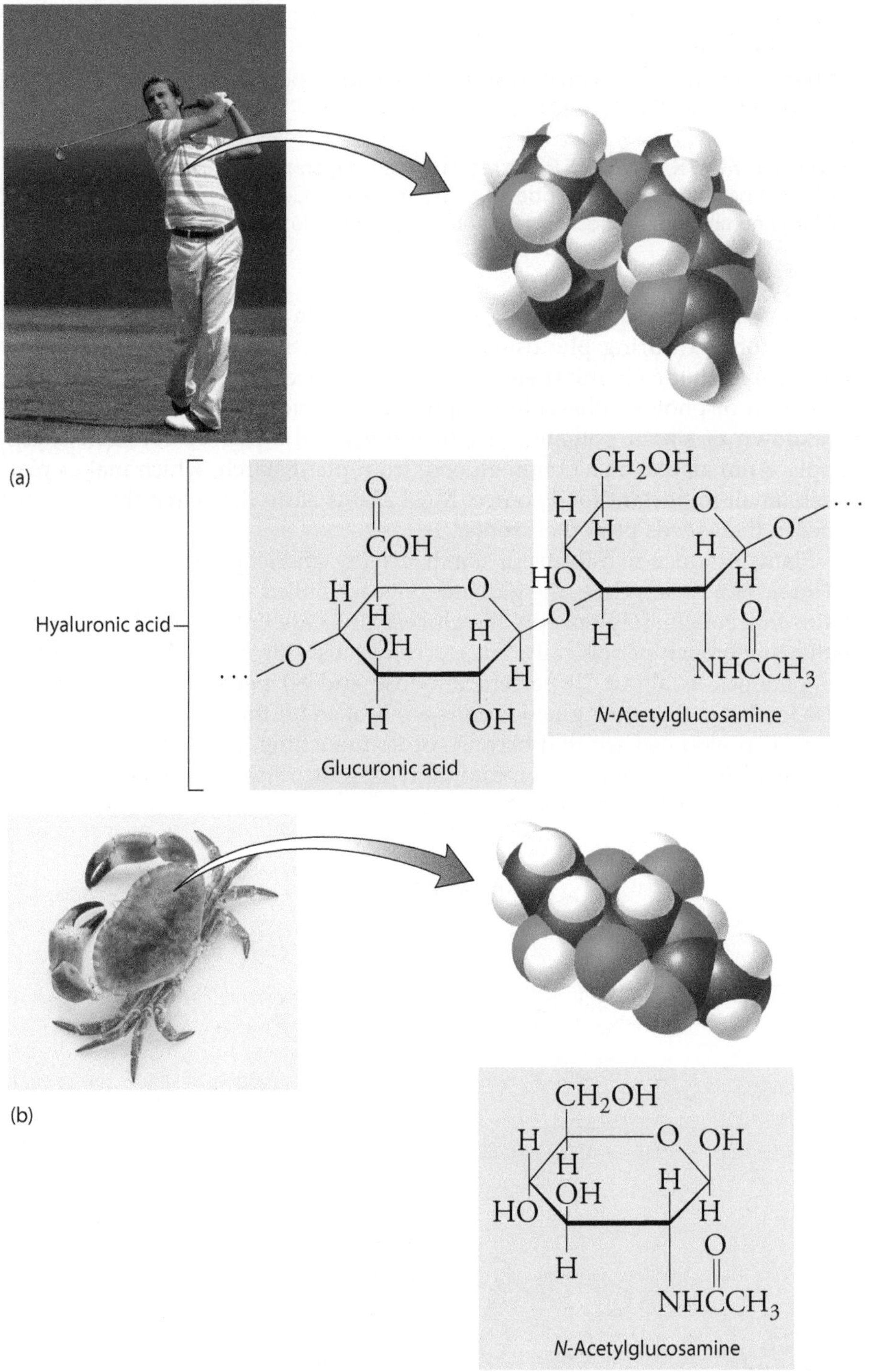

◀ **Figure 13.5**
(a) Hyaluronic acid, a lubricant in bone joints, is a polysaccharide consisting of the monosaccharides glucuronic acid and N-acetylglucosamine. (b) The exoskeletons of insects, crabs, shrimp, and lobsters are made of chitin, a polysaccharide containing only N-acetylglucosamine units.

(a)

Hyaluronic acid

Glucuronic acid

N-Acetylglucosamine

(b)

N-Acetylglucosamine

The exoskeletons (protective shells) of insects and some marine organisms, such as crabs and shrimp, are made of chitin (pronounced kite'-in), a hard, resilient polysaccharide made of the monosaccharide N-acetylglucosamine, shown in **Figure 13.5b**. The wood varnish known as shellac contains the chitin from the exoskeletons of insects. In powdered form, chitin is now finding use as a dietary fiber supplement.

Although a polysaccharide can be made of any type of monosaccharide unit, the polysaccharides in the human diet are made only of glucose. These polysaccharides include *starch*, *glycogen*, and *cellulose*, which differ from one another only in the way the glucose units are connected together. All polysaccharides, but especially the ones in our diet, are known as *complex carbohydrates*. *Complex* refers to the multitude of monosaccharide units linked together.

Carbohydrates	
Simple	**Complex**
Monosaccharides	*Polysaccharides*
Glucose	Hyaluronic acid
Fructose	
	Chitin
Disaccharides	Starch
Sucrose	Glycogen
Lactose	Cellulose

READINGCHECK

What is starch?

Starch is a polysaccharide produced by plants to store the abundance of glucose formed during photosynthesis, the process by which plants convert solar energy to the chemical energy of sugar molecules. (We shall return to a discussion of photosynthesis in Chapter 15.) On cloudy days or at night, the breakdown of starch polymers to glucose gives the plant a constant energy supply. Animals can also obtain glucose from plant starch, which makes plant starch an all-important food source. Most plants store the starch they produce either in their seeds or in their roots.

Plants produce two forms of starch, *amylose* and *amylopectin*, as illustrated in **Figure 13.6**. In amylose, the glucose units are linked together in unbranched chains that coil. In amylopectin, the glucose units are linked together in coiling chains that branch periodically. The starch of most starchy foods, such as bread and potatoes, is about 20 percent amylose and 80 percent amylopectin. As these foods are digested, glucose units are broken off the ends of the chains. In Figure 13.6, you can see that because of its branching, amylopectin has more ends than amylose does. Therefore, amylopectin releases glucose units at a faster rate than amylose does.

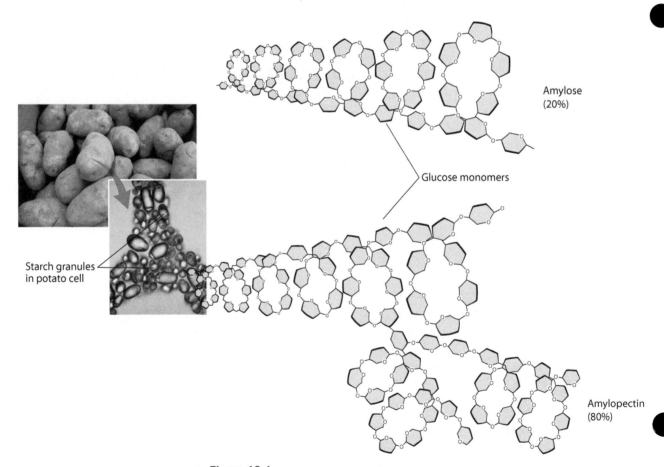

Amylose (20%)

Glucose monomers

Starch granules in potato cell

Amylopectin (80%)

▲ Figure 13.6
Amylose and amylopectin are two forms of plant starch.

CONCEPT CHECK

When amylose and amylopectin are digested, which one supplies glucose at a faster rate?

CHECK YOUR ANSWER There is more branching in amylopectin and therefore a greater number of ends per molecule. The digestion of amylopectin produces more glucose molecules at a faster rate and therefore is a quicker and more immediate supply of glucose.

Animals store their excess glucose by converting it to **glycogen**, a polymer made of hundreds of glucose monomers and sometimes referred to as *animal starch*. Glycogen has a structure much like that of amylopectin but with a greater extent of branching, as shown in **Figure 13.7**. Between meals, when glucose levels drop, the body metabolizes glycogen to glucose. Glycogen therefore serves as our glucose reserve. Most of this glycogen is stored in our livers and muscle tissue.

Cellulose, a structural component of plant cell walls, is also a polysaccharide of glucose. The glucose in cellulose, however, is slightly different from the glucose in starch and glycogen. As highlighted in **Figure 13.8**, the glucose in cellulose is identical to the glucose in starch except for the orientation of one hydroxyl group. Because this particular hydroxyl group is involved in the joining of glucose monomers, the way in which the glucose units join together is affected. The glucose units of starch and glycogen are connected by what is called an alpha (α) linkage, which results in a coiling of the polysaccharide strand. The glucose units of cellulose are connected by a beta (β) linkage, which

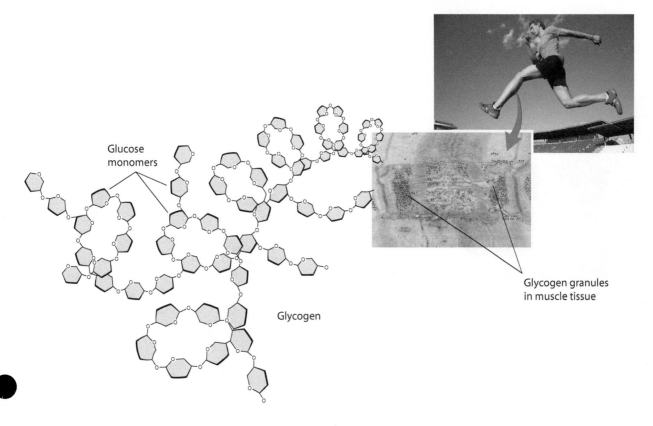

Glucose monomers

Glycogen

Glycogen granules in muscle tissue

▲ Figure 13.7
The complex carbohydrate glycogen, a polymer of glucose, is found in animal tissue.

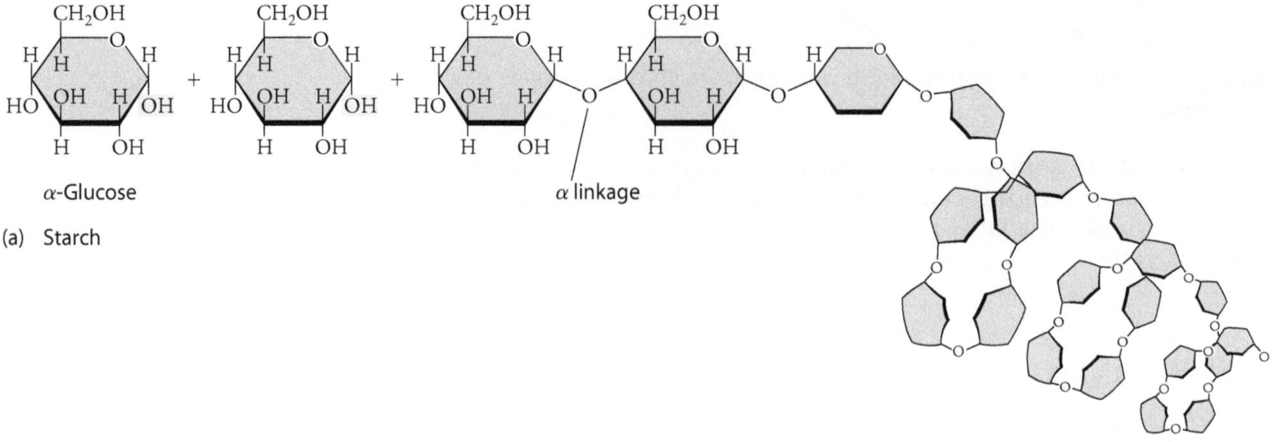

α-Glucose

(a) Starch

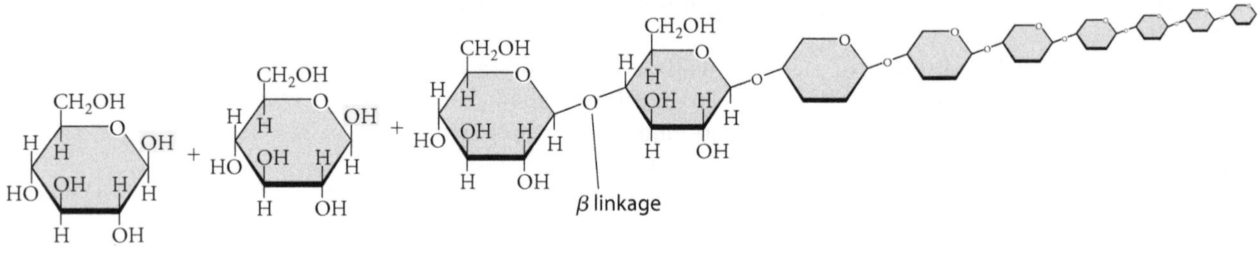

α linkage

β-Glucose

(b) Cellulose

β linkage

▲ **Figure 13.8**
The glucose units of starch and cellulose are bonded in different orientations. (a) In starch, polymerization of α-glucose units results in polysaccharide strands that tend to coil. (b) In cellulose, polymerization of β-glucose units results in polysaccharide strands that do not coil and so can align with one another.

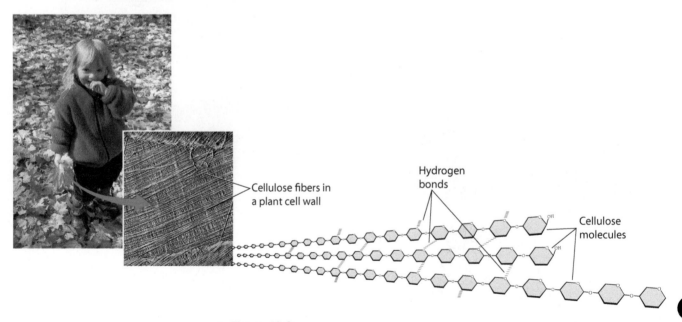

Cellulose fibers in a plant cell wall

Hydrogen bonds

Cellulose molecules

▲ **Figure 13.9**
Strands of cellulose in a plant, including the leaves held by Maitreya, are joined by hydrogen bonds. These microscopic fibers are laid down in a crisscross pattern to give strength in many directions.

results in straight uncoiled, unbranched chains. These two attributes of cellulose molecules allow the polysaccharide strands to align much like strands of uncooked spaghetti. This alignment maximizes the number of hydrogen bonds between strands, which makes cellulose a tough material. Plants produce microscopic cellulose fibers in a crisscross fashion, as shown in **Figure 13.9**, increasing their structural strength.

Cellulose serves as the primary structural component of all plants. Cotton is nearly pure cellulose. Wood, made largely of cellulose, can support trees that are as much as 30 meters tall. Cellulose is by far the most abundant organic compound on the Earth.

Most animals, including humans, are not able to break down cellulose into its monomer units, glucose. Instead, the cellulose in the food we eat serves as dietary fiber that helps in regulating bowel movements. In the large intestine, cellulose-based fiber absorbs water and has a laxative effect. Waste products are therefore moved along faster and, with them, harmful bacteria and toxins, including carcinogens. Microorganisms that live in the digestive tracts of wood-eating termites and grass-eating ruminants, such as cattle, sheep, and goats, can break down cellulose to glucose. Strictly speaking, termites and ruminants do not digest cellulose. Rather, they digest the glucose produced by the cellulose-digesting microorganisms that live inside them, as shown in **Figure 13.10**.

▲ Figure 13.10
(a) A termite digests glucose produced by (b) cellulose-digesting microorganisms living in the termite's digestive tract.

13.3 Lipids Are Insoluble in Water

EXPLAIN THIS

Why would it be a good idea to coat yourself in grease before swimming the English Channel?

Lipids are a broad class of biomolecules. Although structurally diverse, all lipids are insoluble in water because their molecular structures are largely made of hydrocarbons. In this section, we discuss two important types of lipids: *fats* and *steroids*.

Fats Are Used for Energy and Insulation

A **fat** is any biomolecule formed from the reaction between glycerol, $C_3H_8O_3$, and three fatty acid molecules, as shown in **Figure 13.11**. A *fatty acid* is a long-chain hydrocarbon terminating in a carboxylic acid group. Typically, the chains include an even number of carbon atoms (between 12 and 18) and may be either saturated or unsaturated. Recall from Chapter 12 that a *saturated* chain contains no carbon–carbon double bonds and that an *unsaturated* chain contains one or more carbon–carbon double bonds. Note that, like carbohydrates, fats contain only carbon, hydrogen, and oxygen. Fats and carbohydrates have similar elemental compositions because both plants and animals synthesize fats from carbohydrates. There is a significant difference in the structures of these two types of biomolecules, however, as you can readily see by comparing Figures 13.8 and 13.11. Because fats are synthesized from three fatty acids and glycerol, they are commonly referred to as *triglycerides*.

Fats are stored in the body in localized regions known as fat deposits. These deposits serve as important reservoirs of energy. Fat deposits directly under

▶ **Figure 13.11**
A typical fat molecule, also known as a triglyceride, is the combination of one glycerol unit and three fatty acid molecules. Note that this reaction involves the formation of three ester functional groups.

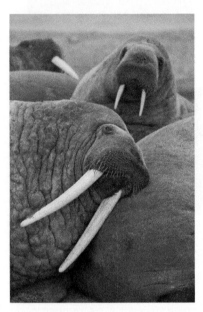

▲ **Figure 13.12**
The walrus and other polar species are insulated from the cold by a thick layer of fat beneath their skin.

 READINGCHECK

What sort of molecular attractions hold the aligned chains of fat molecules together?

Glycerol   Fatty acid molecules   Fat moleocule, a triglyceride

Loss of 3 H_2O molecules

the skin help to insulate us from the cold, good news for the walrus shown in **Figure 13.12**. In addition, vital organs, such as the heart and kidneys, are cushioned against injury by fat deposits.

The digestion of fat is accompanied by the release of considerably more energy than is produced by the digestion of an equivalent amount of either carbohydrate or protein. There are about 38 kilojoules (9 Calories) of energy in 1 gram of fat but only about 17 kilojoules (4 Calories) of energy in 1 gram of carbohydrate or protein. (Recall that the energy content of food is often reported in Calories, with an uppercase C, where 1 Calorie = 1 kilocalorie = 1000 calories.)

CONCEPTCHECK

Give two reasons animals living in cold climates tend to form a thick layer of fat just prior to the onset of winter.

CHECK YOUR ANSWER Fat provides a source of energy during the winter, when food is generally scarce, and it provides insulation from cold winter temperatures.

As shown in **Figure 13.13a**, saturated fatty acid molecules are able to pack tightly together because the linear chains of saturated fatty acids align easily with one another. Induced dipole–induced dipole attractions hold the aligned chains together. This gives saturated fats, such as those found within beef, relatively high melting points, and as a result, they tend to be solid at room temperature. Unsaturated fats—those made from unsaturated fatty acids—have fatty acid chains that are "kinked" wherever double bonds occur, as shown in **Figure 13.13b**. The kinks inhibit alignment, and as a result, unsaturated fats tend to have relatively low melting points. These fats are liquid at room temperature and are commonly referred to as *oils*. Most vegetable oils are liquid at room temperature because they contain a high proportion of unsaturated fats.

The fat from an animal or a plant is a mixture of different fat molecules having various degrees of unsaturation. Fat molecules containing only one carbon–carbon double bond per fatty acid chain are *monounsaturated*. Those containing more than one double bond per fatty acid chain are *polyunsaturated*. Table 13.1 shows the percentage of saturated, monounsaturated, and polyunsaturated fats in a number of widely used dietary fats.

Steroids Contain Four Carbon Rings

Steroids are a class of lipids that have in common a system of four fused carbon rings. Cholesterol, shown in **Figure 13.14**, is the most abundant steroid by far, and it serves as the starting material for the biosynthesis of almost all other steroids, including the sex hormones estradiol and

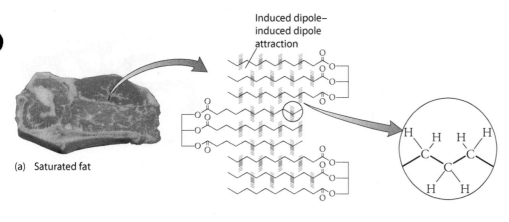

(a) Saturated fat

◀ Figure 13.13
(a) Saturated fats are typically solid at room temperature, because of molecular attractions between fatty acid chains. (b) Unsaturated fats are typically liquid at room temperature, because molecular attractions are inhibited by the kinked nature of the fatty acid chains.

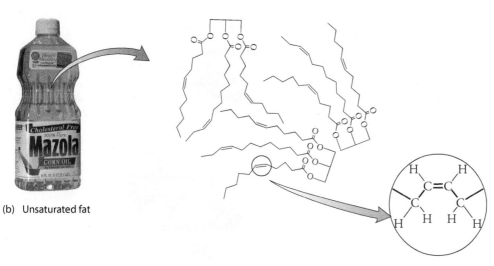

(b) Unsaturated fat

testosterone, also shown in Figure 13.14. *Hormones* are chemicals produced by one part of the body to influence other parts of the body. For example, estradiol, produced by the ovaries, and testosterone, produced by the testes, are responsible for the development of secondary sex characteristics in other parts of the body.

Cholesterol is found throughout the body. In fact, the human brain is about 10 percent cholesterol by weight. Our bodies synthesize cholesterol in the liver. In addition, we obtain cholesterol through a diet of animal products.

Many synthetic steroids having a wide variety of biological effects have been prepared. Prednisone, for instance, is often prescribed as an anti-inflammatory agent for the treatment of arthritis. Synthetic steroids that mimic the muscle-building properties of testosterone are also available. These are the *anabolic steroids*, which physicians prescribe to assist patients suffering from

Lipids

Fats	**Steroids**
Saturated	Cholesterol
Coconut	Testosterone
	Estradiol
Unsaturated	
Olive	

TABLE 13.1 Degree of Unsaturation in Some Common Fats

PERCENTAGE OF TOTAL FATTY ACID CONTENT			
Fat or Oil	Saturated	Monounsaturated	Polyunsaturated
Coconut	93	6	1
Palm	57	36	7
Lard	44	46	10
Peanut	21	49	30
Olive	15	73	12
Corn	14	29	57
Soybean	14	24	62
Canola	6	58	36

▶ Figure 13.14
The sex hormones estradiol and
testosterone are steroids that the
body produces from the most
abundant steroid of all—cholesterol.

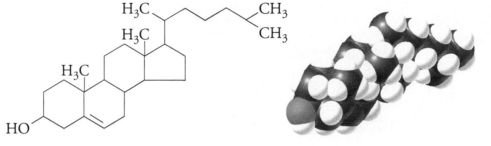

Cholesterol

**CHEMICAL
CONNECTIONS**

**How is vegetable oil
connected to motor oil?**

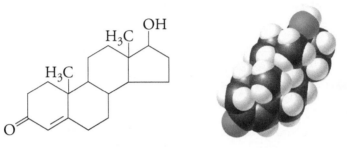

Testosterone

Estradiol

hormone imbalances and those recovering from severe starvation. Athletes
have found that these steroids improve performance, but they have many
negative side effects, including impotence, changes in sexual characteristics,
and liver toxicity.

CONCEPTCHECK

How are fats and steroids similar to each other?

CHECK YOUR ANSWER They are both lipids and, hence, insoluble in water.

13.4 Proteins Are Polymers of Amino Acids

LEARNING OBJECTIVE

Classify the structure of a protein
based upon the organization of
its amino acids and describe how
enzymes work.

EXPLAIN THIS

Why is lysine the most alkaline amino acid?

Proteins are large polymeric biomolecules made of monomer units called
amino acids. An **amino acid** consists of an amine group and a carboxylic acid
group bonded to the same carbon atom, as shown in **Figure 13.15**. A side group
is also attached to the same carbon. All proteins are made from some number

and combination of 20 different amino-acid building blocks. These 20 amino acids differ from one another by the chemical identity of their side groups, as shown in **Figure 13.16**.

Amino acids are linked together by *peptide bonds* formed in a condensation reaction between the carboxylic acid end of one amino acid and the amine end of a second amino acid, as illustrated in **Figure 13.17**. (Recall from Section 12.8 that condensation reactions are marked by the loss of a small molecule, such as water.) A collection of amino acids linked together in this fashion is called a *peptide*. The number of amino acids in a peptide is indicated by a prefix. A dipeptide is made from two amino acids, a tripeptide from three, a tetrapeptide from four, and so on. Peptides of ten or more amino acids are called *polypeptides*. Proteins are naturally occurring polypeptides that have some biological function, and they consist of large numbers of amino acids, often up to several hundred. For example, the molecular formula of one of the proteins found in milk, casein, is $C_{1864}H_{3012}O_{576}N_{468}S_{21}$, which gives you an idea of the size and complexity of some protein molecules.

Plant and animal tissues contain proteins in both dissolved and solid forms. Dissolved proteins reside in the liquid found inside cells and in other liquids in the body, such as blood. Solid proteins form skin, muscles, hair, nails, and horns. The human body contains many thousands of different proteins, such as those in **Figure 13.18**.

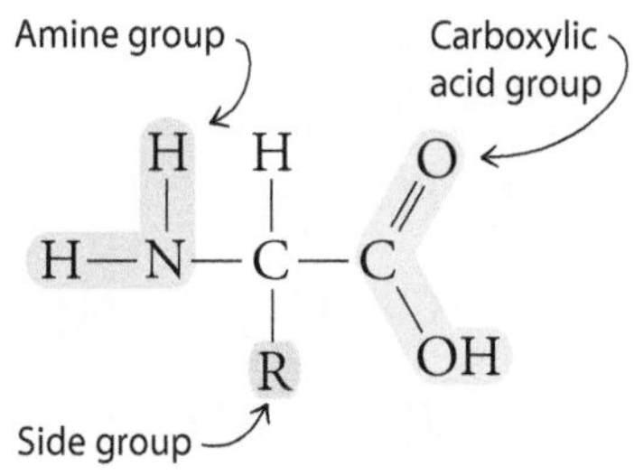

▲ **Figure 13.15**
The general structure of an amino acid, with R representing the side group that makes each amino acid different from all others.

Amino acid: a small biomolecule

Polypeptide: a polymer of amino acids

Protein: a polypeptide having biological function

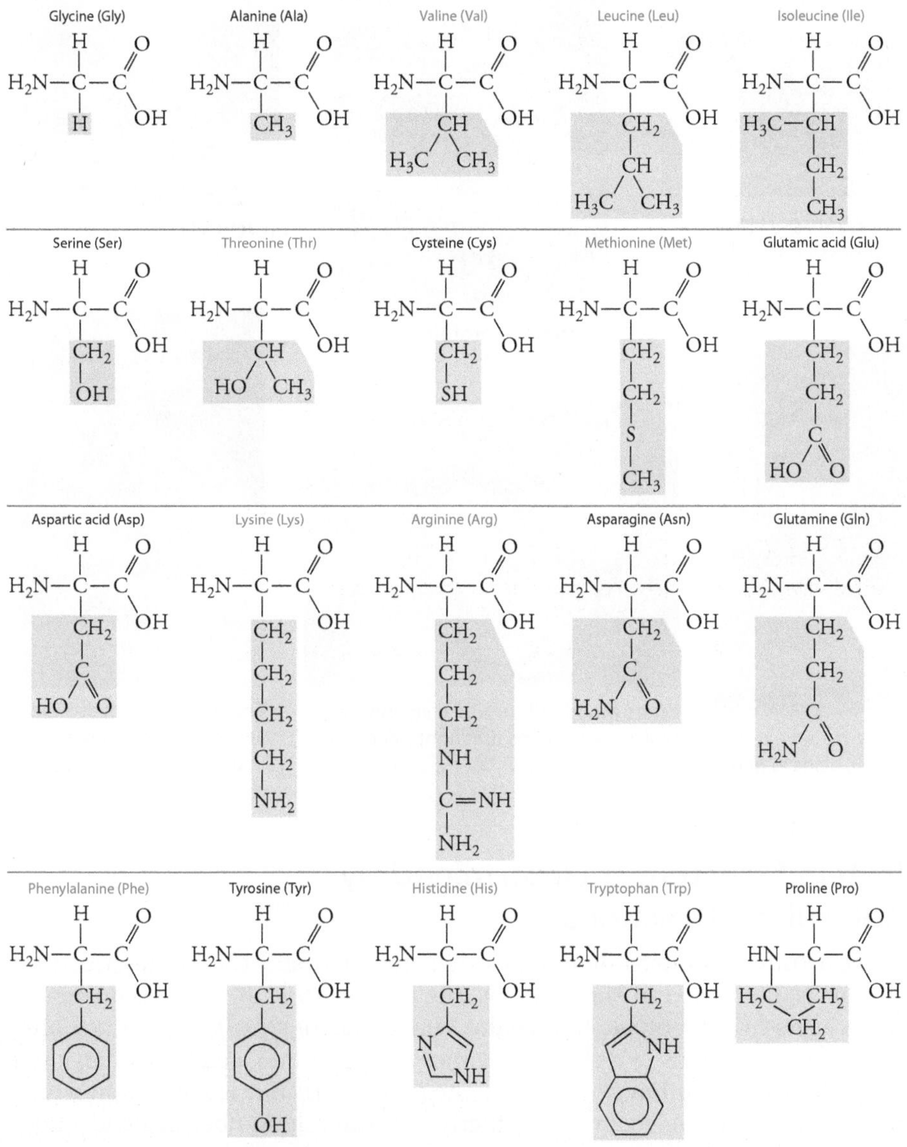

◀ **Figure 13.16**
These 20 amino acids are the building blocks of proteins. The side groups are highlighted in green, and the names printed in red indicate the nutritionally *essential* amino acids, discussed in Section 13.8.

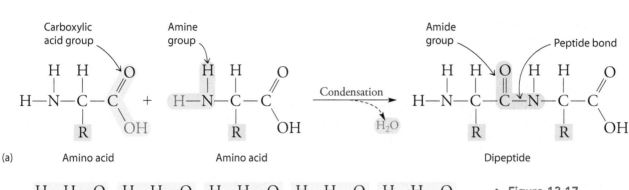

(a)   Amino acid   Amino acid   Dipeptide

(b)   Polypeptide

▲ **Figure 13.17**
(a) Formation of a peptide bond from the condensation of two amino acids. The resulting dipeptide contains an amide group. (b) A polypeptide is ten or more amino acids linked together by peptide bonds.

Some proteins are **hormones** that regulate body metabolism and growth.

Hemoglobin, a **transport protein**, is the component of blood that takes the oxygen we breathe to our cells.

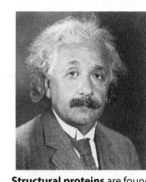

Contractile proteins in muscles allow us to move.

Structural image...

Structural proteins are found in skin, hair, and bones.

Storage proteins serve as a source of amino acids in milk.

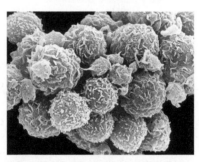

White blood cells produce **antibodies**, which are proteins that fight infections.

Enzymes are proteins that act as catalysts for reactions in the body, including the digestion of food.

▲ **Figure 13.18**
The variety of proteins in the human body.

Protein Structure Is Determined by Molecular Attractions

READING CHECK

What is a protein's primary structure?

The structure of proteins determines their function and can be described on four levels, illustrated in **Figure 13.19**. The *primary structure* is the sequence of amino acids in the polypeptide chain. The *secondary structure* describes how various short portions of a chain are either wrapped into a coil called an *alpha helix* or folded into a thin *pleated sheet*. The *tertiary structure* is the way in which an entire polypeptide chain may either twist into a long fiber or bend into a

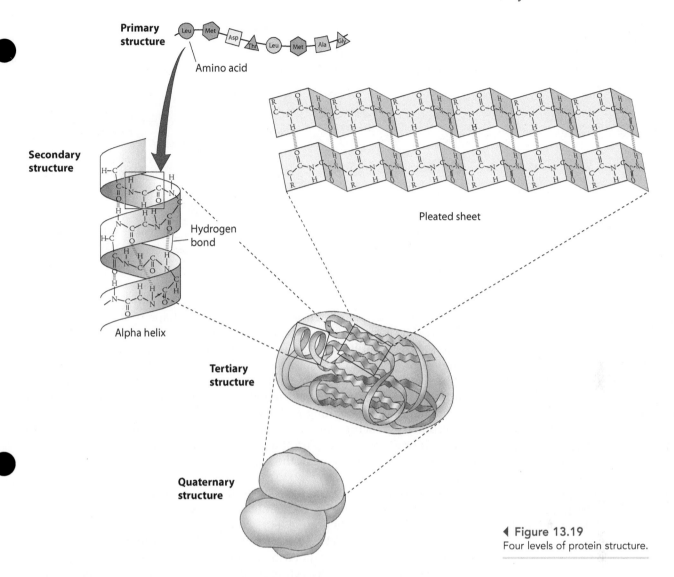

◀ Figure 13.19
Four levels of protein structure.

globular clump. The *quaternary structure* describes how separate protein chains may join to form one larger complex. Each level of structure is determined by the level before it, which means that, ultimately, it is the sequence of amino acids that creates the overall protein shape. This final shape is maintained both by chemical bonds and by weaker molecular attractions between amino acid side groups.

In long polypeptides, the number of possible variations in primary structure is astronomical. For example, the number of 20-unit polypeptides possible from a single set of 20 different amino acids is a whopping 2.43×10^{18}! The number of possible polypeptides when more than 100 units are being combined is seemingly infinite. This diversity is exactly what is needed for building a living organism.

Although functioning proteins have very specific amino acid sequences, slight variations can often be tolerated. In some cases, however, a slight variation can result in a life-threatening condition. For example, some people have a version of hemoglobin—a protein found in red blood cells—that has an incorrect amino acid in about 300. That "minor" error is responsible for sickle-cell anemia, an inherited condition with painful and often lethal effects. The sickle shape of red blood cells characteristic of this disease is shown in **Figure 13.20**.

Attractions between neighboring amino acids in a polypeptide chain are what cause the local contortions that constitute the secondary structure of the polypeptide. This secondary structure takes its alpha-helix form when simpler amino acids, such as glycine and alanine, are grouped together along the

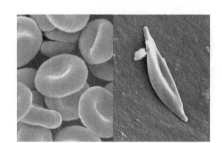

▲ Figure 13.20
On the left is a red blood cell containing the normal hemoglobin protein; on the right, a sickled red blood cell containing hemoglobin that has a single wrong amino acid. The curved shape is reminiscent of the curved-blade harvesting tool known as a sickle.

▶ **Figure 13.21**
Electrical forces of attraction maintain the tertiary structure of a polypeptide.

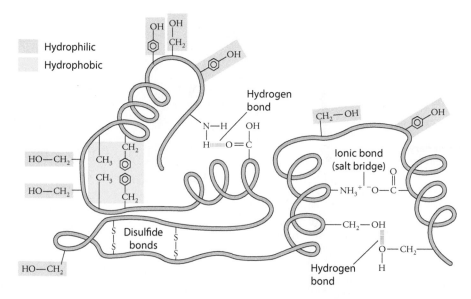

Proteins (polypeptides with biological functions)	
Structure	**Description**
Primary	Sequence of amino acids
Secondary	Alpha helix or pleated sheet
Tertiary	Overall shape of single polypeptide
Quarternary	Complex of two or more polypeptides

 FOR YOUR INFORMATION

A fibrous protein related to keratin is collagen, which is the main protein of connective tissues, such as skin, blood vessels, and muscle tendons. Collagen differs from keratin in that it has far fewer crosslinks between adjacent alpha helices. As we grow older, however, the number of crosslinks within collagen increases, and this causes the collagen to lose some of its elasticity. As a result, our blood vessels become less resilient and our faces become more wrinkled.

polypeptide chain. As shown in Figure 13.19, the shape of the alpha helix is maintained by hydrogen bonds between successive turns of the spiral. Proteins containing many alpha helices, such as wool, can be stretched because of the springlike properties of the alpha helix. The pleated-sheet version of secondary structure forms when predominantly nonpolar amino acids, such as phenylalanine and valine, are grouped together. Proteins containing many pleated sheets, such as silk, are strong and flexible but not easily stretched.

Secondary structure can vary from one portion of a polypeptide chain to another. (For example, one part of a given chain may have a helical secondary structure while another part of the same chain has a pleated-sheet secondary structure).

Tertiary structure refers to the way an entire chain is shaped. Like secondary structure, tertiary structure is maintained by various electrical attractions between amino acid side groups. For some proteins, such as the hypothetical one shown in **Figure 13.21**, these attractions include disulfide bonds between opposing cysteine amino acids. Also important are the ionic bonds (also known as *salt bridges*) that occur between oppositely charged ions. Hydrogen bonding between side groups can also contribute to tertiary structure, as can induced dipole–induced dipole attractions between nonpolar side groups, such as those found in phenylalanine and valine. Because attractions of this last type tend to exclude water, they are also known as *hydrophobic attractions*.

Hydrophilic attractions between a protein and an aqueous medium, such as cytoplasm or blood, also help maintain tertiary structure. In a protein dissolved in an aqueous medium, the polypeptide chain is folded in a compact way so that nonpolar side groups are on the inside of the molecule and polar side groups are on the outside, where they interact with the water. The tertiary structure of this sort of protein, as shown in Figure 13.21, is said to be *globular*. Globular proteins tend to be made of many different types of amino acids.

The two simplest amino acids are glycine and alanine (see Figure 13.16). Proteins made primarily of these two amino acids tend to form alpha helices, which can align to create long fibers. The tertiary structures of these sorts of proteins are said to be *fibrous*. An important example of a fibrous tertiary structure is *keratin*, the protein that is the main component of hair and fingernails, as shown in **Figure 13.22**. The strength of this keratin is established primarily by a number of crosslinks between adjacent alpha helices. In general, thick hair has more disulfide crosslinks than does fine hair. Also, fingernails are hard because of extensive disulfide crosslinking.

Disulfide crosslinks enable hair to hold a particular shape, such as a curl. **Figure 13.23** illustrates a permanent wave, which can modify the degree of curl in hair. The hair is first treated with a reducing agent that cleaves some of the

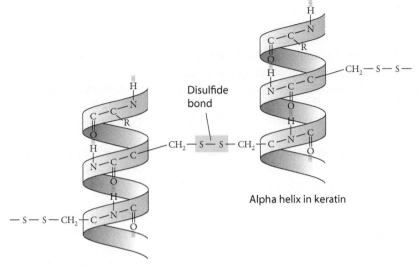

◀ **Figure 13.22**
Parallel polypeptide chains may be cross-linked by a disulfide bond between two cysteine side groups.

Disulfide bond

Alpha helix in keratin

Alpha helix in keratin

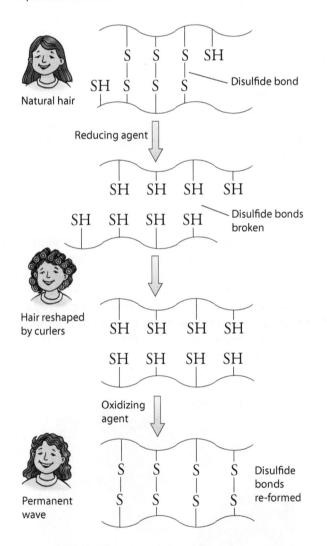

◀ **Figure 13.23**
A permanent wave breaks and re-forms disulfide bonds in hair.

Natural hair

Disulfide bond

Reducing agent

Disulfide bonds broken

Hair reshaped by curlers

Oxidizing agent

Permanent wave

Disulfide bonds re-formed

disulfide crosslinks. This is usually a smelly step, because some of the sulfur is reduced to odorous hydrogen sulfide. Cleaving disulfide crosslinks, however, allows the keratin to become more flexible. The hair is then set into the desired shape, using rollers if the desired shape is curls or a flat surface if the desired shape is straight. An oxidizing agent is then applied to restore the disulfide crosslinks, which hold the hair in its new orientation.

Hydrogen bonds between adjacent alpha helices also play an important role in making keratin a tough material. When keratin gets wet, these hydrogen bonds are disrupted, which is why fingernails soften in water. Hair also softens in water. As water molecules slip between the alpha helices, the polypeptide strands slide past one another to the extent permitted by the disulfide cross-links. As water molecules evaporate, hydrogen bonds between alpha helices are reestablished, and the hair toughens to its original shape—unless, of course, the hair is mechanically held in a different shape by, for example, rollers. Hair fashioned by wetting maintains its new shape only temporarily, however, because disulfide crosslinks ultimately pull the hair back to its natural look. Interestingly, a small amount of water in hair enhances the molecular attractions between alpha helices. Therefore, curls hold longer in humid climates than in dry climates.

Many proteins consist of two or more polypeptide chains. Such proteins have a quaternary structure resulting from bonding and interactions among these chains. A good example is hemoglobin (**Figure 13.24**), the oxygen-holding component of red blood cells. It is a complex of four polypeptides inside which four iron-bearing heme groups are tightly nestled.

CONCEPTCHECK

Distinguish among hemoglobin's primary, secondary, tertiary, and quaternary structures.

CHECK YOUR ANSWER Hemoglobin's primary structure is its sequence of amino acids along each polypeptide. The twisting of each polypeptide into an alpha helix is its secondary structure. The folding up of the full length of each alpha helix into a globular shape is its tertiary structure. The combined four polypeptides is the quaternary structure.

Proteins are functional only under very specific conditions, such as at a particular pH and temperature. Changes in these conditions can disrupt the chemical attractions within a protein and thereby cause a loss of structure, which necessarily means a loss of biological function. A protein that has lost its structure is said to be *denatured*. A hard-boiled egg, for example, consists of denatured proteins that can in no way support the development of a chick. The same atoms are there, but, as usual, it is the arrangement of atoms and their spatial orientations that make all the difference.

▶ **Figure 13.24**
Computer-generated model of the quaternary structure of hemoglobin, a protein consisting of four interlinked polypeptide chains, each shown in a different color. The heme group, shown in dark red, is a disk shaped molecule that bears an iron ion attracting center. Computers are important tools for the study of biomolecules, as they help scientists visualize complex 3-dimensional structures.

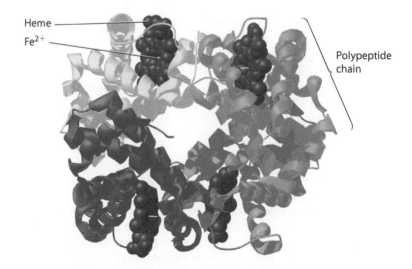

Heme

Fe^{2+}

Polypeptide chain

Enzymes Are Biological Catalysts

Enzymes are a class of proteins that catalyze (speed up) biochemical reactions. Their function has everything to do with their structure. Wad up a piece of paper and you will find many nooks and crannies in the paper wad. In a similar fashion, there are nooks and crannies on the surface of an enzyme. Some of these sites, called *receptor sites*, are special in that reactant molecules, called *substrates*, are able to fit inside them. Like a hand in a glove, a substrate molecule must have the right shape in order to fit a receptor site. Molecular attractions, such as hydrogen bonding, then hold the substrate molecule in the receptor site in an optimal position for reacting. The resulting product molecule or molecules are then released, freeing up the receptor site for other substrate molecules.

Figure 13.25 shows how an enzyme called *sucrase* breaks down sucrose to its monosaccharide units. Once sucrose binds to the empty receptor site on the sucrase, the enzyme facilitates the breaking of the covalent bond that links the glucose and fructose units together. It does so by holding the sucrose molecule in a certain conformation and then changing the electronic characteristics of this covalent bond so that it breaks easily when attacked by water molecules. In the final step, the individual glucose and fructose units are released from the enzyme, which is then free to catalyze the splitting of another sucrose molecule.

While some enzymes, such as sucrase, catalyze the splitting of substrate molecules, others catalyze the joining together of substrate molecules. In all cases, enzymes are so efficient that a single enzyme molecule may act on thousands or even millions of substrate molecules per second. Without enzymes, most biochemical reactions would not occur at rates rapid enough to support life.

A chemical that interferes with an enzyme's activity is called an *inhibitor*. Inhibitors work by binding to an enzyme and thereby preventing a substrate from binding. They are important regulators of cell metabolism. In many instances, an inhibitor is the very product created in an enzyme-catalyzed reaction. Once the enzyme produces a sufficient concentration of product, it begins to shut down because the product molecules function as inhibitors. When the product concentrations diminish enough, the inhibition stops and the enzyme can resume catalyzing the reaction. As we shall see in the next chapter, many drugs act either by inhibiting enzymes or by mimicking an enzyme's natural substrate.

◀ **Figure 13.25**
Upon binding to the receptor site on the enzyme sucrase, the substrate sucrose is split into its two monosaccharide units, glucose and fructose.

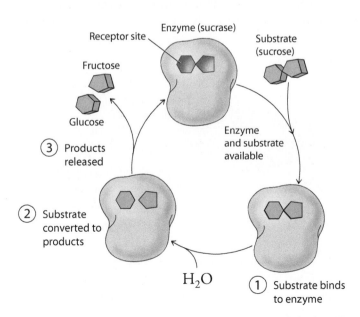

13.5 Nucleic Acids Code for Proteins

Identify nucleic acids as polymers of nucleotides and describe how they code for the building of proteins.

READINGCHECK

What information is held by a nucleic acid?

EXPLAIN THIS

Which is more polar: DNA or RNA?

The number of different ways in which amino acids can be arranged to make proteins is astronomical. Yet somehow our bodies are able to assemble amino acids in just the right order to build proteins that have highly functional structures. How the body builds these proteins begins with the fourth category of biomolecules, which are the *nucleic acids*. **Nucleic acids** hold the information for how amino acids need to be linked together to form the organism. The nucleic acids of a fish, for example, direct the formation of the proteins that come together to make the fish. Likewise, the nucleic acids of a human direct the formation of the proteins that come together to make the human. The details of how this works are something you may learn about in a follow-up course in chemistry or biology. For now, we simply introduce you to the chemical structures of these amazing biomolecules.

A nucleic acid is a polymer. The monomer units of the nucleic acid polymer are **nucleotides,** as shown in **Figure 13.26**. There are two major classes of nucleic acids, which are the *deoxyribonucleic acids*, or simply DNA, and the *ribonucleic acids*, also known as RNA. DNA is found primarily within the cell's nucleus, where it holds all the organism's genetic information. RNA is synthesized from DNA and transports this genetic information to structures outside the nucleus, where the proteins are built from a supply of individual amino acids.

Figure 13.27 illustrates the two major classes of nucleic acids, distinguished from each other by the type of ribose in the nucleotide monomers. Those with an oxygen atom missing on one of the carbon atoms in the ribose ring are **deoxyribonucleic acids,** or simply DNA. These polymers, which are the primary source of genetic information in plants and animals, are found in the cell nucleus as well as in certain organelles known as mitochondria. Nucleic acids that have an oxygen atom on each carbon atom of the ribose, as shown in Figure 13.27, are **ribonucleic acids** (RNA). These polymers are synthesized in the nucleus and then travel outside the nucleus to the cytoplasm, where they direct the synthesis of proteins.

There are four types of nucleotides in a DNA polymer (**Figure 13.27a**). These four nucleotides differ from one another in the nitrogenous base they contain, which may be adenine (A), guanine (G), cytosine (C), or thymine (T). There are

▶ **Figure 13.26**
A nucleic acid is a long polymeric chain of nucleotides, each nucleotide consisting of a nitrogenous base, a ribose sugar, and a phosphate group. (*Nitrogenous* means "containing nitrogen atoms," and ribose is a certain sugar containing five carbon atoms.)

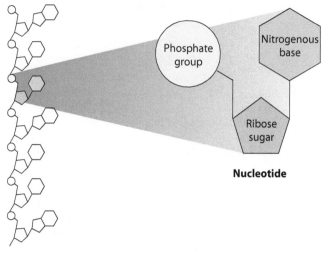

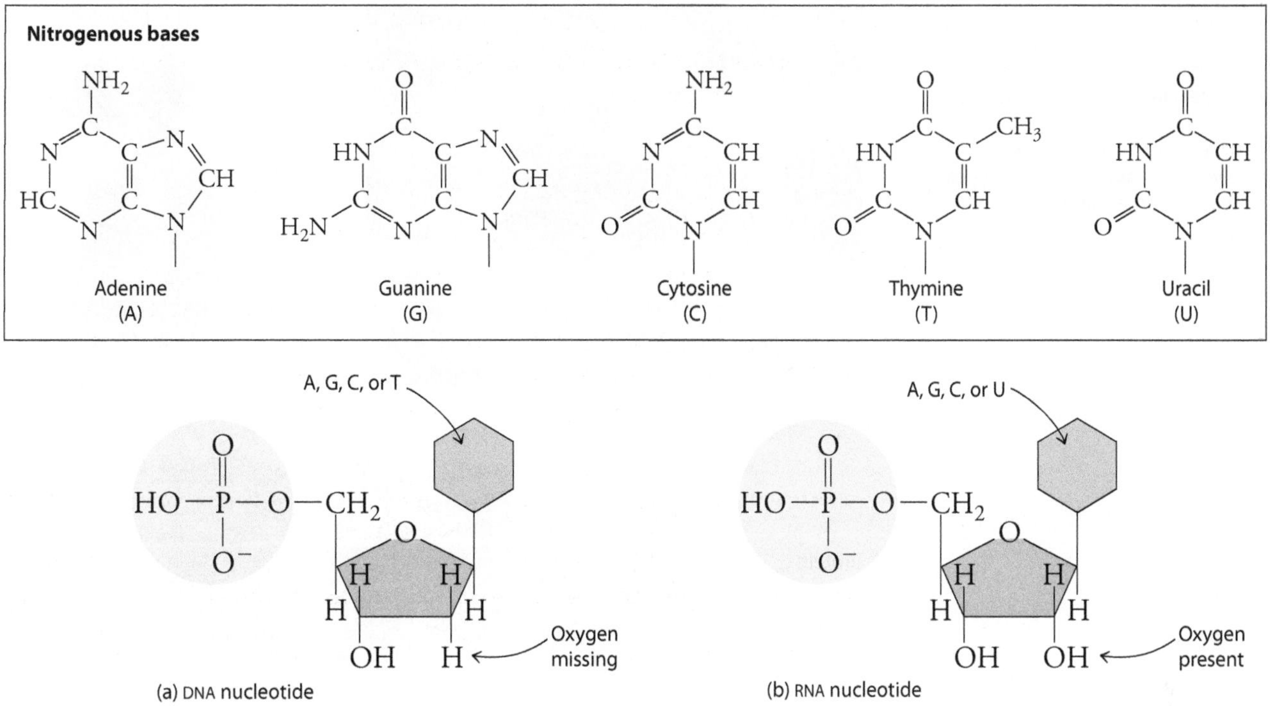

▲ **Figure 13.27**
(a) The ribose in nucleotides of DNA is missing an oxygen on one of the carbons. The nitrogenous bases of DNA are adenine, guanine, cytosine, and thymine. (b) The ribose in nucleotides of RNA is fully oxygenated. The nitrogenous bases of RNA are adenine, guanine, cytosine, and uracil.

also four types of nucleotides used to build RNA polymers (**Figure 13.27b**). The RNA nucleotides contain the same nitrogenous bases found in DNA nucleotides except for thymine. Instead of thymine, RNA nucleotides contain the nitrogenous base uracil (U), whose structure is only slightly different from that of thymine.

CONCEPTCHECK

What are the two ways in which DNA nucleotides differ structurally from RNA nucleotides?

CHECK YOUR ANSWER All DNA nucleotides lack an oxygen atom on the ribose. Also, the nitrogenous base in a DNA nucleotide may be adenine, guanine, cytosine, or thymine. The nitrogenous base in an RNA nucleotide may be adenine, guanine, cytosine, or uracil.

DNA Is the Template of Life

Genetics is the study of how living bodies come into being. The story of our modern-day understanding of genetics began in the 1850s in an abbey garden, where a monk named Gregor Mendel (1822–1884) documented how varieties of sweet peas could pass traits, such as flower color, from one generation to the next. From Mendel's work arose the idea that heritable traits are passed from parents to offspring in discrete units called *genes*. In the early 1900s, researchers correlated Mendel's heritable genes to cellular microstructures known as **chromosomes**, which are elongated bundles of DNA and protein that form whenever a cell is getting ready to divide (**Figure 13.28**). Each gene, it was found, resides at a specific location on a particular chromosome. Offspring inherit genes by receiving replicates of their parents' chromosomes.

Up until the 1940s, it was not known whether the DNA portion or the protein portion of the chromosomes was the carrier of genetic information. Most investigators were inclined to believe that proteins, with their great diversity, were

▶ **Figure 13.28**
Onion cells in the process of dividing show chromosomes that consist of DNA and protein molecules clumped together. During division, the chromosomes dupli-cate themselves in such a way that each new cell receives a full set of chromosomes identical to the set in the parent cell.

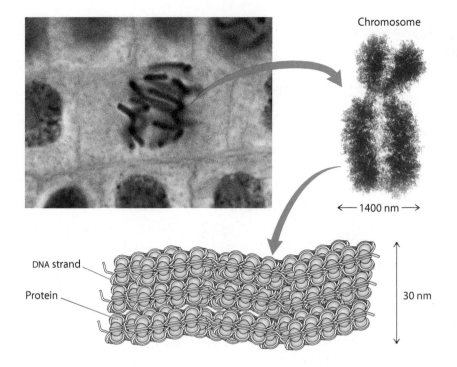

Chromosome

←— 1400 nm —→

DNA strand

Protein

30 nm

the carriers. In the 1940s, however, DNA was found to be a polymer containing adenine, guanine, cytosine, and thymine. Along the DNA chain, these bases could be sorted in any order. This potential variability in base sequencing opened up the possibility that DNA was the carrier of genetic information.

In 1953, James Watson (b. 1928), an American biologist, and Francis Crick (1916–2004), a British biophysicist, together deduced that DNA occurs as two separate strands of nucleotides coiled around each other in a double helix, as illustrated in **Figure 13.29**. The strands are held together by hydrogen bonding between opposing nitrogenous bases.

▶ **Figure 13.29**
A ladder made of rope sides and rigid wooden rungs can be twisted to create a double helix. The ropes are the equivalent of DNA's two sugar–phosphate backbones, which are shown as the yellow and orange atoms in the computer rendition shown on the right. The rungs represent pairs of nitrogenous bases, which are seen in the computer rendition as shades of lavender and blue.

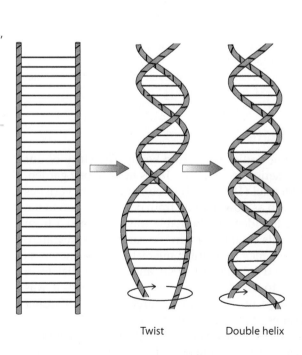

Twist   Double helix

What is most critical about Watson and Crick's model is that hydrogen bonding occurs only between specific bases—guanine pairs only with cytosine and adenine pairs only with thymine, as shown in **Figure 13.30**. This means that if you know the sequence of one strand, you can automatically deduce the sequence of the second strand, also known as the *complementary strand*. For example, if the first strand contains the sequence CTGA, the complementary strand must contain the sequence GACT.

In living tissue, cells divide to create duplicates of themselves. In a maturing organism, cells divide frequently to provide growth. In a mature organism, cells divide at a rate sufficient for the replacement of those that die. Each time a cell divides, genetic information must be preserved. In a process called **replication**, DNA strands are duplicated so that each newly formed cell receives a copy. Replication also allows copies of DNA to pass from parent to offspring.

With their model, Watson and Crick proposed that DNA replication begins with the unraveling of the double helix. Each single strand then serves as a template for the synthesis of its complementary strand. Free nucleotides are coupled to the single strand of DNA according to the rules of base pairing: guanine + cytosine, adenine + thymine. One double helix thus turns into two, as shown in **Figure 13.31**. As a cell divides, one of the two new strands is segregated in each of the two newly formed cells.

For their elucidation of DNA's secondary structure and function, Watson and Crick, along with biophysicist Maurice Wilkins (1916–2004), received the 1962 Nobel Prize in Physiology and Medicine. Subsequent research soon led to an understanding of how the nucleotide sequences in DNA translate into the synthesis of proteins. The Nobel laureates are shown in **Figure 13.32**.

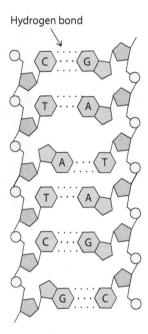

▲ **Figure 13.30**
The two strands of nucleotides in a DNA molecule are held together by hydrogen bonding between complementary nitrogenous bases: adenine with thymine and guanine with cytosine.

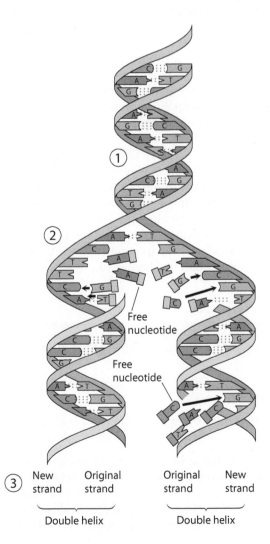

New strand | Original strand | Original strand | New strand

Double helix   Double helix

◀ **Figure 13.31**
The replication of DNA. (1) The double helix of DNA unwinds. (2) Each single strand serves as a template for the formation of a new DNA strand containing the complementary sequence. (3) Two daughter double helices are formed, each containing one of the parent strands.

(a)   (b)   (c)

▲ Figure 13.32
(a) Watson and Crick in 1953 with their model of the DNA double helix. In discovering this model, Watson and Crick relied heavily on the experimental evidence gathered by other researchers; most notably, the research team of (b) Maurice Wilkins and (c) Rosalind Franklin (1920–1958) provided the X-ray images of DNA crystals critical to Watson and Crick's elucidation of DNA's structure. Franklin did not share in the 1962 Nobel Prize because it is not awarded posthumously.

One Gene Codes for One Polypeptide

As we saw in Section 13.4, the shape of a protein molecule is determined by the protein's primary structure, which is the sequence of amino acids. So what, then, controls a protein's amino acid sequence? The answer to this question is the gene.

In modern terms, a **gene** is a particular sequence of nucleotides along the DNA strand. Each gene codes for the synthesis of one or more proteins in an organism. Organisms, of course, need enormous numbers of genes to produce all the proteins they need. The number of genes contained in a single human cell, for instance, is estimated to be on the order of 25,000. To accommodate this many genes, each DNA molecule is very long, containing on the order of 3.1 billion base pairs. Interestingly, genes make up only about 20 percent of a DNA molecule. The other 80 percent of the nucleotides in a DNA strand appear to serve primarily as spacers, their main job being to separate genes on the DNA molecule.

Much has been learned about the chemistry of nucleic acids over the past half century. We now know, for example, many details of how the information in a gene is translated into a protein. This knowledge has given rise to *genetic engineering*, which allows scientists to manipulate the behavior of nucleic acids for a variety of purposes, such as treating human disease and creating new agricultural crops, as described in Chapter 15.

FOR YOUR INFORMATION

Initially, scientists estimated that there were about 100,000 different human genes. Over the past decade, however, this estimate has been gradually reduced to about 25,000 genes, which is only about 5000 more genes than in the DNA of the roundworm.

13.6 Vitamins Are Organic, Minerals Are Inorganic

LEARNING OBJECTIVE

Distinguish vitamins from minerals and the roles they play in our nutrition.

EXPLAIN THIS

Is the salt within food destroyed when the food is burnt?

In addition to carbohydrates, lipids, proteins, and nucleic acids, our bodies require vitamins and minerals in order to survive and function properly. **Vitamins** are organic chemicals that, by assisting in various biochemical reactions, help us maintain good health. **Minerals** are inorganic chemicals that play a variety of roles in the body. Both vitamins and minerals are obtained from our diet (**Figure 13.33**).

Some, such as the iron in hemoglobin, are vital components of biomolecules. Others, such as the calcium in bone, are integral parts of structures. Deficiencies in vitamins or minerals are the causes of certain diseases. Lack of vitamin C, for example, leads to scurvy, a disease marked by a deterioration of the gums. Lack of iron leads to anemia, which results in general fatigue and an irregular heartbeat.

Vitamins are classified as either lipid soluble or water soluble, as shown in Table 13.2. The lipid-soluble vitamins tend to accumulate in fatty tissue, where they may be stored for years. Adults can remain free of a deficiency disease for quite some time because of these vitamin reserves. Children, on the other hand, because they have yet to build up such reserves, are particularly vulnerable to these diseases. In developing nations, for example, many children suffer permanent blindness because of a lack of vitamin A.

A lack of lipid-soluble vitamins can be detrimental, but so can excessive amounts, particularly of vitamins A and D, which can accumulate to dangerous levels. Too much vitamin A causes dry skin, irritability, and headaches. Excessive amounts of vitamin D lead to diarrhea, nausea, and calcification of joints and other body parts. Vitamins E and K are less harmful in large quantities because they are readily metabolized.

The water-soluble vitamins are not retained by the body for long periods of time. Instead, because they are soluble in water, they are readily excreted in urine and must therefore be ingested frequently. It is difficult to harm yourself by taking in too much of the water-soluble vitamins. Your body simply absorbs what it immediately needs and excretes the rest. Foods boiled in water tend to lose their water soluble vitamins, which after dissolving in the water are poured down the drain along with the water. This includes the B vitamins, which is a class of chemically related compounds, as well as vitamin C, which is a common name for the chemical ascorbic acid. For this reason, many people prefer steaming or microwaving their vegetables. Also, foods should not be overcooked, as both lipid-soluble and water-soluble vitamins are destroyed by heat.

All minerals are ionic compounds of various elements. They are classified according to the quantities we need. *Macrominerals*, the ones we need in greatest quantity, make up about 4 percent of our body weight. For the macrominerals listed in Table 13.3, the amounts we need each day are measured in grams. The amounts of *trace minerals* we should ingest daily are measured in milligrams. Finally, there are the *ultratrace minerals*, for which we measure the recommended daily intake in micrograms or even picograms.

Our bodies require balanced amounts of minerals, meaning that ingesting too much is as harmful as ingesting too little. Ultratrace minerals are particularly toxic when taken in large quantities. Cadmium, chromium, and nickel, for example,

READINGCHECK

What are the two classifications of vitamins?

▲ Figure 13.33
Each year people spend more than $20 billion on vitamin and mineral supplements. But are these supplements always beneficial to one's health? Not necessarily. Read on for more detail.

TABLE 13.2   Some Vitamins Needed by the Human Body

VITAMIN	FUNCTION	DEFICIENCY SYNDROME
Lipid Soluble		
Vitamin A (retinol)	Precursor to rhodopsin, a chemical used for vision; assists in inhibiting bacterial and viral infections	Night blindness
Vitamin D (calciferol)	Helps incorporate calcium into the body	Weak bones, rickets
Vitamin E (tocopherol)	Inhibits oxidation of polyunsaturated fats; free radical scavenger; helps maintain circulatory and nervous systems	Diminished hemoglobin
Vitamin K (phylloquinone)	Helps maintain ability to form blood clots	Abnormal bleeding
Water Soluble		
B vitamins	Coenzymes in biochemical reactions for growth and energy production	Various nerve and skin disorders, anemia
Vitamin C (ascorbic acid)	Antioxidant; assists in inhibiting bacterial and viral infections	Scurvy

TABLE 13.3   Some Macrominerals Needed by the Human Body

MACROMINERAL (IONIC FORM)	SOME FUNCTIONS	DEFICIENCY SYNDROME
Sodium (Na^+)	Transportation of molecules across cell membrane, nerve function	Muscle cramps, reduced appetite
Potassium (K^+)	Transportation of molecules across cell membrane, nerve function	Muscular weakness, paralysis, nausea, heart failure
Calcium (Ca^{2+})	Bone and tooth formation, nerve and muscle function	Retarded growth, possible loss of bone mass
Magnesium (Mg^{2+})	Enzyme function	Nervous system disturbances
Chlorine (Cl^-)	Transportation of molecules across cell membrane, digestive fluid, nerve function	Muscle cramps, reduced appetite
Phosphorus ($H_2PO_4^-$)	Bone and tooth formation, nucleotide synthesis	Weakness, calcium loss
Sulfur (SO_4^{2-})	Amino acid component	Protein deficiency

are potent carcinogens, and arsenic is a well-known poison. Yet our bodies need microquantities of these minerals if we are to stay healthy. Eating a well-balanced diet is often the best way to obtain a good balance of minerals. Many people take mineral supplements, but they should monitor the doses they take.

Two of the minerals most prevalent in our diet are potassium and sodium ions. Both are involved in nerve-signal transmission and in the transport of molecules into and out of cells. For good health, we need more potassium than sodium, a situation we share with all other living organisms, both plants and animals. When we eat plants or animals without additives and without excessive processing, the potassium/sodium balance is optimal. When food is boiled or deep-fried, however, both potassium and sodium ions are stripped away along with the liquids in which they are dissolved. Salting the food with sodium chloride then raises the sodium ion content to a level higher than the potassium ion content, which is not healthy.

Phosphorus is another important dietary mineral, which we consume in the form of phosphate ions, such as $H_2PO_4^-$. Look back at Figure 13.26 to see that phosphate ions form the backbone of nucleic acids. In addition, phosphate ions are components of the energy-producing compound adenosine triphosphate (ATP), shown in **Figure 13.34**.

ATP is one of several direct sources of energy for most of the energy-requiring processes in the body, such as tissue building, muscle contraction, transmission of nerve impulses, heat production, and movement of molecules into and out of cells. The human body goes through a great deal of ATP—about 8 grams per minute during strenuous exercise. It is a short-lived molecule and so must

FOR YOUR INFORMATION

Within each of our cells are small organelles called *mitochondria* that produce most of our ATP. Interestingly, mitochondria have their own DNA and their own genetic code, which suggests that mitochondria were once separate living entities with which we now live symbiotically. Mitochondria reproduce themselves by cloning and are then passed down from one generation to the next only through the mother. The fact that nearly all modern humans have mitochondrial DNA of almost the same genetic makeup suggests that we all probably arose from the same single ancestral mother. After studying slight variations in mitochondrial DNA among people of different cultures, scientists estimate that our common mother may have lived somewhere in Africa about 200,000 years ago.

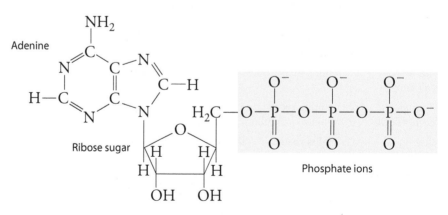

▲ Figure 13.34
Phosphate ions are an important part of the ATP molecule.

be produced continuously. The many chemical pathways by which foods are oxidized to yield ATP have been mapped extensively by biochemists.

Various poisons act by blocking the synthesis of ATP. Carbon monoxide, for example, binds to the iron of hemoglobin, thereby preventing hemoglobin from carrying oxygen. The reason the body needs oxygen, however, is so that it can be used to oxidize carbohydrates, lipids, and proteins to form ATP. So without oxygen, the body becomes starved of energy-yielding ATP and quickly dies. Cyanide also blocks the synthesis of ATP, but does so by incapacitating enzymes that play an important role in ATP synthesis. Interestingly, ATP is also used by the body to allow muscles to relax after contraction. When a body dies, no matter what the cause, ATP synthesis comes to a halt and all the body muscles become stiff—a condition known as rigor mortis.

13.7 Metabolism Is the Cycling of Biomolecules through the Body

EXPLAIN THIS

Is catabolism an example of oxidation or reduction?

Your body takes in biomolecules in the food you eat and breaks them down to their molecular components. Then one of two things happens: either your body "burns" these molecular components for their energy content through a process known as *cellular respiration* or these components are used as the building blocks for your body's own versions of carbohydrates, lipids, proteins, and nucleic acids. The sum total of all these biochemical activities is what we call **metabolism**. Two forms of metabolism are *catabolism* and *anabolism*. **Figure 13.35** shows the major catabolic and anabolic pathways of living organisms.

All metabolic reactions that involve the breaking down of biomolecules are grouped under the heading of **catabolism**. Digestion and cellular respiration are examples of catabolic reactions. Digestion begins with the hydrolysis of food molecules, a reaction in which water is used to sever bonds in the molecules and separate large molecules into their smaller component parts. The small molecules

LEARNING OBJECTIVE

Classify metabolic reactions as either catabolic or anabolic.

 READINGCHECK

What are the two forms of metabolism?

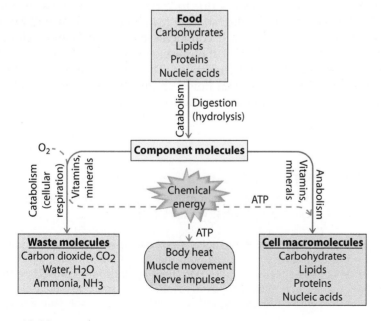

▲ **Figure 13.35**
Metabolic pathways for the food we ingest. Catabolic pathways are indicated by the purple arrows; anabolic pathways, by the blue arrow.

▶ Figure 13.36
The Belgian blue bull, developed through many years of selective breeding, has a defective gene that does not produce myostatin, which is a hormone that inhibits the anabolic formation of muscle. Without myostatin, the muscles of the bull become much more massive. Inhibitors of myostatin in humans may help offset the loss of muscle mass that occurs with conditions such as muscular dystrophy or as occurs in each of us as we age. Interested? Read the Contextual Chemistry essay, "The Genetics of Muscle Fitness," at the end of Chapter 14.

formed in digestion, such as the glucose units of complex carbohydrates, then migrate to all the cells of the body and take part in cellular respiration. There, in a series of steps, the small food molecules lose electrons to the oxygen that was inhaled through our lungs and, as a result, break down to even smaller molecules, such as carbon dioxide, water, and ammonia, which are excreted. Through this process, high-energy molecules, such as ATP, are created. These high-energy molecules are able to drive reactions that produce body heat, muscle movement, and nerve impulses. They also are responsible for fueling **anabolism**, which is the general term for all the energy-requiring chemical reactions that produce large biomolecules from smaller molecules (**Figure 13.36**).

The types of biomolecules produced by anabolism are the same as the types found in food—carbohydrates, lipids, proteins, and nucleic acids. These products of anabolism are, if you will, the host's own version of what the food once was. And if the host ever becomes food, anabolic reactions in the subsequent host will result in different versions of the molecules. Thus, organisms in a food chain live off one another by absorbing one another's energy via catabolic reactions and then rearranging the remaining atoms and molecules via anabolic reactions into the biomolecules they need to survive.

Catabolism and anabolism work together. In healthy muscle tissue, for example, the rate of muscle degradation (catabolism) is matched by the rate of muscle building (anabolism). If you increase your food supply and exercise vigorously, it is possible to favor the muscle-building anabolic reactions over the muscle-destroying catabolic reactions. The result is an increase in muscle mass. Stop eating and exercising, however, and these anabolic reactions lose out to the catabolic reactions. The result is a decrease in muscle mass—you begin to waste away.

CONCEPTCHECK
Anabolic steroids help people gain muscle mass. If there were such a thing as a catabolic steroid, what would be its effect?

CHECK YOUR ANSWER Anabolic? Catabolic? Which is which? Many students recognize the term *anabolic steroids* from the sports news media, which are quick to report on famous athletes caught using these steroids for improved performance. Anabolism, therefore, is muscle-building, so catabolism must be muscle-degrading. A catabolic steroid would cause a loss of muscle mass, which is not generally desirable.

13.8 The Food Pyramid Summarizes a Healthful Diet

EXPLAIN THIS

Why can't your body produce proteins from carbohydrates and fats alone?

LEARNING OBJECTIVE

Describe how the body utilizes carbohydrates, fats, and proteins.

The food pyramid, shown in **Figure 13.37**, summarizes the food intake recommendations of the United States Department of Agriculture (USDA). According to this pyramid, an individual's daily diet should consist mostly of grains (such as bread, cereals, and pastas), fruits, and vegetables, with the amount of dairy products and meats fairly limited and foods high in sugars or fats consumed only sparingly.

We can get some insight into the reasons behind these recommendations by looking at how the body handles the biomolecules contained in these foods.

Carbohydrates Predominate in Most Foods

There are two types of carbohydrates—nondigestible, called dietary *fiber*, and digestible, mainly starches and sugars. As discussed in Section 13.2, dietary fiber helps keep things moving in the bowels, especially in the large intestine. There are two kinds of fiber—water-insoluble and water-soluble. Insoluble fiber consists mainly of cellulose, which is found in all food derived from plants. In general, the less processed the food, the higher the insoluble fiber content (**Figure 13.38**). Brown rice, for example, has a greater proportion of insoluble fiber than does white rice, which is made by milling away the rice seed's outer coating (along with numerous vitamins and minerals).

Soluble fiber is made of certain types of starches that are resistant to digestion in the small intestine. An example is pectin, which is added to jams and jellies because it acts as a thickening agent, becoming a gel when dissolved in a limited amount of water. Soluble fiber tends to lower cholesterol levels in the blood through its interactions with *bile salts*, which are cholesterol-derived substances produced in the liver and then secreted into the intestine. As shown in **Figure 13.39**, one of the functions of bile salts is to carry ingested lipids through the membranes of the intestine and into the bloodstream. The bile salts are then reabsorbed by the liver and cycled back to the intestine. Soluble fiber in the intestine binds to bile salts, which are then efficiently passed out of the body rather than reabsorbed. The liver responds by producing more bile salts, but to do so, it must utilize cholesterol, which it collects from the bloodstream. By this indirect route of binding with bile salts, soluble fiber tends to lower a person's cholesterol level. Foods rich in soluble fiber include fruits and certain grains, such as oats and barley.

Carbohydrates
Nondigestible (dietary fiber)
   Insoluble (cellulose)
   Soluble (certain starches)
Digestible
   Starches and sugars

◀ Figure 13.37
The food pyramid.

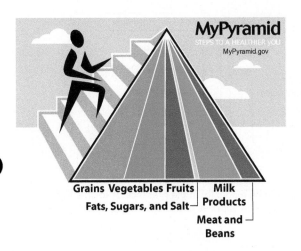

▲ Figure 13.38
The grains, vegetables, and fruits of the pyramid are important sources of food primarily because they contain a good balance of all nutrients—carbohydrates, fats, proteins, nucleic acids, vitamins, and minerals. However, the predominant component of these foods is carbohydrates.

READINGCHECK

What does the American Diabetes Association recommend about the glycemic index?

During digestion, the digestible carbohydrates—both starches and sugars—are transformed to glucose, which is absorbed into the bloodstream through the walls of the small intestine. The body then utilizes this glucose to build energy molecules, such as ATP.

Carbohydrate-containing foods are rated for how quickly they cause an increase in blood glucose levels. This rating is done with what is known as the *glycemic index*. The index compares how much a given food increases a person's blood glucose level relative to the increase seen when pure glucose is ingested, with the latter increase assigned a standard value of 100. In general, foods that are high in starch or sugar but low in dietary fiber are high on the glycemic index, a baked potato being a prime example.

The glycemic index for a particular food can vary greatly from one person to the next. How the food was prepared can also make a big difference. Thus, index values, such as the ones shown in Table 13.4, are merely ballpark figures. Given this qualification, however, the index provides valuable information for those people, such as individuals with diabetes, who need to pay close attention to their blood sugar levels.

A number of problems are associated with eating carbohydrates with high glycemic indices. For example, the rapid spike in blood glucose levels causes the body to produce extra *insulin*, a blood-soluble protein that causes liver, muscle, and fat tissues to take up glucose from the blood. This lowers blood glucose levels. The body responds by releasing glucose-yielding glycogen, but also by triggering a sense of hunger, even if the person just ate. A meal rich in foods high on the glycemic index therefore promotes overeating, which usually leads to obesity.

Many professional organizations, such as the American Diabetes Association, caution that priority should be given to the quantity of carbohydrates ingested rather than to the glycemic index of the food containing those carbohydrates. What really counts is the total number of Calories absorbed, not whether these Calories come from foods high or low on the index. For most people, however, ingesting foods low on the index makes maintaining a healthful caloric intake more manageable by controlling cravings.

Another advantage of eating carbohydrates from foods that are low on the index is that these foods provide energy to the body over an extended period of time. For athletes, a diet rich in foods low on the glycemic index, such as spaghetti, translates to greater endurance. Interestingly, this greater endurance

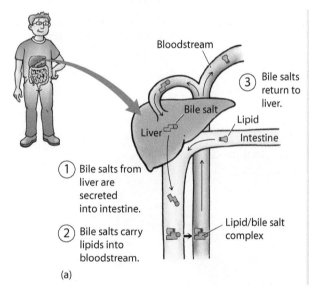

(a)

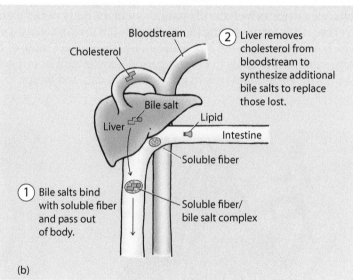

(b)

▲ Figure 13.39
(a) With no soluble fiber present, bile salts are recycled to the liver and no new ones need to be made. (b) In the presence of soluble fiber, bile salts are removed from the body. The liver then must use cholesterol from the blood to make new supplies of bile salts. Thus, by binding with bile salts, soluble fiber indirectly decreases the amount of cholesterol in the blood.

TABLE 13.4   Glycemic Index for Select Foods

FOOD	GLYCEMIC INDEX	FOOD	GLYCEMIC INDEX
Glucose	100	Honey	58
Baked potato	85	Brown rice	55
Cornflakes	83	Popcorn	55
Jelly beans	80	Sweet potato	54
French fries	75	Banana	54
White bread	71	Milk chocolate	49
Life-Savers candy	70	Orange	44
Whole wheat bread	68	Spaghetti, boiled 5 minutes	36
Sucrose	64	Skim milk	32
Raisins	64	Whole milk	27
High-fructose corn syrup	62	Grapefruit	25
White rice	58	Peanuts	15

is just as useful for bodybuilders as it is for marathon runners. Having the energy required for building muscles is far more critical than having the supply of raw materials needed. Furthermore, the body's metabolism is versatile enough to generate proteins out of glucose. (FYI, it can even generate glucose out of proteins.) Thus, the bodybuilder's supply of proteins is ensured. A diet rich in carbohydrates is therefore more effective in allowing a bodybuilder to build muscles than is a diet rich in proteins.

Despite the many advantages of eating carbohydrates low on the glycemic index, foods rich in carbohydrates high on the index, such as sucrose, are now more popular than ever. Many of these foods are highly processed and compose a narrow portion of the food pyramid. Although they are good for providing energy, the USDA recommends that they be consumed only sparingly, because they lack many of the essential nutrients.

Unsaturated Fats Are Generally More Healthful Than Saturated Fats

Because your body uses saturated fats to synthesize cholesterol, eating more saturated fats allows you to synthesize more cholesterol. Unsaturated fats, by contrast, are not ideal starting materials for cholesterol synthesis.

Another reason unsaturated fats are more healthful relates to the way fats are associated with cholesterol. Fats and cholesterol are both lipids, which, on their own, are insoluble in blood. In order to move through the bloodstream, these compounds are packaged with bile salts, as was discussed earlier. Most lipids, however, are made water soluble by being packaged with water-soluble proteins in complexes called *lipoproteins*. Lipoproteins are classified according to density, as noted in Table 13.5. Very low-density lipoproteins (VLDLs) serve primarily in the transport of fats throughout the body. Low-density lipoproteins (LDLs) transport cholesterol to the cells, where it is used to build cell membranes. High-density lipoproteins (HDLs) bring cholesterol to the liver, where it is transformed into a variety of useful biomolecules.

 FOR YOUR INFORMATION

Your blood sugar level needs to remain within a very narrow range. When the blood sugar level gets too high, your body produces the hormone insulin, which causes tissues, such as muscle, to remove glucose from your blood. Interestingly, well-exercised muscles are much more effective at removing blood glucose compared to muscles that are out of shape. Regular exercise, therefore, effectively lowers the glycemic index of the foods you eat.

A candy bar is good for a quick energy fix, but chow down on a spaghetti feast the night before a strenuous workout for long-run energy.

TABLE 13.5   The Classification of Lipoproteins

LIPOPROTEIN	PERCENT PROTEIN	DENSITY (G/ML)	PRIMARY FUNCTION
Very low-density (VLDL)	5	1.006–1.019	Fat transport
Low-density (LDL)	25	1.019–1.063	Cholesterol transport (to cells to build cell walls)
High-density (HDL)	50	1.063–1.210	Cholesterol transport (to liver for processing)

FOR YOUR
INFORMATION

Many people suffer from a condition in which the body produces too much insulin in response to a rapid rise in blood sugar. The excess insulin causes the blood sugar level to drop to abnormally low levels, which creates numerous symptoms, such as fatigue, mood swings, trembling, and fainting. The condition is called *hypoglycemia*, and it is best controlled through a diet that minimizes the ingestion of sugar, white flour, alcohol, caffeine, and tobacco.

A diet high in saturated fats leads to elevated VLDL and LDL levels in the bloodstream. This is undesirable because these lipoproteins tend to form fatty deposits called *plaque* in the artery walls. Plaque deposits can become inflamed to the point where they rupture, releasing blood-clotting factors into the bloodstream. A blood clot formed around the rupture site is released into the bloodstream, where it can become lodged in a blood vessel and block the flow of blood to a particular region of the body. When that region is in the heart, the result is a heart attack. When that region is in the brain, the result is a stroke.

In contrast to saturated fats, unsaturated fats tend to increase blood HDL levels, which is desirable because these lipoproteins are effective at *removing* plaque from artery walls.

CONCEPT CHECK

For what two reasons are unsaturated fats better for you than saturated fats?

CHECK YOUR ANSWER Unsaturated fats are not so readily used by your body to synthesize cholesterol. They also tend to increase the proportion of high-density lipoproteins, which lower the level of cholesterol in your blood and help relieve the buildup of arterial plaque.

Triglycerides with unsaturated fatty acid components, as noted in Section 13.3, tend to be liquid at room temperature. They can be transformed to a more solid consistency, however, by *hydrogenation*, a chemical process in which hydrogen atoms are added to carbon–carbon double bonds. Mix a partially hydrogenated vegetable oil with yellow food coloring, a little salt, and small amounts of the organic compound butyric acid for flavor and you have margarine, which became popular around the time of World War II as an alternative to butter. Many food products, such as chocolate bars, contain partially hydrogenated vegetable oils so that they are of a consistency that sells well in the marketplace. Hydrogenation increases the percentage of saturated fats, however, and therefore makes these fats less healthful. Furthermore, as **Figure 13.40** shows, some of the double bonds that remain are transformed to the *trans-* structural isomer (see Section 12.2). Because carbon chains containing *trans-* double bonds tend to be less kinked than chains containing *cis-* bonds, the partially hydrogenated fat has straighter chains. This means the fat is more likely to mimic the action of saturated fats in the body.

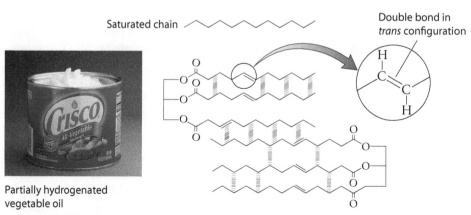

Partially hydrogenated vegetable oil

▲ Figure 13.40
Hydrogenation can lead to *trans-* double bonds in the fatty acid chain.

We Should Monitor Our Intake of Essential Amino Acids

Proteins are useful for their energy content, just as starches, sugars, and fats are, but perhaps the greatest importance of proteins lies in how our bodies use them for building such structures as enzymes, bones, muscles, and skin. Of the 20 amino acids the human body uses to build proteins, adults can produce, in sufficient amounts, 12 of them from carbohydrates and fatty acids. We must obtain the remaining eight, listed in Table 13.6, from our food. These eight amino acids are called *essential amino acids,* in the sense that it is essential we get adequate amounts of them from our food. To support rapid growth, infants and children require, in addition to the eight amino acids listed for adults in Table 13.6, large amounts of arginine and histidine, which they can obtain only from their diets. Therefore, for infants and juveniles, there is a total of ten essential amino acids. (The term *essential* is unfortunate, because in truth, all 20 amino acids are vital to our good health.)

Why our bodies produce ample amounts of some amino acids and not others can be explained by looking at the chemical structures of the amino acid side groups, shown in Figure 13.16 back on page 409. The nonessential amino acids have side groups that tend to be simple and therefore can be produced by the body without much effort. The essential amino acids, however, tend to be biochemically more difficult to create. The body, therefore, can save energy by obtaining these amino acids from outside sources. Over the course of evolution, our capacity to synthesize these amino acids diminished. Similarly, we have lost the capacity to synthesize vitamins, which are also complex molecules more efficiently obtained from dietary sources. In other words, we let other living organisms go through the metabolic expense of synthesizing these biomolecules and then we eat those organisms.

In general, the more closely the amino acid composition of ingested protein resembles the amino acid composition of the animal eating the protein, the greater the nutritional quality of that protein is for that animal. For humans, mammalian protein is of the highest nutritional quality, followed by protein from fish and poultry and then by protein from fruits and vegetables. Plant proteins, in particular, are often deficient in lysine, methionine, or tryptophan. A vegetarian diet provides adequate protein only if it includes a variety of protein sources so that a deficiency in one source is compensated for by an abundance in another source, as shown in **Figure 13.41**.

The old adage "You are what you eat" has a literal foundation. With the exception of the oxygen you obtain through your lungs, nearly every atom or molecule in your body got there by first passing through your mouth and into your stomach. All the biomolecules needed for the energy and development of a fetus growing in the womb, like Maitreya Suchocki in **Figure 13.42**, must first pass through the lungs and mouth of her mother, which is why it is so vital that her mother eat a well-balanced diet and maintain a healthful lifestyle while

CHEMICAL
CONNECTIONS

How is your fingernail
connected to sunshine?

TABLE 13.6   The Essential Amino Acids

Arginine		Essential for children
Histidine		
Isoleucine		
Leucine		
Lysine		
Methionine		
Phenylalanine	Essential for adults	
Threonine		
Tryptophan		
Valine		

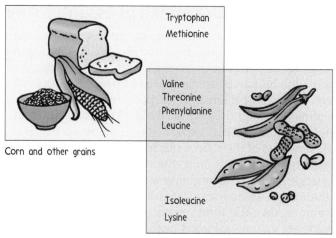

Tryptophan
Methionine

Valine
Threonine
Phenylalanine
Leucine

Corn and other grains

Isoleucine
Lysine

Beans and other legumes

▲ **Figure 13.41**
Sufficient protein can generally be obtained in a vegetarian diet by combining a legume, such as peas or beans, with a grain, such as wheat or corn. Familiar meals containing such a combination include a peanut butter sandwich, corn tortillas and refried beans, and rice and tofu.

she is pregnant. And then, a mere 40 weeks later, her mother's food has been transformed by the actions of Maitreya's DNA into a whole new body ripe for exploring the world around her.

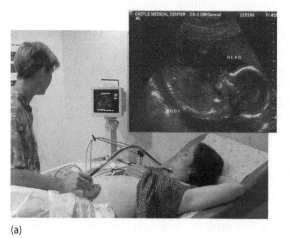

(a)

(b)

▲ **Figure 13.42**
(a) As a fetus, Maitreya is undergoing the most rapid growth rate of her life, and thus her dependence on a healthful diet is as great as it will ever be. (b) As a baby, Maitreya's nutritional needs are still great, as is her ability to imitate her mother during a copyedit session.

Chapter 13 Review

LEARNING OBJECTIVES

Identify the basic components of a cell and the four major classes of biomolecules. (13.1)	→	*Questions 1, 2, 33, 34*
Recognize the molecular structures of simple and complex carbohydrates. (13.2)	→	*Questions 3–6, 26, 28, 35–40*
Compare and contrast the properties of fats and steroids. (13.3)	→	*Questions 7–9, 27, 41–43*
Classify the structure of a protein based upon the organization of its amino acids and describe how enzymes work. (13.4)	→	*Questions 10–13, 29, 44–50*
Identify nucleic acids as polymers of nucleotides and describe how they code for the building of proteins. (13.5)	→	*Questions 14–17, 30, 31, 51–54*
Distinguish vitamins from minerals and the roles they play in our nutrition. (13.6)	→	*Questions 18, 19, 32, 55–59, 68*
Classify metabolic reactions as either catabolic or anabolic. (13.7)	→	*Questions 20, 21, 60–62, 70*
Describe how the body utilizes carbohydrates, fats, and proteins. (13.8)	→	*Questions 22–25, 63–67, 69*

SUMMARY OF TERMS (KNOWLEDGE)

Amino acid The monomers of polypeptides, each monomer consisting of an amine group and a carboxylic acid group bonded to the same carbon atom.

Anabolism A general term for all the energy-requiring chemical reactions that produce large biomolecules from smaller molecules.

Carbohydrate Organic molecules produced by photosynthetic plants, containing only carbon, hydrogen, and oxygen.

Catabolism Chemical reactions that break down biomolecules in the body.

Chromosomes An elongated bundle of DNA and protein that appears in a cell's nucleus just prior to cell division.

Deoxyribonucleic acid (DNA) A nucleic acid containing the sugar deoxyribose and having a double helical structure; it carries genetic code in its nucleotide sequence.

Enzyme A protein that catalyzes (speeds up) bio-chemical reactions.

Fat A biomolecule that packs a lot of energy per gram and consists of a glycerol unit attached to three fatty acid molecules.

Gene A particular sequence of nucleotides along the DNA strand that leads a cell to manufacture a particular polypeptide.

Glycogen A polymer made of hundreds of glucose monomers and sometimes referred to as animal starch.

Lipid A broad class of biomolecules, such as fats and oils, not soluble in water because their structures are largely made of hydrocarbons.

Metabolism The general term describing the sum of all the chemical reactions in the body.

Minerals Inorganic chemicals that play a wide variety of roles in the body and are obtained from our diet.

Nucleic acid A long polymeric chain of nucleotide monomers holding the information for how amino acids need to be linked together to form an organism.

Nucleotide A nucleic acid monomer consisting of three parts: a nitrogenous base, ribose (in RNA) or deoxyribose (in DNA), and a phosphate group.

Protein A polymer of amino acids having some biological function.

Replication The process by which DNA strands are duplicated.

Ribonucleic acid (RNA) A nucleic acid containing a fully oxygenated ribose; it executes protein synthesis based on code read from DNA.

Saccharide Another term for carbohydrate. The prefixes *mono-*, *di-*, and *poly-* are used with this term to indicate the size of the carbohydrate.

Vitamins Organic chemicals that assist in various biochemical reactions in the body and are obtained from our diet.

READING CHECK QUESTIONS (COMPREHENSION)

13.1 Biomolecules Are Produced and Utilized in Cells

1. Is the cell nucleus within the cytoplasm of a cell, or is the cytoplasm within the cell nucleus?

2. What are the four major categories of biomolecules discussed in this chapter?

13.2 Carbohydrates Give Structure and Energy

3. Are all carbohydrates digestible by humans?

4. Why do plants produce starch?

5. Which monosaccharide do starches and cellulose have in common?

6. What is the most abundant organic compound on the Earth?

13.3 Lipids Are Insoluble in Water

7. What are the structural components of a triglyceride?

8. What makes a saturated fat saturated?

9. What do all steroids have in common?

13.4 Proteins Are Polymers of Amino Acids

10. How do various amino acids differ from one another?

11. What do a peptide, a polypeptide, and a protein all have in common?

12. What is the role of enzymes in the body?

13. What holds a substrate to its receptor site?

13.5 Nucleic Acids Code for Proteins

14. What is the difference between a nucleic acid and a nucleotide?

15. Where in a cell is deoxyribonucleic acid found?

16. Which four nitrogenous bases are found in DNA? In RNA?

17. What happens to the DNA double helix during replication?

13.6 Vitamins Are Organic, Minerals Are Inorganic

18. What are two classes of vitamins?

19. Why is it often more healthful to eat vegetables that have been steamed rather than boiled?

13.7 Metabolism Is the Cycling of Biomolecules through the Body

20. What is the general outcome of catabolism?

21. What is the general outcome of anabolism?

13.8 The Food Pyramid Summarizes a Healthful Diet

22. According to the food pyramid, which type of biomolecule should be the primary component of our diets?

23. Are all dietary fibers made of cellulose?

24. Which type of lipoproteins has a greater association with the formation of plaque on artery wall: LDLs or HDLs?

25. Why doesn't the human body synthesize the essential amino acids?

CONFIRM THE CHEMISTRY (HANDS-ON APPLICATION)

26. To each of two drinking glasses, add a teaspoon of potato broth, which you can make by boiling a few slices of potato to mushiness in about a cup of water. Add a tablespoon of fresh water to each glass to dilute the broth. Collect a good volume of saliva in your mouth and gracefully spit into one of the glasses. Swirl to mix. After a few minutes, add a drop of iodine solution (available from stores as a disinfectant) to each glass. Iodine forms a blue complex with starch. The intensity of the color of the solution is proportional to the amount of starch present. Did one solution turn darker than the other? Why or why not?

27. Obtain several different brands of "light" butter or margarine spreads. Place about the same volume of each sample in a test tube or narrow glass. Add enough so that, when melted, the liquid will be at least 2 cm deep. Label each test tube with the brand it contains. Melt all samples in a microwave oven. (Watch carefully because this doesn't take long.) How many layers do you see form in each test tube? Which is more dense: water or fat? Which brand contains the most water? What happens when you cool the test tubes in the refrigerator? Which sample do you suppose contains the greatest proportion of saturated or trans fats?

THINK AND COMPARE (ANALYSIS)

28. Rank the following molecules in order of increasing sweetness: glucose, cellulose, starch.

29. Rank the following amino acids in order of most acidic to least acidic: phenylalanine, tyrosine, histidine.

30. Rank the following molecules in order of increasing molecular mass: cholesterol, glycine, deoxyribonucleic acid.

31. Rank the following in order of increasing size: gene, nucleic acid, nucleotide.

32. Rank the following minerals in order of how much we need each day: calcium, sodium, potassium, chromium.

THINK AND EXPLAIN (SYNTHESIS)

13.1 Biomolecules Are Produced and Utilized in Cells

33. The plasma membrane is made of what kind of biomolecules?

34. Carbohydrates give structure and energy, while lipids are insoluble in water. What is true of proteins and nucleic acids?

13.2 Carbohydrates Give Structure and Energy

35. Does a carbohydrate contain water?

36. Why do photosynthetic plants create carbohydrates?

37. In what ways are cellulose and starch similar to each other? In what ways are they different from each other?

38. What term from Chapter 12 describes the relationship between alpha and beta glucose?

39. Why does starch begin to taste sweet after it has been in your mouth for a few minutes?

40. Why can glycogen be rapidly converted into glucose?

13.3 Lipids Are Insoluble in Water

41. Why are lipids insoluble in water?

42. Why is it important to have cholesterol in your body?

43. Could a food product containing glycerol and fatty acids but no triglycerides be advertised as being fat-free? If so, how might such advertising be misleading?

13.4 Proteins Are Polymers of Amino Acids

44. Silk is more waterproof than cotton. Why?

45. You are a beautician about to apply a reducing agent to a customer with fine hair who wants permanent curls in his hair. Should the reducing agent be regular strength, concentrated, or diluted?

46. When an unknown peptide containing five amino acids is treated with an enzyme that hydrolyzes only the serine–leucine peptide bond, the fragments Leu—Cys, Ser, and Leu—Ser are formed. What was the original amino acid sequence in the peptide?

47. Identify the molecular attractions occurring in this large protein at the locations a, b, and c:

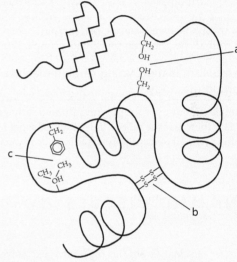

Distinguish among the primary, secondary, and tertiary structures of this protein.

48. Why do changes in pH interfere with the function of proteins? For your answer, consider the acid-base character of many amino acid side chains.

49. What are two common secondary structures in a protein? The two common tertiary structures?

50. Name the four structural levels possible in a protein. Describe the details of each.

13.5 Nucleic Acids Code for Proteins

51. Why did scientists first think that proteins rather than nucleic acids were the holders of genetic information?

52. What do ATP and RNA have in common? What is different about them?

53. A common source of DNA damage is the spontaneous loss of the amine group on cytosine and the formation of an amide. This occurs at a rate of about 100 times a day. Fortunately, we produce enzymes able to detect and repair such degraded cytosines. Given this information, suggest why DNA differs from RNA in possessing the nucleotide thymine rather than uracil.

54. Why is the number of adenines in a DNA molecule always the same as the number of thymines?

13.6 Vitamins Are Organic, Minerals Are Inorganic

55. Both water-soluble and water-insoluble vitamins can be toxic in large quantities. Our bodies are more tolerant of overdoses of water-soluble vitamins, however. Why?

56. The dietary minerals must be in a water-soluble ionic form in order for the body to make use of them. Why?

57. A friend of yours loads up on vitamin C once a week instead of spacing it out over time. She argues the convenience of not having to take pills every day. What advice do you have for her?

58. Which statement is more accurate:

　a. Vitamins are needed by the body to avoid vitamin-deficiency diseases such as scurvy.

　b. Vitamins are needed by the body so that many of its catabolic and anabolic reactions can proceed efficiently.

59. Adenosine triphosphate, ATP, readily loses its outermost phosphate group to become adenosine diphosphate, ADP. When this phosphate group is lost, it rapidly accelerates away from its neighboring phosphate groups. Why?

13.7 Metabolism Is the Cycling of Biomolecules through the Body

60. Mammals cannot produce polyunsaturated fatty acids. How is it then that the tallow (fat) obtained from beef contains up to 10 percent polyunsaturated fatty acids?

61. The human body stores excess glucose as glycogen and excess fat as fatty tissue that can accumulate beneath the skin. How does the body store any excess amino acids?

62. Oxygen, O_2, is certainly good for you. Does it follow that if small amounts of oxygen are good for you, large amounts of oxygen would be especially good for you?

13.8 The Food Pyramid Summarizes a Healthful Diet

63. Suggest why the glycemic index for sucrose is only about 64 percent that of glucose.

64. Suggest why starch takes longer than sucrose to break down to glucose in the intestine.

65. Is it possible to eat a food low on the glycemic index and still experience a significant increase in blood glucose?

66. Peanut butter has more protein per gram than a hard-boiled egg, and yet the egg represents a more nutritious source of protein. Why?

67. Cold cereal is often fortified with all sorts of vitamins and minerals but is deficient in the amino acid lysine. How might this deficiency be compensated for in a breakfast meal?

THINK AND DISCUSS (EVALUATION)

68. A number of scientific studies report that excessive doses of vitamin A increase mortality risk by about 16 percent. Similarly, excessive vitamin E increases one's risk of death by about 4 percent. Dietary supplement manufacturers claim these vitamins have an antioxidant effect, eliminating free radical molecules linked to disease. As any physician knows, however, the body's immune defense mechanisms rely on free radicals to kill off invading infections. Pretend you are a marketing executive for a multimillion-dollar company specializing in vitamin supplements. Write a press release statement in response to a scientific study that shows your product to be harmful. Why might you decide to do nothing in response to the scientific study?

69. Some diets, most notably the Atkins diet, call for large amounts of protein and fat and small amounts of carbohydrate. One of the claims of such diets is that for the same number of calories, a meal high in protein and fat leaves a person with less of an urge to eat later on. One of the arguments against such diets is that they are hard on the kidneys and liver and that they fail to emphasize the importance of regular exercise. What are your thoughts on effective dieting?

70. Some athletes continue to use anabolic steroids despite the negative side effects of these drugs. What legal consequences, if any, should such an athlete face? Should these steroids be banned? Should it be illegal for physicians to prescribe them for athletic use? Is it the right of the individual to decide what goes into his or her body? Should sporting events be segregated into those that allow the use of performance-enhancing chemicals and those that do not?

READINESS ASSURANCE TEST (RAT)

If you have a good handle on this chapter, then you should be able to score at least 7 out of 10 on this RAT. Check Your Answers online at www.ConceptualChemistry.com. If you score less than 7, you need to study further before moving on.

Choose the BEST answer to the following.

1. Which of the following statements is most accurate?

　a. You are undergoing continuous chemical change within your body from moment to moment.

b. You have the exact same molecules in your body as when you were born.

c. You have the exact same types of chemicals in your body as when you were younger.

d. You are chemically the same as you were when you were born.

e. None of the above

2. The monosaccharide is the

   a. fundamental unit of genetic material.

   b. fundamental unit of life.

   c. monomer of the sucrose polymer.

   d. basic repeating unit of a polymer.

   e. fundamental unit of a carbohydrate.

3. Which of these functional groups plays the biggest role in the properties of carbohydrates?

   a. Alcohols

   b. Amides

   c. Esters

   d. Ethers

   e. Phenols

4. Which is **not** a function of fat within the body?

   a. It acts as a cushion to prevent injury.

   b. It acts as an energy reserve.

   c. It acts as a source of glucose.

   d. It acts as insulation.

   e. All of the above are functions of fat.

5. Why do most steroids have chemical structures very similar to that of cholesterol?

   a. Most steroids are synthesized from cholesterol.

   b. Cholesterol is synthesized from saturated fats, and steroids are made from unsaturated fats.

   c. Cholesterol is a fat, which is a lipid, like steroids.

   d. Steroids are made from saturated fats, and cholesterol is made from unsaturated fats.

   e. None of the above is true.

6. How are proteins similar to starches?

   a. Both are polymers.

   b. Both can be consumed and used as an energy source.

   c. Both are made of glucose monomers.

   d. Both are lipids.

   e. Both A and B are true.

7. The structure of ATP is most closely related to that of

   a. a carbohydrate.

   b. a lipid.

   c. an amino acid.

   d. a nucleic acid.

   e. a nucleotide.

8. Vitamins such as the B vitamins and vitamin C are often lost by boiling vegetables. Why?

   a. They are water soluble and are poured down the drain with the water.

   b. The heat of the boiling water causes the vitamin to evaporate.

   c. They are fat-soluble compounds, and vegetables do not have much fat.

   d. The heat of the boiling water causes the vitamins to become locked within the cellulose.

   e. None of the above is true.

9. Which of the following statements is true about metabolism?

   a. Metabolism involves the use of energy to make new biomolecules.

   b. Metabolism involves the rate of cellular respiration.

   c. Metabolism involves the destruction of biomolecules and the production of energy.

   d. All of the above are true.

   e. None of the above is true.

10. Which statement about proteins is true?

   a. The source of a protein (mammal, fish, or poultry) has little effect on nutritional value.

   b. The simpler an amino acid is, the more likely it is to be synthesized by our bodies.

   c. All of the amino acids in the human body can be synthesized by our cells.

   d. All of the above are true.

   e. None of the above is true.

Partially Hydrogenated Fats

Partially hydrogenated. It is a term that seems to be on the ingredients lists of most food products nowadays. There are others, such as *trans-free, low in saturated fats,* and *high in unsaturated fats.* What do these expressions mean? What is supposed to be good for you and what is bad? By studying this textbook, you already have learned much chemistry and the answers to these sorts of mysteries are easily at hand.

One of the building blocks of dietary fats is fatty acid molecules, as shown to the right. Every fatty acid molecule consists of a long chain of nonpolar carbon atoms attached to a polar carbon and an oxygen unit known as a carboxylic acid. As discussed in Chapter 13, there are two types of fats—those made from fatty acids with no double bonds and those made from fatty acids with one or more double bonds. Fatty acids with no double bonds are called *saturated* fatty acids, while those with one or more double bonds are called *unsaturated* fatty acids.

Saturated fat molecules are able to pack tightly together because their fatty acids point straight out and align with one another, much like a bunch of wooden matchsticks in a match box. Induced dipole–induced dipole attractions hold the aligned chains together. This gives saturated fats, such as lard, relatively high melting points, and as a result, they tend to be solid at room temperature. The fatty acid chains of unsaturated fats are "kinked" wherever double bonds occur. The kinks inhibit alignment, and as a result, unsaturated fats tend to have relatively low melting points. These fats are liquid at room temperature and are commonly referred to as *oils.* Most vegetable oils are liquid at room temperature because of the high proportion of unsaturated fats they contain.

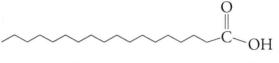

nonpolar carbon chain   polar carboxylic acid

Saturated fatty acid

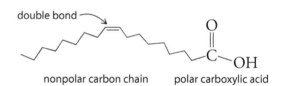

double bond

nonpolar carbon chain   polar carboxylic acid

Unsaturated fatty acid

So which are better for you: saturated or unsaturated fats? The answer is that the body needs both to remain healthy. For example, the body uses saturated fats to produce cholesterol, which is needed to strengthen cell membranes. Numerous biomolecules are also made from unsaturated fats. Furthermore, both saturated and unsaturated fats are required by the body because of their high energy content. A diet high in saturated fats, however, leads to the formation of plaque in artery walls, and this predisposes the body to heart attacks and strokes. A diet high in unsaturated fats, by contrast, can actually work to lower blood cholesterol, thereby reducing the risk of coronary heart disease. For these reason, diets high in unsaturated fats—the ones with the double-bond kinks—are generally considered to be healthier.

Saturated fats tend to be solid, while unsaturated fats tend to be liquid. Think about this from the point of view of, say, a manufacturer of chocolate candies. Your chocolate candies need to be formulated with fats to give them that delicious creamy flavor. If you choose saturated fats, your chocolate bars will take on more of a

solid consistency, which allows you to wrap them and then package them neatly in boxes for display on the grocery store shelf. If you choose the healthier unsaturated fats, however, your chocolate candies will likely melt the moment they leave your air-conditioned factory. Furthermore, compared to saturated fats, unsaturated fats tend to go rancid a lot faster; so your chocolate candies formulated with unsaturated fats will have a much shorter shelf life. What do you do?

It has been known for over 100 years that unsaturated oils can be transformed to a more solid consistency by *hydrogenation*, a chemical process in which hydrogen atoms are added to carbon double bonds. Any desired consistency can be obtained by adjusting the degree of hydrogenation. Complete hydrogenation *saturates* all the double bonds, resulting in a hardened saturated fat. Partial hydrogenation, however, saturates only some of the double bonds, resulting in a mixture of both saturated and unsaturated fats. This mixture is semisolid, which, from a marketing perspective, is most desirable.

One of the first commercial fat products to arise from hydrogenation was margarine, which is a blend of partially hydrogenated vegetable oils, water, salt, butter flavoring, emulsifiers, and preservatives. Initially, margarine's best selling point was its low price. But by the 1970s, it began to be sold as a healthier alternative to butter, which had been revealed to increase the incidence of heart disease. A couple of decades later, however, the good news for margarine and the multitude of other products containing partially hydrogenated fats began to fade when studies showed that hydrogenation also creates what

are known as trans unsaturated fats. In the trans unsaturated fat, the orientation of the trans double bond makes for a linear molecule that resembles a saturated fat.

Trans fats, it was found, increase the incidence of coronary heart disease by mimicking the plaque-forming properties of saturated fats. According to the U.S. Food and Drug Administration, an average candy bar contains 4 grams of saturated fats and 3 grams of trans fats. A serving of stick margarine contains 2 grams of saturated fats and 3 grams of trans fats. French fries contain 7 grams of saturated fats and 8 grams of trans fats. So the proportion of trans fats in many commonly consumed foods is significant. To help the consumer make informed food choices, the FDA requires that nutrition labels display the content of trans fats. From these labels, you will find that "When in doubt, the soft one wins out, but liquid is always better."

CONCEPTCHECK

List the following products in order of increasing proportion of trans fats: stick margarine, soft tub margarine, butter.

CHECK YOUR ANSWER
Interestingly, trans fats occur naturally, but only to a very small extent. The trans fat content of butter is therefore negligible. The soft tub margarine is soft because it wasn't hydrogenated very long, which means that it has fewer trans fats than does the stick margarine. In order of increasing trans fat content: butter < soft tub margarine < stick margarine.

Think and Discuss

1. Today in the United States, margarine outsells butter by a 2:1 ratio. Do you generally use margarine or butter or neither? Why? Also, butter has practically no trans fats. Does this mean that butter is better for you than margarine?

2. In 1886, the U.S. Congress passed the Margarine Act, which added a hefty tax to margarine and required expensive licenses to make or sell it. States also passed laws forbidding manufacturers to add yellow coloring to the naturally pale spread. Why do you suppose these laws were enacted? Do you agree with these sorts of laws? Explain.

3. Some vegetable oils, such as palm oil, are high in saturated fats. Do you think these highly saturated vegetable oils will become more or less popular as the public becomes more mindful of buying products advertised as "trans free"? Why?

4. Trans fats have been added to the nutrition labels of foods. What next? Will nutrition labels eventually become so detailed that they rarely get read by consumers? At what point does a label provide too much information?

5. Processed foods are made of the same kinds of atoms that are in fresh foods. So why all the hype against processed foods?

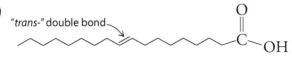

Trans unsaturated fatty acid

Morphine

▲ Slits cut into the pod of the opium poppy plant exude a thick oil that contains a variety of pain-killing analgesics, such as morphine.

14

Medicinal Chemistry

THE MAIN IDEA

Medicines are like keys that unlock various biological responses.

In 78 A.D., the Greek physician Dioscorides wrote *Materia Medica*, a treatise in which he described about 600 plants known to have medicinal properties. Included in this list was the morphine-producing opium poppy. With the development of chemistry in the early 1800s came the understanding that natural products owe their medicinal properties to certain substances they contain. Morphine was first isolated from opium in 1806. Quinine, a drug once used in fighting malaria, was isolated from the bark of the cinchona tree in 1820. Soon, compounds produced in the laboratory were also found to have medicinal properties. In the 1840s, the anesthetic activity of diethyl ether made painless surgery and dentistry possible.

In the 1860s, Louis Pasteur confirmed the germ theory of disease with his discovery of bacteria. This led to the discovery of the antiseptic properties of phenol, as discussed in Chapter 12. The first major advance toward curing bacterial diseases was not made until the 1930s, when sulfur-containing compounds known as sulfa drugs were developed. Next came penicillin. Subsequent research has led to an ever-expanding array of medicines—both natural and synthetic. Today, more than 25,000 prescription and 300,000 nonprescription medicines are available in the United States.

Chemistry

Paper Wad Receptor Site

Most drugs work by binding to tiny nooks and crannies within our bodies, called receptor sites. The biological effect of the drug occurs only upon the binding of the drug to its receptor site. For this activity, the clay represents a drug, which has a receptor site found within the paper ball.

PROCEDURE

1. Wad a sheet of paper into a ball.
2. Soften a teaspoon of craft clay by kneading it with your hands or by heating the clay in a microwave oven for a few seconds.
3. Press the softened clay into one of the nooks or crannies of the paper ball. Let the clay cool (in a freezer, if available) so that the clay retains the shape of that nook or cranny.

ANALYZE AND CONCLUDE

1. Once you pull your clay out of the paper, does the paper have to open up and flex to re-accept the clay? How well does the clay fit in other nooks and crannies?
2. If two drugs have the same biological effect, what might be true about their 3-dimensional shapes?
3. Some drug molecules have rigid structures that do not flex very easily. Others have a more relaxed structure that can contort into a variety of shapes. Which of these types of drug molecules might have a greater variety of biological effects? Which is likely to have a more specific biological effect?

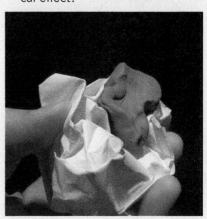

14.1   Medicines Are Drugs That Benefit the Body

EXPLAIN THIS

When might a medicine actually cause you harm?

What is a medicine? A *medicine* is a drug taken for the purpose of improving a person's health. So what then is a drug? Loosely defined, a *drug* is any substance other than food or water that affects the way the body functions. The word *drug* refers to a wide range of chemical substances. All medicines are drugs, but not all drugs are medicines. Many drugs are used for nonmedical purposes, some legal and others illegal. Legal nonmedical drugs include alcohol, caffeine, and nicotine. Illegal nonmedical drugs include LSD and cocaine.

There are a variety of ways to classify drugs. For example, drugs can be classified according to the way they are derived, as is shown in Table 14.1. Drugs that are "natural products" come directly from terrestrial or marine plants or animals. Drugs that are "chemical derivatives" are natural products that have been chemically modified to increase potency or decrease side effects. "Synthetic" drugs are made completely in the laboratory.

Perhaps the most common way to classify drugs is according to their primary biological effects. Note, however, that most drugs exhibit a broad spectrum of activity, which means they have multiple effects on the body and may fall under several classifications. Aspirin, for example, relieves pain, but it also reduces fever and inflammation, thins the blood, and causes ringing in the ears. Morphine relieves pain, but it also causes constipation and suppresses the urge to cough.

LEARNING OBJECTIVE

Classify drugs by their origin and describe the synergistic effect.

READINGCHECK

Do most drugs have a single effect on the body or multiple effects?

Bacterial infections, not cancer or heart attacks, were the leading cause of death in the United States prior to the discovery of antibiotics in the 1930s.

TABLE 14.1 The Origins of Some Common Drugs

ORIGIN	DRUG	BIOLOGICAL EFFECT
Natural product	Caffeine	Nerve stimulant
	Reserpine	Hypertension reducer
	Vincristine	Anticancer agent
	Penicillin	Antibiotic
	Morphine	Analgesic
Chemical derivative of natural product	Prednisone	Antirheumatic
	Ampicillin	Antibiotic
	LSD	Hallucinogen
	Chloroquine	Antimalarial
	Ethynodiol diacetate	Contraceptive
Synthetic	Valium	Antidepressant
	Benadryl	Antihistamine
	Allobarbital	Sedative–hypnotic
	Phencyclidine	Veterinary anesthetic
	Methadone	Analgesic

FOR YOUR INFORMATION

The toxicity of any substance is in the size of the dose. Fresh water, for example, can be lethal if you drink too much of it. Why? Because too much would flush out dissolved ions that are absolutely essential for your health. Similarly, while small amounts of fluoride protect against tooth decay, larger amounts can cause your teeth to be mottled. Worse still, excessive fluoride binds with calcium in your blood, forming lethal calcium fluoride crystals. Again, for emphasis, the toxicity of any substance is in the size of the dose.

At times, the multiple effects of a drug are desirable. For example, aspirin's pain-reducing and fever-reducing properties work well together in treating flu symptoms in adults. In addition, aspirin's blood-thinning ability helps prevent heart disease. Morphine was widely used during the American Civil War for both relieving the pain of battle wounds and controlling diarrhea. Often, however, the side effects of a drug are less desirable. Ringing in the ears and upset stomach are a few of the negative side effects of aspirin, and a major side effect of morphine is its addictiveness. A main goal of drug research, therefore, is to find drugs that are specific in their action and that have minimal side effects.

Although two drugs that are taken together may have different primary activities, they may share a common secondary activity. This secondary effect that both drugs share can be amplified when the two drugs are taken together. One drug's enhancing the action of another is called a **synergistic effect.** A synergistic effect is often more powerful than the sum of the activities of the two drugs taken separately. One of the great challenges for physicians and pharmacists is keeping track of all possible combinations of drugs and potential synergistic effects they might have.

The synergism that results from mixing drugs that have the same primary effect is particularly hazardous. For example, a moderate dose of a sedative combined with a moderate amount of alcohol may be lethal. In fact, most drug overdoses are the result of a combination of drugs rather than the abuse of a single drug.

CONCEPT CHECK

Distinguish between a drug and a medicine.

CHECK YOUR ANSWER A drug is any substance administered to affect body function. A medicine is any drug administered for its therapeutic effect. All medicines are drugs, but not all drugs are medicines.

14.2 The Lock-and-Key Model Guides the Synthesis of New Medicines

EXPLAIN THIS

Receptor sites are made of what sort of biomolecules?

To find new and more effective medicines, chemists use various models that describe the way drugs work. By far, the most useful model of drug action is the **lock-and-key model.** The basis of this model is that there is a connection between a drug's chemical structure and its biological effect. For example, morphine and all related pain-relieving opioids, such as codeine and heroin, have the T-shaped structure shown in **Figure 14.1.**

According to the lock-and-key model illustrated in **Figure 14.2,** biologically active molecules function by fitting into *receptor sites*, where they are held by intermolecular attractions, such as hydrogen bonding. When a drug molecule fits into a receptor site the way a key fits into a lock, a particular biological event is triggered, such as a nerve impulse or even a chemical reaction. In order for a molecule to fit into a particular receptor site, however, it must have the proper shape, just as a key must have properly shaped notches in order to fit into a lock.

Another facet of this model is that the molecular attractions holding a drug to a receptor site are easily broken. (Recall from Section 7.1 that most molecular attractions are many times weaker than chemical bonds.) A drug is therefore held to a receptor site only temporarily. Once the drug is removed from the receptor site, the drug's chemical structure is degraded and the effects of the drug are said to have "worn off."

Using this model, we can understand why some drugs are more potent than others. Heroin, for example, is a more potent painkiller than morphine because the chemical structure of heroin allows tighter binding to its receptor sites.

Why do our bodies have receptor sites? Receptor sites allow different parts of the body to communicate with each other. If a large bear threatened you, for example, your nervous system would generate molecules that would bind to

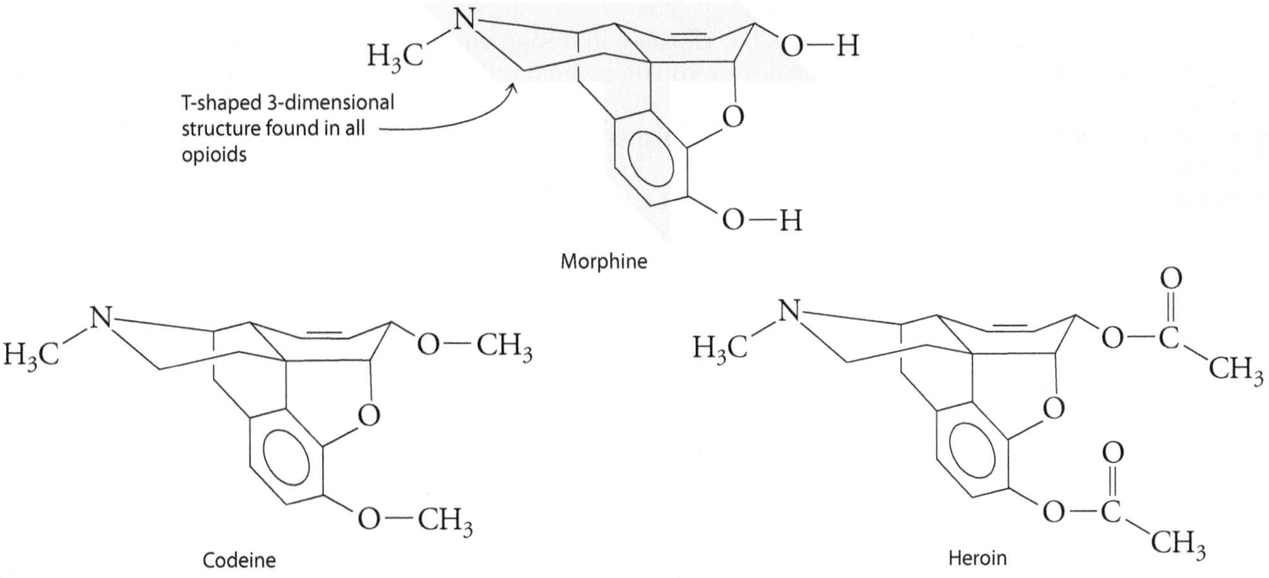

LEARNING OBJECTIVE

Describe the lock-and-key model and explain how this model is used in the development of new medicines.

READINGCHECK

What is the basis of the lock-and-key model?

Morphine

Codeine

Heroin

▲ **Figure 14.1**
All drugs that act like morphine have the same basic 3-dimensional shape as morphine.

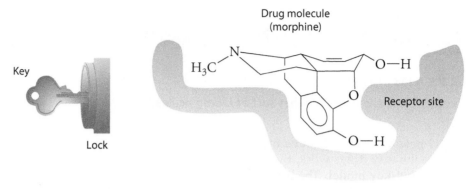

Drug molecule
(morphine)

Key

Lock

Receptor site

▲ **Figure 14.2**
Many drugs act by fitting into receptor sites on molecules, much as a key fits into a lock.

 FOR YOUR INFORMATION

Since 1999, the Bill and Melinda Gates Foundation has granted over $5 billion to global health plans. This includes the Vaccine Fund of $750,000,000 to support the immunization of children in 74 countries through the purchase of new vaccines. A $42 million Gates Foundation grant was also used to help support the Institute for OneWorld Health, the first nonprofit pharmaceutical company in the United States. The mission of OneWorld Health is to develop safe, effective, and affordable new medicines for people with infectious diseases in the developing world.

▲ **Figure 14.3**
Ethnobotanists directed natural products chemists to the yellow coating on the root of the African Bobgunnia tree. Indigenous people have known for many generations that this coating has medicinal properties. From extracts of the coating, the chemists isolated a compound that is highly effective in treating fungal infections. This compound, produced by the tree to protect itself from root rot, shows much promise in the treatment of the opportunistic fungal infections that plague those suffering from AIDS.

receptor sites within your adrenal glands, which would then create the stimulant adrenaline. Once released into the bloodstream, the adrenaline would bind to adrenaline receptor sites on your muscles, which would tell your muscles to power up—so that you could climb a nearby tree.

Many drugs act by binding to the receptor sites of these naturally occurring communication molecules. Some drugs, upon binding, cause the same biological effect as the natural molecule. These drugs are called **agonists.** An example of an agonist is morphine, which mimics molecules the body produces to relieve pain—the endorphins.

Upon binding to a receptor site, other drugs initiate no biological effect except, once bound, they prevent other active molecules from binding. Thus, the receptor site is effectively blocked. Such a drug is called an **antagonist.** A powerful antagonist to endorphins is the drug naloxone. Administering naloxone to someone high on endorphins (or morphine) causes a rapid loss of that high and the immediate start of withdrawal symptoms.

The lock-and-key model has developed into one of the most important tools of pharmaceutical study. Knowing the precise shape of a target receptor site allows chemists to design molecules that have an optimal fit and a specific biological effect.

Biochemical systems are so complex, however, that our knowledge is still limited, as is our capacity to design effective medicines. For this reason, most new medicines are still discovered rather than designed. One important avenue for drug discovery is *ethnobotany*. An *ethnobotanist* is a researcher who learns about the medicinal plants used in indigenous cultures, such as the root of the Bobgunnia tree, shown in **Figure 14.3**. Today, hundreds of clinically useful prescription drugs have been derived from plants. About three-quarters of these came to the attention of the pharmaceutical industry as a result of their use in folk medicine.

Another important method of drug discovery is the random screening of vast numbers of compounds. Each year, for example, the National Cancer Institute screens approximately 20,000 compounds for anticancer activity. One successful hit was the compound Taxol®, shown in **Figure 14.4**. This compound shows significant activity against several forms of cancer, especially ovarian cancer. As was discussed in Section 12.7, the yew tree produces only small amounts of this natural product, which is why the Taxol® used to treat cancer is synthesized in the laboratory.

A drug isolated from a natural source is not necessarily better or more gentle than one produced in the laboratory. Aspirin, for example, is a human-made chemical derivative, and it is certainly more gentle than morphine, which is 100 percent natural. The main advantage of natural products is

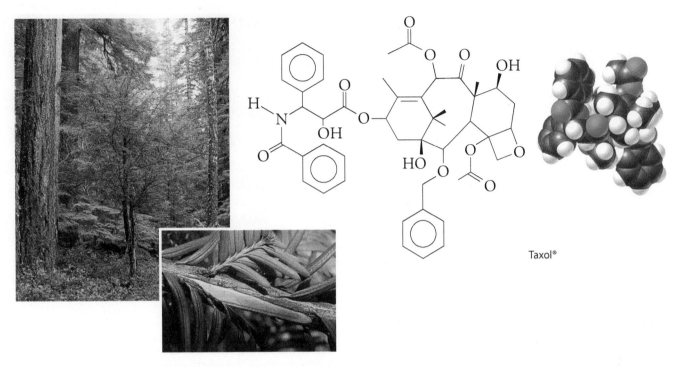

Taxol®

▲ **Figure 14.4**
Originally isolated from the bark of the Pacific yew tree, Taxol® is a complex natural product useful in the treatment of various forms of cancer.

their great *diversity.* Each year more than 3000 new chemical compounds are discovered from plants. Many of these compounds are biologically active, serving the plant as a chemical defense against disease or predators. Nicotine, for example, is a naturally occurring insecticide produced by the tobacco plant to protect itself from insects.

It has been estimated that only 5000 plant species have been exhaustively studied for possible medical applications. This is a minor fraction of the estimated 250,000 to 300,000 plant species on our planet, most of which are located in tropical rainforests. That we know little or nothing about much of the plant kingdom has raised justified and well-publicized concern. For as rainforests are being destroyed, plant species that might yield useful medicines are also being destroyed.

A recent laboratory approach intended to mimic nature's chemical diversity is known as **combinatorial chemistry;** it is a method of generating a large "library" of related compounds. Combinatorial chemistry takes advantage of the many different ways in which a series of reacting chemicals may be combined. Microquantities of reagents are combined in a grid to maximize the number of possible products, as is illustrated in **Figure 14.5a**. The result is a great number of closely related compounds that can be screened for biological activity. The most active derivatives are analyzed for chemical structure and then synthesized on a larger scale for further testing or clinical trials. A typical array is shown in **Figure 14.5b**.

CONCEPTCHECK

Why are chemicals from natural sources so suitable for making drugs?

CHECK YOUR ANSWER Because there are so many naturally derived chemicals with biological effects. Their vast diversity permits the manufacture of the many different types of medicines needed to combat the many different types of human illnesses.

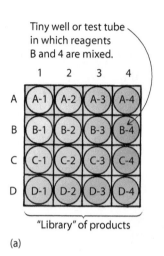

Tiny well or test tube in which reagents B and 4 are mixed.

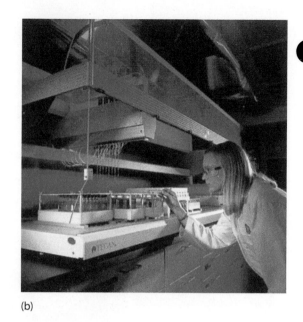

(a)　　　　　　　　　　　　(b)

▲ Figure 14.5
(a) Eight hypothetical starting materials, A through D and 1 through 4, can be combined in various ways to yield 16 products, each of which may have some biological activity not found in any of the starting materials. (b) A multitude of products are thus immediately available to be screened for medicinal activity.

14.3 Chemotherapy Cures the Host by Killing the Disease

LEARNING OBJECTIVE

Describe how chemotherapy protects us from bacterial and viral infections as well as cancer.

READINGCHECK

How does chemotherapy work?

EXPLAIN THIS

How is an insecticide like a chemotherapeutic?

The use of drugs that destroy disease-causing agents without doing excessive harm to the animal host is known as **chemotherapy.** This approach is effective in the treatment of many diseases, including bacterial infections. It works by taking advantage of the ways a disease-causing agent, also known as a *pathogen,* is different from a host.

Sulfa Drugs and Antibiotics

Developed in the 1930s, *sulfa drugs* were the first drugs used to treat bacterial infections. They work by taking advantage of a striking difference between humans and bacteria. Both humans and bacteria must have the nutrient *folic acid* in order to remain healthy. While we humans can obtain folic acid from what we eat, bacteria cannot absorb folic acid from outside sources. Instead, bacteria must make their own supply of folic acid. For this, they possess receptor sites that help make folic acid from a simpler molecule found in all bacteria, para-aminobenzoic acid (PABA). The PABA attaches to the specific receptor site and is converted to folic acid, as shown in **Figure 14.6**.

Sulfa drugs have a close structural resemblance to PABA. When taken by a person suffering from a bacterial infection, a sulfa drug is transformed by the body into the compound *sulfanilamide,* which attaches to the bacterial receptor sites designed for PABA, as shown in **Figure 14.7**. This prevents the bacteria from synthesizing folic acid. Without folic acid, the bacteria soon die. The patient, however, lives on because folic acid is part of his diet.

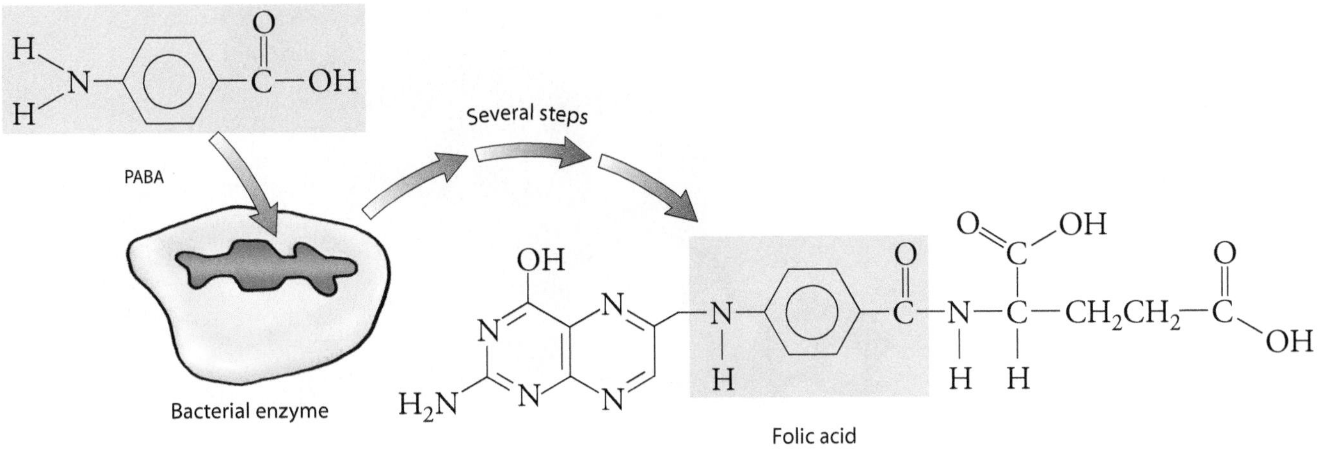

PABA

Several steps

Bacterial enzyme

Folic acid

▲ **Figure 14.6**
Bacterial enzymes use para-aminobenzoic acid (PABA) to synthesize folic acid.

CONCEPT CHECK

How is sulfanilamide poisonous to bacteria but not to humans?

CHECK YOUR ANSWER Sulfanilamide is poisonous to bacteria because it prevents them from synthesizing the folic acid they need to survive. Humans utilize folic acid from their diet, so they are not bothered by sulfanilamide's ability to disrupt the synthesis of folic acid.

Antibiotics are chemicals that prevent the growth of bacteria. They are produced by microorganisms such as molds, fungi, and even bacteria. Penicillin was the first antibiotic discovered. Many derivatives of penicillin, such as the penicillin G shown in **Figure 14.8**, have since been isolated from microorganisms as well as prepared in the laboratory. Penicillins and the closely related compounds known as cephalosporins, also shown in Figure 14.8, kill bacteria by inactivating receptor sites responsible for strengthening the bacterial cell wall. With this receptor site inactivated, bacterial cell walls grow weak and eventually burst.

Chemotherapy Can Inhibit the Ability of Viruses to Replicate

So far, chemotherapy has been more successful in treating bacterial infections than in treating viral infections. Perhaps the greatest obstacle to effective viral treatment lies in the nature of viruses. When not attached to a host, a virus is an

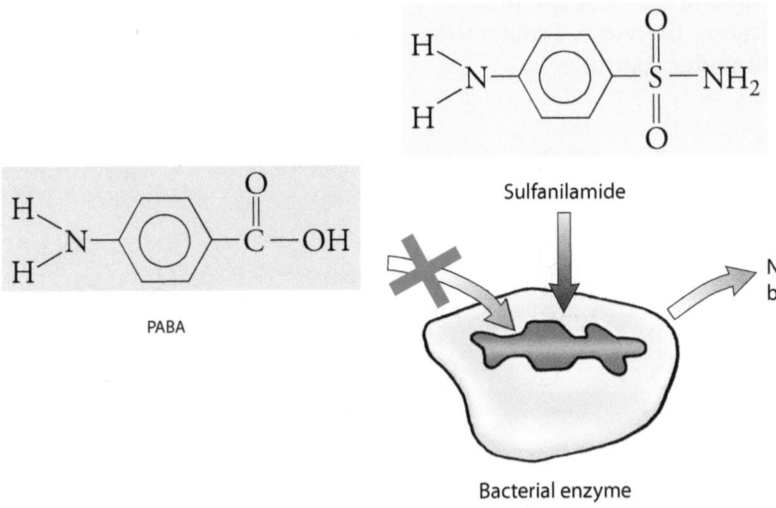

Sulfanilamide

PABA

No folic acid; bacteria die

Bacterial enzyme

◀ **Figure 14.7**
In the body, sulfa drugs are transformed into sulfanilamide, which binds to the bacterial receptor sites and keeps them from doing their job.

▶ **Figure 14.8**
Penicillins, such as penicillin G, and cephalosporins, such as cephalexin, as well as most other antibiotics, are produced by microorganisms that can be mass-produced in large vats. The antibiotics are then harvested and purified.

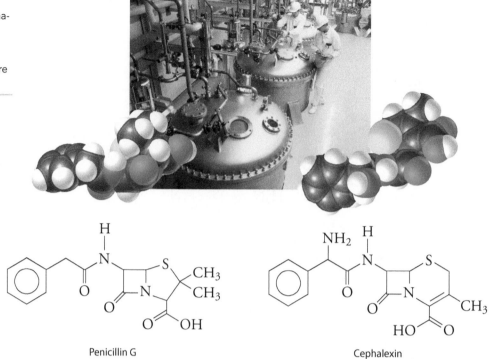

Penicillin G

Cephalexin

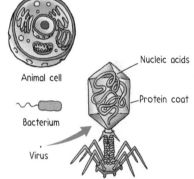

Animal cell

Bacterium

Virus

Nucleic acids

Protein coat

▲ **Figure 14.9**
Viruses are much smaller than bacteria and many times smaller than animal cells. (Notice the small dot representing the virus.) The smallest of all pathogens, viruses consist mostly of nucleic acids enclosed in a protein coat.

▼ **Figure 14.10**
Before a nucleoside such as guanosine can be incorporated into RNA or DNA, it must be activated through the attachment of three phosphate groups.

inert, lifeless bundle of biomolecules—and it's difficult to kill something that's not alive! A typical virus, shown in **Figure 14.9**, consists of only one strand or several strands of either RNA or DNA encapsulated in a protein coat. Some viruses infect by attaching to a cell and then injecting their genetic contents into the cell. Once inside the cell, the virus's genetic information is incorporated into the host DNA and replicated by the host cell. Eventually, the cell bursts because it is overstuffed with a multitude of viral replicates, which then spread to infect other host cells.

Most common antiviral drugs are derivatives of *nucleosides,* which are similar to nucleotides (Section 13.5) but without a phosphate group. Nucleosides roam freely in all cells and are used by the cells to create RNA or DNA. Before being used, however, the nucleosides must first be primed with three phosphate groups, as shown in **Figure 14.10**. Various synthetic derivatives of nucleosides are readily primed by virus-infected cells but not by uninfected cells. Two synthetic nucleoside derivatives, both shown in **Figure 14.11**, are acyclovir, sold under the trade name Zovirax®, and zidovudine, sold under the trade name AZT. Once incorporated in the RNA or DNA of a virus-infected host cell, these nucleoside derivatives disrupt protein synthesis, and the infected cell dies before replicating the virus. Proliferation of the virus, although not halted, is thus brought under control.

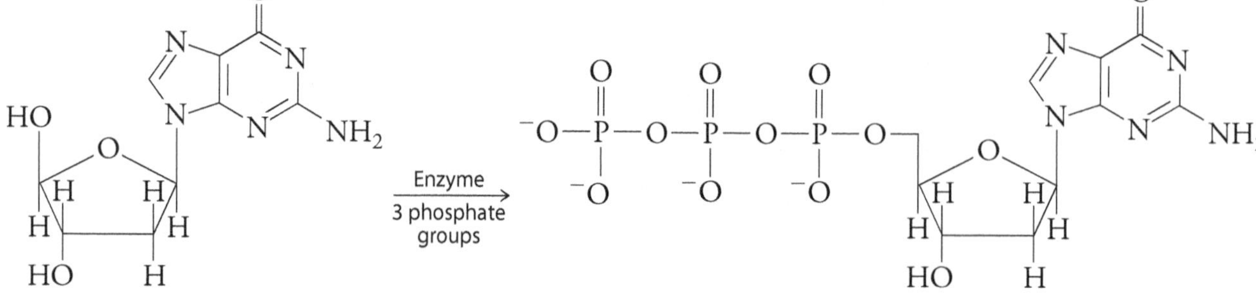

Guanosine

Enzyme
3 phosphate
groups

Guanosine triphosphate

Nucleoside

Nucleoside derivative

Deoxyguanosine

Acyclovir
(Zovirax)

Deoxythymidine

Zidovudine (AZT)

◀ **Figure 14.11**
Acyclovir (Zovirax®) is a derivative of the nucleoside deoxyguanosine, and zidovudine (AZT) is a derivative of the nucleoside deoxythymidine.

Acyclovir is useful in the treatment of herpes. Oral herpes is caused by the herpes simplex virus 1 (HSV-1), and genital herpes is caused by the herpes simplex virus 2 (HSV-2). More than 90 percent of the world's population is infected with the oral herpes virus, although many infected people do not exhibit symptoms. Genital herpes is the most prevalent noncurable sexually transmitted disease. In the United States, about 30 million people are infected with HSV-2 and an estimated 200,000 to 500,000 new cases are seen each year.

Zidovudine is used to suppress the replication of the human immunodeficiency virus (HIV), which is responsible for acquired immune deficiency syndrome, AIDS (**Figure 14.12**). According to the World Health Organization, about 3 million people are infected by this virus each year, while about 2 million people die as a result of AIDS. As of 2010, about 34 million people were living with an HIV infection.

HIV research has led to a new class of antiviral agents known as *protease inhibitors*. The life cycles of many viruses, including HIV, depend on the actions of enzymes known as proteases, which break down proteins. A protease, for

FOR YOUR INFORMATION

An unusual feature of the herpes viruses is their preference for nerve cells. Stress to the nervous system, such as emotional stress or sunburn, can cause the virus to replicate, which leads to an outbreak through the skin. When they are not replicating, these viruses remain in nerve cells, where they are not detected by the body's immune system, which has little activity in nerve cells. While remaining dormant in the nervous system, the viruses evade not only the immune system but also antiviral drugs.

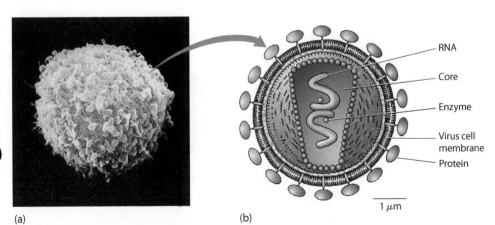

(a)

(b)

1 μm

◀ **Figure 14.12**
(a) The small green bodies covering this white blood cell are human immunodeficiency viruses. (b) The anatomy of HIV. After the initial infection, the infected person's immune response eliminates most of the virus. Some of the virus, however, remains dormant in infected cells and evades the immune response. Over a period of years, HIV reactivates itself, the immune system collapses, and the person succumbs to opportunistic diseases such as cancer and pneumonia.

FOR YOUR INFORMATION

The microorganism, Bugula neritina, is a common marine fouling agent that coats surfaces such as the hull of a boat. This species also produces small quantities of a structurally complex molecule, bryostatin 1, which shows much biological activity. This includes awakening HIV from its defensive dormant phase and thus making this virus susceptible to anti-virals and, potentially, a total eradiation from the human body. Bryostatin 1 is difficult to isolate and difficult to synthesize in the laboratory. Easier to produce analogs, however, are showing even greater promise in the treatment of HIV. This is a classic example of where nature has given us a valuable head start in the development of chemotherapeutics.

example, might be used by a virus to penetrate the proteins on the cell membrane of a host cell or to break down the host's polypeptides to create a supply of amino acids necessary for viral replication. Drugs that block the action of proteases control viral proliferation. Nelfinavir, sold under the trade name Viracept®, is an example of an effective protease inhibitor and is shown in **Figure 14.13**. Patients receiving a "cocktail" of a protease inhibitor and nucleoside antiviral agents may have their HIV counts brought below detectable levels. Although it is unlikely that this regimen can totally eliminate HIV from an infected person, the highly reduced viral counts tend to significantly delay the onset of AIDS and reduce the patient's infectiousness to others.

CONCEPT CHECK

A virus is so much simpler than a bacterium. Why then are viruses so much more difficult to target with chemotherapeutics?

CHECK YOUR ANSWER Chemotherapeutics act by interfering with one or more of the chemical reactions a pathogen needs in order to exist. The more complex a pathogen, the more ways there are to interfere with its life cycle. The fact that viruses are so simple means we have few avenues for a chemotherapeutic approach.

Cancer Chemotherapy Targets Rapidly Growing Cells

Cells periodically lose the ability to control their own growth and begin multiplying rapidly. Normally, the immune system recognizes these renegade cells and destroys them. Occasionally, however, they escape this line of defense and continue to multiply unchecked. The result can be a hard mass of tissue, called a *tumor*, that deprives healthy cells of oxygen and nutrients. Cells from a tumor may break away and be carried to other sites in the body, where they lodge and continue to multiply, forming additional tumors. As tumors multiply, more and more healthy cells are damaged and eventually die. Ultimately, the whole body may die. This is the process of cancer, the second leading cause of death in most developed nations. At present mortality rates, one in six of us will die of this disease.

Chemotherapy is most effective at the early stages of cancer, because drugs work best on cells that are in the process of dividing, a process called *cellular mitosis*, shown in **Figure 14.14**. When a tumor is young, most of its cells are undergoing mitosis. As the tumor ages, however, the fraction of cells in this growth phase decreases, reducing drug sensitivity. Drugs also have a difficult time destroying all cells in a large tumor. A 100-gram tumor, for example, may contain about 100 billion cells. Killing 99.9 percent of these cells would still leave 100 million cells—too many for the patient's immune response to control. The same treatment against a 50-milligram tumor containing about a million

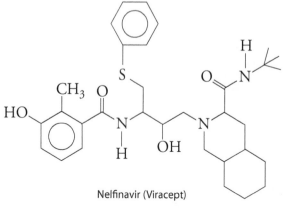
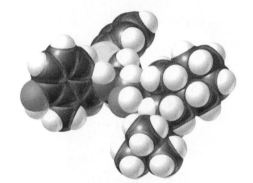

Nelfinavir (Viracept)

▲ Figure 14.13
The protease inhibitor nelfinavir.

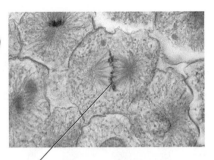
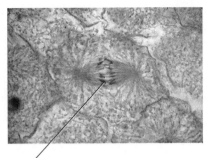
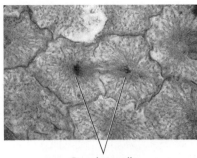

Parent cell with single set of chromosomes

Duplicated chromosomes

Daughter cells

▲ **Figure 14.14**
During cellular mitosis, DNA and certain cellular proteins bundle together into chromosomes, which are visible under a microscope. These chromosomes duplicate themselves and then divide evenly into two separate cells called "daughter cells."

cells would leave only about 1000 cells, which can be controlled by the immune system. Survival rates from cancer are therefore greatly increased by early diagnosis. Hence, you are advised to keep close watch on your body for unusual signs and schedule regular checkups with your physician.

Unfortunately, cancerous cells are not the only cells in the body that divide. Normal cells divide periodically, and some types of cells, such as those in the gastrointestinal tract and in hair follicles, are always in a state of cellular division. As a consequence, cancer chemotherapeutics are noted for their toxicity; patients undergoing treatments often experience gastrointestinal problems and hair loss.

DNA is the target of many anticancer compounds, because during cellular mitosis, strands of DNA are unwound and therefore susceptible to chemical attack. A variety of chemicals may be used to selectively kill cells that are in the process of dividing. The compound 5-fluorouracil, for example, shown in **Figure 14.15**, is mistaken by a cell for the nucleotide base uracil. Once incorporated into the DNA of the cancerous cell, 5-fluorouracil's nonnucleotide structure interferes with the normal DNA workings, and the cell dies. Harsher agents, such as cyclophosphamide and cisplatin, also shown in Figure 14.15, destroy DNA's ability to function by chemically bonding to the DNA or by cross-linking the two strands of the double helix.

Some anticancer drugs kill cancerous cells without acting on DNA. Certain alkaloids, such as vincristine, shown in **Figure 14.16**, and Taxol®, of Figure 14.4, kill dividing cells by preventing the formation of cellular microstructures they need for division.

◀ **Figure 14.15**
These anticancer agents kill dividing cells by targeting the cells' DNA.

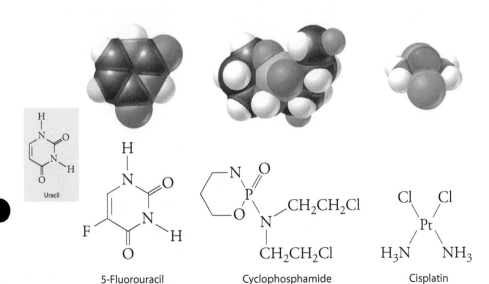

Uracil

5-Fluorouracil

Cyclophosphamide

Cisplatin

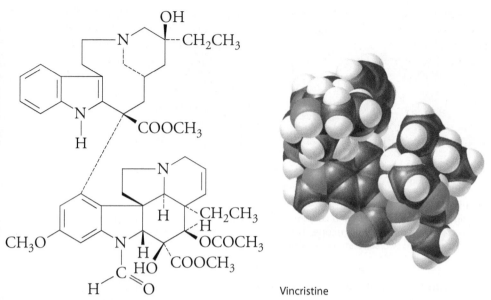

Vincristine

▲ **Figure 14.16**
Vincristine is a naturally occurring alkaloid that shows significant anticancer activity. It is isolated from a plant closely related to the periwinkle, a common ornamental plant of tropical and temperate regions.

As another point of attack, cancerous cells have high metabolic rates, which means they rely heavily on biochemical nutrients, such as the dihydrofolic acid shown in **Figure 14.17**. The anticancer agent methotrexate is structurally very similar to dihydrofolic acid and works by binding to dihydrofolic acid receptor sites in the cancerous cells, thereby interfering with metabolic reactions in the cells.

The high metabolism of cancer cells means they also need to be fed an adequate supply of blood—a source of both nutrients and oxygen. The growth of a tumor, therefore, requires that new blood vessels be allowed to grow alongside new tumor cells. The growth of new blood vessels is called *angiogenesis*. Drugs blocking the growth of new blood vessels are called *angiogenesis inhibitors*. Some powerful angiogenesis inhibitors have been developed that selectively bind to a protein that cancer cells need to promote

▶ **Figure 14.17**
Methotrexate disturbs the metabolism of cancerous cells by substituting for dihydrofolic acid at dihydrofolic acid receptor sites.

Dihydrofolic acid

Methotrexate

angiogenesis. Upon binding, the inhibitor deactivates the protein. This shuts down the growth of new blood vessels into the tumor, which begins to shrink as cancer cells die from a lack of blood. Importantly, angiogenesis inhibitors are also being developed to inhibit angiogenesis in fat cells to help treat people with obesity.

The angiogenesis inhibitors just described are examples of monoclonal antibodies, which have much potential in treating many different kinds of cancer. An antibody is a protein macromolecule the body produces to identify and then destroy an infectious agent, such as a bacterium or virus and even a cancer cell (**Figure 14.18**). Antibodies specific for certain cancer cells, called *monoclonal antibodies*, can be created in the laboratory. The term *monoclonal* means these antibodies are identical clones of each other, which means they can be produced in quantities sufficient for clinical use. The monoclonal antibody rituximab, for example, is effective at killing cancerous cells of certain lymphomas. Although not a cure, the rituximab helps to maintain a state of remission.

Cancer chemotherapy combined with radiation therapy and/or surgery can be effective in minimizing the growth of or even curing many forms of cancer. As our understanding of cellular mechanics continues to grow, so will our ability to increase the overall survival rates of cancer patients. A case in point lies with chronic myeloid leukemia, CML, a form of cancer in which certain blood cells proliferate because of an abnormal protein produced by a DNA mutation. From a mapping of the receptor sites on this protein, scientists were able to design a molecule, called Gleevec®, that binds to these receptor sites, as shown in **Figure 14.19**, and inhibits their functioning. Although not a cure for this once very deadly cancer, Gleevec® stops the progression of this disease and related diseases, with remarkable 5-year survival rates of around 95 percent. There are many forms of cancer, each with its own unique biochemistry. As Gleevec® demonstrates, our best hope lies not with a single magic bullet, but in our ability to learn the details of each cancer's biochemistry. From what we learn, we are then empowered to develop targeted treatments.

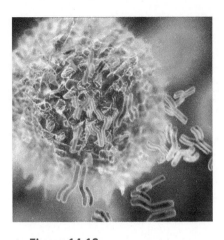

▲ **Figure 14.18**
Antibodies are Y-shaped macromolecules that are a frontline defense in a person's immune system. Once an antibody binds to a pathogen, a cascade of events leads to the destruction of the pathogen.
F. Hoffmann-La Roche Ltd.

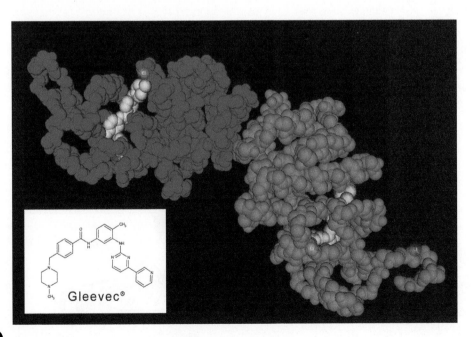

▲ **Figure 14.19**
The amino acid chains of a renegade protein that causes a cancerous proliferation of blood cells are shown in blue and pink. This protein is formed from a certain mutation of DNA. The Gleevec® molecule, shown in gold, binds to receptor sites of this protein, rendering it defunct.

14.4 The Nervous System Is a Network of Neurons

LEARNING OBJECTIVE

Summarize how a nerve impulse travels along a neuron and across the synapse to an adjacent neuron using neurotransmitters.

EXPLAIN THIS

Why is energy necessary for a neuron to remain at rest?

Many drugs function by affecting the nervous system. To understand how these drugs work, it is important to know the basic structure and functions of the nervous system.

Thoughts, physical actions, and sensory input all involve the transmission of electric signals through the body. The path for these signals is a network of nerve cells, or *neurons*. **Neurons** are specialized cells capable of sending and receiving electric impulses. First, in what is called the *resting phase*, a nerve cell primes itself for an impulse by ejecting sodium ions, as shown in **Figure 14.20a**. A higher concentration of sodium ions outside the neuron than inside creates a charge imbalance. The charge imbalance gives rise to an electric potential of around −70 millivolts across the cell membrane. As shown in **Figure 14.20b**, a nerve impulse is a reversal in this electric potential that travels down the length of the neuron to the *synaptic terminals*. The reversal of the electric potential within an impulse occurs as sodium ions flush back into the neuron.

After the impulse passes a given point along the neuron, the cell again ejects sodium ions at that point to re-establish the original distribution of ions and the −70-millivolt potential.

Unlike the wires in an electric circuit, most neurons are not physically connected to one another. Nor are they connected to the muscles or glands on which they act. Rather, as **Figure 14.21** shows, they are separated from one another or from a muscle or gland by a narrow gap known as the **synaptic cleft**.

READINGCHECK

What is a nerve impulse?

▼ **Figure 14.20**
(a) The resting phase of a neuron maintains a greater concentration of sodium ions outside the cell. This results in a voltage of about −70 millivolts. (b) In the impulse phase, sodium ions flush back into the cell through channels to give a voltage of about +30 millivolts.

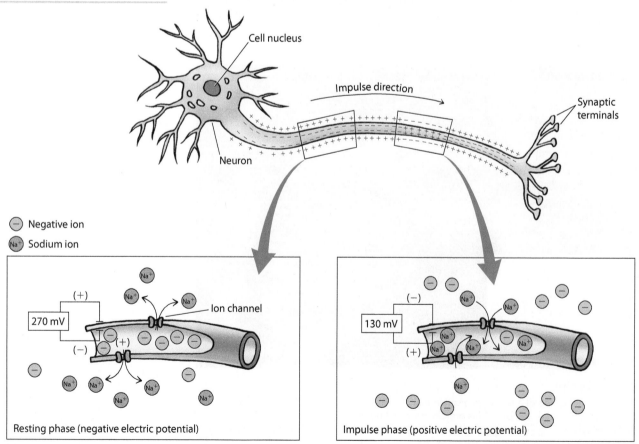

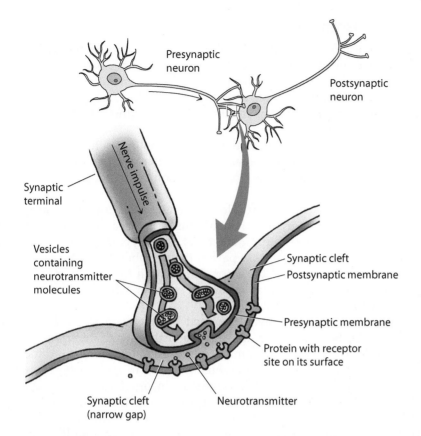

◀ **Figure 14.21**
The passage of neurotransmitters across a synaptic cleft.

Presynaptic neuron

Postsynaptic neuron

Nerve impulse

Synaptic terminal

Vesicles containing neurotransmitter molecules

Synaptic cleft
Postsynaptic membrane

Presynaptic membrane

Protein with receptor site on its surface

Synaptic cleft (narrow gap)

Neurotransmitter

A nerve impulse reaching a synaptic cleft causes bubble-like compartments in the terminal, called *vesicles,* to release neurotransmitters into the cleft. **Neurotransmitters** are organic compounds released by a neuron that are capable of activating receptor sites within an adjacent neuron.

A neurotransmitter, once released into the synaptic cleft, migrates across the cleft to receptor sites on the opposite side. If the receptor sites are located on a *postsynaptic neuron,* as shown in Figure 14.21, the binding of the neurotransmitter may start a nerve impulse in that neuron. If the receptor sites are located on a muscle or an organ, then binding of the neurotransmitter may start a bodily response, such as muscle contraction or the release of hormones.

Two important classes of neurons are the *stress* and *maintenance* neurons. Both types are always firing. But in times of stress, as when facing an angry bear or giving a speech, the stress neurons are more active than the maintenance neurons. This condition is the *fight-or-flight* response, during which fear causes stress neurons to trigger rapid bodily changes to help defend against impending danger: the mind becomes alert, air passages in the nose and lungs open to bring in more oxygen, the heart beats faster to spread the oxygenated blood throughout the body, and nonessential activities such as digestion are temporarily stopped. In times of relaxation, such as sitting down in front of the television with a bowl of potato chips, the maintenance neurons are more active than the stress neurons. Under these conditions, digestive juices are secreted, intestinal muscles push food through the gut, the pupils constrict to sharpen vision, and the heart pulses at a minimal rate.

CHEMICAL CONNECTIONS

How is pumping iron (weight lifting) connected to pumping sodium ions?

My stress neurons are firing more than my maintenance neurons.

My maintenance neurons are firing more than my stress neurons.

CONCEPTCHECK

What is a neurotransmitter?

CHECK YOUR ANSWER A neurotransmitter is a small organic molecule released by a neuron. It influences neighboring tissues, such as nerve membranes, by binding to receptor sites.

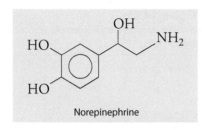

Norepinephrine

Acetylcholine

▲ **Figure 14.22**
The chemical structures of the stress neurotransmitter norepinephrine and the maintenance neurotransmitter acetylcholine.

 FOR YOUR INFORMATION

Recently developed drugs, known as ampakines, have been clinically shown to enhance learning and memory skills. These "smart pills" are being developed as a possible treatment for narcolepsy, attention deficit disorder, and Alzheimer's disease. Once approved by the FDA, physicians can prescribe them for off-label uses, such as jet lag and age-related forgetfulness. These agents act primarily within the central nervous system (brain and spinal cord), and they do not cause the jitteriness commonly associated with caffeine and amphetamines.

Neurotransmitters

On the chemical level, stress and maintenance neurons can be distinguished by the types of neurotransmitters they use. The primary neurotransmitter for stress neurons is *norepinephrine*. The primary neurotransmitter for maintenance neurons is *acetylcholine*. Both are shown in **Figure 14.22**. As we shall see in the following sections, many drugs function by altering the balance of stress and maintenance neuron activity.

In addition to norepinephrine and acetylcholine, a host of other neurotransmitters contribute to a broad range of effects. Three examples are the neurotransmitters dopamine, serotonin, and gamma-aminobutyric acid, shown in **Figure 14.23**.

Dopamine plays a significant role in activating the brain's reward center, which is located in the hypothalamus. This is an area at the lower middle of the brain, as illustrated in **Figure 14.24**. The hypothalamus is the main control center for the involuntary part of the peripheral nervous system and for emotional response and behavior. Stimulation of the reward center by dopamine results in a pleasurable sense of *euphoria,* which is an exaggerated sense of well-being.

Serotonin is the neurotransmitter used by the brain to block unneeded nerve impulses. So that we can make sense of the world, the frontal lobes of the brain selectively block out a multitude of signals coming from the lower brain and from the peripheral nervous system. We are not born with this ability to selectively block out information. In order to have an appropriate focus on the world, newborns must learn from experience which lights, sounds, smells, and feelings outside and inside their bodies must be damped. A healthy, mature brain is one in which serotonin successfully suppresses lower-brain nerve signals. Information that does make it to the higher brain can then be sorted efficiently.

Drugs such as LSD, which modify the action of serotonin, alter the brain's ability to sort information, and this alters perception. While hallucinating, for example, an LSD user rarely sees something that isn't there. Rather, the user has an altered perception of something that does exist.

The control of physical responses ultimately allows us to perform such complex tasks as driving a car or playing the piano. The control of emotional responses allows us to refine our behavior, such as overcoming anxiety in tense social interactions or remaining calm in an emergency. The brain controls both physical and emotional responses by inhibiting the transmission of nerve impulses. The neurotransmitter responsible for this inhibition—*gamma-aminobutyric acid* (GABA)—is *the* major inhibitory neurotransmitter of the brain. Without it, coordinated movements and emotional skills would not be possible.

Dopamine

Serotonin

Gamma-aminobutyric acid (GABA)

▲ **Figure 14.23**
The chemical structures of three neurotransmitters important to the central nervous system.

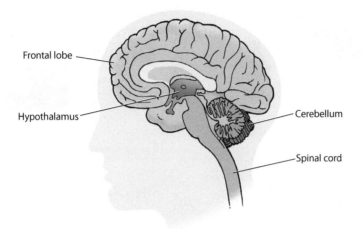

▲ Figure 14.24
The human brain.

CONCEPTCHECK

Match the neurotransmitter to its primary function:

_____ norepinephrine   a. inhibits nerve transmission

_____ acetylcholine   b. stimulates reward center

_____ dopamine   c. selectively blocks nerve impulses

_____ serotonin   d. maintains stressed state

_____ GABA   e. maintains relaxed state

CHECK YOUR ANSWERS   d, e, b, c, a.

14.5 Psychoactive Drugs Alter the Mind or Behavior

EXPLAIN THIS

Structurally speaking, how are THC and ethanol different from most other psychoactive drugs?

Any drug that affects the mind or behavior is classified as **psychoactive.** In this chapter, we focus on two classes of psychoactive drugs: stimulants and depressants.

Stimulants Activate the Stress Neurons

By enhancing the intensity of our reactions to stimuli, *stimulants* cause brief periods of heightened awareness, quick thinking, and elevated mood. They exert this effect by activating the stress neurons. Four widely recognized stimulants are amphetamines, cocaine, caffeine, and nicotine.

Amphetamines are a family of stimulants that include the parent compound *amphetamine* (also known as speed) and such derivatives as methamphetamine and pseudoephedrine. As you can see by comparing **Figure 14.25** with Figures 14.23 and 14.22, these drugs are structurally similar to the neurotransmitters norepinephrine and dopamine. Amphetamines

LEARNING OBJECTIVE

Describe how stimulants, hallucinogens, and depressants work and the problems that occur with the abuse of these chemicals.

▶ Figure 14.25
Amphetamines are a family of compounds structurally related to the neurotransmitters norepinephrine and dopamine.

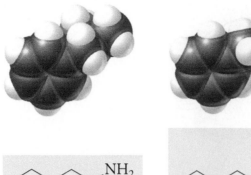

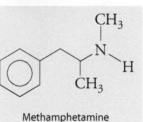

Amphetamine

Methamphetamine

Pseudoephedrine

FOR YOUR INFORMATION

In 1900, about 2 to 5 percent of the U.S. population was addicted to morphine and cocaine, which were the prime and usually secret ingredients of unregulated potions advertised to alleviate practically any illness. This drug addiction was reduced dramatically after the passage of the 1906 Pure Food and Drug Act. Three important aspects of this act were (1) the creation of the Food and Drug Administration, which was given authority to approve all foods and drugs meant for human consumption; (2) the requirement that certain drugs could be sold only by prescription; and (3) the requirement that habit-forming medicines be labeled as such.

bind to receptor sites for these neurotransmitters. So amphetamines mimic many of the effects of norepinephrine and dopamine on the stress neurons, including the fight-or-flight response and a sense of euphoria.

The stimulating and mood-altering effects of amphetamines give them a high abuse potential. Side effects of these drugs include insomnia, irritability, loss of appetite, and paranoia. Amphetamines take a particularly hard toll on the heart. Hyperactive heart muscles are prone to tearing. Subsequent scarring of tissue ultimately leads to a weaker heart. Furthermore, amphetamines cause blood vessels to constrict and blood pressure to rise. These conditions increase the likelihood of heart attack or stroke, especially for the one person out of four whose blood pressure is already high.

Cocaine, a natural product isolated from the South American coca plant, shown in **Figure 14.26**, is one of the more notorious and abused stimulants. Once in the bloodstream, cocaine produces a sense of euphoria and increased stamina. It is also a powerful local anesthetic when applied topically. Within a few decades of its first isolation from plant material in 1860, cocaine was used as a local anesthetic for eye surgery and dentistry. This practice was stopped once safer local anesthetics were discovered in the early 1900s.

▶ Figure 14.26
The South American coca plant has been used by indigenous cultures for many years in religious ceremonies and as an aid to staying awake on long hunting trips. Leaves are either chewed or ground to a powder that is inhaled nasally.

Cocaine

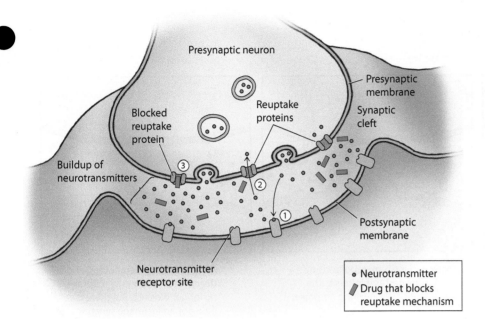

◀ Figure 14.27
(1) Neurotransmitters bind to their postsynaptic receptors. (2) Neurotransmitters are reabsorbed by the presynaptic neuron that released them through proteins embedded in the presynaptic membrane. (3) A drug that interferes with reuptake causes a buildup of neurotransmitters in the synaptic cleft.

Amphetamines and cocaine both have an effect on a process called neurotransmitter **reuptake,** which is the body's way of recycling neurotransmitters, molecules that are difficult to synthesize. As shown in **Figure 14.27**, special membrane-embedded proteins pull once-used neurotransmitter molecules back into the presynaptic neuron so that they can be reused. Amphetamines open the reuptake channels so that neurotransmitters, such as norepinephrine and dopamine, pass freely in and out of the synapse. This increases the concentration of these neurotransmitters in the synapse, which provides a feeling of being high. Cocaine acts a bit differently in that it binds specifically to the dopamine reuptake protein, thereby blocking the reuptake of this neurotransmitter. As a result, the concentration of dopamine within the synapse increases, which leads to a feeling of euphoria.

Amphetamines and cocaine share a similar profile of addictiveness. As with most drugs, the degree of addictiveness is greatly dependent upon how the drug is taken. In general, the quicker the drug is absorbed by the body, the greater the level of addictiveness. Furthermore, this relationship is not linear. A drug absorbed twice as fast, for example, may end up being not twice as addictive, but ten times as addictive. This is explained by the fact that when the drug is absorbed quickly, the concentration of the drug is much greater, which leads to a greater intensity.

Injecting the drug into a vein is the fastest route of administration, followed by inhaling the vapors of the free-base form of the drug. Recall from Chapter 10 that free-base means the nitrogen is "free" of a hydrogen ion, which makes the molecule nonpolar and volatile when heated. The street drug crystal meth, also known as ice, is the free-base form of methamphetamine. Similarly, crack is the free-base form of cocaine. Crystal meth and crack are extremely dangerous drugs of abuse. The intense high is routinely followed by an intense low, which results as the supply of neurotransmitters becomes exhausted. Of course, the low prompts the user to seek more of the drug. The "milder" but still dangerous hydrochloride salts of these drugs are typically sold as a powder. The salts of these drugs are soluble in water, which is why they can be absorbed (in order of increasing rates) by swallowing, snorting, or inserting a suppository.

The mechanism of cocaine's action is illustrated in **Figure 14.28**. The euphoric state induced by cocaine (and amphetamines) is only temporary because enzymes in the cleft metabolize, and hence deactivate, the dopamine. Once the cocaine is metabolized by enzymes, dopamine reuptake is again permitted. By this time, however, there is very little dopamine in the cleft to be reabsorbed. Nor is there an adequate supply of dopamine in the presynaptic

- ○ Dopamine
- ◑ Deactivated dopamine
- 🄴 Enzyme
- ▯ Cocaine
- ◪ Deactivated cocaine

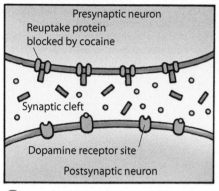

Presynaptic neuron
Reuptake protein blocked by cocaine

Synaptic cleft

Dopamine receptor site
Postsynaptic neuron

① High levels of dopamine remain active in the synaptic cleft as cocaine blocks dopamine reuptake sites. This causes cocaine's euphoric effect.

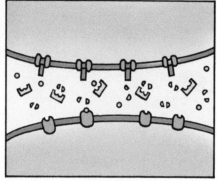

② Dopamine is metabolized and deactivated as it loiters in the synaptic cleft awaiting reuptake.

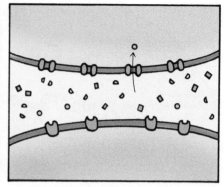

③ After cocaine is metabolized and deactivated, dopamine reuptake is no longer blocked. But there is very little dopamine in the synaptic cleft or in the neurons, and the cocaine user experiences extreme depression.

▲ Figure 14.28
Cocaine affects dopamine levels in the synaptic clefts of the brain's reward center.

CHEMICAL CONNECTIONS

How are any good feelings you've ever had connected to dopamine?

READINGCHECK

What are the two types of dependence involved with drug addiction?

neuron, which is unable to make sufficient quantities of dopamine without the recycling process. The net result is a depletion of dopamine that causes severe depression.

Long-term amphetamine or cocaine abuse leads to a deterioration of the nervous system. The body recognizes the excessive stimulatory actions produced by these drugs. To deal with the overstimulation, the body creates more depressant receptor sites for neurotransmitters that inhibit nerve transmission. A tolerance for the drugs therefore develops. Then to receive the same stimulatory effect, the abuser is forced to increase the dose. And this induces the body to create even more depressant receptor sites. The end result over the long term is that the abuser's natural levels of dopamine and norepinephrine are insufficient to make up for the excessive number of depressant sites. Lasting personality changes are thus often observed. Addicts, even when recovered, often report feelings of psychological depression.

Drug addiction is not completely understood, but scientists do know that it involves both physical and psychological dependence. **Physical dependence** is the need to continue taking the drug to avoid withdrawal symptoms. For amphetamines, drug withdrawal symptoms include depression, fatigue, and a strong desire to eat. **Psychological dependence** is the *craving* to continue drug use. This craving may be the most serious and deep-rooted aspect of addiction. It can persist even after withdrawal from physical dependence, frequently leading to renewed drug-seeking behavior.

To recover, it is imperative that the addict be removed from all cues that bring the experience of drug use to mind. This, however, is not usually possible as it may involve moving to a completely new social environment.

CONCEPTCHECK
What are two ways in which amphetamines and cocaine exert their effects?

CHECK YOUR ANSWER Amphetamines and cocaine in the synaptic cleft both mimic the action of neurotransmitters. They also disturb the reuptake of neurotransmitters, which results in an increase in the concentration of neurotransmitters in the cleft.

Caffeine, depicted in **Figure 14.29**, is a much milder and legal stimulant. Caffeine facilitates the release of norepinephrine into synaptic clefts. Caffeine also exerts many other effects on the body, such as the dilation of arteries, relaxation of bronchial and gastrointestinal muscles, and stimulation of stomach acid secretion.

The caffeine people ingest comes from various natural sources, including coffee beans, teas, kola nuts, and cocoa beans. Kola nut extracts are used for making cola drinks, and cocoa beans (not to be confused with the cocaine-producing coca plant) are roasted and then ground into a paste used for making chocolate. Caffeine is relatively easy to remove from these natural products using high-pressure carbon dioxide, which selectively dissolves the caffeine. This allows the economical production of "decaffeinated" beverages, many of which, however, still contain small amounts of caffeine. Interestingly, cola drink manufacturers use decaffeinated kola nut extract in their beverages. The caffeine is added in a separate step to guarantee a particular caffeine concentration. In the United States, about 2 million pounds of caffeine is added to soft drinks each year. Table 14.2 shows the caffeine content of various commercial products. For comparison, the maximum daily dose of caffeine tolerable by most adults is about 1500 milligrams.

Nicotine is another legal, but far more toxic, stimulant. As noted earlier, tobacco plants produce nicotine as a chemical defense against insects. This compound is so potent that a single dose of only about 60 milligrams is lethal. A single cigarette may contain up to 5 milligrams of nicotine. Most of this is destroyed by the heat of the burning embers, so that less than 1 milligram is typically inhaled by the smoker.

Nicotine and the neurotransmitter acetylcholine, which acts on maintenance neurons, have similar structures, as **Figure 14.30** illustrates. Nicotine molecules are therefore able to bind to acetylcholine receptor sites and trigger many of acetylcholine's effects, including relaxation and increased digestion. This explains the tendency of smokers to smoke after eating meals. In addition, acetylcholine is used for muscle contraction, so the smoker may also experience some muscle stimulation immediately after smoking. After these initial responses, however, nicotine molecules remain bound to the acetylcholine receptor sites. This blocks acetylcholine molecules from binding. The result is that the activity of these neurons is depressed.

Recall that both maintenance neurons and stress neurons are always working. Thus, inhibiting the activity of one type makes the other type more effective. So as nicotine depresses the maintenance neurons, it favors the stress neurons. This raises the smoker's blood pressure and stresses the heart.

In the brain, nicotine affects the stress system directly by enhancing the release of stress neurotransmitters, such as norepinephrine. Nicotine also increases the levels of dopamine in the reward center. Furthermore, when inhaled, nicotine is a fast-acting drug. All of these factors give nicotine a high level of addictiveness. Animal studies show inhaled nicotine to be about

Caffeine

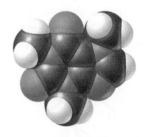

▲ Figure 14.29
A coffee plant with its ripening caffeine-containing beans.

TABLE 14.2 Approximate Caffeine Content of Various Products

PRODUCT	CAFFEINE CONTENT
Energy drink	100–1000 mg/cup
Brewed coffee	100–150 mg/cup
Instant coffee	50–100 mg/cup
Decaffeinated coffee	2–10 mg/cup
Black tea	50–150 mg/cup
Cola drink	35–55 mg/12 oz
Chocolate bar	1–2 mg/oz
Over-the-counter stimulant	100 mg/dose

▶ **Figure 14.30**
Nicotine is able to bind to receptor sites for acetylcholine because of structural similarities.

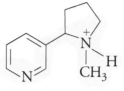

Nicotine

Acetylcholine

In addition to serving as a stimulant, nicotine also has some analgesic properties, which means that it enhances the ability to tolerate pain. A structurally related analog of nicotine, known as epibatidine (shown below), is even better at producing an analgesic effect and has been shown to be over 200 times more analgesic than morphine. Further studies show that epibatidine and nicotine bind to a different analgesia-producing receptor site from that occupied by morphine, which means that nicotine analogs represent a whole new class of analgesics. Epibatidine is isolated from the skin of a poisonous frog of Ecuador. Its toxicity is too great to allow clinical applications. Medicinal chemists, however, are working hard to discover other nicotine analogs that have optimal analgesic effects with minimal toxicity and, ideally, a low level of addictiveness.

Epibatidine

six times as addictive as injected heroin. Because nicotine leaves the body quickly, withdrawal symptoms begin about 1 hour after a cigarette is smoked, which means the smoker is inclined to light up frequently.

Figure 14.31 shows what a smoker's lungs look like. In the United States, about 450,000 individuals die each year from such tobacco-related health problems as emphysema, heart disease, and various forms of cancer, especially lung cancer, which is brought on primarily by tobacco's tar component. Some relief from the addiction can be obtained with nicotine chewing gum and nicotine skin patches. In order for any method to be effective, however, the smoker must first genuinely want to quit smoking.

CONCEPT CHECK

Caffeine and nicotine both add stress to the nervous system, but they do so by different means. Briefly describe the difference.

CHECK YOUR ANSWER Caffeine stimulates the release of the stress neurotransmitter norepinephrine; nicotine enhances the release of norepinephrine too, but it also depresses the action of the maintenance neurotransmitter acetylcholine.

Hallucinogens and Cannabinoids Alter Perceptions

A *hallucinogen*, also known as a *psychedelic*, is any drug that can alter visual perceptions and skew the user's sense of time. Hallucinogens have a pronounced effect on moods, thought patterns, and behavior. Lysergic acid diethylamide (LSD) and mescaline represent two main categories of hallucinogens. *Cannabinoids*, the psychoactive component of marijuana, are in a closely related category of drugs. Cannabinoids do not alter visual perceptions and therefore are not true hallucinogens. They are similar to hallucinogens in other regards, however, such as in their ability to alter a person's sense of time.

▶ **Figure 14.31**
The path of tobacco from the field to a smoker's lungs. About 46 million Americans smoke despite an awareness of the dangers of this habit.

Tobacco field

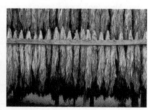

Tobacco curing on racks

Cigarettes being manufactured

User

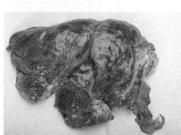

Blackened lungs

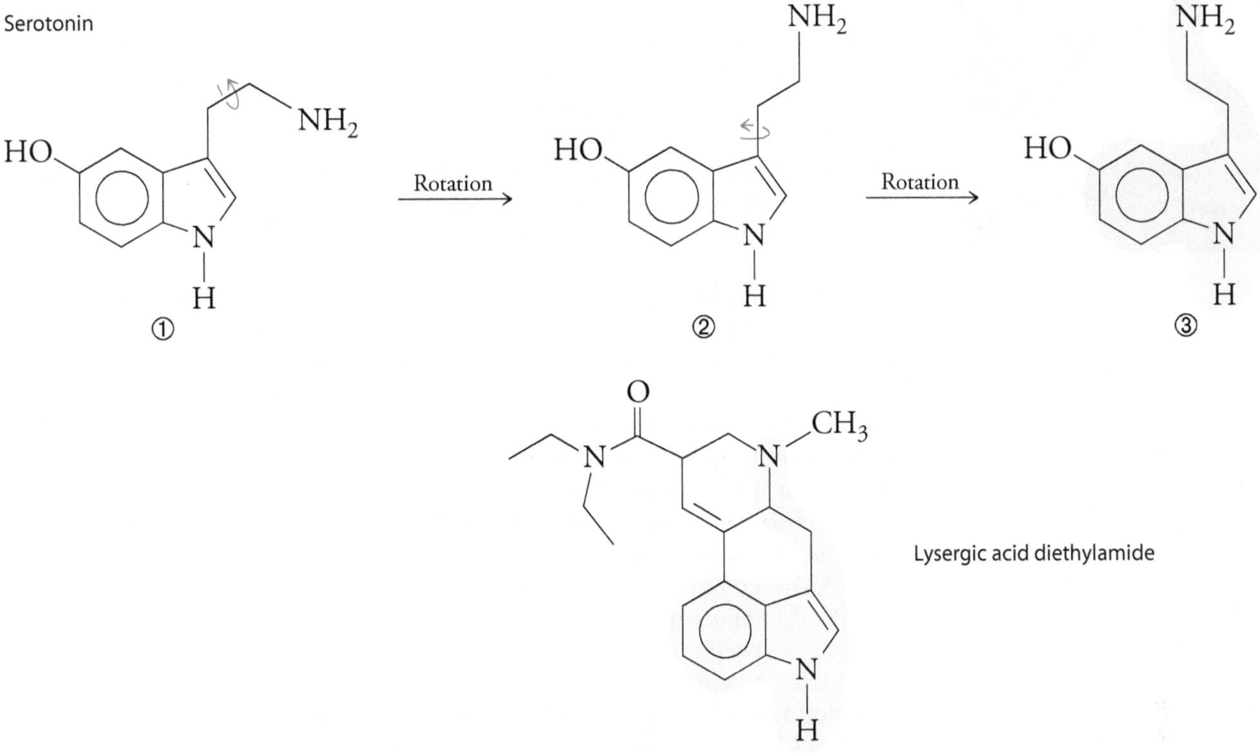

Serotonin

Lysergic acid diethylamide

LSD is the prototypic hallucinogen. Its molecular structure, shown in **Figure 14.32**, is very similar to that of serotonin, and this similarity permits LSD to activate serotonin receptors even more effectively than serotonin. It is LSD's ability to interfere with serotonin's work that causes LSD users to experience an altered sense of reality. Because LSD also stimulates the reward center, the change in sensory organization is usually, but not always, characterized as a favorable experience. LSD also triggers the stress neurons, resulting in enlarged pupils, elevated blood pressure and heart rate, nausea, and tremors. These stress effects can shift the mood of an affected person to panic and anxiety. Because the LSD molecule is nonpolar, quantities may be trapped and hidden away in nonpolar fatty tissue, only to be released months later and result in a mild recurrence of the experience known as a flashback.

In the early 1970s, there was a significant rise in the street use of hallucinogenic derivatives of the compound phenylethylamine, shown in **Figure 14.33**.

▲ **Figure 14.32**
The side chain of serotonin can rotate into a number of conformations. Upon binding to a receptor site, however, the side chain is likely to be held in conformation 3. Note how the LSD molecule can be superimposed on structure 3. LSD may therefore be thought of as a modified serotonin molecule in which the side chain is held in the ideal conformation for receptor binding.

◀ **Figure 14.33**
Hallucinogenic derivatives of phenylethylamine.

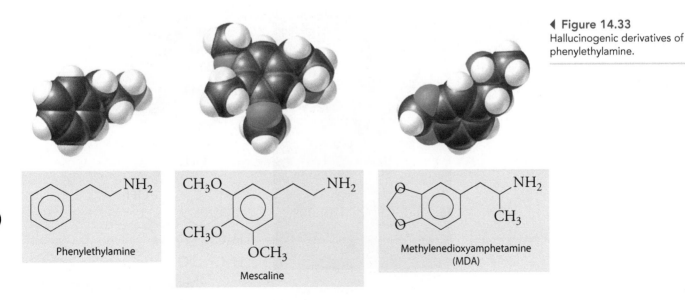

Phenylethylamine

Mescaline

Methylenedioxyamphetamine (MDA)

▲ Figure 14.34
The peyote cactus is a source of the hallucinogen mescaline.

Initially, interest was focused on *mescaline,* the hallucinogenic component of several species of cacti used in religious ceremonies by some Native American tribes in the western United States. One plant yielding this compound is pictured in **Figure 14.34**. Synthetic derivatives, such as methylenedioxyamphetamine (MDA), also became popular. Unlike LSD, these hallucinogens do not exert their effects by binding to serotonin receptor sites. Rather, they stimulate the release of excess quantities of serotonin. This pathway is not as effective as LSD's direct approach, and as a result, these drugs are 1/200 to 1/4000 as potent as LSD. Because larger doses of mescaline and MDA are required, a multitude of other effects are seen, such as a marked stimulation of the stress neurons. In addition, regular use of these compounds causes withdrawal symptoms.

Cannabinoids are the psychoactive components of marijuana, which has the species name *Cannabis sativa.* Concentrations of cannabinoids vary greatly from plant to plant. The original strains of this plant species contain very little of these psychoactive components and have been used for many centuries for their great fiber qualities. On average, strains of *Cannabis* that may be smoked for psychoactive effects contain about 4 percent cannabinoid derivatives. The most active of these derivatives is the compound Δ^9-tetrahydrocannabinol (THC), shown in **Figure 14.35**.

We do not completely understand how THC exerts its psychoactive effect. In 1990, a specific receptor site for the THC molecule was discovered. A few years later, a peptide occurring naturally in the body was found to bind to this receptor and initiate marijuana-like responses. These results suggest that THC functions by mimicking this naturally occurring peptide.

Most notably, cannabinoids accumulate in the area of the brain where short-term memories are sorted. Every experience we ever have goes through this center. Some things—the image of a sidewalk crack you saw on a morning walk, say—get thrown out. Other experiences, such as your first date, get filed away in long-term memory storage. Cannabinoids disrupt this filing system so that memories are not sorted appropriately. In addition, people under the influence of cannabinoids may have a distorted sense of time and unclear thoughts. Another effect of cannabinoids is unrestful sleep. The brain sorts through memories during a phase of sleep marked by rapid eye movement (REM). People who smoke marijuana lose REM sleep time, which results in irritability the following day. Once the memory filing center is cleared of the drug, which may take days or even weeks, the brain makes up for lost time by having extra-long periods of REM.

▶ Figure 14.35
The major psychoactive component of marijuana is Δ^9-tetrahydrocannabinol.

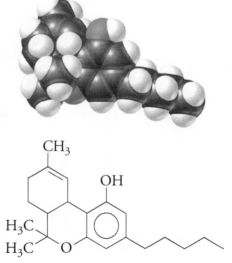

Δ^9-Tetrahydrocannabinol (THC)

Cannabinoids are also known to have a number of medicinal uses. For example, they reduce the symptoms of glaucoma, a disease in which there is a dangerous buildup of pressure within the eyes. They also help to stimulate a person's appetite, which is important for people undergoing cancer chemotherapy and for those suffering from diseases such as AIDS. For these purposes, many states allow marijuana to be prescribed by a physician.

CONCEPTCHECK

Why is marijuana not considered a true hallucinogen?

CHECK YOUR ANSWER The chemicals in marijuana do not significantly alter visual perceptions.

Depressants Inhibit the Ability of Neurons to Conduct Impulses

Depressants are a class of drugs that inhibit the ability of neurons to conduct impulses. Two commonly used depressants are ethanol and benzodiazepines.

Ethanol, also known as *ethyl alcohol* or simply *alcohol*, is by far the most widely used depressant. Its structure is shown in **Figure 14.36**. In the United States, about a third of the population, or about 100 million people, drink alcohol. It is well established that alcohol consumption leads to about 150,000 deaths each year in the United States. The causes of these deaths are overdoses of alcohol alone, overdoses of alcohol combined with other depressants, alcohol-induced violent crime, cirrhosis of the liver, and alcohol-related traffic accidents.

Benzodiazepines are a potent class of antianxiety agents. Compared to many other types of depressants, benzodiazepines are relatively safe and rarely produce cardiovascular and respiratory depression. Their antianxiety effects were identified by chance in 1957. During a routine laboratory cleanup, a synthesized compound that had been sitting on the shelf for two years was submitted for routine testing despite the fact that compounds thought to have similar structures had shown no promising pharmacologic activity. This particular compound, however, shown in **Figure 14.37** and now known as chlordiazepoxide, contained an unexpected seven-membered ring. Chlordiazepoxide showed a significant calming effect in humans and by 1960 was marketed as an antianxiety agent under the trade name Librium®. Shortly thereafter, a derivative, diazepam, was found to be five to ten times more potent than Librium. In 1963, diazepam hit the market under the trade name Valium®.

Alcohol and benzodiazepines exert their depressant effect primarily by enhancing the action of GABA. As shown in **Figure 14.38**, GABA keeps electric impulses from passing through a neuron by binding to a receptor site on a channel that penetrates the cell membrane of the neuron. Figure 14.38a shows that when GABA binds to the receptor site, the channel opens, allowing chloride ions

GABA wabba doo!

It's illegal and very dangerous to drive when your GABA receptor sites are highly activated.

$$CH_3CH_2{-}OH$$
Ethanol

◄ **Figure 14.36**
One of the initial effects of alcohol is a depression of social inhibitions, which can serve to bolster mood. Alcohol is not a stimulant, however. From the first sip to the last, body systems are being depressed.

▶ **Figure 14.37**
The benzodiazepines Librium®
and Valium®.

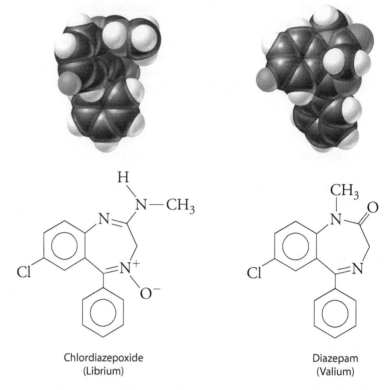

Chlordiazepoxide
(Librium)

Diazepam
(Valium)

to migrate into the neuron. The resulting negative charge buildup in the neuron maintains the negative electric potential across the cell membrane. This inhibits a reversal to a positive potential and prevents an impulse from traveling along the neuron. (For clarification, review Figure 14.20 and the text describing it.)

Ethanol mimics the effect of GABA by binding to GABA receptor sites. This allows chloride ions to enter the neuron, as shown in Figure 14.38b. The effect of alcohol is dose-dependent, which means the greater the amount consumed, the greater the effect. At small concentrations, few chloride ions are permitted into the neuron; these low concentrations of ions decrease inhibitions, alter judgment, and impair muscle control. As the person continues to drink and the chloride ion concentration inside the neuron rises, both reflexes and consciousness diminish, eventually to the point of coma and then death.

Recall from our discussion of cocaine and amphetamines that the body responds to the long-term abuse of these stimulants by creating more depressant receptor sites. Likewise, the body recognizes the excessive inhibitory actions produced by alcohol and tries to recover by increasing the number of synaptic receptor sites that lead to nerve excitation. In this way, an individual can develop a tolerance for alcohol. To receive the same inhibitory effect, the drinker is forced to drink more, which induces the body to create even more excitable synaptic receptor sites. Eventually, an excess of these excitatory receptor sites leads to perpetual body tremors, which can be subdued either by more drinking or, with greater difficulty, by a long-term cessation of alcohol consumption.

▼ **Figure 14.38**
(a) When GABA binds to its receptor site, a channel opens to allow negatively charged chloride ions into the neuron. The high concentration of negative ions inside the neuron prevents the electric potential from reversing from negative to positive. Because that reversal is necessary if an impulse is to travel through a neuron, no impulse can move through the neuron. (b) Ethanol mimics GABA by binding to GABA receptor sites.

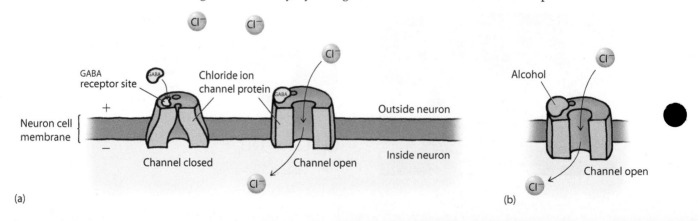

(a)

(b)

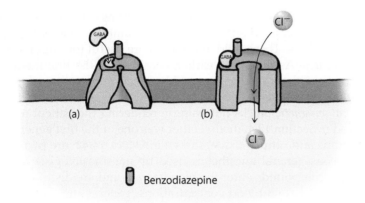

◀ Figure 14.39
The receptor sites for benzodiazepines are adjacent to GABA receptor sites. (a) Benzodiazepines cannot open the chloride channel on their own. (b) Rather, benzodiazepines help GABA in its channel-opening task.

Figure 14.39 illustrates how benzodiazepines exert their depressant effects by binding to receptor sites located adjacent to GABA receptor sites. Benzodiazepine binding merely helps GABA bind. Because benzodiazepine doesn't directly open chloride-ion channels, overdoses of this compound are less hazardous than those of many alternative depressants, making the benzodiazepines the drugs of choice for treating symptoms of anxiety.

CONCEPT CHECK

Does the activity of a neuron increase or decrease as chloride ions are allowed to pass into it?

CHECK YOUR ANSWER Chloride ions inside the neuron help to maintain the negative electric potential. This inhibits the neuron from being able to conduct an impulse (see Figure 14.38). Chloride ions, therefore, decrease the activity of a neuron.

14.6 Pain Relievers Inhibit the Transmission or Perception of Pain

EXPLAIN THIS

Is the placebo effect real or imagined?

Physical pain is a complex body response to injury. On the cellular level, pain-inducing biochemicals are rapidly synthesized at the site of injury, where they initiate swelling, inflammation, and other responses that get your body's attention. These pain signals are sent through the nervous system to the brain, where the pain is perceived. Drugs act at various stages of this process to alleviate pain, as shown in **Figure 14.40**.

LEARNING OBJECTIVE

Compare and contrast how anesthetics, analgesics, and endorphins act to alleviate pain.

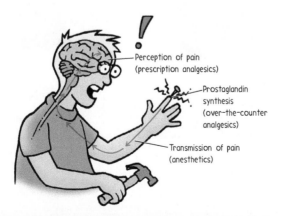

◀ Figure 14.40
Injury to tissue causes the transmission of pain signals to the brain. Pain relievers prevent this transmission, inhibit the inflammatory response, or damp the brain's ability to perceive the pain.

FOR YOUR INFORMATION

The 19th-century chemist Felix Hoffmann created aspirin, acetylsalicylic acid, by adding the acetyl functional group to salicylic acid (see Chapter 12). Two weeks later, he applied the same chemistry to morphine to create diacetylmorphine. The company he was working for, Bayer, named and marketed this product "heroin" for the "heroic" feeling it inspired in users. It was erroneously advertised as being a nonaddictive alternative to morphine. More importantly, however, heroin is a powerful cough suppressant. At the time, tuberculosis and pneumonia were the leading causes of death. Heroin became a much welcomed medicine, as it allowed people sick from these diseases to obtain a restorative night's sleep. By 1913, the negative aspects of heroin became widely known and Bayer stopped producing it, focusing instead on selling aspirin.

READINGCHECK

How does a general anesthetic knock out pain?

▶ **Figure 14.41**
Local anesthetics have similar structural features, including an aromatic ring, an intermediate chain, and an amine group. Ask your dentist which ones she uses for your treatment.

▶ **Figure 14.42**
The chemical structures of sevoflurane and nitrous oxide.

Anesthetics prevent neurons from transmitting sensations to the brain. *Local anesthetics* are applied either topically to numb the skin or by injection to numb deeper tissues. These mild anesthetics are useful for minor surgical and dental procedures. As described earlier, cocaine was the first medically used local anesthetic. Others having fewer side effects soon followed, such as the ones shown in **Figure 14.41**.

A *general anesthetic* blocks out pain by rendering the patient unconscious. As discussed in Section 12.4, diethyl ether was one of the first general anesthetics. Sevoflurane and nitrous oxide, shown in **Figure 14.42**, are two of the more popular gaseous general anesthetics used by anesthesiologists today. When inhaled, these compounds enter the bloodstream and are distributed throughout the body. At certain blood concentrations, general anesthetics render the individual unconscious, which is useful for invasive surgery. General anesthesia must be monitored very carefully, however, to avoid a major shutdown of the nervous system and consequent death.

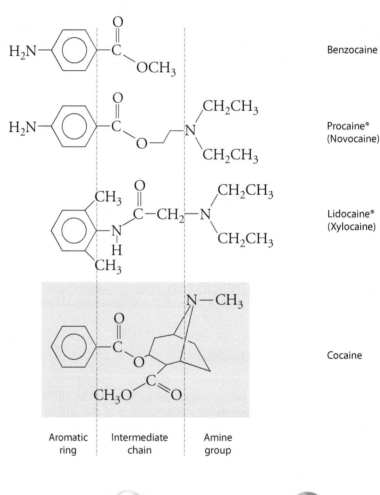

Benzocaine

Procaine® (Novocaine)

Lidocaine® (Xylocaine)

Cocaine

Aromatic ring | Intermediate chain | Amine group

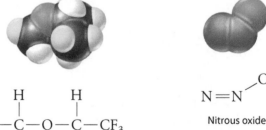

Sevoflurane

Nitrous oxide

Analgesics are a class of drugs that help us to tolerate pain without abolishing nerve sensations. Over-the-counter analgesics, such as aspirin, ibuprofen, and acetaminophen, inhibit the formation of *prostaglandins*. As **Figure 14.43** illustrates, prostaglandins are biochemicals the body quickly synthesizes to generate pain signals. These analgesics also reduce fever, because of the role prostaglandins play in raising body temperature. In addition to reducing pain and fever, aspirin and ibuprofen act as anti-inflammatory agents, because they block the formation of a certain type of prostaglandin responsible for inflammation. Acetaminophen does not act on inflammation. These three analgesics are shown in **Figure 14.44**.

The more potent opioid analgesics—morphine, codeine, and heroin (see Figure 14.1)—moderate the brain's perception of pain by binding to receptor sites on neurons in the central nervous system, which includes the brain and spinal column. Initial discovery of these receptor sites raised the question of why they exist. At the time, some hypothesized that opioids mimic the action of a naturally occurring brain chemical. *Endorphins,* a group of large biomolecules that have strong opioid activity, were subsequently isolated from brain tissue. It has been suggested that endorphins evolved as a means of suppressing awareness of pain that would otherwise be incapacitating in life-threatening situations. The "runner's high" experienced by many athletes after a vigorous workout is caused by endorphins.

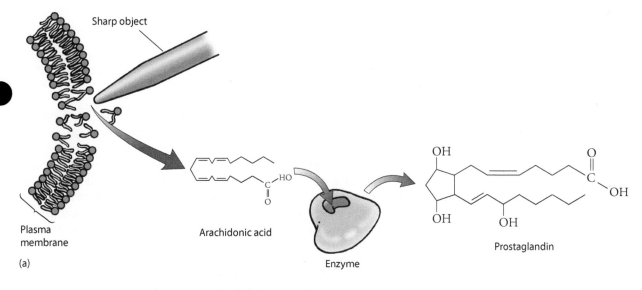

(a)

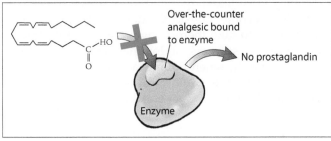

(b)

▲ Figure 14.43
(a) Prostaglandins, which cause pain signals to be sent to the brain, are synthesized in response to injury. The starting material for all prostaglandins is arachidonic acid, which is found in the membranes of all cells. Arachidonic acid is transformed to prostaglandins with the help of an enzyme. There are a variety of prostaglandins, each having its own effect, but all have a chemical structure resembling the one shown here. (b) Analgesics inhibit the synthesis of prostaglandins by binding to the arachidonic acid receptor site. With no prostaglandins, no pain signals are generated.

▶ **Figure 14.44**
Aspirin and ibuprofen block the formation of the prostaglandins responsible for pain, fever, and inflammation. Acetaminophen only blocks the formation of the prostaglandins responsible for pain and fever.

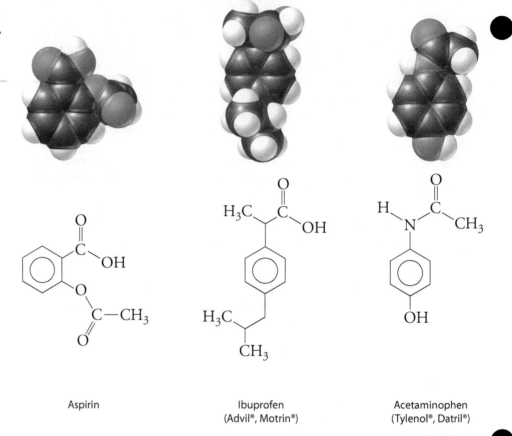

Aspirin

Ibuprofen
(Advil®, Motrin®)

Acetaminophen
(Tylenol®, Datril®)

 **FOR YOUR INFORMATION**

Oxycodone and hydrocodone are two derivatives of morphine available by prescription for pain management. According to the Centers for Disease Control and Prevention, abuse of these prescription painkillers is a fast-growing epidemic currently resulting in more than 15,000 deaths in the United States each year. That makes these agents the second-leading cause of injury death after motor vehicle crashes.

Endorphins are also implicated in the *placebo effect*, in which patients experience a reduction in pain after taking what they believe is a drug but is actually a sugar pill. (A *placebo* is any inactive substance used as a control in a scientific experiment.) Through the placebo effect, it is the patient's belief in the effectiveness of a medicine rather than the medicine itself that leads to pain relief. The involvement of endorphins in the placebo effect has been demonstrated by replacing the sugar pills with drugs that block opioids or endorphins from binding to their receptor sites. Under these circumstances, the placebo effect vanishes.

In addition to acting as analgesics, opioids can induce euphoria, which is why they are so frequently abused. With repeated use, individuals develop a tolerance to these drugs: they must take larger and larger doses to achieve the same effect. Abusers also become physically dependent on opioids, which means they must continue to take them to avoid severe withdrawal symptoms, such as chills, sweating, stiffness, abdominal cramps, vomiting, weight loss, and anxiety. Interestingly, when opioids are used primarily for pain relief rather than for pleasure, the withdrawal symptoms are much less dramatic—especially when the patient does not know he has been on these drugs.

CONCEPTCHECK

Distinguish between an anesthetic and an analgesic.

CHECK YOUR ANSWER An anesthetic blocks pain signals from reaching the brain. An analgesic facilitates the ability to manage pain signals once they are received by the brain.

◀ **Figure 14.45**
The structure of methadone (black) superimposed on that of morphine (blue and black).

Methadone

Methadone/Morphine

The most widely used approach to treating opioid addiction is methadone maintenance. *Methadone*, shown in **Figure 14.45**, is a synthetic opioid derivative that has most of the effects of other opioids, including euphoria, but differs in that it retains much of its activity when taken orally. This means that doses are very easy to control and monitor. The withdrawal symptoms of methadone are also far less severe, and the addict may be slowly weaned off the opioid without excessive stress. An addict may be freed of physical dependence in a matter of months. The psychological dependence, however, usually persists throughout the individual's life, which is why the relapse rate is so high.

14.7 Medicines for the Heart

EXPLAIN THIS

How do cows get their cholesterol?

Heart disease is any condition that diminishes the heart's ability to pump blood. A common heart disease is *arteriosclerosis*, a buildup of plaque on the inside walls of arteries. As discussed in Section 13.8, plaque deposits are primarily an accumulation of low-density lipoproteins, which are high in cholesterol and saturated fats. Plaque-filled arteries are less elastic and have a decreased volume, as shown in **Figure 14.46**. Both of these effects make pumping blood more difficult, and the heart becomes overworked and weakens. Accumulated damage to heart muscle from arteriosclerosis or other stresses can result in abnormal heart rhythms, known as *arrhythmias*. Chest pains, known as *angina*, result from an insufficient oxygen supply to heart muscles. Ultimately, the weakened heart does not adequately circulate blood to the body. People with heart disease have decreased stamina and frequently need to pause to catch their breath during or after exercise.

As discussed in Section 13.8, another danger of arteriosclerosis is the potential for a blood clot around the site of plaque formation. Such a clot can break off and be carried through the bloodstream until it clogs a blood vessel, effectively cutting off the blood supply to tissue, which then begins to die.

LEARNING OBJECTIVE

> Describe how statins, vasodilators, beta blockers, and calcium channel blockers help to protect the heart.

▼ **Figure 14.46**
The heart has to work harder to push blood through vessels narrowed by plaque deposits. Furthermore, inflammation around the plaque can lead to the formation of heart-attack-causing blood clots.

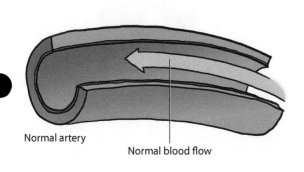

Normal artery

Normal blood flow

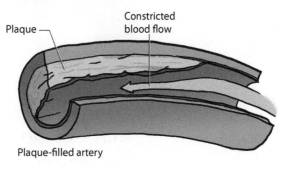

Plaque

Constricted blood flow

Plaque-filled artery

A heart attack occurs when the dying tissue is heart muscle. Some heart attacks progress slowly, allowing the victim time to seek medical assistance, which may involve the administration of a quick-acting clot-dissolving enzyme. Other heart attacks are more rapid, killing the victim within minutes. Surviving a heart attack means living with a heart weakened by dead tissue.

The most effective way to treat heart disease is to do what can be done to prevent it from happening in the first place. This includes avoiding stress; exercising regularly; and eating a well-balanced, low-cholesterol diet. For many people, however, such a healthy lifestyle does little to lower their blood cholesterol levels, because their livers naturally produce more cholesterol than they obtain from their diet. For such people, physicians can prescribe what are known as *statins*, which are drugs that inhibit the synthesis of cholesterol. Two popular statins are atorvastatin (Lipitor®), which is a synthetic drug, and lovastatin (Mevacor®), which is a natural product isolated from the fungus *Aspergillus terreus*.

Vasodilators are a class of drugs that increase the blood supply to the heart by expanding blood vessels. They are useful for treating angina. They also reduce the workload of the heart, because opening up the blood vessels makes pumping the blood easier. Traditional vasodilators include nitroglycerin and amyl nitrite, both shown in **Figure 14.47**. They can be administered by a number of routes: orally or sublingually (under the tongue) or as a transdermal (through the skin) patch. A benefit of the latter two approaches is that they allow the drug to enter the body slowly, in contrast to the effect of taking the drug orally or by injection. These organic nitrates are metabolized to nitric oxide, NO, which has been shown to relax muscles in blood vessels.

Drugs that relax the pumping action of the heart have also been developed. When bound to receptor sites called beta-adrenoceptors in heart muscle, the neurotransmitters norepinephrine and epinephrine stimulate the heart to beat faster. A series of drugs called *beta blockers* slow down and relax an overworked heart by blocking norepinephrine and epinephrine from binding to the beta-adrenoceptors. Propanolol (Inderal®), shown in **Figure 14.48**, was the first beta blocker developed and is useful for treating angina, arrhythmias, and high blood pressure.

Another group of drugs that relax heart muscle are the *calcium channel blockers*. One example is nifedipine, shown in Figure 14.48. Muscle contraction is initiated as a nerve impulse signals calcium ions to enter muscle cells. As their name implies, calcium channel blockers inhibit the flow of calcium ions into muscles, thereby inhibiting muscle contraction. This slows the heart rate, relaxing and dilating the muscles of blood vessels, which lowers blood pressure.

In the United States and in most other developed nations, heart disease is the number-one cause of death for individuals over the age of 65. Because most people in these nations live past this age, heart disease is actually the leading cause of death for all age groups combined, as noted in Table 14.3.

✔ **READINGCHECK**

How do vasodilators increase blood supply?

▶ **Figure 14.47**
The vasodilators nitroglycerin and amyl nitrite.

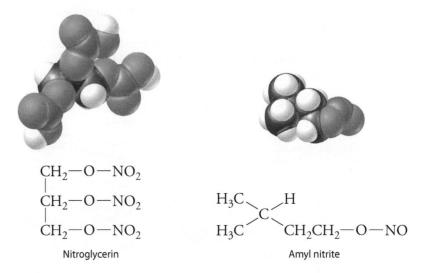

Nitroglycerin

Amyl nitrite

◀ **Figure 14.48**
Propranolol is a beta blocker, and nifedipine is a calcium channel blocker.

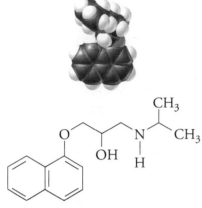

Propranolol (Inderal)

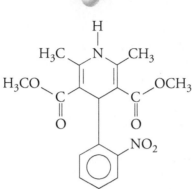

Nifedipine

CONCEPTCHECK

Why do long-time alcoholics require greater doses of a beta blocker in order to relax cardiac muscle?

CHECK YOUR ANSWER As discussed in Section 14.5, long-time excessive drinking leads to an increase in the number of receptor sites for stress neurotransmitters. With more of these receptor sites to block, alcoholics require a correspondingly greater dose of the beta blocker to achieve the desired degree of cardiac relaxation.

Perhaps nowhere is the impact of chemistry on society more evident than in the development of drugs. On the whole, they have increased our life span and improved our quality of living. They have also presented us with a number of ethical and social questions. How do we care for an increasing elderly population? What drugs, if any, should be legally permissible for recreational use? How do we deal with addictive drug use—as a crime, as a disease, or both? As we continue to learn more about ourselves and our ills, we can be sure that more powerful drugs will become available. All drugs, however, carry certain risks that we should recognize. As most physicians would point out, drugs offer many benefits, but they are no substitute for a healthy lifestyle and preventive approaches to medicine.

Remember, seven days without exercise makes one weak!

TABLE 14.3 Causes of Death in the United States

AGE GROUP (YEARS)	CAUSE	TOP TEN CAUSES, ALL AGES COMBINED
15–24	Accident	1. Heart disease
25–44	HIV infection	2. Cancer
45–64	Cancer	3. Stroke
>65	Heart disease	4. Lung disease
		5. Accidents
		6. Diabetes
		7. Alzheimer's disease
		8. Influenza/pneumonia
		9. Kidney disease
		10. Sepsis

Chapter 14 Review

LEARNING OBJECTIVES

Classify drugs by their origin and describe the synergistic effect. (14.1)	→	*Questions 1–3, 26, 27, 31–34, 71, 75*
Describe the lock-and-key model and explain how this model is used in the development of new medicines. (14.2)	→	*Questions 4–5, 35–39*
Describe how chemotherapy protects us from bacterial and viral infections as well as cancer. (14.3)	→	*Questions 6–9, 29, 40–45*
Summarize how a nerve impulse travels along a neuron and across the synapse to an adjacent neuron using neurotransmitters. (14.4)	→	*Questions 10–13, 25, 46–52*
Describe how stimulants, hallucinogens, and depressants work and the problems that occur with the abuse of these chemicals. (14.5)	→	*Questions 14–17, 30, 53–62, 72–74*
Compare and contrast how anesthetics, analgesics, and endorphins act to alleviate pain. (14.6)	→	*Questions 18–21, 28, 63–67*
Describe how statins, vasodilators, beta blockers, and calcium channel blockers help to protect the heart. (14.7)	→	*Questions 22–24, 68–70*

SUMMARY OF TERMS (KNOWLEDGE)

Agonist A molecule that binds to a receptor site and initiates a biological effect.

Analgesic A drug that allows us to tolerate pain without abolishing nerve sensations.

Anesthetic A drug that prevents neurons from transmitting sensations to the brain.

Antagonist A molecule that binds to a receptor site and does not initiate a biological effect except that it blocks other active molecules from binding to the receptor site.

Chemotherapy The use of drugs to destroy pathogens without destroying the animal host.

Combinatorial chemistry A laboratory approach intended to mimic nature's chemical diversity by taking advantage of the many different ways in which a series of reacting chemicals can be combined.

Lock-and-key model A model based upon the idea that there is a connection between a drug's chemical structure and its biological effect.

Neuron A specialized cell capable of sending and receiving electric impulses.

Neurotransmitter An organic compound released by a neuron and capable of activating receptor sites within an adjacent neuron.

Physical dependence A dependence characterized by the need to continue taking a drug to avoid withdrawal symptoms.

Psychoactive drug A drug that affects the mind or behavior.

Psychological dependence A deep-rooted craving for a drug.

Reuptake A mechanism whereby a presynaptic neuron absorbs neurotransmitters from the synaptic cleft for reuse.

Synaptic cleft A narrow gap across which neurotransmitters pass either from one neuron to the next or from a neuron to a muscle or gland.

Synergistic effect One drug enhancing the effect of another.

READING CHECK QUESTIONS (COMPREHENSION)

14.1 Medicines Are Drugs That Benefit the Body

1. What are the three origins of drugs?
2. Are a drug's side effects necessarily bad?
3. What is the synergistic effect?

14.2 The Lock-and-Key Model Guides the Synthesis of New Medicines

4. In the lock-and-key model, is a drug viewed as the lock or the key?

5. What holds a drug to its receptor site?

14.3 Chemotherapy Cures the Host by Killing the Disease

6. Why do bacteria need PABA but humans can do without it?
7. How does penicillin G cure bacterial infections?
8. When is chemotherapy most effective against cancer?
9. How does methotrexate work?

14.4 The Nervous System Is a Network of Neurons

10. What are the symptoms that a person's stress neurons have been activated?

11. What are some of the processes occurring in the body when maintenance neurons are more active than stress neurons?

12. What neurotransmitter is known for its strong activity in the brain's reward center?

13. What is the role of GABA in the nervous system?

14.5 Psychoactive Drugs Alter the Mind or Behavior

14. What neurotransmitter does nicotine mimic?

15. What drugs enhance the action of GABA?

16. What is neurotransmitter reuptake?

17. Which neurotransmitter does LSD mimic?

14.6 Pain Relievers Inhibit the Transmission or Perception of Pain

18. What is an anesthetic?

19. What is an analgesic?

20. Where are the major opioid receptor sites located?

21. What biochemical is thought to be responsible for the placebo effect?

14.7 Medicines for the Heart

22. What is angina? What is its cause?

23. What role does nitrogen oxide play in the treatment of angina?

24. How does a vasodilator reduce the workload on the heart?

CONFIRM THE CHEMISTRY (HANDS-ON APPLICATION)

25. Nerve impulses travel at speeds of up to 100 meters per second, but neurotransmitters travel across the synaptic cleft at a much slower 0.00001 meter per second. Why so slow? Once the neurotransmitters are released into the cleft, no "magnet" forces them quickly to the opposite side. Instead, they are prodded to the other side merely by the random bumping of jiggling molecules in the cleft—a process known as *diffusion*.

Recall from Section 2.6 that molecules slow down with decreasing temperature. The effect of temperature on diffusion can be readily seen by adding food coloring to water. Fill one glass with the ice-cold water, one with the warm water, and one with the hot water. Allow the glasses of water to stand for a couple of minutes so that the water is perfectly still. Add a drop of food coloring to each glass. The drop will sink to the bottom and then begin to diffuse. Observe how long it takes the water to become uniformly colored. Use your observations to suggest why cold-blooded animals become sluggish at colder temperatures.

26. Medicines are organic chemicals that tend to deteriorate over time, especially when subjected to warm and moist conditions, such as that found in a bathroom medicine cabinet. This is why medicines come with expiration dates. After such a date, although the medicine might look fine, much of it has chemically transformed into a variety of decomposition products. Besides the medicine being less effective, some of the decomposition products might be harmful. What might be an ideal place to store your medicines? Expired medicines are no longer effective, and they are often found by children. For those reasons, consider throwing away your expired medicines.

THINK AND COMPARE (ANALYSIS)

27. Rank the following from least ideal to most ideal places for you to throw away your expired medicines:
   a. the toilet
   b. the trash can
   c. your local pharmacy

28. Rank in order of increasing addictive qualities:
   a. benzodiazepines
   b. nicotine
   c. heroin

29. Rank in order of increasing size:
   a. animal cell   **b.** bacterium   **c.** virus

30. Rank in order of its ability to fit within the serotonin receptor site:
   a. serotonin   **b.** LSD   **c.** THC

THINK AND EXPLAIN (SYNTHESIS)

14.1 Medicines Are Drugs That Benefit the Body

31. Why are organic chemicals so suitable for making drugs?

32. Aspirin can cure a headache, but when you pop an aspirin tablet, how does the aspirin know to go to your aching head rather than your throbbing toe?

33. When is a drug overdose most likely to happen?

34. When might two drugs taken together not exert a synergistic effect?

14.2 The Lock-and-Key Model Guides the Synthesis of New Medicines

35. What advantages are there to synthesizing a naturally occurring medicine, such as Taxol®, in the laboratory rather than isolating it from nature?

36. How is the laboratory method called combinatorial chemistry similar to the search for drugs in nature?

37. Which is better for you: a drug that is a natural product or a drug that is synthetic?

38. Classify each of the following drugs as an agonist or antagonist:

a. sulfanilamide   **b.** THC
c. nelfinavir   **d.** ethanol
e. Gleevec®   **f.** ibuprofen

39. Classify each of the following drugs as an agonist or antagonist:

a. amphetamine   **b.** morphine
c. nicotine   **d.** atorvastatin
e. LSD   **f.** propanolol

14.3 Chemotherapy Cures the Host by Killing the Disease

40. How does chemotherapy work to fight a disease or infection?

41. How do some molds or fungi protect themselves from bacterial infections?

42. Why is cancer treated most successfully in its earliest stages?

43. Would formulating a sulfa drug with PABA be likely to increase or decrease its antibacterial properties?

44. Why do protease inhibitors work so well when used in conjunction with antiviral nucleosides?

45. Why do some antiviral agents exhibit anticancer activity?

14.4 The Nervous System Is a Network of Neurons

46. What flushes back into a neuron as a nerve impulse passes through the neuron?

47. What is the main difference between stress neurons and maintenance neurons?

48. Compared to direct connections between neurons, what is an advantage of synaptic clefts?

49. How is dopamine connected to the amino acid tyrosine?

50. What amino acid does the body likely use to create serotonin?

51. What functional group do you add to dopamine to make norepinephrine?

52. Phenylethylamine is a neurotransmitter implicated in the sensation of being in love for the first time. It looks like dopamine, but with no oxygen atoms. Draw the structure.

14.5 Psychoactive Drugs Alter the Mind or Behavior

53. How is psychological dependence distinguished from physical dependence?

54. What symptoms show that a person's stress neurons have been activated?

55. Why do so many stimulant drugs result in a depressed state in a person after an initial high?

56. How is a drug addict's addiction similar to the need for food? How is it different?

57. Nicotine solutions are available from lawn and garden stores as an insecticide. Why must gardeners handle this product with extreme care?

58. Seeds of the morning glory plant contain the natural product lysergic acid, and yet they are only marginally hallucinogenic. Why?

59. Suggest why withdrawal symptoms are observed after a person's repeated use of MDA but not after repeated use of LSD.

60. Why do heavy drinkers have a greater tolerance for alcohol?

61. An excess of which of the neurotransmitters discussed in this chapter would most likely hinder your ability to perform a complicated physical performance, such as a ballet dance?

62. One of the active components of marijuana, Δ^9-tetrahydrocannabinol, is available as a prescription drug under the trade name Marinol®, which is taken orally. What advantage and disadvantage does Marinol® hold for a person suffering from nausea?

14.6 Pain Relievers Inhibit the Transmission or Perception of Pain

63. A variety of gaseous compounds behave as general anesthetics even though their structures have very little in common. Does this support the role of a receptor site for their mode of action?

64. How might the structure of benzocaine be modified to create a compound having greater anesthetic properties?

65. At one time, halothane, $CF_3CHBrCl$, was widely used as a general anesthetic. Suggest why its use is now banned.

66. Which is the more appropriate statement: opioids have endorphin activity, or endorphins have opioid activity? Explain your answer.

67. Why is methadone not very useful in the treatment of cocaine addiction?

14.7 Medicines for the Heart

68. What problems are associated with a buildup of plaque on the inner walls of blood vessels?

69. Distinguish between a beta blocker and a calcium channel blocker.

70. A person may feel more relaxed after smoking a cigarette, but her heart is actually stressed. Why?

THINK AND DISCUSS (EVALUATION)

71. Medicines such as pain relievers and antidepressants are being found in the drinking water supplies of many municipalities. How did these medicines get there? Does it matter that they are there? Should something be done about it? If so, what?

72. Alcohol-free and caffeine-free beverages have been quite successful in the marketplace, while nicotine-free tobacco products have yet to be introduced. Speculate about possible reasons.

73. Would making tobacco illegal help or hurt people trying to kick the habit of using tobacco products such as cigarettes? What unintended consequences might arise from the prohibition of tobacco? How can society best convince people not to pick up the tobacco habit?

74. Why might someone find it more challenging to take a drug for a mental illness than a drug that targets a physical illness?

75. Should you be permitted access to any medicine that might save your life? What if you had no health insurance and could not afford the medicine? What if you were a homeless person living on the streets? What should be the moral, social, and economic responsibilities of a company that develops and produces medicines?

READINESS ASSURANCE TEST (RAT)

If you have a good handle on this chapter, then you should be able to score at least 7 out of 10 on this RAT. Check your answers online at www.ConceptualChemistry.com. If you score less than 7, you need to study further before moving on.

Choose the BEST answer to the following.

1. Which statement is true about a medicine?
   a. It is a drug that provides a euphoric effect.
   b. It is a drug that kills bacteria.
   c. It is a drug that is isolated from a plant.
   d. It is a drug that has therapeutic properties.
   e. None of the above.

2. Which of these would be an example of a synergistic effect?
   a. Penicillin kills both infectious and beneficial bacteria in the intestine.
   b. Antidepressants and cold medicine together can lead to seizures.
   c. Caffeine suppresses the appetite and causes the jitters.
   d. Aspirin reduces fever and thins the blood.
   e. None of the above.

3. The side effects of drugs are
   a. a multiplying effect seen in a drug due to other interactions.
   b. a beneficial aspect of a drug that has not been exploited.
   c. any behavior of a drug that is opposite its primary function.
   d. any behavior of a drug that is not fulfilling its primary function.
   e. none of the above.

4. Which would not be part of the drug discovery process?
   a. Examining folklore for herbs and potions
   b. Changing an old drug a little and testing the new drug for activity
   c. Learning the exact shape of various receptor sites
   d. Randomly testing new compounds for drug activity
   e. All of the above

5. What is the main difference between bacteria and humans that gives sulfa drugs their antibacterial properties?
   a. Bacteria cells can readily absorb sulfa drugs; human cells cannot.
   b. Human cells can readily absorb sulfa drugs; bacteria cells cannot.
   c. Bacteria cells can readily absorb folic acid; human cells cannot.
   d. Human cells can readily absorb folic acid; bacteria cells cannot.
   e. None of the above.

6. How is a set of neurons different from a wire conductor?
   a. The wire can carry more information.
   b. Neurons do not need to physically touch to conduct.
   c. Only the wire conducts electricity.
   d. Two of the above are correct.
   e. All of the above are correct.

7. What is taking place in the synaptic cleft?
   a. Receptor sites are forming due to neuron activity.
   b. Negative ions are accumulating due to neuron firing.
   c. Sodium ions are generating charge.
   d. Neurotransmitters are diffusing between the cells.
   e. None of the above.

8. Which of these neurotransmitters would most likely help you avoid a car accident?
   a. dopamine   b. norepinephrine
   c. acetylcholine   d. GABA
   e. serotonin

9. Physical pain may involve
   a. the synthesis of chemicals at the site of injury.
   b. nerve impulses.
   c. a sensation perceived in the brain.
   d. inflammation.
   e. all of the above.

10. How do over-the-counter analgesics such as aspirin work?
    a. They block nerve impulses.
    b. They alter the perception of pain in the brain.
    c. They block the production of pain-producing chemicals.
    d. All of the above.
    e. None of the above.

The Genetics of Muscle Fitness

There are three types of muscles: smooth, cardiac, and skeletal. Smooth muscles help move food and certain fluids through your body. Cardiac muscles pump blood through your heart. Skeletal muscles attach to your bones. They allow you to run; dance; lift; turn; and, in general, move through your environment. Physical fitness involves resilience in all muscle types. For this essay, however, our focus will be on the more physically apparent skeletal muscles, which can account for up to 40 percent of your total mass.

Muscle tissue is energetically expensive to maintain. Even at rest, your skeletal muscles consume about 25 percent of your energy. Thus, it is not in your body's interest to have more muscle mass than it needs. For this reason, our muscles adapt. If we exercise a lot, our muscles become bigger and stronger. Conversely, muscles wither away without exercise—a fully immobilized muscle loses about one-third of its mass within weeks. From personal experience, we know that muscles are responsive to use. But how is this accomplished? Why do your muscles get stronger when you exercise and get weaker when you don't? The answer is that exercise involves more than your muscles. When you exercise, you are also exercising your DNA.

As discussed in Chapter 13, DNA holds the information for how to build proteins. This includes all the proteins associated with building your muscles, of which there are two kinds. First, there are the proteins that actually become part of your muscles. Second, there are "regulator" proteins that serve to direct the building of your muscles. These regulator proteins act like managers at a construction site, in that they determine when building should speed up or

slow down. If DNA gets a signal that more muscle mass is needed, it creates more regulator proteins that are designed to speed up muscle protein synthesis where needed. Likewise, if DNA gets a signal that muscle production needs to slow down, it creates more regulator proteins designed to inhibit muscle protein synthesis. So DNA is the master controller. It actually produces hundreds of different muscle regulator proteins, and each one has a specific purpose.

Regulator proteins also play a role in how much energy is made available to muscles. An example is the GLUT regulator protein, which sits on the surface of muscle tissue. Its function is to pull glucose, an energy-rich sugar molecule, from the bloodstream and into the muscle. You always have some GLUT proteins to enable muscle movement. As you exercise, however, you stimulate your DNA to create more of these proteins. A single decent exercise routine causes a significant increase in GLUT proteins. What this means is that your muscles

can now pull glucose more efficiently out of your bloodstream, even while you are at rest. If you were to eat a potato, your blood sugar level would not rise as much as it would have if you hadn't exercised.

GLUT proteins generally degrade after 24 hours. With continued exercise, your DNA replenishes the GLUT proteins, and eventually an optimal number of them are retained. If you were to stop exercising, their positive effects would naturally fade away within a day or two. With no exercise over many years, your DNA becomes inefficient at producing GLUT proteins. The result is often unusually high blood sugar levels and a disease called type 2 diabetes.

Most of the energy used by your cells comes from a high-energy molecule known as *ATP* (see Section 13.6). ATP is produced using energy that comes from the oxidation of food. This occurs in small cellular organelles called *mitochondria*. How efficient mitochondria are at producing ATP is a function of the number of regulator proteins they contain. So what produces these regulator proteins? Your DNA. What happens when you exercise? You stimulate your DNA to make more of these regulator proteins within mitochondria; hence, more ATP becomes available to your cells, including your muscle cells. It takes about a week for the number of these regulator proteins to double. After about a month of regular exercise, you reach a plateau. In everyday language you would say you are "in shape." What you really mean, however, is that your DNA is in shape, because it is now producing an optimal number of regulatory proteins that produce an optimal amount of ATP so that you are fit to perform physically demanding tasks, such as running a marathon. So what happens when you don't exercise? ATP production is minimized, and the energy of food is directed to the production of energy-storage tissues such as fat. Why does the body produce fat? So that the precious energy it contains might be available for labor-intensive activities at a later date, when food may not be so abundant.

Our bodies are smart, but the rules of the game have recently changed. For many thousands of years, we relied on our DNA-mediated performance mechanisms to allow us to hunt, gather, and grow food; to build shelters; and to walk long distances. These are all labor-intensive activities. It has only been within the last 100 years that such activities have become mechanized. We walk much less because we have cars. We gather much less of our own food because we have grocery stores. The Centers for Disease Control and Prevention have noted a growing increase in obesity that coincides with the rise in sedentary lifestyles. While obesity can be traced to the intake of food as well as the quality of that food, another half of the equation often gets neglected, which is exercise. Our bodies are designed to be physically active. Harm comes to us when we neglect this calling.

Muscle Doping

Our knowledge of DNA is recent, but we are well on our way to learning how to control and tinker with its mechanisms. The IGF-I regulator protein speeds up muscle-building processes. To build bigger and stronger muscle tissue, why don't we simply inject IGF-I into our muscles? This doesn't work because the IGF-I dissipates within hours. What is needed is a continual source of IGF-I. Toward this end, researchers have successfully implanted the DNA that codes for IGF-I within the muscles of rats, which, without exercise, develop up to 30 percent more muscle mass. With exercise, the genetically "doped" muscles of these "super rats" become nearly twice as strong.

The insertion of DNA into a person to allow the creation of certain proteins is known as *gene therapy*. The idea of gene therapy was dreamed up soon after the discovery of the structure of DNA in the 1950s. Although the idea is straightforward, it is only within the past decade, after years of intensive basic research, that success is starting to be realized. Muscle-enhancement gene therapy is poised to become one of the first gene therapies to be made available to the general population. Such therapy holds great promise for the treatment of muscular dystrophy, as well as for the treatment of muscle weakness that comes with age. But it also raises a number of interesting questions. How safe is the procedure? Might such therapy lead to cancer? Are the inherent risks worth the benefits? Will we be more or less likely to exercise after having our muscles doped with muscle-strengthening DNA? Should athletes be allowed to genetically dope their muscles for better performance? How about military personnel? The ethical issues surrounding genetic enhancement are many and complex.

CONCEPTCHECK

What do 7 days without exercise make?

CHECK YOUR ANSWER Seven days without exercise makes one weak. Got it? Good. Now get to it.

Think and Discuss

1. Should athletes be allowed to genetically dope their muscles for better performance? If so, should there be two sets of Olympics, one for the doped and another for the nondoped? Many Olympians may already benefit from natural mutations that enhance their performance. Why not level the playing field by making such enhancements available to all athletes?

2. A new line of drugs known as nootropics are being developed to help us learn. Consider the social implications. What parallels might there be between sports-enhancing drugs and these intelligence-enhancing drugs? If these drugs are found to be safe, should they be available only by prescription? If they are found to be unsafe, should they be banned or should they be controlled, like alcohol and tobacco?

3. Gene therapy is not passed along to offspring because it does not affect the reproductive cells. For an individual, however, a single dose may last a lifetime. So what restrictions should be placed on muscle-enhancement gene doping? Should it be allowed only for the treatment of medical conditions? Should it be made available only to people over the age of 21?

4. Might gene therapy ever become as routine as vaccinations? By when? What might be some social or political hurdles to the acceptance of such therapies?

5. How much money would you be willing to pay for effective muscle-enhancement gene therapy?

6. Muscle-enhancement gene therapy also interferes with fat deposition. How might the food industry respond to a population of consumers who could eat all they wanted and still remain trim without exercising much?

▲ Genetically engineered golden rice provides nutrients that help prevent blindness.

15

Optimizing Food Production

THE MAIN IDEA

Agriculture employs much chemistry.

Each year, about 500,000 children in developing countries become irreversibly blind because of a deficiency of vitamin A. In an effort to stop this tragedy, scientists created a new strain of rice genetically engineered to produce the orange pigment beta-carotene, which the body uses to make vitamin A. The amount of beta-carotene in this rice gives the rice a golden hue.

Developing a new strain of a crop to meet our nutritional needs is nothing new. Most of our major crops are the result of centuries, if not millennia, of *selective breeding*, a process whereby organisms that offer more value are selectively bred over ones of less value. What is different about golden rice is that it was created over a single generation by inserting genes from a daffodil and a bacterium into the DNA of a strain of rice.

Provided here is a showcase of the chemistry involved in the production of food, primarily the food derived from plants, which feeds both us and our livestock. Along the way, you will be introduced to many of the fundamental concepts of agriculture, such as soil composition, fertilizers, pesticides, and transgenic crops such as golden rice.

Chemistry

Cleaning Your Insects

Perhaps the most environmentally friendly insecticides are solutions of soap or detergent. Insects are perforated with tiny holes, called spiracles, through which atmospheric oxygen migrates directly into cells. These holes are easily penetrated by liquid soap or detergent, which then blocks the exchange of atmospheric gases and suffocates the insect. In general, the larger the insect, the more concentrated a soap solution needs to be in order to kill the insect efficiently. A dilute soap solution, for example, will quickly annihilate aphids, but only a relatively concentrated one will kill cockroaches.

PROCEDURE

Use a measuring spoon to create soap solutions of various concentrations. Use non-bacteriocidal soap to minimize harm to beneficial bacteria in the soil. Stir your solutions gently to avoid excessive foaming. Pour each solution into a pump spray bottle and write the concentration (in teaspoons per cup, say) on each bottle. Test the effectiveness of each solution on an infested plant. Follow with a spray of fresh water to remove any residual soap from the plant.

ANALYZE AND CONCLUDE

1. What are some of the advantages and disadvantages of using soap rather than a commercial insecticide to kill insects?

2. During the time of the dinosaurs, the percentage of oxygen in the Earth's atmosphere was much greater than it is today. What effect might this have had on the potential size of insects?

15.1 Humans Eat at All Trophic Levels

EXPLAIN THIS

Why are fossils of the Tyrannosaurus rex so rare?

The formation of food begins with *photosynthesis*, the biochemical process used by plants to create carbohydrates and oxygen from solar energy, water, and atmospheric carbon dioxide:

$$\underset{\substack{\text{Carbon} \\ \text{dioxide}}}{6\,CO_2} \; + \; \underset{\text{Water}}{6\,H_2O} \; \xoverset{\text{Sunlight}}{\longrightarrow} \; \underset{\text{Carbohydrate}}{C_6H_{12}O_6} \; + \; \underset{\text{Oxygen}}{6\,O_2}$$

Each day, only about 1 percent of the solar energy reaching the Earth's surface is used in photosynthesis. On a global scale, this is enough to produce 170 billion tons of organic material per year. The energy content in this amount of organic matter is the total annual energy budget for virtually all living organisms.

The route food energy takes through a community of organisms is determined by the community's **trophic structure**, which is the pattern of feeding relationships in the community. Trophic structures, also known as *food chains*, consist of a number of hierarchical levels, shown in **Figure 15.1**. The first trophic level is **producers**, most of which are photosynthetic organisms that use light energy to power the synthesis of organic compounds. Plants are the main producers on land. In water, the main producers are photosynthetic organisms known as *phytoplankton* (*phyto-* means "plant").

LEARNING OBJECTIVE

Describe the flow of energy and nutrients between trophic levels.

Off the coast of Newfoundland, Canada, the ocean remains relatively shallow for hundreds of miles in an area known as the Canadian Grand Banks. For centuries, the Grand Banks have been known for thriving fish populations, especially cod, a tertiary consumer with a life span of about 25 years. The cod were once so plentiful that boats had trouble pushing through them. By the 1980s, overfishing had removed the bulk of sexually mature adults. Furthermore, trawling of the floor of the Grand Banks destroyed vital habitats. By 1993, the cod were depleted to the point that the fishery industry there collapsed, throwing some 40,000 people out of work. A moratorium on cod fishing remains in effect, but cod stocks have yet to recover. A similar situation is now occurring in the North Sea of Europe.

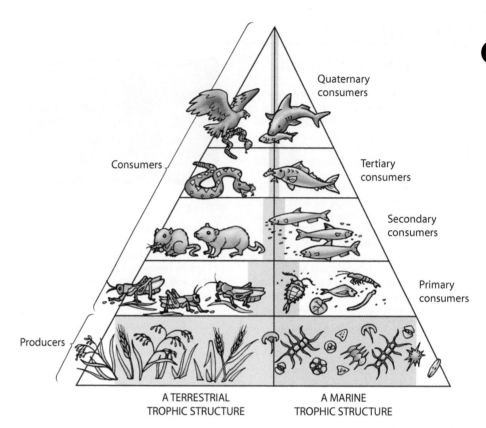

▲ Figure 15.1
Terrestrial and marine trophic structures. Energy and nutrients pass through the trophic levels when one organism feeds on another. The shaded blocks represent the amount of energy transferred from one trophic level to the next.

READING CHECK

What happens to an organism that dies and is not eaten?

▲ Figure 15.2
Most people are primary consumers, with a diet consisting primarily of grains.

Producers support all other trophic levels, collectively called **consumers.** Organisms that consume producers are *primary* consumers. In terrestrial environments, these are *herbivores* ("grass-eaters"), such as grazing mammals, most insects, and most birds. The primary consumers in aquatic environments are the many microscopic organisms collectively known as *zooplankton.* Above the primary consumers, the trophic levels are made of *carnivores* ("meat-eaters"), each level eating consumers from lower levels. Secondary consumers eat primary consumers, tertiary consumers eat secondary consumers, and quaternary consumers eat tertiary consumers. Any organism that dies before being eaten becomes subject to the action of **decomposers,** organisms that break down organic material into simpler substances that then act as soil nutrients. Common decomposers are earthworms, insects, fungi, and microorganisms.

With each transfer of energy from one trophic level to the next, there is a significant loss of energy. Typically, not more than 10 percent of the energy contained in the organic material of one trophic level is incorporated into the next higher level. The availability of food energy is therefore greatest for the organisms lowest on the food chain. A grasshopper, for example, will find many more blades of grass to feed on than a field mouse will find grasshoppers. And the field mouse will find more grasshoppers than a snake will find field mice. This dwindling supply of food resources quickly limits the number of trophic levels, which rarely exceeds the quaternary level. Accordingly, the higher the trophic level, the smaller the possible population of organisms.

We humans eat at all trophic levels. When we eat such things as fruits, vegetables, or the grains shown in **Figure 15.2**, we are primary consumers; when we eat beef or other meat from herbivores, we are secondary consumers. When we eat fish such as trout or salmon, which eat insects and other small animals, we are tertiary or quaternary consumers. Our great and growing numbers, however, are possible only because of our ability to eat as primary consumers.

Eating meat is a luxury. For the people who will eat the chickens shown in **Figure 15.3**, for instance, the amount of biochemical energy they will obtain from eating the chickens is minuscule compared with the amount of biochemical energy used in raising the chickens. In the United States, more than 70 percent of grain production is fed to livestock. Producing meat therefore requires that more land be cultivated, more water be used for irrigation, and more fertilizers and pesticides be applied to croplands. If people in the United States ate 10 percent less meat, the savings in resources could feed 100 million people. As the human population expands, meat consumption will likely become even more of a luxury than it is today.

▲ Figure 15.3
In affluent countries, eating meat is quite common. In the United States, for example, chickens outnumber people 44 to 1.

CONCEPT CHECK
The orca (killer whale) eats both sharks and phytoplankton-feeding gray whales. In which case is the orca eating at a higher trophic level?

CHECK YOUR ANSWER Sharks feed on fish and marine mammals, which make them secondary, tertiary, or quaternary consumers. Phytoplankton-feeding gray whales, however, are primary consumers. When an orca feeds on a shark, it is eating at a higher trophic level than when it feeds on a gray whale.

15.2 Plants Require Nutrients

EXPLAIN THIS

Why is the ratio of sodium to potassium ions greater in the oceans than on land?

LEARNING OBJECTIVE

Distinguish between the macro- and micronutrients needed by plants.

Plant material consists primarily of carbohydrates, which are made of the elements carbon, oxygen, and hydrogen. Plants get these three elements from carbon dioxide and water, but the soil in which the plants live also provides many other elements vital to their survival and good health. Table 15.1 lists these nutrients as *macronutrients*, those nutrients needed in large quantities, and *micronutrients*, those nutrients needed only in trace quantities. Some micronutrients are needed in such trace quantities that a plant's lifetime supply is provided by the seed from which the plant grew.

Plants Utilize Nitrogen, Phosphorus, and Potassium

Plants need nitrogen to build proteins and a variety of other biomolecules, such as *chlorophyll*, the green pigment responsible for photosynthesis. As Table 15.1 shows, plants are able to absorb nitrogen from the soil in the form of ammonium ions, NH_4^+, and nitrate ions, NO_3^-. **Figure 15.4** shows the natural sources of nitrogen for a plant. The two reactions shown there are examples of **nitrogen fixation**, which is defined as any chemical reaction that converts atmospheric nitrogen to a form of nitrogen that plants can use. The two most common forms are ammonium ions and nitrate ions.

Most of the ammonium ions in soil come from nitrogen fixation carried out either by bacteria living in the soil or by microorganisms living in root nodules of certain plants, especially those of the legume family, including clover, alfalfa, beans, and peas (plants generally called *nitrogen fixers*). These microorganisms possess the enzyme *nitrogenase*, which catalyzes the formation of ammonium ions from atmospheric nitrogen and soil-bound hydrogen ions, as Figure 15.4a shows.

TABLE 15.1   Essential Elements for Most Plants

ELEMENT	FORM AVAILABLE TO PLANTS
Macronutrients	
Nitrogen, N	NO_3^-, NH_4^+
Potassium, K	K^+
Calcium, Ca	Ca^{2+}
Magnesium, Mg	Mg^{2+}
Phosphorus, P	$H_2PO_4^{2-}$, HPO_4^{2-}
Sulfur, S	SO_4^{2-}
Micronutrients	
Chlorine, Cl	Cl^-
Iron, Fe	Fe^{3+}, Fe^{2+}
Boron, B	$H_2BO_3^-$
Manganese, Mn	Mn^{2+}
Zinc, Zn	Zn^{2+}
Copper, Cu	Cu^+, Cu^{2+}
Molybdenum, Mo	MoO_4^{2-}

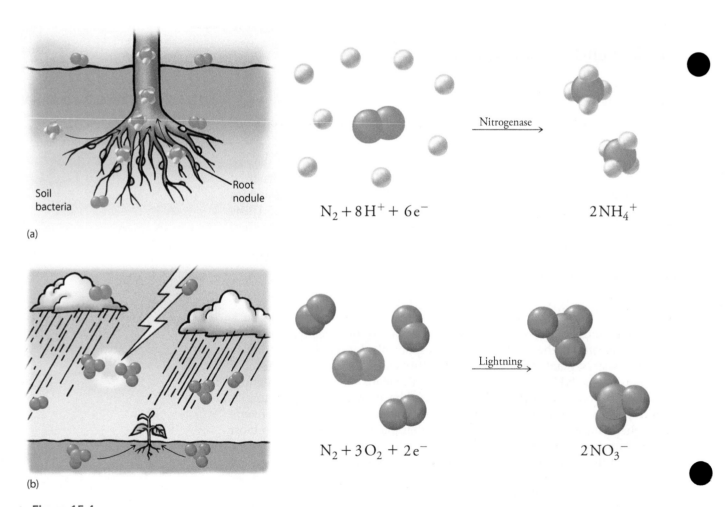

Soil bacteria

Root nodule

(a)

$$N_2 + 8\,H^+ + 6\,e^-$$

Nitrogenase

$$2\,NH_4^+$$

(b)

$$N_2 + 3\,O_2 + 2\,e^-$$

Lightning

$$2\,NO_3^-$$

▲ Figure 15.4
Two pathways for nitrogen fixation, a source of nitrogen for plants. (a) Both free-living bacteria in the soil and microorganisms in root nodules produce ammonium ions. (b) Lightning provides the energy needed to form nitrate ions from atmospheric nitrogen.

Nitrogen fixation also results to a lesser extent from the action of lightning on atmospheric nitrogen, as Figure 15.4b shows. The high energy of the lightning is sufficient to initiate the oxidization of atmospheric nitrogen to nitrate ions, which are then washed into the soil by rain.

In a natural setting, nitrogen fixation is the original source of ammonium and nitrate ions in the soil. Most of this fixed nitrogen, however, is recycled from one organism to the next. After a plant dies, for example, bacterial decomposition of the plant releases ammonium and nitrate ions back into the soil, where these ions become available to plants that are still living.

CONCEPT CHECK

Why are the seeds of nitrogen-fixing plants such as soybeans and peanuts unusually high in protein?

CHECK YOUR ANSWER Plants use nitrogen to build proteins. Nitrogen-fixing plants assimilate a lot of nitrogen, so they produce proteins in large quantities.

Plants deficient in nitrogen have stunted growth. Because nitrogen is needed for making chlorophyll, another symptom of nitrogen deficiency in plants is yellow leaves, as **Figure 15.5a** shows. The yellowing is most pronounced in older leaves—younger ones remain green longer, because soluble forms of nitrogen are transported to them from older dying leaves.

Plants need phosphorus to build nucleic acids, phospholipids, and a variety of energy-carrying biomolecules such as ATP. All phosphorus comes to plants in the form of phosphate ions. The major natural source of these ions is eroded phosphate-containing rock. Significant amounts of phosphates are also recycled as organisms die and become incorporated into the soil. After nitrogen, phosphorus is most often the limiting element in soil. Phosphorus-deficient plants are stunted, as **Figure 15.5b** shows.

Potassium ions activate many of the enzymes essential to photosynthesis and respiration. As with phosphates, the major natural sources of potassium ions are eroded rock and the recycling of potassium ions from decomposing plant material. After nitrogen and phosphorus, soils are usually most deficient in potassium. Symptoms of potassium deficiency include small areas of dead tissue, usually along leaf tips or edges, as shown in **Figure 15.5c**. As with nitrogen and phosphorus, potassium ions are easily redistributed from mature to younger parts of the plant, so deficiency symptoms first appear on older leaves. When cereal grains, such as wheat or corn, are potassium-deficient,

READING CHECK

What macronutrients do plants need for making chlorophyll?

(a)

(b)

(c)

▲ Figure 15.5
(a) The leaves of nitrogen-deficient plants turn yellow prematurely. (b) Phosphorus deficiencies are marked by stunted growth. (c) The leaves of potassium-deficient plants develop dead areas, here seen as the pale yellow region along the leaf perimeter.

they develop weak stalks and their roots become more easily infected with root-rotting organisms. These two factors cause potassium-deficient plants to be easily bent to the ground by wind, rain, or snow.

Plants Also Utilize Calcium, Magnesium, and Sulfur

Both calcium and magnesium are absorbed by plants as the positively charged calcium and magnesium ions, Ca^{2+} and Mg^{2+}, respectively, and sulfur is absorbed as the negatively charged sulfate ion, SO_4^{2-}. Most topsoils contain enough of these ions for adequate plant growth.

Calcium ions are essential for building cell walls. Once absorbed by the plant, calcium ions are relatively immobile; that is, they do not travel well from one part of the plant to another. The plant therefore is not very capable of reallocating calcium supplies in times of need. This is why new-growth zones, such as the tips of roots and stems, are most susceptible to calcium deficiencies. The results are twisted and deformed growth patterns.

Magnesium ions are essential for the formation of chlorophyll, which is the green-pigmented molecule essential for photosynthesis. Chlorophyll houses a magnesium ion at the center of a structure called a porphyrin ring, as shown in **Figure 15.6**. Besides its presence in chlorophyll, magnesium is essential because it activates many metabolic enzymes. Deficiencies in magnesium, which are rare, result in yellow leaves, because the plant is unable to generate chlorophyll.

Most of the sulfur in plants occurs in proteins, especially in the amino acids cysteine and methionine. Another essential compound that contains sulfur is coenzyme A, a compound essential for respiration, for the synthesis and break-down of fatty acids, and for the vitamins thiamine and biotin. Sulfur can be absorbed by leaves as gaseous sulfur dioxide, SO_2, an environmental pollutant released from active volcanoes and from the burning of wood or fossil fuels.

CONCEPT CHECK

Plants require more calcium and magnesium than phosphorus and potassium, yet deficiencies of calcium and magnesium are much more infrequent than deficiencies of phosphorus and potassium. How can this be?

CHECK YOUR ANSWER The potential for nutrient deficiency depends not only on the amount of nutrient needed but also on the nutrient's availability. Calcium and magnesium are abundant in most soils, but phosphorus and potassium are not. Deficiencies of phosphorus and potassium therefore tend to be more frequent.

▼ Figure 15.6
Magnesium ions in the green pigment chlorophyll are vital to photosynthesis.

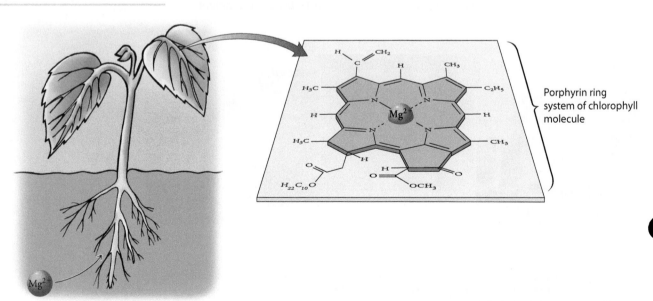

Porphyrin ring system of chlorophyll molecule

15.3 Soil Fertility Is Determined by Soil Structure and Nutrient Retention

EXPLAIN THIS

Why don't leaves accumulate beneath a tree that loses its leaves every year?

Soil is a mixture of sand, silt, and clay. All three of these components are ground-up rock; the differences are in how finely the particles are ground. Sand particles are the largest, and clay particles are the smallest.

Soil often occurs as a series of horizontal layers called **soil horizons,** shown in **Figure 15.7**. The deepest horizon, which lies just above solid rock, is the *substratum,* which is rock just beginning to be disintegrated into soil by the action of water that has seeped down to this level. No growing plant material is found in the substratum. Above the substratum is the *subsoil,* which consists mostly of clay. Only the deepest roots penetrate into the subsoil, which may be up to 1 meter thick. Above the subsoil is the *topsoil,* which lies on the surface and varies in thickness from a few centimeters to 2 meters. The topsoil usually contains sand, silt, and clay in about equal amounts. This is the horizon where the roots of plants absorb most of their nutrients.

Fertile topsoil is a mixture of at least four components—mineral particles, water, air, and organic matter. The mineral particles are the particles of sand, silt, and clay. Many of the nutrients that plants need are released as these particles are formed from the erosion of rock. The size of the particles greatly affects soil fertility. Large particles result in porous soil that has many pockets of space that collect water and air—up to 25 percent of the volume of fertile topsoil consists of pockets. Roots absorb water and the oxygen from the air in these pockets. Small particles pack tightly together so that no or only very few air pockets are present, and as a result, little or no oxygen or water is available to roots. This is why plants do not grow well in clay soils. **Figure 15.8** compares these two extremes of soil.

The organic matter in topsoil is a mixture of fallen plant material, the remains of dead animals, and such decomposers as bacteria and fungi, as **Figure 15.9** illustrates. This organic matter is called **humus,** and it is rich in a variety of plant nutrients. Humus tends to be porous, giving roots access to subterranean water and oxygen. It also binds the soil, helping to prevent erosion.

The flow of water through soil is called *percolation.* The more porous the soil, the greater the rate of percolation. With excessive percolation, flowing water removes many water-soluble nutrients needed to make the soil productive.

LEARNING OBJECTIVE

Identify what makes for healthy soil.

READING CHECK

Fertile topsoil is a mixture of what four components?

◀ Figure 15.7
The vertical structure of soil is a series of layers called soil horizons.

Topsoil

Subsoil

Substratum

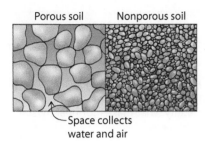

Porous soil   Nonporous soil

Space collects water and air

▲ Figure 15.8
Large soil particles create larger pockets of space than do smaller soil particles.

This process is known as *leaching*. With too little percolation, topsoil becomes waterlogged, choking off a plant's supply of oxygen. Soils with optimal percolation drain water from all but the smallest air pockets.

Soil Readily Retains Positively Charged Ions

Mineral particles play an important role in keeping nutrients in soil. As Table 15.1 shows, many plant nutrients are positively charged ions. The surfaces of most mineral particles, however, are negatively charged. **Figure 15.10** illustrates how the resulting ionic attractions help keep nutrients from being washed away. The degree of nutrient retention is most pronounced in clay soils, because its mineral particles are the smallest and thus have the largest surface area relative to volume.

Decaying matter in humus contains many carboxylic acid and phenolic groups, which at a typical soil pH are ionized to negatively charged carboxylate and phenolate ions. So, like mineral particles, humus helps retain positively charged nutrients.

> ### CONCEPTCHECK
> Soil is able to retain ammonium ions, NH_4^+, better than nitrate ions, NO_3^-. Why?
>
> **CHECK YOUR ANSWER** Mineral particles and bits of humus in soil are negatively charged and therefore hold on to positively charged ammonium ions but repel negatively charged nitrate ions.

The pH of soil is largely a function of the amount of carbon dioxide present. Recall from Section 10.5 that carbon dioxide reacts with water to form carbonic acid, which in turn forms hydronium ions:

$$O{=}C{=}O \;+\; H_2O \;\longrightarrow\; \underset{\text{Carbonic acid}}{HO{-}\overset{O}{\overset{\|}{C}}{-}OH} \;+\; H_2O \;\longrightarrow\; \underset{\text{Bicarbonate ion}}{HO{-}\overset{O}{\overset{\|}{C}}{-}O^-} \;+\; H_3O^+$$

Carbon dioxide   Water   Carbonic acid   Water   Bicarbonate ion   Hydronium ion

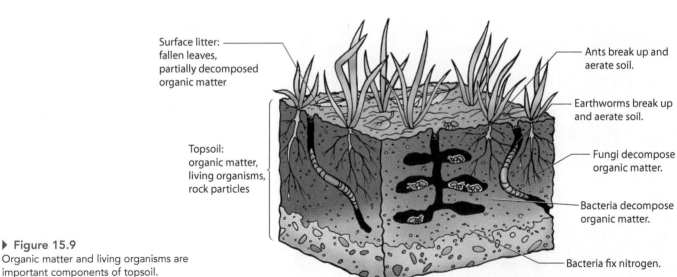

Surface litter: fallen leaves, partially decomposed organic matter

Topsoil: organic matter, living organisms, rock particles

Ants break up and aerate soil.

Earthworms break up and aerate soil.

Fungi decompose organic matter.

Bacteria decompose organic matter.

Bacteria fix nitrogen.

▶ Figure 15.9
Organic matter and living organisms are important components of topsoil.

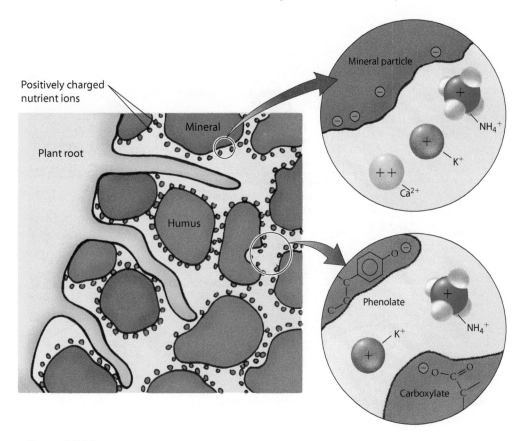

Figure 15.10
The negatively charged surfaces of soil mineral particles and humus help retain positively charged nutrient ions.

The greater the amount of carbon dioxide in soil, the more hydronium ions, and so the lower the pH. Soil that has a low pH is referred to as *sour*. (Recall from Chapter 10 that many acidic foods, such as lemons, are characteristically sour.) Two main sources of soil carbon dioxide are humus and plant roots. The humus releases carbon dioxide as it decays, and plant roots release carbon dioxide as a product of cellular respiration. A healthy soil may have enough carbon dioxide released from these processes to give a pH range from about 4 to 7. If the soil becomes too acidic, a weak base, such as calcium carbonate (known as lime or limestone), can be added.

Hydronium ions are able to displace nutrient ions held to mineral particles and humus. Plants use this fact to great advantage. **Figure 15.11** illustrates how a plant releases carbon dioxide through its root system and in doing so generates nutrient-displacing hydronium ions. The displaced nutrients, no longer attached to soil particles, thus become available to the plant.

CONCEPTCHECK

How might a below-normal pH in soil lead to nutrient deficiencies in plants?

CHECK YOUR ANSWER When the soil pH is below normal, the water in the soil pockets contains an abundance of hydronium ions, which displace large numbers of nutrient ions from soil and humus particles. Most of these nutrients wash away before the plant is able to absorb them. In a short time, the soils are depleted of nutrients and the plants become nutrient-deficient.

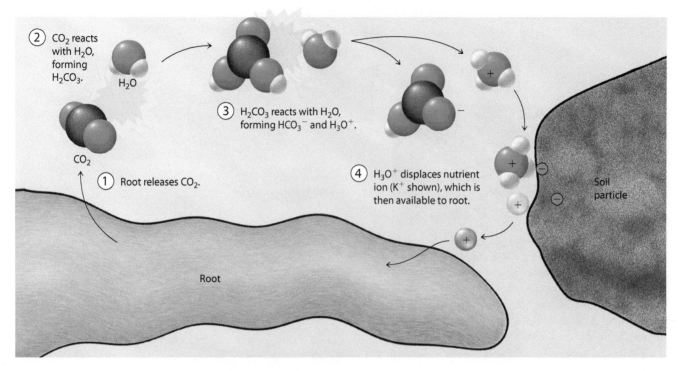

② CO_2 reacts with H_2O, forming H_2CO_3.

H_2O

③ H_2CO_3 reacts with H_2O, forming HCO_3^- and H_3O^+.

CO_2

① Root releases CO_2.

④ H_3O^+ displaces nutrient ion (K^+ shown), which is then available to root.

Soil particle

Root

▲ Figure 15.11
By releasing carbon dioxide, a plant guarantees a steady flow of nutrients from the soil to its roots.

15.4 Natural and Synthetic Fertilizers Help Restore Soil Fertility

LEARNING OBJECTIVE

Describe the origins of straight and mixed fertilizers.

EXPLAIN THIS

Do the roots of a plant need air?

As a soil loses its plant nutrients to harvested crops and to leaching, it loses its fertility. Farmers amend soil by adding *fertilizers*, which are replacement sources for these lost nutrients. Naturally occurring fertilizers are compost and minerals. **Compost** is decayed organic matter, which can be animal manure, food scraps, or plant material. Mineral fertilizers are mined. Saltpeter, $NaNO_3$, for example, was once used extensively as a source of nitrogen, but by the late 1800s, the supply of this nitrogen-containing mineral was almost exhausted. A new source of nitrogen for fertilizers came along in 1913, when Fritz Haber (1868–1934), a German chemist, developed a process for producing ammonia from hydrogen and atmospheric nitrogen:

$$N_2 \quad + \quad 3\,H_2 \quad \rightarrow \quad 2\,NH_3$$

Nitrogen   Hydrogen   Ammonia

READINGCHECK

Most ammonia used today as fertilizer is made from what process?

This technique is now the primary means of producing ammonia, which can be stored in high-pressure tanks as a liquid and injected into the soil. Alternatively, the ammonia can be converted to a water-soluble salt, such as ammonium nitrate, NH_4NO_3, that is then applied to the soil either as a solid or in solution. The mining of other nutrients, such as phosphorus and potassium, still remains an important endeavor.

In times past, mineral fertilizers were used just as they came from the ground. Today, however, chemists have learned how to mix and match minerals to obtain many different formulations, each suitable for a different soil problem or the specific requirements of a particular plant. All these formulated mineral fertilizers are referred to as *chemically manufactured fertilizers* or, more frequently, *synthetic fertilizers*. Don't take the word *synthetic* literally, though, because except for what is produced by the Haber reaction, all the minerals in synthetic fertilizers originally came from the ground.

A fertilizer that contains only one nutrient is called a **straight fertilizer.** Ammonium nitrate, NH_4NO_3, is an example of a straight fertilizer, yielding only nitrogen. Any fertilizer containing a mixture of the three most essential nutrients (nitrogen, phosphorus, and potassium) is called either a *complete fertilizer* or a **mixed fertilizer.** All mixed fertilizers are graded by the N-P-K system, which lists the percent of nitrogen (N), phosphorus (P), and potassium (K) they contain, as **Figure 15.12** shows. A typical mixed fertilizer might be graded 6-12-12. A typical compost, by contrast, might be rated anywhere from 0.5-0.5-0.5 to 4-4-4. Compost N-P-K ratings are much lower because of their high percentage of organic bulk. This organic bulk helps to keep the soil loose for aeration, however, and serves as food for beneficial organisms that live in the soil. Because of the negative electric charges it carries, the organic bulk also attracts positively charged nutrient ions, which are then not so readily leached away.

The effect that nitrogen-containing synthetic fertilizers can have on yields is significant. It requires a lot of energy to mine and refine synthetic fertilizers, however, so they are expensive. For example, of the total energy required to produce corn in the United States, at least one-third is needed to produce, transport, and apply the fertilizer. Nevertheless, synthetic fertilizers are widely used, and our present food supply depends on them.

CHEMICAL
CONNECTIONS

How is a steak dinner
connected to the atmosphere?

CONCEPTCHECK

Which N-P-K rating would you expect for coffee grounds, which contain significant quantities of the alkaloid caffeine: 2-0.3-0.2, 0.3-2-0.2, or 0.3-0.2-2?

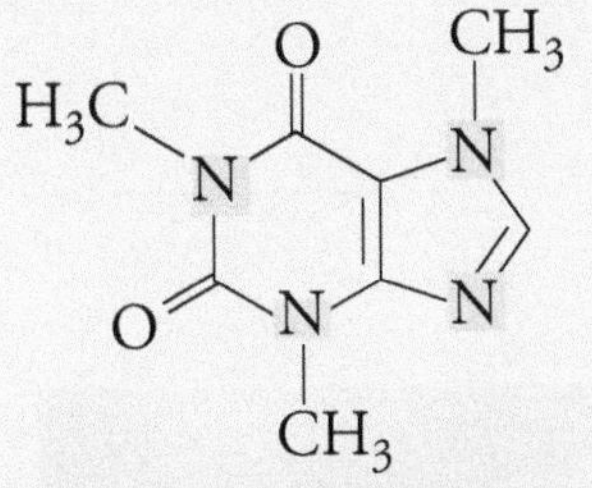

Caffeine

CHECK YOUR ANSWER The fact that caffeine is a nitrogen-containing compound means that coffee grounds must contain a relatively high proportion of nitrogen, as is indicated by their N-P-K rating of 2-0.3-0.2.

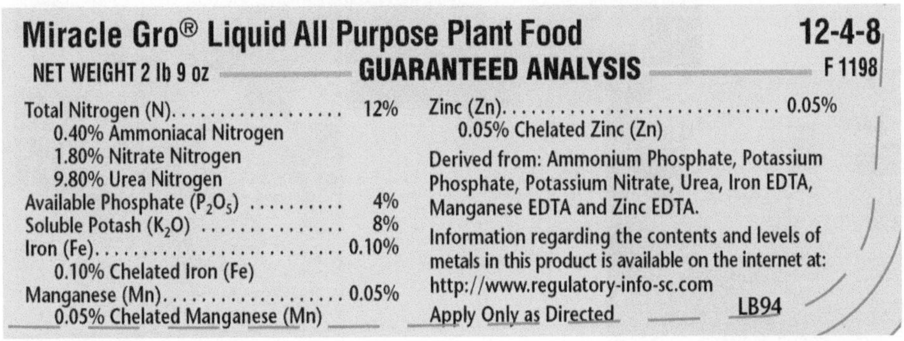

Miracle Gro® Liquid All Purpose Plant Food   **12-4-8**

NET WEIGHT 2 lb 9 oz ——— **GUARANTEED ANALYSIS** ——— F 1198

Total Nitrogen (N).................. 12%	Zinc (Zn)............................ 0.05%	
0.40% Ammoniacal Nitrogen	0.05% Chelated Zinc (Zn)	
1.80% Nitrate Nitrogen	Derived from: Ammonium Phosphate, Potassium	
9.80% Urea Nitrogen	Phosphate, Potassium Nitrate, Urea, Iron EDTA,	
Available Phosphate (P_2O_5) 4%	Manganese EDTA and Zinc EDTA.	
Soluble Potash (K_2O) 8%	Information regarding the contents and levels of	
Iron (Fe)........................ 0.10%	metals in this product is available on the internet at:	
0.10% Chelated Iron (Fe)	http://www.regulatory-info-sc.com	
Manganese (Mn)................... 0.05%	Apply Only as Directed ———— LB94	
0.05% Chelated Manganese (Mn)		

▲ Figure 15.12
Fertilizers are rated by the percentages of nitrogen, phosphorus, and potassium they contain.

15.5 Pesticides Kill Insects, Weeds, and Fungi

LEARNING OBJECTIVE

Describe the benefits and risks of insecticides, herbicides, and fungicides.

 FOR YOUR INFORMATION

Many nations still rely on DDT as an economical method of controlling insects that carry human diseases. DDT can be most effective in this manner, but not without consequences. In Malaysia, for example, thatched roofs were once sprayed with DDT to kill malaria-carrying mosquitos. Wasps that kill straw-eating moths were also killed. The moths prospered, leading to the destruction of the thatched roofs. Furthermore, cats that ate geckos that ate DDT-laden cockroaches died. With no cats, rats multiplied, as did the bubonic-plague-carrying fleas that lived on the rats. Cats were therefore dropped by parachute to remote villages to help prevent outbreaks of the plague. When it comes to ecosystems, you can never change only one thing.

EXPLAIN THIS

Why does the tobacco leaf contain nicotine?

A high-yield crop needs more than adequate nutrition. It also needs defense against a host of natural enemies, a few of which are shown in **Figure 15.13**. To control these pests, farmers can apply substances known as *pesticides*. There are several kinds of pesticides, including insect-killing *insecticides*, weed-killing *herbicides*, and fungus-killing *fungicides*.

Insecticides Kill Insects

Most species of insects are beneficial or even essential to agriculture. Honeybees, for example, are responsible for the pollination of $10 billion worth of produce in the United States. Countless other species take part in nutrient recycling and help maintain soil quality. A small minority of insect species, however, have continually threatened our capacity to grow, harvest, and store crops, and it is against these species that insecticides are used. The most widely used insecticides are chlorinated hydrocarbons, organophosphorus compounds, and carbamates.

The *chlorinated hydrocarbons* have a remarkable persistence, killing insects for months and years on treated surfaces. There are at least two reasons for this persistence. First, chlorinated hydrocarbons tend to be nonbiodegradable, which means there are no natural pathways to break them down chemically. Second, they are nonpolar compounds, which means they are insoluble in water and thus are not washed away by rainwater.

▲ Figure 15.13
Crop pests such as these threaten crop yields.

In 1939, there was a breakthrough in the fight against insect pests with the chemical synthesis of the chlorinated hydrocarbon DDT, shown in **Figure 15.14**. During the 1940s and 1950s, DDT was applied liberally to crops, resulting in markedly greater yields. In addition to protecting plants, DDT protected people from disease. It was applied to rivers, streams, and villages to help control the proliferation of mosquitoes, lice, and tsetse flies, which spread malaria, typhus, and sleeping sickness, respectively. According to the World Health Organization, by protecting against these diseases, DDT has saved an estimated 25 million human lives.

Insect populations began to develop a resistance to DDT within a few years of its first application. Furthermore, DDT was found to be toxic to wildlife, including the natural predators of insects, such as birds. With fewer natural predators, DDT-resistant insects were able to thrive. The early increased crop yields resulting from DDT use were therefore not sustainable.

In the 1950s and 1960s, these and other negative aspects of DDT and other pesticides were brought to the public's attention by a number of publications, including the biologist Rachel Carson's book *Silent Spring*. Carson, shown in **Figure 15.15**, described the importance of understanding the dynamics of ecosystems, most of which are highly sensitive to human activities. She also described a phenomenon known as **bioaccumulation,** whereby a toxic chemical that enters a food chain at a low trophic level becomes more concentrated in organisms higher up the chain, as illustrated in **Figure 15.16**. In bodies of water sprayed with DDT, for example, small amounts of the pesticide were ingested and stored in the nonpolar lipids of aquatic microorganisms. Because these microorganisms serve as food for animals at higher trophic levels, DDT became more concentrated in the body fat of these larger creatures. Predatory birds at the top of the chain accumulated the greatest amounts of DDT. Eventually, the elevated DDT levels affected avian population numbers, because the egg shells of affected birds were too thin and fragile to support the growing chick embryos. DDT contributed to the decline of many bird populations and the near extinction of some species of ospreys, hawks, eagles, and falcons. In the early 1970s, the United States and many other countries banned the use of DDT. Within a matter of years, many wildlife species in these countries were able to recover.

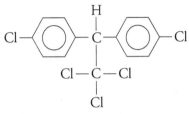

▲ Figure 15.14
The chemical name for DDT is dichlorodiphenyltrichloroethane.

▲ Figure 15.15
Rachel Carson was a pioneer in the fight against the excessive use of pesticides.

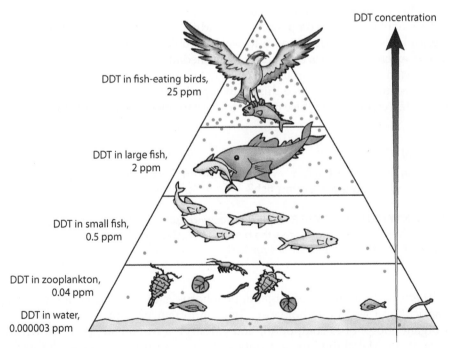

DDT concentration

DDT in fish-eating birds, 25 ppm

DDT in large fish, 2 ppm

DDT in small fish, 0.5 ppm

DDT in zooplankton, 0.04 ppm

DDT in water, 0.000003 ppm

▲ Figure 15.16
The DDT concentration in a food chain can be magnified from 0.000003 parts per million (ppm) as a pollutant in the water to 25 ppm in a bird at the top of the chain.

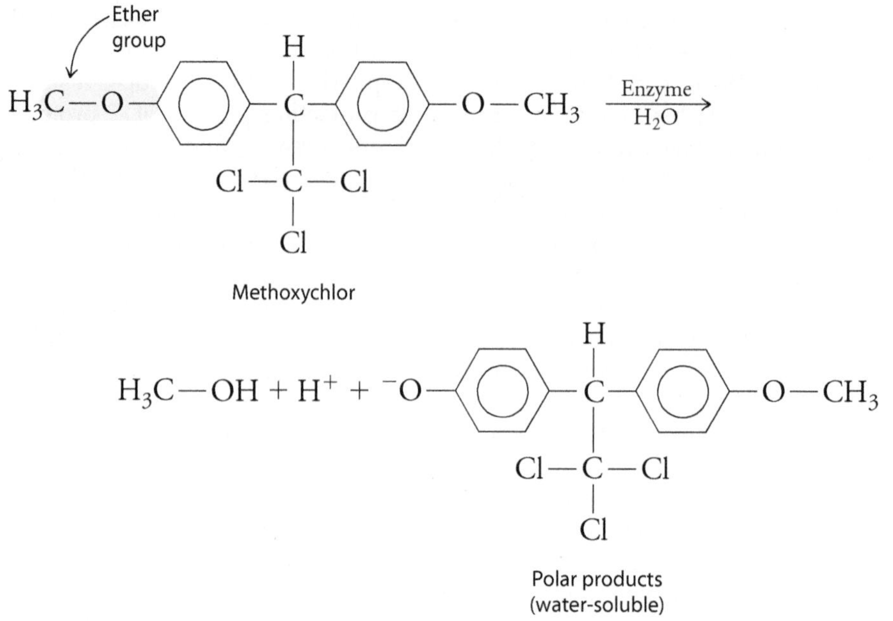

Ether group

H₃C—O—⟨benzene⟩—C(H)—⟨benzene⟩—O—CH₃ $\xrightarrow[H_2O]{\text{Enzyme}}$

Cl—C—Cl
|
Cl

Methoxychlor

H₃C—OH + H⁺ + ⁻O—⟨benzene⟩—C(H)—⟨benzene⟩—O—CH₃

Cl—C—Cl
|
Cl

Polar products
(water-soluble)

▲ **Figure 15.17**
Methoxychlor is one of many alternatives to DDT. Enzymes in the liver can cleave the ether groups to produce polar products. Look back at Figure 15.14, and you will see that DDT lacks ether groups.

Many chlorinated hydrocarbon alternatives to DDT have been developed. One of the earliest substitutes was methoxychlor, shown in **Figure 15.17**. This compound has much lower toxicity in most animals and, unlike DDT, is not readily stored in animal fat. Look carefully at the structures of methoxychlor and DDT and you'll see that they are identical except that methoxychlor has two ether groups whereas DDT has two chlorine atoms. Because the structures are nearly identical, they have nearly the same level of toxicity in insects. In higher animals, however, the oxygen atoms facilitate detoxification. Specifically, enzymes in the liver cleave the ether groups to synthesize polar products that are readily excreted through the kidneys.

Organophosphorus compounds and carbamates, in contrast to chlorinated hydrocarbons, readily decompose to water-soluble components and so do not act over extended periods of time. Their immediate toxicity to both insects and animals is much greater than that of chlorinated hydrocarbons, however. Added safety precautions are required during the application of organophosphates and carbamates, especially because of their toxicity to honeybees (**Figure 15.18**).

▲ **Figure 15.18**
Honeybees do not forage at night. Quick-acting pesticides, such as organophosphorus compounds and carbamates, are therefore best applied in the evening. By the time the bees return the next day, these pesticides have lost much of their toxicity.

CONCEPT**CHECK**
Why are fish-eating birds more susceptible to DDT contamination compared to plankton-eating fish?

CHECK YOUR ANSWER A plankton-eating fish gets DDT from all the plankton it eats. This same DDT then enters the bird that eats this fish. But because the bird eats many fish, its exposure to DDT is amplified.

There are hundreds of organophosphorus and carbamate insecticides in agricultural and household use. Two important examples are malathion, an organophosphorus compound, and carbaryl, a carbamate, both shown in **Figure 15.19**. Malathion kills a variety of insects, such as aphids, leafhoppers, beetles, and spider mites. Carbaryl, like many other carbamates, is relatively selective in the types of insects that it kills.

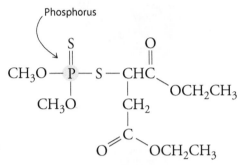

Phosphorus

Malathion
(an organophosphorus compound)

Carbamate group

Carbaryl
(a carbamate)

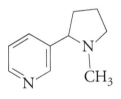

Imidacloprid

The most widely used insecticides, however, are the neonicotinoids, such as the compound imidacloprid, shown in **Figure 15.20**. These compounds are derivatives of the compound nicotine, which is a naturally occurring neurotoxin, both to insects and mammals. The neonicotinoids, however, are selectively more toxic to insects than they are to mammals, making them safer for us to handle. These compounds tend to be water-soluble and can be added to irrigation water or even sprayed on the seeds, which grow into plants that then retain enough neonicotinioids to fight off insects. The neonicotinoids are biodegradable, especially when exposed to sunlight. There is strong evidence, however, that these compounds are having a negative impact on bee populations. Notably, they may be responsible, in part, for a phenomenon called *colony collapse disorder*.

Nicotine

▲ **Figure 15.20**
Imidacloprid is an example of a neonicotinoid, which is a compound that mimics the insecticidal properties of naturally occurring nicotine.

Herbicides Kill Plants

Weeds compete with crop plants for valuable nutrients. The traditional method for controlling weeds is to plow them under the soil, where in decomposing, they release the nutrients they absorbed while they were alive. Plowing also aerates the soil, but it is either labor-intensive or energy-intensive and can lead to topsoil erosion. In the early 1900s, farmers noted that certain fertilizers, such as calcium cyanamide, CaNCN, selectively kill weeds while causing little harm to crops. This prompted a broad search for chemicals that act as herbicides. Today, a farmer can choose from hundreds of herbicides, many tailored for a specific weed. Farmers in the United States apply almost 600 million pounds of herbicides annually, which is about three times more than the amount of insecticides they apply.

Two selective herbicides are the carboxylic acids 2,4-dichlorophenoxyacetic acid (2,4-D) and 2,4,5-trichlorophenoxyacetic acid (2,4,5-T), shown in **Figure 15.21**. Both mimic the action of plant growth hormones and are selective in killing broad-leafed plants but not grass-like crops such as corn and wheat. A herbicide known as *Agent Orange* is a blend of 2,4-D and 2,4,5-T. During the Vietnam War, U.S. military forces applied more than 15 million gallons of Agent Orange and related herbicides in an effort to defoliate jungle areas that could harbor enemy troops. Health problems in Vietnamese troops and civilians, U.S. troops, and others exposed to Agent Orange have since been linked to a minor contaminant of the herbicide—the highly toxic compound 2,3,7,8-tetrachlorodibenzo-*p*-dioxin

READINGCHECK

What disadvantage is there to plowing weeds under the soil?

▶ **Figure 15.21**
The herbicides 2,4-D and 2,4,5-T and the dioxin contaminant TCDD.

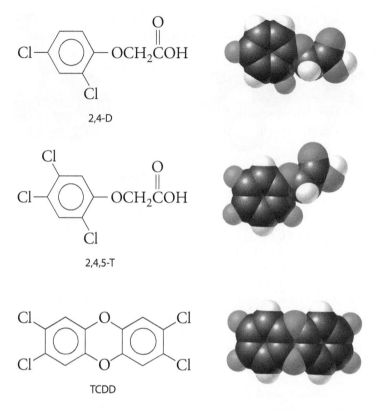

2,4-D

2,4,5-T

TCDD

(TCDD). This contaminant was generated as a side product in the manufacture of 2,4,5-T. In 1985, because of this contamination, the use of 2,4,5-T was prohibited by the U.S. Environmental Protection Agency. TCDD-free methods of 2,4,5-T production, however, have since been developed, which raises the possibility that 2,4,5-T may once again be introduced as an effective herbicide.

Three other commonly used herbicides are atrazine, paraquat, and glyphosate, all shown in **Figure 15.22**. Atrazine is toxic to common weeds but not to many grass-like crops, which can rapidly detoxify this herbicide through metabolism.

▶ **Figure 15.22**
The herbicides atrazine, paraquat, and glyphosate.

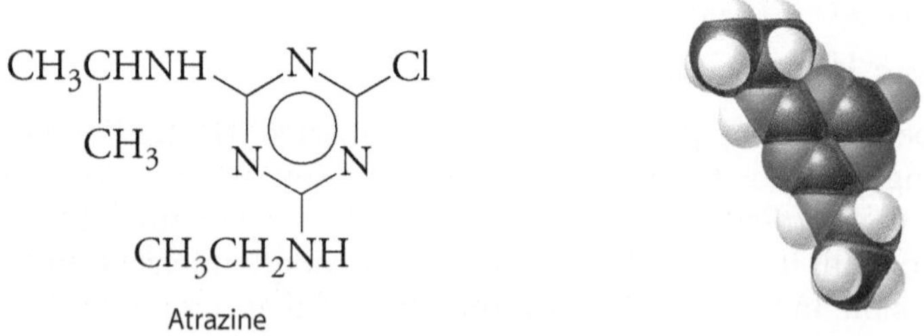

Atrazine

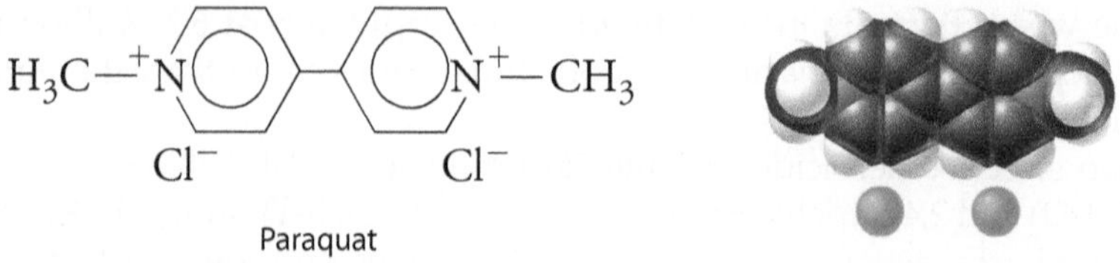

Paraquat

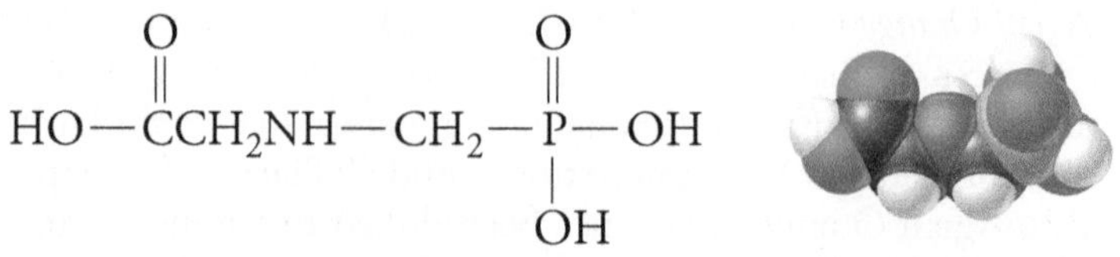

Glyphosate

Paraquat kills weeds in their sprouting phase. During the 1970s and 1980s, this herbicide was sprayed aerially to destroy drug-producing poppy and marijuana fields in the United States, Mexico, and much of Central America and South America. Paraquat residues made their way into the illicit drug products, however, causing lung damage in users. So for ethical reasons, the spraying of paraquat on drug-producing plants is no longer common practice.

Glyphosate is a nonselective herbicide that affects a biochemical process common to all plants—the biosynthesis of the amino acids tyrosine and phenylalanine. Glyphosate has low toxicity in animals because most animals do not synthesize these amino acids, obtaining them from food instead. Glyphosate is the active ingredient of the herbicide Roundup®.

$$(CH_3)_2N - \overset{\overset{\displaystyle S}{\|}}{C} - S$$
$$(CH_3)_2N - \underset{\underset{\displaystyle S}{\|}}{C} - S$$

Thiram

▲ **Figure 15.23**
The fungicide thiram.

Fungicides Kill Fungi

As decomposers, fungi play an important role in soil formation, but they can also harm crops. Most of the harm they cause occurs during a plant's early growth stages. Fungi can also spoil stored food and are particularly devastating to the world's fruit harvest.

In the United States, farmers use about 100 million pounds of fungicides annually, meaning fungicides rank third after herbicides and insecticides in the amounts used. An example of a fungicide is thiram, widely used on fruits and vegetables (see **Figure 15.23**).

During the last 60 years, pesticides have benefited our society by preventing disease and increasing food production. Our need for pesticides will continue, but greater specificity will certainly be demanded. Furthermore, it is becoming increasingly apparent that the benefits of using pesticides must be considered in the context of potential risks.

15.6 There Is Much to Learn from Past Agricultural Practices

EXPLAIN THIS

Does fresh water contain salt?

Over the past 100 years, there have been dramatic increases in crop yields. An acre of U.S. farmland in 1900 yielded about 30 bushels of corn. Today, that same acre yields on the order of 130 bushels of corn. This increased efficiency has meant a significant drop in the number of people needed to farm. In the early 1900s, about 33 million people in the United States lived and worked on farms. Today, only 2 million people are engaged in commercial farming in this country, raising crops and livestock.

Many of the farming methods used to obtain high yields have significant disadvantages. Pesticides and fertilizers, for example, pose certain risks. Pesticides are inherently toxic, and each year thousands of people working in agriculture are poisoned by the mishandling of these dangerous compounds. Fertilizers help plants grow, but major portions of applied fertilizer are washed into streams, rivers, ponds, and lakes, where they upset ecosystems, especially by promoting excessive growth of algae (see Section 16.3). Fertilizer runoff from fields, illustrated in **Figure 15.24**, can also contaminate drinking water supplies and thus affect human health. An ailment known as blue-baby syndrome, for example, results from drinking water containing high concentrations of nitrate ions, a main ingredient of most fertilizers. Nitrate ions in the bloodstream compete with oxygen for the positively charged iron ions of hemoglobin molecules. This leads to a form of anemia known as methemoglobinemia, to which babies are particularly sensitive. Aside from shortness of breath, one of the major symptoms is a bluish color of the skin.

LEARNING OBJECTIVE

Provide examples of poor agricultural practices.

CHEMICAL CONNECTIONS

How is the salinity of the ocean connected to photosynthesis?

▶ **Figure 15.24**
The water running off this farm field contains many pesticides and fertilizers that can be harmful to ecosystems and human health.

Poor maintenance of topsoil is also a major concern. Synthetic fertilizers have no organic bulk and do not provide a food source for soil microorganisms and earthworms. Over time, a soil treated with only these fertilizers loses biological activity, which diminishes the soil's fertility. Soils void of organic bulk become chalky and susceptible to wind erosion. Chalky soils lose their capacity to hold water, which means that more applied fertilizer is leached away. Ever-increasing amounts of fertilizer are thus needed.

Over the past 100 years, damaging farming practices have decreased the amount of topsoil in parts of the United States by as much as 50 percent. During the 1930s, farming practices and drought conditions created giant dust storms, such as the one shown in **Figure 15.25**, that removed major portions of the topsoil in Kansas, Oklahoma, Colorado, and Texas. In one storm, large dust clouds were carried all the way from the Midwest to Washington, D.C., and then into the Atlantic Ocean. Politicians in that city observing the effects of

▶ **Figure 15.25**
Poor soil conservation practices in the early 1900s contributed to the loss of much topsoil to wind storms thick with dust.

poor soil management right outside their windows quickly passed legislation that created the Soil Erosion Service, which became the National Resources Conservation Service; it continues to this day in its efforts to help protect the nation's topsoil for future generations.

Another limited resource required for farming is fresh water. In regions where rainfall is insufficient to support large crops, water is either channeled into fields from lakes, rivers, and streams or pumped from the ground. In many areas, groundwater is the primary source of fresh water, but excessive use of groundwater can lead to land subsidence, illustrated in **Figure 15.26**.

Any source of water other than rainwater requires an irrigation method to deliver water to the fields. Flooding is a common method, but it is not efficient because most of the water is lost in runoff and evaporation. Sprinkler systems are an improvement over flooding because they do not cause soil erosion. Such systems also lose large amounts of water, however, because a significant portion of the airborne water evaporates before reaching the ground.

All liquid water on the Earth's surface, no matter how fresh, contains some salts. After irrigation water evaporates from farmland, these salts are left behind, and over time, repeated irrigation causes the salinity of the soil to increase. This process is known as **salinization**, and it leads to a rapid decrease in productivity. To counteract growing soil salinity, farmers flood the land with huge quantities of water. As the water drains into a river, it washes the unwanted salts—along with significant amounts of topsoil—into the river. Thus, a river passing through farmlands gets saltier and saltier as it runs to the sea, as depicted in **Figure 15.27**.

READINGCHECK

What happens to farmland after repeated irrigation?

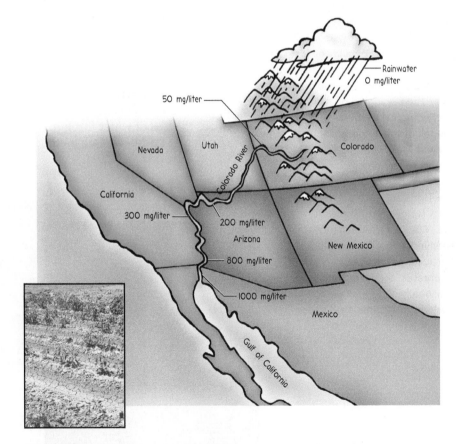

▲ **Figure 15.26**
The San Joaquin Valley of California has subsided by more than 35 feet since the pumping of groundwater began in the 1920s.

▲ **Figure 15.27**
As a river flows along, runoff from agricultural fields can add to the river's natural salinity. By the time the Colorado River reaches the Gulf of California, for example, it is too salty for productive farming. A typical safe drinking water standard for salt content is 500 milligrams/liter. Agricultural damage occurs when soil salinity reaches a concentration of about 800 milligrams/liter. The inset photo shows salt deposits accumulated in an irrigated field.

15.7 High Agricultural Yields Can Be Sustained with Proper Practices

LEARNING OBJECTIVE

Provide examples of sustainable agricultural practices.

 FOR YOUR INFORMATION

Those living in the agricultural regions of the arid southwestern United States are most familiar with the accumulation of visible salty sediments, as shown in the inset of Figure 15.27. Agricultural salinization, however, has been a problem since the beginning of agriculture. Archeological evidence shows that the fall of the Sumerian society of ancient Mesopotamia was likely due in part to the accumulation of salts on farmlands irrigated by waters of the Tigris and Euphrates rivers in what is now Iraq. Civilizations along the Nile river in Africa, however, survived through the millennia because, unlike the Tigris and Euphrates, the Nile river floods seasonally, which allows a seasonal flushing of accumulated salts.

EXPLAIN THIS

Why does organic farming exclude the use of pesticides when the pesticides themselves are organic?

Agriculture is the organized use of resources for the production of food. Whether these resources—mainly topsoil and fresh water—will be available for future generations depends on how well we manage them now. We already know from experience that pesticides and fertilizers cannot be applied liberally without threatening both topsoil quality and our supplies of clean groundwater (not to mention our health and that of the planet).

Over the past several decades, there have been strong movements toward developing methods and technologies that will sustain agricultural resources over the long term. Problems associated with irrigation, for example, can be solved by **microirrigation**, which is any method of delivering water directly to plant roots. Microirrigation not only prevents topsoil erosion but also minimizes the loss of water through evaporation, which in turn minimizes the salinization of farmland. One method of microirrigation is shown in **Figure 15.28**.

▲ Figure 15.28
The microirrigation method known as drip irrigation delivers water through long, narrow strips of punctured plastic tubing. The system provides only as much water as the plants need.

Organic Farming Is Environmentally Friendly

For controlling pests and maintaining fertile soil, the conventional agricultural industry is now looking at the efforts of many small-scale farmers who have demonstrated that significant crop yields can be obtained without pesticides or synthetic fertilizers. This method of farming is known as **organic farming**, where the term *organic* is used to indicate a concern for the environment and a commitment to using only chemicals that occur in nature.

To protect against pests, organic farmers alternate the crops planted on a particular plot of land. Such *crop rotation* works fairly well, because different crops are damaged by different pests. A pest that thrives on one season's crop of corn, for example, will do poorly on the next season's crop of alfalfa. Crop rotation also mitigates the depletion of minerals and nutrients. For fertilizer, organic farmers rely on compost, shown in **Figure 15.29**. They also include nitrogen-fixing plants in their crop rotation schedules.

Many claims are made that food produced organically is better for human consumption. Chemically, however, the molecules that plants absorb from synthetic fertilizers are the same as those they absorb from natural fertilizers. If organically grown produce tastes any better or is more nutritious than conventionally grown produce, the reason likely has to do with the genetic strain of the produce or with the greater attention paid to growing conditions, such as the health of the soil and the amount of water made available to plants.

Organic farming tends to be benevolent to the environment. In addition to avoiding the potential runoff of pesticides and fertilizers, organic farming is energy efficient, using only about 40 percent as much energy as conventional farming. A large portion of the energy savings arises because the manufacture of pesticides and fertilizers is energy intensive. For instance, each year in the United States, about 300 million barrels of oil is consumed for the production of nitrogen fertilizers.

Because much organically grown food, such as that shown in **Figure 15.30**, is grown locally, by purchasing it, you help support local farmers. Ultimately, though, your purchase of organically grown food is a vote in favor of environmentally friendly methods of farming.

▲ Figure 15.29
Odor-free backyard compost bins are easy to build and maintain. The secret is to optimize the exposure to atmospheric oxygen, which favors aerobic decomposition.

CONCEPTCHECK

Which are made of organic chemicals, organically grown foods or conventionally grown foods?

CHECK YOUR ANSWER Regardless of whether they are grown with natural fertilizer or synthetic fertilizer, all foods are made of organic chemicals—carbohydrates, lipids, proteins, nucleic acids, and vitamins. A thoroughly rinsed conventionally grown carrot may be just as good for you—or even better—as one grown without the use of synthetic fertilizers and pesticides. The *organic* in organic farming is a term used to designate a natural method of farming.

FOR YOUR INFORMATION

The outer coat of the naturally occurring rice grain contains bran as well as a number of valuable nutrients. Natural rice is transformed into white rice by milling away this outer coat. But why mill away this outer coat when it is so nutritious? The outer coat also includes fats that, over time, oxidize to make the grain turn rancid. Thus, unprocessed rice, also known as brown rice, is not well suited for long-term storage.

◀ Figure 15.30
(a) Market forces often result in higher prices for organically grown food.
(b) A product can bear this USDA organic seal if at least 95 percent of its ingredients are certified organic.

(a)

(b)

▲ Figure 15.31
Complementary crops such as legumes and corn are grown in alternating strips to enhance soil fertility. The strips follow the contour of the land to minimize erosion from rainwater or irrigation.

Integrated Crop Management Is a Strategy for Sustainable Agriculture

To meet concerns about sustaining agricultural resources over the long term, groups from industry, government, and academia have identified a whole-farm strategy called **integrated crop management** (ICM). This method of farming involves managing crops profitably with respect for the environment in ways that suit local soil, climatic, and economic conditions. Its aim is to safeguard a farm's natural assets over the long term through the use of practices that avoid waste, enhance energy efficiency, and minimize pollution. ICM is not a rigidly defined form of crop production, but rather a dynamic system that adapts and makes sensible use of the latest research, technology, advice, and experience.

One of the more significant aspects of ICM is its emphasis on multicropping, which means growing different crops on the same area of land either simultaneously, as shown in **Figure 15.31**, or in rotation from season to season. Like organic farming, multicropping achieves significant pest control, and it can be used to improve soil fertility. For example, nitrogen-generating crops such as legumes are a good complement to nitrogen-depleting crops such as corn.

An important component of ICM is **integrated pest management** (IPM), one of the aims of which is to reduce the initial severity of pest infestation. This can be accomplished through a number of routes. When a farm is started, for example, only crops that fit the local climate, soil, and topography should be grown. This selectivity makes for crops that are hardy and pest-resistant. Crops should also be rotated as much as possible to reduce pest and weed problems. Another IPM strategy is growing tree crops or hedges either around the perimeter of a farm or interspersed throughout the farm. These trees and hedges provide habitat, cover, and refuge for beneficial insects and such pest predators as spiders, snakes, and birds. As an added benefit, the trees and hedges also protect the land from wind erosion. Yet another IPM strategy is to breed and cultivate plants that have a natural resistance to pests. For centuries, this was accomplished by selectively mating plants that showed the greatest resistance. Today, this age-old method is quickly being supplanted by the techniques of genetic engineering.

Another aim of IPM is to minimize the use of pesticides. For example, many farmers now use the global positioning satellite (GPS) system to target precise pesticide applications. Using infrared satellite photography, illustrated in **Figure 15.32**, and careful walk-through assessment of field conditions, farmers can match pesticide blends with crop needs. Computers link application equipment with the GPS satellites, which "beam" pesticide application adjustments every few seconds as a farmer moves across a field. This same technology also works well with selective delivery of synthetic fertilizers.

▶ Figure 15.32
(a) Infrared satellite images reveal much information about our agricultural lands. Shades of red, for example, indicate crop health, browns show how much pesticide has been applied, and black shows areas prone to flooding. (b) In a practice called precision farming, a computer system using data from high-resolution satellite imagery guides a GPS-equipped tractor to pinpoint nutrient and pesticide placement.

(a)

(b)

Many other methods of pest control can be used in place of chemical pesticides or in combination with them to minimize the need for these agents. Depending on the availability of labor, egg masses or larvae can be hand-picked off plants. Instead of using herbicides, weeds can be tilled under. An insect population can also be controlled through various biological approaches, such as by introducing large numbers of sterile insects into a population or by introducing natural predators, as shown in **Figure 15.33**.

Another way to control the proliferation of insects is to modify their behavior through the use of **pheromones**, which are volatile organic molecules that insects release to communicate with one another. Each insect species produces its own set of pheromones, some as warning signals and others as sexual attractants. Sexual pheromones synthesized in the laboratory can be used to lure harmful insects to localized insecticide deposits, reducing the need for spraying an entire field, as depicted in **Figure 15.34**.

Nature is sophisticated, and if we are to work with nature in a sustainable way, our methods must also be sophisticated. New and improved techniques provide the farmer with a menu of possible actions in response to nature's ever-changing parameters. With each action, however, the farmer must be aware of its potential environmental impact. In this sense, the human who farms sustainably is not dominating nature but rather working with it.

READING CHECK

How are pheromones used to protect crops?

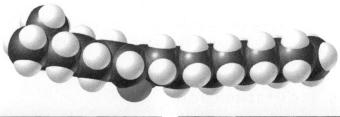

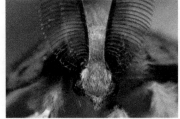 

◀ **Figure 15.34**
Female gypsy moths emit the pheromone disparlure (top) to entice male gypsy moths (bottom left) into mating. The males are so sensitive to this compound that they can detect one molecule in 1017 molecules of air. This astounding sensitivity enables them to respond to a female that may be more than 1 kilometer away. However, they can also be tricked into responding to insecticide traps laced with synthetic disparlure (bottom right).

Chapter 15 Review

LEARNING OBJECTIVES

Describe the flow of energy and nutrients between trophic levels. (15.1)	→	*Questions 1–3, 24, 28, 29, 30, 55*
Distinguish between the macro- and micronutrients needed by plants. (15.2)	→	*Questions 4–7, 25, 31–35*
Identify what makes for healthy soil. (15.3)	→	*Questions 8–10, 23, 36–40,*
Describe the origins of straight and mixed fertilizers. (15.4)	→	*Questions 11–13, 41–43*
Describe the benefits and risks of insecticides, herbicides, and fungicides. (15.5)	→	*Questions 14–16, 26, 44–47, 57*
Provide examples of poor agricultural practices. (15.6)	→	*Questions 17–19, 27, 48–50*
Provide examples of sustainable agricultural practices. (15.7)	→	*Questions 20–22, 51–54, 56, 58–60*

SUMMARY OF TERMS (KNOWLEDGE)

Bioaccumulation The process whereby a toxic chemical that enters a food chain at a low trophic level becomes more concentrated in organisms higher up the chain.

Compost Fertilizer formed by the decay of organic matter.

Consumer An organism that takes in the matter and energy of other organisms.

Decomposer An organism in the soil that transforms once-living matter to nutrients.

Horizon A layer of soil.

Humus The organic matter of topsoil.

Integrated crop management A whole-farm strategy that involves managing crops in ways that suit local soil, climatic, and economic conditions.

Integrated pest management A pest-control strategy that emphasizes preventing, planning, and using a variety of pest-control resources.

Microirrigation A method of delivering water directly to plant roots.

Mixed fertilizer A fertilizer that contains the plant nutrients nitrogen, phosphorus, and potassium.

Nitrogen fixation A chemical reaction that converts atmospheric nitrogen to some form of nitrogen usable by plants.

Organic farming Farming without the use of pesticides or synthetic fertilizers.

Pheromone An organic molecule secreted by insects to communicate with one another.

Producer An organism at the bottom of a trophic structure.

Salinization The process whereby irrigated land becomes saltier as irrigation water evaporates.

Straight fertilizer A fertilizer that contains only one nutrient.

Trophic structure The pattern of feeding relationships in a community of organisms.

READING CHECK QUESTIONS (COMPREHENSION)

15.1 Humans Eat at All Trophic Levels

1. What are the two major chemical products of photosynthesis?

2. In a trophic structure, what distinguishes a producer from a consumer?

3. Why are the number of trophic levels limited?

15.2 Plants Require Nutrients

4. Do plants require oxygen?

5. What effect have terrestrial plants had on the composition of ocean water?

6. Why are calcium and magnesium deficiencies rare in plants?

7. What is the most common form of sulfur absorbed by plants?

15.3 Soil Fertility Is Determined by Soil Structure and Nutrient Retention

8. Which soil horizon is void of life?

9. What are four important components of fertile topsoil?

10. What is one of the great advantages to having humus in soil?

15.4 Natural and Synthetic Fertilizers Help Restore Soil Fertility

11. What is the difference between a straight fertilizer and a mixed fertilizer?

12. What advantages do mixed synthetic fertilizers have over compost?

13. What advantages does compost have over mixed synthetic fertilizers?

15.5 Pesticides Kill Insects, Weeds, and Fungi

14. What are three classes of insecticides?

15. Is DDT still being used today?

16. Glyphosate interferes with the biosynthesis of which two amino acids?

15.6 There Is Much to Learn from Past Agricultural Practices

17. How do pesticides and fertilizers end up in our drinking water?

18. What is missing from synthetic fertilizers that makes them harmful to soil?

19. How is irrigation damaging to topsoil?

15.7 High Agricultural Yields Can Be Sustained with Proper Practices

20. What are the advantages of microirrigation?

21. What is organic farming?

22. How is space technology used to reduce the amount of pesticides applied to farmlands?

CONFIRM THE CHEMISTRY (HANDS-ON APPLICATION)

23. You can measure the pH of soil samples using the red cabbage pH indicator described in Chapter 10. Saturate a soil sample with water, making a thin mud slurry. Stir the slurry and allow it to settle until about 1 centimeter of water appears above the soil. If this top layer of water does not appear, add more water, stir, and allow for further settling.

You will need to filter this soil-treated water. To do so, remove the bulb from a cooking baster and stuff several cotton balls into the top end of the baster. Use a chopstick or skewer to compact the cotton toward the narrow end of the baster. Pour the water from the settled soil into the top of the baster. Attach the bulb and squeeze the liquid through the cotton and into one of the glasses. If the filtered water is still brown or cloudy, repeat the filtering using clean cotton balls each time.

Add to a second glass as much fresh water as there is filtered water in the first glass. Add equal amounts of the pH indicator to the two glasses. Compare colors, recalling from Chapter 10 that a deeper red color means greater acidity and that green indicates alkaline. You might want to test soil samples from several different sites and compare your results.

THINK AND COMPARE (ANALYSIS)

24. Rank the following organisms from lowest to highest trophic level:
   a. beetle
   b. cougar
   c. kangaroo rat
   d. gray fox
   e. cactus

25. Rank the following relative amounts of macronutrient ions in dry plant material from greatest to least:
   a. nitrogen, N
   b. sulfur, S
   c. magnesium, Mg
   d. phosphorus, P
   e. calcium, Ca

26. Rank the following in order of increasing concentration of mercury, Hg:
   a. zooplankton
   b. trout
   c. swordfish

27. Rank the following in order of increasing salt concentration:
   a. rainwater
   b. lake water
   c. stream water
   d. ocean water

THINK AND EXPLAIN (SYNTHESIS)

15.1 Humans Eat at All Trophic Levels

28. What would happen to a forest without the action of decomposers?

29. What happens to most biochemical energy as it passes from one trophic level to the next?

30. Why would a predominately meat-based diet be a severe restriction on the possible size of the human population?

15.2 Plants Require Nutrients

31. Thoroughly dried dead plant material is 44.4 percent oxygen by weight. How is this oxygen held in the plant so that it doesn't escape into the atmosphere?

32. Why do the leaves of plants deficient in nitrogen turn yellow? Consider the structure of the porphyrin ring shown in Figure 15.6.

33. Earthworms are repelled by the high concentrations of nutrients in soils treated with synthetic fertilizers. How might this affect the soil structure?

34. How many carbonyl groups are depicted in the illustration of the porphyrin ring of Figure 15.6?

35. Why is atmospheric nitrogen, N_2, so chemically unreactive?

15.3 Soil Fertility Is Determined by Soil Structure and Nutrient Retention

36. How does humus contribute to the acidity of soil?

37. Why do groundskeepers poke holes in football fields?

38. Pockets of open space in soil are necessary for plant health. If these pockets are too large, however, plant health suffers. Why?

39. Why do soils that contain a high percentage of clay retain plant nutrients so well?

40. Why don't plants grow well in nutrient-rich soils that contain a high percentage of clay?

15.4 Natural and Synthetic Fertilizers Help Restore Soil Fertility

41. What happens to ammonium ions in alkaline soil? How might this reaction favor the loss of nitrogen from the soil?

42. How does the organic bulk of compost help to maintain fertile soil conditions?

43. Which is better for a plant: an ammonium ion from compost or an ammonium ion from synthetic fertilizer?

15.5 Pesticides Kill Insects, Weeds, and Fungi

44. Why do chlorinated hydrocarbons persist longer in the environment compared to organophosphorus compounds?

45. Why does DDT have such a strong affinity for fat tissue?

46. What do the structures of DDT and TCDD have in common? How are these compounds different?

47. Which compound is more acidic: 2,4,5-T or TCDD?

15.6 There Is Much to Learn from Past Agricultural Practices

48. When only synthetic fertilizers are used on a crop, the quantity used needs to be increased over time. Why?

49. How might periodic floods from rainstorms benefit irrigated cropland?

50. Why are the oceans salty?

15.7 High Agricultural Yields Can Be Sustained with Proper Practices

51. Distinguish between organic farming and integrated crop management.

52. Why is it beneficial to grow two or three different crops simultaneously in the same field?

53. Besides lure-and-trap, how else might pheromones be used to decrease insect populations?

54. How is petroleum connected to the food we eat?

THINK AND DISCUSS (EVALUATION)

55. For a person, what are the benefits of eating high on the food chain? What are the benefits of eating low on the food chain? How does this compare with the benefits for the human population as a whole?

56. As stated at the beginning of this chapter, rice has been genetically engineered to contain a precursor to vitamin A. Proponents for golden rice note that this rice is provided as a humanitarian endeavor at no financial cost to the farmers of developing regions. Activist organizations such as Greenpeace, however, are adamantly opposed to golden rice. Why?

57. Should there be an international ban on the production and use of DDT? Why or why not?

58. The caption to Figure 15.30 notes that market forces often result in higher prices for organically grown foods. Identify some of these market forces.

59. Assume you are a Brazilian farmer looking to clear-cut a rainforest to make room for cattle pasture. An environmental activist from the predominately beef-eating United States knocks at your door to convince you not to cut. What are some of the arguments the activist might present? What counter arguments can you think of for proceeding with your plan?

60. Which of the following options should be more effective at sustaining the lives of poor people living in a developing nation? Option 1: Large farms where surplus amounts of food are grown by relatively few owners of large tracts of land. Option 2: Many small farms where sufficient amounts of food are grown by numerous owners of small tracts of land.

READINESS ASSURANCE TEST (RAT)

If you have a good handle on this chapter, then you should be able to score at least 7 out of 10 on this RAT. Check your answers online at www.ConceptualChemistry.com. If you score less than 7, you need to study further before moving on.

Choose the BEST answer to the following.

1. What is acting as the energy storage media in photosynthesis?
   a. chlorophyll
   b. carbohydrates
   c. carbon dioxide
   d. water
   e. oxygen

2. Why would a predominately meat-based diet be a severe restriction on the possible size of the human population?
   a. Not enough biochemical energy is available at that trophic level to support a large population.
   b. The land area taken up by livestock would prevent people from finding new places to live.
   c. Such a diet would shorten the average life span of humans.
   d. The hormones found in beef have a negative effect on human fertility.

3. Why might soil with a very low pH be more susceptible to leaching?
   a. The acid ions exchange with the nutrient ions, which are then rinsed away.
   b. The base ions start to degrade the minerals and lead to insoluble ion salts.
   c. The acid ions decompose the vitamins and minerals stored in the roots.
   d. The base ions decompose the vitamins and minerals stored in the roots and suck the valuable minerals into the water.
   e. None of the above is true.

4. The use of tractors and other heavy farm machinery compacts soil. How might this compacting affect soil fertility?
   a. The increased pressure causes fertilizer molecules to react with one another, which has the effect of harming soil fertility.
   b. The weight of the heavy machinery kills decomposers such as worms, which has the effect of harming soil fertility.
   c. The soil fertility is harmed because the weight of the heavy machinery squeezes out the water.
   d. Once compacted, the soil is not able to hold water and air, which harms the fertility of the soil.

5. What happens to ammonium ions, NH_4^+, in alkaline soil? How might this reaction favor the loss of nitrogen from the soil?
   a. In alkaline soils, ammonium ions have a greater tendency to bind with inorganic particles, thereby becoming unavailable to plants.
   b. Ammonium ions turn into nitrate ions, which are lost through water runoff.

   c. Ammonium ions turn into ammonia, which is a gas that can escape into the atmosphere.
   d. Ammonium ions turn into nitrogen gas, which escapes back into the atmosphere.

6. When only synthetic fertilizers are used on a crop, the quantity used needs to be increased over time. Why?
   a. They are washed away by seasonal rains, but residual fertilizers remain.
   b. Synthetic fertilizers provide no organic bulk.
   c. More and more genetically engineered crops are being planted.
   d. Organisms in the soil develop a resistance to these fertilizers.

7. Shown below are the N-P-K ratings for three fertilizer additives: sawdust, fish meal, and wood ashes. Using what you know about the chemical composition of these substances, assign each to its most likely N-P-K rating: 5-3-3, 0-1.5-8, or 0.2-0-0.2.
   a. sawdust: 5-3-3; fish meal: 0-1.5-8; wood ashes: 0.2-0-0.2
   b. sawdust: 0.2-0-0.2; fish meal: 0-1.5-8; wood ashes: 5-3-3
   c. sawdust: 0-1.5-8; fish meal: 0.2-0-0.2; wood ashes: 5-3-3
   d. sawdust: 0.2-0-0.2; fish meal: 5-3-3; wood ashes: 0-1.5-8

8. Many farmers in the United States are paid by the government not to farm their land. Based on your understanding of the concepts presented in this chapter, cite one reason this might be so.
   a. It keeps the price of food artificially high.
   b. It provides a means of controlling the global population of people.
   c. It conserves the topsoil.
   d. Farmed land has dramatic effects on local weather patterns.

9. Industrial wastes and agricultural pollutants are made of the same kinds of atoms already present in the environment. So why worry about them?
   a. Industrial waste and agricultural pollutant atoms are "contaminated" and in need of purification.
   b. Industrial waste and agricultural pollutant atoms combine to form harmful molecules with markedly different properties from the atoms that make up the molecules.
   c. Industrial waste and agricultural pollutant atoms become "abnormal," in much the same way a cancerous structure forms.
   d. Industrial waste and agricultural pollutant atoms aren't much of a concern. The government makes an issue of them mostly to serve political ends.

10. The specific aim of IPM is to
   a. promote the popularity of organic farming.
   b. minimize the amount of water lost during irrigation.
   c. improve soil fertility.
   d. minimize the use of pesticides.

Genetically Modified Foods

Over the past couple of decades, advances in our understanding of genetics have led to profound developments in agriculture. For centuries, farmers have improved crops and domestic animals by breeding for desirable traits. This uncertain and often lengthy process can now be performed relatively quickly and with great certainty using the tools of modern molecular biology to introduce genes for desired traits into plants and animals. The resulting organisms are called *transgenic organisms* because they contain one or more genes from another species. In the media and marketplace, transgenic organisms are also known as *genetically modified organisms*, or GMOs for short.

As an example, consider that transgenic bacteria have been engineered to mass-produce a variety of valuable proteins, including bovine growth hormone (BGH). When this hormone is injected into dairy or beef cattle, it raises milk production or improves weight gain. It has passed U.S. government-sponsored safety standards and is now being used extensively on dairy herds in the United States. European and Canadian governments, however, do not allow its use in cattle. One reason is that BGH increases the rates of infections among cattle. This prompts farmers to overuse antibiotics that end up being consumed by humans.

Most of the progress in transgenic agriculture has been with plants. Several major crops have been engineered with genes that create proteins having insecticidal properties. The insect pest is killed only when it feeds on the crop. Figure 1 illustrates this technique for corn. With such a mechanism, most—although not necessarily all—nontarget benevolent organisms are left unharmed and the need for pesticide application is reduced. Other major crops have been engineered with genes that make them resistant to the herbicide glyphosate, meaning that the herbicide kills weeds in a field but doesn't threaten the crop planted there. Researchers have also inserted into sweet potato plants a gene coding for a dietary protein. This protein contains significant amounts of the amino acids essential to adult humans. Figure 2 shows these protein-rich sweet potatoes, which are easy to cultivate and hold special value for developing nations, where high-quality protein foods are hard to come by.

Worldwide, about 70 million acres of farmland are cultivated with transgenic crops. As a result, about one-third of the world corn harvest and more than one-half of the world soybean harvest now come from genetically engineered plants.

▲ Figure 2
Gene transfer has made sweet potatoes a better protein source.

The examples just described involve the transposing of only one gene or a couple of genes into the transgenic organism. Many desirable traits, however, involve clusters of genes. An important example is the ability to fix nitrogen. Intense research is currently under way to transpose all the genes necessary for nitrogen fixation into plants that do not naturally fix nitrogen. With such a transgenic species, the expensive production and application of nitrogen fertilizers becomes unnecessary. Because many genes are involved, the system is complicated and currently beyond biotechnical capabilities, but perhaps that won't be true for long.

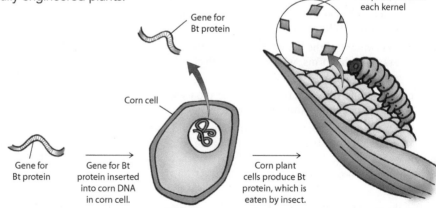

▲ Figure 1
The bacterium *Bacillus thuringiensis* (Bt) produces proteins that are toxic to insects such as the corn borer, a devastating corn pest. The external application of Bt proteins on corn, however, cannot control the corn borer once it is inside the stalk. Corn is made resistant to the corn borer by splicing the gene for the Bt protein into corn DNA. The resulting corn plant produces the Bt protein in its cells and is thus fully resistant to the corn borer.

There is heated debate about genetically engineered agricultural products. Some scientists argue that producing transgenic organisms is only an extension of traditional cross-breeding, the procedure that has given us such new and interesting products as the tangelo (a tangerine–grapefruit hybrid). In general, the Food and Drug Administration has held that if the result of genetic engineering is not significantly different from a product already on the market, testing is not required. On the other side of the argument are scientists who believe that creating transgenic organisms is radically different from hybridizing closely related species of plants or animals. There is concern, for example, that genetically engineered crops might grow too well, ultimately reseeding themselves in areas where they are not desired and thus becoming "superweeds." Transgenic crops might also pass their new genes to close relatives in neighboring wild areas, creating offspring that would be difficult to control.

Stay tuned for developments in the area of transgenic agriculture, such as the promising development of golden rice discussed at the beginning of this chapter. The power of genetic engineering, however, demands that we move cautiously with all necessary safeguards in place. One of the more important safeguards, no doubt, will be a well-informed general public.

But are transgenic organisms really necessary to feed our growing human population? Demographers project that, the human population will begin to stabilize within a century at about 14 billion inhabitants. Will we be able to feed ourselves at that point? The answer is probably yes, but the assumption here is that our food supply expands in a way that does not destroy the natural environment. For agriculture to be sustainable, a steady stream of new technologies that minimize environmental damage must be developed. Transgenic organisms are likely to play a signficant role.

Interestingly, the most critical problems faced by those seeking to counteract world hunger are more likely to be social rather than technical. Above all, efforts toward stabilizing the world population must continue in earnest. Already, most of the world's farmable land is now under cultivation. As the population grows, more food will be required while at the same time more farmable land will be lost to residential and business development. In tropical areas, economic pressures to slash and burn rainforests for the formation of additional farmland will probably continue.

Even with a stable world population, it cannot be assumed that a large-enough food supply will lead to the end of world hunger. Today, the abundance of food is at an all-time high, and yet an estimated 8.7 million individuals, most of them young children, die each year from a lack of adequate nutrition. Amartya Sen, a leader in the fight against world hunger and the 1998 Nobel laureate in Economics, points out that in most circumstances, malnutrition arises not from a lack of food, but from a lack of appropriate social infrastructure, as Figure 3 shows. Backed by strong evidence, Sen argues that "public action can eradicate the terrible and resilient problems of starvation and hunger in the world in which we live." Efforts toward optimizing the food yields from agriculture must therefore be matched by efforts to build social, political, and economic systems that give those people facing starvation the means for survival. World hunger is not inevitable.

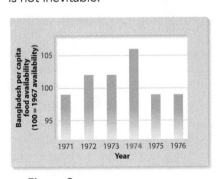

▲ Figure 3
The Bangladesh famine of 1974 occurred during a period when the amount of food available per person in that country was at a peak. It was unemployment, hoarding, and inflated food prices that drove millions to their death.

Think and Discuss

1. How might genetic engineering be used to counteract the negative effects of salinization?

2. Bovine growth hormone was made available to dairy farmers in the early 1990s. At that time, there was no shortage of milk, nor was there any anticipated shortage of milk. Why did the farmers then start using this growth hormone on their cows? Who benefited most? Many physicians are concerned about a possible, although not yet proven, link between growth-hormone-induced milk and certain cancers in humans. Knowing this, are you willing to drink milk from a cow whose milk production has been increased by injections of bovine growth hormone? Why or why not?

3. Should transgenic foods be labeled as such in the stores where they are sold? Would you be reluctant to buy this food? What if the food were processed and in a box, such as a box of crackers? What if the food were fresh produce on display next to organic fresh produce? What if the GMO fresh produce looked healthier than the organic fresh produce and were less expensive?

4. Why might environmentalists welcome the introduction of corn genetically engineered to fix its own nitrogen? Why might they be opposed to it?

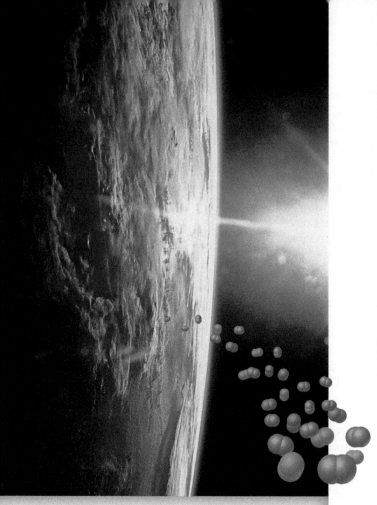

▲ Seen from space, the Earth's atmosphere appears as a narrow blue sliver on the horizon.

Protecting Water and Air Resources

THE MAIN IDEA

The Earth is huge, but so is our ability to transform the environment.

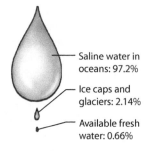

Saline water in oceans: 97.2%

Ice caps and glaciers: 2.14%

Available fresh water: 0.66%

There is a lot of water on the Earth, but about 97.2 percent of this water is saline (salty) ocean water. Another 2.14 percent is fresh water frozen in polar ice caps and glaciers. All the remaining water, less than 1 percent of the Earth's total, comprises water vapor in the atmosphere, water in the ground, and water in rivers and lakes—the fresh water we rely on in our daily lives. The Earth is large, having a diameter of about 13,000 kilometers. The atmosphere that surrounds the Earth, however, is only about 30 kilometers thick. From space, the atmosphere appears only as a narrow band along the horizon. Consider that if the Earth were the size of an apple, its atmosphere would be about as thick as the skin of the apple.

Our planet is a gigantic terrarium, and collectively, we are its caretakers. With this job comes the responsibility to learn how the Earth's resources can be properly managed for the benefit of all its inhabitants. In this chapter, we explore some of the fundamental dynamics of the Earth's water and atmosphere and the impact of human activities.

Chemistry

Rain in a Can

When water vapor condenses in a closed container, very low pressure is created inside the container. The atmospheric pressure on the outside then has the capacity to crush the container. In this activity, you will see how this works for water vapor condensing inside an aluminum soda can. Note: avoid touching the steam produced in this activity—steam burns can be severe.

PROCEDURE

1. Add ice water (at least 1 inch deep) to a saucepan and set aside.
2. Put a tablespoon of water in an aluminum can and heat on a stove until steam can be seen coming from the top.
3. Grasp the can with tongs and invert it (top first) into the ice water in the saucepan. Hold it there for a few moments until the can implodes.

ANALYZE AND CONCLUDE

1. How much air was in the can when it was being heated and steam started coming out of the top? How much water vapor was in the can at that time?
2. What happened to the temperature inside the can when it was inverted in the ice water? What do you suppose then happened to the water vapor within the can?
3. Which occupies more volume: water in a gaseous phase or the same mass of water in a liquid phase?
4. What happened to the pressure within the can as the water vapor within it condensed? How did this internal pressure compare to the external pressure exerted by the atmosphere?

16.1 Water on the Move

EXPLAIN THIS

How can snow on a mountaintop disappear without melting?

The Earth's water is constantly circulating, powered by the heat of the Sun and the force of gravity. The Sun's heat causes water from the Earth's oceans, lakes, rivers, and glaciers to evaporate into the atmosphere. As the atmosphere becomes saturated with moisture, the water precipitates in the form of either rain or snow. This constant water movement and phase changing is called the **hydrologic cycle.** As **Figure 16.1** shows, the route of water through the cycle can be from ocean directly back to ocean or can take a more circuitous route over the ground and even underground.

In the direct route, water molecules in the ocean evaporate into the atmosphere, condense to form clouds, and then precipitate into the ocean as either rain or snow, to begin the cycle anew.

The cycle is more complex when precipitation falls on land. As with the direct route, the cycle "begins" with ocean water evaporating into the atmosphere. Instead of forming clouds over the water, however, the moist air is blown by winds until it is over land. Now there are four possibilities for what happens to the water once it precipitates. It may (1) evaporate from the land back into the atmosphere, (2) infiltrate into the ground, (3) become part of a snowpack or glacier, or (4) drain to a river and then flow back to the ocean.

Water that seeps below the Earth's surface fills the spaces between soil particles until the soil reaches *saturation*, at which point every space is filled with water. The upper boundary of the saturated zone is called the **water table.** The depth of the water table varies with precipitation and with climate. It ranges from zero depth in marshes and swamps (meaning the water table is at ground level at these locations) to hundreds of meters deep in some desert regions.

LEARNING OBJECTIVE

Describe how water circulates through the hydrologic cycle.

 CHEMICAL CONNECTIONS

How are the tears of a crying baby in the United States connected to the Yangtze River in China?

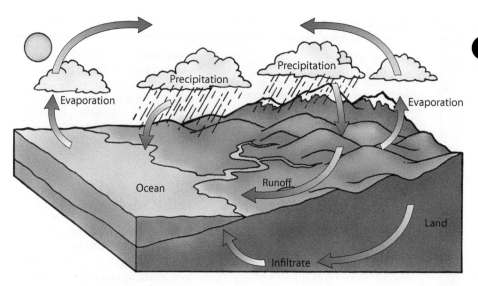

▲ **Figure 16.1**
The hydrologic cycle. Water evaporated at the Earth's surface enters the atmosphere as water vapor, condenses into clouds, precipitates as rain or snow, and falls back to the surface, only to go through the cycle yet another time.

READINGCHECK

Is it possible for the water table to lie above ground?

The water table also tends to follow the contours of the land and lowers during times of drought, as shown in **Figure 16.2**. Many lakes and streams are simply regions where the water table lies above the land surface.

All water below the Earth's surface is called *groundwater*. (Liquid water that is on the surface—in streams, rivers, and lakes—is called, naturally enough, *surface water*.) Any water-bearing soil layer is called an **aquifer,** which can be thought of as an underground water reservoir. Aquifers underlie the land in many places and collectively contain an enormous amount of fresh water— approximately 35 times the total volume of water in fresh water lakes, rivers, and streams combined. More than half the land area of the United States is underlain by aquifers, such as the Ogallala Aquifer, stretching from South Dakota to Texas and from Colorado to Arkansas.

As the human population grows, the demand for fresh water grows. Precipitation is the Earth's only natural source of groundwater recharge. Although the reservoir of groundwater is great, when the pumping rate exceeds the recharge rate, there can be a problem. In wet climates, such as in the Pacific Northwest, extraction is often balanced by recharge. In dry climates,

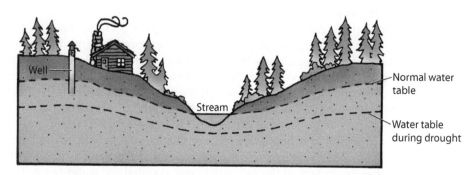

▲ **Figure 16.2**
The water table in any location roughly parallels surface contouring. In times of drought, the water table falls, reducing stream flow and drying up wells. It also falls when the amount of water pumped out of a well exceeds the amount replaced as precipitated water infiltrates the ground and recharges the supply.

however, extraction can easily exceed recharge. To support large populations, these areas must import their water from distant sources, typically through aqueducts. In southern California, for example, most of the fresh water comes from the Colorado River through aqueducts that stretch hundreds of miles.

The Ogallala Aquifer, which is below the dry High Plains, has supplied water to this thirsty agricultural region for more than 100 years. Most of the aquifer is water from the last ice age, some 11,000 years ago, that has been locked underground with few sources of replenishment. If pumping were to cease, it would take many thousands of years for the water table to return to its original level. In this respect, the Ogallala, unlike most other aquifers, is a limited and nonrenewable resource.

As water is removed from the spaces between soil particles, the sediments compact and the ground surface is lowered—it *subsides*. In areas where groundwater withdrawal has been extreme, the ground surface has subsided significantly. In the United States, extensive groundwater withdrawal for irrigation of the San Joaquin Valley of California has caused the water table to drop 75 meters in 20 years, and the resulting land subsidence has been significant.

Probably the most well-known example of land subsidence is the Leaning Tower of Pisa in Italy, shown in **Figure 16.3**. Over the years, as groundwater has been withdrawn to supply the growing city, the tilt of the tower has increased.

▲ **Figure 16.3**
The Leaning Tower of Pisa, built centuries ago, slowly acquired a deviation from the vertical of about 4.6 meters as a result of groundwater withdrawal. The tower's foundation has been stabilized by groundwater withdrawal management, and the tower should remain stable for years to come.

CONCEPT CHECK

An aquifer is a body of underground fresh water. Where does this water come from?

CHECK YOUR ANSWER The source of all natural underground (and aboveground) fresh water is the atmosphere, which gets most of its moisture from the evaporation of ocean water.

16.2 Collectively, We Consume Huge Amounts of Water

EXPLAIN THIS

You don't use water in your household for a 2-hour period, yet your water meter shows otherwise. What's the problem?

The U.S. Geological Survey (USGS) began compiling national water use data in 1950. Since then, this federal agency has been conducting surveys at 5-year intervals. According to their data, the rate at which water enters all U.S. aquifers combined is about 6790 billion liters/day.* In 2005, we were withdrawing water from these aquifers at an average rate of 1319 billion liters/day, meaning we were taking out about 20 percent of this water supply. As shown in **Figure 16.4**, the bulk of this water was used for irrigation and as a coolant in the generation of thermoelectric power.

The numbers in Figure 16.4 tell us that, based on a population of 301 million, the 2005 per-person water usage in the United States was 4380 liters/day (1319 billion liters/day ÷ 301 million persons). According to the USGS, each person's personal use was about 8 percent of that amount, which comes to about 350 liters/day. However, only about one-fourth of that 350 liters is water we

LEARNING OBJECTIVE

Review water consumption trends in the United States.

 READING CHECK

How frequently does the USGS tabulate national water use data?

---

*These reports are made every 5 years, but it takes about 5 years to compile and cross-check the data. Thus, at the time of the printing of this textbook, the report for 2010 was not yet available. To check on the status of the 2010 report, go to http://water.usgs.gov/watuse.

▶ Figure 16.4
Water usage in the United States (2005) in billions of liters per day.

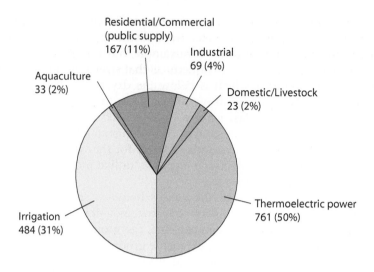

Residential/Commercial (public supply)
167 (11%)

Industrial
69 (4%)

Aquaculture
33 (2%)

Domestic/Livestock
23 (2%)

Thermoelectric power
761 (50%)

Irrigation
484 (31%)

FOR YOUR INFORMATION

Groundwater is found beneath nearly every square foot of land. In some regions, such as deserts, this groundwater is many meters deep and difficult to reach. In moist regions, such as Europe or most of North America, the groundwater is relatively close to the surface and easy to find. Dowsing is a form of pseudoscience in which a person claims to be able to find groundwater using a pronged stick that vibrates wildly upon the discovery of the water. A well is dug at the indicated location, groundwater is revealed, and the dowser is paid for her services. Of course, the dowser is usually successful because groundwater is nearly everywhere. The real test of a dowser would be to find a region where water is *not* located.

either drink or use to water our lawns or gardens. The other three-fourths—about 263 liters!—becomes household wastewater from our bathtubs, toilets, sinks, and washing machines.

It may seem that because we as a population consume only about 20 percent of the fresh water available to us, there is little need to conserve. This percentage, however, is only an average. In many drier regions of the western United States, water usage already exceeds the rate at which aquifers in the region are recharged. In Albuquerque, New Mexico, for example, escalating water consumption has caused the underlying aquifer to drop by about 50 meters over the past 40 years.

The quality of fresh water varies from one region to another. Deep water deposits, for example, are often high in dissolved solids. So even in regions where fresh water is plentiful, it's important to conserve water to help protect that smaller portion of the water supply that is of greatest purity. Furthermore, we are not the only species that relies on fresh water. Many ecosystems, such as lakes and wetlands, are already stressed by our increasing water demands. Water conservation can go far to alleviate this stress even in the face of our growing population.

According to the USGS, there is good news regarding water conservation efforts in the United States. As shown in **Figure 16.5**, total fresh water withdrawal peaked in 1980 at around 1420 billion liters/day. By 2005, however, the

▶ Figure 16.5
Total withdrawals of fresh water increased from 1950 to 1980, largely because of expanded irrigation systems and urban development. After 1980, however, conservation measures reduced water usage even in the face of a growing population.

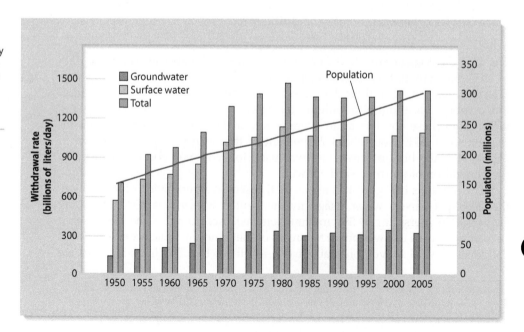

withdrawal rate had declined to 1319 billion liters/day even though the population grew from 230 million to 301 million over the same time period. This remarkable savings in fresh water was largely the result of improved irrigation techniques (Section 15.7), but enhanced public awareness of water resources and conservation programs was also a contributing factor.

16.3 Human Activities Can Pollute Water

EXPLAIN THIS

Why is methane a significant component of the gas emanating from a buried landfill?

Water pollution can arise from either point sources or nonpoint sources. A **point source** is one specific, well-defined location where a pollutant enters a body of water. An example of a point source is the wastewater pipes of a factory or sewage treatment plant, as shown in **Figure 16.6a**. Point sources are relatively easy to monitor and regulate. A **nonpoint source** is a source in which pollutants originate at diverse locations, oil residue on streets being one example. Water becomes polluted as rain washes the oil residue into streams, rivers, and lakes. Agricultural runoff and household chemicals making their way into the storm drains of **Figure 16.6b** are two other common examples of nonpoint sources of water pollution. Because it is difficult to monitor and regulate nonpoint sources, the most effective solutions are often public awareness campaigns emphasizing responsible disposal practices. As **Figure 16.7** illustrates, lawn care in the United States is a major nonpoint source of water pollution.

The rate of water contamination from many point sources has decreased markedly since the passing of the Clean Water Act of 1972 and its subsequent amendments. Prior to 1972, the user of a water supply, such as a municipality, was responsible for protecting the supply. Because it is far more efficient to control water pollutants before they are released into the environment, the Clean Water Act shifted the burden of protecting a water supply to anyone discharging wastes into the water, such as a local industry.

LEARNING OBJECTIVE

Identify sources of water pollution and explain the significance of biochemical oxygen demand.

 READING CHECK

What are two sources of nonpoint water pollution?

(a)

(b)

▲ Figure 16.6
(a) This technician is assessing the clarity of the effluent coming from a wastewater treatment facility, a common point source of pollution. (b) Nonpoint sources are not so easily regulated and depend more on public awareness.

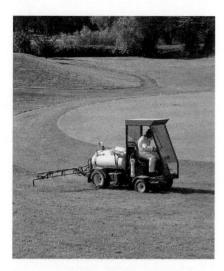

▲ **Figure 16.7**
Lawns cover 25 to 30 million acres in the United States, an area larger than Virginia. People taking care of these lawns use up to two and a half times more pesticides per acre than farmers use on croplands.

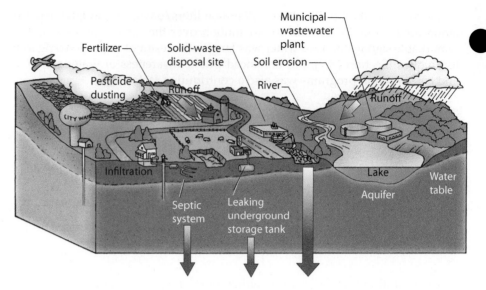

▲ **Figure 16.8**
The arrows indicate some major sources of groundwater contamination.

When rivers and lakes become polluted, they can be cleaned because they are accessible. When groundwater becomes polluted, however, it's a different story. Even after the sources of the pollution have been removed, it can be a lifetime before the contaminants are removed, not only because the groundwater is so inaccessible but also because the flow rate of many aquifers is extremely slow—on the order of only a few centimeters per day! As shown in **Figure 16.8**, groundwater is susceptible to a wide variety of point and nonpoint pollution sources.

Municipal solid-waste disposal sites are a common source of groundwater pollution. Rainwater infiltrating a disposal site may dissolve a variety of chemicals from the solid waste. The resulting solution, known as **leachate,** can move into the groundwater, forming a contamination "plume" that spreads in the direction of groundwater flow, as shown in **Figure 16.9**. To reduce the chances of groundwater contamination, the site can be underlain and capped with layers of compacted clay or plastic sheeting that prevent leachate from entering the ground. A collection system designed to catch any draining leachate may also be used.

▶ **Figure 16.9**
A contaminant plume of leachate spreads in the direction of groundwater flow.

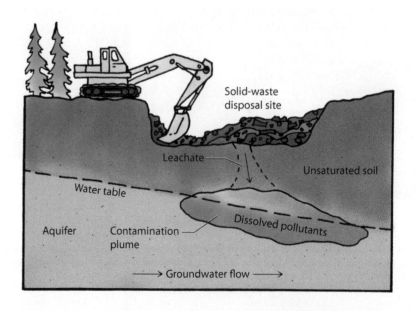

Another common source of groundwater pollution is sewage, which includes drainage from septic tanks and inadequate or broken sewer lines. Animal sewage, especially from factory-style animal farms, is also a source of groundwater (and river water) pollution. Sewage water contains bacteria, which if untreated can cause waterborne diseases such as typhoid, cholera, and infectious hepatitis. If the contaminated groundwater travels relatively quickly through underground sediments in which there are large air pockets, bacteria and viruses can be carried considerable distances. However, if the contaminated groundwater flows through underground sediments in which the air pockets are very small, as is the case with sand, pathogens are filtered out of the water.

CONCEPTCHECK

What is the difference between a point source and a nonpoint source of pollution?

CHECK YOUR ANSWER A point source arises from a specific location—you can pinpoint it on a map. A nonpoint source represents a collection of many sources, each difficult to trace. To specify a nonpoint source on a map, you need to draw a circle.

Microorganisms in Water Alter Levels of Dissolved Oxygen

Naturally occurring water is alive with organisms. At the microscopic level, there are microorganisms, some of them disease-causing and others benign. One of the natural functions of these microorganisms is breaking down organic matter. The body of a dead fish, for example, does not remain at the bottom of a pond forever. Instead, microorganisms such as bacteria digest the organic matter into small compounds of carbon, hydrogen, oxygen, nitrogen, and sulfur.

We identify bacteria as either aerobic or anaerobic. **Aerobic bacteria** decompose organic matter only in the presence of oxygen, O_2. **Anaerobic bacteria** can decompose organic matter in the absence of oxygen. The products of aerobic decomposition are entirely different from the products of anaerobic decomposition. Aerobic bacteria in water utilize oxygen dissolved in the water to transform organic matter to such compounds as carbon dioxide, CO_2, water, H_2O, nitrates, NO_3^-, and sulfates, SO_4^{2-}. All of these products are odorless and, in the quantities produced, cause little harm to the ecosystem. Anaerobic bacteria in water use different chemical mechanisms to decompose organic matter to such products as methane, CH_4 (which is flammable); foul-smelling amines such as putrescine, $NH_2C_4H_8NH_2$; and foul-smelling sulfur compounds such as hydrogen sulfide, H_2S. Cesspools owe their wretched stink to a lack of dissolved oxygen and the resulting anaerobic decomposition.

When organic matter is introduced into a body of water, aerobic bacteria need (or "demand") dissolved oxygen to decompose the organic matter. The term used to describe this demand is **biochemical oxygen demand** (BOD). As more organic matter is introduced, the BOD increases, resulting in a drop in the amount of dissolved oxygen as the bacteria use more and more of it to do their work. If too much organic matter is introduced (say, from the outfall of a sewage treatment plant), dissolved oxygen levels can get so low that aquatic organisms start to die, as shown in **Figure 16.10**. Aerobic bacteria start to work on the bodies of these dead organisms, which lowers the oxygen level even further, killing off even the hardiest aquatic organisms. Ultimately, the dissolved oxygen level reaches zero. At this point, the noxious anaerobic bacteria take over.

CHEMICAL CONNECTIONS

How is a disposable plastic water bottle connected to the middle of an ocean?

READINGCHECK

What do aerobic bacteria utilize to transform organic matter?

FOR YOUR INFORMATION

What is the best way to treat water containing organic wastes? Add air because air contains the oxygen needed to support aerobic decomposition. Conventional secondary wastewater plants use electric energy to blow air into the wastewater. This aeration consumes 60 percent or more of the total electric energy used in wastewater treatment (see Section 16.4). In an advanced integrated pond (AIP) system, good for communities of fewer than 10,000 people, algae and other plants use solar energy and photosynthesis to saturate the wastewater directly with oxygen. AIP systems are particularly applicable in Sun Belt communities, where solar energy is plentiful, and in developing nations, where the supply of electric energy is minimal or nonexistent.

▶ **Figure 16.10**
Sewage entering a river dramatically decreases the level of dissolved oxygen in the water. Because it takes time for aerobic bacteria to decompose organic wastes and because rivers flow, the drop in dissolved oxygen is often most pronounced far downstream. Fish start to die when the concentration of dissolved oxygen dips below 3 milligrams/liter as represented by the dashed line. Once the sewage is consumed, the dissolved oxygen level begins to rise, shown here occurring after 50 km.

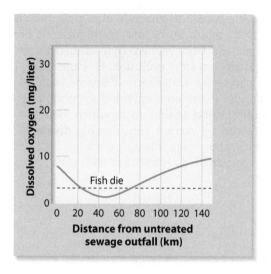

▲ **Figure 16.11**
This algal bloom consumes oxygen dissolved in the water and prevents atmospheric oxygen from mixing into the pond, thereby choking off aquatic life.

CONCEPT CHECK

Which should have a greater capacity to decompose organic matter aerobically: a still pond or a babbling brook?

CHECK YOUR ANSWER The capacity for aerobic decomposition is limited by the amount of dissolved oxygen. In a babbling brook, aeration guarantees that any dissolved oxygen lost to aerobic decomposition is quickly replaced. This is not so with a still pond. Cubic meter for cubic meter, a babbling brook has a greater capacity to decompose organic matter aerobically.

In addition to organic wastes, inorganic wastes such as nitrate and phosphate ions from fertilizers can also cause the level of dissolved oxygen to drop. These ions are nutrients for algae and aquatic plants, which grow rapidly in the presence of the ions, an event called an *algal bloom*. Significantly, the plants and algae in a bloom consume more oxygen at night than they produce through photosynthesis during the day. Also, in some instances, a bloom can cover the surface of a body of water, as shown in **Figure 16.11**, effectively choking off the supply of atmospheric oxygen. As a result, aquatic organisms suffocate and fall to the bottom, along with large amounts of dead algae. Aerobic microorganisms decompose this organic matter to the point that the water loses all its dissolved oxygen and anaerobic microorganisms start functioning. This process whereby inorganic wastes fertilize algae and plants and the resulting overgrowth reduces the concentration of dissolved oxygen in the water is called **eutrophication**, from the Greek word for "well nourished."

16.4 Wastewater Treatment

LEARNING OBJECTIVE

Identify the four stages of wastewater treatment.

EXPLAIN THIS

What chemical holds the record for saving the most lives?

The contents of the sewer systems that underlie most municipalities must be treated before being released into a body of water. The level of treatment depends in great part on whether the treated water is to be released into a river or into the ocean. Wastewater destined for a river requires the highest level of treatment for the benefit of communities downstream.

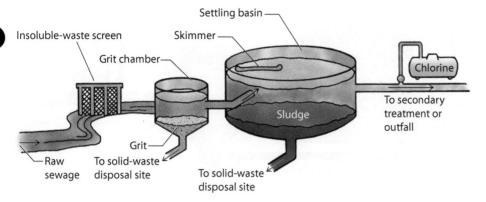

▲ **Figure 16.12**
A schematic for primary-level wastewater treatment. The rotating skimmer on the settling basin is used to remove buoyant materials and artifacts not captured by the screening process.

Human waste loses its form by the time it reaches the wastewater facility, and the wastewater appears as a murky stream. In this stream, however, are many insoluble products—including small plastic items and gritty material such as coffee grounds. Hardened balls of grease from discarded cooking fats are also found. The initial step in all wastewater treatments, therefore, involves the screening out of these insolubles. (You should know that wastewater treatment experts point out that these insolubles—even cooking grease—should be disposed of as solid waste and not be washed down the drain or flushed down the toilet.)

After screening, which may include the use of a grit chamber, the next level of municipal wastewater treatment is *primary* treatment. In primary treatment, screened wastewater enters a large settling basin, where suspended solids settle out as sludge (**Figure 16.12**). After a period of time, the sludge is removed from the bottom of the settling basin and is often sent directly to a landfill as solid wastes. Some facilities, however, are equipped with large furnaces in which dried sludge is burned, sometimes along with other municipal wastes such as paper products. The resulting ash is more compact and takes up less space in a landfill.

Wastewater effluent from primary treatment, as well as higher levels of treatment, is commonly disinfected with either chlorine gas or ozone prior to its release into the environment. A great advantage of using chlorine gas is that it remains in the water for an extended time after leaving the facility. This provides for residual protection against diseases. The chlorine, however, reacts with organic compounds within the effluent to form chlorinated hydrocarbons, many of which are known carcinogens (cancer-causing agents). Also, chlorine kills only bacteria, leaving viruses unharmed. Ozone is more advantageous in that it kills both bacteria and viruses. Also, there are no carcinogenic by-products that result from treating wastewater effluent with ozone. A disadvantage of ozone, however, is that it provides no residual protection for the effluent once it is released. Most facilities in the United States use chlorine for disinfecting, whereas European facilities tend to favor ozone. In a few locations, chlorine and ozone gases have been replaced by strong ultraviolet lamps, which, like ozone, kill both bacteria and viruses but provide no long-term residual protection.

The potential for pathogens to grow in primary effluent is extremely high, and by virtue of the Clean Water Act of 1972, the release of primary effluent is not permitted in most places. A frequently used *secondary* level of treatment, shown in **Figure 16.13**, involves first passing the primary effluent through an aeration tank. This supplies the oxygen necessary for continued decomposition of organic matter by oxygen-dependent aerobic bacteria. The effluent is then sent into a tank where any fine particles not removed in primary treatment can settle. Because sludge from this settling step is high in aerobic bacteria, some of it is recycled back to the aeration tank to increase efficiency. The remainder of the sludge is hauled off to a landfill or an incinerator.

READINGCHECK

What happens to the solids of wastewater during primary treatment?

FOR YOUR INFORMATION

In Hong Kong, about 80 percent of all toilets flush using seawater. Developed since the 1960s, this system now saves the equivalent of about 25 percent of fresh water consumption. Also, effluent from fresh water activities, such as personal hygiene and dishwashing, is treated and reused for watering city trees.

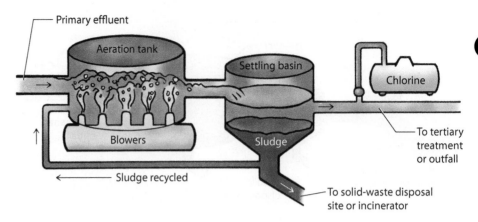

▲ **Figure 16.13**
A schematic for secondary treatment of wastewater from a municipal system.

Many municipalities also require a third level, a *tertiary* level, of wastewater treatment. There are a number of tertiary processes, and most involve filtration of some sort. A common method is to pass secondary-level effluent through a bed of finely powdered carbon, which captures most of the particulate matter and many of the organic molecules not removed in earlier stages. The advantage of tertiary-level treatment is greater protection of our water resources. Unfortunately, tertiary treatment is costly and is normally used only in situations in which the need is deemed vital. Primary and secondary levels of treatment are also not without great cost.

FOR YOUR INFORMATION

For dwellings in remote locations such as summer cabins, many people are opting for composting toilets, which use no water. Rather, they allow human waste to decompose aerobically (with oxygen) as air is vented over the waste, which is buried in peat moss. Dried, odor-free compost, which is removed every few months, is useful as a garden fertilizer.

CONCEPT CHECK

Distinguish among the main functions of primary, secondary, and tertiary wastewater treatment.

CHECK YOUR ANSWER Primary wastewater treatment removes the bulk of solid waste and sludge from the sewage effluent using screening devices and large settling basins. Secondary treatment provides oxygen to oxygen-dependent bacteria that serve to decompose organic matter. Tertiary treatment removes pathogens and wastes not removed by earlier treatments by filtering the effluent through beds of powdered carbon or other fine particles.

16.5 The Earth's Atmosphere Is a Mixture of Gases

LEARNING OBJECTIVE

Describe the formation and composition of the Earth's atmosphere.

EXPLAIN THIS

Why is there no atmosphere on the Moon?

If the Sun no longer provided heat, as represented in **Figure 16.14a**, the air molecules surrounding our planet would settle to the ground—much like popcorn at the bottom of an unplugged popcorn machine. Plug in the popcorn machine and the exploding kernels bumble their way to higher altitudes. Likewise, add solar energy to the air molecules and they, too, bumble their way to higher altitudes. Popcorn kernels attain speeds of 1 meter per second and can rise 1 or 2 meters. But solar-heated air molecules move at about 1600 kilometers per hour, and a few make their way up to more than 50 kilometers in altitude. **Figure 16.14b** shows that if there were no gravity, air molecules would fly into outer space and be lost

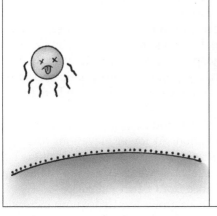

(a) Atmosphere with gravity but no solar heat: molecules lie on the Earth's surface.

(b) Atmosphere with solar heat but no gravity: molecules escape into outer space.

(c) Atmosphere with solar heat and gravity: molecules reach high altitudes but are prevented from escaping into outer space.

▲ Figure 16.14
Our atmosphere is a result of the actions of both solar heat and gravity.

from our planet. Combine heat from the Sun with the Earth's gravity, however, as in **Figure 16.14c**, and the result is a layer of air more than 50 kilometers thick that we call the *atmosphere*. This atmosphere provides oxygen, nitrogen, carbon dioxide, and other gases needed by living organisms. It protects the Earth's inhabitants by absorbing and scattering cosmic radiation. It also protects us from being rained on by cosmic debris, because any material headed toward the Earth burns up before reaching us. It is the heat generated by friction between the flying debris and our atmosphere that causes the debris to burn.

Table 16.1 shows that the Earth's present-day atmosphere is a mixture of gases—primarily nitrogen and oxygen, with small amounts of argon, carbon dioxide, and water vapor and traces of other elements and compounds. This has not always been the composition of the Earth's atmosphere. Oxygen, for example, was not a component until the evolution of photosynthesis in primitive life-forms 3 billion years ago. Carbon dioxide levels have also varied significantly over time.

We have adapted so completely to the invisible air around us that we sometimes forget it has mass. At sea level, 1 cubic meter of air has a mass of 1.18 kilograms. So the air in an average-sized room has a mass of about 60 kilograms—about the average mass of a human.

When you are under water, the weight of the water above you exerts a pressure that pushes against your body. The deeper you go, the more water there is above you and hence the greater the pressure exerted on you. The behavior of

CHEMICAL CONNECTIONS

How is the atmosphere connected to the leaf of a tree?

TABLE 16.1 Composition of Earth's Atmosphere

GASES HAVING FAIRLY CONSTANT CONCENTRATIONS	PERCENT BY VOLUME	GASES HAVING VARIABLE CONCENTRATIONS	PERCENT BY VOLUME
Nitrogen, N_2	78	Water vapor, H_2O	0 to 4
Oxygen, O_2	21	Carbon dioxide, CO_2	0.034
Argon, Ar	0.9	Ozone, O_3	0.000004*
Neon, Ne	0.0018	Carbon monoxide, CO	0.00002*
Helium, He	0.0005	Sulfur dioxide, SO_2	0.000001*
Methane, CH_4	0.0001	Nitrogen dioxide, NO_2	0.000001*
Hydrogen, H_2	0.00005	Particles (dust, pollen)	0.00001*

*Average value in polluted air.

air is the same. Because air has mass, gravity acts upon the air, giving it weight. The weight of the air, in turn, exerts a pressure on any object submerged in the air. This pressure is known as **atmospheric pressure,** and the deeper you go in the atmosphere, the greater this pressure becomes. At sea level, you are at the bottom of an "ocean of air," so the atmospheric pressure is greatest. Climb a mountain so that you are no longer so deep and the atmospheric pressure is less. Venture above the atmosphere and you have entered space, where there is no atmospheric pressure.

If you have ever gone mountain climbing, you have probably noticed that the air grows cooler with increasing elevation. At lower elevations, the air is generally warmer. This is because the Earth's surface radiates much of the heat it absorbs from the Sun. As this heat radiates upward, it warms the air—an effect that decreases with increasing distance from Earth's surface.

You have probably also noticed that the air grows less dense with increasing elevation; that is, there are fewer air molecules to breathe for a given volume. You can understand why this is so by considering a deep pile of feathers. At the bottom of the pile, the feathers are squished together by the weight of the feathers above. At the top of the pile, the feathers remain fluffy and are much less dense. For the same reasons, air molecules close to the Earth's surface are squeezed together by the greater atmospheric pressure. With increasing elevation, the density of the air gradually decreases because of decreasing atmospheric pressure. Unlike a pile of feathers, however, the atmosphere doesn't have a distinct top. Rather, it gradually thins to the near vacuum of outer space. More than half of the atmosphere's mass lies below an altitude of 5.6 kilometers, and about 99 percent lies below an altitude of 30 kilometers.

Scientists classify the atmosphere by dividing it into layers, each layer distinct in its characteristics. The lowest layer is the **troposphere,** which contains 90 percent of the atmospheric mass and essentially all of the atmosphere's water vapor and clouds, as **Figure 16.15** shows. This is where weather occurs. Commercial jets generally fly at the top of the troposphere to minimize the buffeting and jostling caused by weather disturbances. The troposphere extends to a height of about 16 kilometers. Its temperature decreases steadily with increasing altitude. At the top of the troposphere, temperatures average about −50°C.

Above the troposphere is the **stratosphere,** which reaches a height of 50 kilometers. In the stratosphere, at an altitude of 20 to 30 kilometers, lies the *ozone layer.* Stratospheric ozone acts as a sunscreen, protecting the Earth's surface from harmful solar ultraviolet radiation. Stratospheric ozone also affects

READING CHECK

Why is it generally warmer in a valley than at the top of a mountain?

FOR YOUR INFORMATION

Where is the atmospheric pressure greater: at the top or bottom of a 9-inch round balloon? Interestingly, the pressure on the bottom side of the balloon is sufficiently greater to result in a small net force upward. We call this net force upward the *buoyant force,* which for a 9-inch balloon equals about 0.0145 pound. If the balloon weighs more than 0.0145 pound (6.58 grams), it falls. But if it weighs less, it rises, which is the case when the balloon is filled with helium. The air beneath your feet is deeper than the air at your head, so is there a bouyant force acting on you as well? If it weren't for the air, would your measured weight read more or less?

▶ **Figure 16.15**
The two lowest atmospheric layers— troposphere and stratosphere.

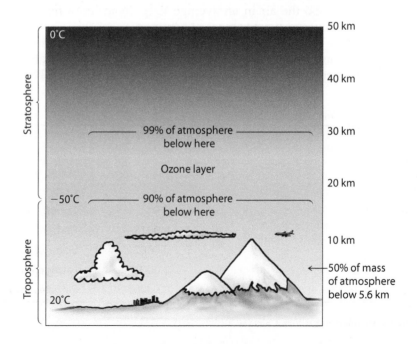

stratospheric temperatures. At the lowest altitudes, the temperature is coolest, because of the solar screening effect of ozone; air at this altitude is literally in the shade of ozone. At higher altitudes, less ozone is available for shading and temperature increases all the way to a warm 0°C at the top of the stratosphere.

CONCEPT CHECK

What effect does the Earth's gravity have on the atmosphere?

CHECK YOUR ANSWER The Earth's gravity pulls molecules in the atmosphere downward, preventing them from escaping into outer space.

CALCULATION CORNER DENSE AS AIR

Knowing the density of air (1.18 kilograms/cubic meter),* it's a straightforward calculation to find the mass of air for any given volume—simply multiply air's density by the volume. In the example given in the text, the volume of the averaged-sized room was assumed to be 4.00 meters × 4.00 meters × 3.00 meters = 48.0 cubic meters. Thus, the mass of the air in the room is

$$1.18 \text{kg/m}^3 \times 48.0 \text{ m}^3 = 56.6 \text{ kg}$$

If you're curious to know how many pounds this is, multiply by the conversion factor 2.20 pounds/1 kilogram:

$$56.6 \text{ kg} \times 2.20 \text{ lb/kg} = 125 \text{ lb}$$

EXAMPLE

What is the mass in kilograms of the air in a classroom that has a volume of 796 cubic meters?

*This assumes a temperature of 25°C and a pressure of 1 atmosphere.

ANSWER

Each cubic meter of air has a mass of 1.25 kilograms, so

$$796 \text{ m}^3 \times 1.18 \text{ kg/m}^3 = 939 \text{ kg}$$

which is as much as the combined mass of 15 students having a mass of about 63 kilograms (138 pounds) each.

YOUR TURN

1. What is the mass in kilograms of the air in an "empty" nonpressurized scuba tank that has an internal volume of 0.0100 cubic meter?

2. What is the mass in kilograms of the air in a scuba tank that has an internal volume of 0.0100 cubic meter and is pressurized so that the density of the air in the tank is 240 kilograms/cubic meter?

The answers for Calculation Corners appear at the end of each chapter.

16.6 Human Activities Have Increased Air Pollution

EXPLAIN THIS

Is the solution to pollution dilution, or should the convention be prevention?

Any material in the atmosphere that is harmful to health is defined as an *air pollutant*. One major source of air pollutants is volcanoes. The largest volcanic blast of the 20th century, for example, was the 1991 eruption of Mount Pinatubo in the Philippines, an eruption that released 20 million tons of the noxious gas sulfur dioxide, SO_2. As **Figure 16.16** shows, this sulfur dioxide managed to travel all the way to India in only 4 days.

In a number of ways, however, humans have surpassed volcanoes as sources of pollution. In the United States alone, for example, industrial and other human activities have been depositing about 20 million tons of sulfur dioxide in the air *every year* since around 1950. By one estimate, human activities account for about 70 percent of all sulfur that enters the global atmosphere.

To stem the human production of air pollutants, the U.S. government passed the Clean Air Act in 1970. This act regulated the gaseous emissions of various industries but was not comprehensive. An amendment in 1977 greatly

LEARNING OBJECTIVE

Differentiate aerosols from particulates and industrial smog from photochemical smog.

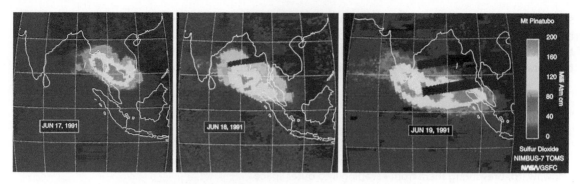

▲ **Figure 16.16**
The cloud of sulfur dioxide generated by the June 15, 1991, eruption of Mount Pinatubo reached India in 4 days. (The black strips are where satellite data are missing.) By July 27, the sulfur dioxide cloud had traveled around the globe.

restricted car emissions, and the most recent amendment, enacted in 1990, overhauled the act by regulating the emissions of nearly all air pollutants, including *aerosols, particulates*, and the components of *smog*.

Aerosols and Particulates Facilitate Chemical Reactions Involving Pollutants

Airborne solid particles, such as ash, soot, metal oxides, and even sea salts, play a major role in air pollution. Particles up to 0.01 millimeter in diameter (too small to be seen with the naked eye) attract water droplets and thereby form **aerosols** that may be visible as fog or smoke. Aerosol particles remain suspended in the atmosphere for extended periods of time and, as **Figure 16.17** shows, serve as sites for many chemical reactions involving pollutants.

Larger solid particles, called **particulates,** tend to settle to the ground faster than the particles that form aerosols and hence do not play as big a role in facilitating atmospheric chemical reactions. While they are airborne, however,

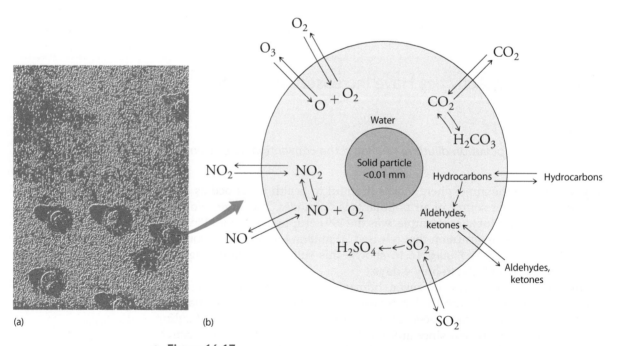

▲ **Figure 16.17**
(a) Micrograph of aerosols in the atmosphere. (b) An aerosol is the site of many chemical reactions involving pollutants. Water surrounding the solid particle attracts airborne molecules that then readily react in aqueous solution before being released back into the atmosphere.

particulates obscure visibility. Atmospheric particulates (and aerosols) also have a global cooling effect because they reflect sunlight back into space. The particulates and aerosols emitted by massive volcanic eruptions have been known to have a profound impact on the weather. A series of eruptions in Indonesia around 1815, for example, led to freezing summer temperatures and brutal winters over mid-latitude regions such as New England and Europe. Crops perished and famines ensued. The effect was most pronounced during 1816, which became known as the "year without a summer."

Industries use a variety of techniques to cut back on emissions of solid particles. Physical methods include filtration, centrifugal separation, and scrubbing, which, as **Figure 16.18** shows, involves spraying gaseous effluents with water. Another method, electrostatic precipitation, shown in **Figure 16.19**, is energy intensive but more than 98 percent effective at removing particles.

There Are Two Kinds of Smog

The term *smog* was coined in 1911 to describe a poisonous mixture of smoke, fog, and air that settled over the city of London and killed 1150 people. Smog has since grown to be a major problem, especially over urban areas, where industrial and human activities abound.

Weather plays an important role in smog formation. Normally, air warmed by the Earth's surface rises to the upper troposphere, where pollutants are dispersed, as shown in **Figure 16.20a**. Parcels of dense, cold air, however, sometimes settle below warm air in a *temperature inversion*, shown in **Figure 16.20b**. Now the air tends to stagnate, which allows a buildup of air pollutants. Temperature inversions may occur just about anywhere, but local geographies make some areas more prone to them than others. The smog of Los Angeles, for example, is trapped by an inversion created when low-level cold air moving eastward from the ocean is capped by a layer of hot air moving westward from the Mojave Desert.

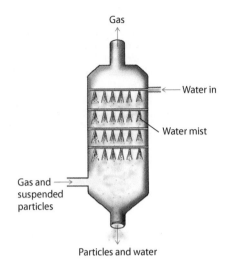

▲ Figure 16.18
During scrubbing of industrial gaseous effluents, a fine mist of water captures and removes solid particles that have diameters as small as 0.001 millimeter.

FOR YOUR INFORMATION

Aerosols and particulates in the atmosphere help to keep the Earth cool by reflecting solar radiation back into space. This effect, known as "global dimming," appears to have masked the full impact of global warming. Efforts to clean the air of aerosols and particulates may have the effect of hastening global warming.

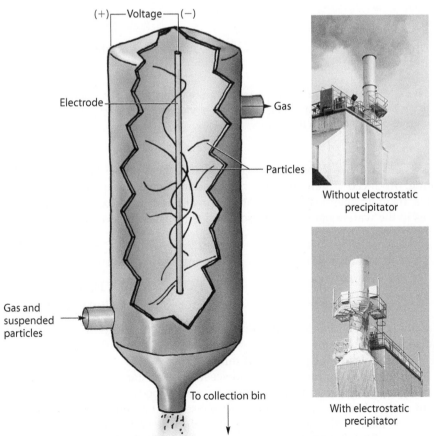

◀ Figure 16.19
(a) Particles in industrial gaseous effluents become negatively charged by an electrode and are attracted to the positively charged wall of the electrostatic precipitator. Once it touches the wall, a particle loses its charge and falls into a collection bin. (b) Smokestacks with and without electrostatic precipitators.

▶ **Figure 16.20**
(a) Smog is removed by rising warm air. (b) In a temperature inversion, smog is trapped as cool air settles below warm air. (The normal scheme of cool above warm is inverted.)

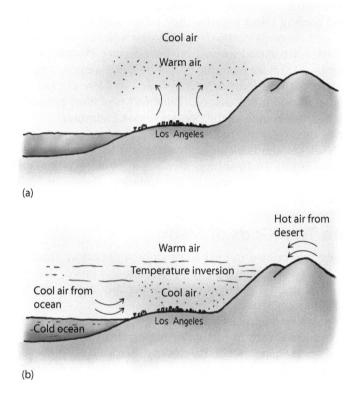

(a)

(b)

Temperature inversions tend to disperse at night because air at higher altitudes cools more quickly than does lower air, which is closer to the Earth's warm surface. This is one reason the skies in many urban areas tend to have less smog in the early morning than in the late afternoon.

CONCEPTCHECK

Why are temperature inversions more common during the day than at night?

CHECK YOUR ANSWER A temperature inversion occurs when a body of warm air sits above a body of denser cold air. The higher air is warmed by the heat of the Sun, which is out only during the day.

READINGCHECK

What are the two types of smog?

There are two types of smog: industrial and photochemical. **Industrial smog,** produced largely from the combustion of coal and oil, is high in particulates. Its main chemical ingredient, however, is sulfur dioxide, which accumulates in the water coating of aerosols and is transformed to sulfuric acid:

$$2 \; SO_2 \quad + \quad O_2 \quad \rightarrow \quad 2 \; SO_3$$
Sulfur dioxide \qquad Oxygen \qquad Sulfur trioxide

$$SO_3 \quad + \quad H_2O \quad \rightarrow \quad H_2SO_4$$
Sulfur trioxide \qquad Water \qquad Sulfuric acid

Breathing aerosols containing even very low concentrations of sulfuric acid can cause severe breathing distress. As discussed in Section 10.5, airborne sulfuric acid is also a leading cause of acid rain.

Although many industries still exceed federal standards regulating sulfur emissions, levels of industrial smog have dropped markedly since the passage of the 1970 Clean Air Act and its subsequent amendments. In the future, however, maintaining low levels of sulfur dioxide emissions will become more difficult as both national economies and world population continue to grow.

Photochemical smog consists of pollutants that participate either directly or indirectly in chemical reactions induced by sunlight. These pollutants are predominantly nitrogen oxides, ozone, and hydrocarbons, and their prime source is the internal-combustion engine. In the combustion chamber, oxygen is mixed with vaporized hydrocarbons for the production of heat, which causes an expansion of gases that drives the piston's power stroke. Atmospheric nitrogen is also present, however, and at the high temperatures characteristic of internal-combustion engines, the nitrogen and oxygen form nitrogen monoxide:

$$heat + N_2 + O_2 \rightarrow 2\,NO$$

Nitrogen monoxide is fairly reactive. Once released from the engine, it reacts rapidly with atmospheric oxygen to form nitrogen dioxide:

$$2\,NO + O_2 \rightarrow 2\,NO_2$$

Nitrogen dioxide is a powerful corrosive agent that acts on metal, stone, and even human tissue. Its brown color is responsible for the brown haze typically seen over a polluted city. Sunlight initiates the transformation of nitrogen dioxide to nitric acid, HNO_3, which, along with sulfuric acid, is a prime component of acid rain. In aerosols, sunlight splits nitrogen dioxide into nitrogen monoxide and atomic oxygen:

$$sunlight + NO_2 \rightarrow NO + O$$

The nitrogen monoxide reacts with atmospheric oxygen to re-form nitrogen dioxide, and the atomic oxygen reacts with atmospheric oxygen to form ozone:

$$O + O_2 \rightarrow O_3$$

Ozone is a pungent pollutant. It causes eye irritation and at high levels can be lethal. Plant life suffers when exposed to even relatively low concentrations of ozone, and it causes rubber to harden and turn brittle. To protect tires from ozone, manufacturers have incorporated paraffin wax, which reacts preferentially with the ozone, sparing the rubber. As was discussed in Section 9.7, ozone is also formed by natural processes in the Earth's stratosphere, where it filters out as much as 95 percent of the Sun's ultraviolet rays. So at the Earth's surface, ozone is a harmful pollutant, while 25 kilometers straight up it serves as a sunscreen and is vital for the good health of all living organisms.

A profile of average urban concentrations of nitrogen monoxide, nitrogen dioxide, and ozone is given in **Figure 16.21**. Early morning rush hour causes a rapid increase in nitrogen monoxide, which by mid-morning has largely been converted to nitrogen dioxide. On a sunny day, following nitrogen dioxide formation, ozone levels begin to peak. In the absence of a temperature inversion, late-afternoon winds clear the pollutants away. After a night of calm, the cycle begins again.

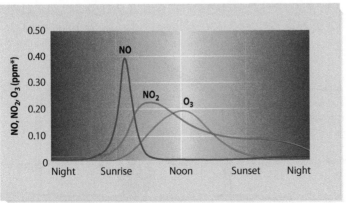

◀ **Figure 16.21**
The average daily concentration of nitrogen monoxide, nitrogen dioxide, and ozone in Los Angeles.

*ppm = parts per million

▲ **Figure 16.22**
Some gasoline nozzles are equipped with a jacket that keeps gasoline vapors from escaping into the atmosphere. Instead, the vapors are directed back to the main tank of the gas station through a secondary hose hidden within the nozzle.

Another class of components in photochemical smog is hydrocarbons, such as those found in gasoline. In the presence of ozone, airborne hydrocarbons are transformed to aldehydes and ketones, many of which add to the foul odor of smog. Also, incomplete combustion of gasoline leads to the release of *polycyclic aromatic hydrocarbons*, which are known carcinogens. Significant amounts of hydrocarbons are also released each time a car is filled with gasoline. Because gasoline is a volatile liquid, any air in a closed tank of gasoline is loaded with gasoline vapors—even when the tank is nearly empty. Every time you fill up at the pump, these vapors, about 10 grams worth, are displaced and vented directly into the atmosphere. Newer gasoline pumps have nozzles designed to trap most of these vapors, as shown in **Figure 16.22**.

CONCEPTCHECK

How does the Sun help disperse air pollutants?

CHECK YOUR ANSWER Sunlight warms the ground, which in turn warms the air, which then rises, carrying with it many air pollutants.

16.7 Carbon Dioxide Helps Keep the Earth Warm

LEARNING OBJECTIVE

Describe the greenhouse effect and potential environmental impacts of increased levels of atmospheric carbon dioxide.

EXPLAIN THIS

How might warmer oceans accelerate global climate change?

Park your car with its windows closed in the bright sun and its interior soon becomes quite toasty. The inside of a greenhouse is similarly toasty. This happens because glass is transparent to visible light but not to infrared, as illustrated in **Figure 16.23**. As you may recall from Figure 4.15, wavelengths of visible light are shorter than wavlengths of infrared. Visible light wavelengths range from 400 nanometers to 740 nanometers, while infrared wavelengths range from 740 nanometers to a million nanometers. Short-wavelength visible light from the Sun enters your car or a greenhouse and warms various objects—car seats, plants, soil, whatever. The warmed objects then emit infrared energy, which cannot escape through the glass; so the infrared energy builds up inside, increasing the temperature.

A similar effect occurs in the Earth's atmosphere, which, like glass, is transparent to visible light emitted by the Sun. The ground absorbs this energy but radiates infrared waves. Atmospheric carbon dioxide, water vapor, and

▶ **Figure 16.23**
Glass acts as a one-way valve, letting visible light in and preventing infrared energy from exiting.

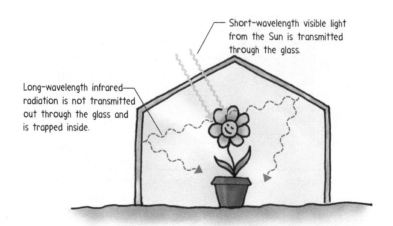

Short-wavelength visible light from the Sun is transmitted through the glass.

Long-wavelength infrared radiation is not transmitted out through the glass and is trapped inside.

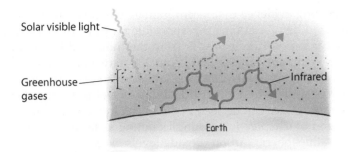

Solar visible light

Greenhouse gases

Infrared

Earth

◀ **Figure 16.24**
The greenhouse effect in the Earth's atmosphere. Visible light from the Sun is absorbed by the ground, which then emits infrared radiation. Carbon dioxide, water vapor, and other greenhouse gases in the atmosphere absorb and re-emit heat that would otherwise be radiated from the Earth into space.

other select gases absorb and re-emit much of this infrared energy back to the ground, as **Figure 16.24** illustrates. This process, called the **greenhouse effect,** helps keep the Earth warm. The greenhouse effect is quite desirable, because the Earth's average temperature would be a frigid –18°C without it. Greenhouse warming also occurs on Venus, but to a far greater extent. The atmosphere surrounding Venus is much thicker than the Earth's atmosphere, and its composition is 95 percent carbon dioxide, which brings surface temperatures to a scorching 450°C.

CONCEPTCHECK

What does it mean to say that the greenhouse effect is like a one-way valve?

CHECK YOUR ANSWER Both the Earth's atmosphere and glass allow incoming visible waves to pass, but not outgoing infrared waves. As a result, radiant energy is trapped.

The role of carbon dioxide as a greenhouse gas is well documented. Core samples from polar ice sheets, for example, show a close relationship between atmospheric levels of carbon dioxide and global temperatures over the past 400,000 years. This relationship is shown in the graph in the Contextual Chemistry essay at the end of Chapter 1. Ancient air in bubbles trapped in the ice core, shown in **Figure 16.25**, can be sampled directly. The age of the air is a function of the depth of the core. Past global temperatures are determined by measuring the deuterium/hydrogen ratio in the trapped air. When global temperatures are relatively high, the ocean is warmer and larger fractions of water containing deuterium evaporate from the ocean and fall as snow. A high deuterium/hydrogen ratio therefore indicates a warmer climate.

READINGCHECK

How do we know of the close relationship between carbon dioxide and global temperatures?

(a)

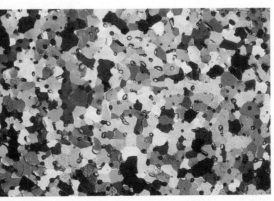

(b)

◀ **Figure 16.25**
(a) Ice cores reveal information about ancient climates. (b) Crystals of ice photographed in polarized light reveal tiny air bubbles containing ancient air.

There is strong evidence that recent human activities, such as the burning of fossil fuels and deforestation, are responsible for some dramatic increases in atmospheric carbon dioxide levels. Prior to the Industrial Revolution, carbon dioxide levels were fairly constant at about 280 parts per million, as shown in **Figure 16.26**. During the 1800s, however, levels began to climb, reaching a level of 300 parts per million in about 1910. Today's level is around 390 parts per million. Interestingly, as can be seen in the graph in the Contextual Chemistry essay at the end of Chapter 1, ice samples dating as far back as 400,000 years do not show atmospheric carbon dioxide levels exceeding 300 parts per million. In step with these increases, average global temperatures since 1860 have increased by about 0.8°C. (Since 1950, the increase has been about another 0.5°C.) Current estimates are that a doubling of today's atmospheric carbon dioxide levels will increase the average global temperature by an additional 1.5°C to 2.5°C.

Carbon dioxide ranks as the number-one gas emitted by human activities. When speaking of atmospheric pollutants such as sulfur dioxide, we talk in terms of millions of tons. The amount of carbon dioxide we pump into the atmosphere, however, is measured in *billions* of tons, as **Figure 16.27** shows. A single tank of gasoline in an automobile produces up to 90 kilograms of carbon dioxide. A jet flying from New York to Los Angeles releases more than 200,000 kilograms (about 300 tons). Above all, our population increases by

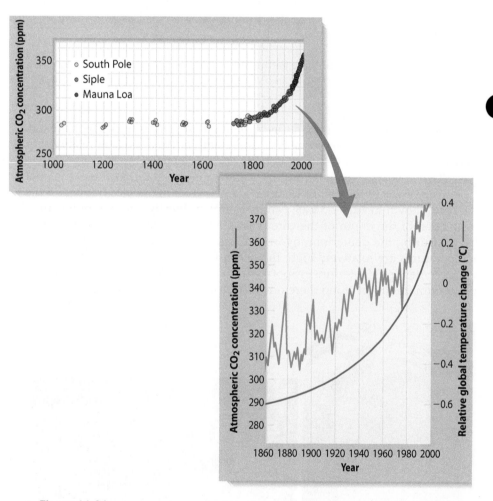

▲ Figure 16.26
Since the Industrial Revolution began in the late 1700s, atmospheric carbon dioxide levels have been increasing at accelerated rates. The yellow and red circles are data from ice samples, and the purple circles are measurements from the Mauna Loa Observatory. (Note: the global temperature changes are shown here relative to the average annual temperature for the years 1950 through 1979.)

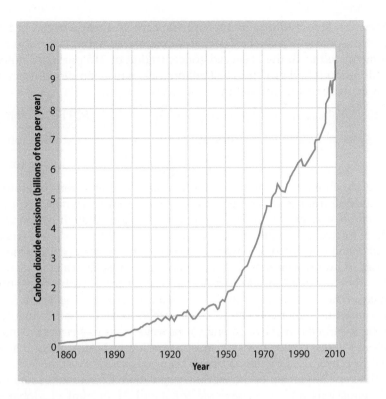

▲ **Figure 16.27**
Carbon dioxide emissions from the burning of fossil fuels have grown dramatically since 1860.

about 236,000 individuals every day, which is about 86 million individuals every year. In 1999, we passed the milestone of 6 billion humans, and a mere 13 years later, in 2012, we surpassed 7 billion humans, each of us responsible for activities that result in the output of carbon dioxide. A satellite view of North America at night, as shown in **Figure 16.28**, quickly reveals the significant impact we humans are having on a planetary scale.

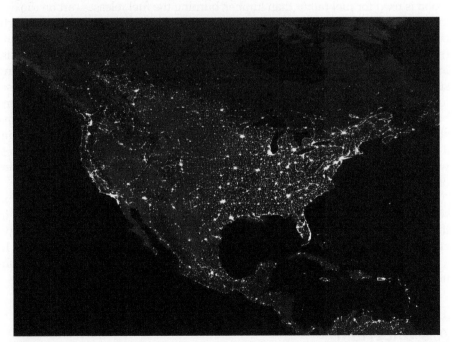

▲ **Figure 16.28**
Our ability to impact the global environment is clearly indicated by a night time view of our planet.

When direct monitoring of atmospheric carbon dioxide began in 1958, the global atmospheric reservoir of carbon dioxide was about 671 billion tons, a figure calculated from the observed concentration of 315 parts per million. By 2010, this amount had grown to 829 billion tons, which simple subtraction tells us is an increase of 158 billion tons:

2010 global atmospheric reservoir of CO_2:	829 billion tons
1958 global atmospheric reservoir of CO_2:	−671 billion tons
Net increase:	158 billion tons

Over the same period, humans released at least 215 billion tons of atmospheric carbon dioxide from fossil fuel emissions alone. From these data, we can get a feel for nature's ability to absorb carbon dioxide. Even though we pumped out 215 billion tons of carbon dioxide, the total quantity in the atmosphere went up by only 158 billion tons. Models suggest that most of the difference was absorbed by the oceans. As we saw in Section 10.6, the ocean, because its water is alkaline, can absorb carbon dioxide. Carbon dioxide can also be absorbed by vegetation during photosynthesis. It has been shown, for example, that trees grow more rapidly when exposed to higher concentrations of carbon dioxide. The fact that levels of atmospheric carbon dioxide are going up, however, tells us that we are exceeding nature's absorbing power. Consider this: according to NASA, more than half of all fossil fuels ever used by humans have been consumed in just the last 20 years.

With about 20 percent of the world's population, China ranks first in carbon dioxide emissions and is responsible for about 22 percent of global carbon dioxide emissions. With about 5 percent of the world's population, the United States is the second leading carbon dioxide emitter, producing about 20 percent of the world's total. The third major emitter of carbon dioxide is the European Union, which, with about 6 percent of the world's population, produces about 14 percent. For these industrialized nations, this adds up to about 56 percent of the global carbon dioxide emissions. Developing nations account for the remaining 44 percent, but their sources of carbon dioxide are evenly split between fossil fuels and deforestation.

Deforestation presents multiple threats to atmospheric resources. If the cut wood is used for fuel rather than lumber, burning the fuel releases carbon dioxide into the atmosphere. Whether the wood is used for fuel or for lumber, though, cutting down any forest destroys a net absorber of carbon dioxide. Furthermore, tropical forests have the capacity to evaporate vast volumes of water, which assist in the formation of clouds. The clouds, in turn, keep regions cool by reflecting sunlight and moist by precipitating rain. Farmers who burn down rainforests for farmland are cutting off their future supply of rainwater. When their farms become desert, they are then spurred to burn even more of the rainforest. So far, about 65 percent of all rainforests have been destroyed. At present rates, within a few decades, remaining rainforests will not be able to sustain regional climates, which will leave more than a billion citizens of the rapidly growing communities of South America, Africa, and Indonesia in the midst of arid land.

As their economies and populations continue to grow over the next several decades, developing nations will likely surpass industrial nations in the amounts of carbon dioxide and other pollutants they emit. New energy-efficient technologies that minimize emissions, however, are now available. In a best-case scenario, developing nations will be able to utilize these new technologies while maintaining needed economic growth.

The Potential Effects of Global Climate Change Are Uncertain

There is consensus among scientists that increased levels of atmospheric carbon dioxide and other greenhouse gases will result in global warming, which will result in changes to global climate systems. How much temperatures may rise,

FOR YOUR INFORMATION

If we were to hold steady and not increase the current rate at which we put CO_2 into the atmosphere, would the level of atmospheric CO_2 also hold steady? The answer is no because at our current rate we already put about twice as much CO_2 into the atmosphere than can be absorbed by plants, soils, and the ocean. To achieve a steady level of atmospheric CO_2, we would need to *decrease* our emissions by about 80%. It's like a slowly draining bathtub. We're pouring water into this tub faster than it can drain, so the water level naturally rises. We've got to close down the spigot—not just stop opening it up—or the tub will eventually overflow.

FOR YOUR INFORMATION

Today, levels of atmospheric carbon dioxide are close to 400 ppm, which is a new level that has likely not been achieved for hundreds of thousands of years. Evidence tells us, however, that about 25 million years ago, carbon dioxide levels of 1000 to 1500 ppm were typical. Of course, there were no polar ice caps way back then, and the average sea level was some 70 meters higher than it is today.

however, is uncertain, as are the potential effects of the temperature increases. This uncertainty is due to the large number of variables that determine global weather. The Sun's intensity, for example, changes over time, as does the ocean's ability to absorb and distribute greenhouse heat. Another variable is the cooling effect of cloud cover, atmospheric dust, aerosols, and ice sheets, which all serve to reflect incoming solar radiation.

A number of mechanisms may ease or even reverse global warming. For example, we may have underestimated the capacity of oceans and plants to absorb carbon dioxide. Greater levels of atmospheric carbon dioxide may simply mean more carbon dioxide in the ocean and more abundant plant life. In addition, warmer global temperatures could mean an increase in cloud cover worldwide and an increase in snowfall in the polar regions. Both of these effects would tend to cool the Earth by increasing the reflection of solar energy. If the cloud cover and snowfall became unusually extensive, continued reflection of solar radiation could even trigger an ice age.

On the other hand, some mechanisms may enhance global warming. Warmer oceans might have a diminished capacity to absorb carbon dioxide because the solubility of carbon dioxide in water decreases with increasing temperature (see Section 7.4). Rapid climatic changes might destroy vast regions of forests and vegetation, meaning those reservoirs for carbon dioxide absorption would no longer exist. Alternatively, more abundant plant life might not be as beneficial for the atmosphere as we had hoped, because although plants absorb carbon dioxide, they also emit other greenhouse gases, such as methane. Warmer global temperatures might also enhance microbial activity in the soil. Microbes decaying organic matter are a significant source of carbon dioxide in dry soils and a source of methane in wet soils. Furthermore, large quantities of methane locked in Arctic permafrost may also be released as a consequence of warmer terrain. As represented in **Figure 16.29**, we just don't know.

An average global temperature increase of only a few degrees would not be felt uniformly around the world. Instead, some places would experience wider fluctuations than others. For example, the number of days temperatures exceed 32°C (90°F) might double in New York City but remain unchanged in Los Angeles. The number of days in polar regions when temperatures rise above 0°C might double or even triple, causing glaciers and polar ice sheets to melt faster. Melting ice combined with the thermal expansion of ocean waters would lead to an increase in sea level. Many climatologists project that a global temperature increase of a few degrees over the next 50 to 100 years may raise sea levels by about 1 meter, enough to inundate many coastal regions and displace millions of people.

▲ **Figure 16.29**
Which weather extreme might become more prevalent as greenhouse gases continue to increase?

Weather is what happens outside on a daily basis. A change in weather is quite common. Climate is different. Climate is the *average* weather that occurs for a particular region. Climate is rather predictable. It may or may not snow in Vermont on January 26, but the climate of Vermont tells the would-be tourists to bring their snow skis, not their water skis.

Small changes in average global temperatures would also change weather patterns. The warming of the equatorial eastern Pacific Ocean during an El Niño, for example, is already known to change local weather patterns throughout the world. If the whole planet were to warm by a few degrees, the impact would be far greater. What is now fertile agricultural land might turn barren, while land now barren might turn fertile. Over the past several decades, for example, average global temperatures have edged upward. In step with this warming trend, the growing seasons of the Great Plains of Canada are up to 2 weeks longer than they were several decades ago. As weather patterns change, one nation's gain may be another nation's loss. Developing nations lacking the resources to make adjustments, however, would be hardest hit.

CONCEPT CHECK

Why are scientists uncertain about the potential effects of global warming?

CHECK YOUR ANSWER The uncertainty is due to the large number of variables that determine global weather patterns. As the debates continue, bear in mind that the issue is not global warming itself but rather its potential effects.

Science tells us that the potential for human-induced global climate change is real but that the degree of impact is uncertain. Actual quantification will come only from a slow but steady accumulation of evidence. What should be done in the meantime is not a scientific issue but rather a societal one.

One societal response to global warming is to adapt to the changes as they occur. Economists argue that large uncertainties in climate projections make it unwise to spend large sums of money trying to avert disasters that may never materialize. Adjusting to immediate changes would be more directed and far less costly. Some measures, however, could be taken now to reduce future difficulties. Irrigation systems, for example, might be made more efficient, because even without a major climatic change, such an improvement would make it easier to cope with normal extremes in weather.

A second societal response is to take preventive measures to minimize global warming and the resulting changes in climate. Emissions of greenhouse gases could be kept low by conserving energy and by shifting to fuels containing lower percentages of carbon, such as natural gas or hydrogen. Cars could be powered not by fossil fuels, but by electricity generated from alternative energy sources such as biomass, solar thermal electric generation, wind power, and photovoltaics. Governments may also come to agree on a set of standards for emissions of carbon dioxide and other greenhouse gases. "Polluting rights" may be granted to each nation based on such factors as population and need for economic growth.

The best public policies will be those that yield benefits even in the absence of global warming. A reduction in fossil fuel use, for example, would curb air pollution, acid rain, and the dependence of many countries on foreign oil producers. Developing alternative energy sources, revising water laws, searching for drought-resistant crop strains, and negotiating international agreements are steps that offer widespread benefits.

We would reduce our greenhouse gas emissions by about 40 percent if all our cars were powered by electricity rather than fossil fuels. Do we have the technology to make the switch? Absolutely, and the change is already upon us. Shai Agassi, founder of Better Place, says, "Trying to save the gasoline car is almost like trying to save the portable CD player three years after the introduction of the iPod."

Chapter 16 Review

LEARNING OBJECTIVES

Describe how water circulates through the hydrologic cycle. (16.1)	→	*Questions 1–4, 46–48*
Review water consumption trends in the United States. (16.2)	→	*Questions 5–8, 34–36, 45, 49–54, 79–83*
Identify sources of water pollution and explain the significance of biochemical oxygen demand. (16.3)	→	*Questions 9–14, 55–58*
Identify the four stages of wastewater treatment. (16.4)	→	*Questions 15–17, 59–61*
Describe the formation and composition of the Earth's atmosphere. (16.5)	→	*Questions 18–23, 37–42, 44, 62–68*
Differentiate aerosols from particulates and industrial smog from photochemical smog. (16.6)	→	*Questions 24–29, 43, 69–75*
Describe the greenhouse effect and potential environmental impacts of increased levels of atmospheric carbon dioxide. (16.7)	→	*Questions 30–33, 76–78, 84–85*

SUMMARY OF TERMS (KNOWLEDGE)

Aerobic bacteria Bacteria able to decompose organic matter in the presence of oxygen.

Aerosol A moisture-coated microscopic airborne particle up to 0.01 millimeter in diameter that is a site for many atmospheric chemical reactions.

Anaerobic bacteria Bacteria able to decompose organic matter in the absence of oxygen.

Aquifer A soil layer in which groundwater may flow.

Atmospheric pressure The pressure exerted on any object immersed in the atmosphere.

Biochemical oxygen demand A measure of the amount of oxygen consumed by aerobic bacteria in water.

Eutrophication The process whereby inorganic wastes in water fertilize algae and plants growing in the water and the resulting overgrowth reduces the dissolved oxygen concentration of the water.

Greenhouse effect The process by which visible light from the Sun is absorbed by the Earth, which then emits infrared energy that cannot escape and thus warms the atmosphere.

Hydrologic cycle The natural circulation of water throughout our planet.

Industrial smog Visible airborne pollution containing large amounts of particulates and sulfur dioxide and produced largely from the combustion of coal and oil.

Leachate A solution formed by water that has percolated through a solid-waste disposal site and picked up water-soluble substances.

Nonpoint source A pollution source in which the pollutants originate at different and often nonspecific locations.

Particulate An airborne particle having a diameter greater than 0.01 millimeter.

Photochemical smog Airborne pollution consisting of pollutants that participate in chemical reactions induced by sunlight.

Point source A specific, well-defined location where pollutants enter a body of water.

Stratosphere The atmospheric layer that lies just above the troposphere and contains the ozone layer.

Troposphere The atmospheric layer closest to the Earth's surface, containing 90 percent of the atmosphere's mass and essentially all water vapor and clouds.

Water table The upper boundary of a soil's zone of saturation, which is the area where every space between soil particles is filled with water.

READING CHECK QUESTIONS (COMPREHENSION)

16.1 Water on the Move

1. In what form does most of the fresh water on our planet exist?
2. What two forces power the water cycle?
3. Where is it possible to see the water table above ground?
4. Is most of the liquid fresh water on our planet located above or below ground?

16.2 Collectively, We Consume Huge Amounts of Water

5. When did the U.S. Geological Survey begin compiling national water-use data?
6. About how many liters of fresh water does the average American consume daily for personal use?
7. Has annual water usage in the United States increased or decreased over the past couple of decades?
8. What human activity consumes most of our fresh water?

16.3 Human Activities Can Pollute Water

9. Why does groundwater take so long to rid itself of contaminants?
10. How can solid-waste disposal sites be designed to minimize the spread of leachates?
11. What type of soil are pathogens not able to pass through?
12. What did the Clean Water Act of 1972 shift?
13. What are some of the main products of aerobic decomposition?
14. What effect does organic matter in water have on the amount of oxygen dissolved in the water?

16.4 Wastewater Treatment

15. What is the first step in treating raw sewage?
16. What happens to the sludge collected from a wastewater treatment plant?

17. Why don't all municipalities require third-level tertiary treatment of wastewater?

16.5 The Earth's Atmosphere Is a Mixture of Gases

18. Why doesn't gravity flatten the atmosphere against the Earth's surface?
19. Which elements make up today's atmosphere?
20. Which chemical compounds make up today's atmosphere?
21. In which atmospheric layer does our weather occur?
22. Does temperature increase or decrease as one moves upward in the troposphere?
23. Does temperature increase or decrease as one moves upward in the stratosphere?

16.6 Human Activities Have Increased Air Pollution

24. What is the difference between an aerosol and a particulate?
25. What is a temperature inversion?
26. What is the difference between industrial smog and photochemical smog?
27. When is ozone useful? When is it harmful?
28. How do unburned hydrocarbons contribute to air pollution?
29. How does a catalytic converter reduce the output of air pollutants from an automobile?

16.7 Carbon Dioxide Helps Keep the Earth Warm

30. The atmosphere, like glass, is transparent to what? Opaque to what?
31. What is the number-one gas emitted by human activities?
32. How do scientists estimate the age of ancient air in bubbles trapped in an ice core?
33. Why do scientists differ in their opinions about the potential effects of global warming?

CONFIRM THE CHEMISTRY (HANDS-ON APPLICATION)

In numbers 34–36, you will be estimating the volume of water you use daily for personal hygiene. You will need a metric ruler, a bucket, a 500-milliliter measuring cup, and a timer that can measure in seconds.

34. Flushing: calculate the volume of water in your toilet's tank by multiplying the height, width, and depth of the water it contains in units of centimeters. Divide by 1000 to convert cubic centimeters to liters. Alternatively, shut off the water valve to the toilet, flush to empty the tank, and then fill to the normal fill line using the measuring cup while keeping track of how much water you add. This is the amount of water used each time you flush. Multiply this number by the avererage number of times you flush each day.

35. Shower/bath: use a measuring cup to add 1 liter of water to the bucket. Mark the water level and then pour out the water, preferably over some plants. Turn on the shower

or bath to a typical flow and time how many seconds it takes to fill the bucket to the marked level. Your volume—1 liter—divided by the number of seconds is the flow rate in liters per second. To convert to liters per minute, multiply this value by 60 seconds/minute. For instance, if it takes 5 seconds to collect 1 liter of water, the flow rate is 1 L/5s × 60s/1min = 12 L/min. The next time you shower or bathe, note how many minutes you run the water and then calculate the volume of water consumed.

36. Bathroom sink activities: turn on your bathroom sink faucet to a typical flow rate and measure the number of seconds it takes to fill the measuring cup. (Recall that 500 milliliters equals 0.5 liter.) Log the number of seconds you run the faucet at this rate over the course of a day as

you wash, brush your teeth, or shave; then calculate the volume of water you've used.

37. Place a card over the open top of a glass filled to the brim with water; then invert the glass. Why does the card stay in place? What happens when the glass is held sideways?

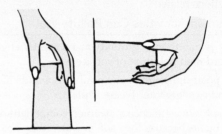

THINK AND SOLVE (MATHEMATICAL APPLICATION)

38. There are 1000 liters in 1 cubic meter and 1000 grams in 1 kilogram. How many grams of air are in 1 liter of air? (Assume a density of 1.18 kg/m^3.)

39. Assume air has an average molar mass of 28.8 grams/mole. Determine how many moles of air molecules are in 1 liter of air. (See Section 9.3 for a review of this calculation.)

40. About 25 trillion (25,000,000,000,000 = 2.5×10^{13}) chlorofluorocarbon (CFC) molecules are in every liter of air you

breathe. How many *moles* of CFC molecules are in every liter of air you breathe? What percentage of air is this?

41. Show that there is 0.041 mole of air molecules in 1.0 liter of air at 1.0 atmosphere of pressure at 25°C, which is 298 K. (You may want to review Section 2.8.)

42. Assuming air has an average molecular mass of 28.8 grams/mole, show that there are about 1.18 grams of air in 1.0 liter of air at 1.0 atmosphere of pressure at 25°C, which is 298 K. (You may want to review Section 9.3.)

THINK AND COMPARE (ANALYSIS)

43. List the order of formation of the following primary constituents of smog starting at dawn in an urban environment.

    a. Nitrogen dioxide, NO_2
    b. Nitrogen monoxide, NO
    c. Ozone, O_3

44. List from highest to lowest percentage the following constituents of the Earth's atmosphere.

    a. Oxygen, O_2
    b. Nitrogen, N_2

    c. Neon, Ne
    d. Argon, Ar
    e. Water, H_2O

45. List from most to least use the four primary uses of fresh water.

    a. Thermolectric power
    b. Industrial
    c. Irrigation
    d. Residential/commercial (public supply)

THINK AND EXPLAIN (SYNTHESIS)

16.1 Water on the Move

46. Look at a map of any part of the world and you'll see that older cities are next to rivers or next to where rivers used to be. Why?

47. The oceans are salt water, and yet evaporation over the ocean surface produces clouds that precipitate fresh water. Please explain.

48. Removal of groundwater can cause subsidence. If the water removal is stopped, will the land likely rise to its original level? Defend your answer.

16.2 Collectively, We Consume Huge Amounts of Water

49. Which consumes more water: people turning on their faucets or people turning on their electricity?

50. Make a rough sketch of a home plumbing system that uses water from an upstairs bathtub to flush a downstairs toilet.

51. How might the water from Southern Asia's wet monsoon season best be captured for use during the dry season?

52. Why is water more effectively stored in the ground than in a large dam reservoir?

53. Gas fills all the space available to it. Why then doesn't the atmosphere go off into space?

54. Research the web to find the major rivers in the world that no longer reach the oceans. What impact does beef consumption have on water use? How are these two questions related?

16.3 Human Activities Can Pollute Water

55. Why is pollution of groundwater a greater environmental hazard than pollution of surface water?

56. Are polar or nonpolar chemical compounds more often found in a leachate? Please explain.

57. Phosphates were once a common component of laundry detergents because they soften water. Why has their use been restricted?

58. How might an air pump be used in the treatment of a small pond affected by eutrophication? What should be done to the pond before the pump is used?

16.4 Wastewater Treatment

59. Is the decomposition of food by bacteria in our digestive systems aerobic or anaerobic? What evidence supports your answer? How do composting toilets work to remove the bad smells of human waste?

60. Where does most of the solid mass of raw sewage end up after being collected at a treatment facility?

61. Why is flushing a toilet with clean water from a municipal supply about as wasteful as flushing it with bottled water?

16.5 The Earth's Atmosphere Is a Mixture of Gases

62. How does the density of air in Death Valley, which is 86 meters below sea level, compare with the density of air at sea level? Please explain.

63. Why do your ears pop when you ride in an airplane that is climbing to higher altitudes?

64. Before boarding an airplane, you buy an airtight foil package of peanuts to eat during your journey. While in flight, you notice that the package is puffed up. Please explain.

65. What are two reasons why there is not much hydrogen gas, H_2, in our atmosphere?

66. Should the atmospheric ratio of nitrogen molecules to oxygen molecules increase or decrease with increasing altitude?

67. We can understand how pressure in water depends on depth by considering a stack of bricks. The pressure at the bottom face of the bottom brick corresponds to the weight of the entire stack. Halfway up the stack, the pressure is half the bottom value because the weight of the bricks above is half the total weight. To explain atmospheric pressure, we could consider a similar scenario, but with compressible bricks made from a material such foam rubber. Why?

68. Would it be slightly more difficult to draw soda through a straw at sea level or on top of a very high mountain? Please explain.

16.6 Human Activities Have Increased Air Pollution

69. Burning coal produces sulfur oxides. Where did the sulfur originate?

70. In a still room, cigar smoke sometimes rises only partway to the ceiling. Why?

71. Airborne sulfur dioxide doesn't remain airborne indefinitely. How is it removed from the atmosphere? Where does most of it end up?

72. Once formed, why is a temperature inversion such a stable weather system?

73. The atmosphere is primarily nitrogen, N_2, and oxygen, O_2. Under what conditions do these two materials react to form nitrogen monoxide, NO? Write a balanced equation for this reaction.

74. A catalytic converter increases the amount of carbon dioxide emitted by an automobile. Is this good news or bad news? Please explain.

75. Why does filling your gas tank in the evening help to minimize photochemical smog?

16.7 Carbon Dioxide Helps Keep the Earth Warm

76. How do greenhouse gases keep the Earth's surface warm?

77. How is the burning of tropical rainforests a triple threat to weather patterns?

78. Why are atmospheric CO_2 levels routinely up to 15 ppm higher in the spring than in the fall? Why are seasonal fluctuations in atmospheric CO_2 much more pronounced in the northern hemisphere compared to the southern hemisphere?

THINK AND DISCUSS (EVALUATION)

79. Many brands of bottled water cost more per liter than gasoline. Why are people are willing to buy such expensive water?

80. Should the federal government place a sales tax on bottled water to help pay for the environmental costs? Might state governments be more likely to create such a tax first?

81. List all the reasons, in order of significance, bottled water has become so popular. Do you foresee bottled water becoming more or less popular within the next five years? Ten years? One hundred years?

82. Pretend you are the president of Egypt, a country whose prime source of fresh water is the Nile river. What actions will you take upon learning that Ethiopia, which is upstream from Egypt, has begun building water projects that will restrict the flow of the Nile?

83. In reference to human nature, Jerome Delli Priscoli, a social scientist with the U.S. Army Corps of Engineers, stated, "The thirst for water may be more persuasive than the impulse toward conflict." Do you agree or disagree with his statement? Might our universal need for water be our salvation or our demise?

84. Which is a more pressing problem: ozone depletion or global climate change? To what degree are the two interrelated?

85. Pretend you are a politician addressing a conference of business leaders from the oil industry. One of the leaders asks you for your stance on global warming. How do you respond so as not to alienate any of these influential executives? What points do you emphasize? What points do you gloss over?

READINESS ASSURANCE TEST (RAT)

If you have a good handle on this chapter, then you should be able to score at least 7 out of 10 on this RAT. Check your answers online at www.ConceptualChemistry.com. If you score less than 7, you need to study further before moving on.

Choose the BEST answer to the following.

1. The oceans are salt water, and yet evaporation over the ocean surface produces clouds that precipitate fresh water. Please explain.

    a. Salt ions are too attracted to the liquid water to evaporate.

    b. Fresh rainwater falls only after the salts have been removed by the action of the clouds.

    c. The salts are too heavy to evaporate with the water.

    d. Two of the above are reasonable answers.

2. Where does most rainfall on the Earth finally end up before becoming rain again?

    a. Groundwater

    b. Lakes and rivers

    c. Snow caps and glaciers

    d. Oceans

3. Why is it important to conserve fresh water?

    a. There is little fresh water available to us on our planet.

    b. As the human population grows, so does our need for fresh water.

    c. It is expensive to purify nonpotable water.

    d. All of the above.

4. A stagnant pond smells worse than a babbling brook because

    a. odors are not transported downstream.

    b. of the type of aquatic life it attracts.

    c. it lacks sufficient dissolved oxygen.

    d. of all of the above.

5. Place the following events in the most likely chronological order:

    1. *The oxygen content drops enough to kill most aquatic animals.*

    2. *A bunch of inorganic ions were released upstream of a pond.*

    3. *Anaerobic bacteria start to dominate the population of living creatures.*

    4. *Aerobic bacteria start to dominate the population of living creatures.*

    5. *Algae start to dominate the population of living creatures.*

       a. 3, 2, 1, 4, 5   b. 2, 4, 1, 5, 3   c. 2, 5, 1, 4, 3

       d. 5, 1, 2, 4, 3   e. None of the above

6. Burning coal produces sulfur oxides. From where, ultimately, did the sulfur originate?

    a. From the photosynthetic plants from which the coal was made millions of years ago

    b. From the atmospheric sulfur oxides that photosynthetic plants use to make their amino acids cysteine and methionine

    c. From volcanoes that belch sulfur into the atmosphere

    d. From nuclear fusion in stars that died out billions of years ago

7. Airborne sulfur dioxide doesn't remain airborne indefinitely. Where does most of it end up?

    a. In the oceans   b. In plants

    c. In metal ore deposits   d. In outer space

8. Does the ozone pollution from automobiles help alleviate the ozone hole over the South Pole?

    a. Yes, because ozone is ozone no matter what the source.

    b. No, because the ozone from automobiles is too dense.

    c. Yes, but automobiles haven't been around long enough for the impact to be significant.

    d. No, because the ozone decomposes before reaching the upper atmosphere.

9. Once formed, why is a temperature inversion such a stable weather system?

    a. Cool air will rise through warm air, but warm air won't rise through cool air.

    b. Warm air is denser than cool air.

    c. Warm air will rise through cool air, but cool air won't rise through warm air.

    d. Temperature inversions are associated with low winds.

10. Geological records indicate that many ice ages were initiated by periods of unusually warm weather. How can warm weather precipitate an ice age?

    a. More sunlight is reflected back into outer space.

    b. More snow accumulates in higher latitudes.

    c. Warm weather increases the rate of evaporation of the oceans.

    d. All of the above.

Pseudoscience

For a claim to qualify as "scientific," it must meet certain standards. For example, the claim must be reproducible by others who have no stake in whether the claim is true or false. The data and subsequent interpretations are open to scrutiny in a social environment where it's okay to have made an honest mistake but not okay to have been dishonest or deceiving. Claims that are presented as scientific but do not meet these standards are what we call *pseudoscience*, which literally means "fake science." In the realm of pseudoscience, skepticism and tests for possible wrongness are downplayed or flatly ignored.

Examples of pseudoscience abound. Astrology is an ancient belief system that supposes there is a correspondence between individuals and the universe—that human affairs are influenced by the positions and movements of planets and other celestial bodies. When astrologers use up-to-date astronomical information and computers that chart the movements of heavenly bodies, they are operating in the realm of science. But when they use the data to produce nontestable astrological revelations, they have crossed over into pseudoscience.

A shaman who studies the oscillations of a pendulum suspended over the abdomen of a pregnant woman can predict the sex of the fetus with an accuracy of 50 percent. Downplaying all the times he was wrong, the shaman can easily collect hundreds of testimonies of success. These testimonies, however, are incomplete evidence for the shaman's ability; hence, they do not qualify as scientific. His claims are pseudoscientific. An example of pseudoscience that has zero success is provided by energy-multiplying machines. These machines are alleged to deliver more energy than they take in. We are told that the designs are "still on the drawing boards and in need of funds for development."

Humans are very good at denial, which may explain why pseudoscience is such a thriving enterprise. Many pseudoscientists do not recognize their efforts as pseudoscience. A practitioner of "absent healing," for example, may truly believe in her ability to cure people she will never meet except through e-mail and credit card exchanges. The pressure to make a decent living in today's fast-paced and often heartless society can be overwhelming. That said, books on pseudoscience greatly outsell books on science in general bookstores. Today, there are more than 20,000 practicing astrologers in the United States. Do people listen to these astrologers just for the fun of it? Many do, but science writer Martin Gardner reported that a greater percentage of Americans today believe in astrology and occult phenomena than did citizens of medieval Europe. Very few newspapers carry a daily science column, but nearly all provide daily horoscopes.

Meanwhile, the results of science literacy tests given to the general public are appalling. Some 63 percent of American adults are unaware that dinosaurs went extinct long before the first humans arose; 75 percent do not know that antibiotics kill bacteria but not viruses. In his book *The Demon-Haunted World*, Carl Sagan wrote that a truer measure of public understanding in science would come from asking deeper questions such as as: "*How do we know that dinosaurs died before the first humans arose?*" "*How do we know that antibiotics kill bacteria and not viruses?*"

▲ Figure 1
Over the past couple thousand years, the Earth's orbit around the Sun has shifted so that the months of the zodiac constellations have also shifted. People born in early May, for example, should now be given the Aries sign, a fact that today's astrologers tend to ignore.

Consider this: According to John Locke, the 17th century philosopher, knowledge must be grounded in physical evidence. In the absence of physical evidence, all that remains is opinion. Thomas Jefferson saw Locke's idea of evidence-based knowledge as the key to making a system of democracy possible. In such a system, differing parties are united by their agreement upon basic facts, which, in turn, allows them to craft effective public policies. The only alternative is some form of authoritarianism where public policy is created based upon the opinions of those with the greatest power.

So democracy and science are natural partners, both arising at about the same time and from the same principles. One unites us in common law. The other unites us in the spirit of discovery. Pseudoscience, by contrast, moves us away from evidence-based thinking. By embracing the methods pseudoscience we risk going back to the good old days of demons and incurable infections. We also leave ourselves open to a form of government that rules by the ideology of the privileged few who know they need not worry about evidence when crafting public policies.

CONCEPTCHECK

The renowned magician James Randi has issued a genuine million-dollar challenge to anyone who can demonstrate, under her own conditions, any psychic, supernatural, or paranormal phenomenon with adequate controls in place. Many have applied, but no one has ever won this challenge. Why?

CHECK YOUR ANSWER We are simply humans who are held firmly within the realm of nature, not above it. The kicker, though, is that nature has many incredible wonders of its own that we have yet to discover. To make these new and exciting discoveries, which would you choose: the tools of science or the tools of pseudoscience?

▶ **Figure 2**
A photograph from the Hubble Space Telescope of colliding galaxies. The tools of science are uniquely suited to helping us understand and appreciate the great wonders of the universe.

Think and Discuss

1. Why do many people trust herbal medicines to prevent disease?

2. Why are health insurance agencies beginning to cover expenses for chiropractic care?

3. Would you rather have a friend or a stranger help you through a sickness? Why? Which health professionals tend to be friendliest?

4. Why are more people afraid of flying than of driving?

5. Why does the general public have so little knowledge or understanding of science?

6. The CEO of a large corporation is told that its operations are leading toward an ecological crisis. Whom might the CEO first contact to confirm the allegation: a politician, a marketing agent, a lawyer, an astrologer, an environmental activist, or a scientist?

7. Who is best situated to persuade a tough-minded business-leader to move his company toward ecologically sound practices: a customer, a politician, a marketing agent, a lawyer, an astrologer, an environmental activist, or a scientist? Rank these people in order of their possible persuasion power.

8. A psychic claims to be able to detect an object randomly concealed within one of several opaque boxes. When tested, the psychic performs no better than what might be expected by chance. What excuses might the psychic give to explain why she was unsuccessful?

9. What advice might you provide to a scientist who was about to appear on a nationally televised program to debate a supernatural or paranormal topic?

10. A concentrated herb solution is found to have a therapeutic effect. Might there still be a therapeutic effect if this solution is diluted to half strength? A quarter strength? What if the solution is diluted repeatedly so that a dose is only one trillion trillionth as strong, meaning that the dose contains practically none of the original herb extract? At what point, if any, does the herb solution stop having a therapeutic effect? Might a dose of pure water mislabeled as containing the extract also have a therapeutic effect?

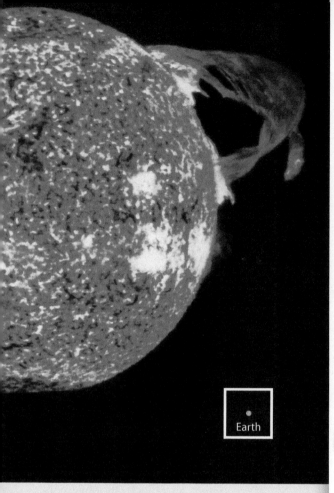

Earth

▲ The Sun is our major source of energy not because it is so hot, but because it is so big.

17

Capturing Energy

THE MAIN IDEA

Dirty energy is most convenient, and clean energy is most abundant.

In our search for energy sources, it is only natural to look to the Sun. The warmth you feel from the Sun, however, isn't so much because the Sun is hot. Indeed, the Sun's surface temperature of 6000°C is no hotter than the flame of some welding torches. Rather, the primary reason you are warmed by the Sun is because it is so big. Look at the lower right of this chapter's opening photograph and you'll see a small blue dot. This dot (painted on the photograph) is the relative size of the Earth. Clearly, when we think about possible energy sources, the enormous energy wealth at the heart of our solar system demands our serious attention.

A number of energy sources that do not depend on the Sun are now available, including nuclear, geothermal, and tidal energies. But whenever we burn plant material, we are releasing solar energy that was captured through photosynthesis. Solar energy is also released when we burn fossil fuels, which are primarily the decayed remains of ancient photosynthetic plants. Electricity-producing hydroelectric dams, windmills, and photovoltaic cells are all driven by solar radiation. Energy abounds. The technical issues we face concern how best to capture this energy.

Chemistry

Solar Pool Cover

Companies claim their "solar pool covers" increase the water temperature by as much as 10°C above the average outdoor temperature. What is the best material from which to make a solar pool cover?

PROCEDURE

1. On a warm, sunny day, fill six identical bowls with tap water and place them outside in direct sunlight. Cover four of the bowls with one of the following materials: aluminum foil, transparent plastic food wrap, colorless bubble wrap, and a black plastic garbage bag. Cut these materials to size and secure one to the top of each bowl with a rubber band, but poke a tiny hole through which you will be able to insert a thermometer. Carefully add several drops of liquid detergent to the fifth bowl and don't do anything to the sixth bowl, which is your control.

2. Allow the bowls to sit in the sun for at least 4 hours. (Good results can also be obtained when skies are lightly overcast.) Take temperature readings every half hour. Stir the water with the thermometer before taking a reading and always rinse the thermometer after dipping it into the detergent-containing water. Record your data.

3. Plot your data on a graph showing temperature on the vertical axis and time on the horizontal axis.

ANALYZE AND CONCLUDE

Is evaporation a cooling or warming process? Which cover becomes the hottest? Which covers allow most solar radiation to enter the water? Which bowl reached the highest temperature? Why?

17.1 Electricity Is a Convenient Form of Energy

EXPLAIN THIS

How is power different from energy?

All usable energy, whatever the source, is delivered to us in the form of either fuel or electricity. The wonder of electricity is the ease with which it can be transmitted to many sites. This property makes electricity one of our most convenient forms of energy. Producing electricity, however, requires the input of some other source of energy, such as the burning of a fuel. We therefore begin this chapter with a brief overview of how electricity is generated and how its consumption is measured.

Electricity is the flow of electric charge. It is generated when a metal wire is forced to move through a magnetic field. By coiling the wire into many loops and rotating the loops through powerful magnetic fields, power companies are able to generate enough electricity to light up cities. **Figure 17.1** illustrates such an *electric generator.*

The many loops of wire wrapped around an iron core form what is known as an *armature.* The armature is connected to an assembly of paddle wheels called a *turbine.* Energy from wind or falling water can cause the turbine and thus the armature to rotate, but most commercial turbines are *steam turbines,* meaning they are driven by steam. To boil the water to create steam requires an energy source, which is usually a fossil fuel or a nuclear fuel.

Still under development are more efficient *gas turbines,* which are driven not by steam, but by the hot combustion products of vaporized alcohols and lightweight hydrocarbons.

LEARNING OBJECTIVE

Provide a brief overview of how electricity is generated and distributed and how its consumption is measured.

READINGCHECK

Why is electricity a convenient form of energy?

▶ **Figure 17.1**
Basic anatomy of an electric generator. Electricity is generated in a looped wire as the wire rotates through a magnetic field. This motion causes electrons in the wire to slosh back and forth. Because the electrons are moving, they possess kinetic energy and so have the capacity to do work.

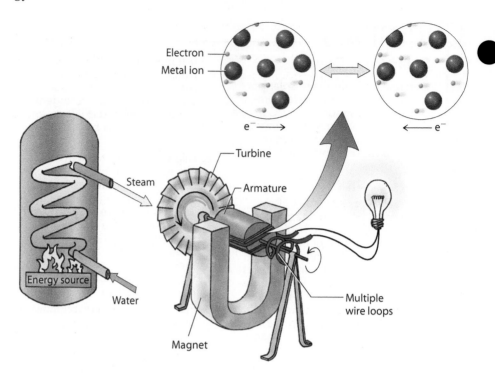

 **FOR Y O U R
INFORMATION**

Steam expands and rises as it forms, and this can be used to drive a turbine. On the opposite side of the turbine, however, the steam condenses back to the liquid phase. In doing so, its volume contracts dramatically. This creates a negative pressure that greatly enhances the flow of gas over the turbines, which is why steam turbines are so energy efficient.

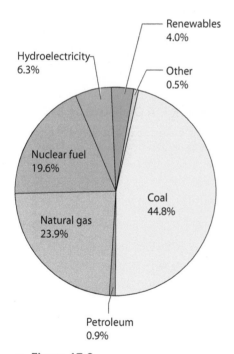

▲ **Figure 17.2**
According to the U.S. Energy Information Agency (EIA), coal, natural gas, and nuclear fuel are the predominant energy sources for the production of electricity in the United States.

C O N C E P T C H E C K

Is electricity more accurately thought of as a source of energy or as a carrier of energy?

CHECK YOUR ANSWER Electricity is energy that is readily transported through wires. In this sense, it is best thought of as a carrier of energy. The energy of electricity is used to run a lightbulb, true, but the source of this energy is not the electricity. Rather, the electricity is merely delivering the energy that was generated by some electric generator, which received energy from some nonelectric source, such as a fossil fuel or a waterfall.

What's a Watt?

Power is defined as the rate at which electrical energy (or any other form of energy) is expended. Power is measured in watts, where 1 **watt** is equal to 1 joule per second:

$$1 \text{ watt} = 1 \text{ joule}/1 \text{ second}$$

A lot of watts means that a lot of energy is being consumed quickly. A 100-watt lightbulb, for example, consumes 100 joules of energy each second, and a 40-watt bulb consumes only 40 joules each second.

The typical U.S. household consumes electrical energy at an average rate of about 800 joules per second, or 800 watts. For a small city of 100,000 households, this adds up to a rate of 80 million watts, or 80 megawatts (MW). This, however, is just the average rate of energy consumption. To meet peak demands, electric power plants must sometimes quadruple their average output. This is why small cities require electric power plants that can produce energy at a power rating of 300 megawatts or higher.

These needs are easily met by present-day power plants. A typical coal-fired plant produces on the order of 500 megawatts of electrical energy, a large nuclear plant can produce on the order of 1500 megawatts, and a large hydro-electric dam can produce more than 10,000 megawatts.

One factor affecting the cost of electricity is the source of the electrical energy. Fossil fuels and nuclear fuels produce hundreds of megawatts of power from a single power plant and are thus able to serve large areas, including cities (**Figure 17.2**). Therefore, economies of scale make electricity from fossil fuels and nuclear fuels relatively inexpensive. Electricity from sources that are not so easily centralized, such as wind energy, have traditionally been more expensive. However, this gap has narrowed significantly as technology has improved and the cost of fossil fuels has increased.

CALCULATION CORNER KILOWATT-HOURS

Take a careful look at your next electric bill. Note that you pay for electrical energy in units of kilowatt-hours. A **kilowatt-hour** (kWh) is the amount of energy consumed in 1 hour at a rate of 1 kilowatt (1000 joules per second). Therefore, if electrical energy costs 15 cents per kilowatt-hour, a lightbulb that has a power rating of 100 watts (0.1 kilowatt) can be run for 10 hours at a cost of 15 cents. The calculation to arrive at this cost is as follows:

Step 1. Calculate the total amount of energy consumed in kilowatt-hours:

power in kilowatts × hours
= energy consumed in kilowatt-hours

0.1 kW × 10 h = 1 kWh

Step 2. Calculate the cost of consuming this much energy:

kilowatt-hours consumed ×
price per kilowatt-hour = cost

1 kWh × \$0.15/kWh = \$0.15

Accordingly, ten 100-watt bulbs can also be run for 1 hour at a cost of 15 cents.

10 × 0.1 kW × 1 h = 1 kWh

1 kWh × \$0.15/kWh = \$0.15

YOUR TURN

How much does it cost to operate ten 100-watt lightbulbs for 10 hours at a cost of 15 cents per kilowatt-hour?

The answers for Calculation Corners appear at the end of each chapter.

Our Aging Electric Power Grid

Today's electric power grid had its beginnings in the late 1800s when Thomas Edison built the first public electric power station to service parts of New York City. He laid copper wires through existing culverts to reach customers within a 2-mile radius of his power station, which could reach no farther because it operated using a form of electricity known as *direct current* (DC). Soon, power stations were developed that served electricity using *alternating current* (AC), which can be efficiently transmitted through wires that are hundreds of miles long. In the United States, large service territories were then developed using alternating currents. Initially, each territory operated independently and was isolated from its neighbors. As the system grew, however, operators started to share resources by connecting their networks (**Figure 17.3**).

FOR YOUR INFORMATION

A moving armature generates electricity. The reverse is also true: electricity (from an external source) can generate a moving armature. In such a case, the armature is called an *electric motor*, which, as you know, has numerous applications. Our standards of living changed markedly after the inventions of electric generators and electric motors.

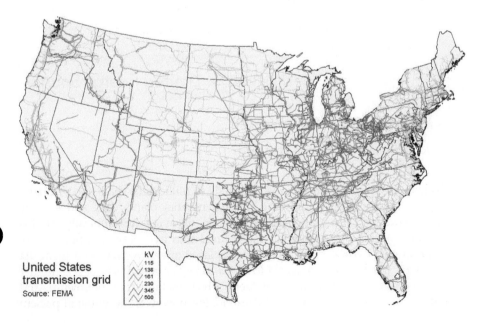

United States transmission grid
Source: FEMA

◀ Figure 17.3
The total length of the North American electric power grid, as developed over the past 100 years, is an estimated 450,000 miles.

It's important to understand that electricity is an on-demand commodity. Flick on a light switch and the nearest power station instantly works a little harder to provide the energy for that light. An advantage of the power grid is that if the demand for electricity at one station becomes excessive, electric power can be taken from the next nearest power station. The grid, therefore, acts as a buffer in that peak demand at one location can be offset by lower demand at another location.

One disadvantage, however, is the potential for regionwide blackouts. When a transmission line goes down, due to either a violent storm or the failure of equipment, the electricity that was to be passing through that transmission line must go somewhere. It then naturally flows into adjacent transmission lines that are already carrying heavy loads. As this happens, the voltage peaks beyond the capacity of the alternative lines, which are then tripped by circuit breakers to shut down. This causes a further buildup of electrical energy in the remaining lines, which also start shutting down. This process continues at an accelerating rate until power stations can shut down their generators. Within a matter of minutes, an entire region has lost power because of a single failure at one location.

Ideally, the power grid would be built using the latest materials. Also, it would be equipped with computerized sensors throughout the system that would monitor the flow of energy. The sensors would allow for a quick and effective response to any potential issues within the grid. In such a case, we would have what electricity advocates call a "smart grid." This, however, is not what we currently have. Instead, our "modern" power grid is a patchwork of wires put together over the past century. Some parts of the grid are new and capable. Most parts, however, are aging and already pushed to their limits. Some sections of the grid are in such desperate need of repair that taking them offline would trigger blackouts.

According to the American Society of Civil Engineers (ASCE), from 2000 to 2010, the United States expended about $63 billion each year on upgrades to the power grid, including the building and upkeep of power stations and transmission lines. This, however, is about $75 billion short of what is actually needed. With continued underinvestment in our electrical infrastructure, the ASCE projects that by 2040, we will need to come up with another $730 billion (in 2010 dollars) to bring the system into good repair. In the absence of this investment, we can expect an electric power grid that becomes less and less reliable. This would be unfortunate given that our economy and lifestyle are becoming more and more dependent upon this most convenient form of energy.

FOR YOUR INFORMATION

Like its electric power grid, the transportation infrastructure of the United States is woefully underfunded. This includes 4 million miles of roads, 600,000 highway bridges, 117,000 miles of rail, and 19,000 airports, most of which were built in the decades after World War II. Congressional studies estimate a shortfall of about $134 billion each year just for maintaining existing systems. If the goal also is to improve upon the systems, then the annual shortfall is about $189 billion.

17.2 Fossil Fuels Are a Widely Used but Limited Energy Source

LEARNING OBJECTIVE

Describe the chemical nature of fossil fuels and their advantages and disadvantages.

EXPLAIN THIS

What do all fossil fuels have in common?

Our fossil fuel supplies were created hundreds of millions of years ago primarily when ancient solar-energy-absorbing (photosynthetic) plants died and became buried in swamps, lakes, and seabeds. Upon decaying anaerobically (in the absence of O_2), this plant material transformed into hydrocarbons. Supplies of these energy-rich hydrocarbons cannot be replaced after we have used them up, which is why they are often referred to as *nonrenewable* energy sources. Estimates vary on the world's supply of fossil fuels, which include coal, petroleum, and natural gas. Even the most conservative estimates, however, show that at present

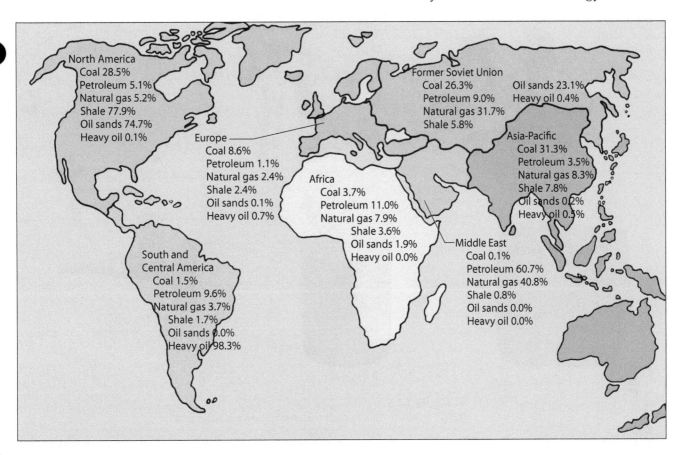

North America
Coal 28.5%
Petroleum 5.1%
Natural gas 5.2%
Shale 77.9%
Oil sands 74.7%
Heavy oil 0.1%

Europe
Coal 8.6%
Petroleum 1.1%
Natural gas 2.4%
Shale 2.4%
Oil sands 0.1%
Heavy oil 0.7%

Former Soviet Union
Coal 26.3%   Oil sands 23.1%
Petroleum 9.0%   Heavy oil 0.4%
Natural gas 31.7%
Shale 5.8%

Asia-Pacific
Coal 31.3%
Petroleum 3.5%
Natural gas 8.3%
Shale 7.8%
Oil sands 0.2%
Heavy oil 0.5%

Africa
Coal 3.7%
Petroleum 11.0%
Natural gas 7.9%
Shale 3.6%
Oil sands 1.9%
Heavy oil 0.0%

Middle East
Coal 0.1%
Petroleum 60.7%
Natural gas 40.8%
Shale 0.8%
Oil sands 0.0%
Heavy oil 0.0%

South and
Central America
Coal 1.5%
Petroleum 9.6%
Natural gas 3.7%
Shale 1.7%
Oil sands 0.0%
Heavy oil 98.3%

▲ **Figure 17.4**
With the exception of coal, fossil fuel deposits are not distributed evenly throughout the world. The data shown here are from the 2010 report of the World Energy Council. The effect of fracking technologies on natural gas reserves has yet to be fully assessed. North America, for example, is now the leading exporter of natural gas because of these technologies. This continent also holds the largest reserves for two of the dirtiest sources of petroleum—from shale (primarily the United States) and oil sands (primarily Canada) deposits.

consumption rates, recoverable petroleum reserves will be depleted within 100 years and recoverable natural gas reserves within 150 years. As depletion approaches, these valuable commodities will become much more costly. Coal reserves, on the other hand, are more abundant and may last another 300 years. Worldwide, nearly all our present energy needs are met by fossil fuels—38 percent from petroleum, about 30 percent from coal, and about 20 percent from natural gas.

Why are fossil fuels so popular? First, they are readily available in many regions of the world, as shown in **Figure 17.4**. Second, gram for gram, they store much more chemical energy compared to other combustible fuels such as wood. Third, they are portable and make excellent fuels for vehicles.

Gases emitted when fossil fuels are burned have negative environmental effects. As discussed in Section 10.5, sulfur and nitrogen oxide emissions lead to acid rain. These gases, along with particulates created from the combustion of fossil fuels, are also a leading cause of urban smog. On a global level, the burning of fossil fuels presents a potentially more devastating disturbance—increased global warming, as discussed in Section 16.7.

The molecular structure of a fossil fuel accounts for its physical phase. As shown in **Figure 17.5**, **coal** is a solid consisting of a tightly bound 3-dimensional network of hydrocarbon chains and rings. **Petroleum**, also called *crude oil*, is a liquid mixture of loosely held hydrocarbon molecules containing not more than 30 carbon atoms each. **Natural gas** is primarily methane, CH_4, which has a boiling point of −163°C. Smaller amounts of gaseous ethane, C_2H_6, and propane, C_3H_8, are also found in natural gas.

READINGCHECK

What accounts for the physical phase of a fossil fuel?

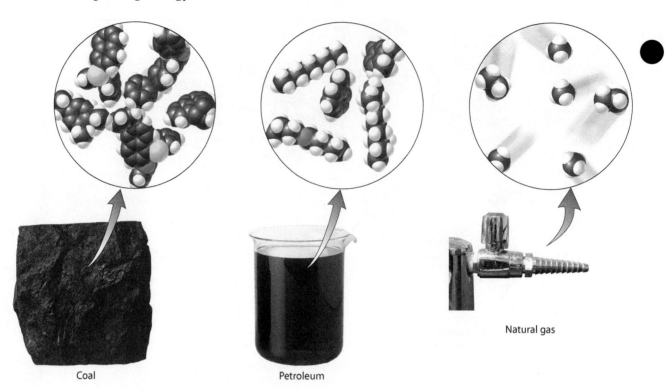

Coal

Petroleum

Natural gas

▲ Figure 17.5
Typical molecular structures of coal, petroleum, and natural gas.

Interestingly, there are other forms of fossil fuels. Canada, for example, is rich in its supply of oil sands, which are geologic deposits resembling a cross between coal and petroleum. These oil sands can be mined and treated to yield crude oil and natural gas. The processes involved, however, are energy intensive and pose environmental hazards. Nonetheless, the development of oil sand fossil fuels in Canada and other nations is progressing rapidly. The highly publicized Keystone pipeline passing from Canada to Houston, Texas, is a product of this industry.

Another example of an unconventional fossil fuel is *methane hydrate*. Most deposits of this material are located kilometers beneath the ocean floor, but in certain locations, the deposits lie just beneath the ocean floor, where researchers can collect samples. Methane hydrate is a white, icy material that is made up of methane gas molecules trapped inside cages of frozen water, as shown in **Figure 17.6**. According to the United States Geological Survey (USGS), the amount of natural gas stored within methane hydrate in the United States alone is about 320,000 trillion cubic feet. By comparison, natural gas reserves within the United States contain about 284 trillion cubic feet. There is no shortage of methane hydrate. The problem is that compared to natural gas, this material is many times more difficult to extract. As a solid, it is not readily pumped from its deep underground or undersea locations. Also, it is usually found within nonporous rock, such as shale, which further helps to lock the methane underground. Researchers are now looking for ways to overcome these obstacles, and the search is on for relatively accessible deposits. Even a small fraction of the vast methane hydrate reserves represents a supply of fossil fuel greater than all the world's coal, petroleum, and natural gas *combined*.

Coal Is the Filthiest Fossil Fuel

Worldwide, the amount of energy available from coal is estimated to be about ten times greater than the amount available from all petroleum and natural gas reserves combined. Coal is also the filthiest fossil fuel, because it contains

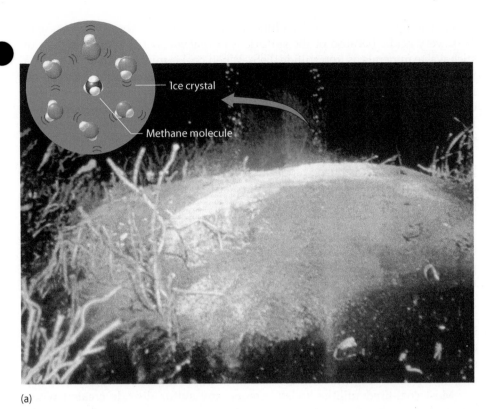
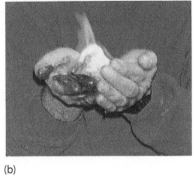

(a)

(b)

▲ **Figure 17.6**
(a) Bubbles of methane gas escaping from a deposit of methane hydrate found on the ocean floor. (b) Once brought to the surface, methane hydrate crystals quickly decompose as the ice melts and the methane gas—seen here having been ignited—is released.

large amounts of such impurities as sulfur, toxic heavy metals, and radioactive isotopes. Burning coal is therefore one of the quickest ways to introduce a variety of pollutants into the air. More than half of the sulfur dioxide and about 30 percent of the nitrogen oxides released into the atmosphere by humans come from the combustion of coal. As with other fossil fuels, the combustion of coal also produces large amounts of carbon dioxide.

Extracting coal from the ground is also harmful to human health and to the environment. Mining coal from underground mines, as the workers in **Figure 17.7** did, is a dangerous job with many health hazards. Local waterways

**CHEMICAL
CONNECTIONS**

How is toast connected to a dinosaur?

◀ **Figure 17.7**
Coal miners in Pennsylvania in the 1930s.

▲ **Figure 17.8**
Pulverized coal floats on water, but impurities sink. This difference in densities allows a simple and efficient means of purifying coal before it is burned.

are contaminated as they receive effluents from the mines. These effluents tend to be very acidic because of the sulfuric acid that forms from the oxidation of such waste minerals as iron sulfide, FeS_2. When coal is mined from the surface, a process called *strip mining,* there are fewer occupational hazards, but the trade-off is that whole ecosystems are destroyed. Although strip mining is initially cheaper than digging, the cost of restoring the ecosystem can be prohibitive. Despite these drawbacks, due to its abundance, about 45 percent of the electric power generated in the United States comes from coal-fired plants.

There are several ways to make burning coal a cleaner process. The coal can be purified before it is burned, pollutants can be filtered out after combustion, or the combustion process can be modified so that it is more efficient and fewer pollutants are produced.

Purifying coal prior to combustion usually involves pulverizing the coal and mixing it with detergents and water. As is demonstrated in **Figure 17.8**, the density of coal is lower than the density of any of its mineral impurities. A proper adjustment of the solution's density therefore allows the coal to float to the surface, where it is skimmed off, while the impurities sink to the bottom. This method of purification, called *flotation,* adds further cost to the coal, which is already expensive because of the high mining and shipping costs. Nonetheless, flotation is successful at removing most of the coal's mineral content, including up to 90 percent of the iron sulfide. Large quantities of sulfur still remain chemically locked in the coal, however. This sulfur can be removed only after combustion.

Most coal-fired utilities today remove any sulfur dioxide created when coal is burned by directing gaseous effluents into a *scrubber,* illustrated in **Figure 17.9**. Within the scrubber, the effluents come in contact with a slurry of limestone, $CaCO_3$. Up to 90 percent of the sulfur dioxide is removed as it reacts with the limestone to form solid calcium sulfate, $CaSO_4$, which is readily collected and sent to a solid-waste disposal site.

As a result of flotation and scrubbing technologies, sulfur dioxide emissions have decreased significantly over the past several decades despite an increase in the use of coal. This is promising, but given our dependence on coal, there is still plenty of room for improvement. For example, nitrogen oxide emissions have remained relatively constant. Also, equipping a coal-fired power plant with a scrubber reduces the efficiency at which the coal energy is converted to electrical energy.

Fewer pollutants and greater efficiencies are achieved by redesigning the combustion process. In conventional power plants, pulverized coal is burned in a combustion chamber, where the heat vaporizes water in steam tubes. Newer chambers send jets of compressed air into the pulverized coal, and as a result, the coal becomes suspended as it burns. This allows the coal to burn more efficiently and provides better transfer of heat from the coal to the steam tubes. Because air-suspended coal burns more efficiently, lower temperatures can be maintained, resulting in a tenfold decrease in nitrogen oxide emissions. (Recall from Chapter 9 that nitrogen oxides form as atmospheric nitrogen and oxygen are subjected to extreme temperatures.) Air-suspended coal can be burned in the presence of limestone, which removes more than 90 percent of the sulfur dioxide as it forms, thus avoiding the need for a scrubber. Cleaned of sulfur and nitrogen oxides, the hot, pressurized effluent can be directed into a gas turbine that generates electricity in tandem with the steam turbine. Overall, this system converts coal energy to electrical energy with an efficiency of about 42 percent.

More efficient combustion is one possible future for coal. Another more hopeful future involves treating coal with pressurized steam and oxygen, a process that produces clean-burning fuel gases such as hydrogen, H_2. All of these technologies are marketed by the coal industry to the general public as "clean coal." Calling it "cleaner coal," however, would be more accurate.

Scrubbed gas to atmosphere

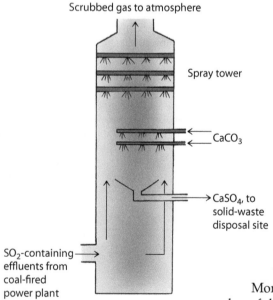

Spray tower

$CaCO_3$

$CaSO_4$, to solid-waste disposal site

SO_2-containing effluents from coal-fired power plant

▲ **Figure 17.9**
A scrubber is used to remove most of the sulfur dioxide created when coal is burned.

CONCEPT CHECK

What is the major advantage of using coal as an energy source?

CHECK YOUR ANSWER Coal is relatively abundant, which means it can be used as an energy source for many years to come.

Petroleum Is the King of Fossil Fuels

The energy content of the coal reserves in the United States far exceeds that of the fossil fuel reserves of all Middle East nations combined. Why, then, does the United States import so much petroleum from these nations? The immediate answer is that petroleum is a liquid, and liquids are far more convenient to handle in bulk quantities.

Consider, for example, that because petroleum is a liquid, it is easy to extract from the Earth. Punch a hole in the ground in the right place and up it comes—no underground mining necessary. Being a liquid also makes petroleum easy to transport. Huge oil tankers easily load and unload liquid petroleum, as shown in **Figure 17.10**. On land, petroleum can be pumped vast distances via a network of pipelines. Coal, on the other hand, must be dug out of the ground with heavy machinery and shipped as solid cargo, usually aboard trucks or freight trains.

Petroleum is also versatile. It contains all the commercially important hydrocarbons, such as those that make up gasoline, diesel fuel, jet fuel, motor oil, heating oil, tar, and even natural gas. Furthermore, petroleum contains much less sulfur than does coal and so produces less sulfur dioxide when burned. So despite its vast coal reserves, the United States has a royal thirst for petroleum, the king of fossil fuels.

Of the 20 million barrels of petroleum consumed each day in the United States, 19 million is burned for energy. The remaining 1 million is used to provide raw material for the production of organic chemicals and polymers. Thus, only one-twentieth of the hydrocarbons consumed daily goes into useful materials. The rest is burned for energy and ends up as heat and smoke.

Natural Gas Is the Purest Fossil Fuel

Natural gas is a component of petroleum, but there are also vast deposits of free natural gas in underground geologic formations. As discussed on page 159, there is currently a boom in the extraction of natural gas from underground shale deposits using a process called hydraulic fracturing. The natural gas can be collected and stored in tanks like the ones shown in **Figure 17.11**.

FOR YOUR INFORMATION

In 2012 dollars, the price of gasoline has remained fairly constant over the past century, hovering around $2.50 per gallon. The recent rise in gas prices resembles the early 1980s, when the price of gasoline rose to an average of $3.37 per gallon.

▲ Figure 17.10
Petroleum is easily and cheaply transported because it is a liquid.

FOR YOUR INFORMATION

In the 1950s, petroleum geologist M. King Hubbert used data about known and projected petroleum reserves in the United States to predict that U.S. oil production would peak in the early 1970s. In 1971, Hubbert's prediction came true. Applying Hubbert's methods to known and projected world oil reserves suggests that world oil production is currently at its peak. This peak in world oil production is known as Hubbert's Peak. It tells us that as world oil production begins to decline, the price of oil and oil-based products, such as plastics and gasoline, will rise exponentially.

◀ Figure 17.11
Natural gas is stored in large spherical tanks, because this shape holds the greatest volume for a given amount of building material.

▲ **Figure 17.12**
If you use gas and it is stored outside your home in a pressurized tank, you are using propane. If there is no tank outside your home, you are using methane.

Natural gas burns more cleanly than petroleum and much more cleanly than coal. This purest of fossil fuels contains negligible quantities of sulfur; hence, burning it produces insignificant amounts of sulfur dioxide. Also, because natural gas burns at lower temperatures, only small amounts of nitrogen oxides are produced. Perhaps most important, however, is that generating energy from natural gas produces less carbon dioxide—about half as much as that produced from burning coal. Because it is a gas, however, this fossil fuel is cumbersome to isolate and transport. Some experts suggest that switching to natural gas as much as possible may buy time and protect the environment until technologies for nonfossil energy sources are perfected and made competitive.

Another advantage of natural gas is that it can be used to generate electricity with great efficiency. With a steam turbine, fossil fuel is burned in a boiler to produce steam, which runs the electricity-generating turbine, as shown in Figure 17.1. Such a system burning natural gas to boil the water produces electricity with an efficiency of about 36 percent, comparable to the 34 percent efficiency attained using coal. However, as mentioned in Section 17.1, the latest development in turbine technology is the *gas turbine,* in which the step of converting water to steam is eliminated. Instead, the hot combustion products of natural gas are what drive the paddle wheels of the turbine. In addition, the exhaust from the gas turbine is sufficiently hot to convert water to steam, which is then directed to an adjacent steam turbine to generate even more electricity. This system of using a gas turbine in tandem with a steam turbine can produce electricity with an efficiency as high as 47 percent. Even higher efficiencies can be attained by chemically converting natural gas to molecular hydrogen, H_2, which can be used to generate electricity in fuel cells, discussed in Section 11.4.

Two types of natural gas are supplied to consumers—one containing primarily methane, CH_4, and the other containing primarily propane, C_3H_8. Methane is lighter than air, which makes it relatively safe to deliver through a network of pipes extending throughout a municipality. If a leak occurs, the methane merely rises skyward, minimizing the fire hazard. Propane is heavier than air and readily liquefies under pressure. Because of these properties, propane is best stored as a liquid in pressurized tanks like the one shown in **Figure 17.12**. Propane tanks are used in areas not connected to municipal gas lines and require periodic filling.

CONCEPT**CHECK**
Of the three forms of fossil fuels, which is most abundant worldwide? Which burns most cleanly? Which is easiest to transport?

CHECK YOUR ANSWER Coal is most abundant. Because it contains few impurities, natural gas burns most cleanly. Because it is a liquid, petroleum is easiest to transport.

17.3 Issues of the Nuclear Industry

LEARNING OBJECTIVE

Describe the benefits and hazards of nuclear energy as generated by nuclear fission.

EXPLAIN THIS

What is the greatest long-term risk associated with nuclear energy?

Figure 17.13 summarizes the two forms of nuclear energy discussed in Chapter 5. One form is nuclear fission, which involves the splitting apart of large atomic nuclei, such as uranium and plutonium. The other is nuclear fusion, which involves the combining of two small atomic nuclei, such as

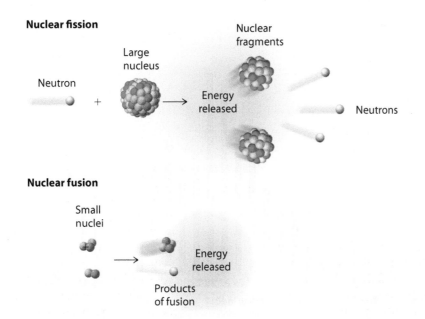

◀ Figure 17.13
Nuclear fission involves the splitting apart of large atomic nuclei. Nuclear fusion involves the coming together of small nuclei.

deuterium and tritium, into a single atomic nucleus, helium. All nuclear power plants to date use nuclear fission. These plants produce electrical energy without emitting atmospheric pollutants.

Nuclear fission energy for the commercial production of electricity has been with us since the 1950s. In the United States, about 20 percent of all electrical energy now originates from about 100 nuclear fission reactors situated throughout the country. Other countries also depend on nuclear fission energy, as is shown in **Figure 17.14**. Worldwide, there are about 436 nuclear reactors in operation and 63 currently under construction. The construction of new facilities, however, has slowed markedly over the past three decades due to negative public sentiments. As shown in **Figure 17.15**, most nuclear power plants are over 25 years old.

Countries turning to nuclear fission energy have decreased their dependence on fossil fuels and have diminished their output of carbon dioxide, sulfur oxides, nitrogen oxides, heavy metals, airborne particulates, and other atmospheric pollutants. Money that would have been spent on foreign oil payments has been saved.

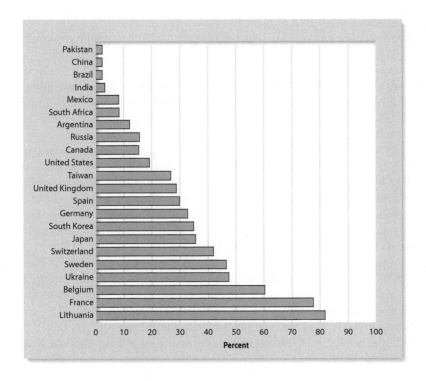

◀ Figure 17.14
Percentage of electricity generated from nuclear fission reactors in selected countries.

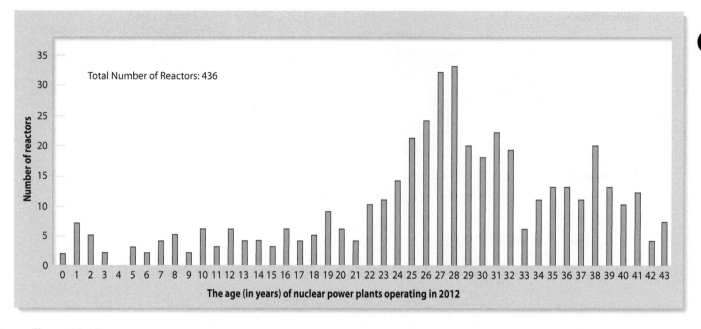

Total Number of Reactors: 436

The age (in years) of nuclear power plants operating in 2012

▲ **Figure 17.15**
The newest nuclear power plants were built before 1990.
Source: IAEA 2012

 FOR YOUR INFORMATION

Since 1983, the federal government has charged a small fee on every kilowatt-hour of electricity generated by nuclear power plants to finance a national disposal site. By 2009, this program had collected about $30 billion, $8 billion of which went to the study of Yucca Mountain.

 FOR YOUR INFORMATION

According to the World Nuclear Association, "During the next 50 years, as Earth's population expands from 7 billion toward 10 billion, humanity will consume more energy than the combined total used in all previous history. With carbon emissions now threatening the stability of the biosphere, the security of our world requires a massive transformation to clean energy. 'Renewables' like solar, wind, and biomass can help. But only nuclear power offers clean, atmosphere-friendly energy on a massive scale."

 READING CHECK

Where is the location of the U.S. long-term radioactive waste repository?

Without an enormous and concerted conservation effort, the world energy demand is going to increase, especially in light of growing populations and the dire need for economic growth in developing countries. Should nuclear energy production come to a standstill, allowing fossil fuels to accommodate this increased energy demand? Or should we continue to operate existing nuclear power plants—and even build new ones—until the alternative sources of energy discussed later in this chapter become feasible on a large scale? Nuclear advocates suggest a fivefold increase in the number of nuclear power plants over the next 50 years. They argue that nuclear fission is an environmentally friendly alternative to increased dependence on fossil fuels.

In the United States, however, the public perception of nuclear energy is less than favorable. There are formidable disadvantages, including the creation of radioactive wastes and the possibility of an accident that would release radioactive substances into the environment. In rebuttal, advocates point out that we cannot insist that nuclear fission energy be absolutely safe while at the same time accept hazards such as tanker spills, offshore drilling, global warming, acid rain, and coal miners' diseases.

So how much radioactive waste is there? According to the U.S. Department of Energy, as of 2002, the most recent year data was made available, about 46,000 tons of spent nuclear fuel rods were stored at reactor sites around the nation. The military is also a significant source of radioactive wastes. Holding tanks at the Hanford nuclear weapons plant in the state of Washington, for example, contain about 200 million liters of highly radioactive wastes. The general consensus among scientists is that our radioactive wastes are best handled by storing them in underground repositories located in geologically stable regions. Water seeping into such a repository, however, could greatly accelerate the corrosion of casks housing the radioactive wastes. Therefore, an effective repository should also be relatively devoid of water and placed hundreds of meters above any water table.

To date, no long-term repositories are in operation anywhere in the world, primarily because few, if any, communities want such a repository in their "backyard." Furthermore, once a potential site gets chosen, extensive and time-consuming evaluations are necessary. For example, tests at a promising site beneath Yucca Mountain in Nevada, shown in **Figure 17.16**, began in 1982. Some 27 years later, in 2009, after $8 billion worth of testing, this site was found to be unacceptable.

In addition to generating radioactive wastes, nuclear power plants pose the risk of having an accident in which radioactive material is released into the environment. The safety design of a nuclear power plant, however, has great bearing on the risks associated with generating nuclear fission energy. In 1979, a nuclear reactor at a facility on Three Mile Island, near Harrisburg, Pennsylvania, heated to the point that the reactor core began to melt. No appreciable radioactivity leaked into the environment, because the core was housed in a containment building (as shown in Figure 5.26). Seven years later, in 1986, a total meltdown occurred at the nuclear power plant shown in **Figure 17.17**, the Chernobyl plant in what is now Ukraine. Notably, the reactor core of the Chernobyl plant was not built and did not operate in accordance with internationally accepted nuclear safety principles. For example, the medium used to control the fission reactions was graphite, which loses its ability to control the reactions as the core temperature rises. Also, the reactor was not housed in a containment building. Because there was no containment building, large amounts of radiation escaped into the environment.

In March 2011, a powerful 9.0 earthquake off the coast of Japan resulted in a massive tsunami 15 m in height (49 feet). The operating nuclear reactors of the Fukushima Daiichi nuclear power plant along the coast of Japan were promptly shut down after the earthquake. The core of any nuclear reactor, however, takes many days to cool upon being shut down. Thus, pumps are needed to pass cooling water over the reactor. The operation of these pumps requires electricity, which comes from the power grid, or in a blackout, from a set of on-site diesel electric generators. Unfortunately, the tsunami that came 50 minutes after the earthquake severed the plant's connection to the grid. Furthermore, the emergency generators were located in the lower levels, which were promptly flooded as the tsunami waves passed over the plant's 5.7 m (19 feet) high seawalls (**Figure 17.18**).

Without power for the cooling system, the Fukushima reactors experienced a full meltdown, which means the uranium fuel turned into a molten phase and dropped to the floor of the containment building. By the next day, however, firefighter crews were able to pump seawater into the containment buildings in an effort to stop the ongoing meltdown. Because of their efforts, less than a meter of the 7-m thick concrete floor was penetrated by the meltdown material, which is to say that most of the radioactivity was contained.

The worst-case scenario would have occurred had the melted fuel penetrated the floor. In such a case, there would have been a massive leak of highly radioactive material into the environment. Radiation, however, was released from the venting of gases and the release of coolant. The extent of this radiation release is estimated to be about 10 percent of the Chernobyl meltdown. At both of these facilities, the efforts to clean up and contain the radioactive wastes will last for many decades.

Recent technological advances hold the promise of increased safety. Early reactors rely on a series of active measures, such as water pumps, that come into play to keep the reactor core cool in the event of an accident. A major drawback is that these safety devices are subject to failure, as was evident at Fukushima. The Generation IV reactor designs discussed in Chapter 5 provide what is called *passive stability*, in which natural processes such as evaporation are used to keep the reactor core cool. Furthermore, the core has a negative

▲ **Figure 17.16**
Yucca Mountain in Nevada was a promising site for a permanent repository for nuclear wastes. Had the site been approved, it would have housed some 70,000 tons of radioactive wastes within a network of tunnels 150 kilometers in combined length.

▲ **Figure 17.17**
In 1986, a meltdown occurred at this nuclear power plant in Chernobyl, Ukraine. Because there was no containment building, large amounts of radioactive material were released into the environment. Three people died outright, and dozens more died from radiation sickness within a few weeks. Thousands who were exposed to high levels of radiation stand an increased risk of cancer. Today, 10,000 square kilometers of land remains contaminated with high levels of radiation.

▶ **Figure 17.18**
Two workers at the Fukushima nuclear power plant died soon after the power station was crippled. Others bravely stayed to help manage the disaster despite high levels of radiation. Across Japan, the earthquake and tsunami directly killed about 20,000 people.

temperature coefficient, which means the reactor shuts itself down as its temperature rises owing to a number of physical effects, such as any swelling of the control rods.

The percentage of the world's electricity produced by fission reactors in 2010 stood at about 15 percent (**Figure 17.19**). Without major efforts to replace aging plants, the International Atomic Energy Agency projects that by 2020, nuclear reactors will account for as little as 9 percent of the world's electricity generation. This poses a tough dilemma because projections are that by 2020, the world's electricity needs will have increased significantly, as indicated in **Figure 17.20**. If nuclear fission energy is phased out, what will replace it?

CONCEPT CHECK

What is the major disadvantage of nuclear fission as an energy source?

CHECK YOUR ANSWER Nuclear fission reactors generate large amounts of radioactive wastes that require permanent large-scale storage facilities.

▼ **Figure 17.19**
World electricity generation by source of energy from 1971 to 2009.

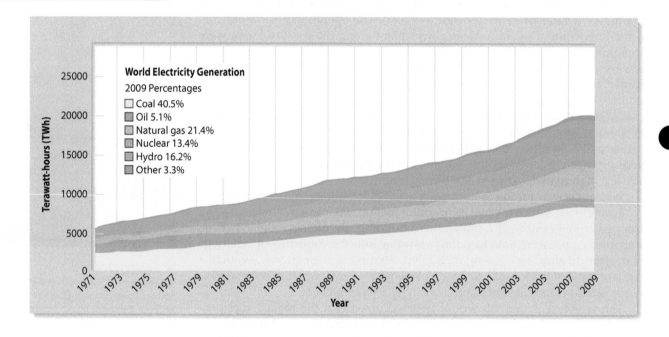

▶ **Figure 17.20**
Total worldwide electricity consumption projected through 2020.

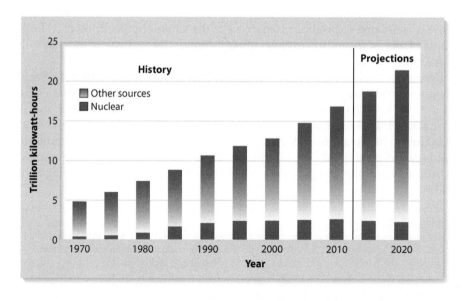

17.4 What Are Sustainable Energy Sources?

EXPLAIN THIS

Trick Question: How can nuclear fusion reactors be an ultimate source of sustainable energy when they generate so much radioactive waste?

LEARNING OBJECTIVE

Describe the ideal sustainable energy source.

The fossil fuels currently available to us are limited. At present rates of consumption, known recoverable oil and gas reserves will disappear by the next century and coal reserves several centuries after that. Furthermore, burning fossil fuels adds undesirable amounts of greenhouse gases to the atmosphere. Nuclear fission reactors do not emit greenhouse gases, but they generate massive quantities of radioactive wastes. Nuclear fusion reactors discussed in Chapter 5 offer many potential benefits, but it may take many decades before both are technologically and economically feasible. So what do we do?

What we ultimately need are *sustainable* energy sources, and we need access to those sources as soon as possible. The ideal sustainable energy source is not only inexhaustible but also environmentally benign. The nuclear fusion reactor has the potential to become an excellent sustainable energy source, but the ultimate source of sustainable energy is the nuclear fusion that goes on within our sun. Other astronomical sustainable energy sources we can tap into include the warm interior of our own planet Earth, as well as tidal forces resulting from the close proximity of the Moon. For the remainder of this chapter, we discuss various technologies that allow us to tap into these solar-system-sized energy sources.

Switching to sustainable energy sources will require commitment from the general public. Perhaps the greatest obstacle to our switching to sustainable energy sources is the present abundance of fossil fuels, which are packed with energy and are incredibly convenient to burn. In a national survey of public attitudes conducted by the U.S. Department of Energy, however, sustainable sources were the most desirable form of energy by far. But can people put their money where their hearts are? Fortunately, technologies are progressing rapidly, and sustainable energy sources will eventually cost the consumer less. This is a critical point, because in a market economy, it is the dollar that speaks. Let's take a look at what some of the major alternative sustainable energy technologies have to offer, as well as some of their potential drawbacks.

 **READINGCHECK**

What is perhaps the greatest obstacle to our switching over to sustainable energy sources?

CONCEPTCHECK

What is a sustainable energy source?

CHECK YOUR ANSWER A sustainable energy source is an energy source that has the potential to be available indefinitely. The technologies that take advantage of these sources should be environmentally benign.

17.5 Water Can Be Used to Generate Electricity

EXPLAIN THIS

How can water that is neither flowing nor boiling be used to move a turbine?

LEARNING OBJECTIVE

Identify three energy sources that allow water to be used to generate electricity.

Water power originates from one of three sources: the Sun, the Earth's hot interior, and the Moon. As you read the following descriptions of methods used to harness energy from water, trace the energy to one of these three sources.

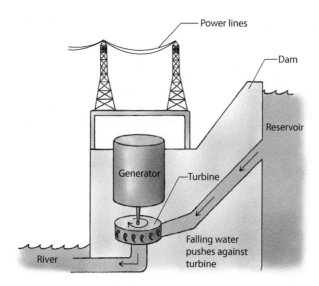

▲ **Figure 17.21**
An overview of a hydroelectric dam.

▲ **Figure 17.22**
The world's largest hydroelectric project, the Three Gorges Dam in China, was completed in 2009. At maximum capacity, the generators of this dam produce about 22,000 MW. This, combined with another 82,000 megawatts from other hydroelectric projects, provides an estimated 30 percent of the country's electricity. The dam extends 2.2 kilometers across, creating a reservoir 600 kilometers long. The flooding of this reservoir displaced about 1.2 million people.

READINGCHECK

OTEC is an acronym for what sustainable energy technology?

Hydroelectric Power Comes from the Kinetic Energy of Flowing Water

Water flowing through a hydroelectric dam rotates a turbine that spins an electric generator to produce electricity, as **Figure 17.21** illustrates. In a modern facility, the efficiency with which kinetic energy is converted to electrical energy can be as high as 95 percent, which translates to low costs for the consumer. Hydroelectric power is clean, producing no pollutants such as carbon dioxide, sulfur dioxide, and other wastes. The source of this power is the Sun, which transports water to mountain altitudes by way of the hydrologic cycle. Hydroelectric power is the kind of sustainable energy that is used most in the United States, supplying about 10 percent of our electricity needs. In developing countries, hydroelectric power supplies about 30 percent of all electricity needs.

There is great potential for increasing the output of hydroelectric power, but not by building more dams. Only 2400 of the 80,000 existing dams in the United States are used to generate power. Many of these "untapped" dams could be fitted with turbines and generators. Furthermore, most hydroelectric dams in the United States were built in the 1940s, when equipment was not as efficient as it is today. New technologies can allow these older plants to operate more efficiently, producing more electricity. The U.S. Department of Energy estimates that a 1 percent improvement in the efficiency of existing U.S. hydropower plants would produce enough power to supply 283,000 households.

Hydroelectric power plants may be pollutant-free, but for the local environment around the dam, there are consequences (**Figure 17.22**). Fish and wildlife habitats are significantly affected. One tragic consequence of some dams is that they prevent fish from reaching their spawning grounds; therefore, fish populations decline. To address this concern, many dams have been retrofitted with fish "ladders" that are supposed to encourage upstream migration to spawning grounds. The success of these ladders has been limited, however. Reservoirs created by dams tend to fill with silt, affecting water quality and limiting the life span of the dam. Furthermore, dams detract from the natural beauty of a river, and in many cases, valuable downstream cropland is lost or disrupted while upstream land habitats are flooded. Dams also need to be well maintained and inspected regularly to minimize the potential of a break.

Temperature Differences in the Ocean Can Generate Electricity

There is always a difference between the temperature of surface seawater (warmer) and the temperature of deep seawater (colder). A process known as *ocean thermal energy conversion* (OTEC) exploits this difference to produce electricity. As shown in **Figure 17.23**, warm surface water is used to boil a liquid that has a low boiling point, such as ammonia. The resulting high-pressure vapor pushes against the turbine, and the movement of the turbine generates electricity. After passing through the turbine, the vapor enters a condenser, where it is exposed to pipes containing cold water pumped up from great ocean depths. As its temperature drops, the vapor condenses to a liquid, which is recycled through the system.

OTEC is limited to regions where differences between ocean surface temperatures and deep-water temperatures are greatest. Fleets of floating offshore rigs have been proposed, with the electricity generated used to produce transportable hydrogen fuel made from seawater. Onshore OTEC plants are best suited for islands such as Hawaii, Guam, and Puerto Rico, where deep waters are relatively close to shore. The world's first OTEC plant, shown

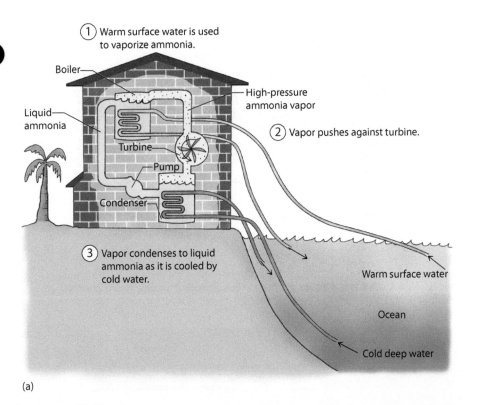

① Warm surface water is used to vaporize ammonia.

Boiler

Liquid ammonia

High-pressure ammonia vapor

② Vapor pushes against turbine.

Turbine

Pump

Condenser

③ Vapor condenses to liquid ammonia as it is cooled by cold water.

Warm surface water

Ocean

Cold deep water

(a)

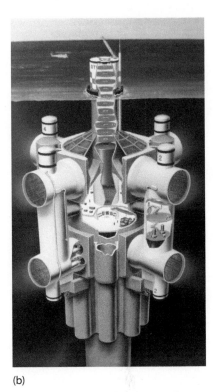

(b)

▲ **Figure 17.23**
An ocean thermal energy conversion operation (a) on land and (b) as a floating rig, which provides electricity to the shore through submerged transmission lines.

in **Figure 17.24**, has been running in Hawaii since 1990, producing about 210 kilowatts of electricity. Most of this electricity is used by a local aquaculture industry, which uses the nutrient-rich deep ocean water piped by the OTEC unit to breed specialty fishes and crustaceans for the U.S. and Japanese markets. Interestingly, the piped cold water is also used to air-condition the OTEC offices and laboratories.

Geothermal Energy Comes from the Earth's Interior

The Earth's interior is quite warm because of radioactive decay and gravitational pressures. In some areas, the heat comes relatively close to the Earth's surface. When this heat pokes through, we see it as lava from a volcano or steam from a geyser. This is *geothermal energy,* and it can be tapped for our benefit. **Figure 17.25** shows some areas in the United States that have geothermal activity.

Hydrothermal energy is produced by pumping naturally occurring hot water or steam from the ground. It is the predominant form of geothermal energy now being used commercially for electricity generation. Californians currently receive about 2600 megawatts from hydrothermal power plants. There are 170 hydrothermal power plants in the United States and many more in Italy, New Zealand, and Iceland. According to the Geothermal Energy Association, the present generating capacity of all hydrothermal power plants worldwide is about 11,000 megawatts. One installation is shown in **Figure 17.26**.

Hydrothermal plants produce electric power at a cost competitive with the cost of power from fossil fuels. Besides generating electricity, hydrothermal energy is used directly to heat buildings. Across the United States, geothermal hot water reservoirs are much more common than geothermal steam reservoirs. Most of the untapped hot water reservoirs are in California, Nevada, Utah, and New Mexico. The temperatures of these reservoirs are not hot enough to drive steam turbines efficiently, but the water is used to boil a secondary fluid, such as butane, whose vapors then drive gas turbines.

▲ **Figure 17.24**
(a) An aerial view of The Natural Energy Laboratory of Hawaii—the world's first OTEC facility—located on the coast near the Kona-Kailua airport. (b) The vertical axis turbine of the OTEC facility. The vapor-to-liquid heat exchanger is seen to the right. A more powerful 1-MW OTEC plant is being constructed at this site to test technologies that are to be deployed on floating OTEC stations.

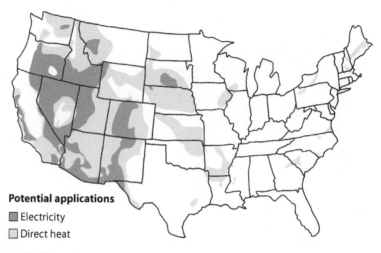

Potential applications

■ Electricity

□ Direct heat

▲ **Figure 17.25**
Regions of the United States where the prospects for geothermal energy are greatest.

CHEMICAL CONNECTIONS

How is the Blue Lagoon of Figure 17.26 connected to radioactivity?

FOR YOUR INFORMATION

As little as 5 meters below the Earth's surface, the temperature remains nearly constant year round. In New England, for example, the ground at this depth remains at about 8°C (46°F). During the winter, ground water can be pumped into a basement heat exchanger, where it causes a low-boiling-point material, such as freon, to evaporate. This cools the water down to about 4°C (39°F) but adds energy to the freon. The gaseous freon is then pumped to a network of baffles, where it condenses, releasing its heat to a second system of water that is piped through the floors of the house. Repeated cycles of the heat exchanger can warm the second system of water to up to 50°C (122°F), which is sufficient to keep the house toasty warm through the winter. This variation of geothermal heating is applicable to nearly any home or business with a backyard. The source of the energy is actually solar. Can you figure out why?

▲ **Figure 17.26**
About 50 percent of the electricity generated in Iceland is from geothermal sources. This is the Blue Lagoon, a warm pool created from the effluent of the hydrothermal power plant visible in the background.

Hot dry rock energy, another form of geothermal energy, involves using hydraulic pressure to open up a large reservoir deep underground. Liquid water is injected into the reservoir, heated by the hot rock, and then brought back to the surface as steam to generate electricity.

Geothermal energy is not without disadvantages. Few gaseous pollutants are emitted, but one of those pollutants is hydrogen sulfide, H_2S, which smells of rotten eggs even at low concentrations. Also, water coming from inside the Earth is often several times more saline than ocean water. This salty water is highly corrosive, and its disposal is a problem. Furthermore, water extracted from deep below is usually contaminated with radioactivity. Withdrawing water from geologically unstable regions also may cause land to subside and possibly trigger earthquakes.

The Energy of Ocean Tides Can Be Harnessed

The Moon's gravitational pull on our planet is uneven. The side of the Earth closest to the Moon experiences the greatest pull, and the side farthest away experiences the weakest pull. The result of this uneven pull is a subtle, planetwide elongation of our oceans. As the Earth spins underneath this elongation, Earth-bound observers notice a perpetual rise and fall of sea level. These are our *ocean tides,* and they can be harnessed for energy.

Tidal power is normally obtained from the filling and emptying of a bay or an estuary, which may be closed in by a dam. When tidal waters flow through the dam (in either direction), they cause a paddle wheel or turbine to rotate, and this motion generates electricity.

The prospects for using tidal power on a large scale are not good. In order to be effective, the tides must be relatively great. This severely limits the number of potential sites worldwide. Furthermore, many of these potential sites are praised for their natural beauty. In such cases, public opposition to building a power plant would be strong. Nonetheless, tidal power is an option some communities may want to consider. One successful site, located in Brittany, a region of France, produces about 240 megawatts of electricity.

CONCEPTCHECK

Of the forms of water power discussed in this section, which are solar in origin? Which is lunar? Which is derived from the Earth's hot interior?

CHECK YOUR ANSWER The energy obtained from dams, from ocean thermal energy conversion, and from a geothermal heat exchanger is solar because they all involve water being warmed by the Sun. Tidal energy is lunar, and other forms of geothermal energy are derived from the Earth's hot interior.

17.6 Biomass Is Chemical Energy

EXPLAIN THIS

Why does ethanol burn more efficiently than gasoline?

Plants use photosynthesis to convert radiant solar energy to chemical energy. This chemical energy comes in the form of the plant material itself—**biomass.** We can use the energy of biomass in two ways: process it to produce transportable fuels or burn it at a properly equipped power plant to produce electricity. Biomass can be grown on demand in regions where water supplies are abundant. If biomass is produced at a sustainable rate, the carbon dioxide released when it is burned balances the carbon dioxide consumed during photosynthesis. Biomass combustion products generally contain about one-third the ash produced by coal and about one-thirtieth the sulfur.

Fuels Can Be Obtained from Biomass

The U.S. transportation industry is 97 percent dependent on petroleum and consumes 63 percent of all petroleum stockpiles. Fuels from biomass, notably ethanol, are natural alternatives to petroleum-based fuels. In fact, ethanol has a higher octane rating than gasoline, which is why it is a preferred fuel for race-car drivers. Ethanol was also the preferred fuel of automotive pioneers Henry Ford and Joseph Diesel, who originally intended their automobiles to run on biofuels. By a curious coincidence, ethanol from fermented grains was about to be used by the automobile industry just prior to the passage of the 18th amendment to the U.S. Constitution prohibiting its use. Prohibition allowed the petroleum industry to take over.

LEARNING OBJECTIVE

Review how biomass can be used to create fuel or electricity.

READINGCHECK

When does the combustion of biomass not contribute to the net production of carbon dioxide, which is a greenhouse gas?

FOR YOUR INFORMATION

Does biomass sound reminiscent of fossil fuels? It should, because biomass is simply a fossil fuel without the fossil—dead plant material that has yet to turn into coal, petroleum, or natural gas. Just about anything you can do with fossil fuels, you can do with biomass, except, of course, deplete it and create as much pollution.

▲ **Figure 17.27**
Gasohol is gasoline containing an alcohol additive. The alcohol provides an octane boost, allowing an engine to run more efficiently with less pollution. If the alcohol is produced from biomass grown within a nation, there is the added benefit of a reduced dependence on foreign oil.

▲ **Figure 17.28**
Brazil's 440 sugar mills produce sugar as well as ethanol.

 FOR YOUR INFORMATION

According to the EPA, in 2011, the United States had about 87 operational power plants using municipal solid waste (MSW) as fuel. Together, these plants generated about 2,500 megawatts, or about 0.3 percent, of total national power generation. Due to economic factors, however, the number of MSW plants to energy plants has decreased by about 15 over the past decade. For some perspective, however, in 2001, we produced more than 229 million tons of municipal waste. Of this, about 34 million tons were burned to generate electricity.

Today in the United States, government programs requiring the addition of 10 percent ethanol to gasoline are becoming widespread. Ethanol, also known as *grain alcohol,* can be prepared from the fermentation of food biomass—any grain will do, but those rich in simple carbohydrates such as sugars work best (**Figure 17.27**).

In 2010, worldwide production of ethanol for fuel was about 87 billion liters. Of that, the United States was the leading producer (49 billion L) followed by Brazil (28 billion L). While ethanol is produced in the United States primarily from the fermentation of corn, in Brazil, it is produced primarily from the fermentation of sugar cane. Brazil started its ethanol for fuel program decades ago in response to the oil crisis of the 1970s. Up until 2005, it was the world's leading producer (**Figure 17.28**). Notably, most vehicles in Brazil are equipped to run using E100, which is all ethanol containing no fossil fuels. Thus, the carbon dioxide produced from these vehicles is offset by the carbon dioxide absorbed by the growing sugar cane.

Ethanol from fermentation is relatively expensive because of the great financial and environmental costs of growing food biomass, a process that requires vast amounts of water and fertilizer. Ethanol derived from petroleum is actually less expensive, but only because crude oil prices are kept artificially low. If taxpayer subsidies, exemptions from paying for the environmental damage from mining and drilling, and the cost of military protection are factored in, the price of crude oil quadruples.

The ultimate starting material for the creation of ethanol is cellulose, which is the world's most abundant organic chemical, found in all plants. The idea is to break cellulose down to its glucose monomers, which can then be fermented to form ethanol. As discussed in Chapter 13, however, the cellulose found in plants is tightly locked. This is important for a plant's structural rigidity, but it makes it very difficult to break down the cellulose into glucose. Nonetheless, much research is being conducted to discover ways in which this can be done in a cost-effective and environmentally friendly manner. One promising approach is to cook the cellulose in liquid ammonia, NH_3. Other approaches involve mimicking the cellulose-digesting action of the microbes that perform this feat within the stomachs of termites and ruminants, such as cows and goats. Ethanol from cellulose is sometimes called *cellulosic ethanol* or *grassohol*. An advantage of cellulosic ethanol is that it does not use up foodstocks such as corn and sugar that could be used instead to help alleviate world hunger.

Biomass Can Be Burned to Generate Electricity

Converting biomass to a transportable fuel is an extra step that decreases energy efficiency. Higher efficiencies are obtained by burning the biomass directly. In the United States, electricity produced from biomass has grown from 200 megawatts in the early 1980s to more than 8000 megawatts in 2000—a 4000 percent increase. Estimates are that this will grow to about 23,000 megawatts by 2020. Most biomass power is generated by paper companies and forest products companies using wood and wood wastes as fuel. Municipalities are experimenting with solid-waste incinerators that provide electricity and waste disposal simultaneously. (On average, about 80 percent of the dry weight of municipal solid waste is combustible organic material.)

The traditional approach to generating electricity from biomass is to burn the biomass in a boiler, where water is converted to steam used to drive a steam turbine. Efficiency can be more than doubled by first converting the biomass to a gaseous fuel, which may be done by applying air and steam at high pressure. Alternatively, gaseous fuels are also produced by mixing the biomass with very hot sand, as is done at the Vermont facility shown in **Figure 17.29**. The gaseous fuel is burned, and the hot combustion products are directed to a gas turbine that generates electricity. In addition, the exhaust gases from the turbine can be used to produce steam for industrial applications or for additional power generation.

> ### CONCEPTCHECK
> What do biomass and fossil fuels have in common?
>
> **CHECK YOUR ANSWER** They both originate from solar energy.

▶ **Figure 17.29**
Since 1984, the Vermont Biomass Gasification Project has supplied more than 50 MW of electricity to the Burlington, Vermont area. The electricity is generated by gas turbines powered by the combustion of a gaseous fuel mixture created as wood chips are mixed with very hot sand (1000°C).

17.7 Energy Can Be Harnessed from Sunlight

EXPLAIN THIS

How might solar energy best be used to dry laundry?

Sunlight can be used directly to heat our homes. Mirrors and lenses can concentrate sunlight onto water to create steam for generating electricity. Sunlight causes winds, which can drive electricity-producing wind turbines. With photovoltaic cells, the energy of sunlight can be converted to an electric current. In this section, we focus on technology that makes use of the effects of direct sunlight.

LEARNING OBJECTIVE

Provide examples of how direct sunlight can be used as a source of useful energy.

Solar Heat Is Easily Collected

Whether you live in a hot or cold climate, you can conserve energy by taking advantage of the Sun's rays. Heating water for bathing and for washing dishes and clothes consumes a considerable amount of energy. A typical household uses about 15 percent of its total energy consumption to heat water. Energy to heat water could just as well come from the Sun as from the local electric power plant or natural gas supplier.

A solar energy collector for heating water is not much more than a black metal box covered with a glass plate, as **Figure 17.30** illustrates. Sunlight passes through the glass into the box, where it is absorbed by the black metal, which becomes hot, emitting infrared radiation. Because glass is opaque to infrared, these rays stay in the box. Water passing through pipes within the box is heated by the trapped warmth.

Water that has passed through a series of collectors can be scalding hot. This water can be stored in well-insulated containers and used for washing purposes. Air passing over coils of solar-heated water becomes warm and may be used to heat buildings.

Solar Thermal Electric Generation Produces Electricity

A variety of technologies collectively known as *solar thermal electric generation* can be used to produce electricity from sunlight. One technology involves pumping a synthetic oil through a pipe positioned near a mirror-coated trough, as

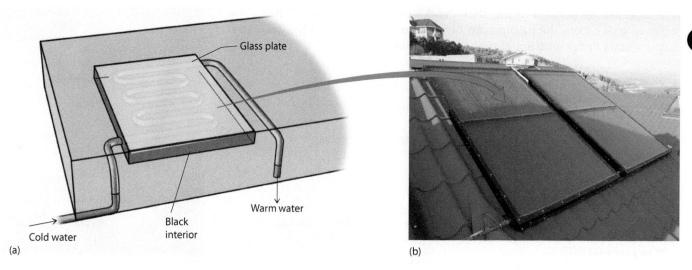

(a)

(b)

▲ Figure 17.30

(a) A solar energy collector is covered with glass to provide a greenhouse effect: sunlight passing into the box is converted to infrared radiation, which cannot escape. (b) Most solar energy collectors are located on rooftops. The collectors are painted black to maximize the absorption of solar heat. The rooftop collector shown here is used for warming an outdoor swimming pool.

shown in **Figure 17.31**. The extremely hot oil is then used to generate steam in a steam turbine used to generate electricity. This design is currently being used in the desert regions of southern California, where more than 360 megawatts of electricity is produced at an operating cost of about 10 cents per kilowatt-hour. A natural gas burner provides supplemental heat during periods of high demand or cloudy weather.

A second technology involves a large array of sun-tracking mirrors reflecting sunlight onto the top of a central tower, where the temperature soars to about 2200°C. Solar heat is carried away by a molten salt, primarily a mixture of sodium nitrate, $NaNO_3$, and potassium nitrate, KNO_3, commonly called *saltpeter*. The hot molten salt is piped to an insulated tank, where the solar thermal energy can be stored for up to a week. When electricity is needed, the salt is pumped to a conventional steam-generating system for the production of electricity. The molten salt recirculates to the central tower for reheating. An important advantage of this approach is that the salt retains heat for a long time, which allows the production of electricity even after the Sun has set or during periods of inclement weather. **Figure 17.32** shows the world's largest commercial central-tower solar thermal energy plant. Located in southern Spain, it started operating in 2011 and generates about 20 megawatts of electricity, which is enough to power about 27,500 homes.

▶ Figure 17.31

A solar thermal electric generation unit. A pipe containing synthetic oil is positioned along a mirror-coated trough. Sunlight hitting the mirror is reflected onto the pipe and heats the oil to 370°C. The hot oil is then pumped out and used to convert water to steam in a turbine in an electric power plant.

◀ Figure 17.32
The Gemasolar thermosolar power plant located in Sevilla, Spain. The central tower is 140 meters tall and is surrounded by 2650 sun-tracking mirrors, each measuring 120 square meters.

Wind Power Is Cheap and Widely Available

Wind power is currently the cheapest form of direct solar energy. This is due, in part, to its simplicity—wind blows on a wind turbine, and the rotation of the turbine blades generates electricity. Most of the early development of wind power took place in California during the 1980s as a result of favorable tax incentives. Costs were relatively high without them, because there was little standardizing of procedures and a lack of experience in mass production. Today, most of these problems have been resolved, and the reliability, performance, and cost of wind power have improved dramatically. In regions where wind speed averages 13 miles per hour, the operating cost of wind-generated electricity now stands at about 4 cents per kilowatt-hour. Unlike conventional energy sources, whose prices keep going up, the cost of wind energy has been coming down. Within a decade, it may well drop below 1 cent per kilowatt-hour, which is significantly cheaper than the cost of electricity from conventional sources such as coal.

In 2011, the worldwide installed capacity for wind energy totaled about 239 gigawatts, which is exponentially greater than the worldwide 1990 total of only 2 gigawatts (**Figure 17.33**). For perspective, however, consider that the United States alone consumes energy at a rate of about 440,000 MW. So wind energy is still only a small percentage of the total, but it is clearly a fast-growing sector of the energy industry. The American Wind Energy Association (AWEA) estimates that by 2020, wind power could supply at least 6 percent of U.S. electric power needs.

As is shown in **Figure 17.34**, most of the wind resources in the United States are concentrated in the northern Great Plains. The U.S. Department of Energy estimates that more than 100 percent of the nation's electricity needs could be

READINGCHECK

Why is wind power the cheapest form of direct solar energy?

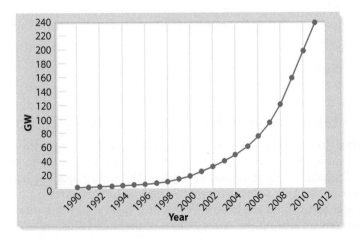

◀ Figure 17.33
China (63 GW) surpassed the United States (47 GW) in 2011 for developing the most power from wind energy. Total world production is expected to exceed 500 GW by 2016.

▶ **Figure 17.34**
Most of the U.S. wind energy potential is in the Great Plains region.

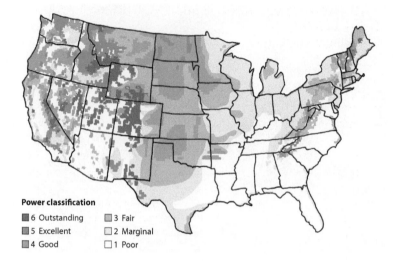

Power classification

■ 6 Outstanding   □ 3 Fair
■ 5 Excellent   □ 2 Marginal
■ 4 Good   □ 1 Poor

met if the wind in these states were fully exploited. At issue, however, is that most of these wind areas are located far away from the more populated areas of the country where demand for electricity is greatest. An expensive infrastructure of new transmission lines would have to be built. Alternatively, electricity from wind could be used to create hydrogen through the electrolysis of water, as we discuss in Section 17.8.

A drawback to wind energy is aesthetic: wind turbines are noisy, and many people do not want to see them on the landscape. Furthermore, the turbines on smaller windmills rotate relatively fast, posing a hazard for birds. Also, all windmills must be manufactured, which requires materials and energy. These drawbacks need to be weighed against the benefit of reduced dependence on fossil fuels and the idea that if people look fondly on the classic windmills of Holland, then perhaps these new-age windmills might one day be looked upon as an attractive sign of prosperity (**Figure 17.35**). Fortunately, much of the wind-rich land in the United States is farmland, and wind power and agricultural activities are compatible. At the Altamont wind farms in California, ranchers lost only about 5 percent of their grazing area. Their land prices, however, quadrupled because of royalties they receive from wind power companies using their land.

For people who purchase small wind turbines for their homes, unused electricity causes their electric meter to run backward. As this happens, the wind-generated electricity becomes available for other people connected to the same grid. Many state laws require utility companies to either purchase this electricity from the turbine owners or provide credit on their electric bill. In this way, the turbine owner is able to "bank" the energy collected on windy days for use on days when the wind speed is too low.

CONCEPTCHECK

What do fossil fuels have in common with the various forms of direct solar energy just discussed?

CHECK YOUR ANSWER They all originate with the Sun.

Photovoltaics Convert Sunlight Directly to Electricity

As described in Chapter 11, photovoltaic cells are the most direct way of converting sunlight to electrical energy. Since their invention in the 1950s, photovoltaics have made remarkable strides. The first major application was in the 1960s in the U.S. space program—space satellites carried photovoltaic cells to power radios and other small electronic devices. Photovoltaics gained further momentum during the energy crisis of the mid-1970s.

▲ **Figure 17.35**
The European Wind Energy Association anticipates the installed wind energy capacity in Europe will reach 180,000 MW by 2020.

The cost of receiving electricity through photovoltaics is primarily a function of the cost of purchasing the equipment and having it installed. As of 2012, a residential 2-kilowatt photovoltaic system is about $14,000. This system would pay for itself in savings from utility bills in roughly 12 years. In sunny climates, however, the payback time would be quicker. Also, photovoltaics, as well as all other sustainable energy sources, become more attractive as the cost of purchasing electricity from fossil fuel–based utility companies continues to rise. Furthermore, keep in mind that the price of photovoltaic systems continues to drop as photovoltaic cell technology improves. Beyond economic reasons, many people are switching to photovoltaic systems simply because they are a clean source of energy and are good for the environment. Today's average residential photovoltaic system has an estimated life span of 25 years.

Worldwide, sales of photovoltaics have grown exponentially over the last several decades. Photovoltaics are now a multi-billion-dollar industry with a strong promise of continued growth. Because they need minimal maintenance and no water, they are well suited to remote or arid regions. Photovoltaics can operate on any scale and are on the verge of being cost-competitive with electricity from fossil and nuclear fuels. More than a billion handheld calculators, several million watches, several million portable lights and battery chargers, and tens of thousands of remote communications facilities are powered by photovoltaic cells. As **Figure 17.36** illustrates, the range of their usefulness is huge.

FOR YOUR INFORMATION

According to the Energy Star efficiency program, if every household in the United States replaced one lightbulb with a compact fluorescent bulb, it would prevent enough pollution to equal removing 1 million cars from the road. Similarly, more than 100 million exit signs are in use throughout the United States. Typically lit by less efficient incandescent bulbs, these signs consume from 30 billion to 35 billion kWh of energy each year. If all U.S. companies switched to Energy Star–qualified exit signs, they would save $75 million in electricity costs. Better still are the even more energy-efficient LED lamps that, unlike fluorescent bulbs, are not a source of mercury pollution.

◀ Figure 17.36
Photovoltaic cells come in many sizes, from those in a handheld calculator to those in a rooftop unit that provides electricity to a house.

17.8 Solar Energy Can Be Stored as Hydrogen

EXPLAIN THIS

What is the difference between hydrogen burning here on the Earth and hydrogen burning in the core of the Sun?

LEARNING OBJECTIVE

Describe the prospects of hydrogen as a carrier of energy.

An ideal fuel is hydrogen, H_2. By weight, hydrogen packs more energy than any other fuel, which is why it is used to launch spacecraft. It is readily produced from the electrolysis of water, and the electrolysis equipment can be driven by solar-derived electricity. In this way, hydrogen provides a convenient means of storing solar energy.

Hydrogen burns cleanly, producing almost nothing but water vapor. The vapor can be used to drive a gas turbine for the production of electricity. The waste heat can be used for industrial purposes or to warm buildings. The water that condenses from the turbine can be used for agriculture or human consumption, or it can be recycled for the production of more hydrogen.

Because it is a gas, hydrogen is easily transported through pipelines. In fact, pumping hydrogen through pipelines is more energy efficient than transmitting electricity through power lines. Thus, hydrogen facilities could be located in regions where production is cheapest—desert regions for solar thermal and photovoltaics, windy regions for wind turbines, and moist regions for biomass—and then transported to meet the energy needs of distant communities.

Hydrogen can even be used as a fuel for automobiles, as shown in **Figure 17.37**. Porous metal alloys have been developed that absorb large quantities of hydrogen gas. Pressing the gas pedal of a car powered by hydrogen fuel sends a warming electric current to the alloy, which is housed in the gas tank. As the alloy heats up, it releases hydrogen, which powers either an internal-combustion engine or an electricity-producing fuel cell. Gas stations of the future may truly be *gas* stations!

C O N C E P T **C H E C K**

Why does the combustion of hydrogen produce no carbon dioxide?

CHECK YOUR ANSWER There is no carbon with which to form carbon dioxide. Hydrogen is an ideal fuel that, when burned, produces almost nothing but water vapor—no carbon dioxide, no carbon monoxide, and no particulate matter.

Fuel Cells Produce Electricity from Fuel

As Section 11.4 discussed, an efficient way to produce electricity from hydrogen (or any other fuel) is with a *fuel cell*. Utilities could produce megawatts of electricity by stacking fuel cells. Present-day fuel cells have an efficiency of about 60 percent, which is well above the 34 percent efficiency of a typical coal-fired power plant. Interestingly enough, one potential source of hydrogen is coal, which when treated with high-pressure steam and oxygen produces hydrogen gas and methane gas. Burning these gases in a gas turbine results in a coal-to-electricity efficiency as high as 42 percent. If the gases are first run through electricity-producing fuel cells, the efficiency is even higher. Meanwhile, because the coal is not burned directly, fewer pollutants are released into the environment.

Photovoltaic Cells Can Be Used to Produce Hydrogen from Water

The cleanest and most abundant source of hydrogen on our planet is water. Creating hydrogen from water is an energy-intensive process, however, that is best done using electrolysis, a technique discussed in Section 11.6. Ideally,

READING CHECK

What is a potential source of abundant quantities of hydrogen?

▶ **Figure 17.37**
A hydrogen-powered test vehicle developed by BMW. The exhaust consists primarily of water vapor. However, many technical and economic hurdles must be cleared for hydrogen-powered vehicles to be commonplace.

the electrical energy required to create hydrogen from water could be provided by a solar-driven photovoltaic cell. The electricity generated by the cell is then sent to a second cell, where water is electrolytically cleaved to produce hydrogen. Also, research that is currently under way seeks to develop photovoltaic surfaces that directly generate hydrogen from water, as shown in **Figure 17.38**.

Over the past several years, there has been great progress toward the realization of a commercially feasible solar-energy-to-hydrogen system. For such a system, the efficiency at which the solar energy is converted to hydrogen needs to be as high as possible. Initial systems had efficiencies of no more than 6 percent. One of the main hurdles was developing a photovoltaic system optimized for 1.23 volts, which is the voltage needed for the hydrolysis of water. Such a system was developed by 1998 and provided an efficiency of 12.4 percent. With newer nanotechnologies now under development, researchers are anticipating efficiencies in excess of 60 percent.

If hydrogen can be efficiently produced using electricity from photovoltaics or wind turbines, then a number of interesting possibilities arise. Photovoltaics and wind turbines are "distributed" sources of energy, as opposed to "centralized" sources. In other words, they can be placed in people's backyards or in neighborhood solar energy parks, in contrast to a centralized coal-fired power plant. The solar electricity and hydrogen can thus be produced at the point of use, thereby avoiding energy losses inherent in shipping energy over long distances. Such a system would not likely put the centralized utility out of business; instead, it would help the utility cope with ever-increasing energy demands. Perhaps most importantly is the fact that the greatest energy growths for the next century will be occurring in developing nations where the infrastructure for centralized energy sources does not exist. Starting people off in these countries with noncentralized and nonpolluting sustainable energy sources would be a worthwhile investment.

▲ **Figure 17.38**
This special photoelectrochemical electrode submerged in water uses the energy of sunlight to generate hydrogen gas, which forms on the electrode's surface. A technical barrier in this case is preventing the electrode from corroding.

But Hydrogen May Not Be the Ultimate Solution

Hydrogen as the fuel of the future has much appeal. There are still, however, numerous drawbacks. According to the National Academy of Sciences, for example, fossil fuels will likely continue to be the primary source of hydrogen for the next several decades. Some argue that funds pushed toward the hydrogen economy would be better spent on increasing current vehicle efficiencies. The demand for hybrid vehicles, for example, is great, but the automotive industry in the United States was late in meeting this demand because the fuel efficiencies of its vehicles were not its top priority. The resulting delay in the implementation of hybrid vehicle technology thus contributed to additional fossil fuel consumption and carbon dioxide output. Major breakthroughs are still required in order for hydrogen fuel cells to become cost-competitive with hybrids. So can we do both? Develop hydrogen cars while advancing hybrid technology at the same time? Some say yes. Others say biofuels, especially cellulosic ethanol, would be more efficient and environmentally friendly. They point to the fact that energy use in the average home contributes more to air pollution and climate change than the average car.

Shai Agassi, founder of BetterPlace.com, strongly believes that the way forward is to develop an efficient infrastructure in support of an all-electric car. In place of gas stations, for example, there will be battery stations where robotics replace your run-down battery with a fresh one quicker than you can swipe your credit card. For the most part, though, drivers charge their cars at pod stations located throughout the community or at home at night when the price of electricity is lowest. In 2012, full-scale trials of this system began in Israel, Denmark, China, and Australia (**Figure 17.39**). Stay tuned for a battery exchanging station near you!

Others point to research indicating that we will soon be able to replicate photosynthesis, whereby we could generate methanol and other organic liquid fuels and feedstocks using sunlight and our increasing concentrations of atmospheric carbon dioxide. Think about that. On the web, use the keywords *artificial photosynthesis* to find that there is no shortage of human ingenuity.

In Perspective

Nearly a quarter of all U.S. facilities that generate electric power will need major renovations, overhauls, or replacement within the next few years. This represents an opportunity for the development of sustainable energy sources. Even greater opportunities exist in developing nations, where there are more than 2 billion people without electricity and many more to come with population growth. Much money will be spent to bring electric power to these people. Immediate choices made about where that energy comes from and how it is produced will affect the environment for a long time to come (**Figure 17.40**).

All regions of the world will benefit from energy conservation, which is the natural partner of sustainable energy. Conservation measures can be quite effective, as was demonstrated after the Arab oil embargo imposed by the Organization of Petroleum Exporting Countries in 1973. From the start of this embargo until 1986, energy consumption in the United States remained constant as a result of increased efficiency while the economy grew by 30 percent. Are we close to our limit in terms of conservation measures? New technologies and materials suggest there is still much we can do. Superinsulated buildings, for example, now require only one-tenth as much heating in winter and cooling in summer. Hybrid combustion–electric vehicles can now drive more than 40 kilometers on 1 liter of gasoline and 25 kilowatts of battery power—this corresponds to about 100 miles per gallon! If you were to replace a single 60-watt incandescent bulb with an equivalent 8-watt LED light, you would be saving the release of about 1800 kg of carbon dioxide each year (assuming a fossil fuel power plant). This LED light would also be about 90 percent cheaper to operate. How many LED light fixtures are in your home?

Our energy needs are increasing primarily because of our increasing population, as shown in Table 17.1. The data in this table assumes that an acceptable standard of living can be had at 0.0030 megawatt per person. On average, however, the 1.2 billion individuals of developed nations consume about 0.0075 megawatt per person. If conservation measures are not set in place and if the far greater number of individuals in developing nations is brought up to 0.0075 megawatt per person, then the world will use 75 million megawatts in 2050 and 112 million megawatts in 2100, six times the amount consumed in 2000.

▲ **Figure 17.40**
Electrical wiring typical along the streets of a densely populated city in India, which reminds us that electrical energy is a highly desired product and vital to modern living. Careful planning for the creation and delivery of this electrical energy is greatly needed.

TABLE 17.1   The World's Increasing Energy Needs

YEAR	WORLD POPULATION (BILLIONS)	×	ENERGY PER PERSON (MW)	=	TOTAL POWER CONSUMED (MW)
2000	6.2	×	0.003	=	19 million
2050	10.0	×	0.003	=	30 million
	10.0	×	0.0075	=	75 million
2100	15.0	×	0.003	=	45 million
	15.0	×	0.0075	=	112 million

As hard as controlling population growth may be, it is likely to be easier than providing increasing numbers of people with energy, food, water, and anything else. For more on this important concept, be sure to read the Contextual Chemistry essay on page 574.

Thomas Jefferson described revolution as an extraordinary event necessary to enable all ordinary events to continue. If a decent standard of living for all human beings is to be held as ordinary, what we need is nothing short of an energy revolution. The days of our using nonrenewable energy sources are numbered. The sooner we make the necessary transitions, the better.

Good chemistry to you.

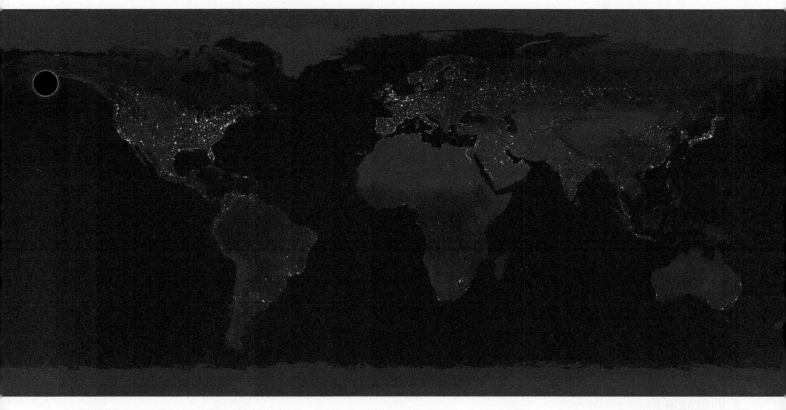

▲ **Figure 17.41**
A composite image from the night skies of the Earth. Where will the energy come from to meet the needs of our growing human population? As our thirst is great, the most likely answer is "all of the sources presented in this chapter."

Chapter 17 Review

LEARNING OBJECTIVES

Provide a brief overview of how electricity is generated and distributed and how its consumption is measured. (17.1)	→	*Questions 1–5, 35–37, 41–44, 75*
Describe the chemical nature of fossil fuels and their advantages and disadvantages. (17.2)	→	*Questions 6–11, 45–48, 71*
Describe the benefits and hazards of nuclear energy as generated by nuclear fission. (17.3)	→	*Questions 12–16, 49–54, 72*
Describe the ideal sustainable energy source. (17.4)	→	*Questions 17–18, 55–56*
Identify three energy sources that allow water to be used to generate electricity. (17.5)	→	*Questions 19–22, 40, 57–61*
Review how biomass can be used to create fuel or electricity. (17.6)	→	*Questions 23–26, 62–63*
Provide examples of how direct sunlight can be used as a source of useful energy. (17.7)	→	*Questions 27–30, 38–39, 64–67, 74*
Describe the prospects of hydrogen as a carrier of energy. (17.8)	→	*Questions 31–34, 68–70, 73*

SUMMARY OF TERMS (KNOWLEDGE)

Biomass A general term for plant material.

Coal A solid consisting of a tightly bound network of hydrocarbon chains and rings.

Kilowatt-hour The amount of energy consumed in 1 hour at a rate of 1 kilowatt.

Natural gas A mixture of methane and small amounts of ethane and propane.

Petroleum A liquid mixture of loosely held hydrocarbon molecules containing not more than 30 carbon atoms each.

Power The rate at which energy is expended.

Watt A unit for measuring power, equal to 1 joule of energy expended per second (J/s).

READING CHECK QUESTIONS (COMPREHENSION)

17.1 Electricity Is a Convenient Form of Energy

1. What does an electric current generate?

2. What does an electrical wire moving through a magnetic field generate?

3. Is electricity a source of energy?

4. What's a watt?

5. What is the estimated total length of the North American electric power grid?

17.2 Fossil Fuels Are a Widely Used but Limited Energy Source

6. Why are fossil fuels such a popular energy resource?

7. Why is coal considered the filthiest fossil fuel?

8. How does a scrubber remove noxious gaseous effluents created in the combustion of coal?

9. Can coal be converted to cleaner-burning fuels?

10. About how many barrels of petroleum are consumed each day in the United States?

11. What chemical is the main component of natural gas?

17.3 Issues of the Nuclear Industry

12. Why has the number of operating nuclear fission plants in the United States decreased over the past two decades?

13. Before what year were most currently operating nuclear power plants built?

14. Why was most of the radiation from the Chernobyl meltdown released into the environment?

15. Why did the backup generators at the Fukushima nuclear power plant fail in March 2011?

16. What was the leading source of energy for the production of electricity throughout the world in 2009?

17.4 What Are Sustainable Energy Sources?

17. What is the ideal sustainable energy source?

18. What is the ultimate source of sustainable energy?

17.5 Water Can Be Used to Generate Electricity

19. Why is Hawaii particularly well suited for generating electricity using OTEC technology?

20. Why are hydrothermal vents often quite smelly?

21. How are tides used to generate electricity?

22. What does OTEC stand for?

17.6 Biomass Is Chemical Energy

23. In what sense is biomass a form of solar energy?

24. Which has a higher octane rating, ethyl alcohol or gasoline?

25. What country is the leading producer of ethanol fuel?

26. What is the most efficient way to convert biomass to electricity?

17.7 Energy Can Be Harnessed from Sunlight

27. Why are solar water heaters painted black?

28. Describe two technologies used to convert solar heat to electricity.

29. What salt is used in the Gemasolar thermal solar power plant?

30. What is a drawback of wind power?

17.8 Solar Energy Can Be Stored as Hydrogen

31. Why is hydrogen such an ideal fuel?

32. How might cars someday store hydrogen fuel?

33. How are fuel cells different from batteries?

34. What is unique about the electric cars produced by the company Better Place?

THINK AND SOLVE (MATHEMATICAL APPLICATION)

35. How much money would you save per hour by replacing a 100-watt incandescent lightbulb with an equally bright 10-watt LED lamp? Assume the cost for electricity to be 15 cents per kilowatt-hour.

36. It is estimated that in 2020, about 200,000,000 MW-hours of electricity will be produced from the combustion of biomass. How much power is this in units of MW? (Hint: there are 8760 hours in a year.)

37. Assuming 1 megawatt of electricity is sufficient to meet the needs of about 350 homes, how many kilowatts of power are needed to meet the needs of a single home?

THINK AND COMPARE (ANALYSIS)

38. Rank the following sources of energy in order of environmental friendliness for the production of electrical energy.

    a. Coal   **b.** Wind
    c. Nuclear fission   **d.** Photovoltaics

39. Rank the following in order of the least number of steps to the most number of steps required to generate electricity.

    a. Solar thermal energy   **b.** Hydroelectric power
    c. OTEC   **d.** Photovoltaics

40. Rank in order of which has the greatest potential, practically speaking, of providing energy.

    a. Hydroelectric power
    b. Geothermal energy
    c. Ocean tides

THINK AND EXPLAIN (SYNTHESIS)

17.1 Electricity Is a Convenient Form of Energy

41. What is the difference between a kilowatt and a kilowatt-hour? Which is a quantity of energy? Which is a measurement of available power?

42. Why are gas turbines more efficient at generating electricity than steam turbines?

43. How does blocking electrical energy from moving into one region quickly block electrical energy from moving into all adjacent regions?

44. Why are airplanes that are flown by the power of electricity rather than the power of gasoline fuel so impractical?

17.2 Fossil Fuels Are a Widely Used but Limited Energy Source

45. How can fossil fuels be considered a form of solar energy?

46. Why is it impractical to burn a fuel in such a way that no nitrogen oxides are formed as combustion products?

47. Why does burning natural gas produce less carbon dioxide than burning either petroleum or coal? Hint: consider the chemical formulas of the compounds making up these materials.

48. How might fracking technology encourage the retirement of many coal-fired power plants?

17.3 Issues of the Nuclear Industry

49. What are some advantages of deriving energy from nuclear fission?

50. Why would a chain-reaction control substance that loses neutron-absorbing power as its temperature increases be dangerous?

51. In 1973, the Organization of Petroleum Exporting Countries stopped exporting its oil for a time. What effect do you suppose this embargo had on the nuclear power industry?

52. What are some advantages of deriving energy from nuclear fusion?

53. Is it possible for a nuclear fission power plant to blow up like an atomic bomb?

54. Why does the graph of Figure 17.19 begin to dip downward around 2008?

17.4 What Are Sustainable Energy Sources?

55. What's the difference between a sustainable energy source and a renewable energy source?

56. Why is it in a country's best national security interests to invest in sustainable energy sources?

17.5 Water Can Be Used to Generate Electricity

57. Why do many environmental groups oppose building more hydroelectric dams when these dams produce virtually no chemical pollutants as they create electric power?

58. At atmospheric pressure, water has a boiling point too high to allow it to be used as the turbine-pushing fluid in an OTEC electric generator. How might the generator be modified to overcome this problem? How might such a modification allow for the production of fresh water from seawater?

59. How is it that here on the Earth we can derive energy from the Moon, which is 400,000 kilometers away?

60. Why is the potential for geothermal energy greater on the western side of the United States than on the eastern side?

61. How is a shallow geothermal heat pump system ultimately a form of solar energy?

17.6 Biomass Is Chemical Energy

62. What are some disadvantages of producing ethanol by fermenting food biomass?

63. What is so special about cellulosic ethanol?

17.7 Energy Can Be Harnessed from Sunlight

64. How does a nonsolar pool cover increase the efficiency of a nonsolar pool heater?

65. What is the utility of south-facing windows in northern climates?

66. Suggest how solar energy might help to destroy pathogens in drinking water.

67. How might solar heat be used to generate colder temperatures?

17.8 Solar Energy Can Be Stored as Hydrogen

68. In the simplified electric generator shown here, a power source is needed at either point A or point B. Where could each of the following power sources be used, at A or at B?

   natural gas _____
   wind _____
   nuclear fusion _____
   hydroelectric _____
   coal _____

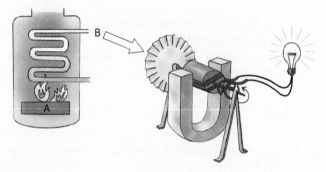

69. What are some major advantages of using seawater to produce hydrogen from electrolysis?

70. A disadvantage of tidal power is that electricity is generated only when tides are coming in or going out. How might a tidal electric power plant be outfitted so that it provides a continuous supply of electricity?

THINK AND DISCUSS (EVALUATION)

71. Why might some people consider it a blessing in disguise that fossil fuels are such a limited resource? Centuries from now, what attitudes about the combustion of fossil fuels are our descendants likely to have?

72. The 1986 accident at Chernobyl, in which dozens of people died and thousands more were exposed to radiation that might lead to cancer in the future, caused fear and outrage worldwide and led some people to call for the closing of all nuclear plants. Yet many people choose to smoke cigarettes in spite of the fact that 2 million people die every year from smoking-related diseases. The risks posed by nuclear power plants are involuntary, risks we all must share like it or not, whereas the risks associated with smoking are voluntary because a person chooses to smoke. Why are we so unaccepting of involuntary risk but accepting of voluntary risk?

73. What are some of the most important scientific and technological challenges facing humanity in the 21st century?

74. List all of the advantages that electricity from wind energy has over electricity from a coal power plant. List all the disadvantages. Assuming your electric bill from each was the same, which would you rather have? Should where you live make a difference in your choice?

75. As an individual, what do you think are some of the most effective ways to conserve energy?

READINESS ASSURANCE TEST (RAT)

If you have a good handle on this chapter, then you should be able to score at least 7 out of 10 on this RAT. Check your answers online at www.ConceptualChemistry.com. If you score less than 7, you need to study further before moving on.

Choose the BEST answer to the following.

1. Construction of the North American power grid began about
   - **a.** 20 years ago.
   - **b.** 40 years ago.
   - **c.** 60 years ago.
   - **d.** 100 years ago.
   - **e.** 150 years ago.

2. Which fossil fuel consists of the largest molecules?
   - **a.** Coal
   - **b.** Petroleum
   - **c.** Natural gas
   - **d.** Oil sands
   - **e.** Methane hydrate

3. Which of the following is most imminent?
   - **a.** Peak coal
   - **b.** Peak oil
   - **c.** Peak natural gas
   - **d.** Peak hydroelectric power

4. What did the Fukushima power plant have that the Chernobyl power plant didn't, resulting in the Chernobyl meltdown being more destructive to the environment?
   - **a.** Backup generators
   - **b.** Containment buildings
   - **c.** A computerized system
   - **d.** Cooling towers

5. The water that enters an OTEC power plant
   - **a.** is cooler than the water that exits the plant.
   - **b.** is warmer than the water that exits the plant.
   - **c.** is about the same temperature as the water that exits the plant.
   - **d.** is less salty than the water that exits the plant.
   - **e.** has two of the above properties.

6. Which fuel is most commonly obtained from *biomass*?
   - **a.** Tritium
   - **b.** Methane
   - **c.** Ethanol
   - **d.** Coal
   - **e.** Hydrogen

7. One megawatt (1.0 MW) of electricity is sufficient to meet the needs of about 350 households in a developed nation. Europe expects to have a wind energy capacity of 180,000 MW by 2020. How many households is this?
   - **a.** 63 million
   - **b.** 6.3 million
   - **c.** 95 million
   - **d.** 630 million

8. Wind power
   - **a.** has become more useful as a source of electricity over the past 20 years because of improvements in equipment.
   - **b.** requires an open area of land that is near a large body of water.
   - **c.** is currently being phased out as a cost-effective method for producing energy.
   - **d.** produces too much air pollution.
   - **e.** requires too many intermediate steps in the process of producing electricity.

9. You order a 2-kW system of photovoltaic cells to be installed on the roof of your house. The cells are guaranteed to last at least 25 years. The cells plus installation plus electrical equipment will cost you $20,000 if you buy now, but $10,000 if you wait 10 years. The average cost of electricity from your utility is 10 cents per kilowatt-hour. What else do you need to know in order to calculate the amount of money you will have spent within 25 years after going with either plan?
   - **a.** The average number of kilowatt-hours you consume each year
   - **b.** The cost of additional storage batteries should you need to store your excess harvested electrical energy
   - **c.** Whether your utility company is willing and able to purchase any excess electricity you produce
   - **d.** All of the above

10. Which of the following is true about hydrogen, H_2?
    - **a.** It is a source of energy.
    - **b.** It is a carrier of energy.
    - **c.** It is a form of energy storage.
    - **d.** It is two of the above.
    - **e.** It is all of the above.

ANSWERS TO CALCULATION CORNER (KILOWATT-HOURS)

Step 1 Total energy consumed in kilowatt-hours:
(10 bulbs)(0.100 kW/bulb)(10 h) = 10 kWh

Step 2 Cost of consuming this much energy:
(10 kWh)(15¢/kWh) = 150¢ = $1.50

Doubling Time

One of the most important things we have trouble perceiving is the amazing effects of consistent doubling. In two doublings, a quantity will double twice ($2^2 = 4$), or quadruple, in size; in three doublings, its size will increase eightfold ($2^3 = 8$); in four doublings, it will increase sixteenfold ($2^4 = 16$); and so on. Continued doubling leads to enormous numbers.

Consider the story of the court mathematician in India who years ago invented the game of chess for his king. The king was so pleased with the game that he offered to repay the mathematician, whose request seemed modest enough. The mathematician requested a single grain of wheat on the first square of the chessboard, two grains on the second square, four on the third square, and so on, doubling the number of grains on each succeeding square until all the squares had been used. At this rate, there would be 2^{63} grains of wheat (about 18 million trillion) on the sixty-fourth square alone. The king soon saw that he could not fill this "modest" request, which amounted to more wheat than had been harvested in the entire history of the Earth!

Steady growth in a steadily expanding environment is one thing, but what happens when steady growth occurs in a finite environment? Consider the growth of bacteria that grow by division, so that one bacterium becomes two, the two divide to become four, the four divide to become eight, and so on. Suppose the number of bacteria grows exponentially with a doubling time of 1 minute (the time it takes for each doubling to occur). Further, suppose one bacterium is put in a bottle at 11 AM and growth continues steadily until the bottle becomes full of bacteria at 12 noon.

CONCEPT CHECK

When was the bottle half full?

CHECK YOUR ANSWER At 11:59 AM because the bacteria doubled in number every minute.

It is startling to note that at 2 minutes before noon, the bottle was only one-fourth full and at 3 minutes before noon is was only one-eighth full. If bacteria could think (and if they were concerned about their future) at what time do you suppose they would sense they were running out of space? Would a serious problem have been evident at, say, 11:55 AM, when the bottle was only 3 percent full (1/32) and had 97 percent open space (just yearning for development)? The point here is that there isn't much time between the moment the effects of growth become noticeable and the time they become overwhelming.

Pretend that at 11:58 AM, some farsighted bacteria see that they are running out of space and launch a full-scale search for new bottles for the thriving population. And further imagine their sense of accomplishment upon finding three new empty bottles. This is three times the space they have ever known. It may seem to the bacteria that their problems are solved—and just in time. If the bacteria are able to migrate to the new bottles and their growth continues at the same rate, what time will it be when the three new bottles are filled to capacity? The disconcerting answer is that all four bottles will be filled to capacity by 12:02 PM! The discovery of the new bottles extends the resource by only two doubling times. In this example, the resource is space—such as land area for a growing population. But it could be coal, oil, uranium, or any other nonrenewable resource.

Although bacteria and other organisms have the potential to multiply exponentially, limiting factors eventually restrict their growth. The number of mice in a field, for example, depends not only on birthrate and food supply but also on the number of hawks and other predators in the vicinity. A "natural balance" of competing factors is normally struck. If the predators are removed, exponential growth of the mouse population can proceed for a while. Remove certain plants from a region and others tend toward exponential growth. All plants, animals, and creatures that inhabit the Earth are in dynamic states of balance—states that change with changing conditions. Hence the environmental adage "You never change only one thing."

The consumption of nonrenewable resources cannot grow exponentially for an indefinite period because the resource is finite and its supply finally expires. This is shown in Figure 1a,

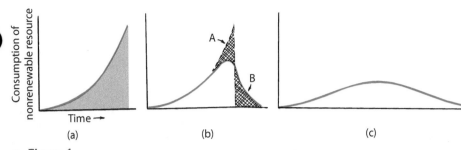

▲ Figure 1
(a) If the exponential rate of consumption for a nonrenewable resource continues until it is depleted, consumption falls abruptly to zero. (b) In practice, the rate of consumption levels off and then falls less abruptly to zero. Note that this curve is made by flipping the top portion of graph(s) (Area A) to the lower right (Area B). (c) At lower consumption rates, the same resource lasts a longer time. In regard to world oil production, the general consensus among geologists is that we reached the peak, also known as "Hubbert's Peak," around 2010. Given this trend, today's "outrageous" gasoline prices will eventually be perceived as outrageously cheap.

where the rate of consumption, such as barrels of oil produced per year, is plotted against time, say, in years. In such a graph, the shaded area under the curve represents the supply of the resource. When the supply is exhausted, consumption ceases altogether. This sudden change is rarely the case, for the rate at which the supply is extracted falls as it becomes more scarce. Furthermore, consumption diminishes as prices escalate. This is shown in Figure 1b. Note that the area under the pink curve is equal to the area under the pink curve in Figure 1a. Why? Because the total supply is the same in both cases. The difference is the time taken to extract that supply. History shows that the rate of production of a non-renewable resource rises and falls in a nearly symmetric manner, as shown in Figure 1c. The time during which production rates rise is approximately equal to the time during which these rates fall to zero or near zero.

The consequences of unchecked exponential growth are staggering. It is very important to ask the following question: is growth really good? Our bodies arise from the doubling of a single fertilized egg, yet we stop growing once we reach adulthood. Continued growth means obesity—or worse, cancer. Might there ever come a day when we value politicians, economists, and businesspeople not for their ability to stimulate growth, but for their ability to diminish growth to rates of zero or less? Is it in us to demand not more, but the same?

CONCEPT CHECK

If world population doubles in 40 years and world food production doubles in 40 years as well, how many people will be starving each year compared to now?

CHECK YOUR ANSWER All things being equal, doubling of food for twice the number of people simply means that twice as many people will be eating as are eating now and twice as many will be starving as are starving now. As you learned in Chapter 15, however, world food production is already great enough to ensure healthy diets for all humans. The chief causes of malnutrition and starvation are political and economical.

Think and Discuss

1. Doubling time approximately equals 70 percent divided by the percent growth rate. For example, if money in the bank grows at a rate of 7 percent per year, then the doubling time equals about 70%/7% = 10 years. In other words, at a rate of 7 percent, a $50 investment will double to $100 in 10 years. Calculate the population doubling times for the following regions (current population in parentheses). Discuss potential reasons for these vast differences in doubling times (see www.prb.org).

World: 1.2% (7.0 billion)
United States: 0.5% (312 million)
Japan: −0.1% (128 million)
India: 1.5% (1.2 billion)
Latin America: 1.2% (596 million)
Europe: 0.0% (740 million)
Nigeria: 2.5% (162 million)
China: 0.5% (1.3 billion)

2. In the U.S. House of Representatives, states are represented according to population—the larger the population of the state, the more representatives the state is allowed to send to Congress to vote on various laws and issues. Should a similar system be set up in the United Nations? How might Europeans feel about such an initiative?

3. The flu epidemic of 1918 killed between 50 million and 100 million people worldwide, which is far more than all the wars in the 20th century put together. Similarly, historians estimate that about 90 percent of Native Americans were killed not by invading Europeans, but by the diseases the Europeans carried with them. Why have pathogens been much more effective at killing us than we have been at killing each other? Why do some consider bioterrorism a greater threat than nuclear terrorism? How might some "farsighted" pathogens feel about jet travel?

4. In all certainty, the human population growth rate on the Earth will eventually reach 0 percent. How soon might this occur? How might this rate be achieved in a worst-case scenario? In a best-case scenario? In a science-fictional scenario?

5. The worldwide trend is that as a country gets wealthier, its rate of population growth declines or even reverses. But which comes first: the wealth or the declining rate? How might an impoverished, highly populated nation reduce its rate of population growth without additional wealth? Why might this nation *not* want to reduce its rate of population growth?

Scientific Notation Is Used to Express Large and Small Numbers

In science, we often encounter very large and very small numbers. Written in standard decimal notation, these numbers can be cumbersome. There are, for example, about 33,460,000,000,000,000,000,000 water molecules in a thimbleful of water, each having a mass of about 0.00000000000000000000002991 gram. To represent such numbers, scientists use a mathematical shorthand called *scientific notation*. Written in this notation, the number of molecules in a thimbleful of water is 3.346×10^{22} and the mass of a single molecule is 2.991×10^{-23} gram.

To understand how this shorthand notation works, consider the large number 50,000,000. Mathematically, this number is equal to 5 multiplied by $10 \times 10 \times 10 \times 10 \times 10 \times 10 \times 10$. (Check this out on your calculator.) We can abbreviate this chain of numbers by writing all the 10s in an exponential form, which gives us the scientific notation 5×10^7. (Note that 10^7 is the same as $10 \times 10 \times 10 \times 10 \times 10 \times 10 \times 10$. Table A.1 shows the exponential form of some other large and small numbers.) Likewise, the small number 0.0005 is mathematically equal to 5 divided by $10 \times 10 \times 10 \times 10$, which is $5/10^4$. Because dividing by a number is equivalent to multiplying by the reciprocal of that number, $5/10^4$ can be written in the form 5×10^{-4}, so in scientific notation 0.0005 becomes 5×10^{-4}. (Note the negative exponent.)

All scientific notation is written in the general form

$$C \times 10^n,$$

where C, called the *coefficient*, is a number between 1 and 9.999 ... and n is the *exponent*. A positive exponent indicates a number greater than 1, and a negative exponent indicates a number between 0 and 1 (*not* a number less than 0). Numbers less than 0 are indicated by putting a negative sign *before the coefficient* (not in the exponent):

	DECIMAL NOTATION	SCIENTIFIC NOTATION
Large positive number (greater than 1)	6,000,000,000	6×10^9
Small positive number (between 0 and 1)	0.0006	6×10^{-4}
Large negative number (less than −1)	−6,000,000,000	-6×10^9
Small negative number (between −1 and 0)	−0.0006	-6×10^{-4}

Table A.2 shows the scientific notation used to express some of the physical data often used in science.

TABLE A.1   Decimal and Exponential Notations

1,000,000	$= 10 \times 10 \times 10 \times 10 \times 10 \times 10 = 10^6$
100,000	$= 10 \times 10 \times 10 \times 10 \times 10 = 10^5$
10,000	$= 10 \times 10 \times 10 \times 10 = 10^4$
1000	$= 10 \times 10 \times 10 = 10^3$
100	$= 10 \times 10 = 10^2$
10	$= 10 = 10^1$
1	$= 1 = 10^0$
0.1	$= 1/10 = 10^{-1}$
0.01	$= 1/(10 \times 10) = 10^{-2}$
0.001	$= 1/(10 \times 10 \times 10) = 10^{-3}$
0.0001	$= 1/(10 \times 10 \times 10 \times 10) = 10^{-4}$
0.00001	$= 1/(10 \times 10 \times 10 \times 10 \times 10) = 10^{-5}$
0.000001	$= 1/(10 \times 10 \times 10 \times 10 \times 10 \times 10) = 10^{-6}$

To change a decimal number that is greater than +1 or less than −1 to scientific notation, you shift the decimal point to the left until you arrive at a number between 1 and 9.999 . . . . For decimal numbers between +1 and −1, you move the decimal point to the right until you arrive at a number between 1 and 9.999 . . . . This number is the coefficient part of the notation. The exponent part is simply the number of places you moved the decimal point. For example, to convert the decimal number 45,000 to scientific notation, move the decimal point four places to the left to get a number between 1 and 9.999 . . . :

$$45,000 = 4.5 \times 10^4$$
4321

The number 0.00045, on the other hand, is converted to scientific notation by moving the decimal point four places to the right to arrive at a number between 1 and 9.999 . . . :

$$0.00045 = 4.5 \times 10^{-4}$$
1234

TABLE A.2   Physical Data

Number of molecules in thimbleful of water	3.346×10^{22}
Mass of water molecule	2.991×10^{-23} *gram*
Average radius of hydrogen atom	5×10^{-11} *meter*
Proton mass	1.6726×10^{-27} *kilogram*
Neutron mass	1.6749×10^{-27} *kilogram*
Electron mass	9.1094×10^{-31} *kilogram*
Electron charge	1.602×10^{-19} *coulomb*
Avogadro's number	6.022×10^{23} *particles*
Atomic mass unit	1.661×10^{-24} *gram*

● Note that because you moved the decimal point to the right in this case, you must put a minus sign on the exponent.

Y O U R T U R N

Express the following exponentials as decimal numbers:

a. 1×10^{-7}

b. 1×10^{8}

c. 8.8×10^{5}

Express the following decimal numbers in scientific notation:

d. 740,000

e. −0.00354

f. 15

WERE THESE YOUR ANSWERS?

a. 0.0000001

b. 100,000,000

c. 880,000

d. 7.4×10^{5}

e. -3.54×10^{-3}

f. 1.5×10^{1}

Significant Figures Are Used to Show Which Digits Have Experimental Meaning

Three kinds of numbers are used in science—those that are *counted*, those that are *defined*, and those that are *measured*. The exact value of a counted or defined number can be stated with absolute certainty. For example, you can count the number of chairs in your classroom or the number of fingers on your hand. Your final count is 100% on the mark. (Of course, this assumes you counted correctly, which is easy when the numbers are small, but not so easy when the numbers run into the millions, as occurs during an election.)

Defined numbers are about exact relationships and are defined as being true. The defined number of centimeters in a meter, the defined number of seconds in an hour, and the defined number of sides on a square are examples. Thus, defined numbers are not subject to error (unless you forget a definition).

Every measured number, however, no matter how carefully measured, has some degree of *uncertainty*. This uncertainty (or *margin of error*) in a measurement can be illustrated by the two metersticks shown in Figure B.1. Both sticks are being used to measure the length of a table. Assuming that the zero end of each meterstick has been carefully and accurately positioned at the left end of the table, how long is the table?

▶ **Figure B.1** Which meterstick offers greater precision?

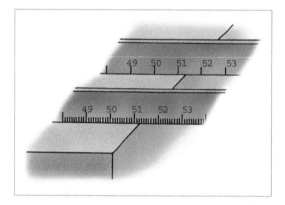

The upper meterstick has a scale marked off in centimeter intervals. Using this scale, you can say with certainty that the length is between 51 and 52 centimeters. You can say further that it is closer to 51 centimeters than to 52 centimeters; you can even estimate it to be 51.2 centimeters.

The scale on the lower meterstick has more subdivisions—and therefore greater precision—because it is marked off in millimeters. With this scale, you can say that the length is definitely between 51.2 and 51.3 centimeters, and you can estimate it to be 51.25 centimeters.

Note how both readings contain some digits that are *certain* and one digit (the last one) that is *estimated*. Note also that the uncertainty in the reading from the lower meterstick is less than the uncertainty in the reading from the upper meterstick. The lower meterstick can give a reading to the hundredths place, but the upper one can give a reading only to the tenths place. The lower one is more *precise* than the top one. So in any measured number, the digits tell us the *magnitude* of the measurement, and the location of the decimal point tells us the *precision* of the measurement. (Figure B.2 illustrates the distinction between *precision* and *accuracy*.)

Significant figures are the digits in any measured value that are known with certainty plus one final digit that is estimated and hence uncertain. These are the digits that reflect the precision of the instrument used to generate the number. They are the digits that have experimental meaning. The measurement 51.2 centimeters made with the upper meterstick in Figure B.1, for example, has three significant figures, and the measurement 51.25 centimeters made with the lower meterstick has four significant figures. The rightmost digit is always an estimated digit, and only one estimated digit is ever recorded for a measurement. It would be incorrect to report 51.253 centimeters as the length measured with the lower meterstick. This five-significant-figure value has two estimated digits (the final 5 and 3) and is incorrect because it indicates a *precision* greater than the meterstick can obtain.

Here are some standard rules for writing and using significant figures.

RULE 1 In numbers that do not contain zeros, all the digits are significant:

4.1327	five significant figures
5.14	three significant figures
369	three significant figures

RULE 2 All zeros between significant digits are significant:

8.052	four significant figures
7059	four significant figures
306	three significant figures

RULE 3 Zeros to the left of the first nonzero digit serve only to fix the position of the decimal point and are not significant:

0.0068	two significant figures
0.0427	three significant figures
0.0003506	four significant figures

RULE 4 In a number that contains digits to the right of the decimal point, zeros to the right of the last nonzero digit are significant:

53.0	three significant figures
53.00	four significant figures
0.00200	three significant figures
0.70050	five significant figures

RULE 5 In a number that has no decimal point and that ends in one or more zeros, the zeros that end the number are not significant. (To scientists, the decimal point is very important. Its presence or absence has great implications in regard to the precision of a measurement.)

3600	two significant figures
290	two significant figures
5,000,000	one significant figure
10	one significant figure
6050	three significant figures

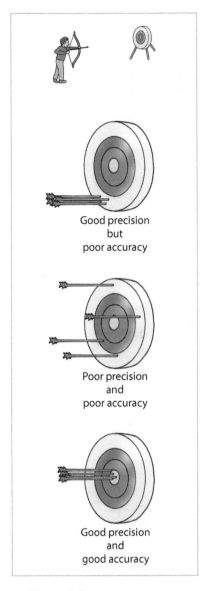

Good precision
but
poor accuracy

Poor precision
and
poor accuracy

Good precision
and
good accuracy

▲ **Figure B.2** Archery as a model for understanding the difference between precision and accuracy. *Precision* means close agreement in a group of measured numbers; *accuracy* means a measured value that is very close to the true value of what is being measured. If you measure the same thing several times and get numbers that are close to one another but are far from the true value (perhaps because your measuring device is not working properly), then your measurements are precise but not accurate. An example would be measuring your weight repeatedly on a broken scale.

RULE 6 When a number is expressed in scientific notation, all digits in the coefficient are taken to be significant:

4.6×10^{-5}	two significant figures
4.60×10^{-5}	three significant figures
4.600×10^{-5}	four significant figures
2×10^{-5}	one significant figure
3.0×10^{-5}	two significant figures
4.00×10^{-5}	three significant figures

YOUR TURN

How many significant figures are in the following?

a. 43,384

b. 43,084

c. 0.004308

d. 43,084.0

e. 43,000

f. 4.30×10^4

WERE THESE YOUR ANSWERS?

a. 5

b. 5

c. 4

d. 6

e. 2

f. 3

In addition to the rules cited above, there is another full set of rules to be followed for significant figures when two or more measured numbers are subtracted, added, divided, or multiplied. These rules are summarized in the appendix of the *Conceptual Chemistry Laboratory Manual*.

Solutions to Odd-Numbered Chapters Questions

CHAPTER 1: About Science

1. An experiment

3. Scientific research usually begins by asking a very broad question.

5. There are 60 carbon atoms in the original buckyball identified by Kroto, Smalley, and Curl. Interestingly, they also discovered larger and smaller buckyballs, but C-60 was most stable.

7. Technology is the application of knowledge of the natural world (usually gained through science) to practical purposes.

9. No. Medical X rays are used because the benefits of their diagnostic powers are judged to be greater than their risk of causing cancer.

11. A hypothesis

13. Chemistry is often called the central science because it touches all of the sciences.

15. Members of the American Chemistry Council have pledged to manufacture their products without causing environmental damage.

17. A prefix is used in the metric system to designate a unit that is larger or smaller than a particular base unit by one or more powers of 10.

19. As noted by Lavoisier, air has mass, which means it also has weight. The weight of this air pushes against us in all directions. The force of this push at sea level is about 14 pounds for every square inch. So how does the card hold up the water in the bottle? Answer: It doesn't. The downward push from the weight of the water in the bottle is less than a pound. The upward (and sideways) push against the outer side of the card from the weight of the air, however, is about 14 pounds. The air wins! Although invisible, air is real stuff. Because of this, birds and airplanes are able to fly.

21. A large risk/benefit ratio such as 100 can be indicated as follows:

$$RISK/benefit = 100$$

This is a risky activity you might choose to avoid. A small risk benefit ratio such as 0.01 can be indicated as follows:

$$risk/BENEFIT = 0.01$$

Because this activity offers much benefit for only little risk, it may be worthwhile. Regarding the purchase of a lottery ticket, the chances of losing a small amount of money (the risk) far exceeds the chances of gaining a large amount of money (the benefit). The risk/benefit ratio, therefore, is quite large. However, if you ignore the probabilities, as many people do, then the risk benefit ratio appears deceptively small.

23. (2000 years)(365 days/1 year)(24 hours/1 day)(60 min/1 hour)(60 sec/1 min) = 63,072,000,000 seconds

25. a < c < b (Traveling by plane is generally much safer than traveling by car, especially over long distances.)

27. The explanations given by science are testable explanations of the natural world.

29. The older, well-seasoned scientist tends to spend more time pondering broad questions and communicating with others. The younger, less-seasoned scientist tends to devote more time on the tedious detailed work, which includes spending a lot of time in the library learning about what is already known and working late at night in the laboratory trying to perform successful experiments.

31. A scientist can develop a hypothesis at any time no matter what she may be doing. She could be barbecuing when suddenly the idea pops into her head. Again, for emphasis, there is no one prescribed path to follow in order to hold to the scientific method.

33. It is a sign of strength for a scientist to change his or her view when faced with evidence inconsistent with that view. Holding to hypotheses and theories that are not testable or that have been shown to be wrong is contrary to the spirit of science.

35. Kroto, Curl, and Smalley were the initial discovers of this molecule, and for this, they received the Nobel Prize. Huffman and Kratschmer's work merely helped to confirm this discovery.

37. During the 1950s nuclear arms race, there was great fear that "the other side," with its different ideological views, would attack. The benefit of the nuclear bombs was their ability to deter either side from being eager to use them. All this obscured the very real risk of placing so much radioactivity in the atmosphere. This radiation made its way into the food chain. The supply of milk, for example, was soon found to be tainted with radioactive strontium. Of course, hindsight is 20/20. It's easy for us to look back on this history and shake our heads. At the time, coming out of the tragedy of World War II, however, the perspective was quite different. Actual risk and our perceptions of that risk are difficult to gauge.

39. Vaccinations prevent sickness, and this is their benefit. While not sick, however, people tend to move on with their lives, often taking their health for granted. In such a case, the benefit of the vaccination becomes invisible and thus difficult to perceive. A vaccination program, however, must continue until the disease has been totally eradicated across the entire population. This is especially true in our modern society, where people travel between different parts of the world so frequently.

41. Science is unable to answer nontestable questions, such as those that are philosophical or religious in nature. Science can, however, generate ideas that have philosophical or religious implications.

43. A scientific theory that *can* be modified to account for new experimental evidence is a theory that is stronger than it was prior to that modification. However, if such modifications to a scientific theory are not possible, then that theory is taken to be wrong and is scrapped.

45. Physics is the most fundamental science as it lays the foundation for chemistry, which is the study of the physics of the atom. Chemistry, in turn, lays the foundation for the most complex science, which is biology.

47. For the United States, the major advantage of using this system is that it is already in use and is familiar to everyone. Instituting a different measurement system would involve considerable disruption in many areas of life, including science, commerce, and industry.

49. Making observations is an activity that occurs on a continual basis, even during the course of other activities. Remember, humans are very good at observing. We do it all the time.

CHAPTER 2: Particles of Matter

1. It would take you 31,800 years to count to a trillion. Do this 125 million times and you would have counted to about the number of atoms in a single grain of sand.

3. The term *atom* was derived from the Greek phrase *a tomos*, which means "not cut" or "that which is indivisible."

5. Mendeleev predicted the existence of elements that had not yet been discovered.

7. Weight can change from one location to the next because it is dependent on gravity.

9. Density is the ratio between the mass of a substance and its volume. Note that as the mass of the substance increases, so does its volume. The ratio of the mass to volume, which is its density, remains the same.

11. The energy due to position is potential energy.

13. A calorie is 4.184 times greater than a joule.

15. The Kelvin scale places zero at the point of zero atomic and molecular motion.

17. The particles in a gas have so much energy that they overcome their attractions to each other and expand to fill all of the space available. In a liquid, the particles tumble loosely around one another. In a solid, the particles are fixed in a 3-dimensional arrangement.

19. Evaporation that occurs beneath the surface of a liquid is called boiling.

21. The volume increases as the temperature increases.

23. The volume of a gas increases as more particles are added to it.

25. No question was asked.

27. No question was asked.

29. The volume of water that gets displaced by a submerged object only depends on the volume of the submerged object.

31. The mass of a 10 kg object anywhere is 10 kg.

33. From Table 1.4, the density of gold is 19.3 g/mL. Use the following formula to find the volume of the sample:

$$V = \frac{M}{D} = \frac{52.3\ g}{19.3\ g/mL} = 2.71\ mL$$

35. There are 0 mL of dirt in the whole, but

$$5\ L \times \frac{1000\ mL}{1L} = 5000\ mL\ of\ air.$$

37. Plug the following values into Charles's Law and solve for the new volume: $V_1 = 1.0$ liters; $T_1 = 298K$; $T_2 = 348K$.

$$\frac{V_1}{T_1} = \frac{V_2}{T_2}$$

$$\frac{(1.0\ liters)}{(298K)} = \frac{V_2}{(348K)}$$

$$\frac{(348K)(1.0\ liters)}{(298K)} = V_2$$

$$1.168\ liters = V_2$$

Round to the appropriate number of significant figures (see Appendix B):

$$1.2\ liters = V_2$$

39. $a < c < b$

41. $c < a = b$

43. $a < b < c$

45. The 50 mL plus 50 mL do not add up to 100 mL because within the mix, many of the smaller water molecules can fit within the pockets of space that were empty in the 50 mL of larger ethanol molecules. This is analogous to the previous question involving the BB's and is yet another example where the existence of molecules helps to explain observed phenomena.

47. That atoms can neither be created nor destroyed in a chemical reaction helps to explain why the mass of all the reacting materials equals the mass of all the products formed.

49. The atoms of the gasoline transform into the atoms of the exhaust fumes, which escape into the atmosphere. The atoms literally go into the gas tank and then out the exhaust pipe. The atoms are conserved, but the gasoline isn't.

51. Yes. A 2 kg iron brick has twice the mass as a 1 kg iron brick. While on the same planet, it also has twice the weight as well as twice the volume.

53. The air inside the car has more inertia than the helium in the helium balloon. As the car starts forward, the greater inertia of the air causes it to pitch backwards, much like the girl's head. As this air holds to the back of the car, the lighter helium balloon moves forward. A similar effect can be seen when sliding a bottle of water on its side across a table. As you accelerate the bottle forward, the water inside the bottle has an inertia that holds it back. Any bubble of air within the bottle thus runs forward in the direction the bottle was pushed.

55. Box A represents the greatest density because it has the greatest number of particles packed within the given volume. Because the particles of this box are packed close together and because they are randomly oriented, this box is representative of the liquid phase. Box C is representative of the gaseous phase, which occurs for a material at higher temperatures. Therefore, this box represents the highest temperature. For most materials, the solid phase is denser than the liquid phase. For the material represented here, however, the liquid phase is seen to be denser than the solid phase. As is explored further in Chapter 8, this is the same case for water, where the solid phase (ice) is less dense than the liquid phase (liquid water).

57. Density is mass divided by volume. If the mass of empty space is zero, then the density of that empty space is also zero.

59. The kinetic energy of a pendulum bob is maximum where it moves fastest, at the lowest point; potential energy is maximum at the uppermost points.

61. Yes. A car burns more gasoline when its lights are on. Lights and other electric devices are run off the battery, which "runs down" the battery. The energy used to recharge the battery ultimately comes from the chemical potential energy of the gasoline.

63. Heat never flows of itself from a lower-temperature substance into a higher-temperature substance.

65. The glass will contract when cold and expand when warmed. Fill the inner glass with cold water while running hot water over the outer glass to help separate the two.

66. At cruising speed (faster than the speed of sound), air friction against the jet raises its temperature dramatically, resulting in this significant thermal expansion.

67. The warmth from the house causes the wood frame of the attic to expand, while the cool of the night causes the frame to contract. As parts of the wood expand and other parts contract, the result is a creaking noise.

69. (b) The trail lines behind each circle in this diagram are meant to indicate that the particles are moving faster. Diagram (a) shows the particles congregated to one side of the box. This is an unlikely scenario because the motion of gaseous particles is randomly oriented. You may have interpreted diagram (a) to mean the gas was hotter, and as everyone knows, hot air rises. Fair enough, but notice how the trail lines in this diagram do not indicate any faster motion. Diagram (c) shows fatter gas particles. Upon the addition of heat, gas particles don't grow any fatter. Instead, they move faster, which is to say they have a greater average kinetic energy.

71. At the cold temperatures of your kitchen freezer, water molecules in the vapor phase are moving relatively slowly, which makes it easier for them to stick to inner surfaces within the freezer or to other water molecules.

73. Gas meters measure the volume of gas that passes through them. As the gas warms, it gains in volume and causes greater measurement readings in the meter. The gas company benefits.

75. Airplane windows are small because the pressure difference between the inside and outside surfaces results in large net forces that are directly proportional to the window's surface area. Larger windows would have to be proportionately thicker to withstand the greater net force—windows on underwater research vessels are similarly small.

77. The surface of the liquid inside the bottle is sealed off from the atmosphere, which can no longer help to push the beverage up through the straw into your mouth.

79. The rubber balloon will expand until it eventually pops. The material of the balloon will fall back to the Earth, and the helium will continue on its way to outer space.

81. To increase the pressure within the cabin of the airplane from 0.743 atm to 1.00 atm would require adding more air molecules to the cabin. This is an application of Avogadro's Law. These added air molecules have weight, which makes the plane heavier. As any pilot knows, the heavier the plane, the more fuel that must be consumed to fly at a given speed. The choice, therefore, is to increase the cabin pressure and slow down or decrease the cabin pressure and speed up. A cabin pressure correspond-ing to around 8000 feet is the happy medium, although it does explain, in part, why flying can be a stress to the body.

83. The concept that matter is made of molecules came about only after many questions about the behavior of matter were asked. If no questions or only a few questions are asked, then the conclusion that matter is made of molecules is not likely to be reached. How you might lead others (or yourself) to accept the idea of molecules is to see how well this idea helps to answer questions and explain observations.

For example, is cinnamon-scented air a single material or a mixture of two materials? If it is two materials, should it not be heavier than the same volume of fresh air? (It is.) Can this air be made fresh by passing it through a filter of activated charcoal? (It can.) Does the charcoal now smell like cinnamon? (It does.) Does the cinnamon smell of the charcoal increase or decrease as it is warmed? (It increases, but eventually tapers off.) Does the charcoal lose or gain weight as the cinnamon smell tapers off? (It loses weight.) Might the charcoal be losing weight as tiny particles (molecules) of cinnamon evaporate from its surface? (That would make sense.) Notice that through this process, you never need to say that molecules exist because someone told you they do. Instead, we come to conclude that molecules exist because they offer the best explanation for what we observe.

85. The scale falls because gravity pulls it downward. An orbiting spacecraft also falls, but the spacecraft is moving sideways so fast (17,500 mph) that it falls *around* the Earth rather than into it. No parking would be possible for a space shuttle alongside the observation deck. Interestingly, astronauts within the orbiting shuttle are as weightless as you are when you jump off a high dive. The only difference is that they're moving sideways super fast, which they can only do above the air drag of the atmosphere. Understanding the rules of nature helps us to appreciate nature. Furthermore, there are many science/technology-oriented decisions that we as citizens need to make—from space travel, to the food we eat, to global warming and more. What if the entire electorate were just as informed about these issues as you are? Would this be a good thing?

CHAPTER 3: Elements of Chemistry

1. Nothing. During a physical change, the chemical identity of a substance remains the same.

3. Determining whether a change is physical or chemical can be difficult because both involve changes in appearance.

5. *Atom* is used to refer to submicroscopic particles in a sample, and *element* is used for microscopic and macroscopic samples.

7. Across any period, the properties of elements gradually change.

9. An element has only one type of atom. A compound has combinations of different types of atoms.

11. The physical and chemical properties of a compound are completely different from the properties of its elemental parts.

13. NO_3^{1-} is a polyatomic ion, so the name of this compound is sodium nitrate.

15. Filtration or distillation can separate mixtures. These methods take advantage of differences in the components' physical properties.

17. Atoms and molecules are very small; so if one atom or molecule out of a trillion is different, then the sample is no longer pure.

19. A centrifuge can be used to determine whether a mixture is a solution or a suspension because it will separate the components of a suspension.

21. The top-down approach in which nanostructures are carved out of larger materials and the bottom-up approach in which nanostructures are pieced together atom by atom.

23. Water vapor condenses into water droplets on the outside of the pot. This water vapor is a product of the reaction of the natural gas with oxygen. Ice water in the pot will favor the condensation of more liquid water from the water vapor. As the pot gets warmer, the condensed water evaporates. Physical changes include the condensation and evaporation of water. Note that the water vapor from the flame results from chemical changes occurring within the flame.

25. The tiny bubbles that form in this activity are from the air that was dissolved in the water. These bubbles contain mostly air, along with some residual water vapor. As we explore in Chapter 7, gases do not dissolve well in hot liquids. Air that is dissolved in room-temperature water, for example, will quickly bubble out when the water is heated. Thus, you can speed up the formation of air bubbles by using warm water. You'll find that boiling *deaerates* the water (that is, removes the atmospheric gases). Chemists sometimes need to use deaerated water, which is made by allowing boiled water to cool in a sealed container. Why don't fish live very long in deaerated water?

27. The difference between these two concentrations is 950 ppm, or 950 milligrams per liter. The amount of salts added to the river water is 950 milligrams per liter.

29. A concentration of 0.16 ppt is 0.16 nanograms per liter. Convert nanograms to milligrams using these equalities: 1 gram $= 10^3$ mg; 1 gram $= 10^9$ ng.

$$(0.16 \text{ ng})(1 \text{ gram}/10^9 \text{ ng})(10^3 \text{ mg}/1 \text{ gram}) = 1.6 \times 10^{-7} \text{ mg}$$
$$= 0.00000016 \text{ mg}$$

31. $c < a < b$

33. $c < a < b$

35. The molecules of the alcohol evaporate into the gaseous phase, which is a physical change.

37. Boiling down the maple syrup involves the evaporation of water. As the syrup hits the snow, it warms the snow, causing it to melt while the syrup becomes more viscous. These are all examples of physical changes. Interestingly, as the maple syrup is boiled, the sugar within the syrup begins to caramelize, which is an example of a chemical change.

39. (a) chemical (b) chemical (c) physical (d) chemical (e) chemical (f) chemical (g) physical

41. There is a practical limit to our ability to purify gold. A sample that is 99.99999999 percent pure still contains 0.00000001 percent impurities. Because atoms are so small, the number of nongold atoms within this sample is quite large. If we define an element as a material made of only one type of atom, then we can argue that even our purest sample of gold is ideally classified not as an element, but as a mixture. On a practical level, however, we may ignore this distinction because we recognize that the number of gold atoms within this super pure sample still far exceeds the number of nongold atoms.

43. Water use to be classified as an element, but that was before people recognized that the basic building block of matter is these tiny particles called atoms. Today, an element is identified as a material consisting of only one kind of atom.

45. The ones that have atomic symbols that don't match their modern atomic names. Examples include iron, Fe; gold, Au; and copper, Cu.

47. Helium is placed on the far right side of the periodic table in group 18 because its physical and chemical properties are most similar to those of the other elements of group 18.

49. Calcium is readily absorbed by the body for the building of bones. Because calcium and strontium are in the same atomic group, they have similar physical and chemical properties. The body, therefore, has a hard time distinguishing between the two, and strontium is absorbed as though it were calcium.

51. The periodic table is a reference to be used, not memorized. You need not memorize the periodic table any more than you need to memorize a dictionary. Both the periodic table and a dictionary should be readily available to you when you need them.

53. Chemical compounds have physical and chemical properties that are different from the elements from which they are made. Oxygen, for example, is a gas at room temperature, as is hydrogen. These two elements combine, however, to make water, which is a liquid at room temperature. Similarly sodium and chlorine, although toxic by themselves, react to form a chemical compound, sodium chloride, that is uniquely different.

55. Each water molecule contains one oxygen atom bound to two hydrogen atoms. This oxygen is not available to form the O_2 that we breathe. Water is uniquely different from the elements oxygen, O_2, and hydrogen, H_2, from which it can be made.

57. Box A: mixture; Box B: compound; Box C: element. Three different types of molecules are shown altogether in the three boxes: one with two light red atoms joined as shown only in Box A, one with a light red and a smaller dark blue atom joined as shown in Boxes A and B, and one with two smaller dark blue atoms joined as shown only in Box C.

59. Strontium phosphate. Note that with polyatomic ion, the convention is to leave out the *tri-* or *di-* suffixes. So to a chemist, the name *tristrontium diphosphate* would sound a little wierd and overdone, but he or she would know what you meant.

61. Nothing. The properties of a compound are uniquely different from the properties of the elements used to make that compound. Copper sulfide, CuS, reveals itself as a dark black powder.

63. Box B appears to contain a liquid as evidenced by the randomly oriented molecules condensed at the bottom of the box. These molecules in the liquid phase of Box B represent a compound because they consist of different types of atoms joined together. There's no way to assume the relative temperatures of the boxes based upon the phases of the materials they contain because these materials are uniquely different from each other.

65. Add the mixture of sand and salt to some water. Stir and then filter the sand. Rinse the sand several times with fresh water to make sure all of the salt has been removed. Collect all the salty water and evaporate the water. The residue that remains will be the salt. After the sand dries, you've got just the sand. For a mixture of sand and iron, take advantage of the fact that only iron is attracted to a magnet.

67. Based upon the differences in physical properties. The iron filings are attracted to a magnet, but the cereal is not. Try this with your next box of iron-fortified cereal.

69. Table salt is generally a heterogeneous mixture of the compound sodium chloride and desiccants that absorb moisture, preventing the salt from clumping. Blood is a suspension, which is an example of a homogeneous mixture. Steel is a solid solution, which is a homogeneous mixture, consisting of mostly iron and smaller amounts of carbon and nickel. Planet Earth is a heterogeneous mixture.

71. The scanning probe microscope provides us with images of the nanoscopic world, which is where individual atoms begin to appear. The optical microscope allows us to see at the microscopic level, which is where individual atoms are too small to be seen. Also, the scanning probe microscope collects images by dragging an ultrasharp needle across the surface. The optical microscope works by collecting and focusing visible light.

73. Chemistry and nanotechnology are similar in that they both focus on the world of the atom. They are different in that chemistry focuses on how to combine atoms in bulk to create novel molecules, also in bulk. Nanotechnology works by manipulating individual atoms or small groups of individual atoms. Great things will no doubt be achieved through nanotechnology, but as a complement to the great things also being achieved through novel chemical reactions.

75. Around the beginning of the 20th century, many people were opposed to the introduction of electricity because of its inherent dangers. Fear, however, may arise from insufficient understanding. As people learned more about electricity— understanding both its dangers and benefits—they came to accept this new technology. The same may be true for nanotechnology. The obligation of the vendors of nanotechnology is to keep us informed. But who will watch over these nanotech companies to make sure greed does not overcome safety? That would be the role of an attentive government and a well-informed consumer.

CHAPTER 4: Subatomic Particles

1. The atoms in the baseball would be the size of Ping-Pong balls if the baseball were the size of the Earth.

3. The ray itself is negatively charged because it is a stream of electrons.

5. Millikan discovered the fundamental increment of all electric charge to be 1.60×10^{-19} coulombs.

7. Rutherford found that a few of the alpha particles were scattered backwards.

9. Elements are listed in the periodic table in order of increasing atomic number.

11. Mass number is the count of the number of nucleons in an isotope. Atomic mass is a measure of the total mass of an atom.

13. A spectroscope separates the light into color components whose frequencies can then be measured.

15. The atoms of each element emit only select frequencies of light. The pattern of these frequencies is unique to that element.

17. No. Bohr's model merely illustrated the different energy levels of an electron in an atom.

19. An electron moves around the nucleus at around 2 million meters per second.

20. Because the electron is moving very fast, its wave nature becomes most pronounced.

21. Two

23. Nitrogen

25. The electrons in the outermost shell of an atom are most responsible for the properties of an atom.

27. Inner-shell electrons diminish the attraction that outer-shell electrons have for the nucleus. The strength of the nuclear charge is also diminished for outer-shell electrons because they are farther away from the nucleus. This diminished nuclear charge experienced by outer-shell electrons is called the effective nuclear charge.

29. The electric force weakens with increasing distance.

31. No questions were asked.

33. The class average where everyone scores an 80 percent is 80 percent. One person scoring higher would slightly raise this average to 81 percent. Similarly, the atomic mass of an element is the average mass of all the various isotopes of that element. The heavier isotopes have the effect of slightly raising the average. Carbon, for example, has an atomic mass of 12.011, which is slightly greater than 12.000 because of the few but heavier carbon-13 isotopes found in naturally occurring carbon.

35. A 50:50 mix of Br-80 and Br-81 would result in an atomic mass of about 80.5; a 50:50 mix of Br-79 and Br-80 would result in an atomic mass of about 79.5. Neither of these is as close to the value reported in the periodic table as is a 50:50 mix of Br-79 and Br-81, which would result in an atomic mass of about 80.0. The answer is c.

37. helium < chlorine < argon

39. A < C < B

41. potassium < sodium < lithium

43. technetium, Tc < indium, In < aluminum, Al

45. Many objects or systems may be described just as well by a physical model as by a conceptual model. In general, the physical model is used to replicate an object or a system of objects on a different scale. The conceptual model, by contrast, is used to represent abstract ideas or to demonstrate the behavior of a system. Of the examples given, the following might be adequately described using a physical model: a gold coin, a car engine, and a virus. The following might be adequately described using a conceptual model: a dollar bill (which represents wealth but is really only a piece of paper), air pollution, and the spread of a sexually transmitted disease.

47. The scanning probe microscope (SPM) only shows us the relative sizes and positions of atoms. It does this by detecting the electric forces that occur between the tip of the SPM needle and the outer electrons of the atom. Recall from Section 3.5 that the atom itself is made of mostly empty space. So the best "image" of the inside of an atom would be a picture of nothing. Thus, it doesn't make sense to talk about taking an "image" of the inside of an atom. Instead, we develop models that provide a visual handle as to how the components of atoms behave.

49. If the particles had a greater charge, they would be bent more because the deflecting force is directly proportional to the charge. (If the particles were more massive, they would be bent less by the magnetic force—obeying the law of inertia.)

51. The fact that only 3 of the 100 marbles bounced back suggests a small obstruction in the cereal box. Perhaps the cereal box is fixed to the floor with a narrow nail sticking up from the floor. This is similar to the line of thinking Rutherford used to conclude that each atom consists of an extremely small, densely packed, positively charged center, which he named the atomic nucleus.

53. The one on the far right where the nucleus is not visible.

55. Yes, as is everyone else's and everything around you. We interact with our environment (for example, bumping our head against a cabinet) because of the repulsive electric fields that prevent atoms from overlapping one another.

57. The remaining nucleus is that of carbon-12.58. The neutron was elusive because of its lack of electric charge. Having no charge, it emits no light, nor is it affected by magnetic fields.

59. From the periodic table, we see that an oxygen atom has a mass of about 16 amu. Two oxygen atoms have a mass of about 32 amu, as does a single oxygen molecule, O_2.

61. A water molecule, H_2O, has a mass of about 18 amu, and a carbon dioxide molecule, CO_2, has a mass of about 44 amu. So a carbon dioxide molecule is more than twice as heavy as a water molecule.

63. White light is not really a color. Rather, it is what we perceive when all the frequencies of visible light come to the eye at the same time.

65. Radio waves are a form of electromangetic radiation, which travels at the speed of light (300,000 km/s).

67. Observe the atomic spectra of each using a spectroscope.

69. Blue light comes from a greater energy transition within an atom.

71. The dimensions of the car and the nature of the materials of the car dictate that certain frequencies will reinforce themselves upon vibration. When the vibration of the tires matches the car's "natural frequency," the result is a resonance, which is self-reinforcing waves. The wave nature of the car, however, is simply due to the back-and-forth vibrations of the materials of the car. The wave nature of the electron, on the other hand, is entirely different. For the electron moving at very high speeds, some of its mass is converted to energy, which is manifested in its wave nature. Given the dimensions of the atom, certain frequencies of the electron will also be "natural" (that is, self-reinforcing). The vibrating car, therefore, is analogous to one of the energy levels of the electron, which is the point at which the electron forms a self-reinforcing standing wave.

73. The diagram on the right shows a greater amount of energy. Note that in going from left to right, an electron from the $2s$ orbital has jumped into the third $2p$ orbital. This process would require the input of energy.

75. An electron in a $3s$ orbital has more potential energy than an electron in a $2p$ orbital. An electron in a $3s$ orbital has more potential energy than an electron in a $2p$ orbital. These answers are obtained by reading the energy-level diagram of Figure 4.29. What do the variations have to do with complications that arise from having more than one electron orbiting around the nucleus? You can learn about that in a more advanced chemistry course.

77. The difference between these two transitions is that the beryllium has a stronger nuclear charge. Boosting beryllium's electron away from the nucleus, therefore, requires more energy.

79. They have similar energy levels and so are grouped within the same shell of orbitals, which is the fourth shell. Elements of the fourth period of the periodic table (potassium, K, through krypton, Kr) all have their outermost electrons within at least one of these orbitals.

81. A shell is a region of space in which electrons may reside. This region of space exists with or without the electrons. The space defined by the shell exists whether or not an electron is to be found there.

83. Both the potassium and sodium atoms are in group 1 of the periodic table. The potassium atom, however, is larger than the sodium atoms because it contains an additional shell of electrons.

85. The approximate effective nuclear charge for any electron can be calculated by subtracting the number of inner-shell electrons from the number of protons in the nucleus. The effective nuclear charge for an outermost-shell electron in fluorine is $+9 - 2 = +7$. The effective nuclear charge for an outermost-shell electron in sulfur is $+16 - 10 = +6$.

87. Effective nuclear charge gives rise to the properties of ionization energy and atomic size.

89. Neon's outermost shell is already filled to capacity with electrons. Any additional electrons would have to occupy the next shell out, which has an effective nuclear charge of zero.

91. Potassium has one electron in its outermost occupied shell, which is the fourth shell. The effective nuclear charge within this shell is relatively weak (+1), so this electron is readily lost. A second electron would need to be lost from the next shell inward (the third shell), where the effective nuclear charge is much stronger (+9). Thus, it is very difficult to pull a second electron away from the potassium atom because this electron is being held so tightly by this much greater effective nuclear charge.

93. Note that each shell has been divided into a series of finer shells known as "subshells." Each subshell corresponds to a specific orbital type. The four of the seventh shell, for example, includes the $7s$ orbital, the $5f$ orbitals, the $6d$ orbitals, and the $7p$ orbitals. Gallium is larger than zinc because it has an electron in three subshells of the fourth shell, and zinc has electrons only in the first inner two subshells of the fourth shell. Thus, what you see here is a refinement on the model presented in Section 4.8. Don't worry about fully understanding this refinement. It is better that you understand that all conceptual models are subject to refinement. We choose the level of refinement that best suits our needs.

95. If you think scientists know all there is to know about the universe, think again. While they certainly know much more than they used to, much still remains unknown. One of the mysteries as of the writing of this textbook is the nature of the second form of matter called dark matter. Astronomers find evidence of massive amounts of this stuff surrounding each galaxy. This form of matter, however, only interacts with the gravitational force. It does not recognize the strong nuclear force (see Chapter 5), which means it cannot clump to form atomic nuclei. Nor does it recognize the electromagnetic force, which is responsible for light and electric charge. Thus, dark matter is invisible to light as well as to our sense of touch. The reason you can't walk through a wall is because of the repulsions between the electrons in your body and the electrons in the wall. If the wall were made of this invisible matter, you would be able to walk through it. Of course, you wouldn't be able to see the wall either. This invisible form of matter that we cannot see or touch is known as *dark matter*. Stay tuned to current developments.

CHAPTER 5: THE ATOMIC NUCLEUS

1. Alpha radiation decreases the atomic number of the emitting element by 2 and the atomic mass number by 4. Beta radiation increases the atomic number of an element by 1 and does not affect the atomic mass number. Gamma radiation does not affect the atomic number or the atomic mass number. So alpha radiation results in the greatest change in atomic number (and hence charge), as well as mass number.

3. Most of the radiation we encounter is natural background radiation that originated in the Earth and in space.

7. A large nucleus requires more neutrons to help overcome the repulsions among the many protons.

9. Most is kinetic energy of the ejected particle, and some is the kinetic energy of the recoiling nucleus.

11. All uranium will eventually decay to lead.

13. 4.5 billion years

15. Carbon-14 lost to decay is replenished with carbon-14 from the atmosphere. When a living thing dies, replenishment stops and the percentage of carbon-14 decreases at a constant rate.

17. An elongated uranium-235 nucleus splits in half because the strong nuclear force between distant nucleons quickly diminishes.

19. A nuclear reactor and a fossil fuel power plant both boil water to produce steam for turbines. The main difference between them is the amount of fuel involved. Nuclear reactors are much more efficient—1 kilogram of uranium-235 yields more than 30 freight car loads of coal.

21. Iron

23. Less after fusion, the mass difference having been converted to energy.

25. Thermonuclear fusion

27. The number of pennies in the mix should not affect the activity of the dimes. The set of coins with only two dimes has gone through a greater number of tosses. The set with 10 dimes would be the more "radioactive" substance. Remember, according to this simulation, radiation is emitted every time a dime lands heads up. The set with only two dimes is analogous to a sample of once-living ancient material.

29. Intensity decreases with distance by the inverse-square law. Twice the distance is one-fourth the intensity and one-fourth the reading. Three times as far is one-ninth the intensity and one-ninth the reading.

31. Count: It's decreased to 40 cps (one half-life); 20 cps (2 half-lives); 10 cps (3 half-lives); and 5 cps, which takes 4 half-lives. If 4 half-lives equals 8 hours, then a single half-life equals $8/4 = 2$ hours.

33. c, b, a

35. $c > a > b$. The longer the half-life, the lower the radioactivity. Of these three isotopes, Ac-225 is most radioactive and U-238 is least radioactive.

37. The strong nuclear force is only effective over extremely short distance. Once the alpha particle leaves the nucleus, its attraction to the nucleus by the strong nuclear force is no longer significant. The electrical repulsion between the alpha particle and the protons of the nucleus, however, is still quite significant. This repulsive force causes the alpha particle to accelerate to high velocities, moving quickly away from the nucleus. Might the surrounding negatively charged electrons cause the alpha particle to slow down? Not significantly. The alpha particle is about 8000 times more massive than an electron. Once it begins moving away from the nucleus, its great mass allows it to plow right through the surrounding electrons.

39. These film badges monitor gamma rays, which are a form of electromagnetic radiation (ultra-high-frequency light). The greater the exposure, the darker the film upon processing.

41. They repel by the electric force and attract each other by the strong nuclear force. The strong force predominates. (If it didn't, there would be no atoms beyond hydrogen.) If the protons are separated to where the longer-range electric force overcomes the shorter-range strong force, they fly apart.

43. Without the strong nuclear force, there would be no elements heavier than hydrogen, which means the periodic table would be much less complicated.

45. All uranium isotopes eventually decay to lead. So any deposit of uranium ore will contain some lead that has been converted from uranium.

47. After beta emission from polonium, the atomic number increases by 1 and becomes 85; the atomic mass is unchanged at 218. However, if an alpha particle is emitted, the atomic number decreases by 2 and becomes 82 and the atomic mass decreases by 4 and drops to 214.

49. If strontium-90 (atomic number 38) emits betas, it should become the element yttrium (atomic number 39); hence, the chemist can test a sample of strontium for traces of yttrium. To verify that the sample is a "pure" beta emitter, the chemist can check to make sure it is emitting no alphas or gammas.

51. No. U-235 (with its shorter half-life) undergoes radioactive decay six times faster than U-238 (half-life of 4.5 billion years), so natural uranium in an older Earth would contain a much smaller percentage of U-235, not enough for a critical reaction without enrichment. Conversely, in a younger Earth, natural uranium would contain a greater percentage of U-235 and would more easily sustain a chain reaction. Interestingly, there is strong evidence that 2 billion years ago, when the percentage of U-235 in uranium ore was greater, a natural reactor existed in Gabon, West Africa.

53. No, not a few years old. Too small a fraction of the carbon-14 has decayed. You couldn't tell the difference between an age of a few years and a few dozen or even a hundred years. To the second question, yes, in a few thousand years, a significant fraction of the carbon-14 has decayed. The method gives best results for ages not so very different from the half-life of the isotope. To the third question, no, not a few million years old. Essentially all of the carbon-14 will have decayed. None will be left to detect. You wouldn't be able to distinguish between an age of a million or 10 million or a hundred million years.

55. Stone tablets cannot be dated by the carbon dating technique. Nonliving stone does not ingest carbon and transform that carbon by radioactive decay. Carbon dating pertains to organic material.

57. The degree of radioactivity coming from this relatively pure uranium ore is high, which makes the mining of this ore particularly hazardous. However, naturally occurring uranium contains more than 99 percent of the U-238 isotope, which does not undergo nuclear fission very readily.

59. When a neutron bounces from a carbon nucleus, the nucleus rebounds, taking some energy away from the neutron and slowing it down so that it will be more effective in stimulating fission events. A lead nucleus is so massive that it scarcely rebounds at all. The neutron bounces with practically no loss of energy and practically no change of speed (like a marble from a bowling ball).

61. Mass per nucleon is greater in uranium than is the mass per nucleon of fission fragments of uranium.

63. Energy would be released from the fissioning of gold and from the fusion of carbon, but by neither fission nor fusion for iron. Neither fission nor fusion will result in a decrease of mass for iron nucleons.

65. The mass of an atomic nucleus is less than the masses of the separate nucleons that compose it. One way to see why is to think about the work that must be done to separate a nucleus into its component nucleons. This work, according to $E = mc^2$, adds mass to the system; so the separated nucleons are more massive than the nucleus from which they came. Notice the large mass per nucleon of hydrogen in the graph of Figure 5.29. The hydrogen nucleus, a single proton, is already "outside" in the sense that it is not bound to other nucleons.

67. Although more energy is released in the fission of a single uranium nucleus than in the fusing of a pair of deuterium nuclei, the much greater number of lighter deuterium atoms in 1 gram of matter compared to the smaller number of heavier uranium atoms in 1 gram of matter results in more energy being liberated per gram for the fusion of deuterium.

69. The radioactive decay of radioactive elements found under the Earth's surface warms the insides of the Earth and is responsible for the molten lava that spews from volcanoes. The thermonuclear fusion of our sun is responsible for warming everything on our planet's surface exposed to the Sun.

71. Energy from the Sun is our chief source of energy, which itself is the energy of fusion. Harnessing that energy on the Earth has proven to be a formidable challenge.

73. Two carbon nuclei would fuse to atomic number 12, and the beta emission would increase it by 1; so the element produced would have atomic number 13, which would be aluminum.

75. Perhaps it's fundamental human behavior to seek control. Perhaps people fear loss of control more than anything else.

77. Obvious changes would occur in the fields of economics and commerce, which would be geared to relative abundance rather than scarcity. Already our present price system, which is geared to and in many ways dependent upon scarcity, often malfunctions in an environment of abundance. Hence, we see instances where scarcity is created to keep the economic system functioning. Gem diamonds, for example, are abundant, but the gem diamond industry works hard to maintain a sense of scarcity. Changes at the international level will likely be worldwide economic reform; changes at the personal level, in a reevaluation of the idea that scarcity ought to be the basis of value. A fusion age could likely see changes that will touch every facet of our way of life.

79. No

CHAPTER 6: How Atoms Bond

1. Two electrons fit in the first shell. Eight electrons fit in the second shell.

3. Electron-dot structures of elements in the same group have the same number of valence electrons.

5. A gain of electrons causes a negatively charged ion.

7. Elements on opposite sides of the periodic table tend to form ionic bonds.

9. An ionic crystal is composed of a multitude of ions grouped together in a highly ordered 3-dimensional array.

11. An alloy is a mixture composed of two or more metallic elements.

13. Elements that tend to form covalent bonds are primarily nonmetallic elements.

15. Oxygen can form two covalent bonds.

17. VSEPR stands for valence-shell electron-pair repulsion.

19. An O atom in a water molecule has four substituents.

21. Fluorine has the greatest electronegativity, and francium has the smallest electronegativity.

23. Dipoles that are equal and opposite each other cancel each other out.

25. Nonpolar molecules tend to have a greater degree of symmetry.

27. The potassium chloride crystals are more rounded at the edges because they are softer crystals. One of the reasons they are softer is because the ionic bond between potassium and chlorine ions is weaker than the ionic bond between sodium and chlorine ions. The reason the bond is weaker is because the potassium ions are larger than the sodium ions, which means the potassium ions cannot get as close to the oppositely charged chloride ions. When it comes to the electric attractions, the farther away, the weaker the force.

29. According to Table 6.1, the carbonate ion carries a 2− charge, which means it has picked up two electrons. These two electrons must have come from the single manganese ion to which it is bound.

31. $MgCl_2$. (Two single negatively charged chlorine ions are needed to balance the one double positively charge magnesium ion.) A shortcut to solving these sorts of problems is to take the charge of one ion and make it the subscript of the opposite ion. For example, take the 2+ charge of the magnesium and make it the subscript on the chlorine. Then take the 1− charge on the chlorine and make that the subscript of the magnesium. Because numeral 1 subscripts are implied when not written, we have not Mg_1Cl_2, but $MgCl_2$.

33. Least polar to most polar: $C—H < N—H < O—H$

   Explanation: The greater the difference in electronegativity between bonded atoms, the greater the polarity of the bond.

35. $c < b < a$

37. Elements in the same period (horizontal row) of the periodic table have valence electrons in the same noble gas shells (See Chapter 4). Their electron dot structures, however, are quite different. Moving from left to right across the period, the number of electrons in the electron-dot structure increases by one for each element.

39. There is room for only one additional electron within the valence shell of a hydrogen atom.

41. There are two inner shells of electrons that shield the valence electron from the nucleus.

43. Neon's outermost electrons are held tightly to the atom by a relatively strong effective nuclear charge (see Sections 4.8 and 4.9). Neon doesn't gain additional electrons because no more room is available in its outermost occupied shell.

45. Atoms with many valence electrons, such as fluorine, F, tend to have relatively strong forces of attraction between their valence electrons and the nucleus. This makes it difficult for them to lose electrons. It does, however, make it easy for them to gain additional electrons.

47. The potassium atom with an additional shell of one electron is larger than the potassium ion, K^+. Both the potassium ion, K^+, and the argon atom, Ar, have the same number of electrons in the same number of shells. The nuclear charge of the potassium ion, however, is greater, and this has the effect of pulling the shells of electrons in closer. The argon atom, Ar, therefore, is the larger of the two.

49. Carefully consider the differences between these two compounds. Where the potassium fluoride contains a potassium ion, the molecular fluorine contains a fluorine. What then are the differences between the potassium ion and the fluorine atom? First, consider size. The potassium ion has three full shells of electrons, which makes it larger. The distance between the two nuclei of the fluorine atoms within F_2, therefore, should be closer together. Second, consider the type of bonding. The F_2 compound is covalent. This involves the overlapping of shells, which allows the nuclei to be closer. So by both considerations, the nuclei of molecular fluorine, F_2, should be closer together.

51. The problem is not whether we have the metal atoms on this planet—we do. The problem is in the expense of collecting them. This expense would be too great if the metal atoms were evenly distributed around the planet. We are fortunate, therefore, that geological formations contain metal ores that have been concentrated by natural processes. Keep in mind that we are able to recycle only the metal atoms that we produce ourselves. If we don't recycle these metal atoms, we'll find substantial shortages of new metal ores from which to feed our ever-growing appetite for metal-based consumer goods and building materials.

53. Hydrogen's electron joins the valence shell of the fluorine atom. Meanwhile, fluorine's unpaired valence electron joins the valence shell of hydrogen.

55. When bonded to an atom with low electronegativity, such as any group 1 element, the nonmetal atom pulls the bonding electrons so closely to itself that an ion is formed.

57. The electron-dot structure for hydrogen cyanide, HCN, is

$$H - C \equiv N\colon$$

Hydrogen cyanide

59. The germanium atom of germanium chloride has a lone pair of electrons that pushes the two substituent chlorine atoms away from a 180° orientation.

61. The compound SF_4 can be expected to be a more polar compound than PF_5 because it is less symmetrical. Note that the three mid-level fluorines in PF_5 are in an orientation such that their electron pulls on the phosphorus balance each other out. For the SF_4, by contrast, the two mid-level fluorines, which are shown to the right, have no fluorine atom on the opposite side of the sulfur to balance their electron pulls. As a result, these two mid-level fluorines will be effective at pulling electrons toward them, thus making that side of the molecule slightly negative.

63. The source of an atom's electronegativity is the positive charge of the nucleus. More specifically, it is the effective nuclear charge experienced within the shell that the bonding electrons are occupying.

65. A selenium-chlorine bond should be more polar. Observe their relative positions in the periodic table. Sulfur and bromine are more equidistant from the upper right-hand corner.

67. Sometimes true and sometimes false. Within any one atomic group, as the number of shells increases, the electronegativity decreases. A group 17 bromine atom, however, has four shells of electrons, yet its electronegativity is greater than that of a group 1 lithium atom, which has only two shells.

69. a. The left compound with the two chlorines on the same side of the molecule is more polar and will thus have a higher boiling point.

b. The chlorine atoms have a relatively strong electronegativity that pulls electrons away from the carbon. In the left molecule, $COCl_2$, this tug of the chlorines is counteracted by a relatively strong tug of the oxygen, which tends to defeat the polarity of this molecule. The hydrogens of the molecule on the right, $C_2H_2Cl_2$, have an electronegativity that is less than that of carbon; so they actually assist the chlorines in allowing electrons to be yanked toward one side, which means this molecule is more polar. A material consisting of the molecule on the right, $C_2H_2Cl_2$, therefore, has the higher boiling point.

71. The single greedy kids ends up being slightly negative, while the two more generous kids are slightly positive (deficient of electrons). The greedy negative kids are actually twice as negative as one of the positive kids is positive. In other words, if the greedy kid had a charge of -1, each positive kid would have a charge of $+0.5$. This is a polar situation where the electrons are not distributed evenly. If all three kids were equally greedy, then the situation would be more balanced (that is, nonpolar).

73. We understand that muscles require exercise in order to stay in shape. What many people don't understand is that the brain is the same way—it too requires exercise in order to stay in shape. Furthermore, similar to the way your muscles become stronger with extra exercise, your mental capacities become stronger with extra mental exercise. This is one of the main reasons we go to school for so many years—we understand the value and benefits of a mind that is well exercised and in shape. And as you will likely discover, learning continues throughout one's life. But learning about the molecular nature of our environment is valuable for more than just the mental exercise. By understanding nature at this level, we gain a deeper appreciation, and with deeper appreciation comes greater respect. More than ever, humans are having a great impact on the environment. Should we do so mindlessly or mindfully? By studying chemistry, you have decided for the latter. We thank you!

75. People face many obstacles to recycling. There's confusion about what can and can't be recycled. There's concern about how clean a container must be before it can be recycled. Some people don't know or understand the value of recycling. Others may be a bit lazy when it comes to recycling. To overcome obstacles such as these, campaigns can be used to educate the general public. An easy-to-read pamphlet describing the do's and don'ts of recycling can be effective. These recycling educational campaigns are typically sponsored by local governments together with local recyling companies. State and federal governments also can play a role in educating the general public and in setting packaging standards for companies that use recyclable material to package their goods. Many states offer refunds for empty containers. Some communities even fine individuals who do not sort plastics correctly. Other communities ask their citizens not to sort at all, under the philosophy that the task of sorting inhibits people's tendency to recycle. Ultimately, this is a global issue such that people of all nations should be encouraged not to waste material resources.

CHAPTER 7: How Molecules Mix

1. A chemical bond is many times stronger than an attraction between molecules.

3. A hydrogen bond is a very strong dipole–dipole attraction involving a hydrogen atom bonded to a highly electronegative atom such as oxygen.

5. The volume of a sugar solution gradually increases as more sugar is dissolved in it.

7. A solute is the component of lesser quantity in a solution (for example, a pinch of salt in a glass of water). A solvent is the component of greater quantity in a solution (for example, the water in a salt water solution).

9. A mole is a very large number: 6.02×10^{23}. For example, a mole of marbles would be enough to cover the entire land area of the United States to a depth greater than 4 meters.

11. The solubility of a gas decreases with increasing temperatures because the gas molecules have more kinetic energy and are more likely to escape from solution. The greater kinetic energy of the solvent molecules also helps the gas molecules to escape.

13. Sugar is very polar, as evidenced by its great solubility in water.

15. Water and soap are attracted to each other by ion–dipole attractions between each water molecule and the polar head of each soap molecule.

17. Large amounts of calcium and magnesium ions are found in hard water.

19. Sodium carbonate (Na_2CO_3) has a 2– charge in the carbonate ion (CO_3)$^{2-}$ to which calcium and magnesium are more attracted than to the 1– charge found in a molecule of soap. The hard water ions calcium and magnesium bind to the carbonate ions, which "softens" the water.

21. People disinfect their water by boiling it or adding disinfecting iodide tablets.

23. The water molecules have both negative and positive ends. Hold a negatively charged balloon up to the water and all the water molecules will rotate so that their positive ends face the balloon. Hold up a positively charged balloon and all the water molecules will rotate so that their negative ends face the balloon. Either way, the balloon attracts.

25. The water level rises just as it would if you were adding sand. It does not matter that what you add dissolves.

27. Divide the number of water molecules by the total number of molecules and multiply by 100 to get the percentage:

999,999 million trillion/1,000,000 million trillion
$$\times\ 100 = 99.9999\%$$

29. mass = (concentration)(volume) = (3.0 g/L)(15 L) = 45 g

31. The total volume of solution should be 20.0g/10.0 g/L = 2.00 L, but this is the volume of solution, not the volume of solvent. Remember that the volume of solution is equal to the combination of the volume of the solute and the solvent and that the volume of water (solvent) is equal to the volume of solution minus the volume of the sodium chloride (solute). Ignoring the rules for significant figures (see Appendix B) and assuming that the 20.0 grams of sodium chloride occupies 7.50 mL (0.00750 L), this volume of water would be:

Volume of solution:	2.00000 L
Volume of solute:	0.00750 L
Volume of water	1.99250 L

But don't waste time measuring out 1.99250 L of water. A far better approach would be to make the solution by adding the sodium chloride (solute) to an empty container calibrated for 10 liters and then adding water (solvent) as needed to make 10.0 liters of solution.

33. These are all nonpolar compounds, but they differ in size. The larger the size of the nonpolar molecule, the easier for that nonpolar molecule to form induced dipoles. Larger nonpolar molecules, therefore, have a greater "stickiness" and hence a higher boiling point. In order of increasing boiling points, these compounds are CF_4, CCl_4, CBr_4, and CI_4.

35. The covalent bonds within a molecule are many times stronger than the attractions occurring between neighboring molecules. We know this because although two molecules can move away from each other (as occurs in the liquid or gaseous phase), the atoms within a molecule remain stuck together as a single unit. To pull the atoms apart requires some form of chemical change.

37. Because the magnitude of the electric charge associated with an ion is much greater.

39. The elements calcium, Ca, and fluorine, F, are on opposite sides of the periodic table, which tells us that the bond between them is ionic. Ionic compounds tend to have high melting points because the attractions between ions extend in all directions. This locks the ions in place, which means a lot of thermal energy is required to break them apart, as occurs during melting. The elements tin, Sn, and chlorine, Cl, are much closer together in the periodic table, which tells us that the bond between them is less ionic and more covalent. Furthermore, the tetrahedral molecular geometry of stannic chloride, $SnCl_4$, is symmetrical, which means the dipole of individual tin–chloride bonds cancels each of them out. This nonpolarity means the stannic chloride molecules are not as attracted to each other, which means they readily evaporate.

41. As described in the FYI in Section 7.2, be it large or small, an individual sugar crystal is transparent. A teaspoon of table sugar appears white because of the way light gets scattered as it passes in and out of the numerous tiny crystals at numerous different angles. Technically speaking, when table sugar dissolves in water, it is this scattering effect that disappears, not the sugar itself.

43. Assuming concentration is given in units of mass (or moles) of solute in a given volume of solution, the concentration necessarily decreases with increasing temperature.

45. To answer this question, you would also need to know the volume of solution. If the volume is 1 liter, then the amount of solute would be equal to (2 mole/liter)(1 liter) = 2 moles. The answer could be expressed simply as 2 moles or as 1.204×10^{24}, which is Avogadro's number times 2.

47. A high boiling point means the substance interacts with itself quite strongly. If you were a water molecule, you would be attracted to both ends of 1,4-butanediol. In fact 1,4-butanediol is infinitely soluble in water.

49. The greater the pressure, the greater the solubility. Recall that the solubility of a gas in a liquid increases with increasing pressure. This principle is used in the manufacture of carbonated beverages.

51. The arrangement of atoms in a molecule makes all the difference in the physical and chemical properties. Ethyl alcohol contains the −OH group, which is polar. This polarity, in turn, is what allows the ethyl alcohol to dissolve in water. The oxygen of dimethyl ether, by contrast, is bonded to two carbon atoms: C−O−C. The difference in electronegativity between oxygen and carbon is not as great as the difference between oxygen and hydrogen. The polarity of the C—O bond, therefore, is less than that of the O—H bond. As a consequence, dimethyl ether is significantly less polar compared to ethyl alcohol and is not readily soluble in water.

53. Air is a gaseous solution, and one of its minor components is water vapor. The process of this water coming "out of solution" in the form of rain or snow is called *precipitation*. The rain or snow is the *precipitate*.

55. Although oxygen gas, O_2, has poor solubility in water, many other examples of gases have good solubility in water. From the concepts discussed in Chapter 6, you should understand that hydrogen chloride is a polar molecule. This gaseous material, therefore, has a good solubility in water by virtue of the dipole–dipole attractions occurring between the HCl and H_2O molecules.

57. The nonpolar molecules have a hard time passing the ionic heads of the fatty acid molecules, which are surrounded by water molecules.

59. This structure is called a *cell*, also known as a *lysosome*. Add a bunch of ions, DNA, organelles, and many other biomolecules (Chapter 13) to the cell and you have a living cell. The bilipid barrier is called a *plasma membrane*, which you will learn all about in your biology classes.

61. The phosphate ions softened the water by binding to calcium ions, which would otherwise interfere with the effectiveness of the soap or detergent. By the 1960s, the damage done by phosphates to fresh water streams and lakes was most apparent. Lake Erie was a notable example. Because the effects were so visible, there was strong support from the public and government for phasing out phosphates from laundry detergent, which occurred during the 1970s and 1980s. In its place arose the carbonate ion, as discussed in this chapter.

63. The net flow of water is reversed in reverse osmosis. Osmosis transfers fresh water into salt water, but reverse osmosis transfers the water molecules in salt water across a semipermeable membrane into a region of fresh water.

65. Distilled water is pure only before you drink it. Once in your stomach, it mixes with everything else to make a nutrient-filled solution. The only difference is that tap water may have contributed a few more milligrams of hard water ions, such as calcium, which your body uses as a mineral. But there's nothing wrong with drinking water that has been distilled. In fact, it is about as pure as any water you can drink.

67. Perfluorocarbons are nonpolar molecules that interact with oxygen molecules by way of induced dipole–induced dipole attractions. A saturated solution of perfluorocarbons contains about 20 percent more oxygen than does the atmosphere we breathe. Lungs absorb this oxygen in much the same way they absorb oxygen from air. Perfluorocarbons also remove carbon dioxide from the lungs. The rodent would drown in the water because of the lack of oxygen there. The goldfish would suffer from oxygen poisoning (too much oxygen for its gills) if it swam in the perfluorocarbons. Potential applications include helping premature babies to breathe, cleaning the lungs of people with lung disorders such as cystic fibrosis, and serving as a backup to the ever-short worldwide blood supply. Many people believe the sacrifice of animal life for research is reasonable only when the animals are treated humanely and compassionately. Many say the same thing about slaughterhouses, where animals are routinely turned into meat.

69. Answers may vary.

CHAPTER 8: How Water Behaves

1. Ice is less dense than water because water expands as it freezes. Each water molecule in the solid phase occupies more space than it does in the liquid phase.

3. As great pressure is applied, the open pockets in the crystalline structure of ice may collapse, which results in the ice melting. This occurs primarily where the pressure is being applied, such as at the bottom of a heavy glacier.

5. When the temperature of 0°C liquid water is increased slightly, it undergoes a net contraction. Water at 4°C is more dense (less volume = more contracted) than it is at 0°C.

7. Cohesive forces result from the molecular attractions within a material; adhesive forces result from the molecular attractions occurring between two different materials.

9. Water will rise higher in a narrow tube because there is more surface area of contact between the water and the wall of the tube relative to the weight of the water contained within the tube.

11. Condensation counteracts evaporation on a hot and humid day. The vapor in the air from the humidity condenses on your skin to make you feel uncomfortably warm.

13. Water can boil at temperatures less than 100°C at higher altitudes, where the atmospheric pressure is less.

15. A substance that heats up quickly will have a low specific heat.

17. Hydrogen bonds

19. The temperature of melting ice does not rise as it is heated because the energy being supplied by the added heat is working to break the hydrogen bonds.

21. It takes much more energy to boil water than to melt it because in a sample of steam, the molecules are relatively free of one another and are not bound together as they are in the liquid phase.

23. As a variation, you can begin your "magic show" having the water in the regular bottle when you demonstrate how atmospheric pressure pushes the card upward so that the water doesn't fall. Ask your friends what will happen when you turn the bottle sideways. (Atmospheric pressure works sideways too.) Then ask them what will happen when you flick off the card. Of course, the water comes rushing out. This builds up an expectation. Ask them if you can repeat the experiment. Pour water back and forth between the two bottles, but this time use the bottle with the mesh screen.

25. A must for every household. Good chemistry to you!

27. Table 8.1 shows that the specific heat of iron is 0.451 J/g°C. So the total amount of heat required is

heat = 100,000 g × 0.451 J/g°C × 30°C = +1,343,000 joules,

which is about 10 million joules less than that required for water.

29. This is a three-part calculation. First, you need to calculate the amount of heat required to raise the temperature of the water from −5.00° to 0.00°C. Then you need to calculate the amount of heat required to transform the 1.00 gram of ice into liquid water. Third, you need to calculate the amount of heat required to raise the temperature of the water from 0.00°C to +5.00°C. Table 8.1 shows that the specific heat of ice is 2.01 J/g°C. Note that this calculation provides three significant figures (see Appendix B).

1) heat = (2.01 J/g°C)(1.0 g)(+5.00°C)

= 10.1 J

2) heat $= (1.00 \text{ g})(+335 \text{ J/g})$

$\qquad = +335 \text{ J}$

3) heat $= (4.18 \text{ J/g°C})(1.00 \text{ g})(+5.00°C)$

$\qquad = +20.9 \text{ J}$

total heat $= 10.1 \text{ J} + 335 \text{ J} + 20.9 \text{ J}$

$\qquad = 366 \text{ J}$

31. Calculate this answer using water's heat of vaporization, which is $+2259 \text{ J/g}$.

heat $= (1.00 \text{ g})(+2259 \text{ J/g})$

$\qquad = 2260 \text{ J}$ (to three significant figures)

33. heat $=$ (specific heat)(mass)(temperature change)

With a little algebra, this transforms into

specific heat $=$ heat/(mass)(temperature change)

specific heat $= (2674 \text{ J})/(575 \text{ g})(5.0°C)$

$\qquad = 0.93 \text{ J/g°C}$ (to two significant figures)

35. $a = b = c$

37. $b < c < a$

39. $a < b < c$

41. The soda can puffs out, and sometimes the lid pops open. This occurs because the water content of the soda freezes and expands.

43. Ice and diamond are both transparent solids having the same six-sided open-pocket crystalline structure. For ice, this crystalline structure is made of water molecules held together by hydrogen bonds. For diamond, this crystalline structure is made of carbon atoms held together by covalent bonds. The covalent bonds of diamond are much stronger than the hydrogen bonds of ice, which is why, compared to ice, diamond is many times harder.

45. Water at 4°C still contains nanoscopic ice crystals. On a macroscopic scale, it behaves as a liquid, but water at 4°C can also be considered a heterogeneous mixture of water and small amounts of ice crystals.

47. Calcium chloride dissolves in water to produce three ions, but sodium chloride dissolves to produce only two. The greater number of ions within a 1 M solution of calcium chloride is more effective at decreasing the number of water molecules entering the solid phase.

49. The water in the ice tray tends to freeze starting at the top followed by the sides, which are directly exposed to the cold of the freezer. As the water freezes into crystalline ice, air molecules are excluded and pushed to the center of the ice cube, where the water is liquid the longest. Eventually, as this central water freezes, the air comes out of solution, forming the opaque bubbles. To make cloudless ice cubes, remove the air from the water by boiling it.

51. If cooling occurred at the bottom of a pond instead of at the surface, ice would still form at the surface, but it would take longer for the pond to freeze. This is because all the water in the pond would have to be reduced to a temperature of 0°C rather than 4°C before the first ice would form. Consider the state where the entire pond is 4°C, as discussed in the chapter. Where cooling occurred at the bottom, the 4°C water at the bottom of the pond would be cooled to 3°C and thus float upward to be replaced by more 4°C water at the bottom. This

will keep happening until the whole pond is 3°C. Then the 3°C water at the bottom would be cooled to 2°C, which would also float upward, and so on. Eventually, the entire pond would be cooled to 0°C. Finally, ice that formed at the bottom, where the cooling process was occurring, would be less dense and would float to the surface (except for ice that may form about material anchored to the bottom of the pond). Of course, this is not how it happens in nature, where cooling typically occurs from the cold air above the water.

53. Mercury sticks to itself (cohesive forces) better than it sticks to the glass (adhesive forces). The cohesive forces are stronger than the adhesive forces. The surface tension of mercury is about eight times that of water.

55. These are the *adhesive* forces between the water and the metal (two different materials). The type of molecular interactions going on is dipole–induced dipole.

57. There are at least two reasons the bubbles get larger as they rise. First, water vapor continues to evaporate into the bubble. Second, as the bubble rises, the external water pressure on the bubble decreases with decreasing depth, which allows the bubble to expand.

59. A bottle wrapped in wet cloth will cool by the evaporation of liquid from the cloth. As evaporation progresses, the average temperature of the liquid left behind in the cloth can easily drop below the temperature of the cool water that wet the cloth in the first place. So to cool a bottle of beer, soda, or some other beverage at a picnic, wet a piece of cloth in a bucket of cool water. Wrap the wet cloth around the bottle to be cooled. As evaporation progresses, the temperature of the water in the cloth drops and cools the bottle to a temperature below that of the bucket of water.

61. No, no, no! When we say boiling is a cooling process, we mean the water left in the pot (and not on your hands) is being cooled relative to the higher temperature it would attain otherwise. Because of the cooling effect of the boiling, the water remains at 100°C instead of getting hotter.

63. A force of attraction between the two marbles causes them to accelerate toward each other just before they collide. Through this acceleration, there is an increase in the kinetic energy of the marbles. Similarly, when two water molecules come together, they accelerate toward each other, which increases their kinetic energy. The water molecules have more kinetic energy after they have come together, and this kinetic energy can be witnessed as an increase in their rates of vibration.

65. The solutes dissolved in the ocean water would lower the oceans specific heat, which is a measure of how much heat a given mass is able to absorb. Consider a kilogram of fresh water and a kilogram of ocean water. Do you understand how the kilogram of ocean water has less water in it? Because it has less water, it is able to absorb less heat. (As you likely discovered in Think and Do #24, the salt solute of ocean water has a rather low specific heat.)

67. As the ocean off the coast of San Francisco cools in the winter, the heat it loses warms the atmosphere with which it comes in contact. This warmed air blows over the California coastline to produce a relatively warm climate. If the winds were easterly instead of westerly, the climate of San Francisco would be chilled by winter winds from dry and cold Nevada. The climate would be reversed in Washington, D.C., because air warmed by the cooling of the Atlantic Ocean would blow over Washington, D.C., and produce a warmer winter climate there.

69. To produce maple syrup requires boiling away most of the water from the sap. The resulting concentrated solution is maple syrup. A lot of energy is required to boil away the water because of water's high heat of vaporization. Interestingly, it takes 40 gallons of sap to produce just 1 gallon of maple syrup. That's a lot of water that needs to be evaporated. Many maple syrup producers get a head start by first removing much of the water using reverse osmosis, which is a technique discussed in Section 7.7.

71. Water is boiling at point A.

73. As pressure increases, the melting temperature of water decreases and the boiling temperature of water increases.

75. When ocean water freezes, forming ice, it excludes salt ions, which remain behind in the aqueous phase. As more ice forms, the concentration of the ions increases. This makes the ocean water saltier as well as denser. Because of its greater density, this saltier water sinks. This process occurs primarily along the perimeter of the ice cap, where the aqueous ocean exposed to frigid Arctic air meets solid sea ice, which, by the way, is primarily fresh. Any water that makes it to a location beneath the center of the ice cap has been stripped of many of its ions after having passed through the perimeter. Oceanographers call this process *brine release* or *salting out*. Any melt from the underside of the ice cap also helps to contribute to the relative freshness of this water.

77. The ice above the Arctic Ocean is already floating. If it were to melt, there would be no rise in the global sea level. The ice over the Greenland subcontinent, however, rests on the land. If it were to melt, this volume of water would run into the oceans, both as liquid fresh water and as new icebergs. Current estimates are that the complete melting of the Greenland ice cap would raise the global sea level by around 7 meters. Will it melt? If so, how soon? Those questions are currently the subject of much scientific research and political debate.

CHAPTER 9: How Chemicals React

1. Coefficients are used to show the ratio in which reactants combine or form in a chemical reaction.

3. A chemical equation must be balanced because the law of conservation of mass says that mass can be neither created nor destroyed. The same number of each atom must be on both sides of the equation.

5. The relative mass of golf balls is greater than that of Ping-Pong balls; therefore, it would take more Ping-Pong balls to equal the same mass of golf balls.

7. The formula mass of NO is 30.006 amu.

9. For water, 18 grams is 1 mole.

11. One mole of water has 6.02×10^{23} water molecules.

13. The amount of energy released when the bond is formed equals the amount of energy needed to break the bond, which is 436 kJ.

15. Energy is consumed by an endothermic reaction.

17. The net entropy of the universe is always increasing.

19. Reactants must collide in a certain orientation with enough energy to react.

21. Assuming the reactant molecules are already mixed together, the first to react are those with sufficient kinetic energy and the proper orientation.

23. Atomic chlorine is a catalyst for the destruction of ozone.

25. A catalyst is unchanged by a chemical reaction.

27. The energy of the lightning or electrostatic sparks passing through the air converts oxygen molecules into ozone molecules. This reaction is endothermic because it requires an input of energy.

29. The balanced chemical equation is $2\ H_2O_2 \rightarrow O_2 + 2\ H_2O$. Bubbles of oxygen gas form upon the mixing of hydrogen peroxide and baker's yeast. The fact that oxygen gas is formed can be demonstrated by inserting a glowing wood splint into the bubbles. The glowing splint will flame up as soon as it makes contact with the oxygen. Students should be directed to do so only under careful supervision.

31. (5.00 g gold)(1 mole gold/197 g gold)
$(6.02 \times 10^{23}$ atoms/1 mole$) = 1.53 \times 10^{22}$ gold atoms

33. The coefficients within this balanced equation tell us the ratio by which reactants react to form products. Accordingly, three moles of oxygen gas are produced for the reaction of every two moles of $KClO_3$ solid. Do you also see that only 1.5 moles of oxygen gas would be produced from the reaction of 1 mole of $KClO_3$ solid?

35. Use the periodic table to find the masses of all the atoms within each molecule. Add these masses together and you'll come up with the given formula masses. Use these masses to help answer the next question.

37. For this reaction, is there enough oxygen to react with all of the methane? Is there enough methane to react with all of the oxygen? The surefire way to find out is to determine how much of one reactant is needed for all of the other reactant to be consumed. According to the following calculation, 16 g of CH_4, would require 64 g of O_2:

$(16\ g\ CH_4)(1\ mole\ CH_4/16\ g\ CH_4)(2\ moles\ O_2/1\ mole\ CH_4)$
$(32\ g\ O_2/1\ mole\ O_2) = 64\ g\ O_2$

But there is only 16 g of O_2, which means not all of the CH_4 is going to be able to react. How much of the CH_4 will react? That can be calculated as follows:

$(16\ g\ O_2)(1\ mole\ O_2/32\ g\ O_2)(1\ mole\ CH_4/2\ mole\ O_2)$
$(16\ g\ CH_4/1\ mole\ CH_4) = 4\ g\ CH_4$

The maximum amount of CO_2 that can be formed is calculated as follows:

$(16\ g\ O_2)(1\ mole\ O_2/32\ g\ O_2)(1\ mole\ CO_2/2\ mole\ O_2)$
$(44\ g\ CO_2/1\ mole\ CO_2) = 11\ g\ CO_2$

39. a.

Energy to break bonds:	Energy released from bond formation:
N—N = 159 kJ	
N—H = 389 kJ	
N—H = 389 kJ	H—H = 436 kJ
N—H = 389 kJ	H—H = 436 kJ
N—H = 389 kJ	N≡N = 946 kJ
Total = 1715 kJ absorbed	Total = 1818 kJ released

NET = 1715 kJ absorbed − 1818 released
     = −103 kJ released (exothermic)

39. b. Energy to break bonds:

O—O = 138 kJ

H—O = 464 kJ

H—O = 464 kJ

O—O = 138 kJ

H—O = 464 kJ

H—O = 464 kJ

Total = 2132 kJ absorbed

Energy released from bond formation:

O=O = 498 kJ

H—O = 464 kJ

H—O = 464 kJ

O—H = 464 kJ

O—H = 464 kJ

Total = 2354 kJ released

NET = 2132 kJ absorbed − 2354 kJ released
= −222 kJ released (exothermic)

41. C < A < B. The endothermic reaction, C, will likely take place more slowly than the exothermic reaction, A, because it requires a decrease in entropy. The fastest of these reactions will be B, which has no energy of activation.

43. Contrary to popular opinion, the entropy of a deck of playing cards has nothing to do with whether it is being shuffled. Historically, there has been a misplaced association between entropy and disorder. But now you know better. Entropy has nothing to do with what our minds perceive as being orderly or disorderly. From the point of view of the molecules within the cards, it makes no difference whether the deck is shuffled. Entropy is merely a measure of the tendency of energy to disperse. The greater the difference between the temperature of the cards and the room, the greater the amount of energy that gets dispersed. Thus, in order of increasing entropy: C < A < B. Of course, a burning deck of cards (at 233°C, which is 451°F) would result in an even greater dispersal of energy.

45. (a) 2, 3, 1 (b) 1, 6, 4 (c) 2, 1, 2 (d) 1, 2, 1, 2

47. Only two diatomic molecules are represented (not three). These are the two shown in the left box, one of which is also shown in the right box. Remember, the atoms before and after the arrow in a balanced chemical equation are the same atoms but in different arrangements.

49. Equation d best describes the reacting chemicals.

51. $Fe_2O_3 + 3\,CO \rightarrow 2\,Fe + 3\,CO_2$

53.

Step 1 (Balance Fe): $2\,FeS_2 + O_2 \rightarrow Fe_2O_3 + SO_2$

Step 2 (Balance S): $2\,FeS_2 + O_2 \rightarrow Fe_2O_3 + 4\,SO_2$

Step 3 (Use a fraction to balance O): $2\,FeS_2 + 11/2\,O_2 \rightarrow Fe_2O_3 + 4\,SO_2$

Step 4 (Multiply entire equation by 2): $2(2\,FeS_2 + 11/2\,O_2 \rightarrow Fe_2O_3 + 4\,SO_2)$

Step 5 (Equation balanced): $4\,FeS_2 + 11\,O_2 \rightarrow 2\,Fe_2O_3 + 8\,SO_2$

55. (a) There is 1 mole of N_2 in 28 grams of N_2. (b) There is 1 mole of O_2 in 32 grams of O_2. (c) There are 2 moles of CH_4 in 32 grams of CH_4. (d) There is 1 mole of F_2 in 38 grams of F_2.

57. A single water molecule has a very small mass of 18 amu.

59. No, because this mass is less than that of a single oxygen atom.

61. There are 69.7 g of gallium, Ga (atomic mass 69.7 amu), in a 145 g sample of gallium arsenide, GaAs. Note that 145 g is the formula mass for this compound.

63. As the carbon-based fuel combusts, it gains mass as it combines with the oxygen from the atmosphere to form carbon dioxide, CO_2, which comes out in the exhaust.

65. The chemical reactions within a disposable battery are exothermic, as evidenced by the electric energy they release.

67. The superheated chamber dampens the ability of the energy from the reaction to be dispersed. This, in turn, makes the reaction less favorable. This is one reason chemists carry out most of their exothermic reactions in a cooled environment, such as within a reaction vessel submerged in an ice bath. Another reason is for safety—they don't want the reaction vessel to explode!

69. The entropy change determines whether the chemical reaction is favorable. If there is an overall increase in entropy, then the reaction will be favorable, which means the reaction can proceed on its own. If there is an overall decrease in entropy, then the reaction will only proceed with the help of a continual source of energy, which will necessarily be coming from some entropy-increasing process such as the combustion of a fuel.

71. The solar energy is more readily dispersed by the water molecules in the gaseous phase.

73. A chemical reaction may be favored and proceed on its own, but often an energy of activation must be overcome first. The flammable red chemicals found at the tip of a match are ready to ignite, but they too need the input of a little energy to overcome their energy of activation. You are providing this when you strike the match against the proper surface. The heat from these burning chemicals then boosts the cellulose molecules in the matchstick to start reacting with the oxygen in the atmosphere. Once a small flame is formed, bigger flames can form. In other words, once initiated, the burning perpetually kick-starts itself and is thus sustainable. Good news for a campfire. Not so good news for an out-of-control brush or forest fire.

75. Endothermic reactions require the input of energy, which can include the input of thermal energy. This gives the molecules greater kinetic energy, which can help their collisions be more effective. The elevated temperature also helps to minimize the unfavorable decrease in entropy due to the heat absorbed by the reaction. Some exothermic reactions are so exothermic that they explode if not run at cold temperatures. The cold temperatures slow down the reactive molecules, which gives the chemist greater control. Also, the heat generated by the reaction is more efficiently dispersed under the colder conditions. This allows for a greater increase in entropy, which helps with the formation of products.

77. In pure oxygen, there is a greater concentration of one of the reactants (the oxygen) for the chemical reaction (combustion). As discussed in this chapter, the greater the concentration of reactants, the greater the rate of the reaction.

79. Photosynthesis produces oxygen, O_2, which migrates from the Earth's surface to high in the stratosphere, where it is converted by the energy of ultraviolet light into ozone, O_3. Plants and all other organisms living on the planet's surface benefit from this ozone because of its ability to shade the planet's surface from ultraviolet light.

81. Hydrogen chloride, HCl, does not stay in the atmosphere for extended periods of time because it is quite soluble in water, as can be deduced from its polarity (see Chapter 6). Thus, atmospheric hydrogen chloride mixes with atmospheric moisture and precipitates with the rain.

83. A timekeeping device, such as a clock, requires a source of energy to keep it running. This energy comes from exothermic reactions that occur because of a net increase in universal entropy. This is true whether the clock is plugged in to a power outlet, runs off a battery, or has solar photovoltaic cells. Even a sundial depends on the light energy coming from the Sun. A living body can also be a timekeeping device (heartbeats and mental counting, for example), but of course, the living body also requires energy that must come from some net increase in universal entropy. The concepts of time and entropy are intricately connected.

85. The cause of the current mass extinction is the destruction of habitats brought on by the exponential rise of the human population. According to some projections, about 20 percent of all plant and animal species on the Earth will be extinct within the next 25 years. On a geologic time scale, this is faster than the blink of an eye. The creatures that do survive will thrive off the waste of humans or off those creatures that don't compete with humans directly or indirectly. Of course, there is also the dark possibility that we humans, like proliferating rabbits on an isolated island, will face extinction due to a depletion of resources. In the words of Harvard biologist E. O. Wilson, "If all mankind were to disappear, the world would regenerate back to the rich state of equilibrium that existed 10 thousand years ago. If insects were to vanish, the environment would collapse into chaos."

CHAPTER 10: Acids and Bases in Our Environment

1. The Brønsted-Lowry definition of an acid and base says that an acid is any chemical that donates a hydrogen ion and a base is any chemical that accepts a hydrogen ion.

3. A chemical that loses a hydrogen ion is behaving as an acid.

5. A solution of a strong acid has more ions in solution, which means it can conduct electricity better than a weak acid.

7. Water is a weak acid.

9. The pH of a solution indicates the acidity of the solution as judged by the concentration of hydronium ions.

11. A buffer solution is any solution that resists changes in pH.

13. A pH of blood that is too high or too low can be lethal.

15. Sulfur dioxide combines with oxygen and water in the air to make sulfuric acid.

17. The ocean absorbs the CO_2, where it is neutralized and not re-released.

19. The pH of the ocean has decreased by about 0.1 pH unit over the past century.

21. The pH of this solution goes up after plain water is added, as evidenced by the change in hue from magenta to purple. This change in hue is easiest to see when the water is added quickly. Understand that pH is merely a measure of the hydronium ion concentration. As you add water, you dilute the concentration of hydronium ions, which has the effect of raising the pH. This is analogous to adding water to a sugar solution. The more water you add, the more dilute the sugar becomes. But will it be possible to bring the pH of the solution to above 7? The answer is no because the plain water you're adding also has hydronium ions and you won't be able to get to a concentration less than that in the plain water. By analogy, imagine you're diluting cherry syrup with gallons and gallons of cherry drink. You will certainly be diluting the cherry syrup, but the concentration of the mixture will never be less than that of the cherry drink you're using for the dilution.

23. The product of the hydroxide ion and hydronium ion concentration is always equal to 1×10^{-14}. So if the hydroxide ion concentration equals 1×10^{-4} M, then the hydronium ion concentration must be equal to 1×10^{-10} M because these two values multiplied together equal 1×10^{-14}.

25. The pH of this solution is 4, and it is acidic.

27. The concentration of hydronium ions in the pH = 1 solution is 0.1 M. Doubling the volume of solution with pure water means that its concentration is cut in half. Therefore, the new concentration of hydronium ions after the addition of 500 mL of water is 0.05 M. To calculate for pH:

$$pH = -\log[H_3O^+] = -\log(0.05) = -(-1.3) = 1.3$$

29. All of these solutions have the same concentration. The difference between them is their acid strength. As discussed in the text, hydrogen chloride is a strong acid, which means nearly all of the hydrogen chloride molecules donate hydrogen ions to form hydronium ions. Acetic acid is a weak acid, which means only a few of the acetic acid molecules in solution donate hydrogen ions to form hydronium ions. The ammonia behaves better as a base than an acid, which means it contributes very few hydrogen ions. The ammonia solution, therefore, has the lowest concentration of hydronium ions. The acetic acid solution has more hydronium ions, but not as many as those found in the hydrogen chloride solution.

31. The pH of the rain has been decreasing (becoming more acidic) with increasing atmospheric concentrations of carbon dioxide. In order of decreasing pH: a > b > c.

33. A salt is the *ionic compound* produced from the reaction of an acid and a base. Water is a covalent compound.

35. It acts as a base. The negative charges on the oxygen atoms are lone pairs, which can accept hydrogen ions. Dissolved in water, the phosphate ion pulls hydrogen ions from the water, thereby creating hydroxide ions, which makes the solution alkaline. The phosphate ion behaves as a base.

37. According to the Lewis definition, the water is behaving as a base because of the positive seeking action of its lone pair of electrons. The aluminum ion is behaving as an acid because of its negative seeking action.

39. The positive sodium ion of sodium hypochlorite combines with the negative chloride ion of hydrochloric acid. Meanwhile, the negative hypochlorite ion combines with the positive hydrogen ion to form hypochlorous acid, HOCl. This is the answer you should give based upon the concepts presented in this chapter. By the way, the hypochlorous acid continues to react with hydrogen chloride to form water and poisonous chlorine gas, Cl_2, which is why bleach and toilet bowl cleaner should NEVER be mixed together.

41. These molecules behave as acids by losing the hydrogen ion from the oxygen atom, which takes on a negative charge. In hypochlorous acid, the chlorine atom pulls the negative charge toward it because of its great electronegativity. This helps to alleviate the oxygen of this negative charge. In other words, the negative charge is spread out over a greater number of atoms, which facilitates the formation of the negative charge. The hypochlorous acid, therefore, is more acidic.

43. As can be deduced from their relative positions in the periodic table, the fluorine atom is much smaller than the iodine atom. The atoms of the H—F bond, therefore, are closer together than are the atoms within the H—I bond. When it comes to electric attractions, closeness wins. Thus, the H—F bond is stronger.

45. As given in the answer to question 43, the H—I bond is weaker; so it is easier to break. This, in turn, means the H—I molecule more readily splits apart to form the hydrogen ion. The H—I, therefore, is a stronger acid. Notably, upon these molecules behaving as acids, the resulting fluoride ion, F^-, is better able to accommodate a negative charge than the resulting iodide ion, I^- (because of fluorine's greater effective nuclear charge). This would favor the H—F being the stronger acid. However, experiments show that the ease of bond breaking is a more significant factor in determining bond strength.

47. The carbonate ion has twice the negative charge as the bicarbonate ion. This makes the carbonate ion more likely to accept the positively charged hydrogen ion. Also, the carbonate ion is a stronger base because it does not "comfortably" accommodate the two negatively charged oxygens. Only one of these oxygens is able to "flip-flop" (spread out) its negative charge to the double bonded oxygen (see question 46). The carbonate ion, therefore, readily accepts a hydrogen ion to help "alleviate" this bounty of negative charge. The negative charge on the oxygen of the bicarbonate ion, however, is accommodated because it can be "flip-flopped" to the adjacent double-bonded oxygen. Its tendency to accept a hydrogen ion, therefore, is less.

49. a. pH is a measure of the hydronium ion concentration. The greater the hydronium ion concentration, the lower the pH. According to the information given in this exercise, as water warms, the hydronium ion concentration increases, albeit only slightly. Thus, pure water that is hot has a slightly lower pH than pure water that is cold.

b. As water warms up, the hydronium ion concentration increases, but so does the hydroxide concentration—and by the same amount. Thus, the pH decreases, yet the solution remains neutral because the hydronium and hydroxide ion concentrations are still equal. At 40°C, for example, the hydronium and hydroxide ion concentrations of pure water are both equal to 1.71×10^{-7} moles per liter (the square root of K_w). The pH of this solution is the minus log of this number, which is 6.77. This is why most pH meters need to be adjusted for the temperature of the solution being measured. Except for this exercise, which probes your powers of analytical thinking, this textbook ignores the slight role temperature plays in pH. Unless noted otherwise, please continue to assume that K_w is a constant 1.0×10^{-14}—in other words, assume that the solution being measured is at 24°C.

51. The sum of the pH and pOH of a solution equals 14. For example, if the pH is 7, then the pOH is also 7. Similarly, if the pH is 5, then the pOH must be 9. Interestingly, the sum of the pH and the pOH is equal to the negative log of K_w, 10^{-14}, which is 14.

53. Yes, provided the added acidic solution is less concentrated with hydronium ions. For example, a 1 M solution of HCl will become less acidic if you add a 0.01 M solution of HCl to it. Both of these solutions are acidic, but the 1 M solution is more acidic than the 0.01 M solution. The 0.01 M solution has the effect of diluting the 1 M solution.

55. The carbon dioxide in your breath reacts with the water to produce carbonic acid, which lowers the pH of the water. If you understand why rainwater is naturally acidic, you'll understand why your breath also is naturally acidic.

57. The hydrogen chloride, which behaves as an acid, reacts with the ammonia, which behaves as a base, to form ammonium chloride. The concentration of ammonium chloride in this system, therefore, increases while the concentration of ammonia decreases.

59. While you are alive, your cells continually produce carbon dioxide. This carbon dioxide is released into your bloodstream, where it is brought to your lungs so that it can be expelled into your breath. As you hold your breath, the amount of CO_2 in your blood increases, causing an increase in the carbonic acid in your body, and the pH of your blood decreases.

61. Beach sand from the Caribbean and many other tropical climates is made primarily of the calcium carbonate remains of coral and shelled creatures. Vinegar is an acid, and the calcium carbonate is a base. The reaction between the two results in the formation of carbon dioxide, which creates bubbles as it is formed. California beach sand comes primarily from the erosion of rocks and minerals, which are made mostly of inert silicon dioxide, SiO_2.

63. The silicon dioxide of granite is inert. The calcium carbonate of limestone, however, reacts with the acids of acid rain and neutralizes them.

65. When fossil fuels burn, they react with the oxygen, O_2, in the air to form gaseous carbon dioxide, CO_2, and water vapor, H_2O. The carbon dioxide reacts with atmospheric moisture to form carbonic acid, which lowers the pH of rainwater. As the acidic rainwater precipitates into the ocean, it neutralizes alkaline ocean salts such as calcium carbonate. This, in turn, lowers the pH of the ocean.

67. Acid rain is typically localized to regions downwind of heavy industry that produces sulfur and nitrogen oxide. Ocean acidification, by contrast, results primarily from the release of carbon dioxide, which is produced in much greater quantities and by a much broader range of sources. Both acid rain and ocean acidification are significant problems. Ocean acidification, however, is more global in scope and therefore should be of greater concern.

69. Is it fair that developed nations mandate through international treaties that developing nations not produce as much carbon dioxide as the developed nations have produced? If the developed nations were allowed to emit all that pollution, why couldn't the developing nations do the same? Imagine corporations, whose primary goal is to make profits, having this same mentality. Yes, countries and industries have the capability to self-regulate, but forces being what they are, this self-regulation is not always ideal for everyone. This is one of the main responsibilities of government, which, ideally, is there to represent the collective will. This gives consumers two voices: one with their pocketbook and the other in the voting booth.

CHAPTER 11: Oxidations and Reductions Charge the World

1. The elements in the upper right of the periodic table (except for noble gases) have the greatest tendency to behave as oxidizing agents.

3. $K \rightarrow K^+ + 1e^-$

5. Electrochemistry is the study of the relationship between electric energy and chemical change.

7. Reduction occurs at the cathode. Remember the "red cat."

9. The lithium donates electrons.

11. Fuel cells also produce the chemical products of the electricity-generating chemical reactions. A hydrogen fuel cell, for example, produces clean water suitable for drinking.

13. Electrons migrate from *n*-type silicon to *p*-type silicon across the junction when these two slices of silicon are pressed close together.

15. Electrolysis uses electric energy to produce a chemical change, whereas a battery uses a spontaneous chemical change to produce electric energy. Both involve oxidation-reduction reactions.

17. Metals located toward the lower left corner of the periodic table, which have the greatest tendency to lose electrons, are the most difficult metals to recover from metallic compounds.

19. Copper (II) sulfide and iron (III) oxide are mixed with limestone and sand and are heated. The iron (III) oxide dissolves in the $CaSiO_3$ formed, and the copper sulfide melts and goes to the bottom of the furnace, where it is removed.

21. After oxidation, both zinc and aluminum form water-insoluble oxidized coats, which are impervious to penetration by air and moisture, that prevent further oxidation.

23. The oxygen atoms become somewhat negatively charged, which means they are gaining electrons.

25. As described in the Hands-On Chemistry activity at the beginning of this chapter, old copper pennies turn dull as the surface copper reacts with atmospheric oxygen to form copper oxides. Dipping the old penny in the salt-vinegar solution causes the copper oxides to react with the acetic acid of the vinegar to form copper acetate, $Cu(CH_3CO_2)_2$. When the copper acetate is exposed to the carbon dioxide (which turns into carbonic acid) of the seltzer water, the blue-green copper carbonate, $CuCO_3$, is formed. Copper roofs tend to turn blue-green as they react with the carbon dioxide of the air and carbonic acid of the rain. Similarly, a penny left in the rain for a long time will turn blue-green.

27. The energy of the battery is causing the electrolysis of water, which generates one volume of oxygen for every two volumes of hydrogen. Note that a greater volume of gas is arising from the negative terminal. Therefore, oxygen gas, O_2, arises from the positive terminal and hydrogen gas, H_2, arises from the negative terminal. Without the pencils, the conditions are not good for the formation of oxygen gas. Instead, hydroxide ions, OH^-, are formed at the positive terminal, and this turns the phenolphthalein pH indicator pink. The battery is quickly ruined because placing it in the conducting liquid short-circuits the terminals, which results in a large drain on the battery.

29. Use the techniques of Section 9.3 to solve this problem. First, convert metric tons of aluminum to grams of aluminum and then to moles of aluminum. Then follow the coefficients of the balanced equation to calculate the moles of carbon dioxide formed. Finally, convert from moles of carbon dioxide to grams of carbon dioxide to metric tons of carbon dioxide:

(1.6 × 10^{13} mt Al)(1000 kg/1 mt)(1000 g/1 kg)
(1 mole Al/27.0 g Al)(3 moles CO_2/4 moles Al)
(44 g CO_2/1 mole CO_2)(1 kg/1000 g)(1 mt/1000 kg)
$$= 2.0 \times 10^{13} \text{ mt of } CO_2$$

31. This reaction is endothermic, requiring the input of electric energy, which is typically produced by the burning of carbon dioxide, which produces fossil fuels.

33. A reducing agent causes other materials to gain electrons. It does so by its tendency to lose electrons. Atoms with low electronegativity tend to lose electrons easily; therefore, they also behave as strong reducing agents. Sodium, Na, has the weakest electronegativity, which means it behaves as the strongest reducing agent of these three elements. (However, it is the weakest oxidizing agent of these three elements). Sulfur, S, is in between, and chlorine, with its greatest electronegativity, behaves as the weakest reducing agent.

35. The iodine atoms, I, gain electrons and are reduced; the bromine ions, Br^-, lose electrons and are oxidized.

37. The sulfur is oxidized as it gains oxygen atoms to form sulfur dioxide.

39. The atom that loses an electron and thus gains a positive charge (the red one) was oxidized.

41. The one that gains a negative charge (the red one) is more likely located closer to the upper right-hand corner of the periodic table. An example is sodium chloride, NaCl, where the sodium (on the left side of the periodic table) loses an electron to acquire a positive charge, Na^+. The chlorine (on the right side of the periodic table) gains an electron to acquire a negative charge, Cl^-.

43. There is a lower ratio of hydrogen atoms in the acetaldehyde product, which tells us that the grain alcohol is being oxidized.

45. The sulfur is losing hydrogen, so it is being oxidized. An oxidizing agent, therefore, is needed. A dilute solution of hydrogen peroxide, H_2O_2, is typically used.

47. Recognize that there is a total charge of 2+ with the reactants, but a total charge of 1+ with the products, which is not balanced. To balance the charges, we need two silver ions, Ag^+. But to account for two silver ions, we also need two silver atoms, Ag, as follows:

$$Sn^{2+} + 2 Ag \rightarrow Sn + Ag^+$$

49. The first step is to balance the atoms by showing two chloride ions, Cl^-, before the arrow:

$$Ce^{4+} + 2 Cl^- \rightarrow Ce^{3+} + Cl_2$$

This brings the total charge of the reactants to 2+, while the total charge of the products is 3+, which is not balanced. To balance the charges, we need two Ce^{4+} ions and two Ce^{3+} ions as follows:

$$2 Ce^{4+} + 2 Cl^- \rightarrow 2 Ce^{3+} + Cl_2$$

51. The anode is where oxidation takes place, and free-roaming electrons are generated. The negative sign at the anode of a battery indicates that this electrode is the source of negatively charged electrons. They run from the anode through an external circuit to the cathode, which bears a positive charge to which the electrons are attracted (see Figure 11.7).

Interestingly, when a battery is recharged, energy is used to force the electrons to move in the opposite direction. In other words, during recharging, the electrons move from the positive electrode to the negative electrode—a place they would never go without the input of energy. Electrons are thus gained at the negative electrode, which is now classified as the cathode because the cathode is where reduction occurs and the gain of electrons is reduction.

53. The fuel-cell power plant produces electricity with greater energy efficiency while also producing fewer pollutants. Built on a smaller scale, the fuel-cell power plant operates on-site, which means there is no need for long transmission lines. Significantly, the heat generated by the fuel-cell power plant is immediately available for heating the water and ventilation systems. This is in contrast to a coal-fired power plant, where excess heat is lost to the environment.

55. A liquid fuel is much denser than a gaseous fuel, which means it occupies much less volume. If the goal is to have the fuel cell miniaturized, then the denser liquid fuel is the way to go. Perhaps the miniaturized fuel cell could tap into a reservoir of gaseous fuel from an adjacent compact bundle of nanofibers. Liquid fuels, however, are handled more conveniently.

57. Electrons dislodged by light in the *n*-type silicon don't cross the junction because the junction is unidirectional. Electrons can only move from the *p*-type to the *n*-type silicon through the junction.

59. The electrolysis of brine (salt water) yields chlorine gas, Cl_2, in addition to sodium hydroxide, NaOH, and hydrogen gas, H_2.

$$2\,NaCl\,(aq) + 2\,H_2O \rightarrow 2\,NaOH\,(aq) + Cl_2\,(g) + 2\,H_2\,(g)$$

61. The active ingredient contains chlorine atoms, which behave as strong oxidizing agents.

63. If copper has a greater tendency to become reduced compared to iron, electrons will preferentially flow from the iron to the copper (and then onto oxygen as indicated in the answer to question 25). Corrosion is thus accelerated around the interface of these two metals. This was one of the prime reasons the Statue of Liberty required a full restoration in 1976.

65. Oxygen, O_2, is blown into the molten cast metal. This causes impurities, such as compounds of phosphorus and sulfur, to become oxidized. These oxidized impurities are less dense than the molten cast metal, which means they float to the surface, where they can be skimmed.

67. The oxygen is chemically bound to hydrogen atoms to make water, which is completely different from oxygen, O_2, which is required for combustion. Another way to phrase an answer to this question would be to say that the oxygen in water is already "reduced" in the sense that it has gained electrons from the hydrogen atoms to which it is attached. Being already reduced, this oxygen atom no longer has a great attraction for additional electrons.

69. The body must oxidize the elemental iron, Fe, into the iron 2+ ion, Fe^{2+}.

71. One of the products of combustion is water vapor.

73. The digestion and subsequent metabolism of foods and drugs tend to make the molecules of foods and drugs more polar. Oxidation is one of the ways the body does this.

75. The volume of the balloon shrinks by about 20 percent because the iron of the steel wool is absorbing the oxygen to form rust.

77. The air we exhale is somewhat more massive than the air we inhale because it contains a greater proportion of the heavier carbon dioxide molecules. Interestingly, the added water vapor in our exhaled breath is less massive than the oxygen it replaced, but not enough to counteract the much greater mass of the carbon dioxide. Breathing causes you to lose weight.

79. With a growing population comes growing energy needs. How should these needs be met? With coal-fired power plants and more nuclear power plants? Both require centralized

infrastructure of power lines, transformers, utility meters and more. Where will the money come from to build this infrastructure? What about the pollution generated by these power sources? Consider the alternative of a decentralized power where each home is equipped with its own source of power in the form of environmentally friendly fuel cells, photovoltaics, or wind turbines. For more on sustainable energy sources, study Chapter 17.

CHAPTER 12: Organic Compounds

1. Structural isomers have different arrangements of carbon atoms, but the number of carbon atoms each has is the same.

3. Boiling points of hydrocarbons are used for fractional distillation.

5. Saturated hydrocarbons have only single bonds. Unsaturated hydrocarbons have multiple bonds.

7. Aromatic compounds contain a six-membered ring system called a benzene ring.

9. Heteroatoms give an organic molecule its "character." They are typically electronegative and affect the physical and electrical properties of the molecule.

11. An alcohol simply contains a hydroxyl functional group, but a phenol has a hydroxyl group attached to a benzene ring.

13. Nitrogen is found in all amines.

15. Alkaloids are alkaline amines found in nature, although they can also be synthesized in the laboratory.

17. Both contain carbonyl groups. Ketones have the carbonyl carbon bonded to two adjacent carbon atoms, but aldehydes have the carbonyl carbon bonded either to one hydrogen and one carbon or to two hydrogens.

19. The quantities of the molecule available in nature may be insufficient to allow for its use by many people.

21. One of the bonds in the multiple bond of a monomer opens to form a new bond with a neighboring monomer molecule.

23. The chlorine atoms of the polyvinylidene chloride are good at forming induced dipole–induced dipole molecular interactions as was discussed in Section 7.1.

25. Teflon® was used to line the valves and ducts of an apparatus they were building to make nuclear weapons.

27. PETE has a T_g of around 69°C, which is why a PETE 2-liter bottle deforms so easily when placed in boiling water. Zipper seal sandwich bags are also commonly made of PETE. People who wash the bags in hot water (rather than throw them away) will note that the bags are much softer and more flexible in hot water than at room temperature. The T_g of polystyrene is around 100°C. This is good news for polystyrene (Styrofoam™) coffee cups, which otherwise would not be able to hold hot water very well.

29. Compound c, C_6H_{12}, has more hydrogens than compound b, C_6H_{10}, which has more hydrogens than compound a, C_6H_8.

31. Recall that the hydroxyl group, −OH, is very polar. The more hydroxyl groups a compound has, the greater its polarity, which means the greater its solubility in water. Compound a is nonpolar and insoluble in water. Compound b has a single hydroxyl group, which makes this compound somewhat soluble in water. Compound c has two hydroxyl groups, which makes it very soluble in water.

33. The order of least to most oxidized is b < a < d < c, where c is most oxidized. Note that this was the order of their presentation in the chapter. The most reduced hydrocarbons were introduced first, followed by the alcohols, followed by the aldehydes, followed by the carboxylic acids.

35. Because of greater induced dipole–induced dipole molecular attractions.

37. Only two structural isomers are shown. Actually, the one in the middle and the one on the right are two conformations of the same isomer.

39. To make it to the top of the fractionating column, a substance must remain in the gaseous phase. Only substances with very low boiling points, such as methane (bp −160°C) are able to make it to the top. According to Figure 12.3, gasoline travels higher than kerosene; so it must have a lower boiling point. Kerosene, therefore, has the higher boiling point.

41. The percent carbon increases as the hydrocarbon gets bigger. Methane's percent carbon is 20%; ethane, 25%; propane, 27%; and butane, 29%.

43. First, find the longest chain of carbon atoms. Notice that this is a five-carbon chain, which is indicated by the *pent-* prefix of *pentane* (*pent-* means 5, as in a pentagon, which is a polygon with five sides). Also notice that the methyl group comes off the second carbon of this chain, which is what the *2* of 2-methyl-pentane indicates. Accordingly, 3-methyl-pentane has the methyl group coming off the third carbon. In organic chemistry, the names of the molecules generally describe their structures.

3-methyl-pentane

45. A is 1,3-cyclohexadiene; B is cyclohexene; C is cyclohexane.

47. (a) C_7H_{16} (b) C_7H_{12} (c) $C_4H_{10}O$ (d) C_4H_8O

49. The first molecule on the far left need not be identified as a cis or trans isomer. The reason for this is because it has two methyl groups attached to the first carbon of the double bond. If the ethyl group attached to the second carbon of the double bond were to flip to the upper side, then we would still have the same molecule.

51. Both of these functional groups are acidic.

53. Replace the hydrogen of a hydroxyl group with a carbon and you have an ether. Similarly, replace the hydrogen of a carboxylic acid with a carbon and you have an ester. Therefore, an ether is to an alcohol as an ester is to a carboxylic acid.

55. The 80 proof vodka is 40 percent ethanol by volume, and hence 60 percent water.

57. Like many natural oils derived from fats, the carbons within cetyl alcohol are arranged in sequence with no branching.

59. Three ethanols surrounding a central nitrogen atom make triethanolamine.

61.

63. No. This label indicates that the decongestant contains the hydrogen chloride salt of phenylephrine but no acidic hydrogen chloride. This organic salt is as different from hydrogen chloride as sodium chloride (table salt) is, which may also go by the name "the hydrogen chloride salt of sodium." Think of it this way: Assume you have a cousin named George. Now you may be George's cousin, but in no way are you George. In a similar fashion, the hydrogen chloride salt of phenylephrine is made using hydrogen chloride, but it is in no way hydrogen chloride. A chemical substance is uniquely different from the elements or compounds from which it is made.

65. 1. ether
2. amide
3. ester
4. amide
5. alcohol
6. aldehyde
7. amine
8. ether
9. ketone

67. The transformation of benzaldehyde to benzoic acid is an oxidation.

69. EDTA is used to help remove lead from people, usually children, suffering from lead poisoning.

71. The female elm bark beetle produces multistriatin, which is a pheromone that attracts the male elm bark beetle. This male beetle, however, carries a fungus that invades and kills the elm tree. Synthetically produced multistriatin can be used to lure the male beetle into traps, thereby protecting the elm tree from disease.

73. It is easier to piece together two large fragments of a molecule than to piece together many tinier fragments. In a retrosynthetic analysis, therefore, it makes the most sense to divide the molecule into the two largest portions possible, which for this molecule means breaking the bond adjacent to the five-membered ring. A working organic chemist would understand how these two fragments could be obtained from the commercially available compounds cyclopentanone and 3-methylbutanal.

cyclopentanone   3-methylbutanal

75. Polypropylene consists of a polyethylene backbone with methyl groups attached to every other carbon atom. This side group interferes with the close packing that could otherwise occur among the molecules. As a consequence, we find that polypropylene is less dense than polyethylene—even low-density polyethylene.

77. Note the similarities between the structure of SBR and polyethylene and polystyrene, all of which possess no heteroatoms. SBR is an addition polymer made from the monomers 1,3-butadiene and styrene mixed together in a 3:1 ratio. Note that the butadiene monomer has two double bonds. One of them is used up in the polymerization. The other one remains after the polymerization. Interestingly, SBR is the key ingredient that allows the formation of bubbles when chewing bubble gum.

79. Initially, the beta-carotene structure looks rather complex. Upon careful examination, however, we find that this molecule is simply the result of the joining together of eight smaller isoprene units. Similarly, many of the molecules you've been studying in this chapter may have looked rather intimidating initially. With a basic understanding of the concepts of chemistry, however, you'll find that you already have a better insight into their properties.

81. Both cellophane and celluloid are produced from cellulose, which is a plant material. Cotton, for example, is a form of pure cellulose. These materials differ in how the cellulose is treated to make the product. For cellophane, the cellulose is treated with concentrated sodium hydroxide followed by carbon disulfide to make a syrupy solution from which the cellophane can be extruded. For celluloid, the cellulose is treated with nitric and sulfuric acids and then made pliable by adding camphor, which serves as a "plasticizer," an agent that makes an otherwise brittle material become soft and supple.

83. The HCl would react with the free base to form the water-soluble but diethyl ether-insoluble hydrochloric acid salt of caffeine. With no water available to dissolve this material, it precipitates out of the diethyl ether as a solid that may be collected by filtration.

85. These are important considerations. Answers may vary.

CHAPTER 13: Nutrients of Life

1. The cell nucleus is within the cytoplasm.

3. Not all carbohydrates can be digested by humans. Cellulose, for example, is a carbohydrate that is not digestible. Many adults also lack the ability to digest the carbohydrate lactose.

5. Starches and cellulose have glucose in common.

7. A triglyceride consists of a glycerol molecule attached to three fatty acid molecules.

9. All steroids have a system of four linked carbon rings.

11. All are made of amino acids.

13. Molecular attractions, including hydrogen bonding and induced dipole–induced dipole attractions, hold a substrate to the receptor site.

15. DNA is found mainly in the cell nucleus.

17. The DNA double helix unwinds, and each strand becomes a template for the formation of new complementary strands. The end result is that one DNA double helix is replicated into two identical DNA double helices.

19. Vegetables boiled in water lose their water-soluble vitamins.

21. Anabolism results in the synthesis of large biomolecules.

23. No. Cellulose is an insoluble fiber. Dietary fibers may also be made of soluble fibers, which are found in oats, barley, and legumes, that slow digestion, thus stabilizing blood glucose levels.

25. It is more energy efficient for the body to obtain these amino acids from other living organisms. The reason is the essential amino acids have side groups that are biochemically difficult to synthesize.

27. Two layers should form. The bottom layer is the denser water, and the top layer is the fat. With the chilled lipid layers, you can assume that, in general, the more solid the sample, the higher its proportion of saturated fats.

As you should have discovered from this activity, the "light" brands of spread contain fewer calories simply because they contain a greater proportion of water. Rather than water, some brands whip air into the spread. Either way, the net result is fewer lipid molecules per serving, which for saturated fats, is not a bad deal. Note that many of the "light" brands are labeled "for spread purposes only, not for cooking." One of the reasons is the excess amount of water they contain.

29. Looking at the side chains of these amino acids from Figure 13.16 and reviewing the discussion of functional groups in Chapter 12: (most acidic) tyrosine > phenylalanine > histidine (least acidic, which is the same as most basic)

31. Nucleotides come together to make a nucleic acid. A gene is a small portion of the nucleic acid sequence of nucleotides. In order of increasing size: nucleotide < gene < nucleic acid.

33. The plasma membrane is made primarily of lipids. How might you know this? Review the concepts of Section 7.5 and then look carefully at questions 58 and 59 at the end of Chapter 7.

35. A carbohydrate is made from water and carbon dioxide, but in no way does it "contain" those two materials. Recall from Chapter 2 that a chemical product such as a carbohydrate is uniquely different from the chemical reactants that were used to make that product.

37. Both cellulose and starch are polymers of glucose. They differ in the way the glucose units are linked together. In cellulose, the beta linkages result in linear polymers that strongly align with each other, giving rise to a tough material useful for structural purposes in plants. In starch, the alpha linkages permit the formation of alpha helices.

39. Enzymes in your mouth break down the starch into individual glucose units, which are sweet.

41. Lipids are made primarily of nonpolar hydrocarbons. Therefore, the lipids do not dissolve in water because they cannot compete with the strong attraction water molecules have for themselves.

43. Technically, a fat molecule is synonymous with a triglyceride. Triglycerides are the fats that are commonly part of one's diet. So if a product contains no triglycerides, there may be some legitimacy in saying that it is "free of fat molecules." Once fat molecules are digested, however, they have been broken down into fatty acids and glycerol molecules. This product, therefore, would still offer your body the same type of molecules that any other type of food does along with the same number of Calories.

45. Your customer's hair is fine because each strand is thin. Because each strand is thin, there is less mass per strand. Because it has less mass per strand, each strand is made of fewer amino acids, which includes fewer cysteine amino acids and disulfide cross-linkings. Adding a concentrated or even a regular strength reducing agent might cause his hair to fall apart completely. Yikes! You should dilute the reducing agent prior to application. You might even skip the reducing step altogether and go directly to the disulfide bond-forming oxidizing step.

47. Location a: dipole–dipole attractions (hydrogen bonding); location b: disulfide bonds; location c: induced dipole–induced dipole (hydrophobic attractions). The primary structure of this protein is the sequence of amino acids that make up the protein. The secondary structure includes the three alpha helix regions (the coils) and the one pleated sheet region (the zigzags). The tertiary structure is the overall shape of the entire polypeptide chain. No quaternary structure is observed because this polypeptide is not associated with an adjacent polypeptide.

49. The two common secondary structures are alpha helices and pleated sheets. Two common tertiary structures are globular proteins and fibrous proteins.

51. There are at least 20 different types of amino acids, which is many more than the different types of nucleotides. With a greater variety of monomers, proteins seemed to offer a better potential for holding a greater wealth of genetic information. Many scientists were surprised to learn of Watson and Crick's structural model for DNA and their proposal to explain how such an elegant structure, with only four nucleotides, could be the holder of our genetic identity.

53. Cytosine with its amine group transformed into an amide is uracil. If uracil were present as a normal component of DNA, it would not be distinguished from those generated by cytosine degradation. The repair enzymes would transform all the uracils—those that came from cytosine and those that didn't— into cytosine, thereby creating only three nucleotides in DNA rather than four. The methyl group on thymine is apparently the "label" that tells the repair enzymes to leave it alone, thus preventing potentially fatal loss of genetic information.

55. Water readily passes through our body, which is itself made primarily of water. Excess quantities of the water-soluble vitamins, therefore, are readily excreted. Water-insoluble vitamins, by contrast, tend to build up in fatty tissues where they can remain for extended periods of time.

57. You should tell your friend that her body needs and will absorb only a certain amount of vitamin C every day. If she loads up with excess amounts only once a week, her body has no way of storing this excess amount. Instead, the excess amount will be excreted, leaving her potentially low on this vitamin for days at a time.

59. All the phosphate groups are negatively charged, which means they repel each other. It is the electric force of repulsion between these phosphate groups that causes the rapid acceleration of the terminal phosphate group when cleaved. In this sense, the ATP molecule is like a compressed spring, which is why it is considered to be a highly energetic molecule.

61. Every muscle in your body is a source of amino acids for yourself (as well as for any organisms that might end up eating you). In times of starvation, your body will access this source of amino acids, resulting in a decrease in muscle mass.

63. Sucrose is a dissaccharide and must be broken into its two saccharide units—only one of which is glucose—before it may be used by the body.

65. You can experience a significant increase in blood glucose by eating a lot of food that is low on the glycemic index. Both the quality and quantity of the food you eat is important.

67. Just add milk and your breakfast of cold cereal will be fully balanced.

69. People's metabolisms vary. A diet that works for one person will not necessarily work for another person no matter how closely the two individuals stick to the regimen. It's interesting that our bodies adapt to our diets. For example, people who have difficulty digesting dairy products can eat more dairy products simply by introducing more dairy products slowly over time. This allows bacterial colonies within a person's gut to adjust to the new conditions. Changing one's diet is easiest and usually most effective when done gradually over time. A person shouldn't start eating large quantities of protein and fat. It is better to slowly build up to larger portions of these foods. Regarding high-protein diets, there is still much debate regarding their risks and benefits. A person should consult a nutritionist or physician. Regarding exercise, remember that seven days without exercise makes one weak!

CHAPTER 14: Medicinal Chemistry

1. Drugs originate from plants or animals or are synthesized.

3. When one drug enhances the effect of another, it is called a synergistic effect.

5. Intermolecular attractions such as hydrogen bonding hold a drug to its receptor site.

7. Penicillin G inhibits the growth of bacterial colonies by binding to receptor sites that the bacteria need for building their cell walls.

9. Methotrexate interferes with the cancer cell's metabolism by binding to the dihydrofolic acid receptor site.

11. Maintenance neurons are responsible for digestion, intestinal muscle function, better vision, and heart rate maintenance.

13. GABA inhibits the output of nerve impulses.

15. Alcohol and benzodiazepines affect the action of GABA.

17. LSD mimics serotonin.

19. Analgesics enhance our ability to tolerate pain without abolishing nerve sensations.

21. Endorphines are thought to be responsible for the placebo effect.

23. Nitrogen oxide, NO, is the metabolic product of vasodilators, which relax muscles found within blood vessels.

25. Because molecules slow down with decreasing temperature, the rate at which neurotransmitters diffuse across the synaptic cleft decreases with temperature. This is one of the reasons cold-blooded animals become sluggish at colder temperatures. As neurotransmitters take longer to diffuse across the synaptic cleft, the rate at which nerve signals reach target muscles slows down.

When outside in the cold without adequate clothing, you may find your extremities becoming numb and your muscles sluggish. This isn't just because of a decrease in the rate of diffusion in your synaptic clefts. Your body responds to cold by diverting blood from your extremities to your internal organs. The speed of neuron transmission, however, depends on blood supply. As the blood supply diminishes, neurons lose out on needed oxygen; thus, nutrients begin to shut down. The result is a numbing effect or loss of muscle control. An ice pack applied to an injury, such as a sprained ankle, puts this principle to good use. The same thing happens to your foot after you've been sitting on it for too long and it "goes to sleep."

27. The least ideal place to throw away your expired medicines is the toilet because it places your medicines in water systems that usually lead to the environment. Note that waste treatment facilities are not equipped to remove pharmaceuticals, many of which end up in people's drinking water. Throwing away your medicines with solid waste is a bit safer, but rain passing through waste disposal sites leach these pharmaceuticals into groundwater. The ideal method of disposal is to take your expired medicines to a pharmacist or a hospital; each has the means of treating the expired medicine as a hazardous waste.

29. In order of increasing size: $c < b < a$

31. It is the vast diversity of organic chemicals that permits the manufacture of the many different types of medicines needed to match the many different types of illnesses.

33. A drug overdose is most likely when two drugs with overlapping effects are taken at the same time. One of the most dangerous combinations of sedatives is with alcohol.

35. In synthesizing the natural product in the laboratory, the chemist is able to create closely related compounds that may have even greater medicinal properties. Also, synthesizing a natural product can be advantageous when the source of the natural product is rare and thus not readily available.

37. Whether a drug is isolated from nature or synthesized in the laboratory makes no difference with respect to "how good it may be for you." A multitude of natural products are harmful, just as many synthetic drugs are harmful. The effectiveness of a drug depends on its chemical structure, not its source.

39. (a) amphetamine (agonist) (b) morphine (agonist) (c) nicotine (agonist) (d) atorvastatin (antagonist) (e) LSD (agonist) (f) propanolol (antagonist)

41. These molds and fungi produce chemicals that inhibit the bacterium's ability to maintain its cell wall. Upon exposure to these chemicals, the cell wall ruptures and the bacterium dies. We can isolate these mold- and fungi-produced chemicals, called antibiotics, and use them as protection against bacteria.

43. Sulfa drugs are transformed into sulfanilamide, which binds to the PABA receptor site on an enzyme that bacteria use to produce folic acid, a vital nutrient. In this sense, the sulfanilamide "competes" with the PABA for the receptor site and is thus said to be an inhibitor (see the discussion on enzymes in Chapter 13). If PABA were provided along with the sulfa drug, there would be a greater chance that PABA, not the sulfanilamide, would bind first, thus helping the bacteria to produce the folic acid that they need. Formulating PABA with a sulfa drug, therefore, is likely to decrease its antibacterial properties.

45. Certain antiviral agents work by upsetting the virus's genetic mechanisms and thereby killing the virus. Certain anticancer agents work by upsetting the genetic mechanism of a cancer cell and thereby killing the cancer cell. Antiviral agents that exhibit anticancer activity are able to upset the genetic mechanisms of both viruses and cancer cells.

47. The main difference between stress and maintenance neurons is the neurotransmitters they produce. Stress neurons use neurotransmitters such as norepinephrine. Maintenance neurons produce neurotransmitters such as acetylcholine.

49. Look carefully at the structure of tyrosine in Figure 13.16 and you will see how this structure is similar to that of dopamine. Add another hydroxyl group to the benzene ring of tyrosine; then take away the carboxylic acid. You now have the structure of dopamine. Interestingly, the amino acid tyrosine is the starting material the body uses to produce dopamine.

51. The hydroxyl group must be added to dopamine to make norepinephrine. Using enzymes, this is, in fact, how the body creates norepinephrine.

53. Physical dependence involves physical symptoms such as depression, fatigue, and a desire to eat. Psychological dependence symptoms can exist long after the physical symptoms have disappeared and can lead to renewed drug-seeking behavior.

55. Many stimulant drugs work by blocking the reuptake of excitatory neurotransmitters. Trapped inside the synaptic cleft, these neurotransmitters are decomposed by enzymes. By the time the reuptake is no longer blocked by the drug, few neurotransmitters are left to be reabsorbed by the presynaptic neuron, which by this point is also deprived of neurotransmitters. With a lack of neurotransmitters, little communication takes place between neurons, which has the effect of making the drug user depressed.

57. Gardeners must exercise extreme care when using solutions of nicotine because nicotine is poisonous to humans.

59. The effective dose of MDA is many times greater than the effective dose of LSD. In other words, while MDA is taken by the milligram, LSD is taken by the microgram. Because so much of the MDA is taken, negative side effects are more likely to appear.

The workings of drug withdrawal, however, are quite complex. Consider the following: A coffee drinker gets a headache when he or she stops drinking coffee. However, as a side effect of the caffeine in the coffee, a headache results, caused by a dilation of blood vessels. Over time, the coffee drinker's body becomes accustomed to this dilation to the point that it is able to counteract the dilation by applying a force that causes constriction. When coffee is no longer consumed, the body doesn't know to stop counteracting the dilation. Blood vessels, therefore, become overly constricted, and that gives rise to a headache, which prompts the coffee drinker to drink more coffee.

Likewise, with repeated use of MDA, the body develops a tolerance to its side effects. Upon stopping the use of MDA, the body is still working hard to tolerate the drug. Thus, it is the body's own counteraction to the drug that drives the user to continue using MDA. It's a vicious cycle driven by some very real chemistry.

61. An excess of any of the neurotransmitters discussed in this chapter has the potential to throw you off balance and hurt your performance. But when it comes to a physical performance, the GABA neurotransmitter would most quickly hinder your coordination.

63. Recall that a molecule must have a certain structure in order to be able to bind to a receptor site. The fact that the structures of these general anesthetics have very little in common suggests that no specific structure is responsible for the general anesthetic effect and, furthermore, that there is no specific receptor site on which they are acting.

65. Halothane, $CF_3CHBrCl$, is a chlorofluorocarbon, which, as described in Chapter 9, has been banned because of the potential danger it poses to stratospheric ozone.

67. The cocaine addict's body is accustomed to a particular kind of high, which is stimulatory. After long-term use, the cocaine addict's body responds by building receptor sites that are depressive in nature to counteract the effect of the stimulation.

Upon withdrawal, the body seeks stimulation to balance the depression the addict is experiencing. Methadone provides only a depression of nerve signals, which is the opposite effect of what the cocaine addict is seeking. Thus, this decreases the chance that the cocaine addict will stick with a methadone maintenance program, which should be reserved strictly for opioid addiction. For a cocaine addict, cessation of all cocaine use, a healthy program of exercise, counseling, and comrade support would be a far better approach.

69. Beta blockers block norepinephrine and epinephrine from binding to beta-adrenoceptors and a calcium channel blocker inhibits the flow of calcium ions into muscles. Both slow down and relax the heart.

71. Many medicines are flushed down drains or toilets directly or after they have passed through the human body into the urine. These drugs end up in downstream water supplies. By the time they reach the downstream consumer, their concentration is at a relatively low level. But this level is easily measured. Perhaps of greatest concern is the effect these low doses will have over the long term. Drinking a single glass of drug-tainted water might not have a noticeable effect. But what if this water is consumed by an individual over his or her lifetime?

73. People would continue to smoke even if tobacco were prohibited. A black market would be established. The government would then need to spend billions of dollars to combat this black market. Prisons would become even more crowded because of the prohibition of many recreational drugs. There's a strong case to be made that educating people about the hazards of drugs and helping users to escape the grip of drugs provides the best long-term solution to any drug problem.

75. The moral, social, and economic responsibility of any person or any corporation is to provide assistance to others to the best of its abilities. A huge part of the success of the human species has come from our ability to work together and help each other in times of need. Life is tough. But when we stick together, we can do amazing things. An overpopulation of humans, however, provides a challenge.

CHAPTER 15: Optimizing Food Production

1. Carbohydrates and oxygen are the two major chemical products of photosynthesis.

3. The number of trophic levels is limited by a dwindling supply of food resources. Each higher trophic level supports successively smaller populations.

5. Terrestrial plants that exclude sodium ions have raised the oceans' levels of sodium ions to more than three times that of the potassium ions.

7. Sulfur is most commonly absorbed by plants as sulfate anions: SO_4^{2-}.

9. Fertile topsoil is composed of organic matter, mineral particles, water, and air.

11. Straight fertilizers contain only one nutrient, but mixed fertilizers contain a mixture of three essential nutrients.

13. Compared to mixed fertilizer, compost has a higher percentage of organic bulk. This keeps soil loose for aeration.

15. DDT was banned in the United States in the 1970s, but it is still used in other countries, primarily to control the spread of malaria-bearing mosquitoes.

17. Pesticides and fertilizers are washed away from fields into streams, rivers, ponds, and lakes to end up in our drinking water.

19. Irrigation is damaging to topsoil because irrigated water contains salts that are left behind once the water evaporates. This process raises the salinity of the soil.

21. Organic farming is farming without the use of synthetic pesticides and fertilizers.

23. No questions are asked.

25. $a > e > c > d > b$

27. $a < c < b < d$

29. The amount of biochemical energy decreases with each passing level because animals being consumed have already lost most of their biochemical energy through their metabolism.

31. This oxygen is not gaseous O_2, but oxygen atoms bound to the cellulose structure.

33. One reason earthworms increase soil fertility is that they help to make the soil more porous as they tunnel through it. The soil structure, therefore, might become more compact as the earthworms retreat from the regions of high nutrient concentration.

35. Nitrogen, N_2, is so chemically unreactive because the two atoms of this molecule are so strongly held to each other by a covalent triple bond. If you want to form a new compound from this material, you must break the two nitrogen atoms away from each other, which takes the energy of a lightning bolt to do.

37. This makes it easier for air to permeate the soil so that oxygen can get to the roots of the grass. Occasionally poking holes also helps the soil to remain loose enough that it can retain water.

39. Clay is a very compact material containing few open spaces. This means water cannot penetrate very well, which means the nutrients are not washed away.

41. The ammonium ion behaves as a weak acid (see Chapter 10) and would donate a hydrogen ion in alkaline soil to form ammonia, NH_3. Although ammonia is soluble in water, it is a gas at ambient temperatures and is thus readily lost to the atmosphere.

43. The plant is unable to distinguish between an ammonium ion from compost or an ammonium ion from synthetic fertilizer. An ammonium ion is an ammonium ion no matter where it is from. The main difference for the plant would be that the compost provides organic bulk, which would be an added benefit to the plant.

45. DDT has such a strong affinity for fat tissue because both it and fat tissue are highly nonpolar substances.

47. The 2,4,5-T is more acidic because of the carboxylic acid functional group it contains.

49. Irrigation inevitably leads to an increase in the salinity of the soil. Although flooding may remove many of the plant nutrients, it would benefit the cropland by removing much of the accumulated salt deposits.

51. Both organic farming and integrated crop management aim to produce crops in an environmentally friendly and sustainable manner. Organic farming, however, tends to avoid the use of technology so that there is a greater reliance on the human touch. Integrated crop management, however, tends to embrace sophisticated technology so that the crops can be produced in large amounts to meet the needs of our expanding human population.

53. Pheromones that insects detect as a warning signal may be used. This may cause the insects to alter their behavior so that they don't feed on the pheromone-treated crops.

55. For the individual, eating high on the food chain means having access to higher concentrations of nutrition. Consider that the Japanese started eating more beef after World War II. As a result, the average height of the Japanese individual increased by several inches. Eating this highly nutritious food, however, also comes with risks, which includes obesity and vulnerability to certain cancers. A well-balanced diet is generally easier to attain if a person focuses primarily on foods that are lower on the food chain. The energy and effort required to raise livestock is many times greater than that needed to grow food from photosynthetic plants. As discussed in this chapter, sufficient resources do not exist for everyone on this planet to live on a meat-intensive diet. The human population gains many benefits from eating primarily foods that are low on the food chain. These benefits include greatly reduced consumption of energy, particularly fossil fuels, and lower outputs of greenhouse gases due to lower fossil fuel consumption and beef consumption (a major producer of methane). Thus, grain that livestock don't eat is available for humans.

57. There are a significant number of alternatives to DDT for fighting mosquito populations. Compared to DDT, these alternatives, although currently more expensive, are just as effective and far less destructive to the environment.

59. If the activist can provide a working model of how to work the land in a more environmentally responsible manner yet still produce agricultural products that can be sold at a greater profit and this model can be sustained over many generations, the rancher is more likely to listen. Such a plan, however, might require an initial capital investment that the rancher cannot afford.

CHAPTER 16: Protecting Water and Air Resources

1. Most of the fresh water on our planet is found in polar ice caps and glaciers.

3. The water table can be seen aboveground in the form of lakes and streams.

5. 1950

7. Annual water use in the United States increased until 1980 and has decreased since then primarily due to improvements in agricultural irrigation methods.

9. Groundwater takes so long to rid itself of contaminants because it is inaccessible and the flow rate of many aquifers is slow.

11. Pathogens cannot pass through sand or underground sediments with small pore sizes.

13. Aerobic decomposition gives off carbon dioxide (CO_2), water (H_2O), nitrates (NO_3^-), and sulfates (SO_4^{2-}).

15. Raw sewage is first treated by screening out insoluble human waste products such as coffee grounds, tiny rocks, pebbles, and balls of oil.

17. A third level of wastewater treatment is expensive, and not all communities are in locations where this level of treatment is necessary.

19. The atmosphere is made up of nitrogen, oxygen, and small traces of argon and other materials such as carbon dioxide and water vapor.

21. Weather occurs in the troposphere.

23. The temperature increases as one goes higher in the stratosphere.

25. Temperature inversion is the reverse of the normal air flow where polluted warm air rises, carrying pollutants away from the surface. Instead, polluted cool air settles below a body of warm air, which traps the cool air and the pollution at the surface.

27. Ozone is useful when found in the stratosphere, where it protects us from harmful ultraviolet rays. Ozone is harmful when it enters the air we breathe here at the bottom of the troposphere.

29. A catalytic converter provides a catalyst to increase the efficiency of the combustion of gasoline and to lower the output of hydrocarbons.

31. Carbon dioxide

33. Scientists differ in their opinions because there are many variables and the outcome is uncertain.

35. Federal regulations mandate that new showerhead flow rates cannot exceed more than 9.5 liters per minute (2.5 gallons per minute). Shorter showers and shallower baths save energy as well as water.

37. The weight of the air pushing against the outer surface of the card is much greater than the weight of the water pushing against the inner surface of the card. This demonstration would not work on the Moon. Do you understand why?

39. $(1.18 \text{ g/L})(1 \text{ L})(1.00 \text{ mole}/28.8 \text{ grams}) = 0.0410$ moles of air

41. Use the ideal gas equation $PV = nRT$ and solve for n, which is the number of moles.

$$(1.0 \text{ atm})(1.0 \text{ L}) = n(0.0821 \text{ L atm/K mol})(298 \text{ K})$$

$$n = 0.041 \text{ mole}$$

43. b, a, c

45. $a > c > d > b$

47. The salts in the ocean do not evaporate along with the water, which is why the moisture that condenses from the sky is fresh water. The reason the salts do not evaporate is they have such a strong affinity for the liquid water to which they are strongly held by many ion–dipole attractions for each ion.

49. According to Figure 16.4, about four times as much water is used to generate electricity as is consumed by domestic and municipal water use combined.

51. Building a dam would allow for the formation of a reservoir that could be tapped in the dry summers. Unfortunately, dams are typically large and expensive projects. They also tend to trap valuable topsoil, which never makes it to downstream farmers. Water from the reservoir must then be piped to where the water is needed. A governmental allocation system would also be required. A far more economical and efficient method is to dig channels that guide monsoon rainwater to wells. Rather than running off to the ocean, the water is retained underground, where there is no loss due to evaporation. The groundwater can then be pumped out of the ground in the summer, using inexpensive hand or foot pumps. No money is spent building an expensive dam. The topsoil is maintained, and farmers are empowered with water right beneath their feet. Such practices have been employed in Bangladesh with much success.

53. The air molecules of the atmosphere are held down by the force of gravity.

55. Groundwater only moves slowly, which means any pollutants added to the groundwater will remain there and not be "flushed away" for quite some time. Furthermore, there is no way to clean polluted groundwater while it is in the ground. Our only choice would be to purifying that which we extract.

57. Phosphates are a nutrient for many plants and microorganisms. In the past, the phosphates from used laundry detergents often made their way into rivers, lakes, and ponds where they caused the growth of plants and microorganisms that grew so rampant, they choked off the natural supply of dissolved oxygen—a process known as eutrophication.

59. The decomposition of food by bacteria in our digestive system is primarily anaerobic simply because little oxygen makes it from our mouths to our intestines, where food decomposition takes place. As a consequence, gas that comes out the other end is frequently of the odoriferous sort. Composting toilets adds air to our wastes, which moves the mode of decomposition from anaerobic to less smelly aerobic microorganisms. It is the same reason fresh dog poop smells worse than week-old dog poop.

61. Our mouths are fairly good at discerning the tastes of residual components of drinking water—so much so that many of us are willing to pay the 1000 percent markup price that water bottlers charge for their products, which are only a fraction of a percentage purer than the water we can obtain from a local water utility. Because their purities are quite comparable, flushing toilets with municipal drinking water is about as wasteful as flushing it with bottled water. At the wastewater treatment facility, human waste is extracted from the water and typically ends up in a landfill. So why not use a composting toilet, skip the waste of water altogether, and send our wastes directly to farmlands rather than to landfills?

63. The pressure of the air on the outside of your eardrums is decreasing faster than the pressure of the air on the inside of your eardrums. Interestingly, commercial airplanes maintain a cabin pressure that is equal to the pressure one experiences at about 2400 meters (8000 feet) up a mountainside.

65. One reason is physical: The gravitational force on the Earth is not strong enough to prevent this relatively light molecule from escaping into space. The second reason is chemical: Hydrogen gas, H_2, is a reactive material that oxidizes when exposed to atmospheric oxygen, O_2, to form water, H_2O.

67. Bricks are solid, which means they are not very compressible. Increasing the weight on the brick, therefore, doesn't cause it to compress by any appreciable extent. The atmosphere, however, is made of a gas, which, like a foam brick, is compressible. Accordingly, the foam bricks at the bottom of the stack will be more squished (denser) than the bricks at the top because the bricks at the bottom bear the greatest weight. Similarly, the air close to the Earth's surface is more squished (denser) than the air high on a mountain because the air close to the surface bears the greatest weight from the air molecules above.

In order to fully understand this answer, you need to understand that air has weight. This may seem counterintuitive because as you hold out your empty hand, you do not experience the air's weight like you might experience the weight of an object, say, a water balloon. But air has weight because it is held down by gravity—if this was not the case, our air would have drifted off into outer space a long time ago. We don't experience the weight of air in a typical way because we are submerged within the air. This weight pushes against our bodies in the form of atmospheric pressure. This is similar to how we can experience the weight of water—not by holding the water, but by submerging ourselves in the water. In such a case, the weight of the water exerts a pressure against our bodies, which is particularly noticeable in our eardrums as we swim to the bottom of a pool.

69. Coal is a fossil fuel, which means it arises from the decomposition of organic matter. Sulfur from organic matter, in turn, is found primarily in the two amino acids—cysteine and methionine (see Figure 13.16). Plants obtain the sulfur to make these amino acids from the sulfur oxides that occur in the atmosphere and were placed there by active volcanoes and by the burning of fossil fuels. Thus, there is no original source of sulfur on the Earth. Rather, like all other elements, the sulfur that is already here recycles through various pathways. Sulfur does, however, have an ultimate origin, and that is with nuclear fusion (see Section 5.9).

71. The airborne sulfur dioxide reacts with oxygen and water to form sulfuric acid, which is carried by rainwater to the planet's surface.

73. Atmospheric nitrogen and oxygen will react with each other under conditions of extreme heat, such as what occurs in an automobile engine or by a lightning bolt (see Chapter 9). The balanced equation for this reaction is $N_2 + O_2 \rightarrow 2\,NO$.

75. The gasoline fumes that leak out from the filling process are not converted to photochemical smog at night. With a steady breeze, they'll be dispersed to other parts of the world by morning.

77. In the burning of forests, CO_2 is released directly into the air. Also, by this burning, an absorber of CO_2 is destroyed. Also, as the forests are destroyed, less water vapor is released into the air to make clouds.

79. Our need for water is great. A person can live weeks without food, but perish without water in a matter of days. People who market bottled water are taking advantage of the fundamental need for water.

81. Marketing, convenience, local water contamination. Perhaps the rise in bottled water use from 1990 to 2008 can be linked to the rapid growth of the economy. In a worsening economic climate, one can imagine the consumption of bottled water leveling off or perhaps decreasing. Once a product has been widely accepted, however, it is not likely to go away anytime soon, especially a product so essential to our health.

83. Perhaps the impulse toward conflict is deeply rooted in the human psyche. Perhaps locally specific conditions are more determinative of likely outcomes in a water management conflict, which is only one dynamic factor among many.

85. You might begin by acknowledging the consensus among scientists that greenhouse gas levels have been rising steadily since the industrial revolution and that this is having the effect of raising the average global temperature. You might also acknowledge that the consequences of such an increase or continued increase are not known with certainty, but the warning flags are there. We would be foolish not to take action that would benefit us even in the unlikely event global warming turns out to be a nonissue. You might continue further with emphasizing the fact that within every crisis is an opportunity and the oil industry is already well seated to lead the country with innovations. For example, the money spent on a new oil refinery might be spent on establishing a wind or solar farm instead. Rather than researching how to optimize fossil fuel yields, company chemists and engineers could build more efficient batteries. The gas station infrastructure the oil industry already owns could also start selling replaceable car batteries that are inserted into an electric vehicle by an automated system. People with gas cars would pull in to buy gasoline. People with electric cars would pull in to swap their spent battery for a fully charged battery. Does the oil industry really expect people still to be driving gasoline vehicles 20 years from now when electric vehicles offer so many more advantages? It would be wise to emphasize the importance of thinking long term.

CHAPTER 17: Capturing Energy

1. An electric current generates a magnetic field.

3. Electricity is a form of energy that requires a source.

5. About 450,000 miles

7. Coal is the filthiest fuel because it contains large proportions of impurities such as sulfur, toxic heavy metals, and radioactive isotopes.

9. Yes. Coal can be converted into cleaner burning fuels by treating it with pressurized steam and oxygen to produce hydrogen gas.

11. Methane, CH_4, is the main component of natural gas.

13. Most currently operating nuclear power plants were built before 1990.

15. The backup generators at the Fukushima power plant failed in March 2011 because they were placed in the lower levels of the power plant where they got flooded by the unusually high tsunami.

17. An ideal sustainable energy source is inexhaustible and environmentally benign.

19. Hawaii is well suited for generating electricity using OTEC technology because of the large differences in surface and deep water temperatures within a relatively short distance from the shore. Interestingly, the developers of OTEC technology in Hawaii are discovering that the "side benefits" of OTEC, such as the production of fresh water, air conditioning, and aquaculture, are proving more worthwhile than is the production of electricity.

21. Bays or estuaries may be closed in by dams, and when tidal waters flow in and out of the dam, they rotate a paddle wheel or turbine to generate electricity.

23. Biomass is a form of solar energy in that it uses photosynthesis to convert solar energy to chemical energy.

25. The United States is currently the world's leading producer of ethanol fuel.

27. Solar water heaters are painted black to absorb and retain sunlight.

29. Salt peter, which is a blend of sodium and potassium nitrates.

31. Hydrogen is an ideal fuel because it packs more energy by weight than any other chemical fuel.

33. Fuel cells are lighter than batteries and can produce water, which batteries cannot do. Fuel cells also may operate continuously, so long as fuel is provided.

35. First, calculate the cost of running the 100-watt lightbulb for one hour, understanding that 100 watts is the same as 0.1 kilowatts:

$$0.1 \text{ kW} \times 1 \text{ h} = 0.1 \text{ kWh}$$

$$0.1 \text{ kWh} \times 15 \text{ cents/kWh} = 1.5 \text{ cents}$$

Then calculate the cost of running the 10-watt LED lamp for one hour, understanding that 10 watts is the same as 0.01 kilowatts:

$$0.01 \text{ kWh} \times 1 \text{ h} = 0.01 \text{ kWh}$$

$$0.01 \text{ kWh} \times 15 \text{ cents/kWh} = 0.15 \text{ cents}$$

The savings for each hour is the difference: 1.5 cents minus 0.15 cents, which equals 1.35 cents per hour. This may not sound like much, but if 50 million households in the United States changed just one lightbulb from a 100-watt incandescent to a 10-watt LED, the total annual savings would be on the order of $1.35 billion.

37. $1.0 \text{ MW} = 1000 \text{ kW}$; $1000 \text{ kW}/350 \text{ homes} = 2.9 \text{ kW per home}$ (two significant figures)

39. $d < b < c < a$

41. The kilowatt is a measurement of the rate at which energy is delivered, which is power. If you have 2 kW available to your home, you can do more than if only 1 kW is available. In both cases, however, you can draw the same amount of energy. With the 2 kW rated system, however, you can do so twice as quickly. Notice that your electric bill is metered in units of kilowatt-hours, which is an amount of energy.

43. Once electric energy is created, it has to go somewhere. If it is blocked from traveling into one region, it will follow all the wires into any adjacent regions, which are already receiving electric energy from some other source. This causes a spike in electric energy, which may trip circuit breakers. This, in turn, results in a buildup of even more energy to all the remaining adjacent regions. The result is a widespread blackout. The power plants need to shut down as soon as possible to minimize the damage.

45. Fossil fuels are made from plants that lived millions of years ago. These plants, in turn, got their energy from the Sun through photosynthesis. Burning fossil fuels, therefore, is an indirect way of using solar energy.

47. In order to produce carbon dioxide, there must be carbon. Accordingly, the lower the proportion of carbon in a fuel, the less carbon dioxide in the exhaust. There is a lower proportion of carbon and a greater proportion of hydrogen in the molecules that make up natural gas. Methane, for example, has one carbon for every four hydrogens. Octane derived from petroleum, however, has eight carbons for every 18 hydrogen atoms, which is the same ratio as one carbon atom for every 2.25 hydrogen atoms.

49. Deriving electric energy from nuclear fission produces almost no atmospheric pollutants, such as carbon dioxide, sulfur oxides, nitrogen oxides, heavy metals, and airborne particulates. As discussed in Chapter 5, there is also an abundant supply of fuel for nuclear fission reactors in the form of plutonium-239, which can be manufactured from uranium-238.

51. The OPEC oil embargo was a boon for the nuclear power industry. The embargo was instrumental in alerting people to the vulnerability of depending on only one source of energy and prompted the building of many new nuclear power plants. Interestingly, there have been no new nuclear facilities constructed in the United States since that time.

53. No. The worst-case scenario for a nuclear power plant is the "meltdown," which occurs when an uncooled nuclear reactor gets so hot that it melts to the floor of the containment building. Nuclear fuel is enriched with fissionable uranium-235 to, at most, 4 percent. The remainder of the fuel is nonfissionable uranium-238. As discussed in Chapter 5, to build a nuclear bomb, uranium-235 must be enriched to a much higher percentage.

55. A sustainable energy source can't be exhausted and should be environmentally benign so that we can use it for a long time without damaging the environment that sustains us. A renewable energy source fits the first half of that definition but not necessarily the second half. A photovoltaic cell or a wind turbine, for example, may provide renewable energy, which is energy that keeps coming without practical end. If, however, the manufacture of these cells and turbines creates excessive pollutants or damages the environment—for example, the mining of rare earth elements—these would no longer classify as sustainable energy sources.

57. While dams themselves produce virtually no chemical pollutants, they do radically alter ecosystems to the detriment of many species, including humans displaced from their homes by the rising flood waters that come behind the dam.

59. Through the gravitational pull between the Earth and the Moon, which results in tidal forces. Interestingly, the energy lost by the Earth–Moon system to the creation of global tides results in a slowing down of the Earth's rotation. During the time of the dinosaurs, days were only about 19 hours long. In the far, far future, the days will be on the order of 47 hours long, such that the Earth's spin will exactly match the Moon's orbit. Thus, the Moon will always appears in the same location in the sky. This Moon will be visible only to creatures on one side of the planet, where it will serve as a useful clock going through all of its phases in a single day.

61. Relative to hot water, cold water doesn't have as much thermal energy, but it still has *some* thermal energy, which, of course, keeps it from freezing. That thermal energy of the cold water not too deep within the ground gets most of this energy from the ground, which gets its energy from sunlight. (Much deeper water gets its heat from radioactivity within the Earth.) Heat pumps that extract some of this thermal energy from underground water, therefore, are utilizing energy that is solar.

63. Cellulose is the most abundant organic chemical found on the Earth. If we can develop a way to easily produce ethanol from cellulose, we will have opened up an almost unlimited energy source.

65. In northern regions of this planet, the Sun shines from the south. As such, houses in these regions are often designed with most of the windows facing south. The houses, therefore, are warmed as the Sun's rays pass through the windows.

67. Use the solar heat to generate electricity, which can be used to run a refrigerator or air conditioner.

69. Seawater already contains the ions that are needed to allow the current to pass through the water. Seawater is also readily abundant.

71. The fact that fossil fuels are a limited resource can be viewed as a positive in the sense that the combustion of fossil fuels results in the production of much pollution. Having realized environmentally benign energy sources, our descendants may scratch their heads as to why we became so reliant on polluting fossil fuels. Also, as discussed in Chapter 12, fossil fuels are an excellent raw material for the production of material goods, such as plastics and pharmaceuticals. Our descendants might be concerned that we burned up most of this valuable raw material.

73. From the point of view of scientists and technologists, there are the challenges of how to cure disease (for example, through genetic engineering) or how to build more powerful computers. In general, however, science and technology is advancing quickly. To keep up with these advances, people must be scientifically and technologically literate. Education, therefore, will be an important challenge. Without a literate general public, science and technology, as funded by the general public, can only move so fast.

75. By using fluorescent or LED bulbs, using low-power laptop computers rather than high-power desktops, driving only when necessary, using high-mileage vehicles, bicycling instead of driving whenever possible, turning off unnecessary lights, conserving water, using an all-electric vehicle instead of a gas-powered vehicle, installing home solar cells or a home wind generator, buying and consuming less, taking vacations closer to home, and teleconferencing rather than attending out-of-town business meetings.

Periodic Table of the Elements, Useful Conversion Factors, and Fundamental Constants

Periodic Table of the Elements

Group																	
1																	**18**
1 **H** 1.0079	**2**											**13**	**14**	**15**	**16**	**17**	2 **He** 4.003
3 **Li** 6.941	4 **Be** 9.012											5 **B** 10.811	6 **C** 12.011	7 **N** 14.007	8 **O** 15.999	9 **F** 18.998	10 **Ne** 20.180
11 **Na** 22.990	12 **Mg** 24.305	**3**	**4**	**5**	**6**	**7**	**8**	**9**	**10**	**11**	**12**	13 **Al** 26.982	14 **Si** 28.086	15 **P** 30.974	16 **S** 32.066	17 **Cl** 35.453	18 **Ar** 39.948
19 **K** 39.098	20 **Ca** 40.078	21 **Sc** 44.956	22 **Ti** 47.88	23 **V** 50.942	24 **Cr** 51.996	25 **Mn** 54.938	26 **Fe** 55.845	27 **Co** 58.933	28 **Ni** 58.69	29 **Cu** 63.546	30 **Zn** 65.39	31 **Ga** 69.723	32 **Ge** 72.61	33 **As** 74.922	34 **Se** 78.96	35 **Br** 79.904	36 **Kr** 83.8
37 **Rb** 85.468	38 **Sr** 87.62	39 **Y** 88.906	40 **Zr** 91.224	41 **Nb** 92.906	42 **Mo** 95.94	43 **Tc** 98	44 **Ru** 101.07	45 **Rh** 102.906	46 **Pd** 106.42	47 **Ag** 107.868	48 **Cd** 112.411	49 **In** 114.82	50 **Sn** 118.71	51 **Sb** 121.76	52 **Te** 127.60	53 **I** 126.905	54 **Xe** 131.29
55 **Cs** 132.905	56 **Ba** 137.327	57 **La** 138.906	72 **Hf** 178.49	73 **Ta** 180.948	74 **W** 183.84	75 **Re** 186.207	76 **Os** 190.23	77 **Ir** 192.22	78 **Pt** 195.08	79 **Au** 196.967	80 **Hg** 200.59	81 **Tl** 204.383	82 **Pb** 207.2	83 **Bi** 208.980	84 **Po** 209	85 **At** 210	86 **Rn** 222
87 **Fr** 223	88 **Ra** 226.025	89 **Ac** 227.028	104 **Rf** 265	105 **Db** 268	106 **Sg** 270	107 **Bh** 269	108 **Hs** 277	109 **Mt** 276	110 **Ds** 281	111 **Rg** 280	112 **Cn** 285	113 **Uut**	114 **Fl** 289	115 **Uup**	116 **Lv** 293	117 **Uus**	118 **Uuo**

Period (left axis)

Lanthanides	58 **Ce** 140.115	59 **Pr** 140.908	60 **Nd** 144.24	61 **Pm** 145	62 **Sm** 150.36	63 **Eu** 151.964	64 **Gd** 157.25	65 **Tb** 158.925	66 **Dy** 162.5	67 **Ho** 164.93	68 **Er** 167.26	69 **Tm** 168.934	70 **Yb** 173.04	71 **Lu** 174.967
Actinides	90 **Th** 232.038	91 **Pa** 231.036	92 **U** 238.029	93 **Np** 237	94 **Pu** 244	95 **Am** 243	96 **Cm** 247	97 **Bk** 247	98 **Cf** 251	99 **Es** 253	100 **Fm** 257	101 **Md** 259	102 **No** 259	103 **Lr** 256

- ☐ Metal
- ☐ Metalloid
- ☐ Nonmetal
- ☐ Not yet confirmed

Atomic masses are averaged by isotopic abundance on the Earth's surface, expressed in atomic mass units. Conventional IUPAC atomic masses are provided as recommended by G. Kaptay in *J. Min. Metall. Sect. BMetall. 48 (1) B (2012)153 159.* For radioactive elements, this is commonly the whole number nearest the most stable isotope of that element.

List of the Elements

NAME	SYMBOL	ATOMIC NUMBER	ATOMIC WEIGHT
Actinium	Ac	89	227.028
Aluminum	Al	13	26.982
Americium	Am	95	243.
Antimony	Sb	51	121.76
Argon	Ar	18	39.948
Arsenic	As	33	74.922
Astatine	At	85	210
Barium	Ba	56	137.327
Berkelium	Bk	97	247
Beryllium	Be	4	9.012
Bismuth	Bi	83	208.980
Bohrium	Bh	107	269
Boron	B	5	10.811
Bromine	Br	35	79.904
Cadmium	Cd	48	112.411
Calcium	Ca	20	40.078
Californium	Cf	98	251
Carbon	C	6	12.011
Cerium	Ce	58	140.115
Cesium	Cs	55	132.905
Chlorine	Cl	17	35.453
Chromium	Cr	24	51.996
Cobalt	Co	27	58.933
Copernicium	Cn	112	285
Copper	Cu	29	63.546
Curium	Cm	96	247
Darmstadtium	Ds	110	281
Dubnium	Db	105	268
Dysprosium	Dy	66	162.5
Einsteinium	Es	99	253
Erbium	Er	68	167.26
Europium	Eu	63	151.964
Fermium	Fm	100	257
Flerovium	Fl	114	289
Fluorine	F	9	18.998
Francium	Fr	87	223
Gadolinium	Gd	64	157.25
Gallium	Ga	31	69.723
Germanium	Ge	32	72.61
Gold	Au	79	196.967
Hafnium	Hf	72	178.49
Hassium	Hs	108	277
Helium	He	2	4.003
Holmium	Ho	67	164.93
Hydrogen	H	1	1.0079
Indium	In	49	114.82
Iodine	I	53	126.905
Iridium	Ir	77	192.22
Iron	Fe	26	55.845
Krypton	Kr	36	83.8
Lanthanum	La	57	138.906
Lawrencium	Lr	103	256
Lead	Pb	82	207.2
Lithium	Li	3	6.941
Livermorium	Lv	116	293
Lutetium	Lu	71	174.967
Magnesium	Mg	12	24.305

NAME	SYMBOL	ATOMIC NUMBER	ATOMIC WEIGHT
Manganese	Mn	25	54.938
Meitnerium	Mt	109	276
Mendelevium	Md	101	259
Mercury	Hg	80	200.59
Molybdenum	Mo	42	95.94
Neodymium	Nd	60	144.24
Neon	Ne	10	20.180
Neptunium	Np	93	237
Nickel	Ni	28	58.69
Niobium	Nb	41	92.906
Nitrogen	N	7	14.007
Nobelium	No	102	259
Osmium	Os	76	190.23
Oxygen	O	8	15.999
Palladium	Pd	46	106.42
Phosphorus	P	15	30.974
Platinum	Pt	78	195.08
Plutonium	Pu	94	244
Polonium	Po	84	209
Potassium	K	19	39.098
Praseodymium	Pr	59	140.908
Promethium	Pm	61	145
Protactinium	Pa	91	231.036
Radium	Ra	88	226.025
Radon	Rn	86	222
Rhenium	Re	75	186.207
Rhodium	Rh	45	102.906
Roentgenium	Rg	111	280
Rubidium	Rb	37	85.468
Ruthenium	Ru	44	101.07
Rutherfordium	Rf	104	265
Samarium	Sm	62	150.36
Scandium	Sc	21	44.956
Seaborgium	Sg	106	270
Selenium	Se	34	78.96
Silicon	Si	14	28.086
Silver	Ag	47	107.868
Sodium	Na	11	22.990
Strontium	Sr	38	87.62
Sulfur	S	16	32.066
Tantalum	Ta	73	180.948
Technetium	Tc	43	98
Tellurium	Te	52	127.60
Terbium	Tb	65	158.925
Thallium	Tl	81	204.383
Thorium	Th	90	232.038
Thulium	Tm	69	168.934
Tin	Sn	50	118.71
Titanium	Ti	22	47.88
Tungsten	W	74	183.84
Uranium	U	92	238.029
Vanadium	V	23	50.942
Xenon	Xe	54	131.29
Ytterbium	Yb	70	173.04
Yttrium	Y	39	88.906
Zinc	Zn	30	65.39
Zirconium	Zr	40	91.224

Useful Conversion Factors

Length
SI unit: meter (m)

1 km = 0.621 37 mi

1 mi = 5280 ft

     = 1.6093 km

1 m = 1.0936 yd

1 in. = 2.54 cm (exactly)

1 cm = 0.393 70 in.

1 Å = 10^{-10} m

1 nm = 10^{-9} m

Mass
SI unit: kilogram (kg)

1 kg = 10^3 g = 2.2046 lb

1 oz = 28.345 g

1 lb = 16 oz = 453.6 g

1 amu = 1.661×10^{-24} g

Temperature
SI unit: Kelvin (K)

0 K = −273.15°C

    = −459.67°F

K = °C + 273.15

$°C = \dfrac{5}{9}(°F - 32)$

$°F = \dfrac{9}{5}(°C) + 32$

Energy
SI unit: Joule (J)

1 J = 0.239 01 cal

1 kJ = 0.239 kcal

1 cal = 4.184 J

Pressure
SI unit: Pascal (Pa)

1 atm = 101,325 Pa

    = 760 mm Hg (torr)

    = 29.9 in. Hg

    = 14.696 lb/in.^{2}

Volume
SI unit: cubic meter (m³)

1 L = 10^{-3} m^3

    = 1 dm^3

    = 10^3 cm^3

    = 1.057 qt

1 gal = 4 qt

    = 3.7854 L

1 mL = 0.0339 fl oz

1 cm^3 = 1 mL

    = 10^{-6} m^3

1 in.^{3} = 16.4 cm^3

Fundamental Constants

Avogadro's number = 6.02×10^{23} = 1 mole

Mass of electron, m_e = $9.109\,390 \times 10^{-31}$ kg

    = 1/1836 of mass of H

Mass of neutron, m_n = $1.674\,929 \times 10^{-27}$ kg

    ≈ mass of H

Mass of proton, m_p = $1.672\,623 \times 10^{-27}$ kg

    ≈ mass of H

Planck's constant, h = 6.626×10^{-34} J · s

Speed of light, c = 3.00×10^8 m/s

Glossary

Absolute zero The lowest possible temperature, which is the temperature at which the atoms of a substance have no kinetic energy: 0 K = −273.15°C = −459.7°F.

Acid A substance that donates hydrogen ions or accepts electron pairs.

Acidic Said of a solution in which the hydronium-ion concentration is higher than the hydroxide-ion concentration.

Activation energy The minimum energy required for a chemical reaction to proceed.

Addition polymer A polymer formed by the joining together of monomer units with no atoms being lost as the polymer forms.

Adhesive force An attractive force between molecules of two different substances.

Aerobic bacteria Bacteria able to decompose organic matter in the presence of oxygen.

Aerosol A moisture-coated microscopic airborne particle up to 0.01 millimeter in diameter that is a site for many atmospheric chemical reactions.

Agonist A molecule that binds to a receptor site and initiates a biological effect.

Alchemy A medieval endeavor concerned with turning other metals to gold.

Alcohol (AL-co-hol) An organic molecule that contains a hydroxyl group bonded to a saturated carbon.

Aldehyde (AL-de-hyde) An organic molecule containing a carbonyl group, the carbon of which is bonded either to one carbon atom and one hydrogen atom or to two hydrogen atoms.

Alkane (AL-kane) A generic word for a saturated hydrocarbon.

Alkene (AL-keen) An unsaturated hydrocarbon containing one or more double bonds.

Alkyne (AL-kine) An unsaturated hydrocarbon containing a triple bond.

Alloy A mixture of two or more metallic elements.

Alpha particle A subatomic particle consisting of the combination of two protons and two neutrons ejected by a radioactive nucleus. The composition of an alpha particle is the same as that of the nucleus of a helium atom.

Amide (AM-id) An organic molecule containing a carbonyl group, the carbon of which is bonded to a nitrogen atom.

Amine (Uh-MEEN) An organic molecule containing a nitrogen atom bonded to one or more saturated carbon atoms.

Amino acid The monomers of polypeptides, each monomer consisting of an amine group and a carboxylic acid group bonded to the same carbon atom.

Amphoteric A description of a substance that can behave as either an acid or a base.

Anabolism A general term for all the energy-requiring chemical reactions that produce large biomolecules from smaller molecules.

Anaerobic bacteria Bacteria able to decompose organic matter in the absence of oxygen.

Analgesic A drug that allows us to tolerate pain without abolishing nerve sensations.

Anesthetic A drug that prevents neurons from transmitting sensations to the brain.

Anode The electrode where chemicals are oxidized.

Antagonist A molecule that binds to a receptor site and does not initiate a biological effect except that it blocks other active molecules from binding to the receptor site.

Applied research A type of research that focuses on developing applications of knowledge gained through basic research.

Aquifer A soil layer in which groundwater may flow.

Aromatic Said of any organic molecule containing a benzene ring.

Atmospheric pressure The pressure exerted on any object immersed in the atmosphere.

Atomic mass The total mass of an atom. The atomic mass of each element presented in the periodic table is the *weighted average* atomic mass of the various isotopes of that element occurring in nature.

Atomic nucleus The dense, positively charged center of every atom.

Atomic number The number of protons in the atomic nucleus of each atom of a given element.

Atomic orbital A volume of space where an electron is likely to be found 90 percent of the time.

Atomic spectrum The pattern of frequencies of electromagnetic radiation emitted by the energized atoms of an element, considered to be an element's "fingerprint."

Atomic symbol An abbreviation for an element or atom.

Atoms Extremely small fundamental units of matter.

Avogadro's Law A gas law that describes the direct relationship between the volume of a gas and the number of gas particles it contains at constant pressure and temperature. The greater the number of particles, the greater the volume.

Avogadro's number The number of particles—6.02×10^{23}—contained in 1 mole of anything.

Base A substance that accepts hydrogen ions or donates electron pairs.

Basic Said of a solution in which the hydroxide-ion concentration is higher than the hydronium-ion concentration. Also sometimes called *alkaline*.

Basic research A type of research that leads us to a greater understanding of how the natural world operates.

Beta particle An electron emitted during the radioactive decay of a radioactive nucleus.

Bioaccumulation The process whereby a toxic chemical that enters a food chain at a low trophic level becomes more concentrated in organisms higher up the chain.

Biochemical oxygen demand A measure of the amount of oxygen consumed by aerobic bacteria in water.

Biomass A general term for plant material.

Boiling Evaporation in which bubbles form beneath the liquid surface.

Bond energy The amount of energy required to pull two bonded atoms apart, which is the same as the amount of energy released when the two atoms are brought together into a bond.

Boyle's Law A gas law that describes the indirect relationship between the pressure of a gas sample and its volume at constant temperature. The smaller the volume, the greater the pressure.

Buffer solution A solution that resists large changes in pH, made from either a weak acid and one of its salts or a weak base and one of its salts.

Capillary action The rising of liquid into a small vertical space due to the interplay of cohesive and adhesive forces.

Carbohydrate Organic molecules produced by photosynthetic plants, containing only carbon, hydrogen, and oxygen.

Carbon-14 dating The process of estimating the age of once-living material by measuring the amount of radioactive carbon-14 present in the material.

Carbonyl group (CAR-bon-EEL) A carbon atom double-bonded to an oxygen atom; found in ketones, aldehydes, amides, carboxylic acids, and esters.

Carboxylic acid (CAR-box-IL-ic) An organic molecule containing a carbonyl group, the carbon of which is bonded to a hydroxyl group.

Catabolism Chemical reactions that break down biomolecules in the body.

Catalyst Any substance that increases the rate of a chemical reaction without itself being consumed by the reaction.

Cathode The electrode where chemicals are reduced.

Chain reaction A self-sustaining reaction in which the products of one reaction event initiate further reaction events.

Charles's Law A gas law that describes the direct relationship between the volume of a gas sample and its temperature at constant pressure. The greater the temperature, the greater the volume.

Chemical bond The force of attraction between two atoms that holds them together within a compound.

Chemical change The formation of new substance(s) by rearranging the atoms of the original material(s).

Chemical equation A representation of a chemical reaction in which reactants are drawn before an arrow that points to the products.

Chemical formula A notation that indicates the composition of a compound, consisting of the atomic symbols for the different elements of the compound and numerical subscripts indicating the ratio in which the atoms combine.

Chemical property A type of property that characterizes the ability of a substance to change into a different substance under specific conditions.

Chemical reaction A term synonymous with chemical change.

Chemistry The study of matter and the transformations it can undergo.

Chemotherapy The use of drugs to destroy pathogens without destroying the animal host.

Chromosomes An elongated bundle of DNA and protein that appears in a cell's nucleus just prior to cell division.

Coal A solid consisting of a tightly bound network of hydrocarbon chains and rings.

Cohesive force An attractive force between molecules of the same substance.

Combinatorial chemistry A laboratory approach intended to mimic nature's chemical diversity by taking advantage of the many different ways in which a series of reacting chemicals can be combined.

Combustion The rapid exothermic oxidation-reduction reaction between a material and molecular oxygen.

Compost Fertilizer formed by the decay of organic matter.

Compound A material in which atoms of different elements are bonded to one another.

Concentration A quantitative measure of the amount of solute per volume of solution.

Conceptual model A representation of a system that helps us predict how the system behaves.

Condensation The transformation of a gas to a liquid.

Condensation polymer A polymer formed by the joining together of monomer units accompanied by the loss of small molecules, such as water.

Configuration A term used to describe how the atoms within a molecule are connected. For example, two structural isomers will consist of the same number and same kinds of atoms, but in different configurations.

Conformation One of a wide range of possible spatial orientations of a particular configuration.

Consumer An organism that takes in the matter and energy of other organisms.

Corrosion The deterioration of a metal, typically caused by atmospheric oxygen.

Covalent bond A chemical bond in which atoms are held together by their mutual attraction for one or more pairs of electrons they share.

Covalent compound A substance, such as an element or a chemical compound, in which atoms are held together by covalent bonds.

Critical mass The minimum mass of fissionable material needed for a sustainable chain reaction.

Decomposer An organism in the soil that transforms once-living matter to nutrients.

Density The amount of mass contained in a sample divided by the volume of the sample.

Deoxyribonucleic acid (DNA) A nucleic acid containing the sugar deoxyribose and having a double helical structure; it carries genetic code in its nucleotide sequence.

Dipole A separation of charge that occurs in a chemical bond because of differences in the electronegativities of the bonded atoms.

Dissolving The process of mixing a solute in a solvent to produce a homogeneous mixture.

Effective nuclear charge The nuclear charge experienced by outer-shell electrons, diminished by the shielding effect of inner-shell electrons and also by the distance from the nucleus.

Electrochemistry The study of the relationship between electrical energy and chemical change.

Electrode Any material that conducts electrons into or out of a medium in which electrochemical reactions are occurring.

Electrolysis The use of electrical energy to produce chemical change.

Electromagnetic spectrum The complete range of waves, from radio waves to gamma rays.

Electron An extremely small, negatively charged subatomic particle found outside the atomic nucleus.

Electron configuration The arrangement of an atom's electrons within orbitals.

Electron-dot structure A shorthand notation of the shell model of the atom, in which valence electrons are shown around an atomic symbol. The electron-dot structure for an atom or ion is sometimes called a Lewis dot symbol, while the electron-dot structure of a molecule or polyatomic ion is sometimes called a *Lewis structure.*

Electronegativity The ability of an atom to attract a bonding pair of electrons to itself when bonded to another atom.

Element A material consisting of only one type of atom.

Elemental formula A notation that uses the atomic symbol and (sometimes) a numerical subscript to denote how many atoms are bonded in one unit of an element.

Endothermic A term that describes a chemical reaction in which there is a net absorption of energy.

Energy The capacity to do work.

Energy-level diagram A schematic drawing used to arrange atomic orbitals in order of increasing energy levels.

Entropy A measure of an amount of energy that has been dispersed. Wherever there is a spreading of energy, there is a corresponding *increase* in entropy.

Enzyme A protein that catalyzes (speeds up) bio-chemical reactions.

Ester (ESS-ter) An organic molecule containing a carbonyl group, the carbon of which is bonded to one carbon atom and one oxygen atom bonded to another carbon atom.

Ether (EETH-er) An organic molecule containing an oxygen atom bonded to two carbon atoms.

Eutrophication The process whereby inorganic wastes in water fertilize algae and plants growing in the water and the resulting overgrowth reduces the dissolved oxygen concentration of the water.

Evaporation The transformation of a liquid to a gas.

Exothermic A term that describes a chemical reaction in which there is a net release of energy.

Fact Something agreed upon by competent observers as being true.

Fat A biomolecule that packs a lot of energy per gram and consists of a glycerol unit attached to three fatty acid molecules.

Formula mass The sum of the atomic masses of the elements in a chemical formula.

Freezing The transformation of a liquid to a solid.

Functional group A specific combination of atoms that behaves as a unit in an organic molecule.

Gamma rays High-frequency electromagnetic radiation emitted by radioactive nuclei.

Gas Matter that has neither a definite volume nor a definite shape, always filling any space available to it.

Gene A particular sequence of nucleotides along the DNA strand that leads a cell to manufacture a particular polypeptide.

Glycogen A polymer made of hundreds of glucose monomers and sometimes referred to as animal starch.

Greenhouse effect The process by which visible light from the Sun is absorbed by the Earth, which then emits infrared energy that cannot escape and thus warms the atmosphere.

Group A vertical column in the periodic table, also known as a family of elements.

Half-life The time required for half the atoms in a sample of a radioactive isotope to decay.

Half-reaction One-half of an oxidation-reduction reaction, represented by an equation showing electrons as either reactants or products.

Hard water Water containing large amounts of calcium and magnesium ions.

Heat The energy that flows from one object to another because of a temperature difference between the two.

Heat of condensation The energy released by a substance as it transforms from gas to liquid.

Heat of freezing The heat energy released by a substance as it transforms from liquid to solid.

Heat of melting The heat energy absorbed by a substance as it transforms from solid to liquid.

Heat of vaporization The heat energy absorbed by a substance as it transforms from liquid to gas.

Heteroatom Any atom other than carbon or hydrogen in an organic molecule.

Heterogeneous mixture A mixture in which the different components can be seen as individual substances.

Homogeneous mixture A mixture in which the components are so finely mixed that any one region of the mixture has the same ratio of substances as any other region.

Horizon A layer of soil.

Humus The organic matter of topsoil.

Hydrocarbon An organic compound containing only carbon and hydrogen atoms.

Hydrogen bond An unusually strong dipole–dipole attraction occurring between molecules that have a hydrogen atom covalently bonded to a small highly electronegative atom, usually nitrogen, oxygen, chlorine, or fluorine.

Hydrologic cycle The natural circulation of water throughout our planet.

Hydronium ion A polyatomic ion made by adding a proton (hydrogen ion) to a water molecule.

Hydroxide ion A polyatomic ion made by removing a proton (hydrogen ion) from a water molecule.

Ideal gas law A gas law that summarizes the pressure, volume, temperature, and number of particles of a gas within a single equation often expressed as $PV = nRT$, where P is presure, V is volume, n is number of molecules, R is the gas constant, and T is temperature given in kelvin.

Impure The state of a material that is a mixture of more than one element or compound.

Induced dipole A temporarily uneven distribution of electrons in an otherwise nonpolar atom or molecule.

Industrial smog Visible airborne pollution containing large amounts of particulates and sulfur dioxide and produced largely from the combustion of coal and oil.

Inner-shell shielding The tendency of inner-shell electrons to partially shield outer-shell electrons from the attractive pull exerted by the positively charged nucleus.

Insoluble Said of a solute that does not dissolve to any appreciable extent in a given solvent.

Integrated crop management A whole-farm strategy that involves managing crops in ways that suit local soil, climatic, and economic conditions.

Integrated pest management A pest-control strategy that emphasizes preventing, planning, and using a variety of pest-control resources.

Ion An atom having a net electrical charge because of either a loss or gain of electrons.

Ionic bond A chemical bond in which there is an electric force of attraction between two oppositely charged ions.

Ionic compound A chemical compound containing ions.

Ionization energy The amount of energy needed to pull an electron away from an atom.

Isotope Any member of a set of atoms of the same element whose nuclei contain the same number of protons but different numbers of neutrons.

Ketone (KEY-tone) An organic molecule containing a carbonyl group, the carbon of which is bonded to two carbon atoms.

Kilowatt-hour The amount of energy consumed in 1 hour at a rate of 1 kilowatt.

Kinetic energy Energy due to motion.

Kinetic molecular theory A theory that explains the properties of solids, liquids, and gases by proposing that they consist of rapidly moving tiny particles, either atoms or molecules or both.

Law of mass conservation The principle that states that matter is neither created nor destroyed during a chemical reaction—atoms merely rearrange, without any apparent loss or gain of mass, to form new molecules.

Leachate A solution formed by water that has percolated through a solid-waste disposal site and picked up water-soluble substances.

Lipid A broad class of biomolecules, such as fats and oils, not soluble in water because their structures are largely made of hydrocarbons.

Liquid Matter that has a definite volume but no definite shape, assuming the shape of its container.

Lock-and-key model A model based upon the idea that there is a connection between a drug's chemical structure and its biological effect.

Mass The quantitative measure of how much matter an object contains.

Mass number The number of nucleons (protons plus neutrons) in the atomic nucleus. Used primarily to identify isotopes.

Matter Anything that has mass and occupies space.

Melting The transformation of a solid to a liquid.

Meniscus The curving of the surface of a liquid at the interface between the liquid surface and its container.

Metabolism The general term describing the sum of all the chemical reactions in the body.

Metal An element that is shiny, opaque, and able to conduct electricity and heat.

Metallic bond A chemical bond in which positively charged metal ions are held together within a "fluid" of loosely held electrons.

Metalloid An element that exhibits some properties of metals and some properties of nonmetals. Six elements recognized as metalloids include boron, B; silicon, Si; germanium, Ge; arsenic, As; antimony, Sb; and tellurium, Te.

Microirrigation A method of delivering water directly to plant roots.

Minerals Inorganic chemicals that play a wide variety of roles in the body and are obtained from our diet.

Mixed fertilizer A fertilizer that contains the plant nutrients nitrogen, phosphorus, and potassium.

Mixture A combination of two or more substances in which each substance retains its properties.

Molar mass The mass of 1 mole of a substance.

Molarity A common unit of concentration equal to the number of moles of a solute per liter of solution.

Mole A very large number equal to 6.02×10^{23} and usually used in reference to the number of atoms, ions, or molecules within a macroscopic amount of a material.

Molecule An extremely small fundamental structure built of atoms.

Molecule The fundamental unit of a chemical compound, which is a group of atoms held tightly together by covalent bonds.

Monomers The small molecular units from which a polymer is formed.

Nanotechnology An area where we engineer materials by manipulating individual atoms or molecules.

Natural gas A mixture of methane and small amounts of ethane and propane.

Neuron A specialized cell capable of sending and receiving electric impulses.

Neurotransmitter An organic compound released by a neuron and capable of activating receptor sites within an adjacent neuron.

Neutral Said of a solution in which the hydronium-ion concentration is equal to the hydroxide-ion concentration.

Neutralization A reaction between an acid and a base.

Neutron An electrically neutral subatomic particle found in the atomic nucleus.

Nitrogen fixation A chemical reaction that converts atmospheric nitrogen to some form of nitrogen usable by plants.

Noble gas shell A graphic representation of a collection of orbitals of comparable energy in a multielectron atom. A noble gas shell can also be viewed as a region of space about the atomic nucleus within which electrons may reside.

Nonbonding pairs Two paired valence electrons that are not participating in a chemical bond.

Nonmetal An element located toward the upper right of the periodic table, with the exception of hydrogen, that is neither a metal nor a metalloid.

Nonpoint source A pollution source in which the pollutants originate at different and often nonspecific locations.

Nonpolar Said of a chemical bond or molecule that has no dipole. In a nonpolar bond or molecule, the electrons are distributed evenly.

Nuclear fission The splitting of the atomic nucleus into two smaller halves.

Nuclear fusion The combining of nuclei of light atoms to form heavier nuclei.

Nucleic acid A long polymeric chain of nucleotide monomers holding the information for how amino acids need to be linked together to form an organism.

Nucleon Any subatomic particle found in an atomic nucleus. Another name for either proton or neutron.

Nucleotide A nucleic acid monomer consisting of three parts: a nitrogenous base, ribose (in RNA) or deoxyribose (in DNA), and a phosphate group.

Ore A geologic deposit containing relatively high concentrations of one or more metal-containing compounds.

Organic chemistry The study of carbon-containing compounds.

Organic farming Farming without the use of pesticides or synthetic fertilizers.

Osmosis The net flow (diffusion) of water across a semipermeable membrane from a region of low solute concentration to a region of high solute concentration.

Oxidation The process whereby a reactant loses one or more electrons.

Particulate An airborne particle having a diameter greater than 0.01 millimeter.

Period A horizontal row in the periodic table.

Periodic table A chart in which all known elements are organized by physical and chemical properties.

Periodic trend The gradual change of any property in the elements across a period of the periodic table.

Petroleum A liquid mixture of loosely held hydrocarbon molecules containing not more than 30 carbon atoms each.

pH A measure of the acidity of a solution, equal to the negative logarithm of the hydronium-ion concentration.

Phenol (fen-ALL) An organic molecule in which a hydroxyl group is bonded to a benzene ring.

Pheromone An organic molecule secreted by insects to communicate with one another.

Photochemical smog Airborne pollution consisting of pollutants that participate in chemical reactions induced by sunlight.

Photoelectric effect The ability of light to knock electrons away from atoms.

Physical change A change in which a substance changes its physical properties without changing its chemical identity.

Physical dependence A dependence characterized by the need to continue taking a drug to avoid withdrawal symptoms.

Physical model A representation of an object on some convenient scale.

Physical property Any physical attribute of a substance, such as color, density, or hardness.

Point source A specific, well-defined location where pollutants enter a body of water.

Polar Said of a chemical bond or molecule that has a dipole. In a polar bond or molecule, electrons are congregated to one side. This makes that side slightly negative, while the opposite side (lacking electrons) becomes slightly positive.

Polyatomic ion An ionically charged molecule.

Polymer A long organic molecule made of many repeating units.

Potential energy Stored energy.

Power The rate at which energy is expended.

Precipitate A solute that has come out of solution.

Probability cloud A plot of the positions of an electron of a given energy over time as a series of tiny dots.

Producer An organism at the bottom of a trophic structure.

Products The new materials formed in a chemical reaction.

Protein A polymer of amino acids having some biological function.

Proton A positively charged subatomic particle found in the atomic nucleus.

Psychoactive drug A drug that affects the mind or behavior.

Psychological dependence A deep-rooted craving for a drug.

Pure The state of a material that consists solely of a single element or compound.

Quantum A small, discrete packet of energy.

Quantum number An integer that specifies the quantized energy level within an atom.

Radioactive Material containing nuclei that are unstable because of a less than optimal balance in the number of neutrons and protons.

Radioactivity The high-energy particles and electromagnetic radiation emitted by a radioactive substance.

Reactants The reacting substances in a chemical reaction.

Reaction rate A measure of how quickly the concentration of products in a chemical reaction increases or the concentration of reactants decreases.

Reduction The process whereby a reactant gains one or more electrons.

Rem A unit for measuring the ability of radiation to harm living tissue.

Replication The process by which DNA strands are duplicated.

Reuptake A mechanism whereby a presynaptic neuron absorbs neurotransmitters from the synaptic cleft for reuse.

Reverse osmosis A technique for purifying water by forcing it through a semipermeable membrane into a region of lower solute concentration.

Ribonucleic acid (RNA) A nucleic acid containing a fully oxygenated ribose; it executes protein synthesis based on code read from DNA.

Saccharide Another term for carbohydrate. The prefixes *mono-*, *di-*, and *poly-* are used with this term to indicate the size of the carbohydrate.

Salinization The process whereby irrigated land becomes saltier as irrigation water evaporates.

Salt An ionic compound commonly formed from the reaction between an acid and a base.

Saturated hydrocarbon A hydrocarbon containing no multiple covalent bonds, with each carbon atom bonded to four other atoms.

Saturated solution A solution containing the maximum amount of solute that will dissolve in its solvent.

Scanning probe microscope A tool of nanotechnology that detects and characterizes the surface atoms of materials by way of an ultrathin probe tip, which is detected by laser light as it is mechanically dragged over the surface.

Science A body of knowledge built from observations, common sense, rational thinking, experimentation, and (sometimes) brilliant insights.

Scientific hypothesis A testable explanation for an observable phenomenon.

Scientific law An experimentally confirmed description of the natural world. Also known as a *principle*.

Scientific theory A well-tested explanation that unifies a broad range of observations within the natural world.

Semipermeable membrane A membrane containing submicroscopic pores that allow passage of water molecules but not of larger solute ions or solute molecules.

Solid Matter that has a definite volume and a definite shape.

Solubility The ability of a solute to dissolve in a given solvent.

Soluble Said of a solute that is capable of dissolving to an appreciable extent in a given solvent.

Solute Any component in a solution that is not the solvent.

Solution A homogeneous mixture in which all components are dissolved in the same phase.

Solvent The component in a solution that is present in the largest amount.

Specific heat The quantity of heat required to change the temperature of 1 gram of a substance by 1 Celsius degree.

Spectroscope A device that uses a prism or diffraction grating to separate light into its color components and measure their frequencies.

Steel Iron strengthened by small percentages of carbon.

Straight fertilizer A fertilizer that contains only one nutrient.

Stratosphere The atmospheric layer that lies just above the troposphere and contains the ozone layer.

Strong nuclear force The attractive force between all nucleons, effective only at very short distances.

Structural isomers Molecules that have the same chemical formula but different chemical structures.

Sublimation The process of a material transforming from a solid directly to a gas without passing through the liquid phase.

Submicroscopic The realm of atoms and molecules, where objects are smaller than can be detected by optical microscopes.

Substituent A term used to describe an atom or a nonbonding pair of electrons surrounding a centrally located atom.

Surface tension The elastic tendency found at the surface of a liquid.

Suspension A homogeneous mixture in which the various components are finely mixed, but not dissolved.

Synaptic cleft A narrow gap across which neurotransmitters pass either from one neuron to the next or from a neuron to a muscle or gland.

Synergistic effect One drug enhancing the effect of another.

Temperature How warm or cold an object is relative to some standard. Also, a measure of the average kinetic energy per molecule of a substance, measured in degrees Celsius, degrees Fahrenheit, or kelvin.

Thermodynamics An area of science concerned with the role energy plays in chemical reactions and other energy-dependent processes.

Thermometer An instrument used to measure temperature.

Thermonuclear fusion Nuclear fusion brought about by high temperatures.

Transmutation The changing of an atomic nucleus of one element into an atomic nucleus of another element through a decrease or increase in the number of protons.

Trophic structure The pattern of feeding relationships in a community of organisms.

Troposphere The atmospheric layer closest to the Earth's surface, containing 90 percent of the atmosphere's mass and essentially all water vapor and clouds.

Unsaturated hydrocarbon A hydrocarbon containing at least one multiple covalent bond.

Unsaturated solution A solution that is capable of dissolving additional solute.

Valence electrons The electrons in the outermost occupied shell of an atom.

Valence shell The outermost occupied shell of an atom.

Valence-shell electron-pair repulsion A model, also known as VSEPR (pronounced ves-per), that explains molecular geometries in terms of electron pairs striving to be as far apart from one another as possible.

Vitamins Organic chemicals that assist in various biochemical reactions in the body and are obtained from our diet.

Volume The amount of space an object occupies.

Water table The upper boundary of a soil's zone of saturation, which is the area where every space between soil particles is filled with water.

Watt A unit for measuring power, equal to 1 joule of energy expended per second (J/s).

Weight The gravitational force of attraction between two bodies (where one body is usually the Earth).

Credits

CHAPTER 1

Chapter Opener NASA Headquarters; **pg 3** John A. Suchocki; **1.1** John A. Suchocki; **1.2** John A. Suchocki; **1.4a** Tommy Lavergne; **1.4b** Harry Kroto; **1.5** American Institute of Physics/Emilio Segre Visual Archives; **1.6** Andy Crawford and Tim Ridley/ Dorling Kindersley; **1.7** Reprinted by permission from Macmillan Publishers Ltd: Nature 318, 162-163 (14 November 1985); **1.8a** Don Huffman; **1.8b** Wolfgang Kratchmer; **1.9** Tyler Boyes/Shutterstock; **1.10** Associated Press/Soren Andersson; **1.11** Jacques-Louis David; **1.12a** Lisa Jeffers-Fabro; **1.12b** John Suchocki; **1.13a** anyaivanova/ Shutterstock; **1.13b** Reuters/Scott Audette; **1.13c** Wes Agresta/Argonne National Laboratory; **1.13d** Francois Gohier/Photo Researchers, Inc.; **1.13e** John O'Meara/John Suchocki; **1.14** Getty Images; **1.16** John Suchocki; **pg 23** vadim kozlovsky/ Shutterstock; **pg 25** Maria-Jose Viñas/NASA Earth Observatory

CHAPTER 2

Chapter Opener NASA; **pg 27** John A. Suchocki; **2.1** Galyna Andrushko/Shutterstock; **2.1 inset** Getty Images; **2.5** Library of Congress; **2.6** Andrew Osipovich Karelin; **2.7** Dmitri Mendeleev; **2.8a** Roy Kaltschmidt/Lawrence Berkeley National Laboratory; **2.8b** IBM Corporate Archives; **2.8c** IBM Corporate Archives; **2.9** John A. Suchocki; **2.10** John A. Suchocki; **2.11** National Bureau of Standards; **2.12a** NASA Earth Observing System; **2.12b** NASA/Goddard Institute for Space Studies; **pg 34** NASA/Goddard Institute for Space Studies; **2.13** Richard Megna/Fundamental Photographs; **2.15** MelissaF84/Shutterstock; **2.16b** Paul G. Hewitt; **2.17** Erich Schrempp/Photo Researchers, Inc.; **2.18** Pearson Education; **2.24a** Paul G. Hewitt; **2.24b** Paul G. Hewitt; **2.24c** Paul G. Hewitt; **2.29a** John Suchocki; **2.29b** John Suchocki; **2.29c** John Suchocki; **2.29d** John Suchocki; **2.31** Hu Meidor; **2.32** John A. Suchocki; **pg 50** Pearson Education; **pg 51 left** Pearson Education; **pg 51 right** Pearson Education; **pg 56 middle** Advanced Image Resources LLC; **pg 56 top** Imagebroker/Alamy; **pg 57 top** Russell Harrison/Tank Service Inc.; **pg 57 bottom** U.S. Department of Agriculture

CHAPTER 3

Chapter Opener John Suchocki; **pg 59** John A. Suchocki; **3.1a** murboy/iStockphoto; **3.1b** Irene Springer/Pearson Education; **3.1c** Getty Images; **3.2a** Russell Illig/Getty Images; **3.2b** Pearson Education; **3.3a** Pearson Education; **3.3b** Eric Schrader/Pearson Education; **3.3c** Netfalls - Remy Musser/Shutterstock; **3.5** Pearson Education; **3.6** Sue Smith/Shutterstock; **3.7** Sharon Hopwood/Paul G. Hewitt; **3.8** Sharon Hopwood/Paul G. Hewitt; **pg 64** Paul G. Hewitt; **3.11a** Rachel Epstein/PhotoEdit; **3.11b** Gary Woodard/Shutterstock; **3.11c** Joseph Sibilsky/Alamy; **3.12a** Getty Images; **3.12b** M. Freeman/Getty Images; **3.12c** SuperStock/Alamy; **3.12d** Steve Cole/Getty Images; **3.12e** MarcelClemens/Shutterstock; **3.12f** Eric Schrader/Pearson Education; **3.12g** GeoStock/ Getty Images; **3.12h** Richard Megna/Fundamental Photographs; **3.13** NASA; **3.17** Pearson Education; **3.21a** Richard Megna/Fundamental Photographs; **3.21b** Pearson Education; **3.21c** Pearson Education; **3.21d** Pearson Education; **3.22** Getty Images; **3.23** Don Geddis; **3.24a** John Suchocki; **3.24b** John Suchocki; **3.26a** Richard Megna/Fundamental Photographs; **3.26b** parema/iStockphoto; **3.27** Aerial Archives/ Alamy; **3.29a** Colin Keates/Dorling Kindersley; **3.29b** Jim Larkin/iStockphoto; **3.29c** Getty Images; **3.29d** BenC/Shutterstock; **3.29e** Hiroshi Sato/ Shutterstock; **3.29f** John Suchocki; **3.30** MikLav/ Shutterstock; **3.32a** Science Photo Library/Alamy; **3.32b** Mondolithic Studios; **3.33a** IBM Corporate Archives; **3.33b** European Communities; **3.33c** Team Nanotec GmbH; **pg 85** Pearson Education; **pg 86 left** Pearson Education; **pg 86 right** Pearson Education; **pg 90** Kzenon/Fotolia; **pg 91 text** Leonard Hayflick, The Prophet of Immortality, Methuselah Foundation, 2004: Retrieved from http://www.mprize.org/ index.php?pagename=newsdetaildisplay&ID=044

CHAPTER 4

Chapter Opener John Suchocki; **pg 93** Pearson Education; **4.1a** CLM/Shutterstock; **4.1b** dundanim/ Shutterstock; **4.3b** National Oceanic and Atmospheric Administration; **4.4a** Karin Hildebrand Lau/ Shutterstock; **4.4b** Richard Megna/Fundamental Photographs; **pg 97** Keystone/Staff/Getty Images; **pg 98 top** Bettmann/Corbis; **pg 98 bottom** Library of Congress; **4.11** John Suchocki; **4.16a** Maxim Kazmin/Fotolia; **4.16b** John Suchocki; **4.17** John Suchocki; **4.18a** Tom Bochsler/Pearson Education; **4.18b** Richard Megna/Fundamental Photographs; **4.18c** Richard Megna/Fundamental Photographs; **4.18d** Richard Megna/Fundamental Photographs; **pg 108** Paul Ehrenfest; **4.23a** John Suchocki; **4.23b** RGB Ventures LLC dba SuperStock/Alamy; **4.24a** John Suchocki; **4.24b** John Suchocki; **4.24c** John Suchocki; **pg 118** Tom Hollyman/Photo Researchers, Inc.; **pg 125 left** John Suchocki; **pg 125 top** John Suchocki; **pg 125 bottom** Pearson Education; **pg 130 left** Jupiter

Images/Alamy Images; **pg 130 top** Collection Roger-Viollet/The Image Works; **pg 130 right** Deco Images II/Alamy; **pg 131** AFP/Getty Images/Newscom

CHAPTER 5

Chapter Opener RelaxFoto.de/iStockphoto; **pg 133** John Suchocki; **5.4** Cordelia Molloy/Photo Researchers, Inc.; **5.6** Steve Cole/Getty Images; **5.7** Mic Smith Photography LLC/Alamy; **5.9** Bionerd via Flickr; **5.16** Saint-Gobain Crystals & Detectors; **5.17** Sgt. Fernando Serna/US Air Force; **5.19** Jack Fields/Photo Researchers, Inc.; **5.26** RelaxFoto.de/iStockphoto; **5.31** NASA; **5.32** International Thermonuclear Experimental Reactor (ITER); w**pg 159** J.B Nicholas/Splash News/Newscom; **pg 160** "Lower 48 states shale plays" Energy Information Agency; 2011

CHAPTER 6

Chapter Opener Charles M. Falco/Photo Researchers, Inc.; **pg 163** Pearson Education; **6.2** Science Source/Photo Researchers, Inc.; **6.8a** Eric Schrader/Pearson Education; **6.8b** John Suchocki; **6.9** iStockphoto.com; **6.10a** lauriek/iStockphoto; **6.10b** marcel/Fotolia; **6.11a** Pearson Education/Eric Schrader; **6.11b** Pearson Education/Eric Schrader; **6.13** US Mint; **6.14** Gary Whitton/Shutterstock; **6.15** Jamie Marshall/Dorling Kindersley; **6.16** Alcoa Inc.; **6.18** Theodore Gray/Element Collection; **6.21a** Pearson Education; **6.21b** Pearson Education; **6.22** Thomas M Perkins/Shutterstock; **6.32** David Taylor/Photo Researchers, Inc.; **6.36** Danny E Hooks/Shutterstock; **pg 188** John Suchocki; **pg 192 top** EPA; **pg 192 bottom** EPA; **pg 193** Stephenson, John "GAO-08-841R Superfund Funding and Costs" United States Government Accountability Office: 2008

CHAPTER 7

Chapter Opener P. Vasiliadis; **pg 195** Pearson Education; **7.6** Pearson Education; **7.10** Jonathan Searle JS/JD/Reuters; **7.11a** Benjah-bmm27; **7.11b** Theodore Gray/Element Collection; **7.12** Pearson Education; **7.14a** Photodisc/Alamy; **7.14b** Pearson Education; **7.14c** Orla/Shutterstock; **7.20** Sergey Peterman/Shutterstock; **7.21** John Suchocki; **7.26** Sheila Terry/Science Photo Library/Photo Researchers, Inc.; **7.30** NEPCCO Environmental Systems; **7.31** SunRay Technologies; **7.32** Saline Water Conversion Corporation; **7.33a** SolAqua; **7.36a** DenGuy/iStockphoto; **7.37** Pearson Education; **pg 221 left** Pearson Education; **pg 221 right** Pearson Education; **pg 224** Gordon Baer/Paul G. Hewitt; **pg 226** Don Farrall/Getty Images

CHAPTER 8

Chapter Opener Kirk Weddle/Getty Images; **pg 229** John Suchocki; **8.1** Eric Schrader/Fundamental Photographs; **8.3** Pearson Education; **8.4** Nuridsany & Perennou/Photo Researchers, Inc.; **8.5** Sportlibrary/Shutterstock; **8.13** Kuzmik/Shutterstock; **8.15a** NASA; **8.15b** Tischenko Irina/Shutterstock; **8.17a** Pearson Education; **8.17b** Pearson Education; **8.18** Pearson Education; **8.23** Lebazele/iStockphoto; **8.24a** John Suchocki; **8.24b** alexford/Fotolia; **8.26** Paul G. Hewitt; **8.27** NASA/Johnson Space Center; **8.29(4)** Roman Sigaev/Shutterstock; **8.31** William Stevenson/Alamy; **8.33** Richard Megna/Fundamental Photographs; **8.35** NOAA; **8.39** koka55/Shutterstock; **8.40** John Suchocki; **8.41** Vladimir Melnik/Shutterstock; **pg 254** Pearson Education; **pg 258 left** Institute of Paper Science & Technology; **pg 258 top** Moreno Soppelsa/Shutterstock; **pg 258 right** Pearson Education; **pg 259** Mike Goldwater/Alamy

CHAPTER 9

Chapter Opener Getty Images; **pg 261** Pearson Education; **9.4a** Pearson Education; **9.4b** Pearson Education; **9.4c** Pearson Education; **9.6a** Creatas/AGE Fotostock; **9.6b** MarKarrass/Alamy; **9.8** NASA; **9.10** John Suchocki; **9.11** Stuart Cupit; **9.19** Tony Campbell/Shutterstock; **9.23** NASA; **9.24** Charles D. Winters/Photo Researchers, Inc.; **9.25** Felix Miozznikov/Shutterstock; **pg 287** Pearson Education; **pg 292** IMAC/Alamy; **pg 293 text** "Mercury and Air Toxics Standards" U.S. Environmental Protection Agency; 2011

CHAPTER 10

Chapter Opener shootsphoto, Germany/Shutterstock; **pg 295** John Suchocki; **10.1a** Lew Robertson/Alamy; **10.1b** Pearson Education; **10.1c** Pearson Education; **10.1d** Pearson Education; **10.2a** Eric Schrader/Pearson Education; **10.2b** Pearson Education; **10.2c** Pearson Education; **10.2d** Pearson Education; **10.5** Pearson Education; **10.9a** Richard Megna/Fundamental Photographs; **10.9b** Richard Megna/Fundamental Photographs; **10.9c** Richard Megna/Fundamental Photographs; **10.12a** Richard Megna/Fundamental Photographs; **10.12b** Eric Schrader/Pearson Education; **10.17** Bilderbox/AGE Fotostock; **10.18a** Simon Pearson; **10.18b** J.L. Carson/Custom Medical Stock Photo/Newscom; **10.18c** Library of Congress; **10.20** Charles D. Winters/Photo Researchers, Inc.; **10.22** Mauna Loa Weather Observatory; **pg 320 top** Nejron/Fotolia; **pg 320 middle** Brooks/Brown/Photo Researchers, Inc.; **pg 320 bottom** Corning Incorporated; **pg 321 top** Stockbyte/Alamy; **pg 321 bottom** Dale Mitchell/Shutterstock

CHAPTER 11

Chapter Opener Andrew Buckin/Shutterstock; **pg 323** John Suchocki; **11.1** Pearson Education; **11.8** Pearson Education; **11.9** Toyota Motor Corporation; **11.11** Ballard Power Systems; **11.13** Keri Enright-Kato/Yale Office of Sustainability; **11.14** Michael C. Liu; **11.15a** Pearson Education; **11.15b** Ed van den Berg; **11.20** John Suchocki; **11.23b** Freeport-McMoRan Copper & Gold; **11.25b** Holly Kuchera/Shutterstock; **11.27** Freeport-McMoRan Copper & Gold; **11.28** John Suchocki; **11.29** Pearson Education; **11.30** Angelika Bentin/Fotolia; **pg 347 left** Pearson Education; **pg 347 right** Pearson Education; **pg 352** John Suchocki

CHAPTER 12

Chapter Opener John Suchocki; **pg 355** John Suchocki; **12.6** Pearson Education; **12.12** Casey Kelbaugh/AGE Fotostock; **12.26a** david hughes/Fotolia; **12.26b** James Stuart Griffith/Shutterstock; **12.27b** Gerald J. Lenhard/Louiana State University, Bugwood.org. Used under a Creative Commons License, http://creativecommons.org/licenses/by/3.0/us/; **12.34** John Suchocki; **12.35** Pearson Education; **12.36a** Pearson Education/Eric Schrader; **12.36b** Sergey Yarochkin/Fotolia; **12.39** Associated Press/Fujitsu Ltd./HO; **12.40** Marco Rametta/Shutterstock; **12.43a** John Suchocki; **12.43b** Richard Megna/Fundamental Photographs; **12.44b** Daisy Images/Alamy; **12.45** Greenwood Communications; **12.46** Pearson Education; **12.47** ATOFINA Chemicals; **12.48** Pearson Education; **pg 388** Pearson Education; **pg 394** Subbotina Anna/Fotolia

CHAPTER 13

Chapter Opener John Suchocki; **pg 397** John Suchocki; **13.1a** John Suchocki; **13.1b** vaivirga/Fotolia; **13.2a** Frank Greenaway/Dorling Kindersley; **13.3a** Geoffrey Jones/Shutterstock; **13.4a** Pearson Education; **13.5a** Gerard Brown/Dorling Kindersley; **13.5b** Frank Greenaway/Dorling Kindersley; **13.6a** GeoM/Shutterstock; **13.6a inset** Jubal Harshaw/Shutterstock; **13.7b** yanlev/Shutterstock; **13.7b inset** SPL/Pearson Education; **13.9a** John Suchocki; **13.9a inset** Biophoto Associates/Photo Researchers, Inc.; **13.10a** Scott Camazine/Photo Researchers, Inc.; **13.10b** Eric V. Grave/Photo Researchers, Inc.; **13.12** Vladimir Melnik/Fotolia; **13.13a** Pearson Education; **13.13b** Pearson Education; **13.18a** CandyBox Images/Fotolia; **13.18b** Janice Carr/CDC; **13.18c** Stefan Schurr/Fotolia; **13.18d** Science Photo Library/Photo Researchers, Inc.; **13.18e** John Suchocki; **13.18f** Science Photo Library/Photo Researchers, Inc.; **13.18g** John Suchocki; **13.20a** Susumu Nishinaga/Photo Researchers, Inc.; **13.20b** Janice Carr/CDC; **13.28a** Pearson Education; **13.29b** Lawrence Livermore National Laboratory; **13.32a** National Cancer Institute; **13.32b** National Library of Medicine; **13.32c** Jenifer Glynn/National Library of Medicine; **13.33** monticellllo/Fotolia; **13.36** David Bagnall/Alamy; **13.37** USDA: http://www.choosemyplate.gov/: 2012; **13.38** Jupiter Images/Alamy; **13.40a** Pearson Education/Eric Schrader; **13.42a** John Suchocki; **13.42b** John Suchocki; **13.42c** John Suchocki; **pg 432 left** Pearson Education; **pg 432 right** Pearson Education; **pg 436** Photodisc/Alamy

CHAPTER 14

Chapter Opener Dr. Jeremy Burgess/Photo Researchers, Inc.; **pg 439** John Suchocki; **14.3** Kurt Hostettmann; **14.4a** David Nanuk/Photo Researchers, Inc.; **14.4b** David Nanuk/Photo Researchers, Inc.; **14.5b** Pfizer/Pharmacia Corporation; **14.8** Maximilian Stock Ltd./Photo Researchers, Inc.; **14.12a** NIBSC/Photo Researchers, Inc.; **14.14a** Michael Abbey/Photo Researchers, Inc.; **14.14b** Michael Abbey/Photo Researchers, Inc.; **14.14c** Michael Abbey/Photo Researchers, Inc.; **14.16a** Scott Camazine/Photo Researchers, Inc.; **14.18** F. Hoffmann-La Roche Ltd; **14.26a** luigipinna/Fotolia; **14.29a** oneclearvision/iStockphoto; **14.31a** IGG digital Graphics Productions GmbH/Alamy; **14.31b** Daniel Dempster Photgraphy/Alamy; **14.31c** Tim Hazel/Getty Images; **14.31d** Jim Mires/Alamy; **14.31e** Christina Pedrazzini/Photo Researchers, Inc.; **14.34** R. Konig/Photo Researchers, Inc.; **14.35** WillSelarep/iStockphoto; **14.36a** Jupter Images/Alamy; **pg 473** Pearson Education; **pg 476** Hoby Finn/AGE Fotostock

CHAPTER 15

Chapter Opener John Suchocki; **pg 479** Pearson Education; **15.2** Searagen/Alamy; **15.3** Glow Images/Alamy; **15.5a** Nigel Cattlin/Photo Researchers, Inc.; **15.5b** Nigel Cattlin/Photo Researchers, Inc.; **15.5c** IPNI, International Plant Nutrition Institute; **15.7** K. W. Fink/Photo Researchers, Inc.; **15.12** Pearson Education; **15.13a** Jack Dykinga/USDA; **15.13b** Rob Flynn/USDA; **15.13c** Nigel Cattlin/Photo Researchers, Inc.; **15.13d** ying/Shutterstock; **15.13e** Garden Picture Library/Getty Images; **15.15** Associated Press; **15.18** Joe Belanger/Alamy; **15.24** U.S. Department of Agriculture; **15.25** U.S. Department of Agriculture; **15.26** U.S. Geological Survey, Denver; **15.27a** U.S. Department of Agriculture; **15.28a** pnason/iStockphoto; **15.28b** Ron Levine/Getty Images; **15.29** Jeff Gynane/Alamy; **15.30a** Pearson Education; **15.30b** U.S. Department of Agriculture; **15.31** Joe Munroe/Photo Researchers, Inc.; **15.32a** NASA; **15.32b** United States Department of Agriculture; **15.33** Marjorie A. Hoy; **15.34a** imagebroker/Alamy; **15.34b** Clearphoto/iStockphoto; **pg 506** C. S. Prakashi

CHAPTER 16

Chapter Opener Science Source/Photo Researchers, Inc.; **pg 509** Dennis Wong; **16.3** 4745052183/Shutterstock; **16.5** U.S. Geological Survey Information Services, Retrieved from http://ga.water.usgs.gov/edu/wateruse-fresh.html; **16.6a** Will McIntyre/Photo Researchers, Inc.; **16.6b** Pearson Education; **16.7** Pearson Education; **16.11** Pearson Education; **16.16a** NASA; **16.16b** NASA; **16.16c** NASA; **16.17a** National Oceanic and Atmospheric Administration (NOAA); **16.19b** PPC Industries; **16.19c** John Suchocki; **16.22** Pearson Education; **16.25a** Reuters/L. Augustin/Laboratoire de Glaciologie et Geophysique de l'Environment (LGGE); **16.25b** US Army Cold Regions Laboratory; **16.28a** NASA/GSFC/Craig Mayhew and Robert Simmon; **16.29a** Galyna Andruchko/Shutterstock; **16.29b** Moodboard/Alamy; **pg 534** Pearson Education; **pg 538** Alex Bartel/Photo Researchers, Inc.; **pg 539** NASA, ESA, the Hubble Heritage (STScI/AURA)-ESA/Hubble Collaboration, and K. Noll (STScI)

CHAPTER 17

Chapter Opener NASA; **pg 541** Pearson Education; **17.2** U.S. Energy Information Administration, US Dept. of Energy; **17.3** United States Power Grid Retrieved from: wikipedia.org/wiki/File:UnitedStatesPowerGrid/jpg; FEMA; **pg 544 text** American Society of Civil Engineers; Failure to Act: The Economic Impact of Current Investment Trends in Electricity Infrastructure: 2011 Adapted from: http://www.asce.org/uploadedFiles/Infrastructure/Failure_to_Act/SCE41%20report_Final-lores.pdf; **17.4** Data compiled from: World Energy Council 2010 Report; World Energy Council; **17.5a** Harry Taylor/Dorling Kindersley, Courtesy of the Natural History Museum, London; **17.5b** Dorling Kindersley; **17.5c** Pearson Education; **17.6a** Robert Sassen/Texas A & M University; **17.6b** GEOMAR Research Center; **17.7** Lawrence Thornton Archive; **17.8** Richard Megna/Fundamental Photographs; **17.10** philipus/Fotolia; **17.11** Stockbyte/Getty Images; **17.12** John Suchocki; **17.14** Nuclear Power Reactors in the World; International Atomic Energy

Agency 2012 Edition Series No. 2, International Atomic Energy Agency 2012; **17.15** Nuclear Power Reactors in the World; International Atomic Energy Agency 2012 Edition Series No. 2, International Atomic Energy Agency 2012; **17.16** U. S. Department of Energy; **17.17** STR/Associated Press; **17.18** Asahi Shimbun/Getty Images; **17.19** OECD Organisation for Economic Cooperation and Development: 2011: Retrieved from: http://www.oecd-ilibrary.org/sites/factbook-2011-en/06/01/04/06-01-04-g1.html?contentType=&itemId=/content/chapter/factbook-2011-49-en&containerItemId=/content/serial/18147364&accessItemIds=&mimeType=text/h; **17.20** AEO 2012 Early Release Overview, U.S. Energy Information Administration, 2012; **17.22** Stringer Shanghai/Reuters; **17.23b** Van D. Bucher/Photo Researchers, Inc.; **17.24a** National Energy Laboratory, Hawaii; **17.24b** National Energy Laboratory, Hawaii; **17.25** National Renewable Energy Laboratory: http://www.nrel.gov/gis/geothermal.html - Geothermal Map; **17.26** Images & Stories/Alamy; **17.27** Saeed Khan/AFP/Getty Images/Newscom; **17.28** b27/ZUMA Press/Newscom; **17.29** Future Energy Resources Corporation; **17.30b** Photos.com; **17.31** SolarWorld Americas; **17.32** Gemasolar solar thermal plant, owned by Torresol Energy ©Torresol Energy; **17.33** D J Milborrow, "Wind energy: a technology that is still evolving" Proceedings of the Institution of Mechanical Engineers, Part A: Journal of Power and Energy Date: Jun 30, 2011 Sage Publication; **17.34** National Renewable Energy Laboratory: http://rredc.nrel.gov/wind/pubs/atlas/maps/chap2/2-01m.html; **17.35** Acciona Energia; **17.36a** Otmar Smit/Shutterstock; **17.36b** Pearson Education; **17.37** Martin Bond/Photo Researchers, Inc.; **17.38** Lewis S. Lewis; **17.39** BetterPlace.com; **17.40** Adam Jones/Danita Delimont/Newscom; **17.41** NASA/GSFC/Craig Mayhew and Robert Simmon

APPENDIX D

pg A34 Kaptay, G: "Journal of Mining and Metallurgy, Section B: Metallurgy vol.48B"; Technical Faculty 2012

Index

Useful Conversion Factors

Length
SI unit: meter (m)

1 km = 0.621 37 mi

1 mi = 5280 ft

      = 1.6093 km

1 m = 1.0936 yd

1 in. = 2.54 cm (exactly)

1 cm = 0.393 70 in.

1 Å = 10^{-10} m

1 nm = 10^{-9} m

Mass
SI unit: kilogram (kg)

1 kg = 10^3 g = 2.2046 lb

1 oz = 28.345 g

1 lb = 16 oz = 453.6 g

1 amu = 1.661×10^{-24} g

Temperature
SI unit: Kelvin (K)

0 K = −273.15°C

    = −459.67°F

K = °C + 273.15

$$°C = \frac{5}{9}(°F - 32)$$

$$°F = \frac{9}{5}(°C) + 32$$

Energy
SI unit: Joule (J)

1 J = 0.239 01 cal

1 kJ = 0.239 kcal

1 cal = 4.184 J

Pressure
SI unit: Pascal (Pa)

1 atm = 101,325 Pa

        = 760 mm Hg (torr)

        = 29.9 in. Hg

        = 14.696 lb/in.^{2}

Volume
SI unit: cubic meter (m^3)

1 L = 10^{-3} m^3

      = 1 dm^3

      = 10^3 cm^3

      = 1.057 qt

1 gal = 4 qt

        = 3.7854 L

1 mL = 0.0339 fl oz

1 cm^3 = 1 mL

          = 10^{-6} m^3

1 in.^{3} = 16.4 cm^3

Fundamental Constants

Avogadro's number = 6.02×10^{23} = 1 mole

Mass of electron, m_e = $9.109\ 390 \times 10^{-31}$ kg

                        = 1/1836 of mass of H

Mass of neutron, m_n = $1.674\ 929 \times 10^{-27}$ kg

                        ≈ mass of H

Mass of proton, m_p = $1.672\ 623 \times 10^{-27}$ kg

                        ≈ mass of H

Planck's constant, h = 6.626×10^{-34} J · s

Speed of light, c = 3.00×10^8 m/s